POINT/COUNTERPOINTS

HUMAN

BIOLOGY

HEALTH, HOMEOSTASIS, AND THE ENVIRONMENT

DANIEL D. CHIRAS
University of Denver

WEST PUBLISHING COMPANY

ST. PAUL   NEW YORK   LOS ANGELES   SAN FRANCISCO

PRODUCTION CREDITS

COPYEDITING AND INDEXING Patricia Lewis
INTERIOR AND COVER DESIGN Diane Beasley
ENVIRONMENT AND HEALTH LOGO Sue Simmons
ARTWORK Wayne Clark, Cyndie C. H.-Wooley, Darwen and Vally Hennings, Georg Klatt, Elizabeth Morales-Denney, and John and Judy Waller. Individual credits follow index.
COMPOSITION Carlisle Communications, Ltd.
PAGE LAYOUT David Farr/Imagesmythe, Inc.
PHOTO RESEARCH John Cunningham/Visuals Unlimited
COVER IMAGE AND FRONTISPIECE Allsport USA/ Vandystadt

COPYRIGHT ©1991 By WEST PUBLISHING COMPANY
50 W. Kellogg Boulevard
P.O. Box 64526
St. Paul, MN 55164-0526

Printed in the United States of America

98 97 96 95 94 93 92 91   8 7 6 5 4 3 2 1 0

Library of Congress Cataloging-in-Publication Data

Chiras, Daniel D.
HUMAN BIOLOGY : HEALTH, HOMEOSTASIS, AND THE ENVIRONMENT / Daniel D. Chiras.

        p.   cm.
    Includes index.
    ISBN 0-314-79498-0 (soft)
    1. Human biology.   I. Title.
    [DNLM: 1. Biology.   2. Physiology.
QH 307 .2 C541h]
QP36.C46   1991
612--dc20
DNLM/DLC
for Library of Congress   90-13063
                                                                CIP

To my son,
Skyler,
with love and affection

ABOUT THE AUTHOR

 Dan Chiras received his Ph.D. in reproductive biology from the University of Kansas Medical School in 1976. During his stay at the medical school, Dr. Chiras was president of both the Graduate Student Council and the All Student Council; in the latter position he was active in sponsoring a university-wide research day in which graduate students presented their research findings. Upon graduation, Dr. Chiras received the Latimer Award for research on ovarian physiology.

In September 1976, Dr. Chiras joined the Biology Department at the University of Colorado in Denver in a teaching and research position. Since 1976, Dr. Chiras has taught numerous undergraduate and graduate courses, including general biology, cell biology, histology, endocrinology, and reproductive biology. Dr. Chiras also has a strong interest in environmental issues and has taught a variety of courses on environmental pollution and its impacts on human health and living systems. Today, Dr. Chiras is an adjunct professor at the University of Colorado in Denver and at the University of Denver; he is also a visiting professor at the University of Washington where he teaches environmental science from time to time. But most of his time is spent writing and lecturing.

Dr. Chiras is the author of numerous technical publications on ovarian physiology and biological education, which have appeared in the *American Biology Teacher* and other journals. He has also written numerous articles for a general audience on pollution, conservation, nuclear energy, and sustainable ethics. Dr. Chiras is the author of a leading textbook, *Environmental Science: Action for a Sustainable Future,* and is the coauthor of *Natural Resource Conservation: An Ecological Approach* (with Oliver S. Owen). Dr. Chiras has also written a high school textbook, *Environmental Science: A Framework for Decision Making*, that is used throughout the nation.

Dr. Chiras's latest book is *Beyond the Fray: Reshaping America's Environmental Response*. Published in 1990, this book offers a host of recommendations for solving environmental problems and building a sustainable society. Professor Chiras is currently at work on two new books.

Besides his active scientific pursuits, Dr. Chiras is an avid kayaker, skier, and hiker. He travels widely, studying the magnificent world we live in. He and his wife and son live in Evergreen, Colorado, overlooking the snowcapped Rocky Mountains.

BRIEF CONTENTS

CONTENTS

IV   REPRODUCTION AND DEVELOPMENT   431

V   EVOLUTION, ECOLOGY, AND BEHAVIOR   529

21 Environmental Issues: Population, Pollution, and Resources   579

PREFACE

Human Biology: Health, Homeostasis, and the Environment was written for the introductory course in human biology, arguably one of the most relevant courses offered in colleges and universities. This book teaches students about the structure and function of their own bodies and describes the delicate internal balance that maintains our health. *Human Biology* also describes how internal imbalances result in disease, such as cancer, stroke, and heart disease. This text also describes new treatments for disease and points out ways to prevent disease in the first place. In many ways, *Human Biology* is a kind of operator's manual for the human body.

Unlike other books, this one takes a broad systems approach. First, it discusses topics such as cells, organs, and organ systems as part of a whole, interdependent system, which is kept in balance by a variety of physiological mechanisms. *Human Biology* also examines human beings in a broader context—that is, as part of the living systems of the planet Earth. It shows how human activities affect the health of the planet and how environmental impacts of human action ultimately affect our own health. The discussions of human health and environmental quality lead to a simple, but inescapable conclusion: what we do to the planet we do to ourselves. Because of this, planet care becomes the ultimate form of self-care.

ORGANIZATION

Human Biology is divided into five parts. Part I provides background information. It describes key biological and chemical principles that apply to all living things—and to our everyday lives. It also shows how humans are both similar to and different from other organisms that share this planet with us. Part I ends with a discussion of science and the scientific method and introduces critical thinking skills that will be useful throughout our lives.

Part II lays additional groundwork by taking a journey through the microscopic world of the cell. It describes the components of cells, how cells function, the nature of chromosomes, and the ways human genes control cells.

Part III outlines the structure and function of the human body. This part shows how organ systems operate and describes common diseases that afflict them. In so doing, it presents a great deal of useful information.

Part IV discusses human reproduction and heredity. It describes the structure and function of the reproductive systems and discusses the dramatic series of events that lead to the formation of a new life. It also explains how characteristics are transmitted from parent to child.

Part V in many ways is the capstone of the book. It provides a larger picture, showing how human beings evolved and how we fit into the Earth's rich biological fabric, the biosphere. The final chapter takes a look at the problems modern society has created and offers solutions for redirecting human society to a sustainable relationship with the life-giving planet.

SPECIAL FEATURES

Human Biology was custom-designed with the student in mind. Written in a friendly, clear style, it presents the basics of human biology. To help keep the book lively, numerous interesting and imaginative analogies have been interspersed to help explain the more difficult concepts. The text also includes numerous real-life examples to show how relevant the study of human biology is to a student's life.

Study Skills

Immediately following the preface is a brief list of study skills—tips on ways students can improve their memory, get more out of what they read, and, in general, study more efficiently. The study skills are designed to help in this and virtually every other course a student will take. Many of the tips will be useful later in life as well.

Health Notes

Many interesting examples are interwoven throughout the text to enliven the subject and make it more relevant. In addition, Health Notes are included in most chapters to present some of the more exciting developments in medicine. Health Notes describe advances in medicine—for example, dramatic advances in treating and preventing diseases or new procedures or discoveries that could someday revolutionize medicine. Students will learn how to reduce stress in their lives and how to avoid unhealthy foods. They will also learn how fetal cell transplants are being used to cure previously incurable diseases. They will see how breast feeding benefits newborns and learn the reasons behind drug and alcohol addiction and much more.

Point/Counterpoints

Many discoveries in biology have profound impacts on our lives. However, the application of new discoveries often results in new and controversial techniques, such as genetic engineering. This book presents a number of modern-day controversies in Point/Counterpoint sections.

Each Point/Counterpoint consists of two brief essays written by distinguished writers and thinkers. These lively essays present opposing views on such important issues of our times as genetic engineering, fetal cell transplantation, cancer, and global warming. Point/Counterpoints offer students a chance to practice critical thinking skills.

Critical Thinking Skills

As noted earlier, Chapter 1 presents a number of rules for improving critical thinking skills. These guidelines will help students become more discerning thinkers, a skill that will prove useful in this and many other college courses.

Additional emphasis is placed on critical thinking throughout the text. Each chapter, for example, has at least one question to help students exercise their critical thinking skills. Critical thinking questions are also included after each Point/Counterpoint.

Environment and Health

As noted earlier, human health and the environment in which we live are closely connected. To help illustrate the connections between the environment and human health, each chapter ends with an Environment and Health section. These brief sections illustrate some of the ways in which the physical and chemical environments affect our health. They also show how our social and psychological environment affects our health.

End-of-Chapter Summaries

Unlike other biology textbooks, this one contains a fairly detailed summary at the end of each chapter. These summaries cover key concepts and ideas. Students can use them for a quick review after reading the chapter and as they prepare for exams. The detailed summaries and extensive questions (discussed below) provide an excellent study guide.

Summary Tables

To help students summarize key concepts, processes, and systems, summary tables have been included in many chapters. Students can use these tables to prepare for exams or to review material after reading the chapter.

Test of Terms

To help students review the key terminology in each chapter, a test of terms has been included at the end of each chapter. Tests of terms contain fill-in-the-blank questions and can be used to assess one's understanding of the material presented in the chapter. They will prove useful when preparing for exams as well. Students can fill in the blanks immediately after reading the chapter or after they have spent some time studying it. Students may find it useful to review the questions before tests.

Test of Concepts

Each chapter contains a number of brief essay questions that will enable students to assess their understanding of the material.

Suggested Readings

Some students are interested in learning more about important subjects they encounter in the study of human biology or want a list of suggested readings to help them write term papers. To satisfy curious students, a list of suggested readings appears at the end of each chapter.

Art Program

This book contains a remarkable collection of drawings and photographs unlike those found in any other book on the market. These colorful illustrations supplement the text, helping to make the more complex concepts and processes understandable.

A Note to Instructors: Supplements

To help you teach this course, the publisher, Professor Robert Hollenbeck at Metropolitan State College, and I have developed a supplement package that includes an instructor's guide, transparencies, transparency masters, slides, and videos.

Instructor's guide. An Instructor's guide with a thousand-question test bank is available from West Publishing Company. For each chapter, the instructor's guide contains lecture outlines with tips and suggested enrichment topics, page-referenced key terms found in the chapter, food-for-thought questions that can be used to provide students with something to think about as they read the next assignment from the book, and a list of films and video sources that might prove valuable in lectures.

Computerized testing. The test bank is also available in a form that allows tests to be computer generated. Contact your West sales representative for details.

Transparency masters, acetates, and slides. A set of transparency masters of important diagrams and acetates of key full-color art pieces will be available

from the publisher for adopters to use in classes. A slide set with other important pieces of art and photographs from the text will also be available.

Videos. West Publishing offers a video library of films that adopters of the text may wish to use in their classes. Contact your local West sales representative for further information.

It should be noted that rather than publishing an accompanying study guide, we have integrated the materials typically found in a study guide into the book itself. This not only saves students time and money, but also makes the book a better learning tool.

ACKNOWLEDGMENTS

Like all textbooks, *Human Biology* summarizes a wealth of information gleaned from the studies of many scientists and writers. Consequently, this book is a monument to the dedication and hard work of many people. I hope I have done justice to their work. To them, I am deeply grateful.

Human Biology is also the product of a great deal of hard work from editors, artists, and others. A mere thanks seems woefully inadequate to express my gratitude to my editor and friend, Jerry Westby, who provided much of the inspiration and creativity for this project. Jerry patiently guided this manuscript from first draft to final draft, skillfully coordinating and commenting on reviews, pointing out strengths and weaknesses, and always pushing for the best possible book. His contribution is immeasurable. I am equally indebted to my production coordinator and friend, Barbara Fuller. Her vigilant attention to detail and uncompromising pursuit of excellence deserve a world of thanks. And special thanks must go to Jerry and Barb for their continued cheerfulness despite a hectic schedule and a sometimes cranky, overstressed author. I am also indebted to Patricia Lewis for her diligence in copyediting. She went through the manuscript with a fine-tooth comb, helping to create a finely polished text.

I am also grateful to the extraordinary team of artists, Judy and John Waller, Georg Klatt, Wayne Clark, Cyndie C. H.-Wooley, Darwen and Vally Hennings, and Elizabeth Morales-Denney, who worked on this book. Words cannot express my appreciation for their talent and hard work. I am honored to have been able to work with them.

I am also deeply indebted to Dr. John Cunningham of Visuals Unlimited for his diligent photo research. His skill and knowledge have helped make this a much better book than I ever dreamed possible.

Special thanks must also go to Kristen McCarthy, the senior promotion manager, for her marketing work; Diane Beasley, who provided the wonderful design for the book; and Denis Ralling, who served as developmental editor for the book and the instructor's supplements. His thoughtful, ever-helpful input was greatly appreciated.

I also extend a special thanks to Professor Richard Shippee of Vincennes University for his detailed and thoughtful comments on the manuscript. Professor Shippee also proofed artist's sketches and offered many helpful comments for improving the art. His untiring assistance throughout the entire project was deeply appreciated. Special thanks also go to my friend Dr. Frank Schumann, who patiently answered a host of medical questions throughout the project.

Many reviewers have commented on the three drafts of this manuscript and offered valuable input. To each and every one, I offer a heartfelt thanks. For those who offered detailed comments above and beyond the call of duty, *muchas gracias*.

Reviewers

Donald K. Alford
Metropolitan State College

David R. Anderson
Pennsylvania State University—Fayette Campus

Jack Bennett
Northern Illinois University

Charles E. Booth
Eastern Connecticut State University

J. D. Brammer
North Dakota State University

Vic Chow
City College of San Francisco

John D. Cowlishaw
Oakland University

Richard Crosby
Treasure Valley Community College

Stephen Freedman
Loyola University of Chicago

Martin Hahn
William Paterson College

John P. Harley
Eastern Kentucky University

Robert R. Hollenbeck
Metropolitan State College

Florence Juillerat
Indiana University—Purdue University

Ruth Logan
Santa Monica College

Charles Mays
DePauw University

David Mork
St. Cloud State University

Lewis Peters
Northern Michigan University

Richard Shippee
Vincennes University

Beverly Silver
James Madison University

David Weisbrot
William Paterson College

Tommy Wynn
North Carolina State University

Finally, many thanks must go to my wife Kathleen, who has helped immeasurably during the two long years it took to write this book. Thanks for typing my reading notes, contacting authors for Point/Counterpoints, hunting through libraries for articles, entering copyeditor's changes on disk, and proofing the galleys and page proofs. Perhaps, more importantly, thanks for being there when I needed you, and thanks for loving and supporting me through it all.

Finally, although he is too young to understand, thanks to my son Skyler for just being alive. He has brought great joy to my life and has recharged my sense of enchantment with the living world.

Daniel D. Chiras
Evergreen, Colorado

STUDY SKILLS

College is a demanding time in the lives of many students. Term papers, tests, reading assignments, and classes require a new level of commitment and intellectual activity. The work load can become overwhelming and frustrating. Fortunately, there are ways to lighten the load, to manage your time efficiently, to increase your chances of getting good grades, and to improve your knowledge and understanding.

This section offers some suggestions to help you manage your studies and improve your mastery of the subjects you study. If you already are adept and efficient at studying and taking tests, you may still benefit from reading this section. Every suggestion you can use to your advantage will help.

To begin, read over the suggestions listed below. Pick a few that seem right for you under each category, then put them into action. Applying these ideas could pay huge dividends—not just in college, but throughout your entire life. Learning is a lifetime endeavor, and those that learn fastest seem to get the most out of life.

Mastering basic study skills will require some work at first and may require that you break some bad habits. In the long run, the additional time investment could save you lots of time, help you get better grades, and become a more efficient learner. Most important, it could help you improve your knowledge and understanding.

GENERAL STUDY SKILLS

- Study in a quiet, well-lighted space.
- Work at a desk or table. Don't lie on a couch or bed.
- Establish a specific time each day to study and stick to your schedule.
- Study when you are most alert. Many people find they retain more if they study in the evening a few hours before bedtime.

- Let your friends and family know when you study and ask them to respect that time.
- Take frequent breaks—one every hour or so. Exercise or move around during your study breaks to help you stay alert.
- Reward yourself after a study session with an ice cream cone or a mental pat on the back.
- Study each subject every day to avoid cramming for tests. Some courses may require more hours than others, so adjust your schedule accordingly.
- Look up new terms or words whose meaning is unclear to you in the glossaries in your textbooks or in a dictionary.

IMPROVING YOUR MEMORY

You can improve your memory by following the PMC method. The PMC method involves three simple learning steps: (1) paying attention, (2) making information memorable, and (3) correlating new information with facts you already know.

Step 1. Paying attention means taking an active role in your education—taking your mind out of neutral. Eliminate distractions when you study. Review what you already know and formulate questions about what you are going to learn *before* a lecture or *before* you read a chapter in the text. Reviewing and questioning help prime the mind.

Step 2. Making information memorable means finding ways to help you retain information in your memory. Repetition, mnemonics, and rhymes are three examples.

- Repetition can help you remember things. The more you hear or read something, the more likely you are to remember it. Scribble notes while you read or study.

- You can also use learning tools, mnemonics, to help remember lists. For example, *keep piling chocolate on for goodness sakes* helps you remember the taxonomic classification scheme: kingdom, phylum, class, order, family, genus, and species.
- Rhymes and sayings are also helpful. If you are having trouble remembering a list of facts, try making up a rhyme.
- If you're having trouble remembering key terms, look up their roots in the dictionary. This helps them stick in your memory. Use the list of prefixes, suffixes, and roots on the back endsheets of this book.
- Draw pictures and diagrams of processes.

Step 3. Correlating with things your know means tying facts together or making sense of the bits and pieces of information you are learning and have learned previously.

- Instead of filling your mind with disjointed facts and figures, try to see how they relate to previous information you have learned. Stop and scan your memory for similar facts. Correlating facts with previous knowledge enables you to comprehend the big picture. The end-of-chapter questions will assist you in this function.
- After studying your notes or reading a chapter in your textbook, determine the main points. How does the new information you have learned fit into your view of life or the general subject under discussion? How can you use the information?

BECOMING A BETTER NOTE TAKER

- Spend 5–10 minutes before each lecture reviewing the material you learned in the previous lecture. This is extremely important for learning.
- Know the topic of each lecture *before* you enter the class. Spend a few minutes *before* each class reflecting on facts you already know about the subject that is to be discussed.
- If possible, read the text *before* each lecture. If not, at least look over the main headings in the chapter, read the topic sentences, and look over the figures.
- Develop a shorthand system of your own. Symbols such as = (equals), w/o (without), w (with), > (greater than), < (less than), ↑ (increase), and ↓ (decrease) can save you time.

- Develop special abbreviations. For example, if you find yourself writing the word human over and over again in your notes, abbreviate it to H. Muscle could be abbreviated as m or mm. Species is sometimes abbreviated sp.
- Omit vowels and abbreviate words to decrease writing time (for example: omt vwls shrten wrtng tme). This takes some practice.
- Don't take down every word your professor says, but be sure your notes contain the main points, supporting information, and important terms.
- Watch for signals from your professor indicating important material ("This is an extremely important point . . . ").
- If possible, sit near the front of the class to avoid distractions.
- Review your notes soon after lecture while they're still fresh in your mind. Be sure to leave room in your notes during class to add material you missed. Recopy your notes if you have the time.
- Compare your notes with those of your classmates to be sure you understood everything and did not miss anything important.
- Attend lecture regularly.
- Use a tape recorder, if necessary and if it's acceptable to your professor, if you have trouble catching all the points.
- If your professor talks too quickly, politely ask him or her to slow down.
- If you are unclear about a point, ask during class. Chances are other students are confused as well. If you are too shy, go up after lecture and ask, or visit your professor during his or her office hours.

HOW TO GET THE MOST OUT OF WHAT YOU READ

- Before you read a chapter or other assigned readings, preview the material by reading the main headings or chapter outline to see how the material is organized.
- Pause over each heading and ask a question or two about each main heading.
- Next, read the first sentence of each paragraph. When you have finished, turn back to the beginning of the chapter and read it thoroughly.
- Take notes in the margin or on a separate sheet of paper. Underline or highlight key points.

- Don't skip terms that are confusing to you. Look them up in the glossary in the back of your text-book or in a dictionary. Make sure you understand each term before you move on.
- Use the study aids in your textbook, including end-of-chapter questions and summaries. Don't just look over the questions and say, "Yeah, I know that." Write out the answer to each question as if you were turning it in for a grade and save your answers for later study. Look up answers to questions that confuse you. This book has questions that test your knowledge of the terms and questions that test your understanding of the concepts and processes. Critical thinking questions are also included to help you sharpen your critical thinking skills.

PREPARING FOR TESTS

- Don't fall behind on your reading assignments and review lecture notes as frequently as possible.
- If you have the time, you may want to outline your notes and your assigned readings. Try to prepare the outline with your book and notes closed. Determine weak areas, then go back to your text or class notes to study those areas.
- Space your study to avoid cramming. One week before your exam, go over all of your notes. Study for two nights, then take a day off. Study again for a couple of days. Take another day off, then make one final push before the exam, being sure to study not only the facts and concepts, but also how the facts are related. Unlike cramming, which puts a lot of information into your brain for a one-time event, spacing will help you retain information for the test and for the rest of your life.
- Be certain you can define all terms and give examples of how they are used.
- Draw key structures over and over until they stick in your memory.
- You may find it useful to write flash cards to review terms and concepts.
- After you have studied your notes and learned the material, look at the big picture—the importance of the knowledge and how the various parts fit together.
- You may want to form a study group to discuss what you are learning and to test one another.

- Attend review sessions offered by your instructor or by your teaching assistant. Study before the review session and go to the session with questions.
- See your professor or class teaching assistant with questions as they arise.
- Take advantage of free or low-cost tutoring offered by your school or, if necessary, hire a private tutor to help you through difficult material. Get help quickly, though. Don't wait until you are way behind. Remember that learning is a two-way street. A tutor won't help unless you are putting in the time.
- If you are stuck on a concept, it may be that you may have missed some important previous material. Look back over your notes or ask your tutor or professor what facts might be missing and causing you to be confused.
- If you have time, write your own tests, including all types of questions.
- Study tests from previous years, if they are available legally.
- Determine how much of a test will come from notes and how much will come from the textbook.

TAKING TESTS

- Eat well and get plenty of exercise and sleep before tests.
- Remain calm during the test by deep breathing.
- Arrive at the exam early or on time.
- If you have questions about the wording of a test question, ask your professor. Don't be shy.
- Look over the entire test first so you can budget your time.
- Skip questions you can't answer right away and come back to them at the end of the session if you have time.
- Read each question carefully and be sure you understand its full meaning before you answer it.
- For essay questions and definitions, organize your ideas on a piece of scrap paper or the back of the test *before* you start writing.

Now take a few moments to go back over the list. Check off those things you already do. Then, mark the new ideas you want to incorporate into your study habits. Make a separate list, if necessary, and post it by your desk or on the wall and keep track of your progress. Good luck!

SUGGESTED READINGS

Atkinson, R. H., and D. G. Longman. 1988. *Reading enhancement and development*. St. Paul, Minn.: West.

Annis, L. F. 1983. *Study techniques*. Dubuque, Iowa: Wm. C. Brown.

Kesselman-Turkel, J., and F. Peterson. 1981. *Study smarts*. Chicago: Contemporary Books.

Longman, D. G., and R. H. Atkinson. 1988. *College learning and study skills*. St. Paul, Minn.: West.

Pauk, W. 1984. *How to study in college*. Boston: Houghton Mifflin.

Rowntree, D. 1983. *Learn how to study*. New York: Scribners.

I
FROM MOLECULES TO HUMANKIND

1 An Introduction to Human Biology

Ascorbic acid crystal.

Preceding page: Cyclists racing on a velodrome.

Three and a half million years ago, our ancestors roamed the grasslands of Africa in search of food and water (Figure 1–1). Neither swift nor strong, with teeth woefully inadequate for capturing and killing prey, our ancestors subsisted on roots, seeds, nuts, fruits, and carrion, animals that had been killed by predators or died from other causes.

By most measures of animal success relevant at the time, our ancestors have to be judged a chancy experiment. They could have easily ended up as a blind alley in evolution. Fortunately, they possessed several characteristics that would aid in survival. Perhaps the most important was the brain. Over the next 3.5 million years, the human brain would help our kind ascend to a place of dominance on Earth, giving us unrivaled power to manipulate the environment to our liking.

Today, we live in a world markedly different from that of our early ancestors. Huge farms produce food for billions of people, many of whom live in towns and cities. Radio waves and microwaves transmit messages throughout the world in fractions of a second, allowing instant communication. Satellites probe the far reaches of our solar system, searching for clues to the origin of the universe. Scientists tinker with the genetic material of the cell, in efforts to alter plants and animals in ways that could increase food production.

In many ways, humans have been an overwhelming success. Over 5 billion of us now live on the planet, making our homes in the jungles, the grasslands, the

FIGURE 1–1 Australopithecus Afarensis Current scientific evidence suggests that *Australopithecus afarensis* was the first humanlike ape. It belongs to a group called hominids, which includes all human forms, living and dead. Skeletal remains indicate that they walked upright on two legs and had humanlike hands and teeth. It and another form much like it survived for two million years by foraging and hunting for food.

mountains, and even the frozen poles. Human achievement, however, has had its price. Polluted skies and streams have come to symbolize our accomplishment (Figure 1–2). Hundreds of species have perished as a result of human habitation and disregard.

This book examines the human organism in its environment, starting with the molecules and cells that make up our bodies and ending with a discussion of our role in the environment. Part I discusses the chemicals and chemical reactions that make life pos-

sible. Part II describes the cells of our bodies that, while functioning independently, provide services vital to the whole. Part III discusses the structure and function of the human body, emphasizing the importance of internal balance for maintaining health and showing ways that balance can be upset. Part IV discusses human reproduction and development, including inheritance and aging. Finally, Part V describes evolution and principles of ecology and outlines major environmental problems facing humankind. It also describes many solutions. Throughout the text, you

(a)

(b)

FIGURE 1-2 The Price of Progress Technological development and economic progress are a double-edged sword. They help make our lives more comfortable and enjoyable, but they also exact a price. Pollution, resource depletion, and the extinction of various species are three examples of the price we pay for progress. (*a*) Urban air pollution, (*b*) clear-cut in the Pacific Northwest.

will encounter many examples of the marvelous and exciting processes taking place in the human body and many examples of how our mental and physical health depends on a healthy environment.

LIFE IN THE BALANCE: HOMEOSTASIS

All life exists as part of a whole, an interdependent network of organisms and their environment. Over time, a complex series of checks and balances has evolved. These occur at all levels of organization—from the cell to the **biosphere**, the thin zone of life on Earth that contains all organisms (Figure 1-3).

What Is Homeostasis?

In cells and in our bodies, these checks and balances create a kind of **dynamic equilibrium**—a state that changes from time to time but remains more or less the same over the long run. Physiologists use the term **homeostasis** (homeo = same; stasis = standing) to describe this state of constancy. *Internal constancy is achieved through a variety of automatic mechanisms. These mechanisms compensate for internal and external changes.* Consider what happens after you eat a meal. Blood sugar levels begin to rise soon after your meal is ingested because sugars from the foods you have eaten pass from the digestive tract into your bloodstream (Figure 1-4). One of the key sugars present in food is glucose. Glucose is an important source of energy for body cells. Although blood glucose levels rise soon after you begin eating, they are kept from rising too high by a chemical substance called insulin. Insulin is a **hormone**, a substance produced in one part of the body that travels to another where it causes a specific effect. As blood sugar levels rise, insulin is released by the pancreas. Insulin enhances glucose uptake by muscle and liver cells, where it is stored for later use. This uptake of glucose lowers its levels in the blood and thus reduces insulin secretion.

This homeostatic mechanism provides at least two benefits. First, it helps maintain relatively constant levels of glucose in the blood. If levels climb too high, a person becomes unconscious. Second, insulin secretion helps prevent the loss of glucose in the urine (Chapter 11).

Insulin release is just one of many homeostatic mechanisms in the body that help maintain internal constancy and ensure normal cellular function. Checks and balances also occur in the biosphere. Consider how rodent populations are kept in balance in Yellowstone National Park. Coyotes are one of the major predators in the park. Along with other predators, they help prevent mice and other rodents from overrunning their food supply. The coyote is, therefore, crucial in maintaining ecological balance. However, coyote populations are also influenced by the size of their prey populations. When mice are abundant, so are coyotes, but when the mouse populations decline, so do predator populations.[1] As a result, a

---

[1]Many other factors also contribute to this balance. Rainfall and sunlight, for example, determine plant growth, which in turn affects the food supply of the mice. Chapter 20 discusses the concept of ecological balance in much more detail.

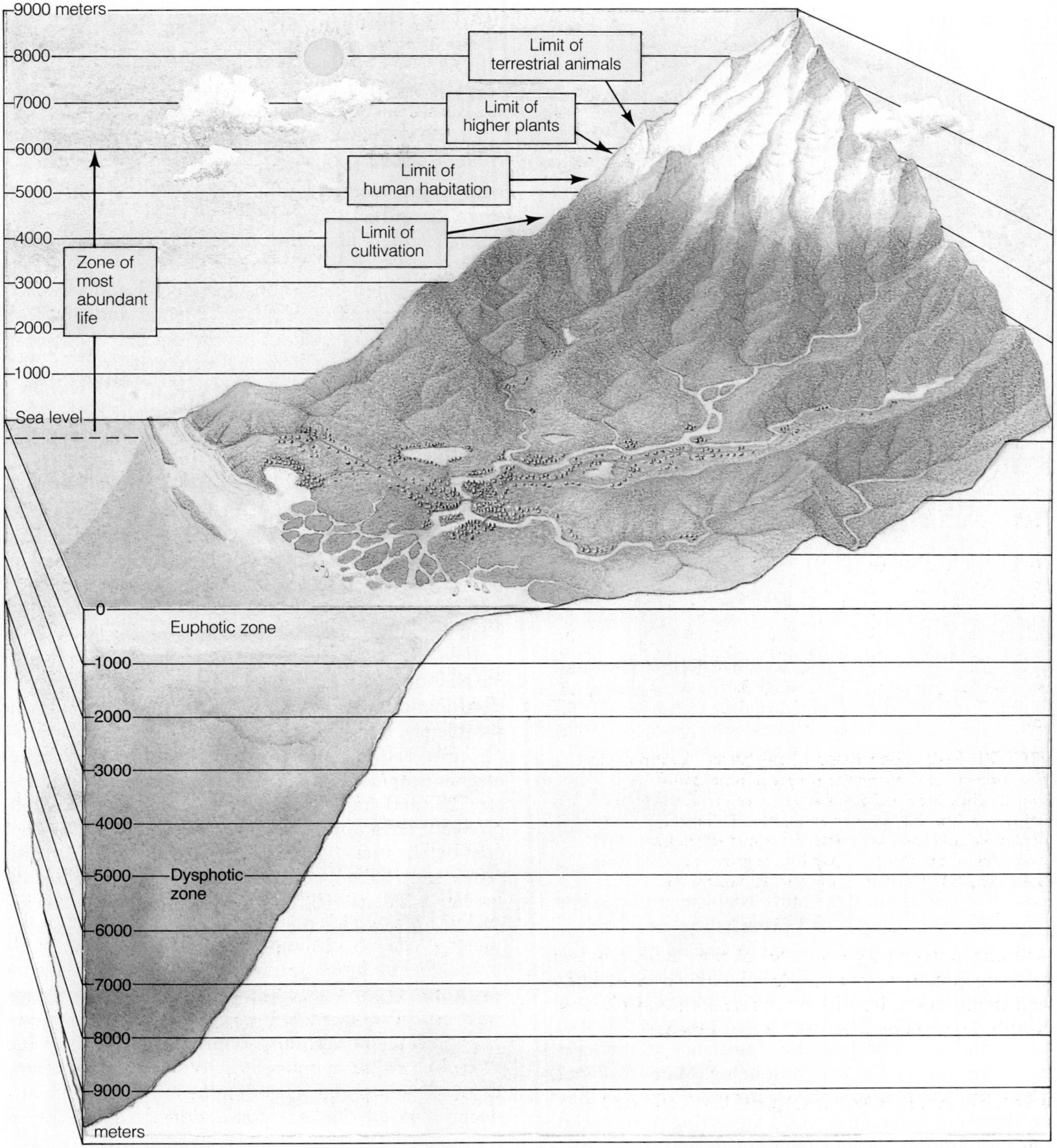
9000 meters
8000
7000
6000
5000
4000
3000
2000
1000
Sea level
Limit of terrestrial animals
Limit of higher plants
Limit of human habitation
Limit of cultivation
Zone of most abundant life
0
Euphotic zone
1000
2000
3000
4000
5000
Dysphotic zone
6000
7000
8000
9000
meters

FIGURE 1–3 The Biosphere The biosphere is the zone of life on Earth. Although the biosphere extends from the tops of the highest mountains to the bottom of the oceans, the zone of most abundant life is a much narrower band.

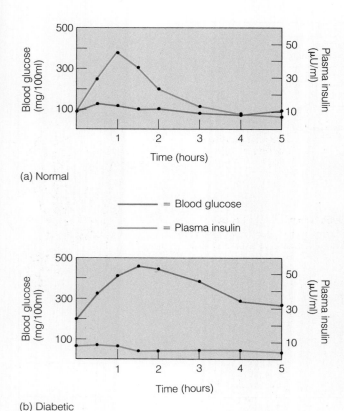

(a) Normal

——— = Blood glucose

——— = Plasma insulin

(b) Diabetic

FIGURE 1–4 Glucose and Insulin Levels Physicians use a test involving the ingestion of 100 grams of glucose to help determine whether an individual has diabetes. (*a*) Normal response. Shortly after ingesting the glucose, the glucose levels in the blood begin to increase, which causes the release of insulin; this, in turn, prevents the glucose concentration from rising further. (*b*) In diabetics who lack insulin, blood glucose levels rise considerably following the administration of glucose.

kind of dynamic equilibrium is established that keeps populations in balance and protects the grasslands.

This book uses the term *homeostasis* to refer to the process of maintaining balance at all levels of biological organization—from cells to the biosphere. Maintaining balance at all levels of organization is essential to maintaining human health. Imbalance at any level can affect our health and well-being. Many forms of pollution, for example, cause trouble because they upset natural ecological cycles. Upsetting these cycles can have impacts on our own health.

What Is Health?

For many years, **health** was defined as the absence of disease (Figure 1–5a). Many people, however, recog-

nized that this definition was much too limited. Today, health has taken on a broader meaning. For our purposes, we will define health as a state of physical and emotional well-being characterized by few risk factors. **Risk factors** are indicators of disease that often precede discernible illnesses. High blood pressure, for example, is a risk factor for heart disease. A person with high blood pressure may feel fine, but could some day suffer a heart attack or stroke. As a result, good health is also marked by the absence of risk factors.

Health experts also consider mental and physical fitness as indicators of health. Mental and physical fitness are measures of our psychological and physical abilities to meet the demands of our daily lives. Fit people, for example, should be able to cope with everyday psychological stresses and should be able to walk up a flight of stairs without becoming short of breath.

As Figure 1–5b illustrates, health is measured on a continuum. The healthiest people have no obvious illness, few, if any, risk factors, and a high level of fitness. The less healthy person has no obvious illness, but some risk factors and so on. Table 1–1 lists healthy habits that have been shown to be related to good health.

Physical well-being depends in large part on homeostatic mechanisms; when these mechanisms are not functioning properly, or when they break down completely, illness results. Persistent stress, for example, can disrupt several of the body's homeostatic mechanisms. Stress is a natural phenomenon in modern life, and most of us cope with it quite well. It probably does not cause harm when it occurs infrequently. If stress is prolonged, however, it can increase the risk of cardiovascular disease (diseases of the heart and arteries). It can also increase the risk of ulcers and weaken the immune system. Stress also

TABLE 1–1 Healthy Habits
Sleep seven to eight hours per day
Eat breakfast regularly
Avoid snacking between meals
Maintain ideal weight
Do not smoke
Avoid alcohol or use alcohol moderately
Exercise regularly

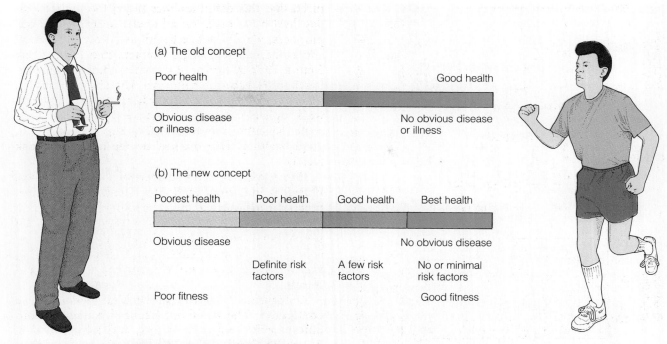

FIGURE 1–5   Old and New Concepts of Health

increases the likelihood of developing mental disorders. Stress causes disease by disrupting homeostatic mechanisms.

Fortunately, stress can be managed or lowered. Exercise, relaxation training, and systematic desensitization all work. These and other measures are discussed in Health Note 1–1.

Ecosystems are communities that consist of organisms and their environment. Ecosystems also possess homeostatic mechanisms that allow them to recover from a wide variety of changes. For instance, a natural landslide in the Pacific Northwest resulting from heavy rain may fill a salmon stream with sediment, burying spawning grounds and killing fish. Over time, however, the sediment will be washed away by the stream, restoring the stream and allowing the fish population to recover. Grasses, shrubs, and trees may take root on the hillside, stabilizing the slope and preventing future landslides. Because of these mechanisms, environments undisturbed by human activity can often persist for thousands of years (Chapter 20). When disturbances occur frequently, when they are severe, or when they overpower the recuperative capacity of the environment, disruption may lead to destruction. Today, pollution, mining, overgrazing, deforestation, and many other human activities

threaten environmental homeostasis and, therefore, the health of the biosphere. Flagrant environmental insults are a threat to human health and welfare. In many ways, they are akin to the risk factors affecting human health.

Human Health Is an Environmental Issue

Human health is generally viewed as a personal issue, but, more and more people are finding that human health and the health of our planet are linked. A healthy person cannot persist without a healthy ecosystem. Toxic pollutants from automobiles, power plants, and factories, for instance, form a thick layer of air pollution in and around many American cities. Air pollution contributes significantly to lung and heart disease. According to a recent study by the Environmental Protection Agency, over 76 million Americans (nearly one of every three men, women, and children) live in cities where ground-level ozone, a pollutant indirectly derived from automobiles, poses a threat to human health. Ozone irritates the eyes and the lungs. In combination with nitrogen dioxide, another common pollutant, ozone destroys the tiny air sacs in the lung, resulting in **emphysema**, a disease that kills its victims slowly and painfully (Figure

HEALTH NOTE 1–1
Reducing Stress in Your Life

Stress is a normal occurrence in everyday life, and not all stress is bad. In some instances, it even helps improve your performance. As you are crossing a busy street, for example, the sound of screeching brakes may start your heart racing and increase your rate of breathing. The signal of danger would result in the release of hormones that elevate your blood levels of glucose. All these changes enable you to perform at an extraordinary level, darting out of the way of the oncoming car.

Other external threats stimulate the stress response as well, but for most of us stress comes from within. That is, we create stress through our thoughts. We worry about our performance on exams, and we worry about embarrassing ourselves at parties. The physical reaction from psychological stress is the same as the reaction that occurs when we are threatened by an outside factor.

Severe stress, if prolonged, can cause great harm. It may interfere with performance at home, at work,

and at school. It may lead to cardiovascular disease, including heart attacks. It increases the risk of developing ulcers and can even weaken the immune system.

Getting rid of stress requires conscious effort, but it can be done. One of the most successful methods of reducing stress is through exercise. A single workout at the gym, a ride on your bicycle, a vigorous hike, or a day of cross-country skiing will reduce tension for two to five hours. Regular exercise may reduce the overall stress in your life. An individual who is easily stressed may find that stress levels decline after several weeks of exercise. Exercise may also relieve depression in some people.

Exercise can be supplemented by relaxation training to reduce stress. As you prepare for a difficult test or get ready for a date that you are nervous about, tension often builds in your muscles. Periodically stopping to release that tension helps you reduce physical stress. For some peo-

ple, getting up and stretching or taking a walk can help reduce the tension. Others find it useful to tighten the muscles forcefully, then let them relax. Still others may need more concentrated efforts. Cassette tapes on relaxation will teach you ways to relax, or you can sign up for stress-reduction classes through a therapist. The more you practice relaxation, the better you will get at relieving tension.

Stress from within is often blown out of proportion. By dealing with the thoughts that exaggerate your stress, you can help eliminate stress before it begins. Start paying attention to the thoughts that provoke anxiety in your life. Are they exaggerated? If so, why? For example, are you nervous before exams? Why? Do you always study adequately? Do you need to prepare better? Would that help reduce your anxiety? Do people asking you questions about a test right before class make you nervous? If so, can you isolate yourself from them? Finding

1–6). Smoking, a form of personal pollution, also contributes to emphysema. Today, emphysema kills more Americans than lung cancer and tuberculosis combined. Cancer-causing chemicals in our drinking water and our food can also alter cells in the body, causing some to proliferate uncontrollably. Unthwarted growth forms tumors that eventually sap our strength and kill us. These examples and others presented in the Environment and Health section in each chapter illustrate that an unhealthy environment results in an unhealthy body.

The health of our environment can also be measured by stress levels. Stress results from crowding, crime, and the pace of modern society and affects our health in many ways. Thus, a healthy environment is not just a clean, pollution-free setting, but an environment with minimal levels of stress. A healthy environment is psychologically and socially stable. If maintaining a healthy planet is essential to maintaining health, it is clear that planet care is the ultimate form of self-care.

THE UNITY AND DIVERSITY OF LIFE

"One touch of Darwin makes the whole world kin," wrote the British dramatist and critic George Bernard Shaw. With typical eloquence, Shaw underscored two important biological principles. First, humans are a product of evolution. Second, we, like all living things, are descendants of the first cells that formed some 3.5 billion years ago.

the source of your anxiety and taking positive action to alleviate it are helpful ways of reducing stress.

Stress reduction is not always so easy, however. Test anxiety, for example, may be deeply rooted in feelings of insecurity and low self-esteem. You can struggle with low self-esteem your entire life or seek professional help. A trained psychologist can help you find the roots of your problem and help you learn to feel better about yourself. Psychological help is as important as medical help these days, given the complexity and pace of our society. There's no shame in seeking counseling. In fact, just asking for help is the first step to recovery. It is a positive step toward healing yourself.

Biofeedback is another form of stress relief. A trained health worker hooks you up to a machine that monitors heart rate, breathing, muscle tension, or some other physiological indicator of stress (see the figure). During a biofeedback session, your trainer will help you to

Biofeedback Student in a biofeedback session.

relax, then perhaps discuss a stressful situation. When one of the indicators shows that you are suffering from stress, a signal is given off. Your goal is to consciously reduce the frequency of the signal. For example, if your heart started beating when you thought about taking an exam, a clicking sound might be heard. By deep breathing and relaxing, you consciously slow down your heart rate, and the clicking sound slows down as well, then disappears. Learning to recognize the symptoms of stress and to counter them when

not hooked to the machine is the goal of biofeedback.

You can also reduce risk by managing your time efficiently and managing your work load. Numerous books on the subject can help you learn to budget your time more effectively. This alone can help you keep from feeling stressed.

It is important to challenge yourself in college, but be realistic in what you expect of yourself. If you must work, for example, sign up for a class load that you can handle. Even if you are not working, take a reasonable class load, and be sure to exercise regularly.

Relieving stress in our lives helps us reduce the risk of cardiovascular disease. It helps us relax and enjoy life. It makes us more pleasant to be around as well. All in all, it is best to start learning early in life how to cope with stress. Lessons learned now will be useful for years to come.

The Characteristics of Life

Humans evolved from the same stock as all other organisms and, therefore, share many of the same biological characteristics. *First, like all forms of life, humans are capable of reproduction and growth.* A new human life begins with the union of an ovum and sperm in the mother's reproductive tract. The cell that develops from fertilization contains genetic information inherited from both parents. This genetic information directs the entire developmental process and probably even influences some of our behavior. As development proceeds, the single cell divides and differentiates, forming tissues, organs, and other body parts.

Human reproduction is called sexual reproduction. **Sexual reproduction** occurs in organisms that pro-

duce offspring by combining sex cells from the mother and the father. Sexual reproduction offers several key advantages over its counterpart, asexual reproduction. **Asexual reproduction** is common in single-celled organisms, such as the amoeba. In these and other single-celled organisms, reproduction occurs by cell division and budding, both of which produce generation after generation of identical offspring (Figure 1–7). Sexual reproduction, in contrast, produces genetically different offspring, for reasons explained in Chapter 19. Genetic differences result in differences in structure, function, and, possibly, behavior. Witness the differences in physical appearance in the offspring of a single set of parents (Figure 1–8).

Biologists call the genetic differences in a group of organisms **variation**. The variation that arises from

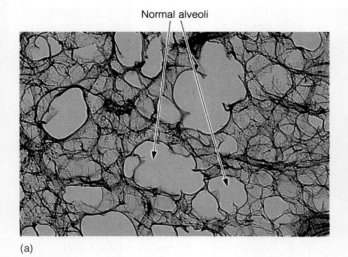

Normal alveoli

(a)

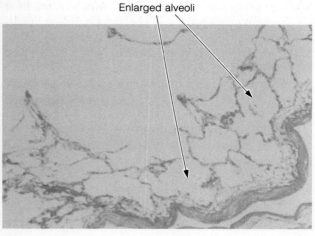

Enlarged alveoli

(b)

FIGURE 1–6 Emphysema Pollutants from cigarette smoke and other sources destroy the tiny air sacs in the lung, which are necessary for the absorption of oxygen into the blood. The walls between the sacs break down, decreasing overall surface area and making it harder for victims of this slow, debilitating disease to breathe. (*a*) Section through a normal lung. (*b*) Section through an emphysemic lung.

sexual reproduction has important evolutionary implications. A genetic difference that leads to a structural, functional, or behavioral difference may provide an animal a survival advantage over others (Chapter 19).

A second characteristic of life is that all organisms consist of the same biochemical substances. Thus, all plants, animals, and microorganisms are remarkably similar. Bacteria and humans, for instance, use the same type of genetic material, and they make protein in similar ways. They even have many of the same enzymes for extracting energy from sugar molecules. This biochemical similarity suggests a common evolutionary origin of all organisms and illustrates an important principle of evolution—notably, that patterns that were successful persisted.

The third feature common to organisms is that the molecules combine to form cells. In many organisms, cells form tissues and organs. To maintain this order, organisms must expend considerable amounts of energy.

FIGURE 1–7 Asexual Reproduction (*a*) Asexual reproduction takes place when an amoeba divides, producing identical offspring. (*b*) Small cellular buds in some organisms like this hydra enlarge to produce genetically identical offspring which eventually break free from the parent.

(a)

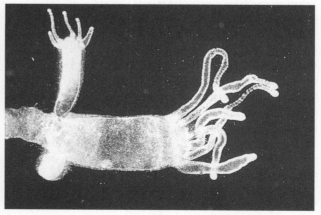

(b)

Without energy, all life would end. Much of the food you eat provides energy and chemical building blocks to maintain the biochemical integrity of our cells, replacing worn-out parts and building new ones. In fact, 70 to 80% of the energy adults need is required to maintain our bodies; the rest is used for activity.

A fourth feature of life is that all living things exhibit metabolism. **Metabolism** is the sum of all of the chemical reactions that occur in an organism. In human cells, thousands of reactions occur each second just to maintain life. Protein is synthesized in cells. Glucose is broken down by cells to release energy needed by cells, and large molecules (glycogen) are made in muscle and liver cells to store excess sugar molecules. The cell is a complex mixture of chemicals; the chemical reactions that occur in the cells, tissues, and organs are the basis of all life.

A fifth characteristic of life is homeostasis, the maintenance of internal constancy, discussed above. Maintaining a constant internal chemical environment is a never-ending task for organisms. Homeostasis, however, involves more than chemical constancy. To stay healthy, blood pressure and body temperature must also be kept fairly constant. Homeostasis requires the maintenance of normal concentrations of red blood cells, which transport oxygen and carbon dioxide. If levels fall substantially, tissues become starved of oxygen, and toxic wastes of cellular metabolism can build up.

A sixth feature common to all forms of life is irritability. **Irritability** is the ability to perceive and respond to stimuli. You are demonstrating irritability when you draw your hands from a hot stove, close your eyes to bright light, or shiver when chilled. The ability to perceive and respond to stimuli allows humans and all other organisms to respond to the environment— that is, to make changes necessary for survival. Irritability, therefore, is an important survival tool. Even single-celled organisms respond to stimuli.

A seventh characteristic is that humans and all other life forms are the products of evolution and subject to change. **Evolution** is a process that leads to structural, functional, and behavioral changes in species— making them better suited to their environment. Two factors are responsible for evolutionary change: variation and environmental conditions. Variation results from natural genetic differences in species and gives rise to differences in structure, function, and behavior. Variations can give some members of species an advantage over others when environmental changes occur. Organisms that have an advantage over others are more likely to survive and reproduce. Thus, they pass

FIGURE 1–8 Diversity in Sexually Reproducing Organisms Sexual reproduction produces incredible diversity in appearance, as the children in this family illustrate.

on the genetic material that gave them an advantage. Over time, the genetic composition of a group of organisms may change. Profound changes can result in the evolution of new species (Chapter 19). The process by which the environment "selects" organisms is called **natural selection**; it is discussed in detail with other aspects of evolution in Chapter 19.

What Makes Us Different?

Humans share many characteristics with the millions of species on Earth. Yet we are a unique form of life. What makes us different? *One of the key differences between humans and other animals is our culture.* Culture has been defined through the years in many ways. In a humorous vein, one critic of modern society said that culture is anything we do that monkeys don't. For our purposes, we will define **culture** as the ideas, customs, skills, and arts of a given people in a given time. Art and literature, for example, are the private domain of humankind (Figure 1–9).

Culture varies from place to place and changes over time. Certainly, modern American culture differs markedly from the early days of American settlement. Our ideas and values have shifted in many ways. And surely, modern American culture differs markedly from the culture of the Japanese, the Chinese, the Iranians, and the British, to name only a few.

Humans differ in other important ways as well. For example, *humans have the power to plan for the future.* While a few other animals seem to share this ability, most of what appears to be planning is probably the result of instinct. A bird's nest-building activities, for

FIGURE 1–9 Elephant Art Art, ideas, literature, and customs are the private domain of human beings . . . well almost. This artwork was produced by an elephant and judged extraordinary by critics who were not aware of the source.

example, are programmed by its genetic material. Nest building is probably not the product of conscious planning. Building skyscrapers and launching rockets, however, are two human activities that require extraordinary planning and knowledge.

On another front, *humans possess an ability, unrivaled in the biological world, to reshape the environment.* The ability of our minds to shape images and the ability of our hands to translate those images into reality have given us a power almost beyond our control. Despite our remarkable technologies and massive efforts, however, attempts to control nature often backfire, creating larger problems than we started out to solve. Attempts to control flooding on the Mississippi River, for instance, have led to more frequent flooding and more widespread damage (Figure 1–10). Attempts to keep the Mississippi River chan-

nel free of silt have reduced the deposition of sediment on the Mississippi River delta. The delta is gradually sinking into the ocean as a result. At one time, sediment deposited by the river offset the natural subsidence (sinking); today, however, because of human efforts, over 1 million acres have been reclaimed by the sea. Much of New Orleans is now below sea level, protected from the sea only by levees and breakwalls.

SCIENCE, CRITICAL THINKING, AND SOCIAL RESPONSIBILITY

Over the years, human societies have accumulated enormous amounts of information about the world. The systematic study of the world and its many parts today falls into the realm of science.

Science and the Scientific Method

Science is both a method of accumulating knowledge and the body of knowledge that results from systematic study. Many people view science as dull and uninteresting—an endeavor best left to a few. Science, however, is an exciting field with many practical applications. The application of science has improved human life in many ways, but the practical application of knowledge gained by scientific research can also lead to problems. Pollution, environmental destruction, and species extinction are partly the result of careless use of scientific knowledge.[2]

Science helps explain many fascinating phenomena around us, such as the weather, heredity, disease, and nutrition. A knowledge of science makes us better voters—better able to discern facts from fiction in a political debate. An understanding of certain aspects of science can help us decide which form of energy our society should develop to meet future energy demands. A knowledge of science can make us smarter consumers, helping us select the best way to control insects in our gardens or the best ways to heat our homes and offices.

The methods of science are known to most of us. We use them on a daily basis. When our cars fail to start or when our computers act up, for example, we very likely engage in the same kind of thinking that scientists use to gather information and test their ideas. This process, called **scientific method**, is shown in Figure 1–11. As illustrated, it begins with obser-

[2]Population growth is also a major contributor to enviromental destruction.

(a)

(b)

FIGURE 1–10 Human Control of Nature (*a*) Levees along the Mississippi River have grown considerably in the past hundred years in a sometimes vain effort to stop flooding. (*b*) Levees and other measures to control the river have led to more frequent flooding.

vation and measurement, often resulting from scientific experimentation.

To understand scientific method, let us use a familiar example. Suppose you sat down at your computer and turned the switch and nothing happened. This observation would lead to a **hypothesis**, a tentative explanation of the cause that is subject to experimentation. Perhaps, you noticed that the lights in the room didn't come on as well. You might hypothesize from these observations that the electricity in the house was off.

The next step in the process is to test your hypothesis by performing an **experiment**. An experiment is a carefully controlled test of a hypothesis. In this case, you would test your hypothesis by trying the lights in the kitchen or the bathroom. If they worked, you would discard your original hypothesis and form a new one. You might, for instance, hypothesize that the circuit breaker to your study had been tripped. To test this hypothesis, you would run another experiment by locating the circuit breaker to see if it was off. If it was, you would conclude that your second hypothesis was valid. To substantiate your conclusion, you would throw the switch, then test your computer and the room light.

This simple process involving observations and measurement, hypothesis, and experimentation is the basis of all science. Today, hundreds of thousands of men and women are performing this cycle of discovery in laboratories, in fields and forests, in space, and in the ocean—all with one purpose in mind: to add to the body of scientific knowledge.

Scientific knowledge, however, is more than just a collection of disconnected facts. Scientific knowledge also consists of numerous theories. **Theories** are the principles of science—the broader generalizations about the world and the operation of its components.[3] Theories are supported by considerable evidence. You cannot test a theory with a single experiment. Darwin's theory of evolution by natural selection is an example. It explains the workings of evolution and fits observations made in many different ways (Chapter 19).

A theory commands respect in science because it has stood the test of time. This does not mean that theories are always correct, however. Key theories, in fact, may be discarded as new scientific evidence is gathered. More often, theories only require modification as new data are made available. Because theories may prove to be wrong or may require modification to fit new observations, scientists must be open-minded about theories and must be willing to examine new evidence that can throw into question their most cherished beliefs.

For the most part, theories are talked about as if they were fact. As a result, some people object to calling a theory a theory for fear that it makes it sound

[3]Unfortunately, the word *theory* is often used loosely in everyday conversation and writing to mean hypothesis.

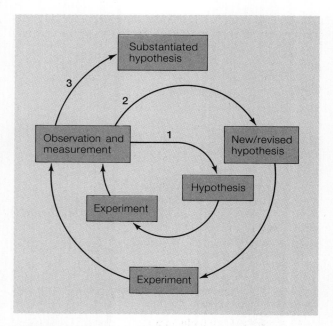

FIGURE 1–11 Scientific Method Scientific study begins with observation and measurement. These activities lead to hypotheses that can be tested by experiments. New and revised hypotheses are derived from experimentation.

more tentative. Good science, however, requires a mind open to change. Even widely held theories that have persisted for many years can be wrong. In A.D. 140, the Greek astronomer Ptolemy hypothesized that the Earth was the center of our solar system. For nearly 1500 years, the geocentric view held sway. Fellow astronomers, for the most part, vigorously defended this position; observations that did not fit the theory were ignored. In 1580, however, Nicolaus Copernicus dared to expound a new theory—the heliocentric view, stating that the sun was the center of the solar system. His work and his calculations created considerable debate, and eventually a new theory emerged.

Will evolutionary theory a hundred years from now resemble the current theory? Will future generations find that biologists have been viewing it incorrectly all along? Probably not. Nonetheless, fundamentally new and different information may come along and change our view of the evolution of life.

Science and Social Values

Science plays a major role in our lives. Today's science, in fact, is often the basis of tomorrow's technology. Lester Milbrath, professor of sociology and political science at the State University of New York at Buffalo, argues that much of our science today is aimed at improving technologies that increase wealth and power. Science and technology, he says, are therefore inadvertently responsible for the environmental problems we face. However, science also provides us with tools that alert us to the dangers of the modern world we have built. And it helps us understand the impacts of technology. Research on the dangers of pollution or the hazards of a technology, for example, helps society steer its course. Scientific knowledge may become the basis of society's action. Thus, science and public policy are linked. Most decisions, however, are not made on the basis of scientific knowledge alone; many are made after weighing human values—what we view as right or wrong.

Science affects the course of human culture and even alters human values. For example, the knowledge that human health and planetary well-being are tightly connected has caused many to change their values. No longer is waste and pollution accepted. In an even broader context, a knowledge of the factors that keep natural ecosystems stable might create a shift in thinking about the way modern society currently carries out its business (Chapter 21).

Science broadens the mind. It allows us to understand the interconnections of living things. It uncovers relationships not obvious to most people, such as the role of bacteria in recycling nutrients. In widening our understanding of the relationships that exist among living organisms and the environment, science helps us understand our dependence on other organisms (Figure 1-12). This view helps us widen our ethical boundaries, so that we may temper our activities, taking into account and seeking to minimize our impacts on other species. Science is also a source of fascination and wonder, which you may experience as you study human biology.

Critical Thinking

Science provides us with a method for systematically analyzing issues. Scientific method also helps us to think more critically. **Critical thinking** is the capacity to distinguish between beliefs (what we believe to be true) and knowledge (facts that are well established and supported by research). In other words, critical thinking is a process by which we separate facts from judgment.

Critical thinking skills provide a way to analyze problems, issues, and information. One of the first requirements of critical thinking is that you take an active part in acquiring information. Don't sit back

FIGURE 1-12 The Gray Whale One of Earth's magnificent creatures, the gray whale migrates annually along the Pacific coast, thrilling onlookers of all ages. The whale, like many other species, is part of a global recycling network that ensures a constant supply of nutrients for all living organisms.

and accept everything you read and hear. Question what you learn, and seek other sources.

Critical thinking, therefore, requires that you first define the problem, issue, or information, then examine the evidence that is available. Your examination of the evidence requires several key steps, which are summarized in Table 1-2.

First, critical thinking requires a clear understanding of terms. As you study human biology, you will encounter many terms that comprise the vocabulary of the subject. Terms are in many ways the bricks of the foundation of our understanding. Without a clear definition of terms, you will be unable to master this or any other subject.

Understanding terms, and making sure others define terms when they talk with you, helps bring clarity to an issue. Socrates destroyed many an argument by insisting on clear definitions of terms. So, as you analyze information, always be certain that you understand the terms.

Second, critical thinking requires an individual to question the methods by which facts are derived. Were the facts derived from experimentation or were they derived from unscientific observations? Experimentation properly carried out usually requires two groups: an experimental group and a control group. The **experimental group** is the one that is tested or manipulated in some way. The control group is treated the same in every respect, except for one, the experimental variable. For example, to test the effect of excess vitamin ingestion, you might start with two groups of

---

TABLE 1-2 Critical Thinking Skills

When analyzing an issue or facts, you may find it useful to follow these steps:

1. Understand and define all terms
2. Question the methods by which facts and information were derived
   - Were the facts derived from experiments?
   - Were the experiments well executed?
   - Did the experiment include a control group and an experimental group?
   - Did the experiment include a sufficient number of subjects?
   - Has the experiment been repeated?
3. Question the source of the information
   - Is the source reliable?
   - Is the source an expert or supposed expert?
4. Question the conclusions
   - Are the conclusions appropriate?
   - Was there enough information to make the conclusions?
5. Uncover assumptions and biases
   - Was the experimental design biased?
   - Are there underlying assumptions that affect the conclusions?
6. Tolerate ambiguity
   - Don't expect all of the facts to be established.
   - Expect controversy.
7. Examine the big picture
   - Look for multiple causes and effects.
   - Look for hidden effects.
   - Look for hidden relationships.

people, each containing a sufficient number of subjects to allow for statistical testing. In a good experiment, both groups would be as similar as possible in age, sex, weight, race, and so on. Both groups would be treated the same throughout the experiment. They would, for example, receive the same diet and would be housed in the same facility. Ideally, they would be exposed to the same level of stress; they would perform similar work and spend similar amounts of time at rest. In a well-run experiment, the only difference in the two groups should be the vitamin supplements given to the experimental group. Any observed differences between the groups, therefore, would be attributed to the treatment.

Good experiments require an adequate number of experimental and control subjects to be sure that any observed differences are real. In many laboratory experiments, at least 10 test animals are required. Groups larger than 10 are even better. For human health studies, much larger groups are generally used.

When you analyze new information, first check to see how it was derived. If the results were obtained from experimentation, were the experiments well planned and executed? Did the experimenters use an adequate number of subjects? Did the experiment have a control group? Were the control and experimental groups treated identically except for the experimental variable?

In science, one experiment is rarely adequate to draw firm conclusions. Thus, caution is advised when the results of new and unique experiments are announced. The media are especially fond of publishing the results and conclusions of new studies, but these studies, while sometimes valid, should be viewed with caution. Before one can be confident of their reliability, the experiment should be repeated, especially by other researchers. If similar results are derived from repetitions of the same experiment, you can have more confidence in the conclusions.

Third, critical thinking requires us to question the source of the facts—that is, who is reporting them. Beware of experts who really aren't what you think they are. For example, you might think that nutritional advice from a physician would be fairly reliable. Unfortunately, most physicians have received little or no training in nutrition. In the 1970s, for example, researchers found that many hospital patients suffered severe malnutrition when under a doctor's care for prolonged periods. Although this resulted in a dramatic reanalysis of nutritional instruction in medical schools, a study of U.S. medical schools in the mid-1980s showed that nutrition training was still inadequate in many schools: many medical students graduate without a full understanding of the role of nutrition in preventing disease and promoting good health.

Beware of experts from all walks of life. Experts from industry are often biased by corporate goals. A drug company expert announces that a new product is safe, but hidden biases may cloud such a person's judgment. Experts from the environmental field may be swayed by a hidden agenda as well.

Fourth, critical thinking requires us to question the conclusions derived from facts. Two of the most important questions you can ask when analyzing an issue

are the following: (1) Do the facts support the conclusions? (2) Are there alternative conclusions? Consider an example. One of the earliest studies on lung cancer showed that people who ate large amounts of refined sugar (sucrose) had a higher incidence of lung cancer. The conclusion was obvious: lung cancer was caused by sugar. Did the facts support this conclusion? Was there an alternative explanation?

A careful reexamination of the patients showed that the lines between cause and effect in this experiment were not what researchers thought they were. It turned out that smokers were more likely to consume more sugar than nonsmokers. Thus, the link between sugar and lung cancer was false. The real link was between smoking and lung cancer.

This example illustrates a key principle of medical research: *a correlation does not necessarily mean causation.* Two factors that appear to be correlated may not be causally related. In 1989, for example, doctors found many patients were suddenly becoming ill. When they and health officials at the Centers for Disease Control (CDC) looked into the problem, they found that the affected individuals had been taking large doses of a chemical called L-tryptophan (an amino acid). L-tryptophan is taken to help people get to sleep or to ease symptoms of premenstrual syndrome: tension, irritability, and bloating that may occur in many women before menstruation. Doctors reported over 1200 cases of L-tryptophan "poisoning" to the CDC; as of February 1990, 12 people had died. Many others were left nearly paralyzed or severely impaired. As a result, L-tryptophan was taken off the market. Headlines in newspapers proclaimed its dangers. Further research, however, suggested that the culprit was not L-tryptophan, but a chemical contaminant in the pills. In this case, the cause and effect were probably incorrectly assigned. Thus, the presumed correlation may prove false. In this case, careful research uncovered what appears to be the real cause of death and illness.

Another problem with conclusions is that they can be influenced by assumptions and bias. *The fifth rule of critical thinking requires us to uncover assumptions and biases that may underlie conclusions.* Many people believe that all radiation exposure is bad for human health. Is this conclusion based on any underlying assumptions? In this case, there is an underlying assumption: the belief that there is no threshold level for radiation—that is, a level below which no harm occurs. As a result, all exposure is believed dangerous. Scientists, however, still debate this assumption. Some claim that there is a threshold level, so the con-

(a)

(b)

FIGURE 1–13   The Changing Face of Agriculture
(*a*) For years farmers rotated crops to ensure soil fertility and to control insects. (*b*) Many modern farms plant acre after acre of the same crop, which is susceptible to disease and insects. When planted on the same field year after year, these crops deplete the soil of important nutrients that are not replenished by synthetic fertilizers.

clusion that all radiation is harmful is not fully accepted.

Critical thinking demands that we examine our own biases and those of the people presenting information to us. Popular health magazines flood readers with advice on the benefits of megadoses of vitamins, but most nutritionists think that much of the advice is not sound. Why? First, it is biased by an underlying assumption—that if a little vitamin is good for you, a lot must be even better. Second, much of the work is often extrapolated from studies on rats and mice, yielding results that cannot always be extrapolated to humans. Research also shows that large doses of some vitamins can cause significant harm (Chapter 7).

Sixth, critical thinking requires a tolerance for ambiguity. Although this may seem contradictory, it is important to remember that ambiguity exists in science. Hard-and-fast answers are not always available. As a result, we must become comfortable with uncertainty. Scientists, for example, believe that the temperature of the Earth is warming as a result of excess carbon dioxide and other gases from human activities that have accumulated in the atmosphere over the past two hundred years. Many scientists are willing to stake their reputations on it, and they quote an impressive body of information in support of their view (for more on this topic, see the Point/Counterpoint in Chapter 21). Not all scientists agree, however. Some think that the conclusions, which are based on computer projections, may be in error. We cannot predict tomorrow's weather. How then can we tell what the

weather will be like in 40 years? Global warming is, therefore, an issue where critical thinkers might reserve opinion. This leads us to the next idea—examining the big picture.

Seventh, critical thinking requires us to examine the big picture. Considering the high stakes of global warming, such as a dramatic shift in world climate that will turn productive farmland to desert and will flood 20% of the world's landmass (as glaciers and the Antarctic ice cap melt), critical thinkers who recognize that the facts are still controversial might choose to err on the conservative side. Destroying the climate of a planet poses such a serious threat that steps to mitigate the problem, even if we are not 100% certain that there is a problem, may be the most intelligent choice. Reducing the threat of global warming can be brought about by marked improvements in energy efficiency in automobiles, factories, and even our own homes (Chapter 21). This is a kind of insurance policy that would have the added benefit of reducing other forms of harmful pollution, stretching our limited supplies of oil and other fuels, and saving substantial amounts of money.

Consider another slightly different example of the benefits of examining the big picture. In 1988, researchers at Monsanto announced that they had discovered a way to alter the genetic material of wheat to make it resistant to a harmful fungus that causes enormous crop damage. To control the fungus now, farmers often rotate wheat from year to year with other crops that do not support the pest (Figure 1–13).

With the new genetically altered strain, say research-ers, farmers will not have to bother rotating crops. They can plant their fields in wheat year after year and can even plant larger crops without worrying about fungus infections.

While this may sound good, it could be an invitation for problems, when one considers the big picture. First, crop rotation helps build soil fertility. Rotating beans, clover, alfalfa and other legumes with wheat, for example, adds nitrogen to the soil and helps maintain soil fertility. Not rotating crops often drains a soil of its nutrients, reducing its productivity over time. Second, crop rotation also helps reduce insect pest populations. By planting a new crop in a field every year, farmers reduce food sources for insects that tend to prefer one crop over another. Because the food supply is not constant from one year to the next, pest populations remain low and manageable. Eliminating crop rotation, therefore, will very likely result in an outbreak of harmful wheat-eating insects. In solving the fungus problem, then, science may contribute to several more. The lesson in this case is that a careful examination of the ecological relationships often throws into question the apparent wisdom of new actions.

As a final note, *critical thinking requires us to examine multiple cause and effect.* This rule is an extension of the last. Far-seeing and intelligent as we are, we are also rather narrow-minded at times. We often fall back on simplistic thinking when analyzing problems. In the 1970s, Paul Ehrlich, a noted ecologist, argued vigorously that the world's environmental problems stem from overpopulation—too many people for the available resources. Another scientist, Barry Commoner, argued that the problems were due to technology and its by-products— pollution. A careful analysis shows that our problems are the result of many factors. Overpopulation and technology are two of the many. Inadequate laws and poor education must be factored into the equation. So must various psychological factors—for instance, our view of the world as something to overcome. Many more could be added to the list.

Critical thinking demands a broader view of cause and effect. Consider all of the contributing factors and their relative contributions to the problem. Avoid simplistic thinking.

As you read this text, you will be presented with examples on which to sharpen your critical thinking skills. The Point/Counterpoints will help you put these important tools into practice. You may also want to use your skills as you study the material.

SUMMARY

LIFE IN THE BALANCE: HOMEOSTASIS

1. Numerous mechanisms exist in our cells and our bodies to maintain internal constancy. These are called homeostatic mechanisms, and the internal balance is called homeostasis.
2. Homeostatic mechanisms are also present in ecosystems, where they help maintain a balance among organisms.
3. Human health is dependent on homeostasis. When homeostatic systems break down, illnesses often result.
4. Health has traditionally been defined as the absence of disease, but a broader definition of health is emerging. Good health includes an absence of risk factors that lead to disease and a high level of mental and physical fitness.
5. Human health and planetary health are tightly linked, making planet care a form of self-care.

THE UNITY AND DIVERSITY OF LIFE

6. Life exists in a variety of forms, but since the Earth's organisms evolved from early cells some 3.5 billion years ago, organisms share many common characteristics, listed below.
7. All life is capable of reproduction and growth.
8. All organisms contain the same types of biochemicals. The molecules are organized to form cells and tissues. To maintain order requires a considerable expenditure of energy.
9. All organisms are capable of metabolism.
10. All organisms exhibit homeostatic mechanisms.
11. All organisms exhibit irritability—the capacity to perceive and respond to stimuli.
12. All organisms are the product of evolutionary development and subject to further evolutionary change.
13. Although humans are similar to organisms in many respects, they also possess unique abilities. One of the key differences is culture. Humans can plan for the future and possess enormous abilities to reshape the Earth through human ingenuity and technology.

SCIENCE, CRITICAL THINKING, AND SOCIAL RESPONSIBILITY

14. Science is both a systematic method of discovery and a body of information about the world around us.
15. Scientists gather information and test ideas through the scientific method. The scientific method begins with observation and measurement, often taken during experiments. Observations and measurements often lead to hypotheses, testable explanations of natural phenomena. Hypotheses are tested in experiments. Results of experiments help scientists support or refute their hypotheses.

16. Scientific knowledge is based on experimentation and observation. Theories emerge from scientific knowledge. Theories are the broad principles of science or generalizations about the way the world works. Theories can change over time as new information becomes available.

17. Science is typically viewed as a value-free endeavor. But values in society sometimes determine the direction of science and can subtly influence the interpretation of data.

18. Scientific discovery can also change social values. New knowledge about our place in the biosphere, for example, may help temper notions of human dominance and help society reach a sustainable relationship with nature.

19. Scientists use critical thinking skills to distinguish knowledge from beliefs or judgment.

20. Critical thinking provides a way to analyze issues and information. It requires that you first define an issue, then study the evidence.

21. Table 1–2 summarizes the critical thinking rules.

EXERCISING YOUR CRITICAL THINKING SKILLS

Find an article on a current medical or environmental problem in a magazine or newspaper and read it. Then, study the facts presented by the author. Do you see places where conclusions are not supported by the facts? Do you see places where bias has entered into the author's writing? Do you see authorities used to support issues who may not be as knowledgeable as you would like or who might be biased? Do you see faulty experiments?

TEST OF TERMS

1. The _____ is the living skin of the planet, consisting of the plants, animals and microorganisms.

2. _____ is the term that describes the internal constancy maintained by automatic mechanisms in plants and animals.

3. The pancreas produces a substance called _____ that lowers blood sugar following a meal to help maintain constancy and prevent waste.

4. The disease caused by smoking and air pollution that results in the breakdown of the tiny air sacs in the lung is called _____ .

5. _____ reproduction occurs when an organism forms offspring by cell division.

6. Genetic _____ is the result of reproduction in which an ovum and sperm unite and is considered the raw material of evolution.

7. The chemical reactions occurring in an organism are collectively referred to as _____ .

8. _____ is the ability to perceive and respond to stimuli and is one of the chief characteristics of all life forms.

9. Changing environmental conditions may weed out certain members of a population of organisms, leaving only those that are genetically capable of surviving and reproducing under the new conditions. This weeding-out process is called _____ .

10. _____ consists of the ideas, customs, skills, and art forms of a given people.

11. Science is a body of _____ and a method of discovering knowledge.

12. The process by which scientists gather and test knowledge about the world around us is called the _____ _____ .

13. _____ are tentative, testable explanations of a observations.

14. A _____ is a principle of science supported by considerable scientific research.

Answers to the Test of Terms are located in Appendix B.

TEST OF CONCEPTS

1. In what ways is planet care a form of self-care? Give as many examples as you can.
2. Describe the concept of homeostasis. How does it apply to humans? How does it apply to ecosystems? Give examples from your experiences if you can.
3. In what ways are humans different from other animals? In what ways are they similar?
4. Describe the scientific method, and give some examples of ways you have used it in the last week in your own life.
5. How do a hypothesis and a theory differ?
6. Find examples of important issues in your daily newspaper where scientific knowledge can help you make decisions.
7. List and discuss the seven critical thinking skills presented in this chapter. Which skills seem to be most important for the kind of thinking you normally do?

SUGGESTED READINGS

Berger, J. J. 1985. *Restoring the Earth: How Americans are working to renew our damaged environment.* New York: Knopf. Excellent and uplifting book on efforts to restore past damage.

Boskin, W., G. Graf and V. Kreisworth. 1990. *Health dynamics: Attitudes and behaviors.* St. Paul, Minn.: West. Overview of health issues of concern to students.

Chiras, D. D. 1991. *Environmental science: Action for a sustainable future,* 3d ed. Redwood City, Calif.: Benjamin/Cummings. Chapter 2 discusses important changes in cultural evolution that have caused a shift in the human- environment interaction.

Gallo, R. C. and L. Montagnier. 1988. AIDS in 1988. *Scientific American* 259(4): 40–48. Illustrates scientific method as it shows how the virus was discovered.

Graham, J. 1989. Sunscreen roulette. *Health* 21(5): 52–57. Evaluates products for sun protection and explains how to assess your own skin type. Helps you practice some of your critical thinking skills.

Hamilton, E. M. N., E. N. Whitney, and F. S. Sizer. 1988. *Nutrition: Concepts and controversies.* St. Paul, Minn.: West. See Controversies sections to exercise your critical thinking skills.

Hardin, G. 1985. *Filters against folly.* New York: Viking. Superb look at some aspects of critical thinking.

Marowitz, H. J. 1989. Roundtable: Models, theory, and the matrix of biological knowledge. *Bioscience* 39(3): 177–79. Looks at the concept of unity within diversity.

McPhee, J. 1989. *The control of nature.* New York: Farrar Straus Giroux. Three case studies showing how humans have attempted to control nature, often with disastrous consequences.

Nontoye, H. J., J. L. Christian, F. J. Nagle, and S. M. Levin. 1988. *Living fit.* Menlo Park, Calif.: Benjamin/Cummings. Highly readable book on health and fitness.

2 Principles of Chemistry

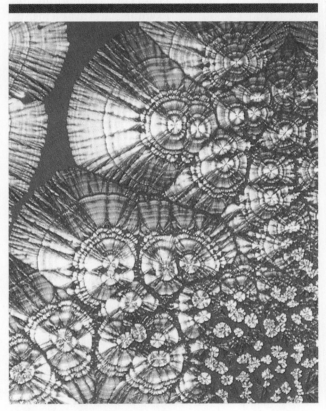

Vitamin C.

Surveys of public knowledge are almost always embarrassing. Survey after survey of geography, for example, show that most Americans are blissfully uninformed about the world around them. If we were to join the ranks of the pollsters and ask questions on chemistry, the responses would display American ingenuity and a sense of humor, but probably little knowledge of the subject. Considering the importance of chemistry to our lives, it is surprising how little most people know about it.

In many ways, the science of chemistry and its practical applications have made our world a better place to live (Figure 2–1). A wide variety of drugs produced in the laboratory, for example, help humanity fight disease. Latex paints, which have replaced toxic lead-based paints, reduce the health hazards to children.

A little knowledge of chemistry is useful to all of us. It can, for example, help us understand the effects of pollution. It can help us remove stains from carpets, and it can help us improve our diets.

There's no question: chemistry is important to our lives. Without a background in chemistry, the study of biology (the study of life) would be as meaningless as the study of medicine without an understanding of the structure and function of the human body. Many

21

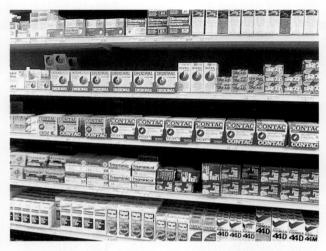

FIGURE 2–1 Better Living Through Chemistry?
Store shelves are packed with drugs to cure everything from warts to ugly facial hair. We are a chemical society, dependent on its benefits and subject to its hazards.

FIGURE 2–2 The Elements Elements are the purest form of matter. They consist of tiny particles called atoms.

biological processes you will study are chemical reactions.

Biologists rely heavily on chemical techniques and frequently study chemical processes in biological systems. Reproductive biologists, for instance, measure hormones in the blood with complex biochemical procedures. Immunologists rely on an arsenal of chemical techniques to better understand how the body protects itself from bacteria and viruses. Neurobiologists probe individual nerve cells to measure internal chemical activity. They also study the chemical substances in the brain that influence behaviors such as depression and alcoholism. Many evolutionary biologists use sophisticated equipment to analyze blood proteins to determine evolutionary relationships among species.

This chapter introduces chemistry and lays the foundation for the study of human biology. It discusses atoms, molecules, and some important principles of chemical reactions. The chapter concludes with an overview of the major biological molecules you will encounter in your study.

ATOMS AND SUBATOMIC PARTICLES

The physical world we live in is composed of matter. In the language of science, **matter** is anything that has mass and occupies space. Matter exists in a multitude of forms. Wood, water, plastic, metal, air, and food, for

example, are all forms of matter. Why are there so many different forms? Part of the answer lies in the fact that matter is made from many different building blocks called elements.

Elements are the purest form of matter. They are substances that cannot be separated into different substances by chemical means. Gold, aluminum, hydrogen, and carbon, for example, are all elements. Science has discovered 92 naturally occurring elements (Figure 2–2). Another dozen or so can be made in the laboratory. Of the 92 naturally occurring elements, only about 20 are found in the human body. Four elements in this group: carbon, oxygen, hydrogen, and nitrogen make up the bulk of all living things. Elements consist of tiny particles of matter invisible to the naked eye known as **atoms**. A bar of gold consists of many gold atoms.

Atoms bond to other atoms, forming a wide variety of different molecules. As Figure 2–3 illustrates, atoms are composed of smaller units, called **subatomic particles**: electrons, protons, and neutrons. Subatomic particles can be separated from an atom by physical methods—for instance, by smashing an atom with a particle traveling at high speed.

Of the three subatomic particles, **protons** are the heaviest. Protons are concentrated in a central, dense region of the atom called the **nucleus** (Figure 2–3). Each proton has a positive charge. Also found in the nucleus of all but one element (hydrogen) are uncharged particles called **neutrons**. They are nearly as heavy as protons. Protons and neutrons are so heavy and so tightly packed in the nucleus that a single cubic centimeter (about ⅓ teaspoon) of protons and neutrons would weigh 100 million tons. Because protons are positively charged, the nucleus has a positive charge.

Surrounding the nucleus is a region called the **electron cloud**. Here small, negatively charged **electrons** orbit the positively charged nucleus at nearly the speed of light. The more energetic electrons tend to orbit the nucleus in the outer regions of the electron cloud (the outer shell), while the less energetic ones remain close to the nucleus in the inner shell. Electrons are held in orbit in large part by the electrostatic attraction between them and the positively charged protons of the nucleus. Since the number of electrons in an atom equals the number of protons, all atoms are electrically neutral.

Carl Sagan, one of America's leading astronomers, once remarked that atoms are mostly space—or electron fluff. If you could enlarge an atom to the size of Mount Everest (29,028 feet), the nucleus would be about the size of a football. The electrons would spin around the nucleus in the vast electron cloud, a space 29,000 feet in diameter.

Elements are listed on a chart called the **periodic table of elements** (Table 2–1; Appendix A). The periodic table lists the vital statistics of the atoms of each element. As illustrated in Table 2–1, each element is represented by a one- or two-letter symbol. Table 2–1

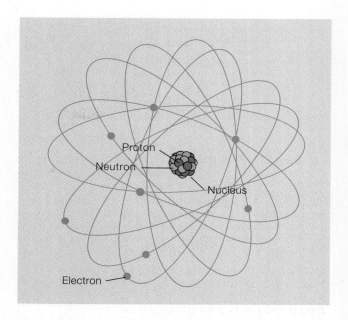

FIGURE 2–3 The Atom The atom consists of two regions. The central nucleus contains protons and neutrons and comprises 99.9% of the mass. Surrounding the nucleus is the electron cloud where the electrons spin furiously around the nucleus. (Figure not drawn to scale.)

TABLE 2–1 A Simplified Periodic Table

I	II	III	IV	V	VI	VII	VIII
1 H Hydrogen 1.0	Atomic number Atomic symbol Atomic weight						2 He Helium 40
3 Li Lithium 7.0	4 Be Beryllium 9.0	5 B Boron 11.0	6 C Carbon 12.0	7 N Nitrogen 14.0	8 O Oxygen 16.0	9 F Fluorine 19.0	10 Ne Neon 20.2
11 Na Sodium 23.0	12 Mg Magnesium	13 Al Aluminum 27.0	14 Si Silicon 28.1	15 P Phosphorus 31.0	16 S Sulfur 32.1	17 Cl Chlorine 35.5	18 Ar Argon 40.0
19 K Potassium 39.1	30 Ca Calcium 40.1						

Note that each element is represented by a one- or two-letter symbol. Elements are listed according to their atomic number (the number of protons). Their atomic mass is also shown.

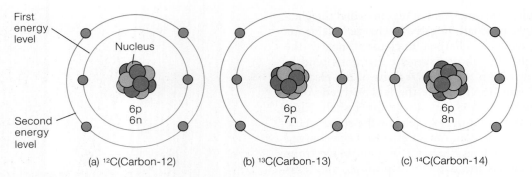

Nucleus

6p
6n

(a) ^{12}C(Carbon-12)

6p
7n

(b) ^{13}C(Carbon-13)

6p
8n

(c) ^{14}C(Carbon-14)

FIGURE 2–4 Isotopes of Carbon (*a*) Carbon-12 is the main isotope of carbon; it constitutes 98.89% of the naturally occurring carbon in the world. (*b*) Carbon-13 forms 1.1% of the world's carbon. (*c*) Carbon-14 is found only in trace amounts. All three isotopes have six electrons; carbon-14 is radioactive, but carbon-12 and carbon-13 are not. These atoms are drawn according to the Bohr model with the nucleus in the center surrounded by concentric circles. Each circle represents an energy level, and all of the electrons within a given energy level are drawn on the circle.

also lists the atomic number of each element. The **atomic number** of an element is equal to the number of protons in the nucleus. Since all atoms are electrically neutral, the atomic number also equals the number of electrons in each atom. The periodic table also lists the **atomic weight**, or the average mass of the atoms of each element.[1] Hydrogen (atomic number 1) with one electron and one proton has an atomic weight of about 1. Helium (atomic number 2) has two protons, two neutrons, and two electrons. Its atomic weight is about 4. Atomic weight is measured in **atomic mass units**. One atomic mass unit is $\frac{1}{12}$ the weight of a carbon atom. Protons and neutrons each have an atomic weight of 1 atomic mass unit. Electrons have almost no mass, and, therefore, barely figure into the atomic weight of an element.

The number of protons in the atoms of any given element is constant. For example, all carbon atoms contain six protons. However, the number of neutrons varies. Most carbon atoms, for example, contain six neutrons. A small percentage of carbon atoms, however, have seven neutrons (Figure 2–4). Another subgroup contains eight. All forms of an atom are known as **isotopes**. Additional neutrons make some nuclei unstable. Therefore, even though the carbon atoms with seven and eight neutrons are still part of the carbon family, they are slightly heavier and less stable. Those with eight neutrons also emit radiation.

Radiation consists of tiny bursts of energy or high-energy particles emitted from the nuclei of radioactive isotopes. Radiation carries energy and mass from an atom's nucleus and is the atom's way of achieving stability.

Many of the benefits of radiation were discovered by scientists at the turn of the century. Since then, modern society has found many additional uses for it in medicine and science, including X-rays and radioactive tracers used in medical tests. The exploitation of radiation, however, has not been without cost, for radiation can damage molecules in cells, leading to changes (mutations) in the genetic material that can result in birth defects and cancer. The Point/Counterpoint examines the pros and cons of a new use of radiation—the irradiation of food.

CHEMICAL BONDS AND MOLECULES

Atoms, by themselves, are rarely encountered in nature because atoms tend to bind to one another, forming molecules. A **molecule** is a chemical composed of two or more atoms.

Ionic and Covalent Bonds

Atoms form two types of chemical bonds: the ionic and the covalent bond. **Ionic bonds** are rather weak links that form between ions. **Ions** are formed when atoms gain or lose electrons from their outermost shells. Figure 2–5 illustrates how an ionic bond forms between a chlorine atom and a sodium atom. As illustrated, the chlorine atom accepts an electron from the sodium atom. As a result, the sodium atom becomes positively charged and is represented by the symbol Na^+. Chlorine, in turn, gains an electron and

---

[1]Atomic weight is approximately equal to the number of protons and neutrons.

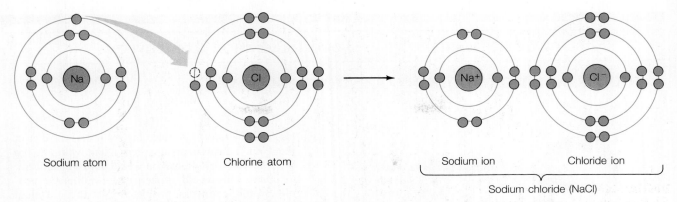

Sodium atom   Chlorine atom   Sodium ion   Chloride ion

Sodium chloride (NaCl)

FIGURE 2–5 **Ions and Ionic Bonds** Sodium and chlorine atoms differ in their electron affinity. Sodium has a weaker affinity and tends to give up its electron to atoms like chlorine. As a result, sodium becomes a positively charged ion, and chlorine becomes a negatively charged ion. The oppositely charged ions attract each other, forming an ionic bond.

is converted to a negatively charged ion, the **chloride ion**, represented by the symbol Cl^-. The oppositely charged ions formed by this electron exchange are attracted to one another, and the electrostatic attraction between them constitutes an ionic bond.

Single molecules of sodium chloride (ordinary table salt) do not exist in nature. Instead, sodium chloride forms crystals containing numerous sodium and chloride ions arranged in a lattice—a three-dimensional structure in which each sodium ion is surrounded by six chloride ions and each chloride ion is surrounded by six sodium ions, as shown in Figure 2–6a. The ions in crystals arrange themselves in such a way that attractive forces between unlike charges are

FIGURE 2–6 **A Sodium Chloride Crystal and Its Dissolution** (*a*) Sodium chloride crystal. In the sodium chloride crystal, each sodium ion is surrounded by six chloride ions and each chloride ion is surrounded by six sodium ions. (*b*) Dissolution of sodium chloride. Water molecules are polar and are attracted to both positively and negatively charged ions. When the sodium chloride crystal is dropped in water, the water molecules pull ions from the salt and tend to push themselves between the positively and negatively charged ions of the crystal, breaking it apart. When this occurs, sodium chloride is said to be dissolved.

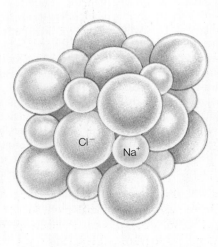

(a) Arrangement of the Na^+ and Cl^- in a sodium chloride crystal.

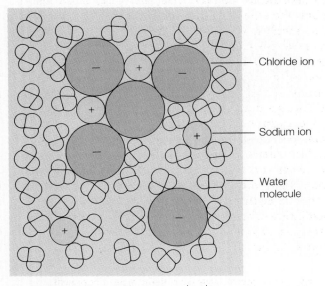

Chloride ion

Sodium ion

Water molecule

(b) Breakup of salt crystal by water molecules.

FOOD IRRADIATION: TOO MANY QUESTIONS TO KNOW

Donald B. Louria

Donald B. Louria, M.D., teaches at the University of Medicine and Dentistry of New Jersey—New Jersey Medical School where he is the chairman of the Department of Preventive Medicine and Community Health.

The debate about the safety of irradiating foods with large doses of cobalt 60 or cesium 137 raises five salient issues:

1. **The safety issue.** Irradiated food does not become radioactive, but the radiation causes chemical changes. The major concerns is over possible deleterious effects on the elderly and on malnourished persons. Some animal experiments raise concerns about the former. A highly controversial, but flawed study on small numbers of children in India suggested that freshly irradiated wheat could produce chromosomal abnormalities. A study of a much larger group of healthy adults showed no such abnormalities. The proponents of food irradiation have derided the Indian study and pointed to the larger, negative study. But that does not answer the concerns. If food irradiation becomes an accepted technique, millions of malnourished people will eat irradiated foods. What is needed is a careful study with adequate numbers of malnourished children and adults. I am prepared to believe that consumption of irradiated foods is unlikely to produce significant harm, but such a belief must be supported by proper data.

2. **The nutrition issue.** Irradiated food loses some of its nutritional value; the extent of loss of vitamin content depends on the type of food and the dose of radiation—the higher the radiation dose, the greater the loss. Furthermore, some evidence suggests that when irradiated foods are processed (frozen, thawed, heated), there is an accelerated loss of vitamins. The irradiation proponents suggest that the effects of irradiation on vitamin content are not different from the effects of processing foods in conventional ways (heating, freezing, and the like). They also maintain that the diet in the United States contains redundant vitamins, so some loss through irradiation would not be of concern. Of course, this would not be true for millions of people in other countries, people over 60 in the United States, and people with various diseases. Arguing that our diet provides adequate vitamins and that destroying the vitamin content of foods by irradiation is of no importance is not likely to be viewed with equanimity by a large proportion of the American public.

3. **The necessity of this technology.** Proponents say that food irradiation will reduce diarrheal illness from infected poultry. However, proper cooking of the poultry would be just as effective. Proponents say irradiating meats will reduce the dangers of trichinosis, a bacterial disease, but trichinosis occurs infrequently. Will irradiating foods prolong their shelf lives and thus feed the world? Shelf lives will be increased: that is an important benefit for the less-developed world, but the proponents have provided no adequate data on the extent of the benefit. It is likely that the foods will be sold primarily in affluent countries where the shelf life issue is of less concern.

4. **The issue of safer competing technologies.** Food irradiation could reduce the use of toxic post-harvest fumigants, chemicals applied to foods in storage. During the next decades, scientist will develop food crops that resist pests, grains that do not require fumigants, and foods that have longer shelf lives. Biotechnology advances are likely to give us much safer alternatives to food irradiation, if we will only have a little patience.

5. **The environmental pollution issue.** It seems a bit incongruous that as our society is becoming increasingly concerned about our inability to get rid of nuclear wastes, we appear to be endorsing food irradiation, a technology that will result in numerous food irradiation plants in the United States. The few irradiation plants operating in the United States have contaminated their workers and the environment. Imagine the potential for contamination if there were hundreds of such plants using radioactive materials.

Food irradiation does have potential benefits: it also raises substantial concerns. Whether we should adopt the technology is obviously a matter for continuing debate: certainly the technology should not be adopted until the issues have been resolved.

FOOD IRRADIATION: SAFE AND SOUND

George G. Giddings

George G. Giddings, Ph.D., had been involved with food irradiation research and development, for the past 28 years since earning a B.Sc. in Food Science and Technology in 1963. In recent years, he has written and spoken extensively on the subject.

Irradiated foods are safe and wholesome. The ionizing radiation process applied to foods offers certain proven public health and economic benefits without significant public health risks when carried out according to well-established principles and procedures. Decades of worldwide research and testing by competent, knowledgeable, objective, and responsible scientists, led to this conclusion. This conclusion is also supported by some 30 years of experience in the sterilization of medical devices and other health care products to prevent infections, plus a growing list of other industrial and consumer products, including foods and their raw materials, ingredients, and packaging materials.

There is organized political opposition to food irradiation by a network of special interest activists serving various political agendas, notably the antinuclear/antiradiation one. This network includes a handful of scientists and medical professionals from other fields who act as "expert witnesses" against food irradiation to serve their 'hidden' agendas. Despite some short-term political successes, their campaign is doomed to failure in the longer term in the face of the unshakable facts, including a growing appreciation for public health and other proven benefits of food irradiation, and its growing worldwide regulatory approval, industrial usage, and public acceptance.

Ionizing radiation processing is undoubtedly the most versatile physical process yet applied to food materials in terms of the range and variety of objectives it can accomplish.

Radiation can:

Inhibit the sprouting of foods, such as potatoes and onions, and delay spoilage.

Rid fruits, vegetables, and grains of insect pests.

Prevent parasites from consuming fish and meats.

Delay microbial spoilage of a wide variety of animal and plant products by reducing microbe levels.

Sterilize food and nonfood packaging materials before filling and sealing, eliminating microorganisms that would otherwise contaminate products.

Sterilize a wide variety of prepared or cooked foods such as meat, poultry, and fishery products. These have already been used to feed astronauts, cosmonauts, and immune-compromised patients.

All of these beneficial effects and more can be readily accomplished by the application of ionizing (gamma, electron, and X-ray) radiation according to well-established principles and procedures. Further, all of these benefits are already being realized in one country after another as food irradiation gradually increases worldwide. Nevertheless, irradiation must compete with a number of other new technologies. As a result, it is not likely to be applied to all, or even a high percentage, of the national and world food supply. It will therefore be used in cases in which it is clearly the best all-around choice.

SHARPENING YOUR CRITICAL THINKING SKILLS

1. What is food irradiation? Why is it used?
2. List and summarize the key points of each author.
3. Using your critical thinking skills analyze each author's position. Are you inclined to agree with either one on all issues? Why or why not?
4. In your opinion, what is the best course for the development and implementation of this technology?

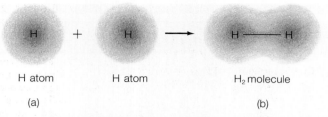

H atom + H atom → H₂ molecule

(a)   (b)

FIGURE 2−7   Covalent Bond   Like a good marriage or friendship, most covalent bonds are characterized by more or less equal sharing of electrons. (*a*) An artist's rendition of two separate hydrogen atoms. The shaded regions represent the atoms' electron clouds. (*b*) Two atoms bound together by a covalent bond. The electrons spin around both nuclei, holding them together to create a molecule.

maximized and repulsive forces between like charges are minimized. Consequently, discrete molecules are not present. In water, sodium chloride crystals break apart, releasing sodium and chloride ions into solution (Figure 2−6b). In solution, the sodium and chloride ions exist independently.

Ionic bonds form between atoms with markedly different affinities for electrons. The bonds form because one of the atoms gives up its outer shell electron(s) to another atom. In contrast, a **covalent bond** occurs when there is a sharing of outer shell electrons between atoms. The "sharing" of electrons results in a stronger bond.

Covalent bonds form between two atoms of approximately equal electron affinity. Consider an example. Two hydrogen atoms, each with one electron, bond to each other by sharing their electrons thus forming a molecule of hydrogen gas, H_2 (Figure 2−7). The electrons in the newly formed molecule circle around both nuclei, holding them together.

The union of two atoms by a covalent bond results in a stable configuration because covalent bond formation allows atoms to fill their outermost shells. When an atom's outermost shell is filled, the atom is in the most stable state possible. Consider a few examples. The carbon atom has four outer shell electrons. A full outer shell would contain eight electrons. Carbon can fill its outer shell by covalently binding to four hydrogen atoms, each with a single electron. When carbon binds to four hydrogen atoms, it forms a molecule of methane whose chemical formula is CH_4. Each of the four hydrogen atoms shares one electron with the carbon atom, giving the carbon a full outer shell. But what about the hydrogen atoms? Are they "satisfied" by this arrangement? The answer is yes. Unlike carbon, the outermost shell of a hydro-

gen atom can contain only two electrons. In methane, each of the hydrogen atoms shares an electron with the carbon atom. Thus, each of the four hydrogen atoms reaches a stable state.

The atoms of nitrogen, another biologically important element, share three electrons. Thus, nitrogen can combine with three hydrogen atoms, forming NH_3, ammonia. Oxygen, which can share two electrons, can combine with hydrogen to form H_2O, water.

Some atoms can share two or three electrons with another atom, forming double and triple covalent bonds, respectively (Figure 2−8). Oxygen, for example, can share two electrons with a carbon atom, forming a double covalent bond (C=O). In this bond, two electrons from the carbon atom orbit around the oxygen atom—in other words, two electrons of the carbon atom are shared with the oxygen atom. This fills the oxygen atom's outer shell. Oxygen also shares two electrons with carbon, helping to fill its outermost shell. The four shared electrons bind the two atoms together and constitute a double covalent bond. Triple covalent bonds result when two atoms share three pairs of electrons (Figure 2−8b).

Not all covalent bonds result in 50−50 sharing. When two atoms with similar but slightly different affinities unite by covalent bonds, the electrons are shared somewhat unequally. For example, oxygen and hydrogen atoms have similar electron affinities, although oxygen's affinity is slightly higher. If oxygen and hydrogen join to form a single covalent bond, the electrons tend to orbit around oxygen more than hydrogen. Because the electron "shared" by hydrogen in this covalent bond tends to spend more time circling around the oxygen atom, the oxygen atom will have a slightly negative charge. The hydrogen atom is visited less frequently by the electron and bears a slightly

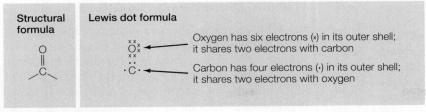

(a) Double covalent bond

FIGURE 2–8 Double and Triple Covalent Bonds (*a*) When atoms share two pairs of electrons, a double covalent bond is formed. Chemists have devised several ways to draw the bonds. In the formula on the left, each line represents a pair of shared electrons. On the right electrons are indicated by xs and dots. (*b*) On rare occasions, atoms share three pairs of electrons, creating triple covalent bonds.

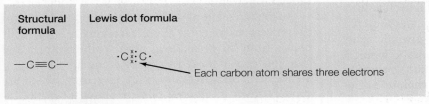

(b) Triple covalent bond

positive charge (Figure 2–9a). The result is a **polar covalent bond**, a covalent bond between two slightly charged atoms.

The presence of polar covalent bonds sometimes makes molecules polar. A polar molecule is one that contains slightly charged atoms. One of the best examples of a polar molecule is water. As shown in Figure 2–9, water molecules are attracted to one another because of the charged atoms they contain. This attraction gives water many of its characteristic features (discussed later).

Hydrogen Bonds

Ionic and covalent bonds form between atoms, creating molecules or crystals. A third type of bond, called the hydrogen bond, can also be found. The **hydrogen bond** is a relatively weak electrostatic attraction that occurs between polar molecules, such as water. As Figure 2–9 illustrates, the hydrogen bond forms between the slightly positive hydrogen atom of one water molecule and the slightly negative oxygen atom in another molecule. Unlike the ionic and covalent bonds that link atoms together to form molecules or crystals, hydrogen bonds form *between* the atoms of molecules or between atoms located at different parts of very large molecules, such as proteins.

Hydrogen bonds also form between hydrogen atoms and nitrogen atoms on different molecules. For instance, they are found in DNA, the hereditary molecule, where slightly negative nitrogen atoms attract slightly positive hydrogen atoms, helping to hold this complex molecule in a double-stranded spiral. Hydrogen bonds are also found in proteins where they help determine and maintain the three-dimensional shape.

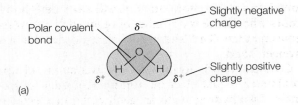

(a)

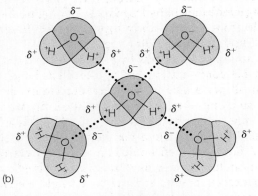

(b)

FIGURE 2–9 Hydrogen Bond (*a*) Slightly unequal sharing of electrons in the water molecule creates a polar molecule. Electrons tend to spend more time around the oxygen nucleus, making it slightly negative. The hydrogen atoms are, therefore, slightly positive. (*b*) Because of the polarity, hydrogen atoms from one molecule are attracted to oxygen atoms of another. The attraction is called a hydrogen bond.

CHEMICAL REACTIONS

Chemical substances form the structures of the human body—our bones, muscles, hearts, brains, and the billions of cells that comprise the body. Many chemicals in the cells and tissues of our bodies also participate in reactions that are essential for growth, reproduction, homeostasis, and development. These metabolic reactions fall into two major categories: (1) synthetic or **anabolic reactions**, in which chemicals combine with others to form new molecules; and (2) **catabolic reactions**, in which chemicals are split apart or dissociate. In anabolic reactions, new chemical bonds are formed; in catabolic reactions, existing chemical bonds are broken. All of the chemical reactions in the body constitute **metabolism**.

FIGURE 2–10   Ben Johnson, Canada's Premier Track Star   Officials at the 1988 Olympics in Seoul, South Korea, found traces of an anabolic steroid in Johnson's blood. Anabolic steroids stimulate protein synthesis in muscle tissue, making for an extraordinary physique and astounding athletic performance. Steroids are banned for use by Olympic athletes and may have adverse psychological and physiological side effects.

Anabolic and Catabolic Reactions

Ben Johnson, Canada's extraordinary track star (Figure 2–10) was stripped of a gold medal in the 1988 Summer Olympics because he allegedly took anabolic steroids—hormones that help build muscle protein, which makes for exceptional performance (Chapter 14). Protein synthesis is one of many anabolic reactions you will encounter in your study of human biology.

Anabolic reactions all require energy, and most of the energy needed for these reactions in your cells comes from the breakdown of the carbohydrates and fats you consume. A runner in a marathon, for instance, eats carbohydrates, such as starch, which are broken down to glucose (a type of sugar) in the intestinal tract, then absorbed by the blood. Glucose molecules are absorbed by muscle cells where they are broken down in a series of reactions that release energy.

Where does that energy in glucose come from? The energy in glucose and other food molecules comes from the sun via plants. Wheat, for example, produces carbohydrates, amino acids, and other food molecules from carbon dioxide (CO_2) in the air and water from the soil in a process called **photosynthesis**. During photosynthesis, plants capture sunlight energy and use it to synthesize organic molecules. Sunlight energy ultimately is used to form the covalent bonds joining the atoms of these molecules. Wheat and other grains are used to produce cereals and other food products. Thus, the cereal you eat for breakfast contains food molecules produced by plants from the sun's energy. The energy stored in the chemical bonds of these molecules is released by your cells when they catabolize carbohydrates (and other molecules) produced by plants.

Virtually all life on Earth is powered by the sun.[2] Even the energy that powers industrial societies was once sunlight. Take coal, for example (Figure 2–11). Coal was produced from plants that lived 200 to 350 million years ago. When these plants died, they were buried by sediments, then slowly converted to coal. The energy released from the burning of coal is sunlight energy trapped by plants several hundred million years earlier. Oil and natural gas have similar origins.

[2]A small number of species of bacteria can derive the energy they need from molecules such as hydrogen sulfide and methane. These chemosynthetic organisms are rare, however.

FIGURE 2–11 Coal This rich deposit of coal was produced from plant matter that grew here several hundred million years ago. The energy of coal is ultimately derived from sunlight.

Coupled Reactions

In the cells of your body, energy that is released from one reaction is often used to drive other energy-requiring reactions. These reactions are therefore said to be **coupled** (Figure 2–12). The energy released by the breakdown of glucose in muscle cells, for instance, is used in the synthesis of a molecule called **adenosine triphosphate (ATP)**. ATP stores the energy it absorbs for later use—for example, to power muscle contraction, protein and carbohydrate synthesis, plasma membrane transport, and many other processes. (ATP is discussed in more detail at the end of this chapter and in Chapter 3.) During coupled reactions, the transfer of energy is never 100% efficient. Some energy is lost as heat.

Chemical Equilibria

Most chemical reactions are a two-way street—traffic proceeds in both directions. In the chemist's parlance, chemical reactions are said to be reversible. In the blood, carbon dioxide released from cells combines with water to form a weak acid, called carbonic acid:

$$H_2O \quad + \quad CO_2 \quad \rightleftarrows \quad H_2CO_3$$
water carbon dioxide carbonic acid

The double arrows indicate that this reaction proceeds in both directions. By convention, the chemicals on the left are the reactants. The chemical on the right, which is formed in the reaction, is called the product.

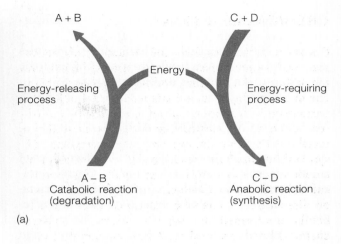

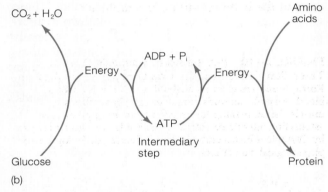

FIGURE 2–12 Coupled Reactions (*a*) In the body's cells, energy-releasing reactions are coupled to energy-requiring reactions. This close association provides energy to drive many important chemical processes. (*b*) An intermediary molecule absorbs energy and transfers it to regions of energy demand, where it liberates the energy. The transfer of energy is never 100% efficient.

This chemical reaction, like all others, reaches a **chemical equilibrium**, a point at which the speed of the forward reaction is equal to that of the reverse reaction. Because the forward and reverse reactions occur at the same rate, the concentration of chemicals on both sides of the reaction remains constant. It is important to note, however, that chemical equilibrium does not mean that the concentration of the products and the reactants is the same, only that the rates of the forward and reverse reactions are equal.

Chemical equilibria can be upset by changes in the concentration of molecules on either side of the reaction. For instance, additional carbon dioxide entering the blood from your muscle cells during vigorous exercise drives the reaction to the right, causing more carbonic acid to form. Other factors, such as temperature, may also affect chemical equilibria.

In the body, chemical equilibria drive molecules across membranes. As a result, they may prevent toxic substances from building up inside cells.

WATER, ACIDS, BASES, AND BUFFERS

Atoms bond together to form molecules. Two types of molecules exist: organic and inorganic. **Organic molecules** are best described as compounds made primarily of carbon and hydrogen and are sometimes referred to as the compounds of carbon. The atoms of organic molecules are joined by covalent bonds, and many organic compounds are large molecules. Those compounds that are not organic are known as **inorganic molecules**. Inorganic molecules are generally smaller than organic molecules and consist of atoms that are frequently joined by ionic bonds. Sodium chloride and magnesium chloride are examples. Not all inorganic compounds are formed by ionic bonds,

however. Water, for instance, is an inorganic compound whose atoms are joined by covalent bonds.

Water, Life, and the Hydrogen Bond

Water performs several key roles in human biology. First, water is a major component of all cells, not just in humans, but in all organisms. Nearly two-thirds of the human body is water. If you weigh 100 pounds, nearly 70 pounds of your body weight is water. Water is also an important solvent in the human body. A **solvent** is a fluid that dissolves other chemical substances. Blood, for example, is about 55% water. The water in our blood dissolves and transports nutrients, hormones, and wastes throughout the body. Water also bathes our cells, providing a route by which wastes and nutrients move between the cells and the blood. Water also participates in many chemical reactions—for example, the breakdown of protein. Water also serves as a lubricant. Saliva, which is largely water, lubricates food in the mouth and esophagus of animals, easing its passage to the stomach. A watery fluid in the joints called **synovial fluid** enables bones to slide over one another, thus facilitating body motion. Finally, water helps us regulate body temperature. Perspiration, for example, rids the body of heat (Figure 2–13).

The Dissociation of Water

Although water molecules are stable, they can dissociate, forming two oppositely charged ions:

$$H_2O \rightleftarrows H^+ + OH^-$$

water   hydrogen ion   hydroxide ion

Hydrogen ions can react with the hydroxide ions, reforming water molecules. Although water dissociates freely, the ratio of water molecules to H^+ and OH^- in the human body is about 500 million to one. Nevertheless, even the slightest changes in the hydrogen ion concentration can alter a biological system, shutting down biochemical pathways and sometimes killing organisms.

Acids, Bases, and Buffers

In pure water, the concentrations of H^+ and OH^- are equal. Thus, a solution of pure water is said to be **neutral**. Chemical neutrality can be upset by adding or removing H^+ and OH^-. Hydrochloric acid (HCl), for instance, dissociates into hydrogen ions (H^+) and

FIGURE 2–13 Perspiration Cools the Body
Perspiration is an automatic reaction that helps rid the body of heat.

chloride ions (Cl^-) when added to water, thus increasing the hydrogen ion concentration. A substance that adds hydrogen ions to a solution of water is an **acid**. Solutions with proportionately more H^+ than OH^- are said to be acidic.

In contrast, **bases** are substances that release hydroxide ions upon dissociation. Sodium hydroxide, for example, dissociates in water forming Na^+ and OH^-. The OH^- reacts with H^+ in solution to form water molecules. Bases, therefore, reduce the concentration of H^+ in solution.

Acidity is measured on the **pH scale** (Figure 2–14). On the pH scale, neutral substances are assigned a pH of 7. Those substances that have a pH greater than 7 are basic; and those with a pH less than 7 are acidic. Thus, the lower the pH, the greater the acidity.

The pH scale ranges from 0 to 14 and is designed to allow scientists to condense a wide range of values into a relatively small scale. Don't be fooled by a small change in pH, however, for a difference of one pH unit represents a tenfold change in acidity. A solution with a pH of 3, for example, is 10 times more acidic than one with a pH of 4 and 100 times more acidic than one with a pH of 5.

Most biochemical reactions occur at pH values between 6 and 8. The blood, for instance, has a pH of 7.4. A slight shift in pH, say, to 7.8, for even a short period can be fatal. The only place in the body where high acidity (pH 1.5–3.5) is tolerated and beneficial is in the stomach. Hydrochloric acid secreted by cells lining the stomach coagulates protein and activates an enzyme that helps break down protein molecules (Chapter 7).

Acids can have profound impacts on organisms and their environment. In the eastern United States, sulfuric acid draining into lakes and streams from abandoned coal mines has disrupted aquatic systems, killing fish and other organisms (Figure 2–15). Acids are also produced in the atmosphere from gaseous pollutants given off by automobiles, power plants, and factories. These acids fall in rain and snow and can increase the acidity in lakes and streams, killing fish and other aquatic organisms (Chapter 21). When the pH level of a lake or stream falls too low (usually below 4.0–4.5), virtually all life disappears.

As a rule, biological systems operate within a narrow pH range, maintained by a simple, but important homeostatic mechanism, chemical buffers. **Buffers** are substances that protect against changes in pH; they help maintain a constant pH by removing hydrogen ions from solution when levels increase or by adding

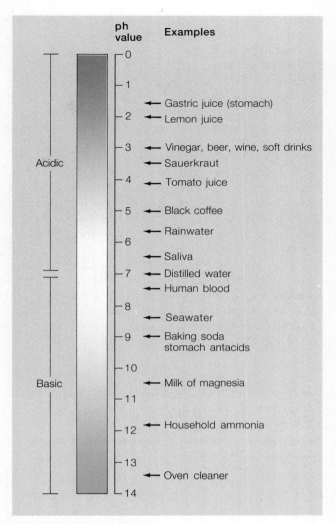

FIGURE 2-14   pH Scale

hydrogen ions when levels fall. One of the most important buffers in the blood is **carbonic acid**, H_2CO_3, which is formed from water and the cellular waste product, carbon dioxide. In an aqueous solution, carbonic acid dissociates into bicarbonate and hydrogen ions:

$$H_2CO_3 \leftrightharpoons H^+ + HCO_3^-$$

carbonic acid   hydrogen ion   bicarbonate
(weak acid)   (weak base)

This reaction shifts back and forth in response to changing levels of hydrogen ions. For example, when hydrogen ions are added to the blood, they combine with bicarbonate driving the reaction to the left. When hydrogen ion concentrations fall, however, the reaction is driven to the right, keeping the pH constant.

FIGURE 2–15 **Acid Mine Drainage** More than 7000 miles of U.S. streams, mostly in Appalachia, are contaminated by abandoned coal mines that leak sulfuric acid.

Molecule	Formula	Uses
Methane	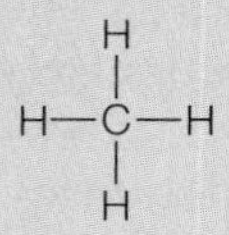 H—C—H with H above and H below	Component of natural gas; burned to generate heat and electricity
Acetylene	H—C≡C—H	Mixed with oxygen and burned to create high-temperature flame for welding
Ethanol	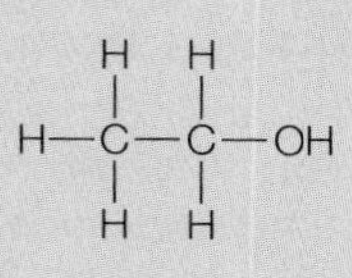 H—C—C—OH (with H's)	Active ingredient of beer, wine, and hard liquors
Amino acid	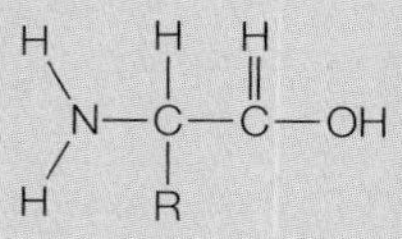 N—C—C—OH (with H, R)	Builds proteins; an essential nutrient
Glucose	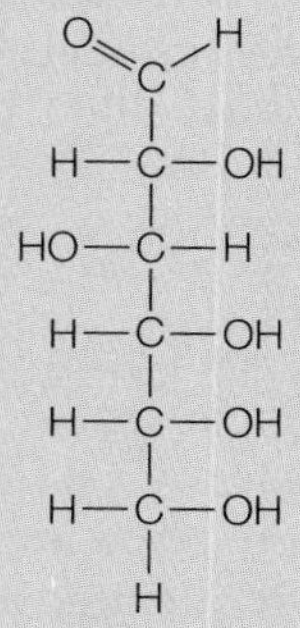 (chain structure)	Chief source of energy in humans; links with many others to form glycogen, a storage form of glucose in muscle and liver cells

FIGURE 2–16 **Carbon's Bonding Options Create Molecular Diversity** Carbon can covalently bond to many other atoms, creating a diverse array of small organic molecules. Some of these molecules, in turn, can bond together to form large chains or polymers.

Many lakes and streams in the United States contain buffers that protect them from acid rain and snow. Others, such as those in high mountain regions, have little buffering capacity. These lakes are, therefore, highly vulnerable to acids in rain and snow (for more on acid deposition, see Chapter 21).

MORE BIOLOGICALLY IMPORTANT MOLECULES

The diversity of organic matter results in large part from the carbon atom. As noted earlier, carbon has four electrons in its outermost shell that it can share with other atoms, including oxygen, hydrogen, nitrogen, and even other carbon atoms. The carbon atom can enter into single, double, and triple covalent bonds as well (Figure 2–16).

Further adding to molecular diversity, many smaller organic molecules, such as amino acids and sugars, can combine to form large organic molecules, called **polymers**. The individual molecules comprising a polymer are called **subunits** or **monomers**. Starch, for example, is a polymer made from hundreds of glucose molecules. Protein is a polymer of amino acid molecules.

The organic molecules found in organisms fit into four classes: (1) carbohydrates, (2) lipids, (3) amino acids and proteins, and (4) nucleic acids. The following sections discuss some of the main features of these biologically important organic molecules.

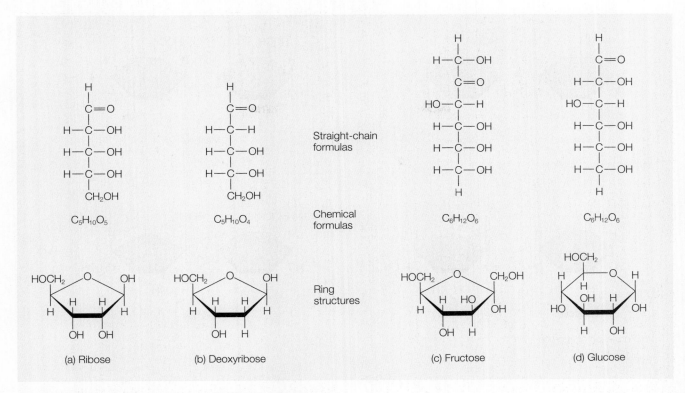

FIGURE 2–17 **Four Monosaccharides** (a) Ribose.
(b) Deoxyribose. (c) Fructose. (d) Glucose.

Carbohydrates

Most organic material consists of carbohydrate. **Carbohydrates** are a group of organic compounds, such as glucose, starch, glycogen, and cellulose, all of which have the general formula $(CH_2O)_n$—hence the name "hydrates of carbon" or carbohydrates.[3] Carbohydrates provide the energy needed by a human at rest and at work. They can also be broken down, forming smaller molecules that may be used to synthesize protein and fats (Chapter 7). Eating too many sweets and starchy foods (rice, potatoes, and bread), therefore, can cause a person to gain weight. Carbohydrates fall into three groups: monosaccharides, disaccharides, and polysaccharides.

Monosaccharides. Monosaccharides are often called simple sugars. These water-soluble molecules usually contain three to seven carbons. Many monosaccha-

rides form ring structures when dissolved in water, as illustrated in Figure 2–17. Many monosaccharides can unite to form disaccharides and polysaccharides (Figure 2–18).

One of the most common monosaccharides is glucose, an important biological molecule. Glucose is a six-carbon sugar whose structure is shown in Figure 2–17. Made by plants during photosynthesis, glucose molecules can be joined to form starch. Starch is stored in the roots (potatoes) and the seeds (wheat and rice) of plants and is a major nutrient for humans and other animals. In the human digestive tract, starch is broken down into glucose molecules, which enter the bloodstream and are distributed to body cells where they are either stored or broken down to release energy.

Disaccharides. Disaccharides are carbohydrates that consist of two monosaccharides covalently bonded to each other. Thus, as Figure 2–18b illustrates, the two monosaccharides glucose and fructose combine to form the disaccharide **sucrose** (table

---

[3]The word *hydrates* refers to the water in the formula. Hydrogens and oxygens do not exist as hydrates (H_2O molecules).

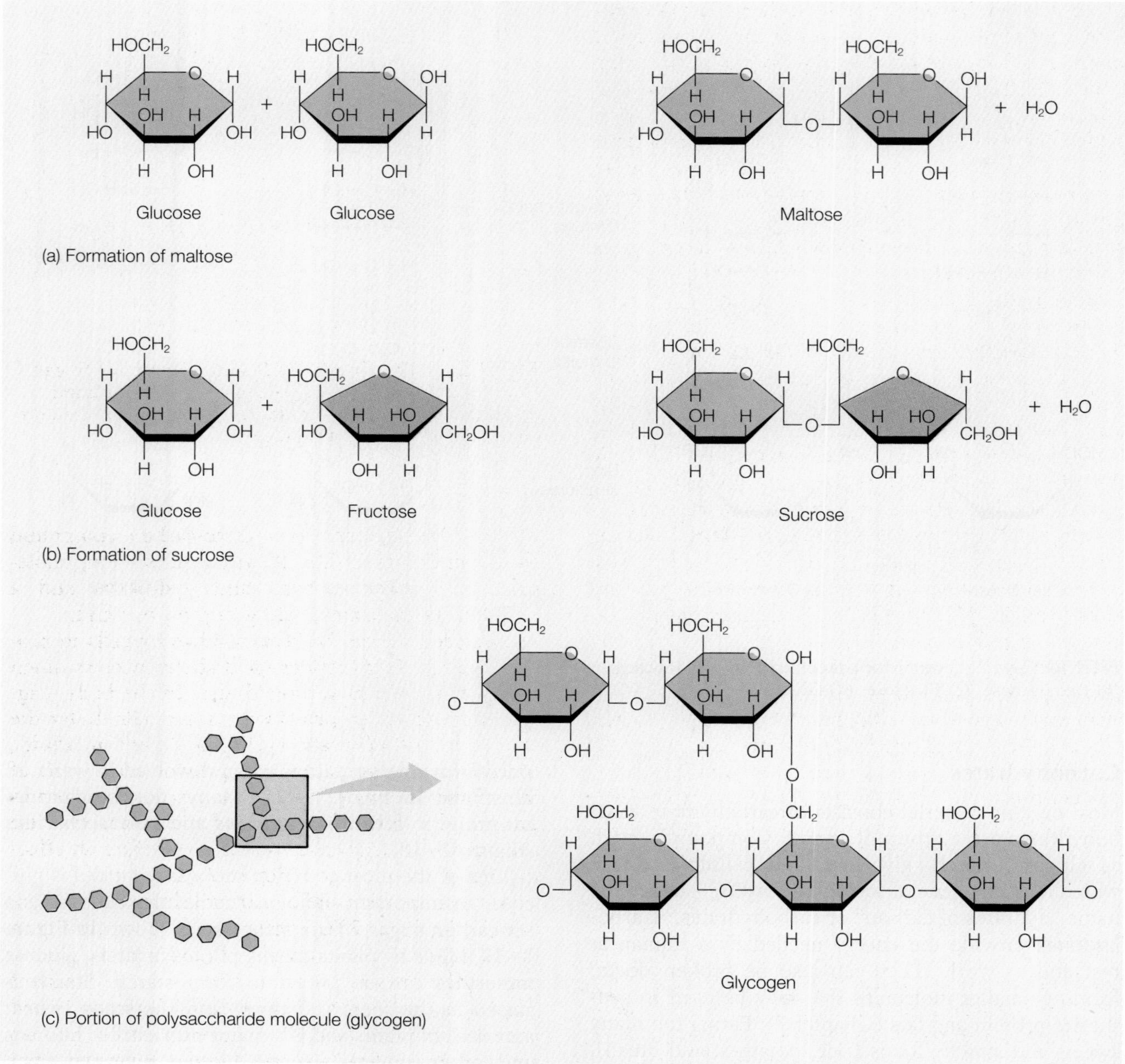

FIGURE 2–18 Disaccharides and Polysaccharides
Simple sugars like glucose can combine to form disaccharides, such as (*a*) maltose and (*b*) sucrose, and polysaccharides, like (*c*) glycogen.

sugar). Sucrose is produced in the leaves of plants and stored in fruits. It is also transported down the tree trunk to nourish cells in the roots. In the fall, for example, trees transport large amounts of sucrose to their roots for use over the long winter. In the early spring, the sugar-rich sap runs upward, providing nourishment to growing buds. New Englanders have long tapped the sugary fluid of maples to obtain the raw material for maple syrup (Figure 2–19).

Polysaccharides. In humans and plants, most monosaccharides react to form long polymers, called **polysaccharides**. The most common building block of polysaccharides is glucose. Plants, for example, use glucose to synthesize two polysaccharides: (1) **cellulose**, one of the materials found in the walls around plant cells that gives wood its rigidity and protects the delicate cells of leaves and other plant tissues, and (2) **starch**, a molecule that stores these energy-rich molecules for later use. Glucose is an important source of energy for humans as noted earlier. Some glucose we ingest is stored in muscle and liver cells as **glycogen**, a polymer of glucose (Figure 2–18c). The stored glycogen is broken down into glucose molecules when needed.

Starch and glycogen are a major source of glucose for humans, but cellulose, though prevalent in our diet, cannot be digested by humans for two reasons: First, the covalent bonds that hold the glucose molecules to one another in cellulose are different from those found in glycogen and starch. Second, humans lack the enzyme needed to break the bonds that hold the glucose molecules together. Consequently, cellulose passes through the digestive system unchanged. Nondigestible cellulose is a principal form of dietary fiber that creates bulk in the diet. Fiber, discussed in more depth in Chapter 7, facilitates the passage of feces through the large intestine and may reduce colon cancer and cholesterol levels.

Not all species are incapable of digesting cellulose. Cattle, sheep, and other grazers, for example, eat plant material rich in cellulose. Although they do not produce the enzyme necessary to digest this material, their digestive tracts contain bacteria with the necessary enzymes. The breakdown of cellulose by bacteria in the digestive tracts of these animals liberates glucose molecules, which can be absorbed by the intestine and used by the animal.

Lipids

Lipids are a diverse group of organic molecules that have one characteristic in common—they are insoluble in water but soluble in organic solvents, such as benzene. In chemistry, like substances dissolve one another. Nonpolar lipids, therefore, do not dissolve in water, which is a polar compound. Biologically important lipids fall into three main categories: (1) triglycerides, (2) phospholipids, and (3) steroids.

Triglycerides. Triglycerides are known to most of us as fats and oils. Cooking oil, for example, is a triglyceride, as are butter and the fat on a steak. Body fat is mostly triglyceride.

Triglycerides are made from four molecules: a molecule of **glycerol** and three **fatty acid** molecules (Fig-

FIGURE 2–19   The Early Spring Harvest of Sugar New Englanders tap the sugary sap of a sugar maple and boil it to make maple syrup. Tubes driven into the tree draw off the sap without harming the trees.

ure 2–20a). Glycerol is a three-carbon compound shown in Figure 2–20a. A fatty acid is a long molecule with many carbons and hydrogens and a −COOH, a carboxylic acid group on one end.

Triglycerides can be broken down by cells to produce energy. Triglycerides yield about twice as much energy per gram as carbohydrates. In the body, triglycerides are stored principally in fat cells under the skin and in other locations (Figure 2–21). In adults, triglycerides from fat reserves provide about half of the cellular energy consumed at rest. Carbohydrates provide the remainder. Moderate (aerobic) exercise also consumes triglycerides and is therefore an effective way to lose weight (Figure 2–22).

At room temperature fats are solid whereas oils are liquid (Figure 2–23). The reason for this difference lies in their chemical makeup. In fats, the carbon atoms of the fatty acids are joined by single covalent bonds (Figure 2–20b). The resulting fatty acid is said to be **saturated** because all of the remaining bond sites are taken up by—or "saturated" with—hydrogens. Because the bonds are saturated, the side chains of the fatty acid molecules are fairly straight; this allows the molecules to pack together. In contrast, the fatty acids in oils contain a number of double covalent bonds and are said to be **unsaturated** (Figure 2–20c). The presence of the double bonds causes the molecules to bend, which prevents tight packing. The looser arrangement of molecules results in a liquid at room temperature.

When numerous double bonds exist, fatty acids are said to be **polyunsaturated**. Studies show that a diet

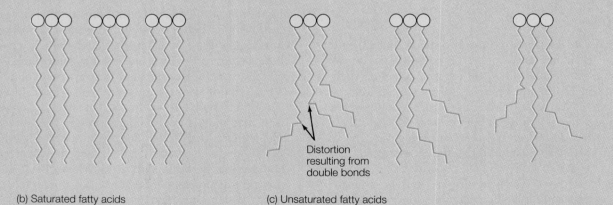

(a) Formation of a triglyceride

Fatty acid

Glycerol

+ 2 Fatty acids

Neutral fat or triglyceride

(b) Saturated fatty acids

(c) Unsaturated fatty acids

Distortion resulting from double bonds

FIGURE 2–20 Triglycerides (left) The triglycerides are the fats and oils. (*a*) Triglycerides consist of glycerol and three fatty acids, covalently bonded as shown. (*b*) Saturated fatty acids are principally derived from animal fats. The side chains lie flat and allow the molecules to pack tightly together so that fats are solid at room temperature. (*c*) Double bonds in unsaturated fatty acids in oils cause the fatty acid chains to bend and prohibit tight packing. Oils, derived chiefly from plants, are therefore liquid at room temperature.

FIGURE 2–21 Fat Cells (*a*) A light micrograph of fat tissue. The clear areas are regions where the fat has dissolved during tissue preparation. Notice that the cytoplasm is reduced to a narrow region just beneath the plasma membrane. (*b*) A scanning electron micrograph of a fat cell.

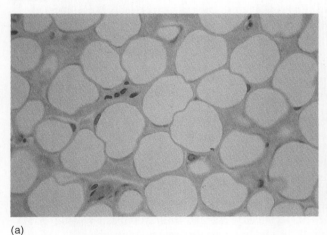

(a)

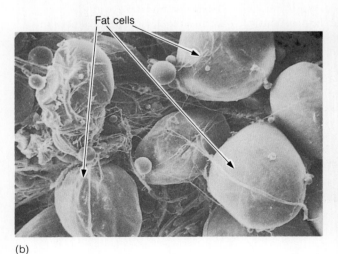

Fat cells

(b)

rich in saturated fat (animal fats) increases the risk of heart disease, resulting from a disease called **atherosclerosis**. It results from the build up of cholesterol deposits on the walls of arteries called **atherosclerotic plaque** (Figure 2–24). Atherosclerosis is associated with three major risk factors: consumption of excessive fat, lack of exercise, and smoking. Plaque can restrict blood flow to the heart and other vital organs, leading to heart attacks and strokes. For reasons not yet well understood, lowering the level of saturated fat in the diet reduces the concentration of cholesterol in the blood. Thus, by replacing butter (high in saturated fat) with margarine (high in unsaturated fat), or using vegetable oil instead of animal fat to make french fries, individuals can reduce their blood cholesterol levels and reduce their risk of heart attack and other problems (For more on cholesterol, see the discussion of steroids below and Health Note 7–1).

FIGURE 2–22 Aerobic Exercise Besides being good for your heart, aerobic exercise can burn off calories and help you lose weight.

HEALTH NOTE 2-1
New Treatments for Cancer

Cancer is a deadly disease that results from the uncontrolled division of body cells, producing a rapidly growing mass of cells or tumor. Cancer cells often leave the site of origin, invading other body tissues and organs where they form secondary tumors.

One of every three Americans will contract cancer in his lifetime, and one of every four will die from it. For most patients, tumors are removed surgically. This procedure is followed by radiation treatments or chemotherapy. Radiation kills remaining cancer cells at the tumor site and can also kill cancer cells in secondary tumors. Radiation can also be used to destroy tumors in hard-to-reach locations, as can chemotherapy, the administration of chemical agents that kill rapidly dividing cells. Unfortunately, radiation and chemotherapeutic drugs also kill normal body cells, especially those that divide rapidly, such as the cells of the hair follicles and the cells lining the intestine. As a result, hair loss, nausea, and diarrhea are common side effects of these treatments. Patients may be sick for several days after a treatment.

Thanks to a relatively simple technique, physicians may have a way to eliminate the harsh chemical treatments for cancer patients. The potential new treatment uses tiny lipid spheres— called liposomes — which are packed with cancer-killing drugs.

Liposomes can be made from synthetic molecules or natural products, such as egg yolks and soy

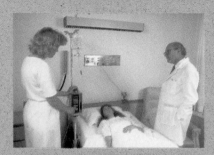

Chemotherapy Patient receives chemotherapy treatment for cancer.

beans. Each tiny fat-soluble sphere contains a watery interior that traps water-soluble drugs. Fat-soluble drugs can be trapped in the fatty exterior. The microspheres act as a drug reservoir that releases its chemicals as the liposome dissolves or after it is ingested by a cell.

Medical researchers are optimistic about the potential use of liposomes for cancer therapy. By packaging chemotherapeutic drugs in the liposomes, medical scientists may be able to protect the body's cells from harm. Experiments with an anticancer drug, doxorubicin, have yielded some interesting results. This highly toxic drug damages heart muscle when administered intravenously. Injection in liposomes, however, reduces its harmful effects on the heart by 80% because much smaller doses can be used.

Researchers have also successfully packaged cytochalasin B, a potential cancer drug, in liposomes. This drug halts the spread of malignant cells but, to be effective, must be present at all times. Liposomes release the

drug very slowly, thus maintaining low levels that are nontoxic to normal cells but fatal to tumor cells.

Researchers are now experimenting with ways to get liposomes to home in on specific targets. One approach that may prove useful involves the use of antibodies— proteins that react with foreign substances or foreign cells, such as cancer cells. By coating drug-containing liposomes with antibodies, medical researchers hope that they can produce site-specific bullets that bind only to cancer cells. Liposomes engulfed by the cell would release their contents into it, thus killing it.

Targeting may also prove beneficial in treating other diseases where the drug is toxic to many cells. Target-specific liposomes may be useful in delivering altered genes to diseased cells. Scientists at the University of Tennessee in Knoxville have injected mice with pieces of DNA encapsulated in antibody-coated liposomes. The liposomes attach to and are ingested by mouse cells. Inside the cells, the DNA is released from the liposomes and may be incorporated into the cell's DNA. There it stimulates the production of an enzyme that the cell had been unable to make.

This startling finding could prove extraordinarily helpful in inserting normal genes into defective cells, a major barrier in the use of genetic engineering to reverse genetic disease in adults. Researchers believe that liposomes loaded with genes that trigger cell death could also be used to treat cancer.

FIGURE 2–23 Fats and Oils At room temperature, fats are solid whereas oils are liquid. The reason for the difference is explained in the text and in Figure 2–20.

Oils can be converted to solids by adding hydrogen atoms. Hydrogen atoms bind to the carbon atoms involved in the double bonds in the unsaturated fatty acids of the triglycerides. Margarine is produced by chemically adding hydrogen to vegetable oil.

Phospholipids. Phospholipids are a class of lipids that contain phosphate, PO_4. The most common type of phospholipid is **phosphoglyceride** (Figure 2–25a). Phosphoglycerides are a major component of the plasma membrane. Like triglycerides, phosphoglycerides are comprised of a molecule of glycerol attached to two fatty acids. Covalently bonded to the third carbon, however, is a phosphate group containing a phosphorus atom and four oxygen atoms. Attached to the phosphate is a group designated by the letter R, a symbol that represents one of a half dozen or more molecules.

The R group is frequently polar or charged. Its presence creates a polar region (the phosphate head) on the molecule, as shown in Figure 2–25b. The long hydrocarbon chains of the fatty acids form a large nonpolar region. In water, phosphoglycerides form tiny globules called micelles, shown in Figure 2–26. As illustrated, the polar region of the phosphoglycer-

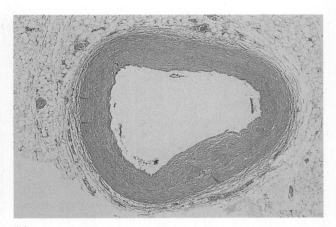

(a)

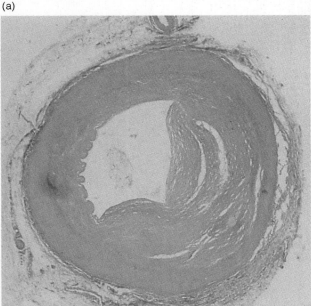

(b)

FIGURE 2–24 Atherosclerosis These cross sections of (*a*) a normal artery and (*b*) a diseased artery show how atherosclerotic plaque can obstruct blood flow.

ide orients outward toward the polar water molecules. The nonpolar ends orient inward, avoiding the polar water molecules.

Phosphoglycerides may also form microspheres or liposomes with a watery core (see Health Note 2–1). The structure of the **plasma membrane**, the layer of protein and lipid surrounding cells, is largely determined by the interaction of phosphoglycerides and water.

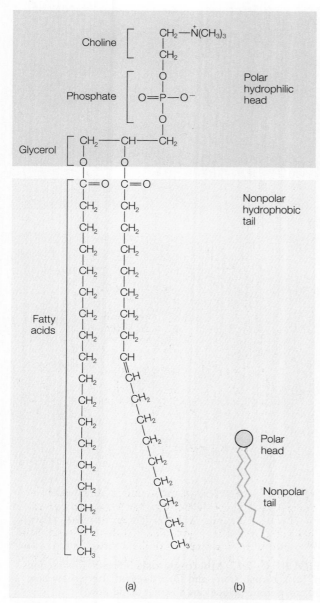

FIGURE 2–25 **Phosphoglyceride** Phosphoglycerides are the predominant phospholipid in the body. (*a*) Each phosphoglyceride consists of a glycerol backbone, two fatty acids, a phosphate, and a variable R group, in this case choline, which is polar. (*b*) Because of the R group, the molecule has a polar head. The nonpolar tail region is formed by the two fatty acid chains.

Steroids. Steroids are quite different from the triglycerides and phospholipids. As shown in Figure 2–27, steroids consist of four fused rings. One of the best known steroids is **cholesterol** (Figure 2–27a). Cholesterol is a component of the plasma membrane

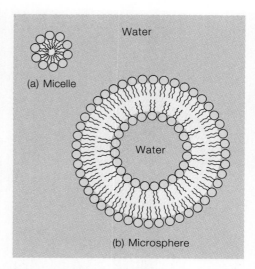

FIGURE 2–26 **Micelle and Microsphere** (*a*) In a solution of water, phosphoglyceride molecules form tiny globules called micelles. Notice how the nonpolar tails of the molecules bunch together while the polar heads stick out. (*b*) Phosphoglycerides may also form microspheres, tiny spheres containing an aqueous core. Again, notice how the polar and nonpolar ends of the molecule seek a chemically similar environment.

and is also used in the body as a raw material to produce other steroids, such as the sex hormones estrogen and testosterone. Cholesterol also deposits in arteries forming atherosclerotic plaques.

Amino Acids, Peptides, and Proteins

Amino Acids. **Amino acids** are the building blocks of proteins. Each amino acid contains a central carbon attached to four functional groups: an **amino group** (NH_2), a **carboxyl group** (COOH), a hydrogen, and a variable group, indicated by the letter R (Figure 2–28a).[4] Twenty different amino acids are common in nature, differing only in the R group; from this seemingly limited pool, the cells build thousands of different proteins in much the same way that musicians create unique and exciting music from a relatively small number of notes.

Proteins are polymers of amino acids that serve many functions in the human body. One class of proteins, the **enzymes**, accelerates chemical reactions in the body. Other proteins have a structural role. **Keratin**, for example, is a protein that makes up hair and

---

[4]The R groups found in amino acids are different from those found in phospholipids.

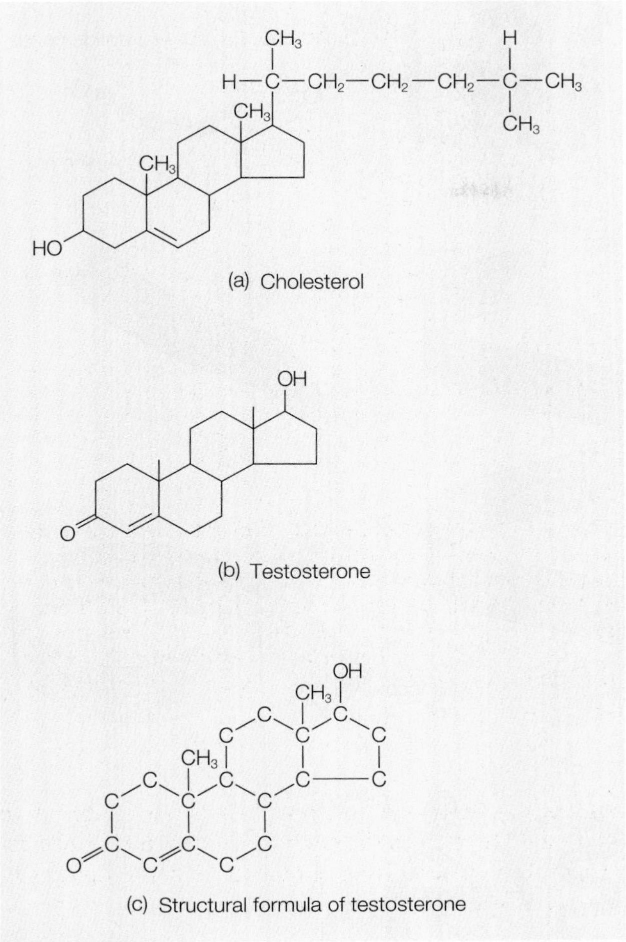

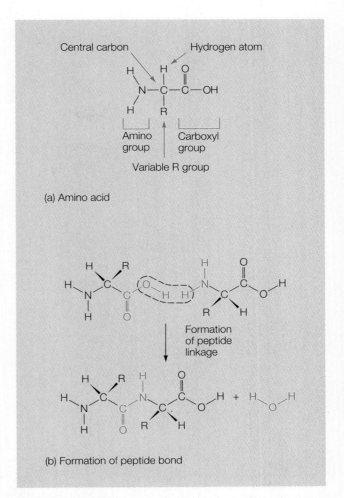

(a) Cholesterol

(b) Testosterone

(c) Structural formula of testosterone

FIGURE 2–27 Steroids Steroids, like (a) cholesterol and (b) testosterone, consist of four rings joined together. The drawings in (a) and (b) are a shorthand way of drawing (c) the structural formula.

(a) Amino acid

(b) Formation of peptide bond

FIGURE 2–28 Structure of Amino Acids and Formation of Peptide Bonds (a) The amino acid is a small organic molecule with four groups. (b) The carboxyl and amino groups react in such a way that a covalent bond is formed between the carbon of the carboxyl group of one amino acid and the nitrogen of the amino group of another. This bond is called a peptide bond.

nails. **Collagen** is the most abundant protein in the body. Collagen fibers help hold many tissues together and are the major component of bone. If the calcium is dissolved from the bone, a rubbery collagenous replica of the bone remains.

Proteins and Peptides. Proteins are synthesized in the cells in reactions in which one amino acid is added at a time, thus forming proteins with hundreds, even thousands, of subunits. The bond linking one amino acid to another is called a **peptide bond** (Figure 2–28b).

The length of amino acid chains in nature varies considerably from the simplest **dipeptides** (a molecule made of two amino acids) to the largest proteins containing 3000 amino acids. By convention, biolo-

gists use the term *protein* to apply to amino acid chains with 20 or more amino acids; smaller chains are called **peptides** or **polypeptides**.

Protein Structure. Although thousands of different types of proteins are found in human cells, each type of protein is structurally and functionally unique. A protein's unique shape results from the sequence in which the different amino acids are joined.

Biochemists recognize four levels of organization in protein structure. The first is the **primary structure**, the order or the sequence of amino acids. Each type of protein has a unique primary structure. As

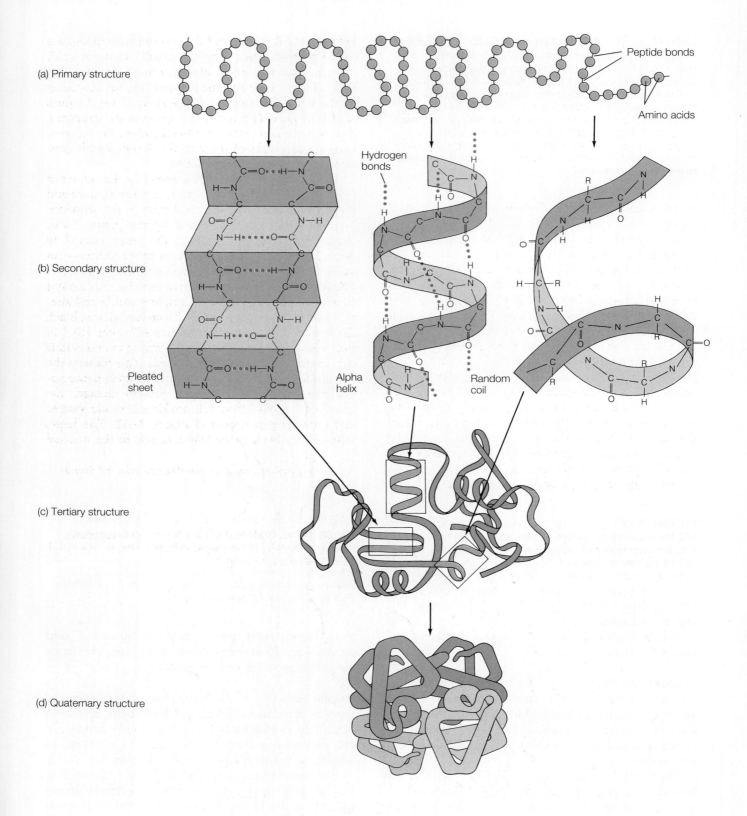

(a) Primary structure

Peptide bonds

Amino acids

(b) Secondary structure

Hydrogen bonds

Pleated sheet

Alpha helix

Random coil

(c) Tertiary structure

(d) Quaternary structure

FIGURE 2–29 Protein Structure (left) (*a*) Primary structure is the sequence of amino acids. It determines to a large extent the complex three-dimensional shape a protein assumes. (*b*) Secondary structure results from a bending or coiling of the primary structure. Three general types are found: pleated sheet, alpha helix, and random coil. (*c*) Tertiary structure results when the secondary structure of a protein is compacted to form its three-dimensional shape upon which its function is dependent. (*d*) Quaternary structure results when two or more globular proteins unite to form a "superprotein."

proteins are produced in human cells, the chain of amino acids begins to bend and fold, creating the **secondary structure**. The chain may form a spiral (alpha helix), random coil, or, in some cases, a complex pleated sheet. The bending and coiling of the primary structure is brought about by hydrogen bonds that form between parts of the chain (Figure 2-29b). Some proteins have all three types of secondary structure in different parts of the molecule.

Proteins that form a pleated sheet generally remain in that form. Further bending is not possible. However, those proteins whose secondary structure is characterized by random coils and/or spirals (helices) can compact further. This compaction converts the protein chain into a globular protein. The three-dimensional shape of this type of protein is called its **tertiary structure** (Figure 2–29c). The tertiary structure of a protein results primarily from interactions between the R groups of the amino acid chain or between the R groups and the predominant molecules in the protein's environment—usually water or lipid.

Some proteins consist of two or more globular subunits (Figure 2–29d). The subunits are held together by the mutual attraction of oppositely charged amino acid R groups. The result is a **quaternary structure**. The best-studied example is hemoglobin, the oxygen-carrying molecule of the red blood cell, which contains four polypeptide subunits.

The primary structure of a protein—the sequence of amino acids—determines its ultimate structure and its function. As described in Chapter 5, the sequence of amino acids is determined by the genes. Thus, **mutations**—minute changes in the genes, caused by radiation, chemical substances, or other factors—can severely alter a protein's three-dimensional structure. Mutations can result in defective proteins that cannot function normally. A good example is **sickle-cell anemia**, which primarily afflicts African Americans, black Africans, and several other groups (Chapter 18).[5] In individuals with the disease, one wrong amino acid is substituted during protein synthesis. This impairs the protein folding, resulting in an odd-shaped hemoglobin molecule. The misshapen protein changes the shape of the red blood cell, making it sickle shaped and inflexible, as shown in Figure 2–30. The large, inflexible cells clog tiny blood vessels in the internal

[5]People from parts of Italy, Greece, Arabia, and India and their descendents.

FIGURE 2–30 Sickle-Cell Anemia Transmission electron micrograph of (*a*) normal blood cell and (*b*) blood cell taken from patient with sickle-cell anemia. Because a single amino acid is incorrect, the structure of the entire hemoglobin molecule and the red blood cell is disrupted, creating inflexible, sickle-shaped cells that clog in tiny blood vessels and lead to death.

(a)

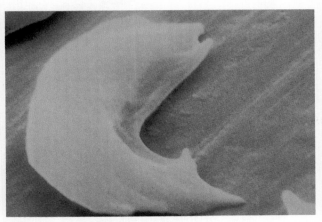

(b)

organs (especially the brain and heart) of its victims. Severe blockage can result in death.

Changes in the physical and chemical environment can also change the three-dimensional structure of a protein. An increase in acidity, for instance, can cause a protein to unfold. The loss of a protein's native shape is called **denaturation**. Buffer systems, described earlier in the chapter, help prevent the denaturation of proteins. Heat also denatures protein. If you drop an egg in a skillet, it solidifies because the heat causes the protein in the egg to unfold, entangling with chains of other molecules. Some denaturation is necessary. For example, in humans and other animals, acids in the stomach denature ingested proteins, which facilitates enzymatic digestion that occurs in the small intestine.

Nucleic Acids

Nucleic acids are a group of organic molecules consisting of three members: RNA, DNA, and nucleotides. Both DNA and RNA are polymers of smaller

FIGURE 2–31 **Nucleotides** (*a*) Nucleotides are the building blocks of RNA and DNA. Nucleotides consist of three molecules: (*b*) a phosphate, (*c*) a simple sugar, and (*d*) a nitrogen-containing base. RNA and DNA nucleotides contain different sugars and bases. Uracil replaces thymine in RNA.

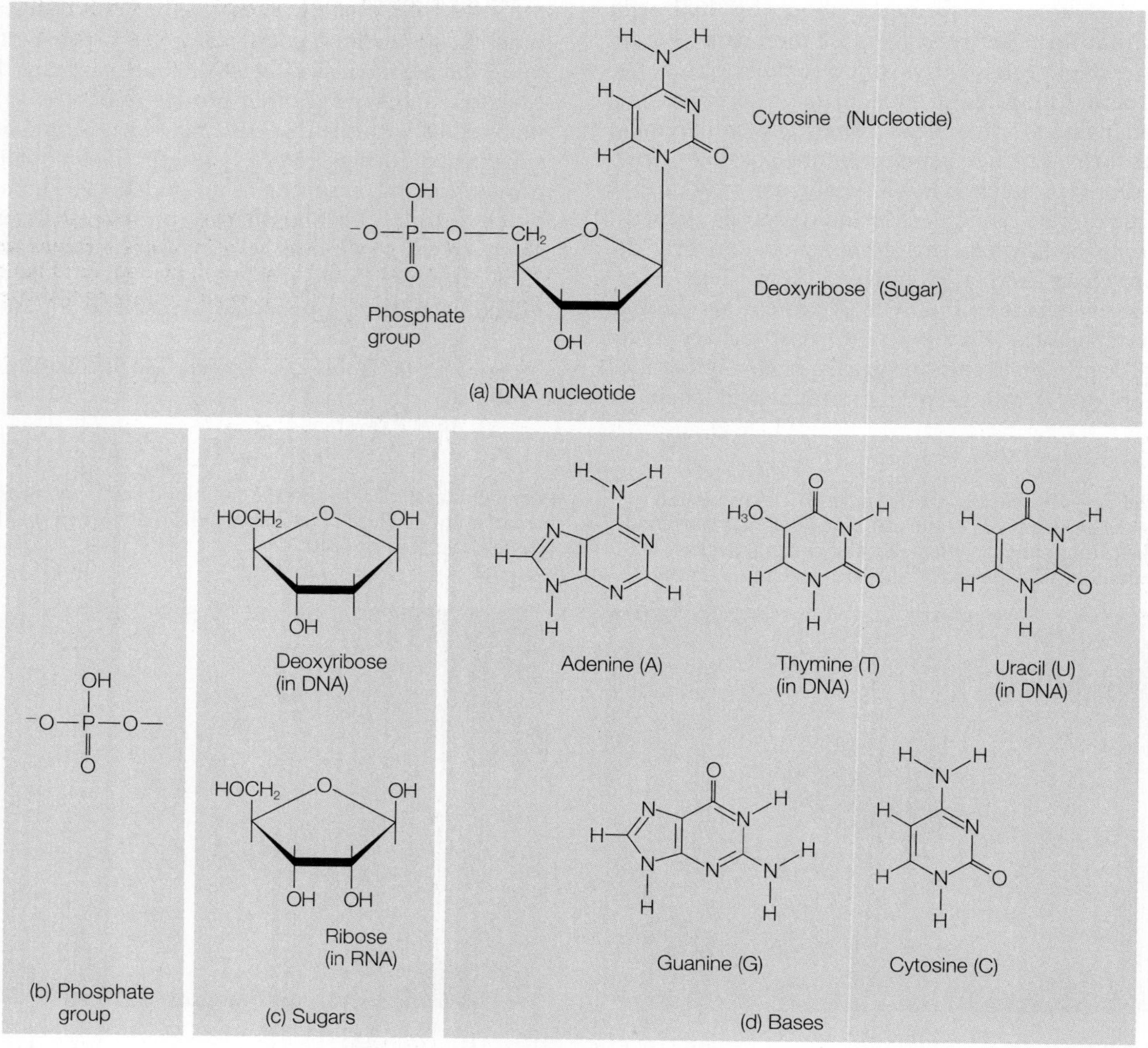

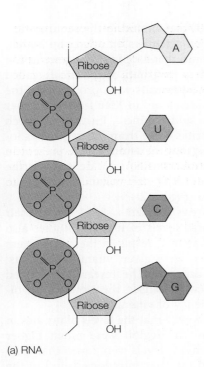

(a) RNA

molecules called **nucleotides**. Each nucleotide consists of a phosphate, a sugar molecule, and a nitrogenous base as shown in Figure 2–31. The nucleotides are covalently bonded to one another to form long chains, known as polynucleotide chains.

DNA or **deoxyribonucleic acid** forms the genes and carries the instructions needed to build and operate a human being. **RNA** or **ribonucleic acid** in humans is a key player in protein synthesis (Chapter 5). RNA is a single strand of nucleotides containing the sugar ribose (Figure 2–32a). DNA, on the other hand, is a double-stranded chain, or **helix**. Its nucleotides differ from those of RNA because they contain **deoxyribose** instead of ribose (Figure 2–31c). The two strands of the DNA molecule are held together by hydrogen bonds (Figure 2–32b). The sequence of nucleotides in DNA is the genetic information that controls all cell functions, (see Chapters 4 and 5).

FIGURE 2–32 RNA and DNA (*a*) RNA is a single-stranded polynucleotide chain. It is sometimes twisted into unusual shapes. (*b*) DNA is a double-stranded molecule containing two polynucleotide chains joined by hydrogen bonds between the bases.

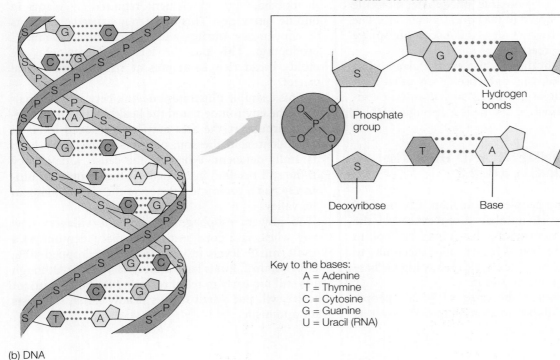

(b) DNA

Key to the bases:
 A = Adenine
 T = Thymine
 C = Cytosine
 G = Guanine
 U = Uracil (RNA)

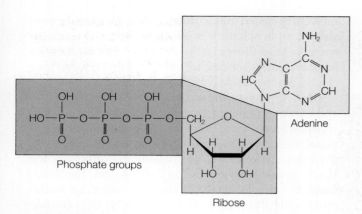

FIGURE 2–33 Molecular Structure of ATP

ATP. Earlier in the chapter we saw that chemical reactions in the body that give off energy are linked to reactions that require energy by an energy carrier, a molecule called ATP. ATP is a nucleotide with three phosphate groups (Figure 2–33). ATP is formed in an energy-absorbing reaction as follows:

$$ADP \quad + \quad P_i + energy \rightleftarrows ATP$$

| Adenosine diphosphate | Inorganic phosphate | Adenosine triphosphate |

The energy absorbed during this process is used to covalently bond the inorganic phosphate molecule (P_i) to ADP and is stored in the bond for later use. The stored energy is released when ATP loses a phosphate group (the reverse reaction).

Organisms cannot directly use energy given off in the catabolic breakdown of glucose or any other molecule to power cellular processes, such as muscle contraction. Energy must be channeled through ATP.

ENVIRONMENT AND HEALTH: TRACKING A KILLER

For 2000 years, the people in Lin Xian, China, have been dying in record numbers from cancer of the esophagus, the muscular tube that transports food to the stomach. One of every four people succumbs to this mysterious killer, whose incidence is higher here than anywhere else in the world.

In 1959, scientists in the valley of 70,000 people began to study the disease in an attempt to discover its origins and hopefully put an end to the scourge. The mysterious deaths led scientists on a lengthy search. Researchers first found that esophageal cancer in Lin Xian was the result of a group of chemicals called nitrosamines. These chemicals were produced in the stomachs of the residents of Lin Xian from two other chemicals: nitrites and amines. Further research showed that the nitrites came from the vegetables local residents ate and that the amines were present in moldy bread, which is considered a delicacy in the region. The complex set of circumstances that led to cancer is outlined in Figure 2–34.

Studies of the area's residents showed that their bodies contained high levels of nitrites. Studies also showed that the vitamin C levels in their diets were low. When residents were given vitamin C tablets, their nitrite levels dropped. The researchers probed further to find out what caused the elevation in nitrites and the deficiency of vitamin C.

A careful study showed that the levels of nitrates in vegetables grown in the region were much higher than expected. Vitamin C levels were lower than normal. That explained the concentrations found in the residents, but why were the levels in plants different, say, from areas a few hundred miles away? The answer, scientists found, lay in the soil. Chemical analyses showed that the topsoil in Lin Xian was deficient in molybdenum, an element required by plants in minute quantities. This deficiency caused the plants to concentrate nitrites and reduce their vitamin C production. This put the local residents who ate moldy bread rich in amines at risk for esophageal cancer.

This example illustrates one link between the quality of the environment and the health of people. Long-term farming in the area may have reduced levels of molybdenum in the soil, or the soil may have been naturally deficient. Whatever the cause, a deficiency in the soil resulted in changes in the plants that ultimately had a profound impact on the people living in the valley.

To prevent esophageal cancer, the villagers now coat wheat and corn seeds with molybdenum. As a result, nitrite levels in vegetables have dropped 40%, and vitamin C levels have increased by 25%. Although it is still too early to tell if restoring the soil's nutrient health will put a halt to this deadly killer, scientists are optimistic.

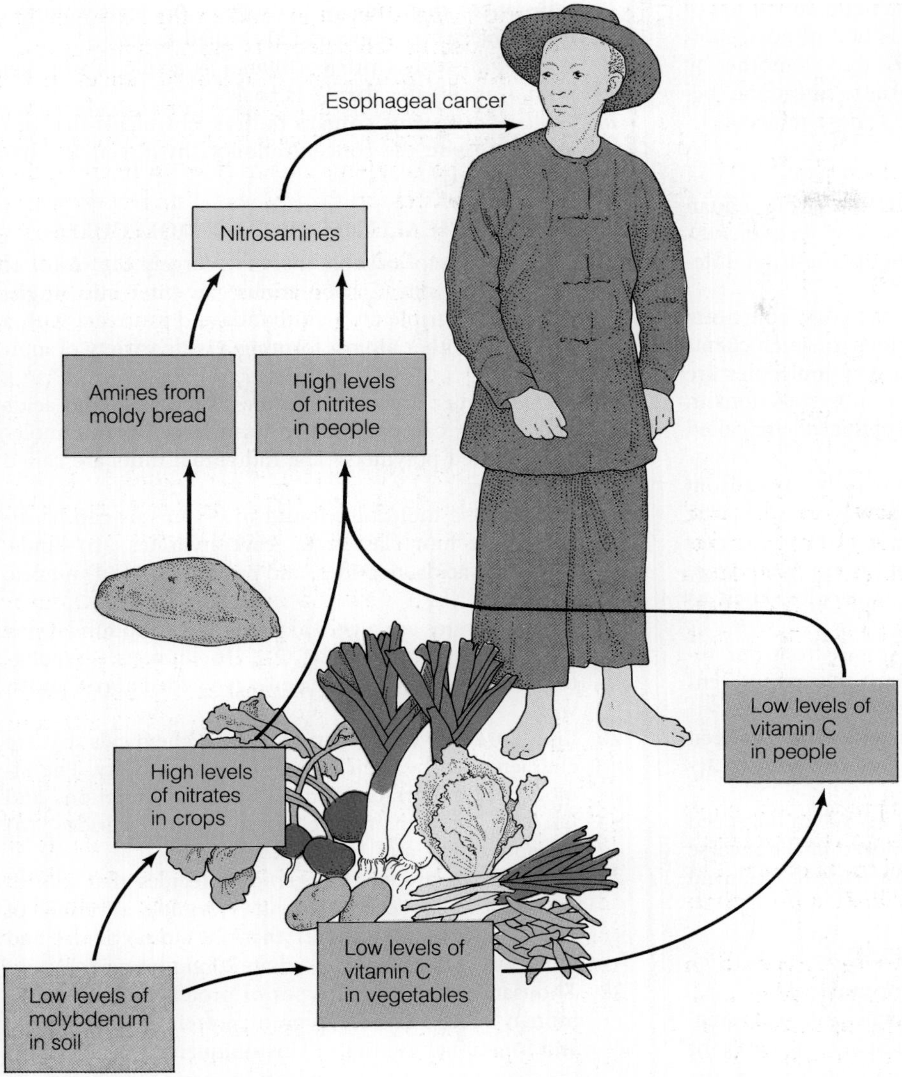

Esophageal cancer

Nitrosamines

Amines from
moldy bread

High levels
of nitrites
in people

High levels
of nitrates
in crops

Low levels of
vitamin C
in people

Low levels of
molybdenum
in soil

Low levels of
vitamin C
in vegetables

FIGURE 2–34   Tracking a Deadly Killer   In Lin Xian, China, the link between esophageal cancer and the soil was uncovered by the diligent work of medical researchers.

SUMMARY

1. An understanding of chemistry is essential to understanding biology.

ATOMS AND SUBATOMIC PARTICLES

2. All organisms and the nonliving components of our world are composed of matter. Matter is anything that has mass and occupies space.
3. All matter is composed of elements, the purest form of matter. Elements, in turn, are composed of atoms.
4. Each atom contains three basic subatomic particles: electrons, protons, and neutrons. Electrons, the smallest of the three subatomic particles, orbit in the electron cloud around the dense, central nucleus. Protons and neutrons are the most massive particles and are found in the nucleus.
5. Elements are listed on the periodic table by atomic number. The atomic number is the number of protons in the nucleus.

CHEMICAL BONDS AND MOLECULES

6. Atoms bond to one another forming molecules.
7. The two principal bonds are covalent and ionic. Covalent bonds form between atoms with similar electron affinity that share one or more pairs of electrons. This sharing holds the atoms together.
8. Ionic bonds form between two oppositely charged ions and are electrostatic attractions.

9. Unequal sharing of electrons results in the formation of a polar covalent bond and the formation of polar molecules. Polar molecules are attracted to one another by hydrogen bonds, a weak electrostatic attraction between the slightly charged atoms of these molecules.

CHEMICAL REACTIONS

10. Chemical substances form the structures of the human body. Many chemicals also participate in reactions essential for growth, reproduction, homeostasis, and development.

11. All of the chemical reactions in the body constitute metabolism. Those metabolic reactions in which chemicals combine with others to form new molecules are synthetic or anabolic reactions, and those reactions in which chemicals are split apart or dissociate are called catabolic reactions.

12. Most of the energy required by body cells comes from the catabolic breakdown of carbohydrates and fats. The energy released from these chemicals ultimately comes from the sun via plants, which capture sunlight energy and use it to synthesize organic molecules, such as glucose.

13. In the cells of the body, energy released from one reaction is often used to drive reactions that require energy. Such reactions are said to be coupled.

14. ATP (adenosine triphosphate) shuttles energy between energy-yielding reactions and energy-requiring reactions.

15. Most chemical reactions are reversible. When the speed of the forward reaction equals the speed of the reverse reaction, the reaction is said to have reached chemical equilibrium.

WATER, ACIDS, BASES, AND BUFFERS

16. Atoms bond to one another to form molecules.

17. Two types of molecules exist: organic and inorganic. Organic molecules are compounds made primarily of carbon and hydrogen. Covalent bonds hold the atoms of organic molecules together, and many organics are large molecules. The rest of the molecules are inorganic. They are frequently much smaller molecules whose atoms are often joined by ionic bonds.

18. Water is an important inorganic molecule and a major component of all body cells. It is an important solvent, which transports many substances in the blood, body tissues, and cells. Water also participates in many chemical reactions, serves as a lubricant, and helps regulate body temperature.

19. Although fairly stable, water molecules can dissociate, forming hydrogen and hydroxide ions. In pure water, the concentration of these ions is equal and water has a pH of 7. Adding acidic substances increases the concentration of hydrogen ions, causing the pH to fall.

Substances that remove hydrogen ions from solution cause the pH to climb and are called bases.

20. A solution with a pH less than 7 is acidic; a solution with a pH greater than 7 is basic.

21. Biological systems contain buffers, chemicals that offset changes in the concentration of hydrogen and hydroxide ions.

MORE BIOLOGICALLY IMPORTANT MOLECULES

22. The diversity of organic matter is largely the result of the carbon atom. Carbon atoms can enter into single, double, and triple covalent bonds and can react with a number of other atoms, forming a wide variety of molecules.

23. Many smaller organic molecules, such as amino acids and sugars, can combine to form large organic molecules, called polymers. The individual units are called monomers.

24. The organic molecules found in organisms can be divided into four classes: (a) carbohydrates, (b) lipids, (c) amino acids, peptides, and proteins, and (d) nucleic acids.

25. Carbohydrates are a group of organic compounds with the general formula $(CH_2O)_n$. In humans, carbohydrates provide much of the energy needed at rest and at work.

26. Lipids are a diverse group of organic chemicals that are characterized by their lack of water solubility. The biologically important lipids serve many functions and fall into three main categories: (a) triglycerides, (b) phospholipids, and (c) steroids.

27. Amino acids are small organic molecules that join to one another by peptide bonds, forming a variety of peptides (containing fewer than 20 amino acids) and proteins (containing more than 20 amino acids).

28. Thousands of different types of proteins are found in human cells, and each type of protein is structurally and functionally unique. This uniqueness results from the sequence of amino acids in a protein. The sequence of amino acids determines the ultimate shape and the function of the protein. A change of even one amino acid can severely disrupt the shape and function of a protein.

29. Nucleic acids are a group of organic molecules consisting of three members: nucleotides, RNA, and DNA, polymers of smaller molecules called nucleotides. A nucleotide consists of a phosphate, a sugar molecule, and a base (either a purine or a pyrimidine).

30. DNA, or deoxyribonucleic acid, forms the genes of our cells and carries all of the genetic information required for cell structure and function.

31. RNA, or ribonucleic acid, plays a key role in protein synthesis.

Concern over the environment and health in recent years has led business to introduce a number of new products, which have been labeled environmentally safe. One of the principal rules of critical thinking is to understand and define all terms. In this case, manufacturers are calling many products environmentally safe because they are biodegradable or photodegradable. Trash bags, shopping bags, and even diapers have been sold under the biodegradable banner. What does *biodegradable* mean? What does *photodegradable* mean?

Now that you have the definitions, you could ask if the manufacturers are using the terms appropriately. Is a biodegradable plastic really biodegradable? Is a photodegradable plastic bag really good for the environment?

Critical thinking on this issue will require that you spend a little time in the library, reading both the industry and environmental perspectives on the issue. Critical thinking requires that you become an active participant in the process: that you seek out additional information and differing viewpoints. Critical thinking also requires that you distinguish between facts and values or judgments.

TEST OF TERMS

1. Anything that occupies space and has mass is called _____ .

2. _____ are the purest form of matter and cannot be separated into other substances by chemical means.

3. Atoms consist of three subatomic particles: _____ , _____ , and _____ .

4. Protons are found in the _____ of the atom and have a _____ charge.

5. Elements are listed on the periodic table of elements by ascending _____ _____ .

6. An isotope is an alternative form of an atom containing one or more additional _____ .

7. _____ released from isotopes help stabilize their nuclei.

8. An _____ is formed when an atom either gains or loses electrons.

9. A _____ bond forms when two atoms share a pair of electrons.

10. A _____ bond forms between the charged ends of two water molecules.

11. Metabolism consists of two broad groups of reactions: the synthetic or _____ reactions, and the breakdown or _____ reactions.

12. Energy-releasing and energy-absorbing reactions are said to be _____ in the cell so that the energy given off by one reaction can be used to drive the other.

13. _____ is a short-term energy storage molecule in the cell that takes up energy and releases it to energy-requiring processes.

14. At _____ _____ the rate of the forward reaction equals the rate of the reverse reaction.

15. Organic molecules are principally made of the elements _____ and _____ and contain _____ bonds.

16. The _____ molecule is a storage form for glucose in human liver and muscle cells.

17. Water dissociates into two ions _____ and _____ .

18. An _____ is any chemical substance that adds hydrogen ions to solution, and a _____ is any chemical substance that removes them.

19. On the pH scale substances with pH values lower than 7 are _____ . The lower the pH reading, the _____ the concentration of hydrogen ions.

20. A substance that helps maintain a constant pH is called a _____ .

21. Simple sugars are also called _____ and are used to synthesize long-chained molecules, such as starch and glycogen, which are in a group of carbohydrates called _____ .

22. A triglyceride like margarine, vegetable oil, or animal fat consists of three _____ _____ molecules and one molecule of _____ .

23. The principal lipid in the cell membrane is _____ .

24. _____ is the lipid that deposits in the walls of arteries causing atherosclerosis.

25. Proteins are _____ of amino acids linked by _____ bonds.

26. _____ are a class of proteins that speed up chemical reactions.

27. The sequence of amino acids in a protein is called its _____ structure. The final three-dimensional shape of a protein is its _____ structure.

28. The _____ _____ , DNA and RNA, are polymers of molecules called _____ .

29. ATP is formed by ADP, _____ and _____ .

Answers to the Test of Terms are located in Appendix B.

TEST OF CONCEPTS

1. Why is the matter on Earth so varied in its appearance?
2. Describe the structure of an atom, using a diagram to further your explanation.
3. Define the following terms: atomic weight, atomic number, and isotope.
4. How do ionic and covalent bonds form? Describe what holds the atoms together in both cases. In what ways are the bonds different?
5. Describe how polar covalent bonds can result in the formation of hydrogen bonds.
6. Temperature is a measure of the speed of molecules: the higher the temperature, the higher the speed. With this knowledge and your knowledge of water and the hydrogen bond, why would you think water has a higher boiling point than a nonpolar liquid like alcohol?
7. Define the following terms: anabolic reaction, catabolic reaction, and coupled reaction. Give an example of each and explain the importance of these terms.
8. Draw the chemical reaction for the formation of carbonic acid. How does this reaction participate in homeostasis?
9. Describe the biological significance of water from a human perspective.
10. How does the body regulate H^+ levels in the blood? Describe the process.
11. List the four major kinds of biological molecules described in the chapter and explain why each one is important.
12. Define the following terms: monosaccharide, disaccharide, and polysaccharide. Give an example of each one.
13. Describe the biological importance of each of the following lipids: triglycerides, steroids, and phosphoglyceride.
14. List some of the functions proteins serve in the cell.
15. Why do changes in the primary structure of a protein alter its tertiary structure and function?
16. How are DNA and RNA similar? How are they different?

SUGGESTED READINGS

Campbell, N. A. 1990. *Biology*. 2d ed. Menlo Park, Calif.: Benjamin/Cummings. See Chapters 2–5 for a more detailed look at important principles of chemistry.

Slabaugh, M. R., and S. L. Seager, 1988. *Chemistry: An introduction*. St. Paul, Minn.: West. An excellent introductory text that elaborates on concepts discussed in this chapter.

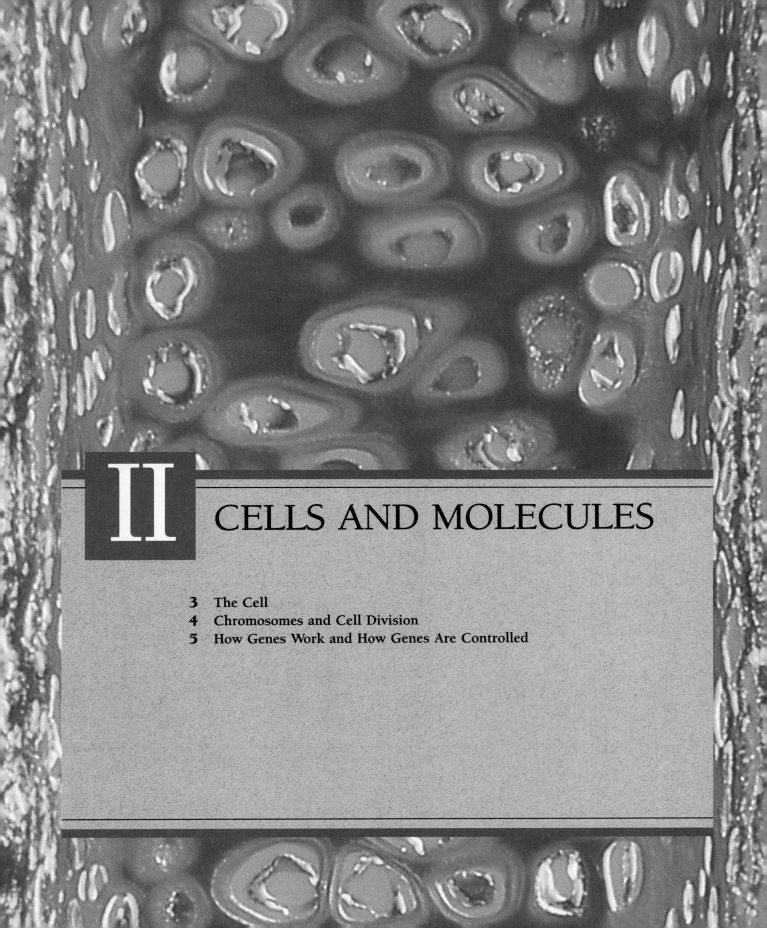

II CELLS AND MOLECULES

3

The Cell

Cells in monolayer culture.

Preceding page: Mammal ear hyaline cartilage.

I n Sweden and Mexico, medical researchers are trying a bold new experiment. In the hopes of finding a cure for several chronic diseases, scientists are transplanting cells from human fetuses into adults with chronic, debilitating diseases. The researchers hope that the fetal cells will survive and reproduce in the new location, replacing defective cells in the brain and other organs—and ultimately reducing the pain and suffering of tens of thousands of people each year.

Like many areas of research in medical science and technology, this work has generated a variety of technical problems, as discussed in Health Note 3–1. Perhaps more important, however, are the moral issues it raises. Some of these issues are discussed in the Point/ Counterpoint in this chapter.

This chapter will help you understand the cell—and a little more about the controversy over cell transplantation. It looks at the structure and function of the cell's parts, providing information that will help you better understand human biology.

AN OVERVIEW OF THE CELL

The fundamental unit of all living things is the cell. In fact, all organisms are comprised of cells. The first living organisms appeared on Earth approximately 3.5 billion years ago. Resembling modern-day bacteria, these organisms lived in the oceans and were the dominant form of life for over 2 billion years (Figure 3–1). The first cells and their modern relatives, the bacteria, are called prokaryotes. **Prokaryotes** are very small single-celled organisms, whose genetic information is contained in a single circular strand of DNA. The DNA of prokaryotes is not membrane bound, and the cells have very little internal cellular differentiation (Figure 3–2a).

Structurally more complex cells are known as eukaryotic cells. **Eukaryotes** contain DNA within a well-defined region, the nucleus, which is delimited by a membrane (Figure 3–2b). The earliest eukaryotic cells lived independently, as do their modern descendants, aquatic amoebae and paramecia. From these early eukaryotes arose the plants, fungi, and animals.

As the evolution of plants, fungi, and animals proceeded, two developments took place. First, organisms became multicellular, comprised of many cells. Second, as evolution proceeded, the cells specialized—that is, they became modified to perform specific functions. Multicellularity and specialization led to an increase in the overall complexity of the Earth's organisms. Organisms evolved complex motor sys-

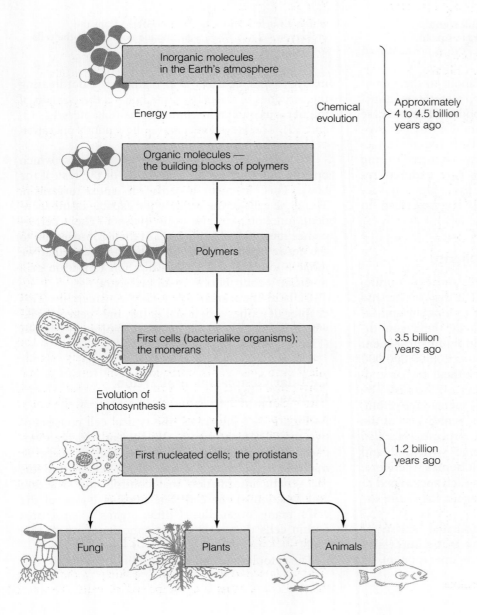

FIGURE 3–1 Evolution of the Cell

Inorganic molecules in the Earth's atmosphere

Energy — Chemical evolution — Approximately 4 to 4.5 billion years ago

Organic molecules — the building blocks of polymers

Polymers

First cells (bacterialike organisms); the monerans — 3.5 billion years ago

Evolution of photosynthesis

First nucleated cells; the protistans — 1.2 billion years ago

Fungi   Plants   Animals

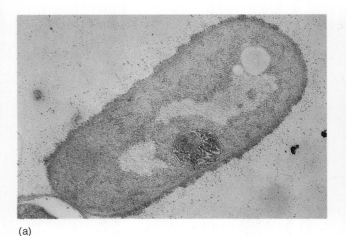

(a)

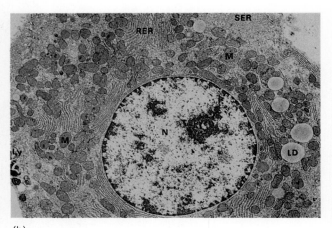

(b)

FIGURE 3–2 Prokaryotes and Eukaryotes
(*a*) Prokaryotes are relatively simple organisms like this bacterium. They contain DNA, but the DNA is not contained within a nucleus. (*b*) Eukaryotes evolved from the prokaryotes. Their DNA is membrane bound, and the cells contain numerous cellular organelles.

tems that enabled them to move about in their environment and complex circulatory systems to transport nutrients throughout their bodies. The increase in organismic complexity was paralleled by the evolution of systems that coordinate body functions. Despite the dramatic changes that have occurred during evolution, the cell's basic features have endured, reinforcing the basic evolutionary principle that successful patterns persist, which were described in Chapter 1.

Coordinating Growing Complexity

Human life begins when the sperm and ovum unite, forming a single cell, the **zygote**. The zygote's genetic information is derived from both its mother and father. During embryonic development, the zygote divides, and the resultant cells multiply in subsequent divisions. A newborn infant contains 20,000,000,000 (20 billion) cells. If you could stack them end on end, they would stretch 200 kilometers (125 miles). By adulthood the number of cells has increased fivefold.

Multiplication of cells, however, is only part of the development process. During that process, cells also differentiate—that is, they undergo structural and functional modifications. The adult human, therefore, contains many different cell types, each specialized to perform specific functions that benefit the entire organism.

Homeostasis requires the coordinated activity of these cells in much the same way that a successful football team requires the smooth operation of its many players, coaches, and managers. Coordinating the activities of the cells of the body is the domain of the nervous and endocrine systems (Chapters 12, 13, and 15). These systems ensure the smooth operation of the entire body.

All cells have their own internal controls, which operate more or less independently of the rest of the body. This, in turn, benefits the entire organism. Thus, the cells of the body, while capable of independent function, are also team players. Their performance must be judged in light of their contribution to the entire organism, in much the same way that we judge a football player's contribution to a team or a musician's contribution to an orchestra.

Cells in the human body vary considerably. This chapter describes a fictional entity, the "general cell," which exists only in the pages of biology textbooks (Figure 3–3).

Cellular Compartmentalization: Increasing Efficiency

As Figure 3–3 illustrates, the typical cell consists of two major compartments, the nuclear and the cytoplasmic compartments. The nuclear compartment, or **nucleus**, is the control center of the cell. It contains the genetic information that controls the structure and function of the cell. The nucleus is one of the cell's many **organelles** ("little organs"), structures within cells that carry out specific functions (Table 3–1).

The cytoplasmic compartment lies between the nu-

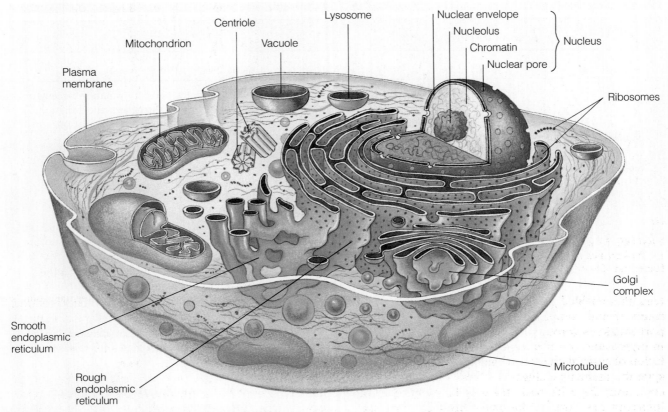

FIGURE 3–3   Structure of the General Eukaryotic Cell

cleus and the outermost structure of the cell, the plasma membrane. The **plasma membrane**, which is made of lipid, protein, and carbohydrate, controls the movement of materials into and out of the cell and performs a number of other important functions as well. The cytoplasmic compartment contains a material called the **cytoplasm.** The cytoplasm, in turn, contains many dissolved substances, such as ions, nutrients, proteins, waste products, vitamins, and dissolved gases. It is from this pool that the cell draws chemicals needed for metabolism.

The cytoplasm also contains numerous organelles. As shown in Figure 3–3, many organelles are membrane bound, forming distinct compartments. This permits the cell to isolate or compartmentalize key functions, much like an efficient factory. Non-membrane-bound organelles, such as ribosomes and microtubules, are also present in the cytoplasm. These organelles perform specific functions as well and will be described later.

Giving shape to the cell and helping to organize the cytoplasm is a network of protein tubules and filaments, called the **cytoskeleton** (Figure 3–4). Many of the organelles and many protein molecules in the cytoplasm are attached to the cytoskeleton. The cytoskeleton, like other organelles, helps organize cellular activities, increasing cellular efficiency. One way it does this is by binding to **enzymes**, proteins that speed up the rate of chemical reactions, as described below. Many chemical reactions in the body occur as part of **metabolic pathways**. Each reaction in the path requires an enzyme. Metabolic pathways are the cellular equivalent of the assembly line. A single molecule enters a pathway and is modified along the way in a series of reactions (Figure 3–5). Enzymes are the agents of change. Like the workers on an assembly line, they alter the product (molecule) as it proceeds down a metabolic pathway. The efficiency of metabolic pathways is greater when their enzymes are fixed to the cytoskeleton.

TABLE 3–1 Overview of Cell Organelles

ORGANELLE	STRUCTURE	FUNCTION
Nucleus	Round or oval body in the center of the cell; surrounded by nuclear envelope	Contains the genetic information necessary for control of cell structure and function; DNA contains hereditary information
Nucelolus	Round or oval body in the nucleus consisting of DNA and RNA	Produces ribosomal RNA
Endoplasmic reticulum	Network of membranous tubules in the cytoplasm of the cell; smooth endoplasmic reticulum contains no ribosomes; rough endoplasmic reticulum is studded with ribosomes	Smooth endoplasmic reticulum (SER) is involved in the production of phospholipids and has many different functions in different cells; rough endoplasmic reticulum (RER) is the site of the synthesis of lysosomal enzymes and proteins for extracellular use
Ribosomes	Small particles found in the cytoplasm; made of RNA and protein	Aid in the production of proteins on the RER and polysomes
Golgi complex	Series of flattened sacs usually located near the nucleus	Sorts, chemically modifies, and packages proteins produced on the RER
Secretory vesicles	Membrane-bound vesicles containing proteins produced by the RER and repackaged by the Golgi complex; contain protein hormones or enzymes	Store protein hormones or enzymes in the cytoplasm awaiting a signal for release
Food vacuole	Membrane-bound vesicle containing material engulfed by the cell	Stores ingested material and combines with lysosome
Lysosome	Round, membrane-bound structure containing digestive enzymes	Combines with food vacuoles and digests materials engulfed by cells
Mitochondria	Round, oval, or elongated structures with a double membrane; the inner membrane is thrown into folds	Complete the breakdown of glucose, producing NADH and ATP
Cytoskeleton	Network of microtubules and microfilaments in the cell	Gives the cell internal support, helps transport molecules in the cell, and binds to enzymes of metabolic pathways
Cilia	Small projections of the cell membrane containing microtubules	Propel materials along the surface of a cell
Flagellae	Large projections of the cell membrane containing microtubules	Provide motive force for cells

Enzymes and Homeostasis

Each chemical reaction in the cell requires a specific enzyme. All told, there are approximately 2000 different enzymes in each cell and an estimated 10,000 different enzymes in the human body. Enzymes are essential to cellular metabolism. A missing enzyme in a metabolic pathway, for example, will shut down the process—in much the same way that a missing worker on an assembly line could interrupt production.

Each enzyme is a large, globular protein with a small region called the active site (Figure 3–6). The **active site** is a small indentation or pocket where the chemical reaction occurs. Its shape corresponds to that of the **substrate(s)**, the molecule(s) undergoing reaction. Substrates fit into the active site in much the same way that a hand fits into a glove. Each enzyme has its own uniquely shaped active site capable of binding to one, or a few, chemical substrates. Thus, enzymes are said to be **specific.**

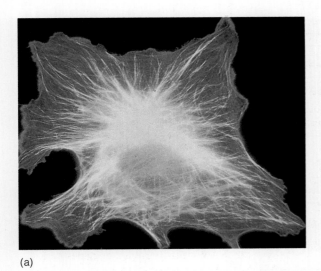

(a)

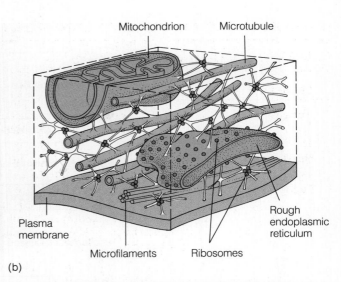

(b)

FIGURE 3–4 The Cytoskeleton (*a*) Photomicrograph of the cytoskeleton of a human fibroblast (connective tissue cell). Microtubbles are green and microfilaments are red. (*b*) Artist's rendition of the cytoskeleton showing its two major components: microtubules (hollow tubules made of protein) and smaller microfilaments (solid fibers made of actin and myosin proteins).

Specificity helps the cells regulate chemical reactions and maintain homeostasis. You can imagine the biochemical disorder that would result if one enzyme regulated all of the thousands of chemical reactions occurring in the cell. By producing many different types of enzymes, each specific for one, or—at most—a few, reactions, the cell can control metabolism with an extraordinary degree of precision. By controlling its enzymes, the cell produces only what it needs when it needs it. This helps conserve energy and chemical substances.

Enzymes belong to a class of compounds called **catalysts**. Catalysts speed up chemical reactions. Although they take an active role in the process, catalysts are left unchanged by the reaction and can therefore be used over and over again. The catalytic converter in your car, for instance, contains an inorganic catalyst that converts carbon monoxide, a poisonous gas in the car's exhaust, into carbon dioxide. Carbon monoxide is an odorless, colorless but highly poisonous gas; carbon dioxide, though poisonous at high levels, is not harmful to human health at levels typically encountered in urban environments.[1]

Enzymes can be regulated—switched on or off—by other molecules, often their own products. This pro-vides a means of controlling metabolic pathways and is a kind of cellular homeostatic mechanism. The end products of metabolic pathways often bind to specific control regions of one of the enzymes in a metabolic pathway, shutting off the enzyme and thus shutting down the entire metabolic pathway (Figure 3–6b). The binding sites, called **allosteric sites**, are molecular switches. The binding of a chemical to an enzyme's allosteric site causes changes in the structure of the enzyme that alter the shape of the its active site. This, in turn, renders the site inactive, making it unable to bind to substrates. When excess product builds up, the chemical reaction shuts down, and the entire metabolic pathway is blocked, in much the same way that an assembly line might shut down if a company were producing more of a product than it could sell. In other enzymes, products of metabolic pathways bind directly to the active site and physically block substrates from entering, blocking the enzyme and shutting down the metabolic pathway.

Control processes such as these are called feedback mechanisms. A **feedback mechanism** is any process that is regulated by its own end product(s). Furnaces in homes operate by a feedback mechanism. The thermostat sends a signal to the furnace that turns the furnace on when the temperature in the house drops below the desired setting. When the room temperature increases to the desired level, the thermostat shuts the furnace off. Heat, the product of the process, therefore "feeds back" on the system, shutting it down.

[1]Even though current levels of carbon dioxide gas in the environment are not harmful to human health, the buildup of carbon dioxide in the atmosphere may be partly responsible for global warming, a gradual increase in global temperature, discussed in Chapter 21.

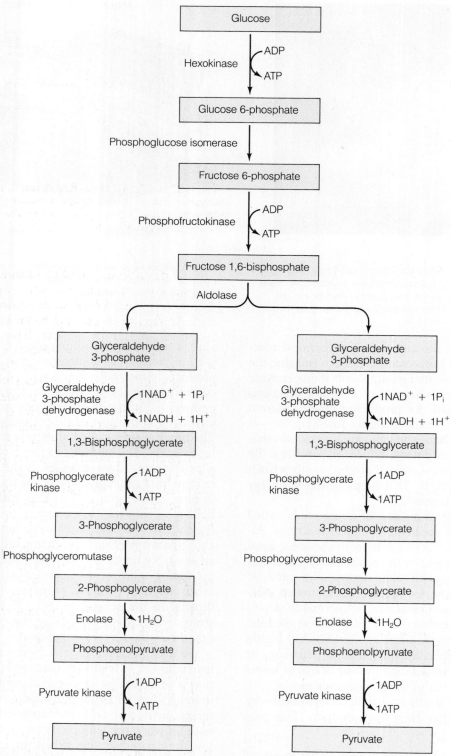

FIGURE 3–5 Metabolic Pathway Reactions in the cell occur as part of larger pathways where the product of one reaction becomes the reactant of another, as in this complex biochemical pathway. This illustrates the steps in glycolysis, the breakdown of glucose in the cytoplasm of human cells. Note that each reaction has its own enzyme.

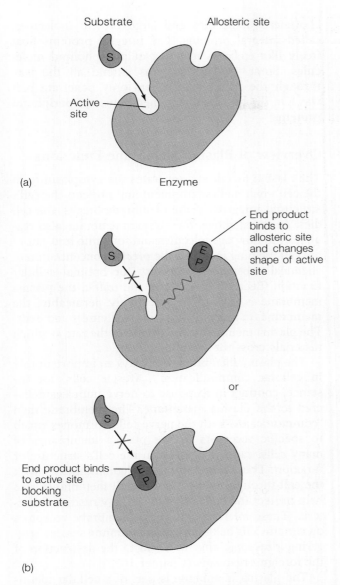

(a)

(b)

FIGURE 3–6   Enzyme Structure and Function
(*a*) Three-dimensional drawing of an enzyme. The active site conforms to the shape of the reacting molecule(s). The allosteric site regulates the enzyme. When products bind to the allosteric site, they can turn the active site on or off, depending on the enzyme. (*b*) End products of a reaction can inhibit the enzyme by binding to the active site or by binding to the allosteric site.

Feedback mechanisms are an essential component of homeostasis. The levels of hormones, glucose, sodium, calcium, and water in the blood, for example, are all controlled by feedback mechanisms. Without them, the body would cease functioning in a matter of seconds.

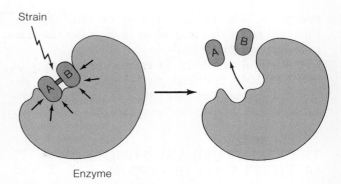

FIGURE 3–7   Enzyme-Regulated Cleavage   As the substrate molecule fits into the active site, it distorts the site slightly, which, in turn, puts strain on the molecule, enabling bonds to break.

How do enzymes speed up chemical reactions? To answer this question, we will consider catabolic (breakdown) reactions first. As shown in Figure 3–7, the substrate that is about to be broken down fits into the active site of its enzyme. The insertion of the molecule into the active site causes a slight distortion of the site, in much the same way that a hand distorts a glove when inserted into it. The change in the shape of the active site strains the covalent bonds of the substrate molecule. If the strain is great enough, a bond may break and the molecule will split apart. For the bond to break without an enzyme, however, a considerable amount of energy is required. If the molecule collided with another molecule, for instance, the energy imparted by the collision might be sufficient to break the bond. Enzymes, therefore, offer a place where intermolecular stresses needed to break bonds can be generated. This greatly reduces the amount of energy needed to break the bond. Reducing the energy required to drive reactions increases the rate at which anabolic reactions occur.

Enzymes also reduce the energy required to drive anabolic reactions. When two molecules react to form a new molecule in the absence of enzymes, the molecules must first be energized. In the chemistry laboratory, anabolic reactions can be initiated by heating reactants in test tubes. Heat accelerates the molecules, increasing the rate and force of collisions. Forceful collisions bring the atoms of the molecules close enough to react.

Active sites of enzymes involved in anabolic reactions provide a place for two molecules to react. The sites may also orient the reacting molecules in ways that increase the likelihood of chemical reaction. Thus, enzymes reduce the need for large amounts of energy to form bonds. Because of this, enzyme-catalyzed reactions occur many times faster than uncatalyzed reactions.

A single molecule of an enzyme can catalyze as many as 1000 reactions per second! Without enzymes, life could not exist; most of the chemical reactions in the human body would occur too slowly.

THE STRUCTURE AND FUNCTIONS OF THE PLASMA MEMBRANE

The cell performs hundreds of tasks. Perhaps one of the most important is the control of molecular traffic across its outermost boundary, the plasma membrane. Controlling what goes in and out of the cell determines the cell's internal chemical composition and helps determine the composition of the fluids lying outside the cell, called **extracellular** or **interstitial fluids**. Thus, the plasma membrane helps regulate the constancy of the internal and external microenvironments of the cell (Table 3–2). The plasma membrane is, therefore, also a cellular homeostatic structure.

Plasma Membrane Structure

The plasma membrane and all of the other membranes of the cell consist of lipids, proteins, and carbohydrates (Figure 3–8). As Figure 3–8 illustrates, most of the lipid molecules in the plasma membrane are phospholipids (Chapter 2). The phospholipid molecules form a double layer. The polar heads of the outer layer of phospholipid molecules protrude outward toward the aqueous extracellular fluid, and the polar heads of the inner layer of molecules protrude inward toward the aqueous cytoplasm (Figure 3–8). Interspersed in the lipid bilayer are a small number of

---

TABLE 3–2 Overview of Plasma Membrane Functions

Ensures the cell's structural integrity

Regulates the flow of molecules and ions into and out of the cell

Maintains the chemical composition of cytoplasm and extracellular fluid

Participates in cellular communication

Forms a cellular identification system

---

cholesterol molecules and large protein molecules, called integral proteins. The **integral proteins** float freely like icebergs in a sea of phospholipid molecules. Some integral proteins extend all the way through the membrane; others only penetrate part way. On the outside of the membrane are **peripheral proteins**.

Overview of Plasma Membrane Functions

The plasma membrane separates the cytoplasm from the cell's external environment and protects the cell's structural integrity; if the membrane breaks, the cell dies. As noted above, the plasma membrane also regulates the flow of molecules and ions into and out of the cell, helping maintain the precise concentration of chemical substances necessary for optimal cellular function. Because it regulates the traffic, the plasma membrane is said to be **selectively permeable**: the membrane, in a sense, "selects" what enters and exits. The plasma membrane also regulates the rate at which materials cross the membrane.

The plasma membrane also plays an important role in cellular communication.[2] Muscle cells, for instance, contract in response to nerve impulses delivered to the plasma membrane. The membrane then "communicates" with the nerve cell. Hormones attach to specific receptors in the plasma membranes of many cells, causing changes in the cell's structure or function. Thus, some hormones "communicate" with the cell through its membrane. Cell membranes also help protect the body from invading viruses and bacteria. These microbes attach to membrane receptors on certain cells belonging to the immune system, triggering a response that will lead to the destruction of the foreign organisms (Chapter 10).

The plasma membrane is part of a cellular identification system. Each person has a unique "cellular fingerprint," which is determined by the composition of integral proteins in the plasma membrane. The integral proteins are **glycoproteins**; that is, protein molecules that have carbohydrate molecules attached to them. Because the glycoprotein composition of the plasma membrane of the cells of an individual is unique, the body can recognize its own cells.[3] More importantly, it allows the body to distinguish foreign cells, including bacteria and even tumor cells. The cellular identification system no doubt evolved to pro-

---

[2]Communication is defined very broadly here to include any kind of message transmission or receival.

[3]The only people with identical cells are identical twins.

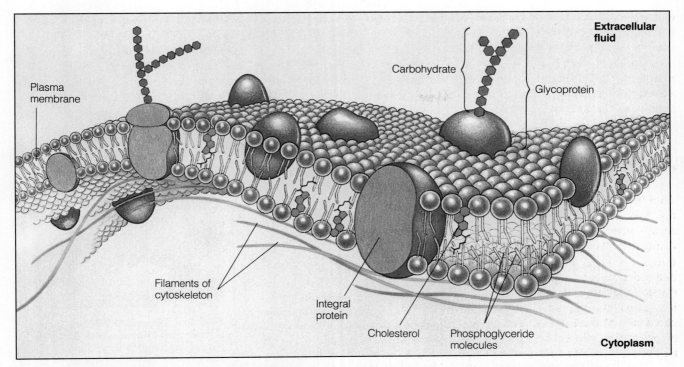

Plasma membrane

Filaments of cytoskeleton

Integral protein

Cholesterol

Phosphoglyceride molecules

Cytoplasm

Carbohydrate

Glycoprotein

Extracellular fluid

FIGURE 3–8 Fluid Mosaic Model of the Plasma Membrane The fluid mosaic model is the most widely accepted theory of the plasma membrane. Phosphoglycerides are the chief lipid component. They are arranged in a bilayer. Integral proteins float like icebergs in a sea of lipid.

vide the body with protection. The advent of tissue and organ transplantation presents some special challenges to medical science because the same system that protects us from cancer, bacteria, and viruses also destroys tissue and organ grafts. The immune system recognizes cells of the transplant as foreign (unless they come from an identical twin) and rejects them. To prevent rejection, patients must be treated with drugs that suppress the immune system. Repressing the immune system, however, makes a person more vulnerable to bacterial and viral infections. Early heart transplants, though heralded as a major breakthrough, all ended in failure. Almost all of the patients who died succumbed to bacterial and viral infections. Thankfully, new drugs are available to suppress graft rejection but permit some degree of immune protection.

Fetal cell transplantation, discussed at the opening of the chapter and in Health Note 3–1, could help doctors avoid tissue rejection. Fetal cells have not fully developed their glycoprotein fingerprint and can therefore be transplanted without triggering an immune reaction.

Controlling the Molecular Traffic across Plasma Membranes

The cytoplasm and the interstitial fluid outside the plasma membrane contain a variety of materials dissolved or suspended in water. Some of these can pass through the membrane; others cannot. Ultimately, the membrane itself determines which molecules can pass through and how quickly. But how does the plasma membrane "control" the traffic?

Control over plasma membrane traffic is achieved in a number of ways (Figure 3–9; Table 3–3). In chemistry you may have learned that "likes dissolve likes." In other words, chemically similar substances readily mix. A polar compound, for example, will readily dissolve in water. The converse is also true: chemically dissimilar substances do not mix. Thus, oil will not dissolve in water. Since the plasma membrane is made mostly of lipid, lipid-soluble materials, such as steroid hormones, oxygen, and carbon dioxide, pass directly through the lipid bilayer of the plasma membrane with ease.

Water-soluble materials cannot and must travel via other avenues. Water and many small water-soluble

HEALTH NOTE 3–1

Fetal Cell Transplants: The Allure of a New Cure

Treating and curing diseases are two different things. Consider diabetes. This disorder afflicts an estimated 6 to 8 million Americans and is characterized by an inability to control blood glucose levels. For years, diabetes has been treated by diet or by daily injections of insulin, a hormone lacking in many victims of the disease.

Despite the widespread use of insulin, a diabetic's life is never easy. Too much exercise or too much insulin can cause blood glucose levels to plummet. Victims are left feeling dizzy and weak. Excess blood sugar, caused by overeating or a failure to take an insulin shot on time, can cause irritability, depression, and fatigue—and even death. Short-term fluctuations in blood sugar are mostly an annoyance to diabetics, but diabetics are also plagued by several serious long-term complications. Damage to blood vessels and

nerves, for instance, are common. Blindness and kidney damage are two common results of the damage at the cellular level (Chapter 15).

Today, thanks to new developments in fetal cell transplantation, a cure may be possible for some victims of this disease. Fetal cell transplantation can be used to insert healthy fetal pancreas cells into diabetics who cannot produce insulin hormone on their own. Here, research suggests, the fetal cells will establish a permanent residence and will begin secreting insulin, ending a patient's dependency on insulin injections.

Fetal cell transplantation offers promise in other areas as well. Medical researchers are hoping that fetal brain cells transplanted into patients with Parkinson's disease will cure this disorder. In 1988, U.S. researchers announced the successful transplantation of human fetal brain cells

Fetal Cell Transplant Patient Once suffering from Parkinson's disease, Donald Nelson now leads a fairly normal life thanks to fetal cell transplantation.

into monkey brains. Seventy days after transplantation, the grafted cells had formed a dense tangle of maturing nerve cells. Two of three grafts showed signs of producing dopa-

---

molecules and ions, for example, are thought to pass through **pores** in the membrane believed to be formed by integral proteins (Figure 3–9b).[4] Some protein pores apparently remain open and allow the free movement of small molecules and ions. Water-soluble materials pass through these pores rather freely, from regions of high concentration to low concentration, in a process called **diffusion**.

The third avenue is provided by **carrier proteins**, molecules that transport the smaller molecules and ions across the membrane in ways not completely understood. Carrier proteins shuttle molecules across membranes from regions of high concentration to low

concentration—movement often said to be "down the concentration gradient." This process is called **facilitated diffusion** to distinguish it from forms of diffusion that do not require carrier proteins (Figure 3–9c).

The fourth transport mechanism is active transport. **Active transport** is the movement of molecules across membranes with the aid of transport molecules and energy supplied by ATP (Figure 3–9d). Active transport often moves molecules and ions from regions of low to high concentration—movement "up the concentration gradient." Active transport occurs in cells that must concentrate chemical substances to function properly. For example, thyroid cells require large quantities of the iodide ion (I^-) to manufacture the hormone thyroxine. Thyroid cells actively transport iodide ions from the bloodstream, where they are present in fairly low concentrations, into the cyto-

---

[4]The evidence for pores is fairly circumstantial. No one has ever seen a pore in electron micrographs of cells, so scientists only hypothesize their existence on the basis of physiological research.

mine, a chemical lacking in the brains of Parkinson's victims. Researchers hope that these cells will restore normal brain levels of dopamine, which could eliminate tremors and rigidity.

Since their announcement, American and Swedish scientists have transplanted human fetal brain tissue into human subjects. Early results have been disappointing, but new studies show more promise, raising hopes that the technique can help millions of people worldwide.

On another front, researchers are looking to fetal cell transplantation to provide a way to reverse the neurological effects of long-term alcohol abuse in humans. Rats treated with alcohol suffer memory loss similar to that which occurs in chronic alcoholics. Recently, medical researchers announced the successful transplantation of fetal brain tissue into the brains of alcohol-treated rats. About nine weeks after receiving transplants of fetal brain tissue in certain areas of the brain, alcohol-treated and non-alcohol-treated rats were subjected to performance tests that measured memory. Twelve of fourteen alcohol-treated rats performed remarkably well on trials, almost as well as control rats who had received no alcohol. Researchers do not know why the brain transplants work but note that similar transplants could improve the lives of many recovering alcoholics whose brains are severely damaged after years of alcohol abuse.

Fetal cell tissue transplantation offers great hope to humankind. The technique could provide physicians with an enormous opportunity to cure diseases long thought to be incurable. Fetal cell tissue transplantation has already stirred considerable controversy in the United States. (See the Point/Counterpoint on page 70.)

In most states, the use of fetal tissue for experimentation is legal, and a federal panel convened to debate the issue concluded that clinical trials should be conducted. The possibility of relieving suffering and saving life, the panel said, "cannot be a matter of moral indifference to those who shape and guide public policy." Unfortunately, say some, the Department of Health and Human Services, which funds such research, has decided to withhold research money, forcing researchers to seek outside money.

To avoid controversy, some researchers are working to develop cell lines of human fetal cells that could be used in place of tissues taken from aborted fetuses. Unfortunately, many researchers believe that tissue culture techniques suitable for fetal cells are a decade away from perfection.

---

plasm, where the iodide concentration is many times higher. If the thyroid cell relied on diffusion, it would not be able to produce enough hormone to meet the body's demand.

How do active transport proteins function? The simplest active transport proteins have two binding sites, as illustrated in Figure 3–9d. One of them attaches to the molecule that is to be moved across the membrane; the other binds to ATP. Biologists hypothesize that the breakdown of the ATP molecule releases energy that causes the protein molecule to change its shape in such a way that the protein propels the molecule attached to it across the membrane—moving it in or out of the cell.

Most active transport molecules transport two substances across the membrane at the same time, often in opposite directions. For example, in nerve cells an active transport protein pumps sodium ions out of the cytoplasm while pumping potassium ions inward. This pump ensures the proper functioning of the nerve cell (Chapter 12).

The fifth, and final, mechanism that enables cells to ingest large molecules, such as proteins, or even other cells, is called **endocytosis**. Endocytosis (literally "into the cell") is illustrated in Figure 3–9e. It requires ATP. Endocytosis consists of two different processes: phagocytosis and pinocytosis. **Phagocytosis** (cell eating) occurs when cells engulf larger particles, such as bacteria and viruses. **Pinocytosis** (cell drinking) occurs when cells engulf extracellular fluids and dissolved materials. In humans, phagocytosis is limited to a relatively few types of cells: those involved in protecting the body against foreign invaders (Chapter 8). However, most, if not all, cells, are capable of pinocytosis.

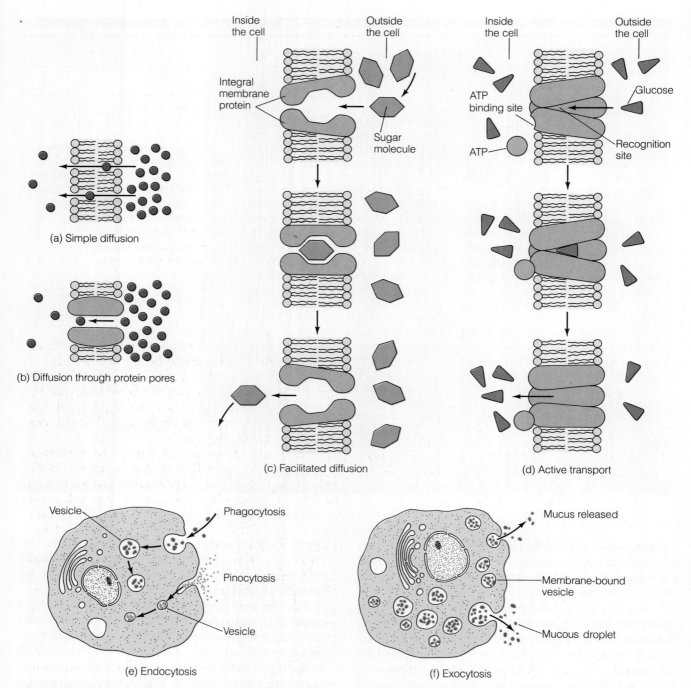

FIGURE 3–9 Membrane Transport Molecules move through the plasma membrane primarily in five ways. (a) Lipid-soluble substances pass through the membrane directly. (b) Water-soluble molecules may diffuse passively through pores formed by protein molecules. (c) Water-soluble molecules may also diffuse through membranes with the assistance of proteins in facilitated diffusion. (d) Other proteins use energy from ATP to move against concentration gradients. (e) Finally, cells may engulf large particles, cell fragments, and even entire cells via endocytosis. (f) Exocytosis, the reverse process, rids the cell of large particles.

TABLE 3–3   Overview of Plasma Membrane Transport

PROCESS	DESCRIPTION
Diffusion	Flow of ions and molecules from high concentrations to low. Water-soluble ions and molecules probably pass through pores; water-insoluble molecules pass directly through the lipid layer
Facilitated diffusion	Flow of ions and molecules from high concentrations to low concentrations with the aid of protein carrier molecules in the membrane
Osmosis	Diffusion of water molecules from regions of high water (low solute) concentration to regions of low water (high solute) concentrations
Active transport	Transport of molecules from regions of low concentration to regions of high concentration with the aid of transport proteins in the cell membrane and ATP
Endocytosis	Active incorporation of liquid and solid materials outside the cell by the plasma membrane. Materials are engulfed by the cell and become surrounded in a membrane
Exocytosis	Release of materials packaged in secretory vesicles

Cells release large molecules, such as hormones, by a process called **exocytosis** ("out of the cell"); the reverse of endocytosis (Figure 3–9f). In the hormone-producing cells of the pituitary gland, for instance, protein hormones are packaged internally into tiny membrane-bound vesicles, called **secretory vesicles** (discussed later). Secretory vesicles migrate to the plasma membrane and fuse with it. At the point of fusion, the membrane breaks down, and the protein hormone is released into the extracellular fluid.

Osmosis

Like any other small molecule, water moves from one side of a plasma membrane to the other by diffusion. The diffusion of water across a selectively permeable membrane is given a special name, **osmosis**. To understand osmosis, consider a simple example. Suppose we had a large bag made of a selectively permeable material that permits water to flow in and out, but will not allow dissolved substances in the water to pass through. Suppose also that we filled the bag with water and sodium chloride (table salt), then submerged it in a large container of distilled water (Figure 3–10). Distilled water has no ions in it. In other words, it is 100% water.

As soon as the bag containing salt water hit the water, it would begin to swell because of the inward flow of water molecules—osmosis. Why does water flow inward? As Figure 3–10 illustrates, water flows from regions of high water concentration to areas of low water concentration in the same way that any other substance diffuses across a membrane. But remember that the salt cannot move out through the membrane. Since water molecules enter, but salt ions remain inside, the bag swells.

Water concentration is a measure of the number of water molecules per milliliter. Pure water obviously has a higher water concentration than salt water; that is, there are more water molecules per unit volume in pure water than in salt water. This concentration difference drives the movement of water across selectively permeable membranes.

Whenever two fluids with different solute (dissolved substance) concentrations are separated by a selectively permeable membrane, water tends to flow from one to the other. The driving force is called **osmotic pressure**. The greater the difference in concentration, the greater the osmotic pressure, and the more quickly water moves.

Osmotic pressure is responsible for the movement of water across tiny blood vessels called capillaries in body tissues (Chapter 8). Thus, osmosis helps equalize water concentrations on opposite sides of membranes and is an important homeostatic mechanism. Because of osmosis, the fluid surrounding body cells generally has the same concentration as the cytoplasm of the cells. Therefore, the extracellular fluid is said to

FIGURE 3–10 Osmosis
Osmosis is the diffusion of water molecules from a region of high water concentration to low water concentration across a semi-permeable membrane. (*a*) To demonstrate the process, immerse a bag of salt water in a solution of pure water. (*b*) Water diffuses into the bag causing it to swell.

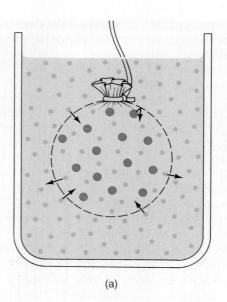

(a)

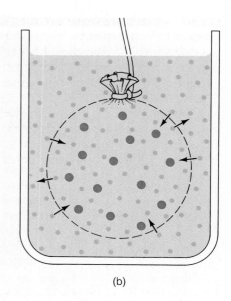

(b)

● Ions of salt　　• Water molecules

be **isotonic** ("having the same strength") to the cytoplasm. Fetal cells used in tissue transplants must be cultured in the laboratory in isotonic solutions to survive. If these cells were immersed in a solution that is more concentrated than the cytoplasm—that is, having more solute and less water—water would move out of the cells in an attempt to equalize the water concentrations on both sides of the membrane.

A solution with a higher solute concentration than the cell's cytoplasm is said to be **hypertonic** ("having a greater strength"). Hypertonic solutions cause cells to shrivel. A solution with a solute concentration lower than the cell's cytoplasm is said to be **hypotonic** ("having a lesser strength"). If fetal cells were placed in a hypotonic solution, water would rush in, causing the cells to swell and burst.

CELLULAR COMPARTMENTALIZATION: CELLULAR ORGANELLES

Cell compartmentalization has evolved and persisted over time because it offers many advantages, efficiency being one of the most important, as explained below. This section examines the structure and function of cell organelles (Table 3–1).

The Cell's Command Center: The Nucleus

The nucleus is usually the cell's largest organelle (Figure 3–11a). It contains DNA, the genetic material that controls the structure and function of the entire cell. The nucleus is the command center of the cell, for its DNA contains the cell's operating instructions (Chapters 4 and 5). Each cell contains the same genetic information, but different portions are used by different cells. Thus, a muscle cell uses a different set of commands than a liver cell. The selective repression of part of a cell's genetic material occurs during cellular differentiation.

The nucleus consists of (1) the nuclear envelope, (2) the chromatin, (3) the nucleoplasm, and (4) the nucleolus. The **nuclear envelope** is a double membrane that isolates the nuclear material from the cytoplasm. Minute channels, the **nuclear pores**, allow materials to pass to and from the nucleus (Figure 3–11b).

The bulk of the nucleus contains long, threadlike fibers of DNA and protein. These fibers are called **chromatin** and appear as fine granules in the nucleus in Figure 3–11a. Proteins, water, and other small molecules and ions are also found in the nucleus, forming a semifluid material called the **nucleoplasm**. Just before cell division, the chromatin fibers coil to form short, compact bodies, known as chromosomes (Figure 3–12). This compaction makes the chromosomes easier to separate when the cell and its nucleus divide (Chapter 4).

Nucleoli ("little nuclei") are temporary structures found in the nuclei of cells between cell divisions. They appear as small, clear, oval structures in light micrographs and dense bodies in electron micrographs. Nucleoli are regions of the DNA that actively produce RNA (Chapter 5). The RNA produced at each nucleolus is called **ribosomal RNA** (rRNA) because it combines with certain proteins to form **ribosomes**, non-membrane-bound organelles that appear as small dark granules in electron micrographs. Ribosomes play an important part in protein synthesis, described in Chapter 5. Ribosomes consist of two subunits, each containing RNA and protein. These subunits enter the cytoplasm through the nuclear pores.

FIGURE 3–11   The Nucleus (*a*) The nucleus houses the genetic information that controls the structure and function of the cell. The nuclear envelope, made of lipid and protein like the plasma membrane, actually consists of two membranes, separated by a space. Pores in the membrane allow the movement of molecules into the nucleus, providing raw materials for the synthesis of DNA and RNA. They also allow RNA molecules to travel into the cytoplasm where they participate in the production of protein. (*b*) Colorized scanning electron micrograph of the nuclear membrane showing numerous pores.

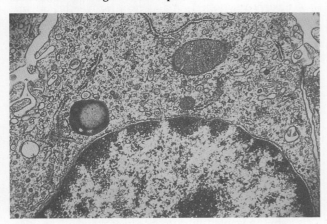

(a)

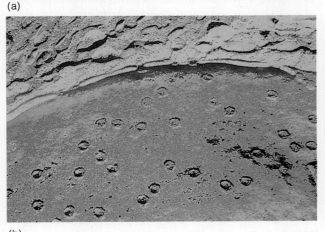

(b)

Energy Production in the Cell: The Mitochondrion

Cells use energy for a great many purposes: to synthesize chemical substances, to transport molecules across membranes, to divide, to contract, and to move about. As noted in Chapter 2, cellular energy is stored in ATP molecules. Most of the ATP produced by cells is synthesized in the **mitochondrion** (plural, **mitochondria**) (Figure 3–13). Mitochondria vary considerably in form and number from cell to cell. Nevertheless, all mitochondria have several common characteristics. For example, all mitochondria contain two membranes. The inner membrane, shown in Figure 3–13b, is thrown into folds, or **cristae**. The inner membrane creates two distinct compartments within the mitochondrion: the inner compartment and the outer compartment. Each compartment has a special function in energy production, described below.

In the human body, glucose is the principal source of cellular energy. (Remember that all cellular energy ultimately comes from the sun.) In cells, glucose is broken down in a series of chemical reactions, each regulated by its own enzyme. A portion of the energy released during the catabolism of glucose is captured by the cell to produce ATP, which, as explained in Chapter 2, is used to power cellular functions.

Cellular energy production is a four-part process.[5] The first occurs in the cytoplasm. Steps 2, 3, and 4

[5]Use of the term *cellular energy production* is not meant to imply that cells create energy—only that they release energy stored in chemical bonds.

FIGURE 3–12   Chromosomes The threadlike chromosomes, made of chromatin fibers consisting of protein and DNA, must condense before the nucleus can divide.

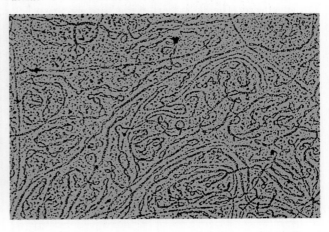

FETAL TISSUE TRANSPLANTS: AUSCHWITZ REVISITED

Thomas J. Longua

Thomas J. Longua has been an educator for 29 years and is currently instructor of anatomy and physiology at the Denver Academy of Court Reporting. He has been active in the pro-life movement for 17 years and is currently serving as president of the Colorado Right to Life Committee.

Opposition to the practice of using tissue from aborted human babies for medical purposes starts with opposition to abortion itself. For at least two and a half millennia, abortion has been universally recognized as the destruction of human life.

In the mid-1960s, Western society decided that it needed an "acceptable" way to rid itself of the unwanted babies it was creating in its new-found sexual permissiveness. Legal abortion was its answer. We who opposed legalization predicted that it would lead to other evils, including infanticide and euthanasia. What most of us did not foresee, however, was the horror of cannibalizing aborted human fetuses for medical purposes. Even when this practice began to be discussed, we still could not believe that "civilized" people could take it seriously. Auschwitz was only a generation away.

When the Nazi barbarities were fully realized, a civilized world convened the Nuremberg Tribunal. The prosecutor of the Doctors' Trials opened his case with this statement: "The defendants in this case are charged with . . . atrocities committed in the name of medical science. . . . The wrongs which we seek to condemn and punish have been so calculated, so malignant, and so devastating, that civilization cannot tolerate their being ignored, because it cannot survive their being repeated."

Yet here we are again, committing the same "malignant wrongs," and with the exact same rationale, including the pretense that the victims are "sub-human." The only difference between the Nazi "atrocities" and modern medical practiceis the age of the victims.

Advocates of such transplants pretend that the procedure is acceptable if it is "separated" from the abortion itself. But this "separation" charade won't wash. First, it is exactly the same defense used by the doctors at Nuremberg: "We caused no deaths. They were all consigned to death by legal authorities. . . we tried to salvage some good from their plight." Furthermore, new abortion techniques have been—and are being—developed, solely ensure that fetal tissue is more usable.

The fact is, the entire practice of fetal tissue transplants rests on the "acceptability" of killing some humans. Indeed, a National Institutes of Health panel that approved such transplants based its findings on the fact that "abortion is legal."

Besides the inherently ghoulish nature of such transplants, this practice will certainly

- Further legitimize—and even sanctify—the original killing. What woman contemplating abortion will not be swayed by the argument that the decision to kill her baby will help others? Dr. James J. Parks, a Denver abortionist bragged, "[My patients] say, 'Thank God, some good is going to come out of this.'" (*New York Times,* November 19, 1989)

- Increase the number of abortions. Some medical journals claim that such tissue could be used to treat a vast array of diseases. In a guest editorial in the November 7, 1988, *Wall Street Journal,* Dr. Emanuel D. Thorne estimated that the current 1.6 million abortions in America every year will not be enough to keep up with the demand for such tissue.

- Increase the pressure to legalize euthanasia. If harvesting unwanted humans becomes standard practice, who can doubt that there will be further demand for organs from older humans whose lives, like those of aborted babies, are determined to be "meaningless"?

- Lead to trafficking in human "spare parts"—even to pregnancy planned specifically for that purpose, as has already happened.

Advocates of fetal tissue transplants depict themselves as "humanitarians," as did the doctors at Nuremberg. But, in fact, those who condone—and even encourage—the slaughter of society's most helpless members are not "humanitarians," but barbarians.

HUMAN FETAL TISSUE SHOULD BE USED TO TREAT HUMAN DISEASE

Curt R. Freed

Curt R. Freed, M.D., is a professor of medicine and pharmacology at the University of Colorado School of Medicine in Denver, Colorado. He has written some 50 articles as well as numerous abstracts, chapters, and reviews on medical topics.

Despite its legalization, abortion remains a controversial issue and will continue to stir debate in the future. In the United States, the debate centers on whether a woman has the right to control her reproduction. The future developments of this political debate are uncertain, but recent elections suggest that pro-choice candidates have been victorious when elections are based on the abortion issue.

Currently, over one million legal abortions are performed in the United States each year; most abortions are performed in the first trimester. For nearly all women, having an abortion is an anguishing choice filled with regret and ambivalence. Nonetheless, the difficult personal decision to terminate a pregnancy is made. After the abortion, fetal tissue is usually discarded.

As an alternative to throwing this tissue away, research has shown that fetal tissue may be useful for treating patients with disabling diseases. For over 50 years, research in animals has demonstrated that fetal tissue has a unique capacity to replace certain cellular deficiencies and so may be useful for treating some chronic diseases of humans. These diseases include Parkinson's disease, diabetes, and some immune system disorders.

Cadaver fetal tissue offers the promise of helping large numbers of Americans with crippling diseases. Parkinson's disease, for example, affects hundreds of thousands of Americans. By reducing the ability to move, the disease can end careers and turn people into invalids. The disease is caused by the death of a small number of critically important nerve cells that produce a chemical called dopamine. Experiments in animals and early experiments in humans indicate that fetal dopamine cells transplanted into the brains of these patients may restore a patient's capacity to move and may eliminate the disease. Patients whose minds work perfectly well and whose bodies are otherwise normal may become healthy and productive citizens once again.

Concern about using cadaver tissue to treat humans has been debated for nearly 40 years. As kidney, cornea, and other organ transplants were developed in the 1950s, many objected to recovering organs from cadavers. In the intervening decades, opinion has changed so that the practice of recovering kidneys, hearts, livers, skin, and corneas from cadavers has gone from a provocative and controversial practice to an accepted policy endorsed by most states. In fact, in most states a check-off box on the back of driver's licenses is used to give permission for organ donation in the event of the death of the driver. Because abortions are induced, some argue that fetal tissue should be regarded differently than other cadaveric tissue. Given the facts that abortion is legal and that fetal tissue is ordinarily discarded, there should be no moral dilemma in using fetal tissue for therapeutic purposes. As with the use of all human tissue for transplant, specific informed consent by the woman donating the tissue must be obtained.

Some have proposed that using fetal tissue for therapeutic purposes will increase the number of abortions. This is preposterous. It strains the imagination to think that a woman would get pregnant and have an abortion simply on the chance that the aborted fetal tissue might be used to treat a patient unknown to her. An unwanted pregnancy is an intimate and deeply personal crisis; it is inconceivable that the pregnancy would be seen primarily as a philanthropic opportunity.

Politics and medicine have frequently mixed in the past and will continue to do so in the future. As a physician, I think it is important to try to improve the health of patients with serious diseases. Legally acquired fetal cadaver tissue that would otherwise be discarded should be used to treat humans with disabling diseases.

SHARPENING YOUR CRITICAL THINKING SKILLS

1. Summarize the positions of each author.
2. Using your critical thinking skills, analyze the view of each author. Are they well substantiated? Do author's biases play a role in each argument?
3. Each position is based on at least one key factor. What are they?
4. Which viewpoint do you agree with? Why? What factors (biases) affect your decision?

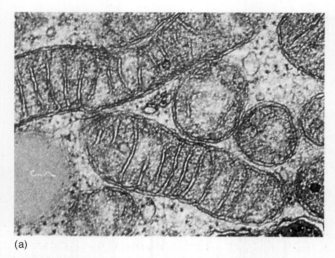

(a)

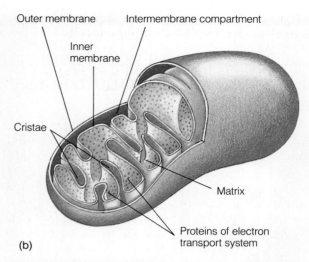

(b)

FIGURE 3–13 The Mitochondrion (a) Like the nucleus, the mitochondrion is delimited by a double membrane made of protein and lipid. The inner membrane, however, is infolded, forming cristae. (b) The infolding creates two distinct compartments: an inner compartment filled with a material called the matrix, and an outer compartment lying between the two membranes. Compartmentalization is essential for ATP synthesis.

occur inside the mitochondrion and are described in more detail in the following section.

During the breakdown of glucose, the cell captures only about a third of the energy contained in the molecule. The rest of the energy is given off as heat. This "waste" product of human metabolism creates body heat, which is necessary for maintaining normal enzymatic function. Thus, "waste" heat is not really wasted.

The importance of body heat is evidenced when body temperature drops. Immersion in the icy water of a spring snowmelt, for instance, can quickly sap a person's strength, leaving victims too weak to swim even a few feet to shore (Figure 3–14). Prolonged exposure to cold without adequate protection may result in **hypothermia**, a steady decrease in body temperature that can lead to death. Too much heat can also be dangerous. Exercise, for example, accelerates the breakdown of glucose in muscle cells and, consequently, increases the amount of heat generated by the body. Increased heat output explains why cross-country skiers can stay warm in the sub-zero winters of Maine or Minnesota—as long as they keep moving (Figure 3–15). Periodic elevations in body temperature are not dangerous, but troubles can arise when body temperature is prolonged. Male long-distance runners, for example, who practice every day for many months, may become temporarily sterile. Raised body temperature temporarily impairs sperm production (Chapter 16). A prolonged elevation of body temperature, resulting from a bacterial or viral infection, can result in death.

Mitochondria play a key role in homeostasis by maintaining energy supplies and body temperature. Their function is necessary for individual cells, but also benefits the entire organism.

During early cellular evolution, the ancestors of mitochondria were probably free-living organisms, similar to bacteria. As explained in more detail in Chapter 19, because of evolutionary pressures, these organisms probably took up residence inside other cells. There they remained, living off the food supply and providing heat and ATP in return. Biologists believe that this relationship persisted over many millions of years, and the once free-living organisms became a permanent fixture of the new cells, the eukaryotes.

How can biologists make such claims? For one thing, mitochondria resemble bacteria in many respects. Moreover, they are capable of dividing and can make many of their own proteins. Cells, in fact, cannot make mitochondria from scratch; all mitochondria come from the division of existing ones. Finally, mitochondria contain their own DNA. These facts and others, described in Chapter 19, suggest that mitochondria were once free-living organisms. Biologist and philosopher Lewis Thomas called them the "essentially foreign creatures" within our cells.

Manufacturing in the Cell: Endoplasmic Reticulum, Ribosomes, and the Golgi Complex

Cells synthesize many chemical substances needed for growth, reproduction, and day-to-day maintenance. Many of these materials are used within the cell; others, such as hormones, are released from the cell and travel to neighboring or distant cells where they elicit some response. This section looks at the cellular manufacturing capabilities and describes the organelles involved in this process.

Endoplasmic Reticulum. Coursing throughout the cytoplasm of many cells is a branched network of channels called the **endoplasmic reticulum** (Figure 3–16a). These channels are formed from flattened sheets of membrane, derived chiefly from the nuclear envelope. In some cells, the endoplasmic reticulum is coated with ribosomes, small darkly stained granules (Figure 3–16b). Ribosome-coated endoplasmic reticulum is called **rough endoplasmic reticulum (RER)**. The RER produces protein for use outside the cell.

Protein hormones and digestive enzymes, for example, are both produced on the RER. Digestive enzymes are manufactured in cells in the pancreas. These enzymes are released from the cell by exocytosis and travel by a system of ducts to the small intestine where digestion takes place.

The RER also produces some digestive enzymes that remain within the cell. These enzymes are bound by a membrane and a part of a cellular organelle called the lysosome. Lysosomes are used to digest materials phagocytized by the cell, a process described in more detail soon.

The RER produces proteins on the outside of the membrane, then transfers them into the channel, or **cisterna** (plural, cisternae) where they are protected from the cytoplasm. In the cisterna, resident enzymes may add small carbohydrate units to some proteins, producing glycoproteins. Thus, the endoplasmic reticulum not only produces proteins, but also chemically modifies some of them.

Most cells also contain some endoplasmic reticulum lacking ribosomes, which is called **smooth endoplasmic reticulum** or **SER** (Figure 3–16c). The SER produces phosphoglycerides used to make the plasma membrane of cells. The SER also performs a variety of additional functions in different cells. In the liver, for

FIGURE 3–14 Canoeist in Trouble Hypothermia sets in quickly when a swimmer is exposed to icy spring snow melt.

FIGURE 3–15 Cross-country Skier Despite the cold, this skier stays comfortably warm thanks to body heat generated by active muscles.

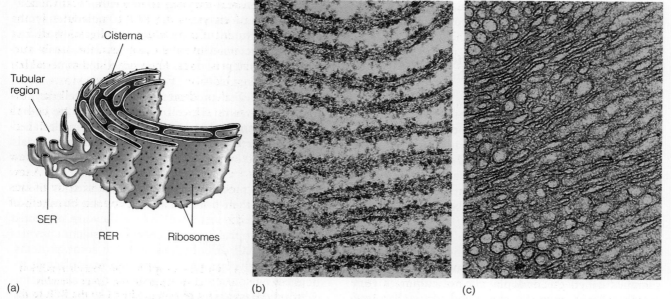

Cisterna

Tubular region

SER

RER

Ribosomes

(a)

(b)

(c)

FIGURE 3–16   The Endoplasmic Reticulum
(a) Created by a network of membranes in the cytoplasm, the endoplasmic reticulum is often studded with small particles, the ribosomes, forming rough endoplasmic

reticulum. This is the site of the synthesis of protein for lysosomes and extracellular use (digestive enzymes and hormones). (b) Electron micrograph of RER and (c) SER.

instance, SER detoxifies certain drugs, such as barbiturates, a type of sedative. The SER is the site of hydrochloric acid production in cells located in the glands of the stomach lining. In the adrenal glands, the SER is the site of steroid hormone synthesis.

Ribosomes.   Ribosomes, mentioned earlier, are tiny particles made of protein and ribosomal RNA (rRNA). They are located along certain sections of the endoplasmic reticulum, but are also found free in the cytoplasm. On the RER and in the cytoplasm, ribosomes attach to molecules of RNA, known as messenger RNA. **Messenger RNA (mRNA)** is produced in the nucleus of the cell and contains the information needed to make proteins. Messenger RNA shuttles information needed to make enzymes and structural proteins from the nucleus to the cytoplasm—the site of all protein synthesis. Ribosomes that bind to the mRNA aid in building the protein.

In the cytoplasm several ribosomes usually attach to a single free mRNA strand, forming an organelle called a **polyribosome** or **polysome**. Polysomes synthesize proteins that are used inside the cell—for example, the structural proteins of the cytoskeleton. The role of the mRNA and the ribosomes in protein synthesis is described in more detail in Chapter 5.

The Golgi Complex.   The rough endoplasmic reticulum transfers the proteins it manufactures into its cisterna where they may be chemically modified. The RER then transports these proteins to the terminal ends of the cisternae (Figure 3–17). As the protein builds up inside the RER, the ends of the cisternae enlarge. Eventually, they grow so large that they pinch off, forming membrane-bound vesicles. These vesicles then migrate to another organelle, the Golgi complex, for further processing (Figure 3–17).

The **Golgi complex** consists of a series of flattened membranes lying on top of one another, forming channels (Figure 3–18). The Golgi complex performs three principal functions. First, it sorts the molecules it receives by destination, in much the same way that postal workers sort outgoing mail. Thus, digestive enzymes that will remain in the cell in lysosomes are segregated from hormones. Second, the Golgi complex, like the RER, contains internal enzymes that chemically modify some of the proteins, adding carbohydrates or other molecules. Finally, the Golgi complex repackages its proteins into two types of membranous vesicles, secretory vesicles (also called secretory granules) and lysosomes.

As illustrated in Figure 3–17, **secretory vesicles** are released from the Golgi complex and stored in the cytoplasm until needed. When the signal for release comes, secretory vesicles travel to the plasma membrane. Here, they fuse with the plasma membrane. The point of fusion breaks down, and the contents of the secretory vesicle are released by exocytosis.

Cellular Digestion: Lysosomes

Lysosomes ("digestive bodies") are membrane-bound organelles that contain digestive enzymes (Figure 3–19a). These enzymes are used for two purposes: to break down materials that enter the cell by endocytosis and to destroy aged or malfunctioning cellular organelles. As noted above, the digestive enzymes of lysosomes are produced by the RER and packaged by the Golgi complex.

The function of lysosomes is shown in Figure 3–19b. As illustrated, material engulfed by the cell is enclosed by a segment of the plasma membrane. The membrane and its enclosed material are called a **food vacuole.** Lysosomes attach to food vacuoles. At the point of fusion, the membranes break down, thus permitting lysosomal enzymes to enter the food vacuole. The lysosomal enzymes digest the molecules in the food vacuole, and the products of digestion diffuse out of the vacuole into the cytoplasm for use in various metabolic processes. The undigested material left behind is expelled from the cell by exocytosis.

Lysosomes also destroy defective cellular organelles, such as mitochondria, helping the cell to maintain its structure and function. How the cell recognizes defective organelles remains a mystery.

Most cells in the human body contain only a few lysosomes to recycle malfunctioning organelles. A few cell types, which scavenge the blood and body tissues for bacteria and viruses, however, contain hundreds of these organelles.

FIGURE 3–17 Protein Synthesis and Secretion
Protein packed in lysosomes and secretory granules (for later export) is synthesized on the RER and transferred in tiny transport vesicles to the Golgi complex. Protein is sometimes chemically modified in the cisternae of both the RER and the Golgi complex. The Golgi complex separates protein by destination and repackages it into secretory granules, which remain in the cytoplasm until secreted by exocytosis.

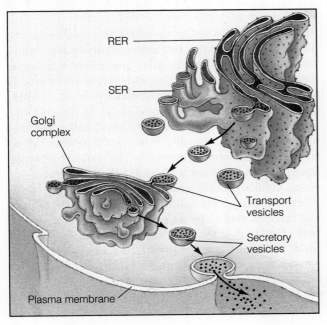

FIGURE 3–18 Golgi Complex (*a*) Artist's rendition of the three-dimensional structure of the Golgi complex. Transport vesicles carry protein produced by the RER to the Golgi complex and fuse to it, releasing the protein into the cisternae of the Golgi complex. Here it is sometimes chemically modified and repackaged. (*b*) Electron micrograph of the Golgi complex.

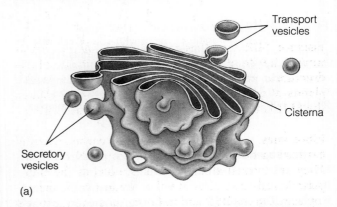

(a)

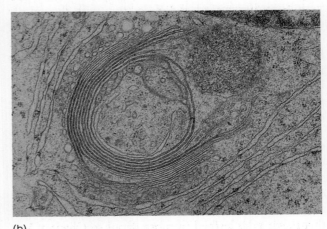

(b)

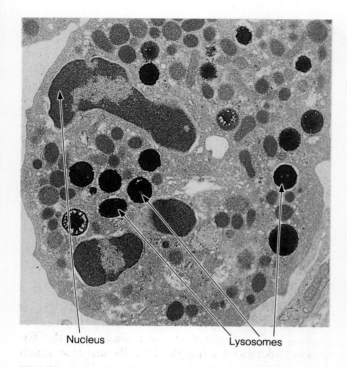

(a)

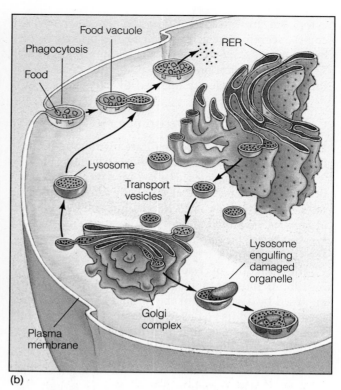

(b)

FIGURE 3–19 The Lysosome (*a*) Electron micrograph of a neutrophil, a type of white blood cell that phagocytizes bacteria and cellular debris in wounds. The dark-staining bodies are lysosomes, filled with digestive enzymes that are used by these cells to break down ingested material. (*b*) The digestive enzymes of the lysosome are produced on the RER and transported to the Golgi complex for repackaging. Lysosomes fuse with food vacuoles, which allows their enzymes to mix with the contents of the food vacuole. The enzymes digest the contents, which diffuse through the membrane into the cytoplasm where they are used. The membrane surrounding the lysosome helps protect the cell from digestive enzymes.

Lysosomes also play an important part in human embryonic development. Initially, during development, for instance, the fingers of the human fetus are webbed. The cells of the webbing, however, are genetically programmed to self-destruct with the aid of enzymes released from lysosomes.

Lysosomal analysis can be a useful diagnostic tool for physicians. Since lysosomal enzymes are released into the bloodstream when heart cells die during a heart attack, blood enzyme measurements can help physicians diagnose heart attacks. Several other diseases can also be detected by blood enzyme levels.

Cell Movement: Cilia and Flagella. Most cells in the body are fixed. The cells that are mobile, however, are usually transported passively in the blood and other body fluids like rafts in a stream. Red blood cells and white blood cells, for instance, are pumped throughout the body in the branching network of arteries and veins. One group of white blood cells, the neutrophils, however, can escape the **capillaries**, thin-walled vessels in body tissues, and migrate through body tissues on their own accord. Neutrophils propel themselves by **ameboid movement**, which is illustrated and explained in Figure 3–20. In infected tissues, neutrophils engulf bacteria, dead cells, and cellular debris. Neutrophils, however, die when their supply of lysosomes gives out. The yellowish white liquid emanating from a wound, called **pus**, contains water, cellular debris, and dead neutrophils.

The human sperm cell utilizes another kind of motive force provided by an organelle called the flagellum (Figure 3–21). The **flagellum** (plural, flagella; Latin for whip) is a long, whiplike extension of the plasma membrane. It contains numerous small tubules, called **microtubules**, that produce and transmit the motive force.

The flagellum resembles another cellular organelle

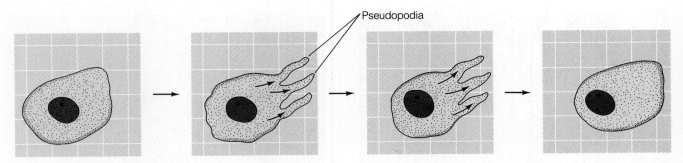

Pseudopodia

FIGURE 3–20 Ameboid Movement Some cells in the human body move by ameboid movement. The cell sends out minute extensions called pseudopodia (false feet), which attach to solid surfaces. Cytoplasm then flows into the pseudopodia, moving the cell forward.

found in several locations in the human body—the cilium (plural, cilia). **Cilia** are much smaller extensions of the cell. In humans, cilia move materials and fluids over the surface of stationary cells (Figure 3–22a).

As Figures 3–22b and 3–22c illustrate, cilia and flagella contain nine pairs of microtubules. The microtubules are arranged in a circle around a central pair. Biologists typically refer to this as the "9 + 2 arrangement." At the base of cilia and flagella is the **basal body** (Figure 3–22d). The basal body gives rise to cilia and flagella and contains nine sets of micro-

tubules (with three microtubules each) and no central pair.

Although cilia and flagella are similar in some ways, they are quite different in others. Flagella, for example, have additional fibers outside the central 9 + 2. It is thought that these fibers give the flagellum additional strength. Flagella are also much longer than cilia and less numerous. Each human sperm cell, for example, has only one flagellum, while ciliated cells lining the respiratory system and parts of the reproductive system contain thousands of cilia each (Figure 3–22a). Flagella and cilia also beat dif-

FIGURE 3–21 The Sperm Cell (*a*) The sperm cell is a marvel of architecture, uniquely "designed" to streamline the cell for its long journey to fertilize the ovum. The nuclear material is compacted into the sperm head. A flagellum propels the sperm through the female reproductive tract. (*b*) Transmission electron micrograph of sperm cells.

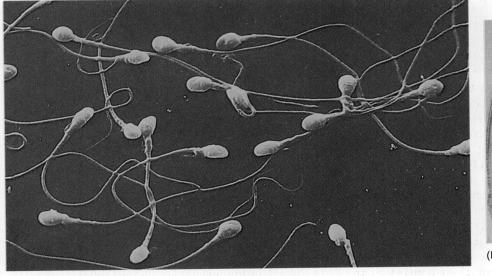

(a)

Flagellum   Acrosome

(b) Mitochondria   Sperm cell nucleus

(a)

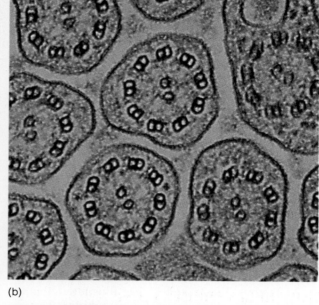
(b)

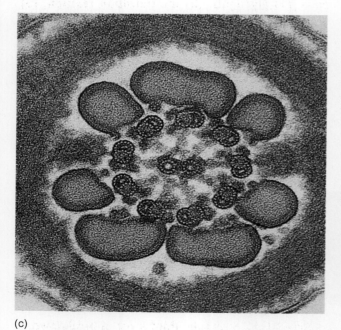
(c)

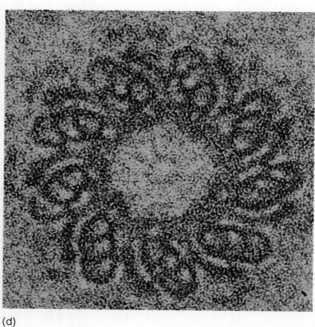
(d)

FIGURE 3–22   Cilium and Flagellum   (*a*) Cilia on cells lining the trachea of the human respiratory tract. (*b*) Numerous cilia in cross section showing the 9 + 2 arrangement of fibers. (*c*) A sperm flagellum showing 9 + 2 arrangement and additional fibers thought to provide support and strength to the vigorously beating tail. (*d*) A basal body is found at the base of each cilium. It consists of nine sets of triplets that give rise to the microtubular fibers of the cilium.

ferently. Cilia perform a stiff rowing motion with a flexible return stroke, whereas flagella beat more like whips, undulating with a continuous motion (Figure 3–23).

Cilia and flagella often beat continuously and, consequently, require a more or less continuous supply of energy, provided by ATP. ATP is produced from en-

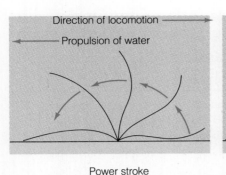

Direction of locomotion →
← Propulsion of water

Power stroke

Return stroke

(a) Cilium

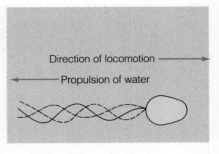

Direction of locomotion →
← Propulsion of water

Continuous propulsion

(b) Flagellum

FIGURE 3–23 Beating of Cilia and Flagella
(*a*) Cilia beat with a stiff power stroke—something of a rowing motion. During the return stroke, the cilium is flexible. (*b*) The flagellum beats like a whip, over and over, propelling the sperm cell forward.

ergy liberated by the breakdown of glucose in mitochondria, which are usually located nearby.

HOW CELLS PRODUCE ENERGY

Cells capture energy contained in carbohydrates (primarily glucose) and triglycerides and, only when absolutely necessary, from protein (Chapter 7). This section briefly explains how cells acquire energy from glucose molecules.

Cellular Respiration

Glucose breakdown occurs in a series of chemical reactions that begins in the cytoplasm and is completed inside the mitochondrion. During these reactions, glucose containing six carbon atoms is broken down into six molecules of carbon dioxide and water. The overall reaction is

glucose + oxygen → carbon + water
                           dioxide

$$C_6H_{12}O_6 + 6\ O_2 \rightarrow 6\ CO_2 + 6\ H_2O \quad \text{(catabolic reaction)}$$

38 ADP   + 38 P_i →   38 ATP   (anabolic reaction)

adenosine   inorganic   adenosine
diphosphate   phosphate   triphosphate

As the reaction shows, the breakdown of glucose is coupled to the synthesis of ATP. Carbon dioxide released from the cells travels in the bloodstream to the lungs, where it is exhaled. Oxygen required by the reaction comes from the air we breathe.

The complete breakdown of glucose is called **cellular respiration**, so named because cells take in oxygen and give off carbon dioxide. Cellular respiration is comprised of four separate but interconnected parts: (1) glycolysis, (2) the transition reaction, (3) the citric acid cycle, and (4) the electron transport system (Table 3–4). Figure 3–24 provides an overview of the three steps. Take a moment to study each of them before reading further.

To understand how the cell acquires energy from glucose, we will begin with the glucose molecule in the cytoplasm, then trace it through the four steps. The first phase, **glycolysis** (sugar breakdown) is a metabolic pathway in the cytoplasm of the cell. As Figure 3–24 shows, during glycolysis, glucose is split in half, forming two three-carbon molecules of pyruvic acid, or pyruvate. The energy released during the reaction nets the cell two molecules of ATP. All in all, it is not a very impressive gain.

Pyruvic acid undergoes a reaction inside the mitochondrion, which is called the transition reaction. The **transition reaction** is an intermediate step in which on one carbon atom is cleaved off each pyruvic acid molecule, forming a two-carbon compound, acetyl CoA.

Acetyl CoA enters the third stage of breakdown, a metabolic pathway located inside the inner compartment of the mitochondrion. This pathway is sometimes called the Krebs cycle, after the scientist who discovered it, but more often it is referred to as the citric acid cycle, after the very first product formed in the reaction sequence (Figure 3–24).

The **citric acid cycle** is a series of reactions that starts and ends at the same place. During this stage,

TABLE 3-4 Overview of Cellular Energy Production

REACTION	LOCATION	DESCRIPTION AND PRODUCTS
Glycolysis	Cytoplasm	Breaks glucose into two three-carbon compounds, pyruvate. Nets two ATPs. Nets two NADH molecules
Transition reaction	Mitochondrion	Removes one carbon dioxide from pyruvate, producing two acetyl CoA. Produces two NADH molecules
Citric acid cycle	Inner compartment of mitochondrion	Completes the breakdown of acetyl cycle CoA. Produces two ATPs per glucose. Produces numerous NADH and FADH molecules
Electron transport	Inner membrane of mitochondrion	Accepts electrons from NADH and FADH, generated by previous system reactions. Produces 34 ATPs

the two-carbon compound produced in the transition reaction binds to a four-carbon resident of the reaction sequence, called **oxaloacetate**. The result is a six-carbon product, **citric acid**. Citric acid undergoes a series of reactions that regenerates the starting material, oxaloacetate, which is used again and again in this cycle.

The citric acid cycle is an elaborate chemical pathway that cleaves off the two carbon atoms introduced at the beginning by each acetyl CoA. The chemical reactions that occur during the citric acid cycle yield two ATPs, one for each acetyl CoA (derived from the original glucose molecule). The reactions of the citric acid cycle also remove electrons and protons from the chemical reactants. Electrons are energetic subatomic particles whose energy is used to produce ATP.

The electrons removed during the chemical reactions of the citric acid cycle are first passed to one of two electron acceptor compounds found in the mitochondrial matrix: **nicotinamide adenine dinucleotide (NAD)** and **flavine adenine dinucleotide (FAD)**. These molecules shuttle the energetic electrons to the fourth and final stage of cellular respiration, the electron transport system. It is in the electron transport system that ATP is produced.

The **electron transport system** is a series of protein molecules found in the inner membrane of the mitochondrion. The protein molecules accept the electrons from FAD and NAD, then pass the electrons from one to another, eventually giving them to oxygen molecules. During their rapid journey along this chain of proteins, the electrons lose energy, much like a hot potato that is passed from one person to the

next. The energy lost along the way is used to make ATP. How does the cell make ATP?

In 1961 British biochemist Peter Mitchell formulated an hypothesis called **chemiosmosis** that describes the way in which ATP is made. In 1978, Mitchell received a Nobel Prize for his work. Chemiosmotic theory states that the proteins of the electron transport chain are dual-purpose molecules: they convey electrons from one to another, but also serve as hydrogen ion pumps. They use the energy lost along the chain to pump hydrogen ions from the inner compartment of the mitochondrion into the outer compartment (Figure 3-25). This causes a buildup of H^+ in the outer compartment, as illustrated in Figure 3-25. Hydrogen ions do not stay put, however. Many of them flow back into the inner compartment through pores in the inner mitochondrial membrane. It is this flow of hydrogen ions that actually powers ATP synthesis. The mechanism is similar to that which occurs when a wire is hooked to a battery. In a battery, the flow of electrons through a wire provides power to light a bulb. According to the hypothesis, the flow of hydrogen ions through the pores of the membrane provides power to drive the synthesis of ATP.

All told, the electron transport system produces 34 ATP molecules per glucose molecule. Since two ATPs are gained from glycolysis and two are produced in the citric acid cycle, the total output for cellular respiration is 38 per glucose molecule.

As shown in Figure 3-25, oxygen molecules in mitochondria combine with "de-energized electrons" from the electron transport system and hydrogen

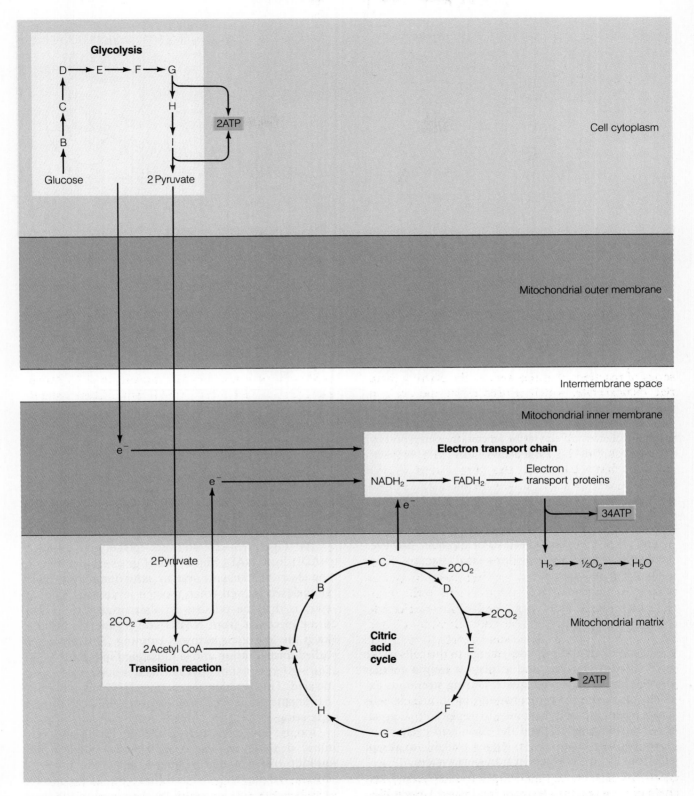

FIGURE 3–24   Cellular Respiration   This process involves four steps: glycolysis, the transition reaction, the citric acid cycle, and the electron transport system. Most of the ATP is produced in the electron transport system from the energy stripped from electrons given off by glycolysis and the citric acid cycle reactions.

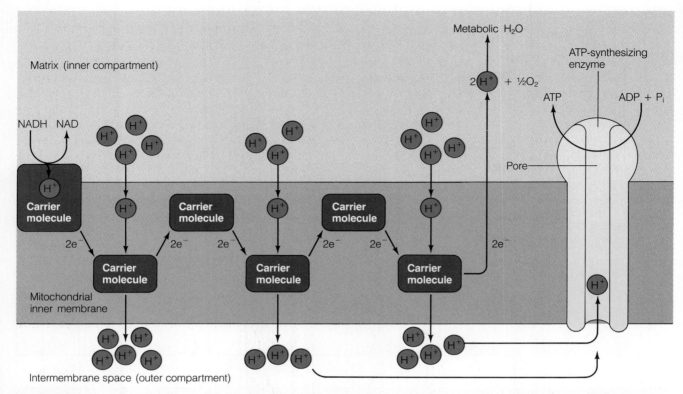

FIGURE 3–25 Chemiosmosis The electrons given over to the electron transport system travel from carrier protein to carrier protein. In the process, the energy lost by the electrons is used to pump protons (hydrogen ions) into the outer compartment. The protons then flow back through a special protein pore. This flow powers the synthesis of ATP production.

ions, forming water, called **metabolic water**. Interestingly, for some animals living in desert environments, the water produced in cellular respiration provides nearly all of their body needs. For humans, however, metabolic water only provides a small fraction of our daily water requirement.

Fermentation: Sore Muscles, Cheese, and Wine

Cellular respiration provides energy to the cells of the body at rest, at work, and at play as long a cellular oxygen levels remain adequate. During strenuous exercise, however, oxygen consumption in muscle cells can exceed the body's ability to replace it. Oxygen levels inside cells fall, and the citric acid cycle shuts down because oxygen is no longer available to accept electrons from the electron transport system.

Rather than shutting down completely, the cell continues to break down glucose through glycolysis.

This process, however, is exceedingly inefficient, netting only two ATPs for each glucose molecule, compared to 38 during cellular respiration.

As Figure 3–24 showed, glycolysis produces NADH from NAD. Since NADH gives its electrons to the electron transport system, why doesn't glycolysis shut down as well when oxygen levels fall? The answer is that the cell has an alternative way to strip those electrons from NADH and regenerate NAD to keep the glycolytic pathway running. This process is called **fermentation** and is illustrated in Figure 3–26. During fermentation, pyruvic acid is converted to lactic acid. This reaction regenerates NAD, allowing it to participate in the glycolytic pathway and keep the cell operating.

Intense exercise, such as weight lifting and running, depletes muscle oxygen and results in the buildup of lactic acid in muscle cells, which causes muscle fatigue. Lactic acid diffuses out of the muscle within hours and is carried by the blood to the liver.

Here lactic acid is converted to pyruvic acid. In the liver, in a series of chemical reactions that is essentially the reverse of glycolysis, pyruvic acid is converted into glucose. Glucose can then be used to synthesize glycogen or can be released into the blood and may be redistributed to body cells where it is used to generate energy.

Fermentation is an emergency source of energy in the human cell when oxygen levels fall. For many bacteria that live in oxygen-free environments, such as the deep mud of a river or lake bottom, fermentation is the major source of energy. Many food products, such as cheese and sauerkraut, are produced by fermentation. Most cheeses, for example, are made from milk combined with certain bacteria. Alcoholic beverages are also produced by mixing yeast (single-celled fungi) with grain. Beer is produced by a yeast that breaks down the carbohydrates in barley and other grains, converting pyruvic acid into ethanol (Figures 3–26 and 3–27).

ENVIRONMENT AND HEALTH: PARKINSON'S AND POLLUTION?

In 1983, Dr. J. W. Langston, a California neurologist, was confronted with a medical puzzle unlike any he had ever seen. Several relatively young men and women were admitted to the hospital in a catatonic stupor. Confined to their beds, the patients lay immobile day after day. They could neither talk nor feed themselves and could not move their limbs. It was as if they had been frozen.

Langston began an intensive study of the victims and found that each of the patients was a drug addict. He found that each of them had taken a synthetic form of heroin that had inadvertently been contaminated with a paralyzing chemical MPTP. Made in the basement of one of California's small-time drug pushers, the synthetic heroin is one of the latest in a long list of "designer" drugs on the market. Sloppy chemistry resulted in the contamination. What makes this

FIGURE 3–26 Fermentation Pyruvic acid from glycolysis is converted to lactic acid in human cells when oxygen levels fall. Fermentation can produce a variety of products in bacteria and yeast cells.

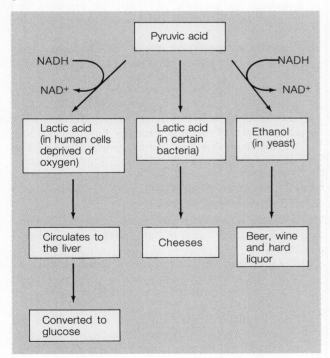

FIGURE 3–27 Products of Fermentation Beer and cheese.

story even more interesting, though, is that MPTP is extraordinarily similar to some common industrial chemicals and widely used pesticides.

Researchers have shown that MPTP attacks cells in a part of the brain called the substantia nigra. Degeneration of the brain cells in substantia nigra has long been implicated in Parkinson's disease, a disorder that afflicts older individuals and is characterized by tremors—involuntary rhythmic shaking of the hands, head, or both. If the disease worsens, patients have difficulty talking and writing. Walking becomes awkward and clumsy. In later stages, memory and thinking can be severely impaired.

What is striking is that individuals exposed to even minute amounts of MPTP develop symptoms similar to those seen in Parkinson's patients. A Swiss researcher who had worked briefly with the chemical was hospitalized in a mental institution for years. Doctors did not know why his body had locked up. They had assumed he was suffering from catatonic schizophrenia.

Some scientists believe that MPTP-like chemicals in the air, water, and soil may be the cause of Parkinson's disease. Opponents argue against this assertion, however, since several studies have failed to show a link between Parkinson's disease and industrial pollution. At a 1985 meeting at the National Institutes of Health, researchers agreed: Parkinson's disease was not produced by environmental contamination.

No sooner had the researchers arrived home than a Canadian researcher announced the results of a study in Quebec Province. The study showed a remarkable correlation between the use of pesticides and the incidence of Parkinson's disease. The researchers argued that pesticides were probably not the only chemical villains. Many neurotoxins could have been present in the area from the nearby chemical factories.

In support of the hypothesis that Parkinson's disease is linked to industrial pollution and pesticide use, researchers note that Parkinson's disease was unheard of before the Industrial Revolution. As the Industrial Revolution and environmental pollution spread, the incidence of the disease rose sharply, reaching a plateau in the early 1900s. No increase has been seen since 1940. However, some researchers warn that an increase in the use of paraquat and other pesticides and an increase in industrial pollution could substantially increase the incidence of the disease in years to come, underscoring the fact that to protect our health we must also protect the quality of our environment.

SUMMARY

AN OVERVIEW OF THE CELL

1. The fundamental unit of all living organisms is the cell. Two types of cells are found: prokaryotes and eukaryotes.
2. Prokaryotes lack nuclei and include the bacteria. Eukaryotes are cells with true nuclei.
3. As evolution proceeded, organisms became multicellular and more complex. The increasing complexity of organisms was paralleled by the development of complex systems of control that help ensure homeostasis.
4. Eukaryotic cells are compartmentalized, a feature that helps isolate various functions and helps promote the smooth and efficient operation of the cell.
5. The two major compartments of the cell are the nucleus and the cytoplasm. Additional compartments are made in the cytoplasm by membrane-bound organelles, such as the Golgi complex, the mitochondrion, and the endoplasmic reticulum.
6. The cytoplasmic compartment contains cytoplasm. Within the cytoplasm are numerous cellular organelles, specialized to carry out specific functions.
7. The shape of the cell is maintained by the cytoskeleton, a network of tubules and filaments to which the enzymes of some metabolic pathways attach.
8. A metabolic pathway is a series of reactions, each of which is catalyzed by its own enzyme.
9. Enzymes are protein molecules that lower the energy required for anabolic and catabolic reactions and thereby greatly accelerate the rate of reactions in the body.
10. Enzymes are specific for one or a few reactions. Specificity is conferred by the shape of the active site, a region of the enzyme that binds to the reactants.
11. Enzyme activity is often regulated by levels of byproducts of metabolic pathways and other cellular chemicals. The active sites can be physically blocked by the products of chemical reactions. Enzymes also contain allosteric sites that bind to products. Binding may activate or deactivate the enzyme.

STRUCTURE AND FUNCTIONS OF THE PLASMA MEMBRANE

12. The outermost boundary of the cell is the plasma membrane, which is made of lipid, protein, and carbohydrate.
13. The plasma membrane ensures the structural integrity of the cell, regulates the flow of materials into and out of the cell, and participates in cellular communication and cellular identification.
14. The membrane controls traffic into and out of the cell in several key ways. Because it is mostly lipid, the membrane is a natural barrier to water-soluble molecules. Lipid-soluble materials pass through with ease.

15. Water-soluble molecules and ions are believed to pass through pores in the membrane by diffusion, the flow of a substance from an area of higher concentration to an area of lower concentration.

16. Carrier proteins may also help transport water-soluble molecules across the membrane. These molecules provide for facilitated diffusion.

17. Some molecules are actively transported across the membrane. Active transport mechanisms use energy to pump materials into or out of the cell. A special class of membrane proteins, the transport molecules, is required for active transport.

18. Active transport allows cells to transport substances from regions of low concentration to regions of high concentration.

19. Liquids and their dissolved materials are engulfed in a process called pinocytosis. Large molecules and cells may be engulfed by phagocytosis. Pinocytosis and phagocytosis are collectively referred to as endocytosis.

20. During endocytosis, the plasma membrane indents and surrounds the material to be ingested. The material engulfed by the cell is incorporated in a membranous vesicle, which is taken into the cell.

21. Movement of water through a membrane from a region of higher water concentration to a region of lower water concentration is known as osmosis.

CELLULAR COMPARTMENTALIZATION: CELLULAR ORGANELLES

22. The nucleus is the largest organelle in most cells. It houses the DNA, which contains the genetic information that determines the structure and function of the cell.

23. The nucleus is bounded by a double membrane, the nuclear envelope, which contains numerous pores that allow many materials (but not the DNA) to pass freely between the nucleus and the cytoplasm.

24. In the cell, energy is produced principally by the breakdown of glucose. Glucose catabolism takes place in the cytoplasm and in the mitochondrion. Most of the ATP produced during this process is produced in the mitochondrion.

25. All mitochondria have two membranes, an outer membrane and an inner membrane, which is thrown into folds, called cristae.

26. Cellular energy production only captures about one-third of the energy contained in glucose; the rest is liberated as heat.

27. The cell synthesizes a variety of chemical substances. The endoplasmic reticulum, a network of membranous channels in the cytoplasm of the cell, is a major site of cellular protein synthesis.

28. Endoplasmic reticulum may be smooth or rough. Smooth endoplasmic reticulum has no ribosomes. Rough endoplasmic reticulum contains many ribosomes on its outer surface.

29. Smooth endoplasmic reticulum produces phospholipids needed to make the cell membrane and has many additional functions in different cells. The rough endoplasmic reticulum is involved in the synthesis of protein for extracellular use and for lysosomes.

30. Proteins produced on the surface of the RER enter the cavity, where they may be chemically modified. Proteins are transferred from the RER to the Golgi complex, where they are sorted, chemically modified, and repackaged into secretory vesicles or lysosomes.

31. Secretory vesicles accumulate in the cytoplasm and are released when needed. Lysosomes remain in the cell where they bind to food vacuoles (material phagocytized or pinocytized by the cell).

32. Most cells in the body are fixed. Many of the mobile cells, however, are transported passively in the blood. Still other cells can move about in tissues via ameboid movement.

33. The human sperm cell utilizes another kind of motive force, the flagellum, a long, whiplike structure.

34. In the human, some cells contain numerous smaller extensions, called cilia, which propel fluids and materials along their surfaces.

35. Both cilia and flagella contain nine pairs of microtubules arranged in a circle around a central pair.

HOW CELLS PRODUCE ENERGY

36. Cells capture energy from carbohydrate, lipid, and occasionally protein catabolism. Glucose, a carbohydrate, is the chief source of energy.

37. Glucose's energy is liberated in a four-part process: glycolysis, the transition reaction, the citric acid cycle, and the electron transport chain.

38. The complete breakdown of a molecule of glucose yields six carbon dioxide molecules, six water molecules, and 38 ATP molecules.

39. Glycolysis occurs in the cytoplasm and produces two molecules of pyruvic acid (pyruvate) for every molecule of glucose. It also nets two ATP molecules.

40. The transition reaction and the citric acid cycle occur in the inner compartment of the mitochondrion. They complete the breakdown of glucose and yield a few ATP molecules and a large number of high-energy electrons. These electrons are passed to the electron transport system via NAD and FAD.

41. In the electron transport system, the electrons are passed down a series of transport proteins. As the electrons move along the series, they give off energy, which is used to pump hydrogen ions in the inner compartment to the outer compartment. The hydrogen ions then flow back into the inner compartment through pores, which is believed to provide the energy needed to produce ATP.

42. Cellular respiration provides energy to the cells of the body at rest, at work, and at play. During strenuous exercise, however, oxygen consumption in muscle cells

can exceed the body's ability to replace it. When oxygen levels inside the cells fall, the citric acid cycle shuts down.

43. When oxygen levels fall, the cell continues to break down glucose through glycolysis. This process is exceedingly inefficient, netting only 2 ATPs for each glucose molecule, compared to 38 during cellular respiration.

44. In the absence of oxygen, the product of glycolysis, pyruvate, undergoes further reaction called fermentation, producing lactic acid.

45. Intense exercise, therefore, can result in the buildup of lactic acid in muscle cells, which causes muscle fatigue. Lactic acid diffuses out of muscle cells and is carried by the blood to the liver where it is converted back into pyruvic acid and then into glucose.

EXERCISING YOUR CRITICAL THINKING SKILLS

Garbage from modern society contains a great deal of organic matter: grass clippings, leaves, kitchen wastes, paper, cardboard, and so on. This waste is dumped in landfills, where it is covered with dirt and supposedly decomposed by bacteria over time. Recently, however, U.S. newspapers and magazines have featured stories about researchers who have recovered newspapers and food buried in an Arizona landfill 35 years ago and still not decomposed. This story has been repeated many times, and many people are now under the impression that our garbage is not decomposing in landfills. One person, in fact, recently suggested that it does not matter whether plastic cups or paper cups are used because neither decompose in a landfill.

This is an issue where critical thinking skills come in handy. Critical thinking requires us to be an active participant: to take some time to research and study matters more carefully. In keeping with this, you might want to read a section on landfills in an environmental science textbook. Did you find any useful information that makes you think that the conclusion that our trash is not decomposing is false?

Reread the first paragraph again carefully. Can you find any clues that would suggest why the conclusions from the study on garbage decomposition might not apply to landfills in Massachusetts or Michigan?

TEST OF TERMS

1. The first cells formed approximately _____ billion years ago.
2. The first cells probably resembled modern-day _____ .
3. _____ are the building blocks of body tissues.
4. The _____ contains the cell's hereditary information contained in molecules of _____ .
5. The network of protein tubules in the cytoplasm of a cell is called the _____ . It binds to organelles and to _____ , protein molecules that increase the rate of chemical reactions.
6. A series of chemical reactions in the cell is called a(n) _____ .
7. The region on the enzyme where the substrate binds is called the _____ site. This site usually only binds to one substrate and therefore makes the enzyme _____ for one chemical

reaction.
8. Enzymes belong to a class of compounds called _____ , which speed up chemical reactions. Unlike inorganic members of this group, the enzyme can be controlled or regulated. One way this happens is by the binding of an end product from a reaction or reaction sequence to the _____ site, located some distance from the region of the protein where the reaction takes place.
9. The control mechanism in biological systems in which a product alters the process is called a _____ mechanism.
10. The plasma membrane consists of a _____ layer of lipid, mostly _____ . Proteins in the membrane can be embedded in the lipid layer of the membrane; these are called _____ proteins.

11. A membrane that is permeable to some substances and impermeable to others is called a _____ _____ membrane.
12. A protein that has carbohydrate units attached is called a _____ . Carbohydrate units are added to protein in the cell in two organelles, the _____ and the _____ .
13. The movement of a molecule (other than water) across the plasma membrane from a region of high concentration to low concentration is called _____ .
14. Facilitated diffusion requires a(n) _____ protein in the membrane to shuttle molecules and ions from one side to the other down a concentration gradient.
15. The process of engulfing another cell, such as a bacterium, is called _____ .
16. Water moves across membranes

through pores. It is driven by concentration differences. The driving force is called _____ pressure.

17. When the concentration of a fluid is the same as that of a cell, the fluid is said to be _____. When the fluid has a solute concentration lower than the cell's cytoplasm, the fluid is said to be _____. Dropping cells into this type of solution will cause them to _____.

18. The double membrane surrounding the nucleus is called the nuclear _____. It is pierced by small openings called nuclear _____.

19. Strands of DNA with associated protein are called _____. They condense before cell division to produce dark-staining bodies called _____.

20. _____ is synthesized at the nucleoli. It combines with _____ to form ribosomes.

21. The _____ is the site of the bulk of the energy production occurring in a cell.

22. Protein for extracellular use and digestive enzymes found in lysosomes are produced on the _____. The instructions for these proteins come from the nucleus in the form of _____ molecules.

23. Secretory vesicles are packages of protein and other cellular products. Secretory granules are produced by the _____ _____.

24. Digestive enzymes are held in a membrane-bound organelle called the _____. It fuses with engulfed material contained in a(n) _____ _____.

25. The long whiplike organelle on human sperm cells is called a _____. The motive force is created by microtubules arranged as _____ pairs surrounding a central doublet. The is the _____ arrangement.

26. Small motile projections on the cells lining much of the respiratory tract are called _____.

27. The partial breakdown of glucose in the cytoplasm of a cell is called _____; it results in the production of _____ molecules of _____ for each molecule of glucose broken down.

28. The complete breakdown of glucose in the cell is called _____; it produces _____ ATP molecules and takes place both in the cytoplasm and the mitochondria.

29. The bulk of the ATP molecules produced during the breakdown of glucose are formed during the _____ _____ _____. The energy needed to make ATP comes from _____ stripped off the reactants during the _____ _____ cycle.

30. When cells of the human organism lack oxygen, energy can still be achieved through _____.

Answers to the Test of Terms are located in Appendix B.

TEST OF CONCEPTS

1. Describe cellular and organismic homeostasis. How does cellular homeostasis affect organismic homeostasis?

2. Cellular compartmentalization allows for greater efficiency. Why? Give an example to support your argument.

3. What is specificity as applied to enzymes? Why is it beneficial to living things?

4. How are enzymes controlled? Explain the term *feedback mechanism*.

5. Draw a diagram of the plasma membrane and describe the five routes by which molecules pass through the membrane.

6. In what ways does the plasma membrane participate in cellular communication?

7. Define the terms *osmosis, isotonic, hypotonic, and hypertonic*. In what way(s) is osmosis similar to diffusion and in what way(s) is it different?

8. Describe the organelles involved in the synthesis, storage, and release of protein for extracellular use, starting with the basic instructions needed to make the protein.

9. In what ways are cilia and flagella similar and in what ways are they different?

10. Describe the complete breakdown of glucose in a human cell. Be sure to include the following terms: glycolysis, transition reaction, citric acid cycle, electron transport system, ATP and NAD, and chemiosmosis.

SUGGESTED READINGS

Becker, W. M. 1986. *The world of the cell.* Menlo Park, Calif.: Benjamin/Cummings. A clearly written but fairly detailed account of cell biology. Good reference material.

DeDuve, C. 1984. *A guided tour of the living cell.* New York: Scientific American Books, Inc. Profusely illustrated and entertainingly written.

Weiss, R. 1988. Forbidding fruits of fetal-cell research. *Science News* 134(19): 296–98. Examines the controversies surrounding the use of fetal cells for medical treatments.

4 Chromosomes and Cell Division

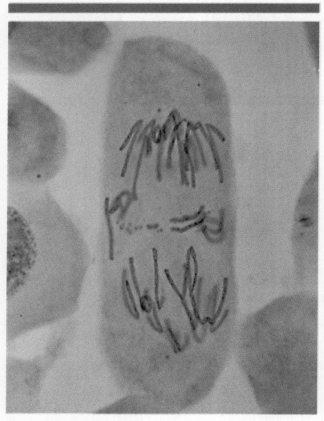

X-ray–disrupted mitosis in an onion cell. Chromosomes have failed to separate completely.

I n 1979 two researchers from the University of Colorado reported a disturbing finding on the effects of extremely low frequency (ELF) magnetic waves produced when electricity flows through high-current electric power lines. Their studies showed that ELF magnetic waves increased the incidence of childhood leukemia in families living near high-voltage power lines. The death rate from cancer in these children was twice what is expected in the general public.

In November 1986, researchers from the University of North Carolina announced results of a similar study. They found an alarming fivefold increase in childhood cancer (particularly leukemia) in families living within 7.5 to 15 meters (25–50 feet) of wires carrying electricity from power substations to neighborhood transformers. Furthermore, a Texas researcher found that ELF fields increased the growth rate of cancer cells in tissue culture and made cancer cells 60 to 70% more resistant to the body's naturally occurring killer cells.

Fortunately, few Americans live close to high-tension wires and power substations. However, these findings are still of concern to health experts, because similar magnetic fields are generated by waterbed heaters and electric blankets, both common in American homes. Research suggests that ELF fields given off by these products increase the likelihood of miscarriage—spontaneous abortion of the embryo or fetus. Some researchers believe that ELF fields may cause birth defects in addition to cancer.

Cancer is the uncontrolled replication of cells in an animal. One of every three Americans will contract cancer, and one of every four Americans will die from it. Despite years of research and billions of dollars spent studying cancer, science is not much closer to finding a cure than it was a decade ago. Modern science may, in fact, be losing the war against many types of cancer. New drugs have helped boost the survival rate of some cancers—for instance, childhood leukemia, a cancer of the blood system. Surgery also works effectively for many skin cancers. Early detection has also helped; the earlier a cancer is detected, the greater the chances of survival. Despite increased efforts to detect and fight the disease, long-term survival rates for cancer patients have only increased 4.2%, compared to an overall increase of 5.1% in survival rates for all other diseases, notes John C. Bailar, a biostatistician at McGill University. Because of the slight increase in survival through medical intervention, some experts believe that the greatest hope today lies in prevention (see Health Note 4–1).

What causes cancer? Why does it kill so effectively? These and other questions are answered in this chapter. Before we can address them, however, we must understand more about cells, especially chromosomes and cell division. As you shall see in this and the next chapters, it is the chromosomes and the process of cell division that are altered by cancer-causing agents.

THE CELL CYCLE

Human life begins as a single cell, formed by the union of the sperm and ovum (Chapter 16). The new cell divides many times during development. During division, each cell goes through a precisely choreographed series of changes, constituting the **cell cycle** (Figure 4–1). Understanding the cell cycle helps us understand cancer and may help scientists find cures for this prevalent disease.

As illustrated in Figure 4–1, the cell cycle consists of two major phases: cell division and interphase. **Cell division** is a period of great change during which the nucleus and the cytoplasm split, dividing the contents of the cell more or less equally and forming two daughter cells. **Interphase** is the period between cell divisions.

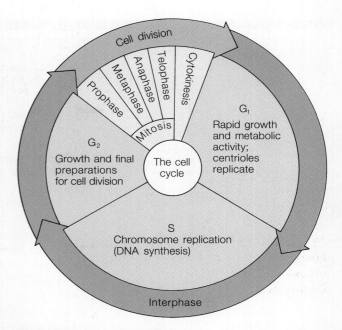

FIGURE 4–1 The Cell Cycle

Interphase

Once thought to be a resting period, interphase is now known to be a period of intense cellular activity. Molecular synthesis and cellular growth take place during interphase and are necessary to prepare cells for cell division. During interphase, the cell duplicates its DNA, cytoplasm, and organelles. Thus, when the cell divides, its offspring, or **daughter cells**, receive an adequate supply of materials and organelles for survival. Cells in interphase, however, are not simply building for the final act—cell division. The things they make and the functions they perform are essential to the whole organism.

Interphase is divided into three parts: G_1, S, and G_2 (Figure 4–1 and Table 4–1). G_1 (gap 1) begins immediately after a cell divides. During G_1, the cell synthesizes RNA, protein, and other molecules necessary for normal cellular function and also stockpiles materials for the two daughter cells that will soon form.

During G_1, the chromosomes in the nucleus consist of a single strand of DNA and protein; these strands are called **chromatin fibers**. Except for the sperm and ovum, all human cells contain 46 chromosomes. A **chromosome** consists of one or two chro-

TABLE 4-1   Phases of the Cell Cycle

Interphase		
	G₁	*Gap 1* Occurs immediately after mitosis Marked by synthesis of RNA, protein, and other molecules
	S	*Synthesis phase* DNA replication Chromosomes become double stranded
	G₂	*Gap 2* Relatively inactive period
Mitosis		
	Prophase	Chromosomes condense Nuclear envelope disappears Centrioles divide and migrate to opposite poles of the dividing cell Spindle fibers form and attach to chromosomes
	Metaphase	Chromosomes line up on equatorial plate of the dividing cell
	Anaphase	Chromosomes begin to separate
	Telophase	Chromosomes migrate or are pulled to opposite poles New nuclear envelope forms Chromosomes uncoil

matin fibers, depending on the stage of the cell cycle. The 46 human chromosomes carry an estimated 100,000 genes. **Genes** are segments of the DNA that control cell structure and function (Chapter 5). Genes in the cells of the iris of the eye, for instance, control the production of pigment, which helps determine eye color. Genes in cells of the pituitary gland control the amount of growth hormone produced on the rough endoplasmic reticulum. The amount of growth hormone available during childhood, in turn, helps determine a person's height.

DNA duplication occurs during the next part of interphase, the **S**—or synthesis—**phase**. After the single-stranded chromosomes replicate, each chromosome consists of two chromatin threads (Figure 4–2). Each thread is called a **chromatid**. The chromatids remain attached to each other until the nucleus divides. Chromosome structure is described in the following section.

The period after DNA synthesis is called G₂. G₂ is a brief period before cell division. Some cellular manufacturing and growth occur during this period, but far less than in G₁.

Cell Division

The final act of the cell cycle is cell division. During cell division, both the nucleus and cytoplasm divide. Nuclear division, or **mitosis**, involves a series of dramatic morphological changes. In order for the nucleus to divide, however, the chromosomes must first become condensed or tightly coiled. Condensation or coiling results in the formation of blocklike structures that are visible in ordinary light microscopic preparations. These structures did not escape early cytologists who coined the name, chromosomes ("colored bodies").

Condensation is only one of the prerequisites of mitosis. Another prerequisite is that the chromosomes be lined up in the center of a cell's nucleus. From this position, chromosomes can be easily split. This alignment, therefore, helps to ensure an even distribution of chromosomes and ensures that each daughter cell will receive a complete set of the parent cell's genetic material.

Cytoplasmic division, or **cytokinesis** (literally "cell movement"), is an independent process that occurs toward the end of mitosis. Both mitosis and cytokinesis are described in more detail below. Before we study these processes, however, we must take a closer look at the chromosome.

THE CHROMOSOME

Located in the nucleus, the chromosome changes substantially during the cell cycle. As noted above, during G_1, each chromosome consists of a single DNA molecule (double helix) and associated protein. The 46 single-stranded chromosomes in the human cell are packed loosely into the nucleus in an apparently random fashion. During the S phase, the DNA strands duplicate (Figure 4–2). Consequently, the single-stranded chromosomes are converted into double-stranded structures. Each strand has exactly the same genetic information, much as a photocopy and an original have the same information.

Early in cell division, the chromosomes begin to coil, compacting in much the same way that a stretched phone cord shortens and compacts as the tension is removed. Chromosomes in the condensed state are metabolically inactive.

The structure of a condensed chromosome is shown in Figure 4–3. As illustrated, each chromatin fiber consists of a strand of DNA and associated proteins, called histones. The **histones** are globular proteins, which are thought to play a role in regulating the DNA's activities (Chapter 5). As illustrated in Figure 4–3, loops of the double helix of DNA encircle the globular histones.

Chromosomal condensation is essential to cell division. Without it, each cell division would no doubt be a chaotic event. The condensation of chromosomes also offers medical scientists an opportunity to examine the chromosomal makeup of an individual. Chromosomal analysis is particularly useful when an older couple (in their mid-to-late thirties or older) is considering having a child, but is worried about the possibility of giving birth to a child with genetic defects. Their concern is well founded, for some genetic defects are known to occur much more frequently in the offspring of mothers (and even fathers) who are 35 years of age or older (Figure 4–4a). One of the most common defects is Down syndrome (Figure 4–4b), which occurs in one of every 700 births. Down syndrome is caused by an error in cell division during gamete formation, which results in an additional

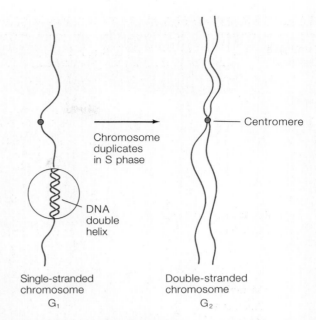

FIGURE 4–2   DNA Replication   During the S phase of the cell cycle, the DNA molecule unwinds and duplicates, forming two strands, each containing a DNA double helix.

chromosome 21 in the female gamete, the ovum. As a result, an additional chromosome 21 is present in all of the body cells of a woman's child (Chapter 18). The additional chromosome disrupts normal physical growth and mental development.

To test for this and other genetic defects, physicians extract fluid from the **amnion**, a sac, surrounding the fetus (Figure 4–5). This fluid contains fetal cells, which can be cultured in the laboratory. In this procedure, called **amniocentesis**, a needle is directed into the amnion using an ultrasound device to avoid injuring the fetus. Ultrasound waves are inaudible, but bounce off dense objects and are reflected back to a detector, much like sonar. The pattern of deflected waves creates an image on a screen, which allows a doctor to direct the needle into the amnion, missing the fetus.

Fetal cells withdrawn in the amniotic fluid are separated and grown in culture dishes for several days. The cultured cells are allowed to divide freely. After the number of fetal cells has increased, the tissue culture is treated with a chemical substance that arrests cell division. The cells are then flattened on a microscope slide, forcing the chromosomes to spread out

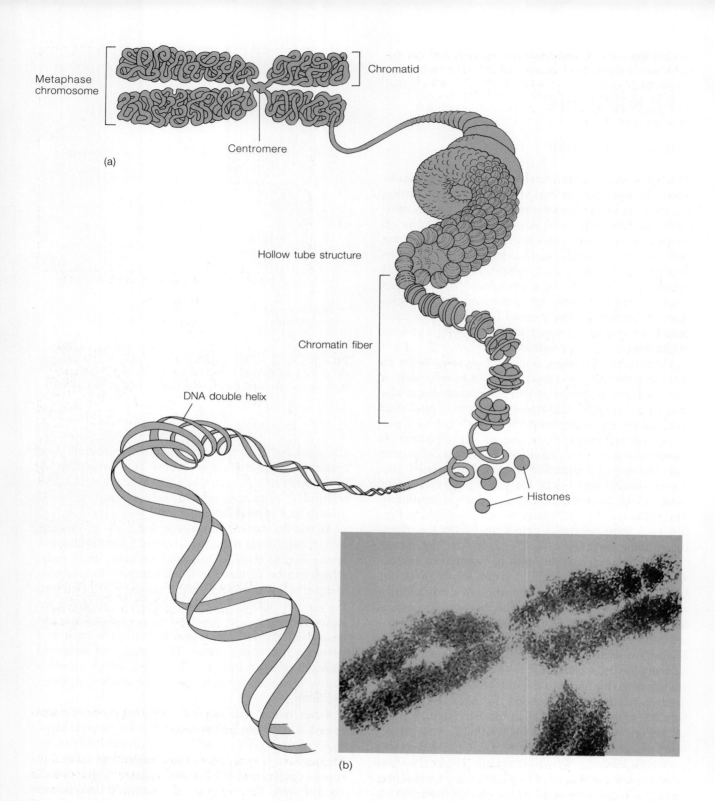

Metaphase chromosome

Chromatid

Centromere

(a)

Hollow tube structure

Chromatin fiber

DNA double helix

Histones

(b)

FIGURE 4–3   Chromosome Structure (left)   (a) The DNA helix wraps around histone molecules forming a chromatin fiber. During the cell cycle, the chromatin fiber coils tightly forming the densely packed chromosome, which first appears during prophase. (b) Scanning electron migrograph of chromosome.

(Figure 4–6a). Technicians photograph the chromosome spread through the microscope, then cut out each chromosome from the picture, lining them up by number. This produces a karyotype, a precise alignment of chromosome pairs (Figure 4–6b).

A computerized device that attaches to the microscope can also be used to produce a karyotype. It photographs the chromosomes through the microscope and projects the image on a computer screen. Technicians isolate the chromosomes on the screen, and the computer creates a karyotype.

As we have seen, human cells contain 23 pairs of chromosomes, making a total of 46 chromosomes. By studying the karyotype, geneticists can determine if there are any defects in the number of chromosomes (either too many or too few) or if there are any structural abnormalities, such as missing segments. Chromosomal defects can have serious consequences. In many cases, chromosome analysis assures parents that their children will be healthy, alleviating much worry during a period of heightened stress. In addition, it lets doctors and parents determine the sex of the baby.

CELL DIVISION: MITOSIS AND CYTOKINESIS

One of the most astounding of all cellular events is cell division. During cellular division, the cell divides its chromosomes equally and distributes its cytoplasm and organelles more or less evenly between two daughter cells. In adults, cell division helps repair tissue damage. It also helps to replace cells lost through normal wear and tear. In children, cell division is a key cause of body growth and serves to repair the many cuts and bruises that are a part of growing up.

Cell division is a series of events. To simplify matters, however, biologists discuss nuclear and cytoplasmic divisions separately. We begin our discussion of cell division with mitosis, or nuclear division.

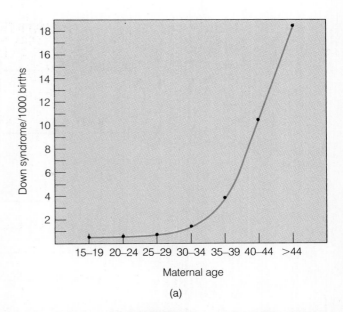

(a)

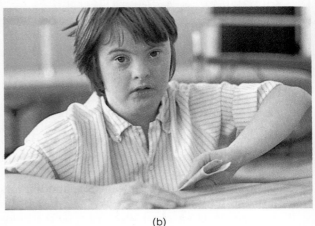

(b)

FIGURE 4–4   Down Syndrome   (a) Caused by an additional chromosome 21 in the cells of the body, this disease increases markedly in pregnancies in women over age 35. (b) All children born with Down syndrome appear quite similar.

Mitosis

Mitosis is divided into four stages: prophase, metaphase, anaphase, and telophase.

Prophase.   Immediately after interphase, the chromosomes begin to thicken and shorten (Figures 4–7a and 4–7b). In prophase, the nucleoli, which were visible during interphase, disappear because the cell

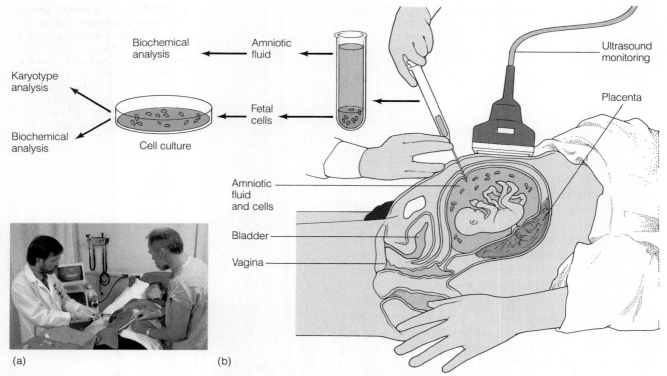

FIGURE 4–5   Amniocentesis   (*a*) A patient undergoing amniocentesis. (*b*) A needle is inserted into the fluid-filled space surrounding the fetus to withdraw fluid containing fetal cells that can be subjected to chromosomal and biochemical analyses. Ultrasound monitors are used to direct the needle to prevent injury to the fetus.

terminates rRNA production. Changes also occur in the cytoplasm during prophase. Two of particular importance are the division of the centriole and the formation of the spindle.

The **centriole** is an organelle found in the cytoplasm (Figure 4–8). It consists of two small, cylindrical structures that are identical to the basal bodies (described in Chapter 3) and, in fact, give rise to basal bodies during cellular development. Like most other organelles, centrioles duplicate during interphase. During prophase, the centrioles separate, migrating to opposite ends of the nucleus.

In animal cells, the centrioles are associated with a transient organelle, the **mitotic spindle**, an array of microtubules that forms in the cytoplasm during prophase (Figure 4–9). The microtubules of the mitotic spindle connect to the chromosomes and later help draw them apart. Also associated with the centrioles is a star burst of microtubules called the **aster**. Its role in cell division remains unclear.

Late in prophase, the nuclear envelope begins to break down. Spindle fibers attach to the chromosomes. As illustrated in Figure 4–9, microtubules that attach to the chromosomes are called **chromosome-to-pole fibers**. The microtubules of the spindle that extend from one centriole to the other are called **pole-to-pole fibers**. Each chromatid of a replicated chromosome contains a region called the **centromere**, a point on each chromatid to which the sister chromatid attaches (Figure 4–2; Table 4–2). The centromere is also the site of spindle fiber attachment.

Metaphase.   During metaphase the chromosomes line up in the center of the cell (Figure 4–7c). The chromosomes align themselves an equal distance from the two poles along the equatorial plane of the cell, the **metaphase plate**. This precise alignment permits the cell to divide each chromosome in two—so that one chromatid from each chromosome can be delivered to each daughter cell.

Anaphase.   When the chromatids of each chromosome begin to separate, the cell enters anaphase (Figure 4–7d). The chromosomes are pulled in opposite

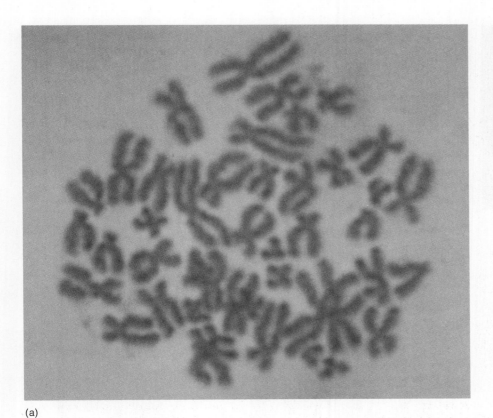

(a)

(b)

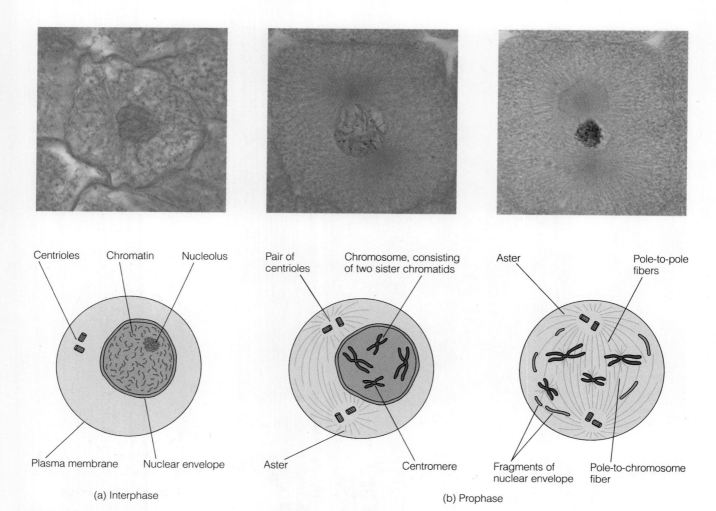

Centrioles   **Chromatin**   **Nucleolus**

Pair of centrioles

Chromosome, consisting of two sister chromatids

Aster

Pole-to-pole fibers

Plasma membrane   **Nuclear envelope**

Aster

Centromere

Fragments of nuclear envelope

Pole-to-chromosome fiber

(a) Interphase

(b) Prophase

FIGURE 4−7 Mitosis These light micrographs and schematic drawings show cell division in an animal cell. After interphase (*a*), mitosis occurs in four stages: (*b*) prophase; (*c*) metaphase; (*d*) anaphase; and (*e*) telophase. Cytokinesis, the division of the cytoplasm, begins in late anaphase or telophase.

directions, and thus move toward opposite poles with the aid of the mitotic spindle. At the end of anaphase, 46 single-stranded chromosomes are present in each daughter cell.

Several mechanisms have been proposed to explain the movement of chromosomes. One theory holds that the pole-to-chromosome fibers slide along the pole-to-pole fibers, thus pushing the poles further apart, causing the cell to elongate, and dragging the chromosome away from the metaphase plate. ATP is probably required. Elongation of the cell may also result from growth of the pole-to-pole fibers. As the cell elongates, the chromosomes are separated. These hypotheses may explain why the chromosomes separate initially, but what causes the chromosomes to

migrate all of the way to the poles remains a mystery. One possibility is that after initial separation of the chromosomes, the pole-to-chromosome fibers may shorten, in effect drawing the chromosomes to the poles of the cell.

Telophase. The final stage of nuclear division is telophase (Figure 4−7e). It is, in many respects, the reverse of prophase. Thus, during telophase the nuclear envelope re-forms. The new nuclear envelope is produced by the endoplasmic reticulum. During telophase, the chromosomes also uncoil, regaining their threadlike appearance. Telophase ends when the daughter cell nuclei appear to be in interphase.

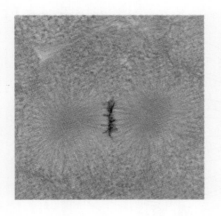
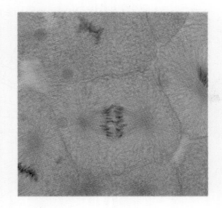
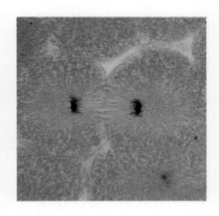

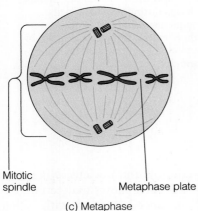
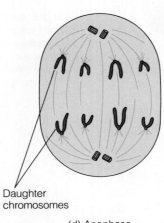
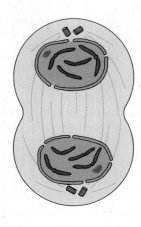

Mitotic spindle

Metaphase plate

Daughter chromosomes

(c) Metaphase

(d) Anaphase

(e) Telophase and cytokinesis

Cytokinesis

Cytokinesis, the division of the cytoplasm, begins late during anaphase or early in telophase. Cytoplasmic division is brought about by a dense network of contractile fibers, or **microfilaments**, that lie beneath the plasma membrane. Microfilaments are composed of proteins that contract when stimulated. The microfilamentous network is part of the cytoskeleton, discussed in Chapter 3.

When the fibers in the midline of the cell contract, they pull the membrane inward (Figure 4–10). The membrane furrows, in much the same way it would if a string were tied around it and tightened. The furrow deepens and eventually pinches the cell in two, thus producing two daughter cells.

Control of Cell Division

Cells divide continually in the developing embryo and fetus. Eventually, though, mature tissues are formed, and cell division slows down and sometimes stops altogether. The cells that stop dividing are highly specialized cells such as muscle and nerve cells. These cells become arrested in various stages of interphase, usually either G_1 or G_2. Although such specialized cells can produce proteins and other molecules necessary for the functions they perform, they cannot divide and replace themselves. A damaged muscle cell, therefore, will not be replaced by other cells. Body muscles grow after exercise by increasing the amount of contractile protein they contain (Chapter 14).

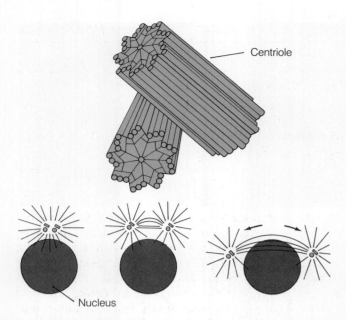

FIGURE 4–8 **Centriole** The centriole, like the basal body, consists of nine sets of microtubules with three in each set. Centrioles are formed in the cytoplasm and replicate during interphase and migrate to opposite poles of the nucleus during mitosis.

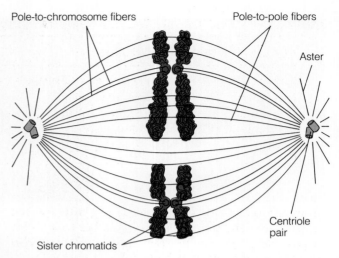

FIGURE 4–9 **Mitotic Spindle** The mitotic spindle, made of microtubules, forms during prophase in the cytoplasm of the cell. Some fibers connect to the chromosomes, while others extend from pole to pole. In ways not clearly understood, the mitotic spindle helps separate each double-stranded chromosome during mitosis.

Other cells stop dividing, but retain the ability to undergo cellular division should the need arise—for instance, to replace damaged tissue. Still other cells continue to divide more or less regularly throughout adulthood. Skin cells, for example, divide quite rapidly, replacing the tens of thousands of cells that are sloughed off every day. It is these actively dividing cells that are most vulnerable to carcinogens (cancer-causing agents) and are, therefore, more likely to become cancerous.

Despite years of intensive research, the control of the cell cycle remains one of the more puzzling aspects of biology. Some studies suggest that chemical messages from the cytoplasm may play a major role in choreographing the events of the cell cycle. This conclusion is based on a simple yet elegant experiment in which cell nuclei in G_1 were transplanted to the cytoplasm of actively dividing cells. In their new location, the G_1 nuclei quickly became active, entering the S phase. The previously inactive chromosomes soon replicated, and the nuclei began to divide. Researchers believe that in such cases one or more chemical factors produced in the cytoplasm of the host cell activated the nuclei.

If the cytoplasm controls the nucleus, what controls the cytoplasm? Research shows that certain external signals—for instance, contact with other cells—may exert some control over cytoplasmic events. Research in **cell cultures**, cells grown in the laboratory under precise conditions in glass containers, for example, indicates that normal cells divide until they spread evenly across the dish (Figure 4–11). At this point, they stop, even though nutrients may be plentiful. Thus, the contact of cells with one another inhibits further growth; this process is called **contact inhibition**. Contact inhibition also occurs in the tissues of the body.

One of the reasons cancer cells grow uncontrollably is that they lose contact inhibition. In a culture dish, cancer cells proliferate wildly, growing on top of one another and quickly utilizing nutrients in the culture dish. In the body, cancer cells grow in a similar kind of insatiable frenzy, forming large growths, called **tumors** that may eventually destroy vital organs and cause death.

Hormones can also affect the cell cycle. During each menstrual cycle, for example, estrogen released

TABLE 4–2 Review of Terminology

Chromatin	General term referring to a strand of DNA and associated histone protein
Chromatin fiber	Strand of chromatin
Chromosome	Structure consisting of one or two chromatin fibers
Chromatid	Generally used to refer to one of the chromatin fibers of a double-stranded chromosome
Centromere	Region of each chromatid to which a sister chromatid attaches

from a woman's ovaries stimulates growth of the breast tissue and the uterine lining in preparation for a possible pregnancy. Several other hormones also stimulate cell division of body cells.

FIGURE 4–10 Cytokinesis (*a*) The cytoplasm of a cell is divided in two by the contraction of a ring of contractile microfilaments. Cytokinesis begins late in anaphase or early in telophase. (*b*) Cell undergoing cytokinesis.

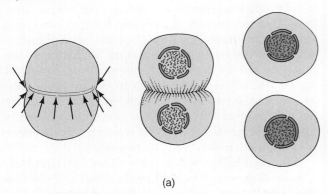

(a)

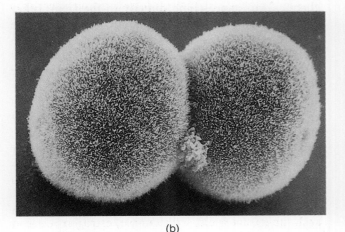

(b)

Growth-promoting and growth-inhibiting factors also play a role in controlling the cell cycle. Skin cells, for example, produce a growth-inhibiting factor that prevents cell division. Damage to the skin, say, a cut or burn, reduces the number of skin cells and presumably reduces the amount of inhibiting factor at the site of injury. Cells released from inhibition proliferate and repair the damaged area. The rate of cell division returns to normal when the tissue is fully repaired.

Obviously, there is no easy answer to the question, what controls the cell cycle? We know many factors play a role, and others may exist as well. An understanding of these mechanisms could aid efforts to treat cancer.

FIGURE 4–11 Contact Inhibition In culture dishes, normal cells grow to form a single layer, as shown. When a cell makes contact with neighboring cells, it stops dividing, even though plenty of nutrients are available. Cancer cells continue to grow, forming multiple layers because they have lost contact inhibition.

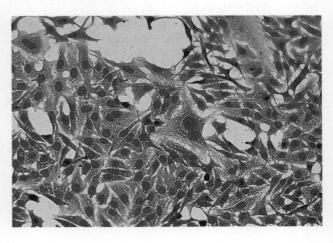

CANCER: UNDERSTANDING A DEADLY DISEASE

In the early 1970s, Americans were shocked by reports of hazardous wastes seeping into their drinking water, pesticides contaminating the food they ate, and pollution in the air they breathed. Some experts estimated that 90% of all cancers were caused by environmental and workplace pollution and other agents such as X-rays, ultraviolet light, and viruses. That estimate caused a storm of controversy and a flurry of research into cancer, a disease that kills nearly 500,000 Americans each year.

More recent studies suggest that the early claims were probably in error. Today, in fact, many health experts believe that only 20 to 40% of all cancers arise from environmental and workplace pollutants. Most of the rest are caused by smoking, diet, and natural causes. No matter what the cause, cancer is a disease that most of us will have close experience with.

There is some encouraging news. Studies show that the mortality rate of all cancers, except lung cancer, has leveled off or decreased since 1950.

This section takes a closer look at cancer. It will familiarize you with some terms and concepts that you will encounter in your lifetime.

The Spread of Cancer

Cancer is a disease in which body cells divide uncontrollably, invading other parts of the body. Not all abnormal cellular proliferation is cancerous. Some cells form small growths that reach a certain size, then stop growing. These are called **benign tumors**. As a rule, benign growths do not pose a significant medical problem, except when they put pressure on nerves or block blood vessels.

Of great concern are the **malignant tumors**, which continue to grow and often spread to other parts of the body, invading vital organs. The term *cancer* is reserved for malignant tumors.

Malignant tumors release individual cells or clusters of cells that can spread throughout the body in the **circulatory system**, the blood vessels, and in the **lymphatic system**, a supplementary network of vessels that drains excess fluid from body tissues (Chapter 8). The spread of cells from one region of the body to another is known as **metastasis**. In distant sites, the cancerous cells form **secondary tumors**. The **primary tumor** from which they arose may be removed surgically or may be destroyed with radiation, but treatment will generally fail unless the secondary tumors are located and destroyed. Unfortunately, finding secondary tumors is extremely difficult, for there are literally thousands of places where the cells can become established. Cancer cells from the breast, for example, may lodge in the brain or lungs or in **lymph nodes**, organs that are interspersed along the lymphatic system and filter the fluid flowing through the vessels. Doctors can take tissue samples or **biopsies** of lymph nodes to assess the spread of the cancer. Cancerous nodes may be removed surgically. Patients can also be treated with **chemotherapeutic agents**, highly toxic drugs that attack rapidly dividing cells (see Health Note 2–1: New Treatments for Cancer). This shotgun approach has many side effects, including nausea, hair loss, and general sickness, but may be a patient's only hope. Some new developments in cancer treatment are described in Health Note 4–1.

What Causes Cancer?

No question interests science more than what causes a cell to begin dividing without control. In other words, what unleashes a cell from its normal controls? The conversion of a normal cell to a cancerous one is called **transformation**. In most cases, transformation is the result of a **mutation**, a change in the DNA caused by a chemical, physical, or biological agent. Not all mutations result in cancer. Some mutations have no effect whatsoever, and still others may be beneficial to the cell. Even when the change is harmful, some damage can be repaired. It's therefore a rare mutation that leads to cancer.

Cancer-causing agents are called **carcinogens**. Chemical carcinogens bind to the DNA and alter its structure, a topic discussed in the next chapter. Some chemical substances bind directly to the DNA; other chemicals are harmless, but are converted in the body to carcinogens, which are capable of binding to the DNA. Nitrites, for example, are fairly harmless by themselves, but in the cells of the body, they are converted to carcinogenic nitrosamines (see the Environment and Health section in Chapter 2). The enzymes responsible for this conversion reside in the smooth endoplasmic reticulum (SER) of liver cells, and possibly others. Although the liver's SER is, for the most part, an ally in detoxifying chemical substances, in this instance, it turns a relatively harmless substance into a carcinogen. The formation of nitrosamines is, unfortunately, only one of many such examples.

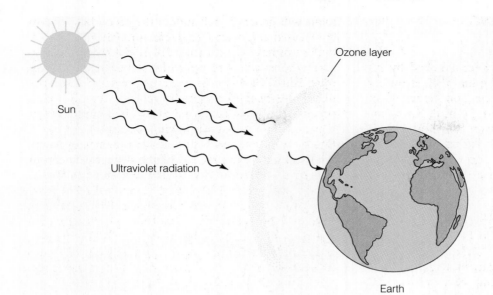

FIGURE 4–12   The Ozone
Layer   High above our heads lies
the ozone layer, a part of the
atmosphere containing ozone. The
ozone layer shields the Earth from
the ultraviolet radiation in sunlight.

Sun

Ultraviolet radiation

Ozone layer

Earth

Mutations do not manifest themselves immediately in cancer. In fact, most tumors do not generally form until 20 or 30 years after the mutation. A few cancers such as leukemia occur within 5 years. This is called the **latent period**, and is one reason why it is difficult for medical researchers to determine the causes of cancer.

In addition to the chemical carcinogens, there are many physical carcinogens. X-rays that a doctor gives to diagnose disease, ultraviolet radiation from the sun or a tanning salon, radon gas given off by soils in some parts of the country, and cosmic rays from the sun all can cause mutations and chromosome breaks that may lead to cancer. Of the physical agents, radiation and ultraviolet light are probably of greatest concern. Radiation interests most people because of the recent disasters at the Chernobyl and Three Mile Island nuclear power stations. Ultraviolet is of interest because of growing concern over the gradual depletion of the Earth's ozone layer, a shield against ultraviolet light from the sun. This topic is discussed in the next section.

Biological agents can also contribute to cancer. In 1911, Peyton Rous discovered that a virus caused cancer in chickens. Since that time, many viruses have been found to cause cancer in a wide variety of animals, including humans. Many of these viruses carry genes that cause cancer in the cells they infect.

ENVIRONMENT AND HEALTH: DEPLETING THE OZONE LAYER

Encircling the Earth, some 12 to 30 miles above your head, is a layer of air rich in ozone (O_3), called the **ozone layer** (Figure 4–12). The ozone layer forms an invisible shield that filters out most of the ultraviolet radiation emitted from the sun. The ozone layer protects all living things—from plants to humans—from harmful ultraviolet radiation. The formation of the ozone layer probably was necessary for the colonization of land many thousands of years ago (Chapter 19). Without it, terrestrial life could probably not exist or would have evolved thick, protective shields.

Today, however, the ozone layer is slowly disappearing. In 1974, two chemists warned that chemical substances used as spray can propellants and refrigerants, the **chlorofluorocarbons (CFCs)**, were eroding the ozone layer. Their conclusions stunned the scientific world, because CFCs had previously been thought to be chemically inert—that is, unreactive. Subsequent research showed that these compounds, while fairly stable, migrate from ground level into the upper atmosphere where they are broken down by sunlight. Chlorine atoms released in the process react with ozone molecules, destroying them, thus slowly depleting the ozone layer.

Cancer: Ultimately, You May Hold the Key to a Cure

Hardly a week goes by without headlines announcing some promising new cure for cancer.

ANTICANCER VACCINE

In 1988, for example, researchers reported preliminary results of a clinical study using an anticancer vaccine to fight lung cancer, which annually kills 150,000 Americans. The new vaccine could double survival rates in patients whose lung cancers are diagnosed early.

Although it is referred to as a vaccine, it is not one in the traditional sense, because it does not prevent the disease, but rather boosts the immune system to fight off an existing disease. In the study, 34 patients with early-stage lung cancer were treated with a unique plasma membrane protein from lung cancer cells. When given to patients in early stages of lung cancer, the protein appears to stimulate the immune system. The immune system mounts an assault on the protein and, in the process, attacks and kills lung can-

Sources of Dietary Fiber Eating food containing dietary fiber may help to reduce the chances of getting colon cancer.

cer cells, especially those that have spread to distant sites.

The therapy increased the five-year survival rates remarkably. The normal five-year survival rate for lung cancer is one in ten. In other words, only one patient in ten is alive five years after diagnosis and conventional treatment—surgery followed by chemotherapy. The immune system boost, however, increased the survival rate to five in ten.

Although this work is encouraging, medical scientists caution that it must be repeated before they can be sure of the effect. If the treatment lives up to its promise, medical researchers believe other cancers might be treatable by the same technique.

MICROSPHERES AND MONOCLONAL ANTIBODIES

In recent years pharmaceutical manufacturers have become excited over the potential for using lipid microspheres containing cancer-killing chemicals (oncotoxins) to treat cancer. As you may recall from Health Note 2–1, liposomes deliver smaller doses of the toxic cancer-killing drugs to cancer cells, helping protect the rest of the body's cells from these toxic chemical substances. Antibodies in the liposome membrane allow the toxic bullet to find its way to its target.

Oncotoxins may also be attached to monoclonal antibodies. Monoclonal antibodies are proteins pro-

In response to the threat, the United States and Sweden soon banned the use of CFC spray can propellants but not refrigerants. Much of the rest of the industrial world remained skeptical and continued to use CFCs for spray cans, refrigerants, blowing agents (to make styrofoam), and cleaning agents. Over the next decade and a half, numerous estimates of ozone depletion were made, but because of conflicting results and because the ozone concentrations in the ozone layer fluctuate naturally from year to year, atmospheric scientists could not tell for certain whether the claims were true.

Then, in 1988 a group of 100 scientists, working under the auspices of NASA, gathered to review the data of nearly 20 years of atmospheric monitoring.

They agreed that the ozone layer was indeed declining, and that levels over the poles had fallen by as much as 10% since 1969. The decline over the heavily populated regions of North America and Europe over the same period was 3%.

Scientists predict that each 1% decrease in the ozone layer will increase the incidence of skin cancer by at least 2%.[1] If this proves correct, the 3% decline will increase the skin cancer rate by 6%. This could result in an additional 20,000 to 60,000 cases of skin cancer each year in the United States alone. Since 4% of all skin cancers are fatal, that means an additional

---

[1] More recent estimates suggest that each 1% decline in the ozone layer will result in a 4% increase in skin cancer.

duced in the laboratory when human cancer cells are injected into an animal. The antibodies with their attached oncotoxin can then be administered to patients. There the antibodies bind to the cell surfaces of cancer cells and destroy them.

PREVENTION

Despite the promising new developments in cancer treatment, many experts stress the importance of prevention. By quitting smoking, or not taking it up in the first place, by limiting exposure to X-rays and ultraviolet light, and by choosing a safe occupation and dwelling, the likelihood of your contracting cancer can be greatly reduced.

Diet may also prove to be helpful in controlling cancer. Dietary fiber in grains, fruits, and vegetables, for example, may help reduce the chances of getting colon cancer, a leading cancer in men (see the figure). Other substances in food may also help reduce the risk of cancer. Recent research suggests that an acid (ellagic acid) found in strawberries and Brazil nuts may destroy carcinogenic chemicals in the body, reducing the likelihood of cancer.

Scientists studying a number of carcinogenic agents found in tobacco smoke, auto exhaust, and foods noted that ellagic acid in strawberries and nuts reduced DNA damage caused by one prevalent carcinogen by 45 to 70%. Using cultures of mouse and human lung tissue, the researchers also found that ellagic acid inhibited the formation of cancer caused by nitrosamines.

Ellagic acid has to be added prior or during exposure to a carcinogen, and researchers have not yet identified the mechanism by which this substance works—or its effectiveness in people. They suspect that the acid competes for binding sites on the DNA molecule to which carcinogens attach.

Clearly, prevention is the best medicine. The critical thinking skills presented in Chapter 1 suggest that individuals should beware of "miracle" cures to cancer. The research on ellagic acid, for example, is new and must be repeated in humans before scientists can be certain of its validity.

EMOTIONS AND CANCER

Studies released very recently indicate that there may also be a link between one's emotional state and the likelihood of developing cancer. Researchers at Johns Hopkins University studied medical students subjected to personality tests between 1948 and 1964, and compared the results of the tests with subsequent health records. They found that over a 30-year period students characterized as "loners" who suppressed emotions beneath a bland exterior were 16 times more likely to develop leukemia and several other cancers than a group that gave vent to emotions and, at times, took active measures to relieve anger and frustration.

A study of nearly 7000 residents of Alameda County, California, for 17 years showed that two types of social isolation (having few close friends and feeling alone even in the presence of friends) elevated the risk of certain cancers in women. A study in Yugoslavia also shows that repression and denial of emotions on a regular basis and in response to stress are related to an increase in cancer and heart disease.

800 to 2,400 deaths each year from skin cancer. Increased ultraviolet light may also harm other animals and plants.

In September 1987, about a year before reports of ozone depletion were endorsed by the atmospheric scientists, a group of 24 nations met in Montreal to hammer out a treaty to curb the use of most CFCs and a few related chemicals also thought to destroy the ozone layer. The nations agreed to cut their use to one-half of the 1986 levels by 1999. Given the accumulation of CFCs in the upper atmosphere since the 1950s, many critics warned that a 50% reduction would fall short of the cutback needed to save the ozone layer. Estimates showed that an eventual 10% reduction of the ozone layer was probable under the Montreal agreement.

Much to the surprise of the critics, in 1988 Dupont, a world leader in CFC production, agreed that further cuts were necessary and called for a total worldwide ban on CFC production. The company argued that stronger regulations could eliminate CFC production by the end of the century. Manufacturers are now actively pursing ways to reduce CFC emissions by using materials more conservatively and by finding safer alternatives.

Shifting to new CFCs will cost the U.S. economy an estimated $5.5 billion between 1990 and 2010, according to manufacturers. The Environmental Protection Agency projects that the cost of complying with the Montreal agreement will equal $27 billion dollars through the year 2075. This price, however, is insignificant compared to the savings from reducing skin cancer deaths alone. In dollars and cents, the cutbacks called for by the Montreal accord would amount to a health cost savings of $6,500 billion by 2075. Add to that the incalculable decrease in human pain and suffering and it becomes clear that good health and a healthy planet are economic as well as social and scientific concerns.

SUMMARY

1. Researchers have found a disturbing correlation between extremely low frequency (ELF) magnetic radiation and leukemia in children. At least one study shows that ELF radiation increases the resistance of cancer cells to the body's natural killer cells.
2. ELF magnetic fields are also generated by waterbeds and electric blankets and may increase miscarriages or spontaneous abortions.

THE CELL CYCLE

3. The life cycle of a cell is called the cell cycle and consists of two basic parts: interphase and cell division.
4. Interphase is a period of active cellular synthesis and is divided into three phases: G_1, S, and G_2.
5. During G_1, the cell produces RNA, proteins, and other molecules.
6. During the S phase, the DNA replicates. After replication, each chromosome in the nucleus of the cell contains two chromatin fibers or chromatids.
7. G_2 is a much shorter period and is relatively inactive.
8. During interphase, the cell replicates its organelles and molecules needed by the two daughter cells. Interphase also contributes to the economy of the organism.
9. Cell division follows interphase. Cell division requires two separate but related processes: mitosis, or nuclear division, and cytokinesis, or cytoplasmic division.

THE CHROMOSOME

10. The chromosome changes considerably during the cell cycle. After mitosis, each of the 46 chromosomes in human cells consists of a single chromatin fiber, composed of a strand of DNA and associated protein.
11. Chromosomes duplicate during the S phase, and each chromosome consists of two chromatin fibers or chromatids attached at their centromeres.
12. The chromosomes are loosely arranged in the nucleus during interphase, but condense during prophase of mitosis. Condensation facilitates chromosome separation. In the condensed state, chromosomes are metabolically inactive.

13. Cells that are about to divide can be isolated and flattened on a microscope slide. This procedure causes the chromosomes to spread out, allowing technicians to count the chromosomes and locate any abnormalities in number or chromosomal shape.
14. Chromosomes can be photographed and aligned according to number, forming a karyotype.

CELL DIVISION: MITOSIS AND CYTOKINESIS

15. During cellular division, the cell divides its chromosomes equally and distributes its cytoplasm and organelles more or less equally between two daughter cells.
16. In adults, cell division helps repair tissue damage and replaces cells lost through normal wear and tear. In growing children, it is a key element of growth and helps to repair damaged tissue.
17. Mitosis is divided into four stages: prophase, metaphase, anaphase, and telophase. Table 4–1 lists the major changes in each stage.
18. The mitotic spindle plays a role in the separation of chromosomes during mitosis. The spindle is an array of microtubules that forms during prophase. The spindle fibers attach to the chromosomes.
19. Several mechanisms have been proposed to explain the movement of chromosomes. According to one theory, the pole-to-chromosome fibers slide along the pole-to-pole fibers, thus pushing the poles of the cell further apart and dragging the chromosome away from the metaphase plate.
20. Elongation of the pole-to-pole fibers may also draw the chromosomes apart. The final transit may be accomplished by a shortening of the pole-to-chromosome fibers.
21. Cytokinesis begins late during anaphase or early in telophase. Cytokinesis results from the contraction of microfilaments beneath the plasma membrane.
22. Research suggests that the cell cycle is controlled at least in part by chemical substances produced in the cytoplasm. These chemicals signal changes in the nucleus of the cell.
23. External controls, such as hormones, growth regulators, cell contact, and other factors, are also imposed on the cell.

CANCER: UNDERSTANDING A DEADLY DISEASE

24. Many health experts believe that 20 to 40% of all cancers arise from environmental and workplace pollutants. Most of the rest are caused by smoking, diet, and natural causes.
25. No matter what the cause, cancer is a disease that most of us will have close experience with.
26. Studies show that the mortality rate of all cancers except lung cancer, has decreased since 1950.

27. Cancer is a disease in which cells divide uncontrollably, forming a tumor. Tumors that fail to grow and spread are called benign tumors and rarely cause medical problems.

28. Malignant tumors are those that continue to grow, often spreading to other parts of the body and invading vital organs. The term *cancer* is reserved for malignant tumors.

29. Malignant tumor cells spread through the body in the circulatory and lymphatic systems, often settling in distant sites where they form secondary tumors.

30. Tumors can be removed surgically or may be destroyed with radiation, but treatment generally fails unless all of the secondary tumors are located and destroyed. Finding secondary tumors is extremely difficult.

31. Patients can also be treated with chemotherapeutic agents, highly toxic chemical substances that attack rapidly dividing cells.

32. The transformation of a normal cell to a cancerous one often results from a mutation, a change in the DNA resulting from a chemical, biological, or physical agent. Not all mutations result in cancer.

33. Cancer-causing agents are called carcinogens. Chemical carcinogens bind to the DNA and alter its structure.

34. Generally, most tumors do not form until 20 or 30 years after the mutation, making it difficult to track their source.

ENVIRONMENT AND HEALTH: DEPLETING THE OZONE LAYER

35. The ozone layer is found in the upper atmosphere. It filters out most of the incoming ultraviolet light, protecting many life-forms from harmful burns and damaging mutations.

36. Unfortunately, certain chemicals called chlorofluorocarbons (CFCs) are destroying the ozone layer. CFCs are used as spray can propellants, refrigerants, coolants, cleaning agents, and blowing agents for foam.

37. According to a recent study, CFCs have caused a 3% decline in ozone levels in the atmosphere over North America.

38. Many nations have agreed to cut CFC production and release by 50% by 1999, but efforts are underway to phase them out completely.

⌘ EXERCISING YOUR CRITICAL THINKING SKILLS

Using the critical thinking skills presented in Chapter 1, describe the link between extremely low frequency electromagnetic waves given off by high-tension wires and cancer. What questions would you want answered before you can support the conclusions?

TEST OF TERMS

1. The cell cycle consists of two major phases, the _____ _____ phase, in which the cell divides, and _____, the period of growth and preparation between divisions.

2. The cell replicates its DNA during the _____ phase. In humans, each of the _____ chromosomes duplicates and consists of _____ strands of DNA.

3. The segment of the DNA that controls a trait is called a(n) _____ .

4. Division of the nucleus during the cell cycle is called _____ , and division of the cytoplasm is called _____ .

5. The unlimited growth of cancer cells is aided by the loss of _____ _____ , a signal moderated by the cell membrane, which stops cell division.

6. During G_2, the chromosomes consist of two _____ strands, each made of DNA and associated protein called _____ .

7. Chromatids of the chromosomes are attached at the _____ . These detach during _____ of nuclear division.

8. The withdrawal of fluid surrounding the developing embryo or fetus is called _____ . A chromosome spread in which the chromosomes are aligned in pairs is a(n) _____ and can be used to find missing segments or extra or missing chromosomes.

9. The _____ _____ appears in prophase and is made of microtubules that attach to the chromosomes.

10. Chromosomes line up in the middle of the cell during _____ .

11. A _____ _____ lying just beneath the plasma membrane contracts and draws the

plasma membrane in, creating a furrow that eventually divides the cytoplasm in two.

12. The spread of cancer cells through the lymphatic and circulatory systems is called _____ . This process results in the establishment of _____ tumors at different sites.

13. The conversion of a normal cell to a cancerous one is called _____ .

14. A chemical or physical agent that causes cancer is a(n) _____ .

15. The chemical substances believed responsible for destroying the _____ layer, a protective region in the atmosphere that blocks ultraviolet radiation, belong to a group of compounds called _____ .

Answers to the Test of Terms are located in Appendix B.

TEST OF CONCEPTS

1. Describe the cell cycle and explain what happens during each part.

2. Discuss the factors that control the cell cycle. Is it likely that many factors play a role in determining when and how often a cell divides?

3. The term *chromosome* applies to both single- and double-stranded structures containing DNA. Draw a diagram of the cell cycle and note how many chromosomes are found in the cell at each stage and whether they are single- or double-stranded.

4. List the stages of mitosis and describe the major cytoplasmic and nuclear changes occurring in each stage. When does cytokinesis occur?

5. In what ways are cancer cells different from normal cells? What events lead to a cancer? Why don't more cells become cancerous? What's the difference between a benign and a malignant tumor?

6. The Environment and Health perspective argues that chemical substances produced by modern society, notably the chlorofluorocarbons, may be making our environment hazardous to human health. Describe how.

SUGGESTED READINGS

Becker, W. M. 1986. *The world of the cell.* Menlo Park, Calif.: Benjamin/Cummings. See Chapter 23 for an interesting discussion of the cellular events of cancer.

Boly, W. 1989. Cancer inc. *Hippocrates* 3(1): 38–48. An assessment of the war on cancer that describes innovative treatments and lists the most common cancers with a comparison of conventional and experimental treatments.

Brodeur, P. 1989. Annals of radiation: The hazards of electromagnetic fields. Part 1: Power lines; Part 2: Something is happening; Part 3: Video display terminals. *The New Yorker* 65 (June 12): 51–88; (June 19): 47–73; (June 26): 39–68. A series for the general public on the hazards of electromagnetic radiation. Part 3 looks at the computer terminal and the evidence that suggests problems may arise from exposure to electromagnetic fields generated by the screen.

Chiras, D. D. 1991. *Environmental science: Action for a sustainable future.* 3d ed. Redwood City, Calif.: Benjamin/Cummings. See Chapter 14 for more on chemical substances and their effects on cells.

Cummings, M. R. 1988. *Human heredity: Principles and issues.* St. Paul, Minn.: West. Covers many topics discussed in this chapter in more detail.

Feldman, M., and L. Eisenbach. 1988. What makes a cell metastatic? *Scientific American* 259(5): 60–68. Discusses the problem of metastasis and how it might be controlled.

McIntosh, J. R., and K. L. McDonald. 1989. The mitotic spindle. *Scientific American* 261(4): 48–56. A survey of recent studies on the structure and behavior of the spindle.

5 How Genes Work and How Genes Are Controlled

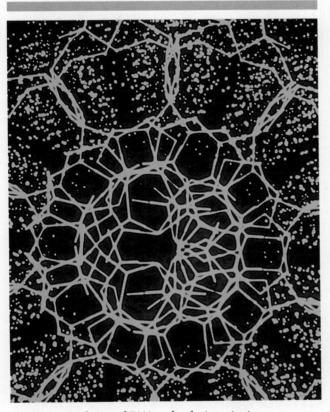

Computer simulation of DNA molecule (top view).

I n 1988, a U.S. military court sentenced a serviceman in Korea to 45 years in prison for rape and attempted murder. Ten days later, a Florida court convicted a man on two counts of first-degree murder. In the trials of both men, prosecutors relied on a new technique called DNA fingerprinting. Criminologists analyze the composition of the DNA molecules found in samples of hair, semen, or blood left at the scene of the crime. Then they compare the DNA in these samples with the DNA of the accused. Without this procedure, prosecutors believe that these two convictions would have been impossible.

Many experts believe that DNA fingerprinting will revolutionize criminal investigation. King County, Washington, in fact, has already begun taking DNA samples from all convicted sex offenders to keep on file. The FBI is also actively developing the technique. The method is promising because of the low chance for error—only one in 4 or 5 trillion. The best conventional methods, such as blood typing and blood enzyme analysis, run the risk of error in one of every 1000 cases or greater, making them of questionable value.

DNA fingerprinting is only one of several exciting new developments in biology. Another promising advance is genetic engineering. **Genetic engineering**

107

emerged in 1973 as the result of research by Stanley Cohen at Stanford University and Herbert Boyer at the University of California at San Francisco. These scientists cut segments of DNA from one bacterium and spliced them into the genetic material of another bacterium. The cutting and splicing of DNA is called genetic engineering, gene splicing, or recombinant DNA technology. Today, these techniques are used for a great many purposes. For example, researchers transfer human genes that control the production of insulin and growth hormones into bacteria. The bacteria, in turn, become hormone factories. This advance has already greatly increased the supply of these hormones, which are used to treat medical disorders. Other new developments may someday permit physicians to insert genes into defective body cells, curing genetic diseases. Genetic engineering could also have profound impacts on agriculture, creating faster-growing and more disease-resistant crops and livestock. Disease-resistant crops could help reduce the use of pesticides worldwide. This new technology may also have adverse impacts, however. Development should proceed with caution.

This chapter describes the gene—how it works and how it is controlled. It will help you better understand genetic engineering and other important concepts of genetics that will be useful in your lifetime. Health Note 5–1 describes some of the real and potential benefits of genetic engineering; the Point/Counterpoint at the end of the chapter introduces you to an important debate over the ethical, social, and environmental issues surrounding genetic engineering.

BUILDING TWO MACROMOLECULES: DNA AND RNA

In 1953 two biologists, James Watson, an American, and Francis Crick, a British scientist, proposed a model for the structure of the DNA molecule. In 1962, Watson and Crick, along with British biophysicist Maurice Wilkins, received the Nobel Prize in Physiology and Medicine.[1] The discovery of the structure of DNA led to a whole new realm of genetics—molecular genetics—with many practical applications, such as DNA fingerprinting and genetic

---

[1]Maurice Wilkins began studying images of molecules produced by X-rays after World War II. One of Wilkin's colleagues, Rosalind Franklin, had made an X-ray image of DNA, which Watson felt suggested that the DNA molecule was a double helix. Unfortunately, Franklin died four years before the Nobel Prize was awarded, and the award is given only to living scientists.

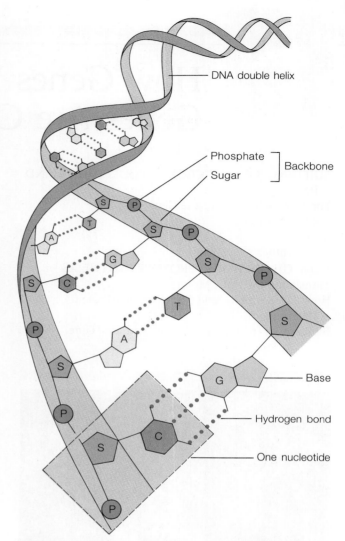

FIGURE 5–1   DNA   DNA consists of two polynucleotide chains wrapped around each other to form a double helix. The sugars and phosphates form the backbone of each chain with the bases projecting inward. The bases on opposite strands are connected by hydrogen bonds.

engineering. This section reviews the structure of DNA and RNA and describes how these molecules are synthesized.

The Molecular Basis of the Gene

DNA is an intriguing molecule that consists of two long chains, each containing millions of nucleotides—hence the name polynucleotide chain (Figure 5–1). The polynucleotide chains twist around one another, forming the **double helix**, a structure that resembles a spiral staircase. The two chains are held

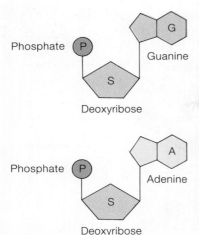

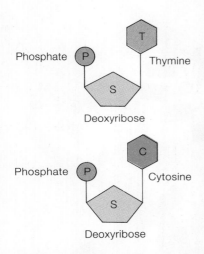

(a) DNA nucleotides containing purine bases

(b) DNA nucleotides containing pyrimidine bases

FIGURE 5–2 Purine and Pyrimidine Bases All nucleotides consist of three parts: a phosphate group, a base, and a five-carbon sugar. DNA nucleotides contain the sugar deoxyribose. Two types of bases are found: purines and pyrimidines. (*a*) The purines are adenine and guanine. (*b*) The pyrimidines are thymine and cytosine.

together by hydrogen bonds that form between the bases, shown in Figure 5–1.

Chapter 2, you may recall, noted that each DNA nucleotide consists of three parts: a base, a phosphate group, and a monosaccharide, deoxyribose. The sugar molecule and the phosphate of the nucleotide join by covalent bonds, forming the backbone of the polynucleotide chains (Figure 5–1). The bases of each chain lie in the inside of the double helix where they

are attracted to bases on the opposite strand by hydrogen bonds. The bases form the steps of the spiral staircase.

As illustrated in Figure 5–2, two types of bases are found in the DNA molecule: purines and pyrimidines. **Purines** consist of two fused rings, while **pyrimidines** consist of only one ring. As shown in Figure 5–2, there are two types of purines and two different pyrimidines in DNA. The purines are adenine (A) and guanine (G).[2] The pyrimidines are cytosine (C) and thymine (T). Purines and pyrimidines are found in both strands of the DNA double helix.

As illustrated in Figure 5–1, purines on one strand always bond (via hydrogen bonding) to pyrimidines on the opposite strand. Adenine, for example, always bonds to thymine. Guanine always bonds to cytosine. Adenine and thymine are, therefore, said to be complementary bases, as are guanine and cytosine. This unalterable coupling, called **complementary base pairing**, ensures the accurate transmission of the genetic information during cell division.

DNA Replication

As you learned in Chapter 4, before a cell divides, it must first make a copy of all of its DNA, thus ensuring that the daughter cells can function independently. Cells replicate their DNA by first splitting each DNA double helix along the hydrogen bonds, then using each strand as a template to produce a new strand (Figure 5–3). The new strand is called a complementary strand.

DNA replication begins when special enzymes begin to pull apart or "unzip" the DNA double helix. As the two strands are separated, the bases on each strand are exposed and thus free to bind to complementary nucleotides in the nucleoplasm. Because adenine only binds to thymine and guanine only binds to cytosine, the new strand is accurately built. The old strand "directs" the synthesis of a new strand one base pair at a time.

Complementary base pairing ensures the accurate replication of information. Accurate replication, in turn, ensures that the daughter cells receive a copy of the information contained in the parent cell's DNA. In turn, the daughter cells will give rise to other cells that receive the same information—barring mutations—creating a line of inheritance, which, for

[2]To remember this, you can use the mnemonic "pure Ag" offered by a professor who taught my genetics course. He was from the agricultural college at my university.

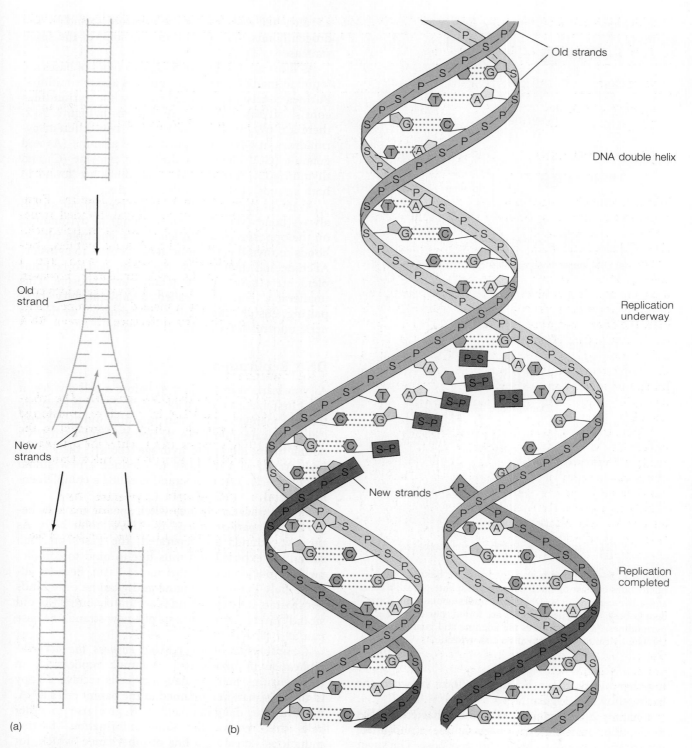

(a)

(b)

Old strands

DNA double helix

Replication underway

New strands

Replication completed

Old strand

New strands

FIGURE 5–3 Semiconservative Replication DNA
replication is semiconservative. (*a*) Each double helix
unwinds, and each half of the helix serves as a template for
the production of a new strand of DNA. When replication is
complete, each new helix contains one old and one new
strand. (*b*) Nucleotides attach to the template one at a time
and are joined to others with the aid of enzymes.

110   PART II   *Cells and Molecules*

TABLE 5–1   Role of RNA Molecules

MOLECULE	ROLE
Messenger RNA (mRNA)	Carries the information needed to make proteins from the nucleus to the cytoplasm
Transfer RNA (tRNA)	Binds to specific amino acids, transports them to the mRNA, and inserts them in the correct location on the mRNA
Ribosomal RNA (rRNA)	Component of the ribosome

humans, stretches back several million years.[3] As noted in Chapter 1, mutations arise spontaneously. If a mutation is beneficial, it will persist and become part of the genetic legacy of an animal or plant.

For a new DNA strand to form, however, the incoming nucleotides must first be aligned properly on the template. Proper alignment permits the hydrogen bonds to form between complementary bases and allows the sugar and phosphate groups to join. Alignment and attachment are aided by an enzyme, **DNA polymerase** (Figure 5–4). DNA polymerase slides along the template, first aligning nucleotides, then linking the nucleotides together by joining phosphates of one to the deoxyribose molecules of its neighbor.

When DNA synthesis is finished, two new DNA molecules exist, each containing one strand from the original double helix and one new strand. (This process is called semiconservative replication.) The chromosome now consists of two chromatids. Each chromatid consists of a DNA double helix with associated histone proteins, described in Chapter 4. The two chromatids of each chromosome are joined at their centromeres after synthesis, but separate when the nucleus divides.

RNA

DNA contains the genetic information, but it is RNA that carries the instructions into the cytoplasm where protein is manufactured. Three types of RNA exist: ribosomal RNA, messenger RNA, and transfer RNA. Each has a unique function in protein synthesis, discussed in the next section and summarized in Table 5–1. Despite the differences in these molecules, all

three share some common structural features. First, all RNAs in human cells are single-stranded molecules (Figure 5–5a). Second, all RNAs are polynucleotides. RNA nucleotides, like DNA nucleotides, consist of three molecules: a sugar, a base, and a phosphate group. In the RNA nucleotide, however, the sugar is ribose. In addition, RNA nucleotides contain the pyrimidine uracil instead of thymine. Table 5–2 summarizes the key differences between RNA and DNA.

RNA Synthesis

RNA is synthesized on the DNA molecule. The information coded in the DNA is, therefore, transferred from the DNA molecule, which is restricted to the nucleus, to RNA, which is free to enter the cytoplasm. The process of RNA production is called **transcrip-**

FIGURE 5–4   Role of DNA Polymerase DNA polymerase binds loosely to the DNA template and to the nucleotide, aligning it for bonding to the previous nucleotide. The enzyme also catalyzes the formation of the bond between the phosphate and sugar.

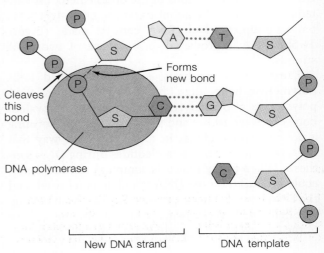

[3]Mutations that become a permanent part of the genetic code may arise during evolution. These mutations are also passed from generation to generation. The emergence of mutations results in a change in the genetic information.

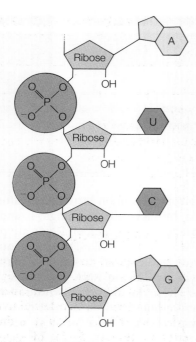

(a) Segment of RNA

FIGURE 5–5 RNA (*a*) RNA is a single-stranded molecule consisting of many RNA nucleotides. (*b*) RNA is synthesized on a DNA template. As shown here, the DNA double helix unwinds and one strand serves as a template for the production of a strand of RNA. Complementary base pairing determines the exact sequence of bases in the RNA molecule.

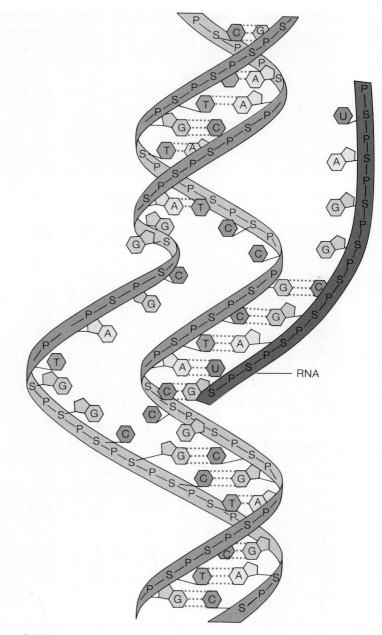

(b) A strand of RNA produced from DNA template

tion. Like note taking, it is merely a transfer of information from one medium (speech) to another (written form). RNA is produced in the cell cycle during the G_1 phase, before the chromosomes duplicate. During G_1, small sections of the double helix unwind temporarily with the aid of enzymes, creating a DNA template upon which RNA can be made (Figure 5–5b). RNA is produced on only one of two DNA strands.

During RNA synthesis, an enzyme called **RNA polymerase** helps align the RNA nucleotides. RNA polymerase also catalyzes the formation of covalent bonds between ribose and phosphate groups, forming a polynucleotide chain, in much the same way that DNA polymerase unites nucleotides during DNA synthesis. When the synthesis is complete, the RNA molecule produced on the DNA template is released, and the DNA strands reunite, re-forming the double helix. The helix, however, unwinds again during the S phase of the cell cycle, during which time the cell replicates

its DNA. During the synthesis of RNA, base pairing ensures the proper sequence of nucleotides on the RNA strand. Figure 5–6 illustrates how base pairing operates.

In summary, the DNA helix is a dual-purpose molecule. It serves as a template for the synthesis of DNA and RNA. By accurately replicating itself, DNA en-

TABLE 5–2 Differences between RNA and DNA in Human Cells	
DNA	RNA
Double stranded	Single-stranded
Contains the sugar deoxyribose	Contains the sugar ribose
Contains adenine, guanine, cytosine, and thymine	Contains adenine, guanine, cytosine, and uracil
Functions primarily in the nucleus	Functions primarily in the cytoplasm

sures the precise transmission of genetic information from cell to cell and, ultimately, from generation to generation. By making RNA, DNA provides instructions used by the cell to make protein in the cytoplasm. Using RNA, in ways described shortly, cells make structural proteins that give them their unique structural and functional characteristics. The cell also makes enzymes and other proteins (like antibodies and hormones) that carry on cellular and organismic functions. In the chain of cellular command, the DNA rules its domain through the RNA. The daily activities of the cell are ultimately controlled by proteins.

HOW GENES WORK: PROTEIN SYNTHESIS

The genes in a cell's nucleus contain a wealth of information. What form is that information in, and how is it used?

FIGURE 5–6 RNA/DNA Base Pairing The sequence of bases on the DNA template determines the sequence of bases on the DNA and RNA molecules synthesized in the nucleus.

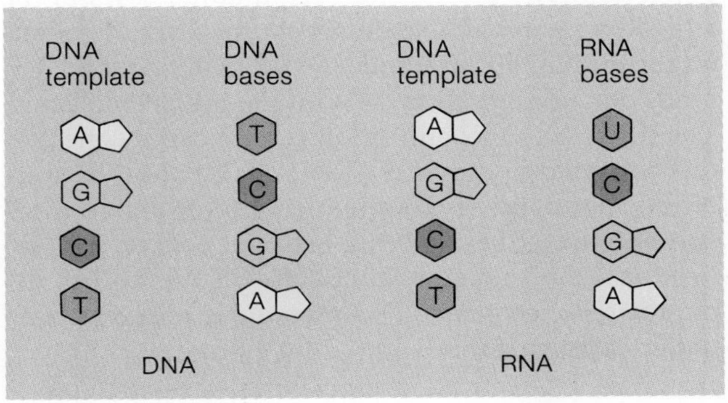

Translating the DNA Message

The answer to these questions represents one of the most fascinating stories of molecular biology. So far, you have learned that the genetic information contained in the DNA molecule is transferred to the RNA, in particular, messenger RNA (mRNA) in the process called transcription (Figure 5–7). Messenger RNA molecules, in turn, carry this information to the cytoplasm.

Protein is synthesized on mRNA molecules located on the endoplasmic reticulum or free within the cytoplasm. The synthesis of protein on an mRNA molecule or template is called **translation** (Figure 5–7). In effect, the RNA message is translated into a new "molecular language"—that of the protein.

FIGURE 5–7 The Central Dogma of Molecular Genetics The DNA controls the cell through protein synthesis. RNA serves as an intermediary, carrying genetic information to the cytoplasm where protein is synthesized.

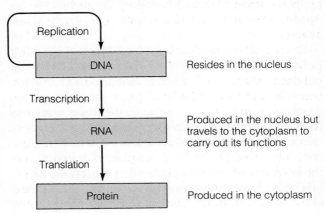

DNA template	DNA bases	DNA template	RNA bases
A	T	A	U
G	C	G	C
C	G	C	G
T	A	T	A
DNA		RNA	

TABLE 5–3 Codons on Messenger RNA and their Corresponding Amino Acids

CODON	AMINO ACID	CODON	AMINO ACID	CODON	AMINO ACID	CODON	AMINO ACID
AAU AAC	Asparagine	CAU CAC	Histidine	GAU GAC	Aspartic acid	UAU UAC	Tyrosine
AAA AAG	Lysine	CAA CAG	Glutamine	GAA GAG	Glutamic acid	UAA UAG	(Stop)* Stop to "Terminator"
ACU ACC ACA ACG	Threonine	CCU CCC CCA CCG	Proline	GCU GCC GCA GCG	Alanine	UCU UCC UCA UCG	Serine
AGU AGC	Serine	CGU CGC	Arginine	GGU GGC	Glycine	UGU UGC	Cysteine
AGA AGG	Arginine	CGA CGG		GGA GGG		UGA UGG	(Stop)* Stop to Terminator Tryptophan
AUU AUC AUA	Isoleucine	CUU CUC CUA	Leucine	GUU GUC GUA	Valine	UUU UUC	Phenylalanine
AUG	Methionine	CUG		GUG		UUA UUG	Leucine

*Stop codons signal the end of the formation of a polypeptide chain.

How Is the Amino Acid Sequence of a Protein Determined?

Research shows that *the sequence of bases in the mRNA codes for the precise sequence of amino acids in proteins and peptides.* Three adjacent bases in the mRNA code for a single amino acid; these three bases are called a **codon** (Table 5–3).

To understand how the codons determine the sequence of amino acids in a protein requires an understanding of two additional players: transfer RNAs (tRNA) and ribosomes. **Transfer RNA** molecules are relatively small molecules of RNA that bind to specific amino acids in the cytoplasm and transport them to the mRNA molecule, where the amino acids are incorporated into the protein molecule under construction. Ribosomes, described in Chapter 3, are made of ribosomal RNA (rRNA) and protein and consist of two subunits. They play a key role in building proteins on the mRNA template. How do mRNA, tRNA, and ribosomes build proteins?

As illustrated in Figure 5–8, the mRNA in the cytoplasm (either free or on the surface of the rough endoplasmic reticulum) consists of many nucleotides. Its bases lie exposed to the cytoplasm. Also found in the cytoplasm are many tRNA molecules. Transfer RNAs bind to specific amino acids and deliver them to the mRNA.

How do tRNA molecules "know" where to bind to the mRNA? On each tRNA is an **anticodon loop**, a segment of the molecule bearing three bases called the **anticodon**. The purine and pyrimidine bases of the anticodons are complementary to the bases of codons on the mRNA. For example, the tRNA anticodon UUU binds only to the codon AAA on the mRNA molecule. The tRNA with this anticodon (UUU) carries one particular amino acid, lysine (Table 5–3). Thus, because of the specificity of tRNA molecules for amino acids and because of base pairing between codons and anticodons, the sequence of codons on the mRNA determines the sequence of amino acids in the protein under construction.

Building a Protein: One Amino Acid at a Time

Proteins are synthesized by adding one amino acid at a time. Each strand of mRNA contains codons marking where protein synthesis should begin and end. These are called the **initiator** and **terminator codons**, respectively.

To understand protein synthesis, let's examine how a cell builds a short peptide. As illustrated in Figure 5–9a, the ribosome first attaches to the mRNA at the initiator codon. Ribosomes contain two binding sites for tRNA molecules. Soon after the ribosome attaches to the mRNA, a tRNA bearing its amino acid binds to the first binding site. Its anticodon binds by hydrogen bonds to the codon. A second tRNA–amino acid then enters the second binding site. The ribosome also contains an enzyme that catalyzes the formation of a peptide bond between the two amino acids. After the peptide bond is formed, the first tRNA (minus its amino acid) leaves the binding site and is free to retrieve another amino acid for use later on. As illustrated in Figure 5–9c, the dipeptide formed in this process is attached to the second tRNA and is held in the second binding site.

The ribosome also contains a contractile protein that allows this structure to move down the mRNA. After the first tRNA is released, the ribosome slides down the mRNA. In the process the tRNA and its dipeptide are shifted to the first binding site. This opens up the second site for another tRNA–amino acid.

A tRNA bound to an amino acid will move into the site. When the correct amino acid moves in, a peptide bond will form between the new amino acid and the previous amino acid, forming a tripeptide. Next, the tRNA in the first binding site detaches and the ribosome shifts down one more codon.

This process repeats itself many times, adding hundreds of amino acids very rapidly. The hemoglobin molecule, for example, is synthesized in 3 minutes. In the bacteria, many proteins are synthesized in 15 to 30 seconds.

As the peptide chain is formed, it begins to bend and twist, forming the secondary structure of the protein or peptide (Figure 5–10). When the ribosome reaches the terminator codon, the entire peptide chain is released. If this process occurs on the rough endoplasmic reticulum, the peptide chain is generally transferred into its cisterna, where it will be chemically modified and packaged into transport vesicles, which are transported to the Golgi complex. If this process occurs in the cytoplasm on a polysome, the protein is released into the cytoplasm.

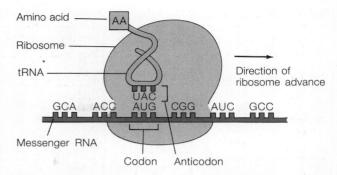

FIGURE 5–8 Messenger RNA Messenger RNA, either free within the cytoplasm or bound to the endoplasmic reticulum, serves as a template for protein synthesis. The codons, each consisting of three bases, determine the sequence of amino acids by binding to anticodons on the tRNA molecules. Each tRNA, in turn, delivers a specific amino acid to the mRNA template.

Inhibiting Protein Synthesis to Kill Bacteria and Cancer Cells

RNA and protein synthesis are vital to cellular function—so important, in fact, that chemical agents that impair these processes can kill a cell. Some antibiotics, for example, kill bacteria by inhibiting bacterial protein or RNA synthesis. Slight differences in the machinery of RNA and protein synthesis in human and bacterial cells "spare" human cells. Antibiotics are, therefore, far less toxic to human cells than to bacteria.

Viruses also invade the human body, but unlike bacteria, viruses do not have their own enzymes or the organelles needed to produce proteins and nucleic acids independently.[4] In order to reproduce, then, viruses must invade other cells. Inside their host, viruses commandeer the cell's protein synthetic machinery, using it to make more viruses. Since viruses reproduce using the enzymes of the body cells and because these enzymes are not affected by antibiotics, viral infections cannot be treated by antibiotics. Patients suffering viral infections (colds, bronchitis, and

[4]Because viruses cannot metabolize, they are not considered living organisms. Microbiologists refer to them as "agents."

FIGURE 5-9   Protein Synthesis

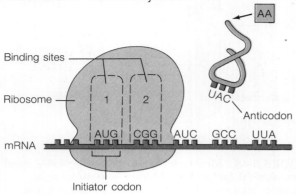

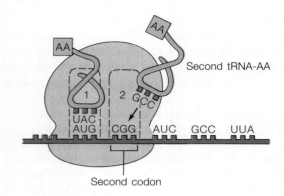

(a) The ribosome binds to the mRNA template. The tRNA binds to a specific amino acid in the cytoplasm.

(b) The tRNA–amino acid (tRNA-AA) complex binds to the first codon and is held in place by the first binding site. A second tRNA-AA complex binds to the second codon.

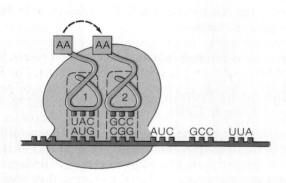

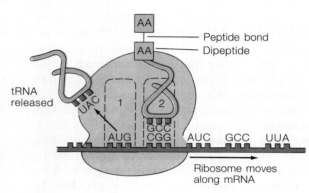

(c) The ribosome contains an enzyme that catalyzes the formation of a peptide bond between the two amino acids. The dipeptide is then attached to the second tRNA. This frees up the first tRNA, which vacates the first binding site.

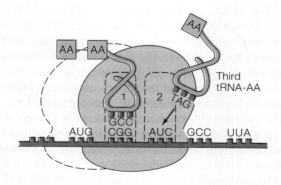

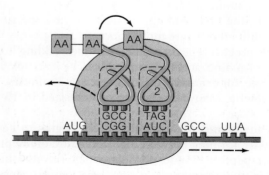

(d) The ribosome next slides down the mRNA, transferring the tRNA-dipeptide to the first binding site and opening up the second binding site to a third amino acid.

(e) The dipeptide is then linked by a peptide bond to the third amino acid, forming a tripeptide. This frees what was the second tRNA. The ribosome next slides down one more codon, exposing the second binding site and freeing it up for the addition of another tRNA-AA. This process repeats itself until the terminator codon is reached.

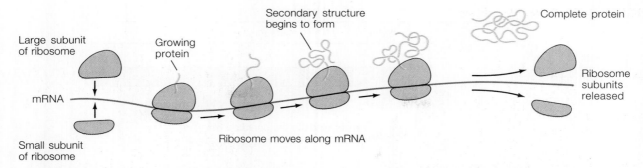

Large subunit
of ribosome

mRNA

Small subunit
of ribosome

Growing
protein

Secondary structure
begins to form

Complete protein

Ribosome moves along mRNA

Ribosome
subunits
released

FIGURE 5–10 Protein Synthesis As the protein is synthesized on the mRNA, it begins to coil and bend, forming its secondary structure. Several ribosomes may "work" a single strand of mRNA simultaneously.

so on) are generally sent home with only a prescription that treats some of the symptoms, such as congestion. The task of curing a viral infection, however, is left to the immune system (Chapter 10).

CONTROLLING THE GENES

In science it is one thing to describe a process, but quite another to understand how a process is controlled. This is certainly true of gene regulation. Although our understanding of gene regulation in bacteria has grown considerably, human gene control remains something of an incomplete puzzle. Understanding the control mechanisms is important, however, because it could help scientists find new ways to treat or even cure diseases, such as cancer. The following section describes gene regulation in bacteria, where the process is relatively well understood.

Control of Genes in Bacteria

In 1961, two French researchers, Francois Jacob and Jacques Monod, who had been studying the genes of a common bacterium, announced a breakthrough in the scientific understanding of bacterial gene control. Jacob and Monod found that bacterial genes consist of three parts: structural genes, regulator genes, and control regions.[5] **Structural genes** code for the production of enzymes and other proteins. **Regulator genes** and **control regions** control the production of mRNA on the structural genes. Jacob and Monod also found that, in bacteria, the control regions and structural genes of a metabolic pathway are all located

[5]They originally labeled the functional gene unit an operon and noted that it consisted of two parts, the structural genes and the regulatory genes (including the regulator gene and control regions).

side-by-side on the chromosome, that is, in a linear cluster. The regulator genes are located nearby. Together, the structural and regulator genes and the control regions form a functional multi-gene unit, called the **operon** (Figure 5–11). For their discoveries, the two scientists were awarded the Nobel Prize in Physiology and Medicine.

Induction of Gene Action. The French researchers discovered two types of operons in bacteria. The first is an **inducible operon**, consisting of a set of genes that remain inactive until needed. Jacob and Monod worked out the details of the inducible operon in the intestinal bacterium with a tongue-twisting name, *Escherichia coli,* or *E. coli* for short (Figure 5–12). *E. coli* live in the human large intestine, digesting left-over materials from food. In the process, they produce vitamin K, which is absorbed into the bloodstream.

E. coli are capable of producing many enzymes needed to digest leftover foodstuffs, but they produce these enzymes only on demand. For example, when lactose (milk sugar) is present, *E. coli* produce three different proteins that allow the bacterium to transport the molecule across its membrane and break it down into its monosaccharide units. What causes the bacterium to activate the synthesis of these proteins?

The signal to begin producing these proteins is lactose itself. Small amounts of lactose diffuse from the large intestine into the bacterium and activate the lactose operon. Lactose is, therefore, said to be an **inducer**, a substance that activates genes and protein production.

To understand how this molecule "turns on" the structural genes, consider Figure 5–11. This diagram shows the location of the structural genes, control regions, and the regulator gene in the lactose operon of *E. coli*. As illustrated, each of the three structural genes codes for a single protein necessary for lactose

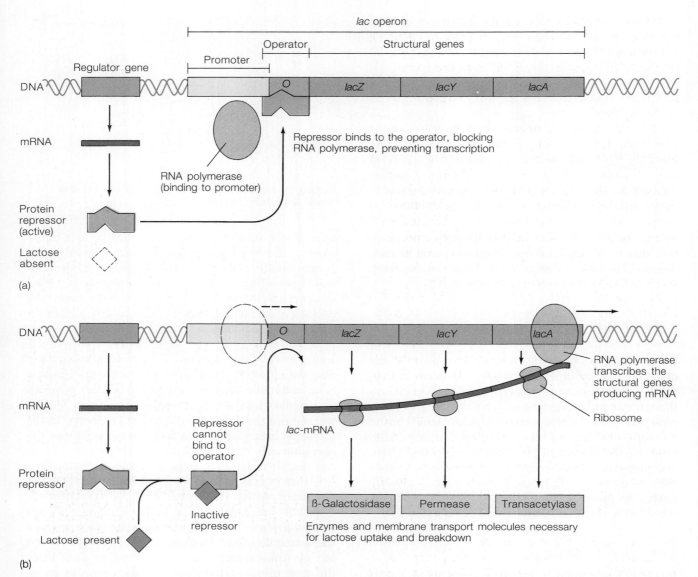

FIGURE 5–11 Inducible Operon (a) The lactose or *lac* operon of bacteria is an inducible operon. It is shut off until activated. Note that the regulator gene of the DNA produces mRNA that codes for a repressor protein. It binds to the operator site of the operon, blocking RNA polymerase from transcribing the structural genes. (b) When lactose is present, it binds to the repressor protein, releasing it from the operator site. This allows RNA polymerase to transcribe the structural genes.

metabolism. The structural genes produce messenger RNA with the assistance of RNA polymerase, described earlier in the chapter. Lying nearby on the DNA is the regulator gene, which controls the activity of the structural genes. The regulator gene codes for the production of a protein called a **repressor**. The repressor protein binds to the control region adjacent to the structural genes, called the **operator site**. The operator site does not produce RNA, but is a kind of switch on the DNA. When the repressor protein is bound to the operator site, the operon is switched off, as illustrated in Figure 5–11. When the repressor protein vacates the site, the operon is switched on. What causes it to leave?

Lactose molecules entering the bacterium bind to the allosteric site on the repressor protein, changing the protein's shape, and causing it to detach from the operator site. Lactose, therefore, turns the operon on. Why does the detachment of the repressor protein lead to the transcription of the three structural genes in this operon?

When the repressor protein binds to the operator site, it physically blocks the transcription of RNA by blocking RNA polymerase, an ezyme that catalyzes the formation of RNA on DNA. RNA polymerase binds to another region of the operon, called the promoter site. The promoter site lies "upstream" from the operator site and the structural genes. When the repressor protein is bound to the operator site, RNA polymerase cannot get to the structural genes to transcribe them. Only when the repressor protein is removed can the structural genes be transcribed.

This complex, but elegant, system provides the bacterium with an on-off switch for some of its genes. Being able to keep genes inactive until needed prevents the synthesis of unnecessary proteins and saves energy and raw materials. Only when substrates are available does the bacterium gear up to deal with them.

Repression of Gene Action. Evolution has provided another strategy for control, the **repressible operon.** Unlike the inducible operon, this one remains active unless turned off. This strategy provides constant amounts of enzymes that bacteria require to produce molecules, such as amino acids. A steady supply of amino acids ensures adequate protein synthesis.

Figure 5–13 illustrates the repression of gene action. As shown in the diagram, the repressible operon is similar in many ways to the inducible operon. In the repressible operon, however, the regulator gene produces a repressor protein that binds to the operator site only when it binds to another molecule, the **corepressor.** Thus, the structural genes remain active until a corepressor is available.

When the corepressor binds to the repressor protein, it presumably changes the shape of the molecule, thus allowing it to bind to the operator site. When the corepressor-repressor complex binds to the operator site, RNA polymerase is blocked and cannot transcribe the structural genes.

Corepressors are generally the end products of metabolic pathways. When these chemical substances are produced faster than they can be used, they build up inside the cell. These molecules then bind to the

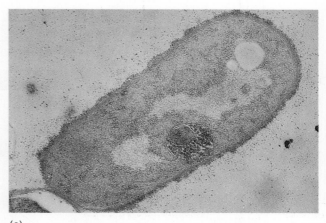

(a)

(b)

FIGURE 5–12   E. coli   (a) A transmission electron micrograph of the bacterium *E. coli,* a common inhabitant of the large intestine of animals. Note that the region of the bacterium containing the circular DNA is not enclosed by a membrane and is therefore called a nucleoid. (b) A circular molecule of DNA from *E. coli.*

repressor proteins. The repressor proteins, in turn, bind to the operator site, shutting down the operon and the production of enzymes in a wonderful example of negative feedback (Chapter 1).

Control of Human Genes

The discoveries of Francois Jacob and Jacques Monod created incredible enthusiasm among biologists, who hoped that human genes would be regulated similarly. Unfortunately, this was not the case. Control of the human **genome**, the entire genetic composition of the human organism, has turned out to be much more complex.

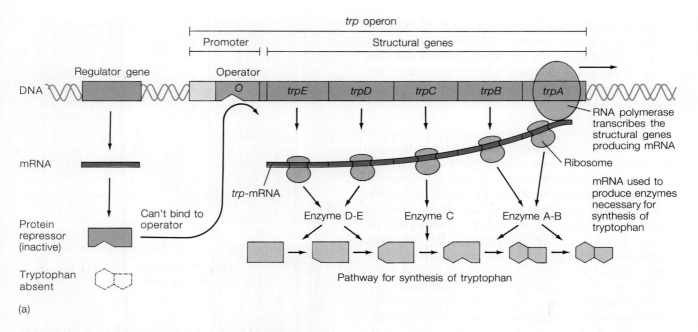

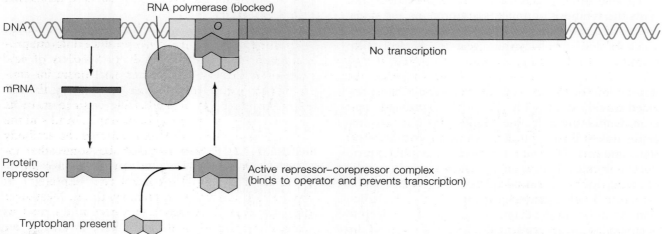

(a)

(b)

FIGURE 5–13 Repressible Operon (a) Structurally similar to the inducible operon, this package of genes is permanently on, until turned off. The regulator gene codes for a repressor protein that cannot bind to the operator site unless corepressor is present. Thus, in the absence of a corepressor, RNA polymerase is free to transcribe the structural genes. (b) When tryptophan, the corepressor, is present, it binds to the repressor protein. This allows the repressor protein to bind to the operator site, blocking RNA polymerase and turning off the operon.

To begin with, gene expression is regulated at several levels (Figure 5–14). These can be divided into at least three major categories: control at the chromosome level, control at transcription, and control after transcription.

Control at the Chromosome. Chapter 4, you may recall, noted that chromatin fibers in the nucleus con- dense and become inactive during prophase of mito- sis. In the condensed state, they cannot produce RNA or new DNA. The molecule is essentially shut down in preparation for nuclear division. During interphase, the cell also selectively inactivates some of its chro- matin. Why is this so?

Each cell in the body contains all of the genetic information needed to operate every single cell in the body. However, a muscle cell, for example, does not need all of the genes that a liver cell needs, and vice versa. Instead of getting rid of the genes it does not need, the muscle cell simply inactivates them.

Inactive chromatin is tightly coiled or compacted in the interphase nucleus and appears as dark clumps called **heterochromatin**. It stays in this state throughout the cell cycle. The cell also packs up some of the chromatin that is not needed at the present moment and unravels it when it is needed. The unwound chromatin is called **euchromatin** ("true chromatin").

The compaction or tight coiling of chromatin partially regulates access to the genes. How the cell condenses a segment of the chromatin is not clearly understood. Chemical modification of the DNA may play a role, but the factors that control the inactivation remain a mystery.

Control of the Gene at Transcription. Research shows that some inducible and repressible systems are present in the human cells. Some geneticists believe that most human genes are inactive and must be switched on by inducers.

Human DNA has an additional mechanism for control at the level of transcription. Certain segments of the DNA, called **enhancers**, greatly increase the activity of nearby genes. In other words, they do not turn genes on and off, but rather step up their activity. An enhancer is a little like the accelerator in your car—it doesn't turn the engine on, it speeds it up. Geneticists think that protein molecules bind to enhancer regions of the chromosome; these proteins somehow make it easier for RNA polymerase to attach to a DNA strand.

Control after Transcription. Another level of control occurs after the production of RNA on the DNA template. Human DNA contains far more genetic material than it needs. A given segment, in fact, has what appear to be many useless or noncoding stretches of DNA interspersed with functional genes. As shown at the top of Figure 5–15, DNA has two types of segments, introns and exons. The *expressed* segments—that is, those that will eventually be used to produce RNA that will be used to make protein—are called **exons**. The *intervening* segments of noncoding DNA are called **introns**.

Both the introns and exons are transcribed during the cell cycle, producing an **RNA transcript**, a complete "read out" of introns and exons. The cell, however, processes the transcript, cutting out the introns

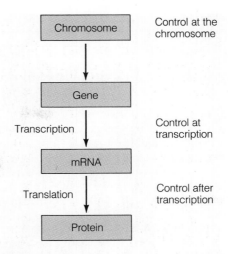

FIGURE 5–14 Gene Expression In humans, gene expression is regulated at many levels.

and then joining the exons to produce a functional mRNA molecule.

The exons can be spliced together in different ways, and the resulting mRNAs produce different proteins. Splicing, therefore, produces a variety of new proteins and represents another mechanism for controlling the expression of genes. In this way, the cell alters the blueprint in predictable ways to meet its immediate needs. Splicing, for example, results in the production of many different varieties of antibody molecules. Health Note 5–1 describes some other applications of gene splicing, or genetic engineering.

Human cells can also control gene expression at the level of translation, protein synthesis. Messenger RNA, for example, which is produced in the nucleus may be transferred to the cytoplasm in an inactive state. The mRNA is said to be masked. Masking allows the cell to build up large supplies of mRNA in preparation for a sudden burst of protein synthesis. When needed, the mRNA is activated and produces large quantities of protein. This phenomenon occurs in the ovum, the female sex cell produced in the ovaries. The ovum produces enormous amounts of masked mRNA as it awaits fertilization. When the sperm arrives and fertilizes the ovum, the unmasking of the mRNA that has accumulated in the cytoplasm results in an explosion of protein synthesis.

Human gene expression is a complex phenomenon. Like the cell cycle, it is no doubt under the influence of many controlling factors. The interplay of these factors regulates gene expression.

Genetic Engineering: Curing the Incurable?

Genetic engineering is really a popular name for something geneticists prefer to call **recombinant DNA technology**. Recombinant DNA technology is so named because it allows scientists to take segments of DNA from one organism and combine them with another organism's DNA. The technique evolved from the discovery of a group of enzymes that snip off segments of the double-stranded DNA molecule. These enzymes, **restriction endonucleases**, cut through both polynucleotide strands of the human DNA and produce numerous fragments, as illustrated in the figure.

To transplant the fragments into other cells, however, the DNA fragments must first be inserted into small circular strands of DNA isolated from bacteria. These strands, called **plasmids**, are separate from the bacterium's main DNA. They can replicate independently and have important genetic functions. Some genes on the plasmid, for example, provide bacteria with resistance to antibiotics. Plasmids provide a vehicle to transfer genes to other cells, such as bacteria and yeast cells (single-celled fungi). In a cell culture, plasmids are rapidly taken up by bacteria.

Bacteria or yeast cells with their newly acquired plasmids can be cultured in the laboratory. Researchers can isolate bacteria containing the human genes they want to study or use and grow those bacteria. As the bacteria duplicate, so do the plasmids, producing many copies of the gene. The gene is then said to be **cloned**.

APPLICATIONS OF GENETIC ENGINEERING

Genetic engineering offers hope in three areas: (1) the mass production of commercially important biochemicals, such as hormones; (2) the transfer of genes to plants or other animals to improve their disease resistance, growth rate, or other desirable traits; and (3) gene therapy, the deliberate introduction of genes into human cells. Critical thinking suggests, however, that these applications may not be problem-free. As you read this section, think of some of the problems that might arise.

Mass Producing Gene Products

The first practical application of genetic engineering was the isolation and cloning of genes that produce human insulin and growth hormone. These genes are now routinely inserted into bacteria and grown in massive quantities, creating large amounts of these proteins for commercial use. Before the advent of genetic engineering, insulin for diabetics was extracted from the pancreases of animals—usually pigs or cattle. This task was costly and time-consuming and produced a hormone that was not chemically identical to human insulin. As a result, the body would recognize these molecules as foreign substances, and mount an attack on them through the immune system (Chapter 10). Although this attack was generally low keyed, it did require physicians to switch their patients to other insulin preparations from time to time.

Companies also use genetic engineering to produce growth hormone, a protein needed to treat children whose pituitary glands are defective. Before the advent of this technique, growth hormone was difficult to find and prohibitively expensive.

Gene Transfer in Plants and Animals

For years, livestock and plant breeders have been selectively breeding hardy animals to produce genetically superior livestock and crops. Thus, people have been performing a kind of genetic engineering for tens of thousands of years. Selective breeding is slow and time-consuming, however. Genetic engineering is much faster.

Scientists recently implanted a gene that codes for human growth hormone into cattle embryos in hopes of producing marketable cattle much faster than by conventional means. Faster-growing cattle could produce more meat per pound of grass. Growth hormone also increases milk production in diary cattle.

Genetic scientists are working to improve plants as well. Already, scientists have introduced genes that allow oats to tolerate salty soil, a growing problem in irrigated farmland throughout the world. Transplanting these same genes into other commercial crop species could allow farmers to use the vast acreage now idle because of the buildup of salts.

Researchers are also working on ways to provide resistance to **herbicides**, chemicals applied to crops to control weeds. Although herbicides generally act only on weed species, they sometimes impair the growth of

Endonuclease cleavage sites

Human DNA

Restriction endonuclease cleaves segment of plasmid

Foreign gene for cloning

Cleaved plasmid

Gene to be cloned

E. coli chromosome

Foreign gene inserted in plasmid

Bacterial host

Plasmid taken up by bacterium

Cloning

Plasmid replicates in bacteria

Restriction Endonucleases Enzymes that snip the double helix have allowed researchers to cut up the DNA of humans and other organisms and insert the segments in plasmids, circular strands of DNA that are separate from the bacterial chromosomes. Plasmids are taken up by bacteria and, therefore, provide an avenue to insert human genes into bacteria for experimentation and other applications.

crops. Thus herbicide-resistant crops might help farmers increase crop yields. One problem with this development, however, is that making plants resistant to herbicides might increase a farmer's dependency on chemicals that could have harmful effects on other species.

Many agricultural improvements have been made by transferring genes to bacteria. For instance, genes found in certain bacteria code for the production of proteins toxic to certain insects. These genes have been transplanted to bacteria that grow on the roots of crop plants, protecting the plant against root-eating pests. Seeds treated with these bacteria provide enough of them to colonize an entire root system.

Genetic researchers have also developed a bacterium that retards the formation of frost on plants. The genetic variant is found naturally in the environment, but not in sufficient quantities to protect crops. Thus, researchers have cloned the bacterium and have begun testing its efficacy in the field. Should it prove successful and safe, the bacterium could save farmers millions of dollars a year by protecting crops against frost damage.

Gene Therapy
Approximately one of every 100 children born in the United States suffers from a serious genetic defect, such as sickle cell anemia or hemophilia. Thanks to advances in genetic engineering, medical science may soon find ways of replacing defective human genes with normal genes, curing previously incurable diseases that cost our society millions of dollars.

(continued on next page)

The largest obstacle, however, is getting the genes into body cells. Bone marrow transplants may provide an avenue. Scientists recently announced a successful experiment in mice to treat a rare genetic condition called Krabbe's disease. This genetic disease results from a deficiency in one enzyme. The absence of the enzyme causes fat to accumulate in the nervous system, causing nerve cells to degenerate. Seizures and visual problems occur early in humans, and most victims die within the first two years of life.

Scientists have found a strain of mice that suffers from a similar condition and have successfully injected cells from bone marrow of genetically normal mice into the defective mice. These cells became established in the lungs and liver and restored enzymatic activity. The bone marrow cells even found their way into the brain. Brain cells of patients with Krabbe's disease are severely affected by this enzyme deficiency, so any cure must replace the genetic material in these cells as well.

Even more promising are microspheres—tiny lipid spheres that can be coated with antibodies that allow them to deliver their contents, in this case, a replacement gene or

two, to body cells (see Health Note 2–1).

Researchers have also developed a transplantation technique that may help surgeons introduce genetically engineered cells into the human body. In a recent experiment, scientists injected genetically altered liver cells into a foamlike material. The foam was then transplanted into rats after being impregnated with a hormone that stimulates the growth of blood vessels from nearby larger vessels. Within a week after implantation, the partially artificial tissue was riddled with a network of blood vessels.

This technique could be used to introduce genetically altered cells (or even fetal cells) into patients with genetic disorders. It could even provide a way to replace insulin-deficient pancreas cells in diabetics or dopamine-producing cells in individuals with Parkinson's disease.

CONTROVERSY OVER GENETIC ENGINEERING

Although genetic engineering has spawned a great deal of enthusiasm, the procedure has its critics. Many safety questions still remain unanswered. Of greatest concern is the possibility of unleashing genetically

altered bacteria or viruses. Critics fear that a new strain might spread through the environment, wreaking havoc on natural ecosystems and, possibly, human populations. Once unleashed, it would be impossible to retrieve. The genetically altered bacterium that retards frost formation, for example, could enter the atmosphere on dust particles, reducing cloud formation, altering global climate.

Other critics object to genetic tinkering, especially the transfer of genes from one species to another. Do humans have the right, they wonder, to interfere with the entire course of evolution? Will genetic engineering have adverse effects on the survival of the human species?

Unfortunately, human experience with genetic engineering is too limited to answer the concerns of critics. Preliminary work suggests that the dangers have been exaggerated and that genetically engineered bacteria are not a threat to ecosystem stability. Still, further research is needed to be certain that in our zeal to make life better we don't do ourselves in. (For a discussion of the pros and cons of genetic engineering, see the Point/Counterpoint in this chapter.)

The Seeds of Cancer within Us: Oncogenes

Chapter 4 described cancer and noted a causal link between cancer and numerous biological, chemical, and physical carcinogens. It also noted that most cancers arise from mutations, alterations of parts of the genetic material. Broadly defined, **mutations** include three conditions: (1) alterations of the DNA itself, such as the deletion or addition of a base; (2) alterations of the chromosome, such as missing segments; and (3) alterations in the chromosome number—too

many or too few chromosomes. Mutations are discussed in more detail in Chapters 18 and 19. For now, consider a few important facts.

In most instances, mutations in DNA are rapidly repaired by enzymes in the nucleus. Some mutations escape repair, but not all of these are harmful. Some could actually improve cellular function. Other body cell mutations are neutral. They have neither a positive nor a negative effect. Mutations in the introns, for instance, would have no effect on the cell. Still other mutations affect vital sections of the genome and may

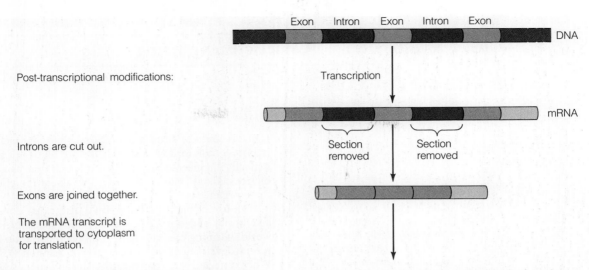

Post-transcriptional modifications:

Introns are cut out.

Exons are joined together.

The mRNA transcript is transported to cytoplasm for translation.

FIGURE 5–15 Posttranscriptional Control of Human Genes One key mechanism of control occurs after transcription of the DNA. DNA contains non-coded segments (introns) and useful segments (exons). A mRNA transcript is produced from the DNA, and the segments produced by introns are removed. The final product is a strand of mRNA containing only RNA copies of the exons. The exon copies can be linked together differently, producing slightly different products.

kill a cell outright or release a cell on a rampage of growth. It is these mutations that may lead to cancer and will concern us here.

Within the last 10 years, researchers have made a startling discovery about mutations and cancer that may explain the underlying cause of most, if not all, cancers. Scientists have found that humans and other organisms contain the seeds of cancer in their own genes. Each human being, for example, has genes that, when mutated, lead to cancerous growth. These genes are called **proto-oncogenes**. Proto-oncogenes are normal genes that code for important cellular structures and functions—for example, for mitotic proteins and plasma membrane receptors for growth factors or hormones. Radiation, ultraviolet light, and chemical carcinogens can mutate these genes, resulting in uncontrolled cellular proliferation—cancer.

Thus, mutagens (substances that cause genetic mutation) cause cancer when they strike the right target within a cell: genes that, under normal conditions, play a key role in cellular growth.

Viruses can also cause cancer. Some viruses carry cancer-causing genes, true **oncogenes** that enter cells and migrate to the nucleus. Here, viral oncogenes may become incorporated in the cell's genome. Not all viruses that cause cancer contain oncogenes, however. Those viruses that do not contain oncogenes are thought to stimulate cancer by activating proto-oncogenes in body cells. For instance, they may in-

troduce a viral control gene into human body cells. Viral control genes regulate the viral genome. If inserted into the human genome near inactive proto-oncogenes, the control gene may lead to cancer.

The discovery of proto-oncogenes, viral oncogenes, and viral control genes has resulted in a quantum leap in the understanding of cancer. Some scientists believe that the key to treating cancer may lie in techniques to turn off activated proto-oncogenes. If medical researchers can find a way to turn off the proto-oncogenes in cancer cells, for example, they may indeed have a cure for cancer. (For more on cancer treatment, see Health Note 4–1.)

ENVIRONMENT AND HEALTH: TOXIC WASTES

The last two chapters have illustrated some important principles of homeostasis. They have also shown that *metabolic pathways that perform beneficial functions can be turned against us when harmful chemicals are present.* Second, *the genome itself can be altered by mutagens, turning cells against the body.*

Life exists in a delicate balance and the most remarkable thing is not that about 500,000 Americans die from cancer each year, but that so many of us escape cancer. Thousands of mutations occur in each cell of your body every day. You are protected from

PLAYING GOD WITH GENETIC CODE

Jeremy Rifkin

Jeremy Rifkin is the president of the Foundation on Economic Trends and president of the Greenhouse Crisis Foundation. Mr. Rifkin's books include Entropy, Algeny, *and* Declaration of a Heretic.

With recombinant DNA technology, it is now possible to snip, insert, recombine, rearrange, and even produce genetic material. The short-term benefits of this extraordinary new power are seductive. Yet, if history has taught us anything, it is that every technological revolution brings with it both benefits and costs. The full-scale use of biotechnology in military research, agriculture, industry, and human reproduction and health raises environmental, economic, and ethical concerns that are without parallel.

The biotechnology industry is preparing to release scores of genetically engineered viruses, bacteria, plant strains, and transgenic animals into the environment over the next few years. A question that must be answered before and such large-scale releases is what risks such organisms pose to human health and to the environment.

Because they are alive, genetically engineered products are inherently more unpredictable than chemical products. Genetically engineered products can reproduce, mutate, and migrate. Once they are released, it is virtually impossible to recall them. A survey of a hundred top U.S. scientists acknowledges the potential benefits of genetic engineering, but warns that "its imprudent or careless use . . . could lead to irreversible, devastating damage to the ecology."

Environmental scientists have compared the risks of releasing biotechnology products to those we have encountered in introducing exotic organisms to North American habitats. Although most of these organisms have adapted to our ecosystems, several, including the chestnut blight and the gypsy moth, have wreaked havoc.

Thus, the long-term cumulative environmental impact of the release of thousands of genetically engineered organisms could be devastating. The sensible approach is to impose a moratorium on the deliberate release of all genetically engineered organisms until such time as a "predictive ecology" can be developed to assess the effect of these organisms.

The use of biotechnology could also cause considerable social and economic dislocation, especially in the world's farming communities. A timely illustration is the recent development of bovine growth hormone (BGH). When injected into dairy cattle, this hormone, cloned through genetic engineering, can increase milk production. Since the U.S. milk market is already flooded, BGH poses a serious threat to dairy farmers. It has been estimated that milk prices may fall 10 to 15% within the first three years after its introduction. Furthermore, the number of dairy farmers may have to be reduced by 25 to 30% to restore market equilibrium. A congressional report has concluded that the use of BGH could cause a shift in U.S. milk production from the traditional smaller dairy farms of the Northeast to the larger dairy farms of the West. Other biotechnology products may lead to similar dislocations with dramatic social, economic, and cultural effects.

The use of biotechnology also raises profound ethical questions. Most worrisome are efforts to cross species boundaries by inserting human genes into animals and animal and plant genes into animals and plants of different species. For example, scientists have fused sheep and goat cells, creating the gheep—a half-sheep, half-goat chimera. Cross-species genetic transfers go far beyond any traditional breeding of plant and animal species.

Before we rush headlong into the biotechnology revolution, we must thoroughly scrutinize the long-term environmental, economic, and ethical issues raised by this new technology. Society failed to consider the "hard" questions about the impacts of the nuclear and petrochemical technologies before it allowed their commercialization. As a result, we are now confronted with a huge environmental and societal bill, including undisposable nuclear wastes, toxic waste dumps, acid rain, the "greenhouse effect," and ozone depletion.

Let us hope that society has learned from these mistakes and this time will raise the important questions before, rather than after, the damage is done.

ETHIC FOR A SMALL PLANET

Nina Federoff

Nina Fedoroff is a plant molecular biologist and geneticist. She is a staff scientist in the Department of Embryology of the Carnegie Institution of Washington and a professor in the Biology Department of the Johns Hopkins University, both in Baltimore. She is a member of the National Academy of Sciences, has served on the Recombinant DNA Advisory Committee of the National Institutes of Health, and is a member of the Biotechnology Advisory Committee of the National Research Council.

As we enter the last decade of the twentieth century, we rejoice in the success of the human cultural enterprise. Spectacular improvements in health, longevity, and living standards have emerged from the exuberance of science. Yet with barely a moment to gain perspective, we must confront the real possibility that the planet cannot continue for long to support our expanding industrial and agricultural activities.

Our protective ozone layer is thinning. The Earth's forests are disappearing ever more rapidly. Flooding worsens. Plant and animal extinction rates are soaring. The atmospheric accumulation of carbon dioxide and other gases may already have begun to affect our global climate. Toxic agricultural and industrial chemicals continue to accumulate in soil, water, and air.

We are beginning to understand that these problems are the price of our comfort: cars, warm houses, food, electricity, the many things we use and throw away. And our very successes in health care have helped bring about an unprecedented growth in our numbers. The existing human population already harvests, diverts, or wastes almost half of the Earth's annual primary photosynthetic yield. Although agricultural productivity has reached extraordinary heights, it is becoming increasingly apparent that present levels cannot be sustained for long. As topsoil is lost and agricultural chemicals lose their effectiveness, the land's productivity declines.

We know that we must learn to live a different way. Fundamental changes must occur in our use of energy, water, and land. Yet short-term profitability remains the central ethical paradigm of the world's dominant human cultures. We have barely begun to construct an ethical framework for living within our resources. Have we time to evolve a new ethic, an ethic for a small planet? And what must it be?

The principle is simple: we must measure what we do for ourselves against the cost, the consequences for the planet's future productivity and habitability. Many would have us turn back the clock and return to a pretechnological world. But we cannot. That road is blocked by our very numbers: the human population has already surpassed 5 billion and is projected to double by the middle of the next century. If we are to avoid the calamity of widespread malnutrition and famine, we have no choice but to use more wisely what we know already. And we must learn more.

Our growing knowledge of biology is critical for our future and that of our planet. The body of genetic knowledge underlying the development of recombinant DNA techniques is one of the towering accomplishments of twentieth-century science. Although no technology holds all of the answers to our very real and mounting environmental problems, it is already evident that recombinant DNA techniques and other biotechnologies can make substantial contributions to putting our human activities on an environmentally sustainable basis.

Recent advances have opened the possibility of using genetic engineering techniques to decrease our reliance on toxic agricultural chemicals, reduce crop losses to diseases and pests, increase milk and meat yields, and ameliorate a variety of contemporary pollution problems. It would be profoundly unethical for us not to use all that we have learned during this extraordinary scientific revolution to ensure that we leave our children the inhabited and habitable planet that we inherited.

SHARPENING YOUR CRITICAL THINKING SKILLS

1. Summarize the dangers Jeremy Rifkin sees in the use of genetic engineering. How does Rifkin support each of his contentions?
2. Summarize Federoff's argument for genetic engineering. What is the basis of her acceptance of this technology?
3. Faced with these opposing viewpoints on genetic engineering, what is your view? Do you need more information?

cancer in part by normal cellular repair mechanisms, which fix damaged DNA. When a mutation occurs, repair enzymes in the nucleus snip off the damaged part, then rebuild the molecule. In a way, evolution has provided us with our own internal medical care system.

Remaining healthy, though, requires that we not stretch that resiliency to the breaking point. And yet, some individuals fear that modern industrial society may be doing just that. Today, some 60,000 chemical substances are in commercial use. The National Academy of Sciences recently noted that few of these substances have been adequately tested for their ability to cause cancer, birth defects, and mutations. Of the pesticides in use, for example, only about 10% have been adequately tested for their ability to cause cancer and birth defects. The extensive use of chemicals and publicity over the ill effects of some have created a chemical paranoia in our society. New research suggests that people may be overreacting to the threat of chemicals. Naturally occurring chemicals in our food, in fact, may cause more cancer than pesticide residues. (For a debate over chemical contamination and the relative importance of pesticides and other industrial chemicals on human health, see the Point/Counterpoint in Chapter 15.)

Some critics believe that modern societies the world over are heading blindly into the future with little concern for the long-term impacts of their actions. Nowhere has the impact been more noticeable than around toxic waste dumps, where for decades some chemical companies have disposed of highly toxic substances, often in cardboard containers or steel barrels that rust within a few years of burial (Figure 5–16). Love Canal in the city of Niagara Falls, New York, was the scene of one of the nation's worst toxic nightmares. Here, in an abandoned canal, Hooker Chemical Company dumped over 20,000 metric tons of highly toxic and carcinogenic wastes from 1947 to 1952.[6] In 1952, the city of Niagara Falls began condemnation proceedings—not to shut down the operations, but to take the land to build a school and housing development. Hooker sold the land to the city for $1 in exchange for a release from future liability.

A few years later, however, the trouble began. Bulldozers preparing the site for construction of the school removed the clay lid that Hooker had placed over the dump site to protect it. In the late 1950s,

[6]A metric ton is 2,240 pounds.

FIGURE 5–16 Hazardous Wastes The careless dumping of hazardous materials in rivers and lakes, in abandoned fields, along highways, and in abandoned warehouses creates a health hazard to humans and many other species and results in costly cleanup efforts. Thankfully, new laws have put tighter controls on hazardous waste disposal, although some industries and waste disposal companies still violate the law.

after the construction was complete, rusty barrels began to work their way up through the ground. Toxic chemicals oozed to the surface, killing trees and gardens and causing chemical burns in children who played in the ooze. Some children even died.

The problem came to a head in the 1970s. After a period of heavy rainfall, toxic wastes began to leak into basements of local residents, and the chemical stench became unbearable. Over 80 different chemical substances turned up in studies of water, air, and soil. Many of the chemicals were known or suspected carcinogens. A New York State Health Department study showed that one of every three pregnant women who lived in the area miscarried. Birth defects were present in one of every five children, far in excess of the expected rate. The chemical substances irritated lungs, gave residents headaches, and brought on convulsions in some people. Genetic studies showed mutations in chromosomes of residents.

Nearly 1000 families were evacuated from the site over the years, and now the homes sit idle, boarded up, in solemn tribute to careless practices. New laws to regulate hazardous waste and concerted efforts of industry have cut back on the reckless disposal of hazardous wastes, but the practice still occurs. Some companies have sought waste disposal sites abroad in the cash-poor Third World.

Cleaning up past mistakes has proven costly. Over $40 million has been spent so far to clean up the Love Canal area, and millions more may have to be spent. What is striking, though, is that some experts estimate that there are nearly 10,000 hazardous waste sites, some worse than Love Canal, in the United States in need of cleanup. Government estimates put the cost of cleaning up the mess at $100 billion. The U.S. Environmental Protection Agency (EPA) has begun the process and is moving slowly but steadily ahead in this costly task. Adding to the problem, however, are the more than 250 million metric tons of hazardous materials that American industry generates annually—that's over a ton of hazardous waste for every man, woman, and child each year!

Improper hazardous waste disposal now pollutes groundwater in many areas—and that concerns public health officials because half the people in the United States get their drinking water from wells. By one estimate, more than 10 million Americans now use tap water contaminated with chemical pollutants in excess of EPA standards. Because of years of careless waste disposal, groundwater contamination is expected to grow worse. Even if we stop polluting now, the problem will linger for another 30 years or more.

A healthy population requires clean water. But how do we get it? Many changes are needed in American society. Especially important are ways to reduce hazardous waste production. By redesigning chemical processes, for example, manufacturers can cut back on their wastes. Manufacturers can also reuse and recycle wastes. A purified waste product from one process may be the raw material for another. Individuals can help reduce hazardous waste by reducing unnecessary consumption. Since Americans consume far more than they need, modest 10 to 20% reductions, which would have little impact on our already high living standard, could have profound effects on hazardous waste generation—and the health of our environment. Just using energy and other resources more frugally and recycling household products, paper, aluminum, glass, and plastics can achieve enormous savings. Beyond that, wastes can be incinerated in high-temperature furnaces, designed to eliminate harmful emissions. Wastes can be chemically neutralized or treated in other ways to render them less harmful. After reductions and chemical modifications come new and improved disposal techniques. However, these should be a last resort, say many experts. Even new landfills lined with thick clay bottoms or synthetic liners may eventually leak, releasing their toxic brew into the ground. Ultimately, most people believe that maintaining a healthy population with a healthy genome requires marked reductions in the waste poured into the environment.

SUMMARY

BUILDING TWO MACROMOLECULES: DNA AND RNA

1. The DNA molecule houses the genetic information. DNA consists of two polynucleotide strands joined by hydrogen bonds between purine and pyrimidine bases on the opposite strands.
2. Each nucleotide in the DNA molecule consists of a purine or pyrimidine base, the sugar deoxyribose, and a phosphate group.
3. Complementary base pairing is an unalterable coupling in which adenine on one strand of the DNA molecule always binds to thymine on the other, and guanine always binds to cytosine.
4. Complementary base pairing ensures accurate replication of the DNA and accurate transmission of genetic information from one cell to another.
5. To replicate, DNA must first unwind. After unwinding, the strands provide templates for the production of complementary DNA strands.
6. The synthesis of protein in the cytoplasm of the cell requires three types of RNA: transfer RNA, ribosomal RNA, and messenger RNA.
7. All three RNA molecules are single-stranded polynucleotide chains. RNA nucleotides contain the sugar ribose instead of deoxyribose, which is found in DNA. RNA nucleotides contain four bases: adenine, guanine, cytosine, and uracil.
8. RNA is synthesized in the nucleus on a template of DNA. The synthesis of RNA is called transcription.

HOW GENES WORK: PROTEIN SYNTHESIS

9. The genetic information contained in the DNA molecule is transferred to messenger RNA. Messenger RNA molecules carry this information to the cytoplasm where proteins are synthesized.
10. Messenger RNA serves as a template for protein synthesis.
11. Transfer RNA molecules deliver amino acid molecules to the mRNA and insert them in the growing chain. Each tRNA binds to a specific amino acid and delivers it to a specific codon, a sequence of three bases on the mRNA. Thus, the sequence of codons determines the sequence of amino acids in the protein.
12. Proteins are synthesized by adding one amino acid at a time.
13. During protein synthesis, the ribosome first attaches to the mRNA at the initiator codon. Soon after the ribo-

some attaches to the mRNA, a tRNA bound to an amino acid enters the first binding site of the ribosome. A second tRNA–amino acid then enters the second site.

14. An enzyme in the ribosome catalyzes the formation of a peptide bond between the two amino acids. After the bond is formed, the first tRNA (minus its amino acid) leaves the first binding site.

15. The ribosome moves down the mRNA, shifting the tRNA bound to its two amino acids to the first binding site and opening the second site for another tRNA–amino acid. This process repeats itself many times.

16. As the peptide chain is formed, it begins to bend and twist, forming the secondary structure of the protein or peptide. When the ribosome reaches the terminator codon, the peptide chain is released.

17. Some antibiotics selectively inhibit protein or RNA synthesis in bacteria and are, therefore, useful in treating bacterial infections. Unlike bacteria, viruses are not affected by antibiotics.

CONTROLLING THE GENES

18. Bacterial DNA contains functional units called operons, consisting of three parts: structural genes, regulator genes, and control regions.

19. Structural genes code for the production of enzymes and other proteins and are controlled by the regulator genes and the control regions.

20. Two types of operons are present: inducible and repressible. In an inducible operon, the structural genes are switched off unless an inducer substance is present. The inducer activates the structural genes by binding to the repressor protein. The repressor protein is produced by the regulator gene and binds to the operator site next to the structural genes, thus preventing the transcription of the structural genes by physically blocking RNA polymerase.

21. When an inducer substance is present, it binds to the repressor protein, causing it to release from the operator site and thus allow RNA polymerase to access the structural genes.

22. Repressible operons are permanently activated but can be switched off by the presence of a chemical substance called a corepressor, usually an end product of a metabolic pathway.

23. The corepressor binds to the repressor protein, thus permitting it to bind to the operator site and block RNA polymerase.

24. Gene control in humans occurs at several levels: at the chromosome, at transcription, and after transcription.

25. Condensation or coiling of the chromatin fibers, for instance, inactivates genes. How the cell condenses a segment of the chromatin is not clearly understood.

26. Research shows that inducible and repressible systems are present in human cells.

27. Research also shows that certain segments of the DNA, called enhancers, can greatly increase the activity of nearby genes. Geneticists think that protein molecules bind to enhancer regions of the chromosome and facilitate gene activity.

28. Another level of control occurs after the production of RNA. Human DNA contains far more genetic material than it needs. Those segments of the DNA used to produce RNA that will be used to make protein are called exons. The intervening segments of noncoding DNA are called introns.

29. Both the introns and exons are transcribed during the cell cycle. The cell, however, removes the introns and joins the exons to produce a functional mRNA molecule.

30. The exons can be spliced together in different ways; the resulting mRNAs produce different proteins.

31. Human cells can also control gene expression at the level of translation, protein synthesis. Messenger RNA may be transferred to the cytoplasm in an inactive or masked state. Masking permits the cell to build up large supplies of mRNA in preparation for a sudden burst of protein synthesis.

32. Most cancers arise from mutations. Mutations include three basic changes: (a) alterations of the DNA itself; (b) alterations of the chromosome; and (c) alternations in the chromosome number.

33. Humans and other organisms contain specific genes, called proto-oncogenes, which when mutated lead to cancerous growth. Proto-oncogenes are normal genes that code for cellular structures and functions, such as cell adhesion, mitotic proteins, and the production of plasma membrane receptors for growth factors or hormones.

34. Radiation, ultraviolet light, and chemical carcinogens can mutate these genes, resulting in uncontrolled cellular proliferation—cancer.

35. Viruses can also cause cancer. Some viruses, for example, possess oncogenes, which enter human body cells and are incorporated in the cell's genome, stimulating uncontrolled cellular division. Other viruses carry genes that stimulate cancer by activating human proto-oncogenes.

ENVIRONMENT AND HEALTH: TOXIC WASTES

36. Numerous chemicals are released into the environment from factories, farms, automobiles, and other sources. Many chemicals now in common use have not been tested to determine their ability to cause cancer and birth defects.

37. New research suggests that people may be overreacting to the threat of chemicals. Naturally occurring chemicals in our food, in fact, may cause more cancer than pesticide residues.

38. In some locations, such as toxic waste dumps, chemicals are present in extremely high concentrations. Decades of improper waste disposal have left a legacy of pollution, ill health, and contaminated groundwater.

39. Reducing the contamination of our environment by hazardous wastes will require concerted efforts on the part of governments, businesses, and individuals. Especially important are ways to reduce hazardous waste production by redesigning processes, recycling and re-using wastes, and finding substitutes.

EXERCISING YOUR CRITICAL THINKING SKILLS

One of America's leading drug companies is spending about $300 million dollars to construct a facility that will produce synthetic growth hormone to be sold to dairy farmers. The hormone will be produced by genetically engineered bacteria that contain growth hormone genes isolated and transplanted from dairy cattle. Given to milk cows, the hormone dramatically increases milk production. Business economists believe that an increase in milk production will reduce the cost of milk to the consumer. Based on this information, does introducing synthetic growth hormone seem like a good idea?

After you have thought about it for a while, consider some additional facts. The American dairy industry already produces an excess of milk. The federal government currently buys the surplus, dehydrates some of it, and makes cheese out of the rest. The government stockpiles the dehydrated milk and cheese at considerable cost to taxpayers. Now what do you think?

Before you draw a firm conclusion, however, remember that cheese and dehydrated milk are given to the needy. The food is not going to waste. Will increasing the surplus produce more food for the needy? Do we need more cheese and milk for indigent Americans?

Before you draw any conclusions consider another factor: the effect increasing the surplus of milk will have on small dairy operations. Many owners of small dairy herds believe that the large producers who use the hormone will increase their milk production. As noted above, this will probably drive down the cost of milk and milk products, but could also put many small dairy farmers out of business. This could adversely affect the economies of many rural regions.

Some consumer groups are concerned about the potential health effects of using synthetic growth hormone. A trace of growth hormone is present in milk produced normally. Will the use of synthetic growth hormone increase the concentration in milk? What effect will that have on your health or the health of children?

Make a list of the pros and cons for the use of synthetic growth hormone. Make a list of questions—those that you can answer and those that you cannot answer. Do you need more information to decide?

If you are able to make up your mind, what factors swayed your opinion? Did the concerns of one group outweigh the concerns of another? What critical thinking rules did you use in this exercise?

TEST OF TERMS

1. DNA consists of two polynucleotide chains held together by _____ bonds, forming a spiral staircase or _____ _____ .

2. A purine molecule is joined to a sugar molecule and a phosphate group, forming a _____ _____ .

3. In the DNA molecule, adenine forms hydrogen bonds only with _____ on the opposite strand, and _____ binds only with cytosine.

4. The production of RNA on a DNA template is called _____

whereas the production of protein on messenger RNA is called _____ _____ .

5. A series of three bases on the RNA molecule codes for a specific amino acid and is called a(n) _____ .

6. _____ molecules bind to amino acids in the cytoplasm and deliver them to the peptide chain being synthesized on the mRNA.

7. _____ pairing is responsible for the accurate production of protein on mRNA and the accurate replication of a strand of DNA.

8. The _____ genes, the regulator gene, and the control region

form a unit called the _____ .

9. An inducible _____ is a set of genes that remains inactive until needed. In order for the structural genes to produce mRNA, RNA polymerase must be able to bind to the _____ site. It can't if the _____ protein is bound to the _____ site.

10. Densely packed chromatin is metabolically inactive and is called _____ .

11. _____ are segments of the DNA that increase the activity of nearby genes in humans.

12. The human genome contains segments of DNA that are not coded, called _____, and segments that code for protein, called _____.
13. Viruses carry cancer-causing genes called _____.
14. Human cells also have genes that, when mutated, can result in cancer. These are called _____.

Answers to the Test of Terms are located in Appendix B.

TEST OF CONCEPTS

1. What is the genetic code of DNA? How is the genetic code, housed in the DNA, translated into instructions the cell can understand?
2. Describe how DNA is synthesized. How is the accurate replication of DNA ensured by the cell?
3. List the three types of RNA and briefly describe their function in protein synthesis.
4. In what ways are RNA and DNA similar? In what ways are they different?
5. Describe, in detail, the production of protein on mRNA. Be sure to note the enzymes involved and the role of the ribosome.
6. Discuss how bacterial cells control their genes. In what way does an end product of a metabolic pathway control genes? In what way does a starting material control the genes?
7. Discuss the various levels at which human genes are controlled. Give an example of each.
8. What is a proto-oncogene? How is it affected by a mutagen, a chemical or physical agent that causes mutation?

SUGGESTED READINGS

Campbell, N. 1987. *Biology*. Menlo Park, Calif.: Benjamin/Cummings. For a more detailed look at genetics, see Chapters 15–19.

Cummings, M. R. 1988. *Human heredity: Principles and issues*. St. Paul, Minn.: West. Chapters 7–9 cover this material in more detail.

Moody, M. D. 1989. DNA analysis in forensic science. *Bioscience* 39(1): 31–35. Looks at genetic tools used to solve violent crimes.

Ptashne, M. 1989. How gene activators work. *Scientific American* 260(1): 40–47. Discusses how regulatory proteins switch genes on and off.

Radman, M., and R. Wagner. 1988. The high fidelity of DNA duplication. *Scientific American* 259(2): 40–46. Looks at the way three enzyme systems work together to make sure DNA is replicated correctly.

Watson, J. D. 1980. *The double helix*. New York: Norton. Delightful account of the discovery of the structure of DNA.

Weiss, R. 1989. DNA takes the stand. *Science News* 136(5): 74–76. Looks at the legal and ethical issues regarding the use of DNA evidence to help convict criminals.

Wickelgren, I. 1989. DNA's extended domain. *Science News* 136(15): 234–37. This article examines compelling evidence to support the theory that additional DNA could exist anchored to the plasma membrane.

III

THE HUMAN ORGANISM
Structure and Function of the Human Body

Principles of Structure and Function

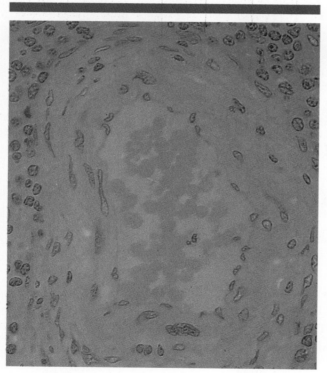

Blood vessel artery in spleen.

Preceding page: Nuclear scan of normal heart.

David Eastland was driving home from work late one evening when he fell asleep at the wheel. His car swerved off the road and struck a tree, crushing the vehicle and his right leg; a large segment of his femur (thigh bone) was destroyed.

In earlier times, this accident would have cost Eastland his leg for surgeons had no way of repairing such extensive bone damage. Even bone grafts are insufficient to repair severe injuries.

Today, surgeons can remake the human skeleton using a material called demineralized bone matter (DBM). Demineralized bone matter consists primarily of the protein collagen and is produced by immersing the bones of human cadavers and animals in a weak acid that dissolves away the mineral matter of the bone, leaving behind the rubbery DBM. After rebuilding the missing bone with DBM, surgeons inject bone cells into it. Bone cells are derived from small fragments taken from a patient's own bone. Transplanted bone cells grow and divide in the DBM, converting the implant into a functional bone, practically indistinguishable from normal bone.

Demineralized bone matter can be used to repair birth defects and to replace bone segments removed because of bone cysts, tumors, and infections. One of

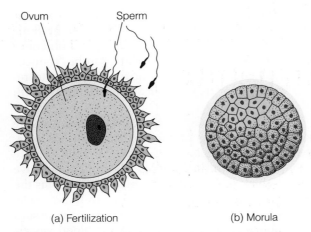

(a) Fertilization   (b) Morula

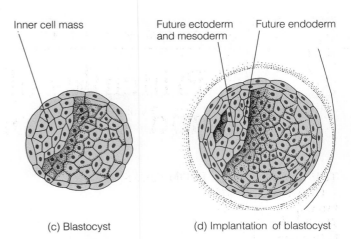

(c) Blastocyst   (d) Implantation of blastocyst

FIGURE 6–1 Synopsis of Embryonic Development
The fertilized ovum (*a*) divides mitotically, eventually forming a solid ball of cells, the morula (*b*), which later becomes a hollow blastocyst (*c*). The inner cell mass shown here becomes the embryo, differentiating into ectoderm, endoderm, and mesoderm in later stages of development (*d*).

the most striking success stories is that of a boy born with a disease known as cloverleaf syndrome. In this disease, growth in the bones of the skull fails to keep pace with the brain's growth, eventually killing its victims. Dr. John Mulliken, a surgeon at Boston's Children Hospital, removed the skull of a six-year-old boy and fashioned a new and larger one from DBM. Four years later, the boy is alive and doing well.

Surgeons are using similar techniques to form artificial skin and build new tissues containing genetically engineered cells. Using cells and supporting materials, such as collagen, physicians can reconstruct body parts lost or damaged in accidents or by disease. To understand the difficulty of this task, we turn our attention to the structure of the human body to see how cells combine to form tissues and how tissues, in turn, form organs.

THE BODY HUMAN: FROM CELLS TO ORGAN SYSTEMS

To begin our study of the structure and function of the human body, we look back once again at the beginning of human life. Chapter 3 pointed out that the life of a human being begins as a single fertilized ovum. That cell contains all of the information needed to develop into a human being and all of the information needed to control the complex life functions, such as growth, reproduction, and homeostasis.

From Cells to Tissues

During embryonic development, the fertilized ovum divides many times, producing first a ball of cells called the **morula** (Figure 6–1). Soon thereafter, the cells undergo a process called **differentiation**, structural and functional specialization. Three types of cells emerge in the human embryo: ectoderm, mesoderm, and endoderm. **Ectoderm** lies on the outside of the embryo and gives rise to the skin, the eyes, and the nervous system. **Mesoderm** lies in the middle and forms muscle, bone, and cartilage. **Endoderm** is the innermost layer and forms the lining of the intestinal tract and several digestive glands.

During embryonic development, ectodermal, mesodermal, and endodermal cells give rise to a variety of highly differentiated cells. Some of these cells produce fibers and other extracellular materials that often bind cells together, forming **tissues** (from the Latin word "to weave"). Extracellular materials may be liquid (as in blood), semisolid (as in cartilage), or solid (as in bone).

Tissues consist of cells and their extracellular products and combine in a variety of ways to form **organs**, discrete structures in the body that carry out specialized functions.

The Primary Tissues

Four major tissue types are found in adults: epithelial, connective, muscle, and nervous. These are called pri-

TABLE 6–1	The Primary Tissues and Their Subtypes	
Epithelial tissue	Connective tissue	
Membranous	Connective tissue proper	
Glandular	Specialized connective tissue	
Muscle tissue		
Cardiac	Blood	
Skeletal	Bone	
Smooth	Cartilage	
Nervous tissue		
Conductive		
Supportive		

tiring. Skeletal muscle cells tire fairly easily when exercised. Despite this and other differences, the three types of muscle tissues are structurally and functionally similar. Skeletal muscle, in fact, can be "trained" to contract repeatedly and is now being used to replace worn out or damaged heart muscle (see Health Note 6–1).

The primary tissues exist in all organs in varying amounts. The lining of the stomach, for example, consists of a single layer of epithelial cells (Figure 6–2). Just beneath the lining is a layer of connective tissue. A thick sheet of smooth muscle cells forms the bulk of the wall of the stomach. Smooth muscle cells are also found in blood vessels supplying the tissues of the stomach. Nerves enter with the blood vessels and control the flow of blood.

mary tissues. Table 6–1 lists the primary tissues and their subtypes; as you can see, each primary tissue has two or three subtypes. Muscle tissue, for instance, comes in three varieties: cardiac, skeletal, and smooth. Cardiac muscle cells are found exclusively in the heart and resemble the skeletal muscle cells, located in body muscles. Unlike skeletal muscle cells, cardiac muscle cells can contract repetitively without

Epithelium

Epithelial tissue exists in two basic forms: glandular and membranous. **Membranous epithelium** consists of sheets of cells tightly packed together, forming the external coverings or linings of organs. Figure 6–3 shows some of the remarkable variety in epithelial membranes and illustrates the presence of two basic types of epithelium: **simple epithelia**, consisting of a

FIGURE 6–2 Human Stomach (*a*) The stomach, like all organs, contains all four primary tissues. (*b*) These are shown here in cross section.

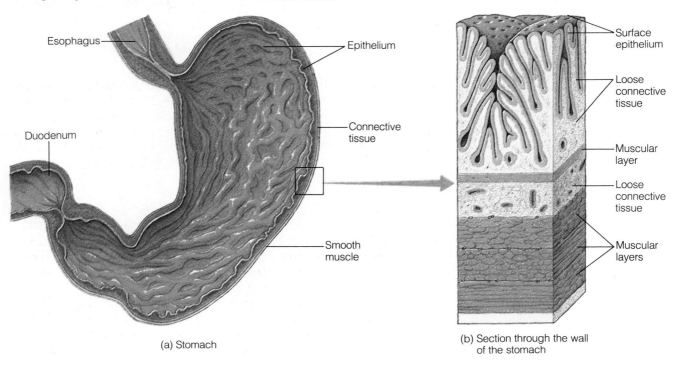

Esophagus

Duodenum

Epithelium

Connective tissue

Smooth muscle

(a) Stomach

Surface epithelium

Loose connective tissue

Muscular layer

Loose connective tissue

Muscular layers

(b) Section through the wall of the stomach

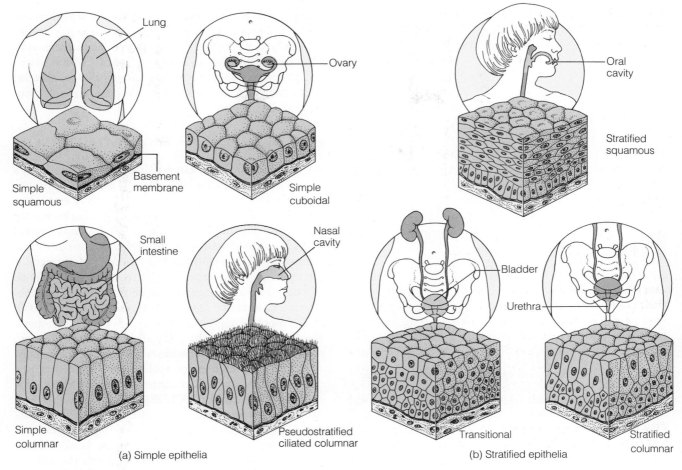

Lung

Ovary

Oral cavity

Basement membrane

Simple squamous

Simple cuboidal

Stratified squamous

Small intestine

Nasal cavity

Bladder

Urethra

Simple columnar

Pseudostratified ciliated columnar

Transitional

Stratified columnar

(a) Simple epithelia

(b) Stratified epithelia

FIGURE 6–3 Membranous Epithelia Single-celled (simple) epithelia (*a*) and stratified epithelia (*b*) exist in different parts of the body.

single layer of cells, and **stratified epithelia**, consisting of many layers of cells. Blood vessels, for example, are lined by a single layer of flattened cells. The lining of the intestinal tract consists of column-shaped (columnar) epithelial cells.[1]

Glandular epithelia are clumps of cells that form the glands of the body. Epithelial glands arise during embryonic development from tiny outpocketings of epithelial membranes, as illustrated in Figure 6–4. Some glands remain connected to the epithelium by hollow ducts and are called **exocrine glands** (glands of external secretion); products of the exocrine glands

flow through ducts into some other body part. Sweat glands in the skin, for example, produce a clear, watery fluid that is released onto the surface of the skin by small ducts. This fluid evaporates from the skin, helping cool the body. Salivary glands, located around the oral cavity, produce saliva, a fluid that is released into the mouth.

Some glandular epithelial cells break off completely from their embryonic source, forming **endocrine glands** (glands of internal secretion), as illustrated in Figure 6–4. The endocrine glands produce hormones that are released from the cell and diffuse into the bloodstream where they travel to other parts of the body (Chapter 15).

In architecture, form (the structure of a building) follows function—in other words, architects design with function in mind. The human body exhibits a similar relationship between form and function. The

[1]You will also note the presence of two other types: pseudostratified ciliated columnar and transitional. The pseudostratified epithelium appears stratified because the nuclei are located at different levels, but this epithelium consists of a single layer of cells. Transitional epithelium is a special type of stratified epithelium.

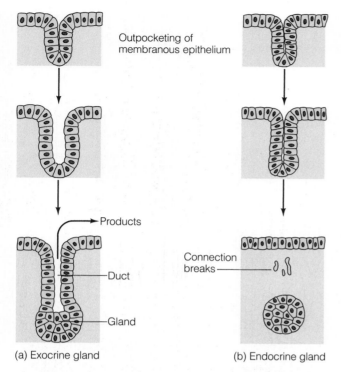

Outpocketing of membranous epithelium

Products

Duct

Gland

Connection breaks

(a) Exocrine gland

(b) Endocrine gland

FIGURE 6–4 Formation of Endocrine and Exocrine Glands (*a*) Exocrine glands arise from outpocketings of membranous epithelia that retain their connection. (*b*) Endocrine glands lose this connection and must secrete their products into the bloodstream.

epithelia provide many examples of this correlation. The membranous epithelia, for example, serve a variety of functions. Some protect underlying tissues. Others absorb and transport materials from one location to another. Still others are involved in secreting products. In each case, the structure of the membrane suggests its function. The outer layer of the skin, the **epidermis**, for example, is an epithelial membrane that protects underlying tissues not only from microorganisms prevalent in the environment, but also from moisture loss. The epidermis consists of numerous cell layers (Figure 6–5). The cells flatten toward the surface and are tightly joined together by special connections reducing the skin's permeability. As the epidermal cells move outward, they become isolated from the underlying blood supply and die, creating a dry protective layer that impedes water loss and presents a formidable barrier to microorganisms.

Another example of the relationship between structure and function is the epithelium of the small intestine. It consists of a single layer of columnar cells, uniquely suited to absorb food materials. The

columnar epithelial cells of the small intestine are also structurally modified to enhance food absorption. As illustrated in Figure 6–5c, the surfaces of these cells are thrown into folds, called **microvilli**, which markedly increase the surface area of the cell available for absorption. The more surface area, the more efficient is food absorption.

Connective Tissue

As the name implies, **connective tissue** is the body's glue. Connective tissues bind cells and other tissues together and can be found in all organs. Connective tissue in the form of bone and cartilage also makes up the skeleton (Chapter 14). The body contains several types of connective tissue, all of which consist of two components: cells and varying amounts of extracellular material. Two types of connective tissue will be discussed here: connective tissue proper and the specialized connective tissues: bone, cartilage, and blood.

Connective tissue proper is an important structural component of the body and is composed of two major types: dense connective tissue and loose connective tissue. **Dense connective tissue** consists primarily of densely packed connective tissue fibers—for example, those found in ligaments and tendons (Figure 6–6a). Ligaments join bones to bones at joints and provide support for joints.[2] The layer of the skin underlying the epidermis, the **dermis**, is also dense connective tissue (Figure 6–5a). It binds the epidermis to underlying muscle and bone.

Loose connective tissue is a packing material and contains numerous connective tissue cells in a loose network of collagen and elastic fibers (Figure 6–6b). These fibers are both made of protein. Loose connective tissue forms around blood vessels in the body and in skeletal muscles where it binds the cells together. It also lies beneath epithelial linings of the intestine and trachea, anchoring them to underlying structures.

The chief difference between dense and loose connective tissues lies in the ratio of cells to extracellular fibers. Dense connective tissue has far more fibers than loose connective tissue. The fibers found in dense and loose connective tissue are produced by **fibroblasts**, cells that reside in connective tissue.

Fibroblasts repair damage created by cuts or tears in body tissues. When skin is cut, for example, fibroblasts in the dermis migrate into the injured area where they begin producing large amounts of colla-

[2]Use the mnemonic LBJ to remember that ligaments connect bones to bones at joints.

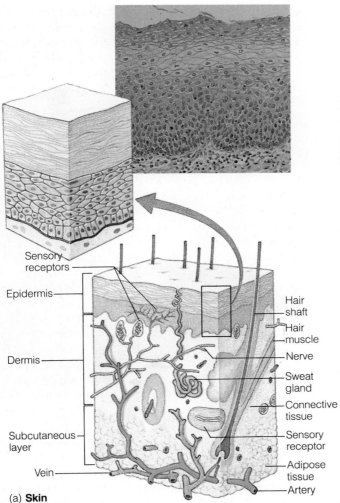

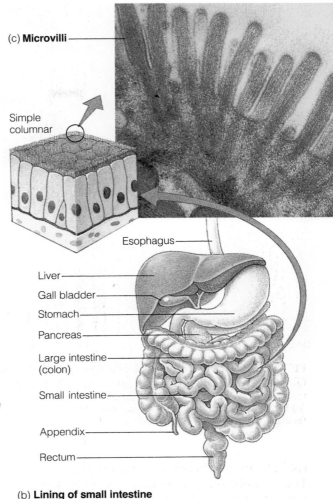

(c) **Microvilli**

Simple columnar

Esophagus

Liver

Gall bladder

Stomach

Pancreas

Large intestine (colon)

Small intestine

Appendix

Rectum

(b) **Lining of small intestine**

Sensory receptors

Epidermis

Dermis

Subcutaneous layer

Vein

(a) **Skin**

Hair shaft

Hair muscle

Nerve

Sweat gland

Connective tissue

Sensory receptor

Adipose tissue

Artery

FIGURE 6–5 Comparison of Two Epithelia with Markedly Different Functions (a) A cross section of the skin showing the stratified squamous epithelium of the epidermis (above), which protects underlying skin from sunlight and desiccation, and (b) the simple columnar epithelium of the lining of the small intestine, which is specialized for absorption. (c) The surfaces of the cells lining the intestine are thrown into folds (microvilli) that greatly increase the surface area for absorption.

gen. The collagen fibers fill the wound, closing it off. The epidermis grows over the damaged area helping repair the damage and restoring the integrity of the skin. When the cut is small, the epidermis covers the damaged area completely, leaving no scar, but in larger wounds the epidermal cells are often unable to grow over the entire wound, leaving some of the underlying collagen exposed and producing a scar.

Loose connective tissues contain several cells that protect against bacterial and viral invasion. One of the most important is the **macrophage** ("big eater"). This cell is capable of phagocytosis and contains numerous lysosomes used to digest phagocytized material. Microorganisms that penetrate skin that has been cut or injured enter the underlying connective tissue where they are engulfed by macrophages. This prevents bacteria from spreading to other parts of the body. Macrophages also play a role in immune protection (Chapter 10).

Some loose connective tissues contain large, conspicuous fat cells. The fat cell is one of the most distinguishable of all body cells. In fully formed cells, a huge fat globule occupies virtually the entire cell, pressing the cytoplasm and the nucleus to the periph-

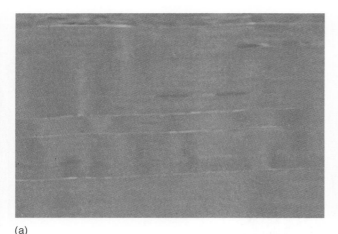

(a)

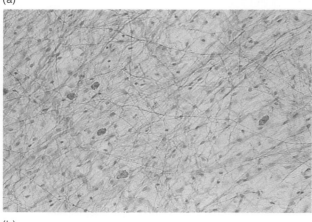

(b)

FIGURE 6–6 Light Micrographs of Connective Tissue (*a*) Dense connective tissue. (*b*) Loose connective tissue.

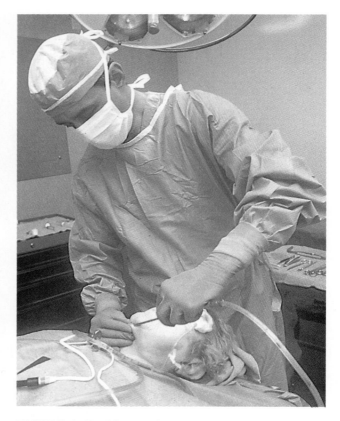

FIGURE 6–7 Liposuction During liposuction surgery, the physician aspirates fat from deposits lying beneath the skin, helping to reduce unsightly accumulations.

ery. Fat cells occur singly or in groups of varying size. Large numbers of fat cells in a given region form a modified type of loose connective tissue known as **adipose tissue** or, simply, fat. Adipose tissue is an important store for lipids, particularly triglycerides, which are used as an energy source (Chapters 2 and 7).

Reducing weight often means reducing the deposits of body fat. Exercise and reduced intake of food are essential elements of any weight-loss plan, but surgical measures can also be used. Surgeons, for instance, can remove excess fat via a technique called **liposuction** (Figure 6–7). In this procedure, surgeons make small incisions in the skin through which they insert a suction device. Fat is aspirated from deposits under the skin in various locations, such as the thighs, buttocks, and abdomen. The fat extracted from one re-

gion can even be transferred to other regions, such as the breast, to resculpt the human body. Liposuction is a relatively safe technique, but it is not free from risk.

Loose connective tissues also contain many mobile cells that are derived from the bloodstream. Cells, such as the lymphocyte and neutrophil, play an important role in protecting the body from foreign invaders and are described in Chapters 8 and 10.

Specialized Connective Tissues

The body contains three types of specialized connective tissue: cartilage, bone, and blood.

Cartilage. Cartilage consists of cells embedded in an abundant extracellular material, the **matrix**. Surrounding virtually all types of cartilage is a layer of dense, irregularly packed connective tissue, the **perichondrium** ("around the cartilage"). This layer contains blood vessels, which supply nutrients to cartilage cells through diffusion. Because cartilage cells are

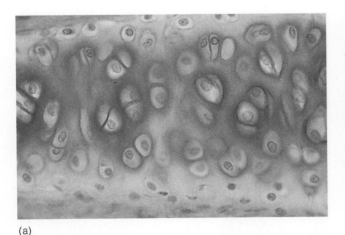

(a)

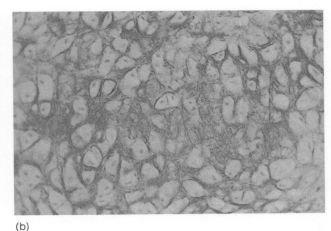

(b)

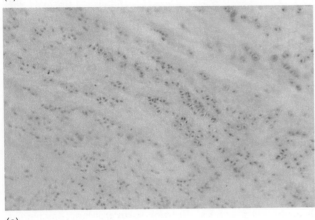

(c)

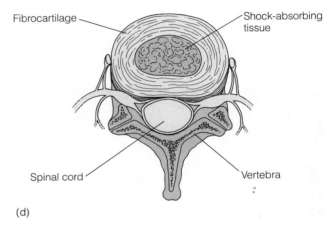

(d)

FIGURE 6–8 Light Micrographs of Cartilage
(*a*) Hyaline. (*b*) Elastic. (*c*) Fibrous. (*d*) Intervertebral disk

showing arrangement of fibrocartilage in a protective ring around the soft, spongy part of the disk that absorbs shock.

nourished by diffusion from outlying capillaries, damaged cartilage heals very slowly. Damage to cartilage in a joint, therefore, may take years to repair.

Three types of cartilage are found in the body: hyaline, fibrous, and elastic. The most prevalent type of cartilage is **hyaline cartilage** (Figure 6–8a). Hyaline cartilage contains numerous collagen fibers, which appear white to the naked eye. Hyaline cartilage is located in many areas. It is found on the ends of many bones, where it reduces abrasion that occurs during movement. It is also the flexible material in the nose and is found in the larynx and the rings of the trachea that you can feel below the larynx. The ends of the ribs that join to the sternum (breast bone) are hyaline cartilage. In embryonic development, the first skeleton is hyaline cartilage, much of which is later converted to bone.

Elastic Cartilage is similar to hyaline cartilage, but contains many wavy elastic fibers, which give it much greater flexibility (Figure 6–8b). Elastic cartilage is found in regions where support and flexibility are required—for example, the external ears and eustachian tubes, which help equalize pressure in the inner ear (Chapter 13).

Fibrocartilage is the rarest type of cartilage. The extracellular matrix of fibrocartilage consists of numerous bundles of collagen fibers and fewer cells than other types (Figure 6–8c). Fibrocartilage is found in the **intervertebral disks**, the shock-absorbing tissue between the vertebrae, the bones of the spine (Figure 6–8d). An intervertebral disk consists of a soft, cushiony central region that absorbs shock. Fibrocartilage forms a ring around the central portion of the disk, holding it in place. Over time, the fibrocartilage ring

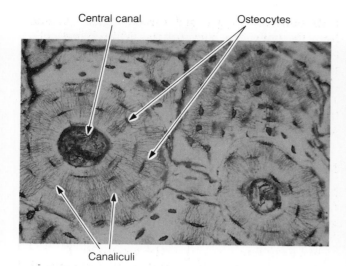

Central canal | Osteocytes

Canaliculi

(a)

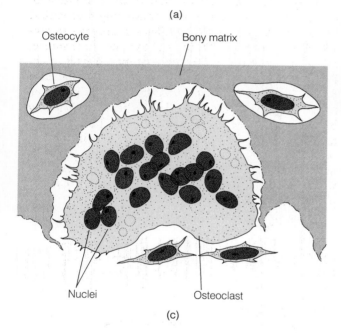

Osteocyte | Bony matrix

Nuclei | Osteoclast

(c)

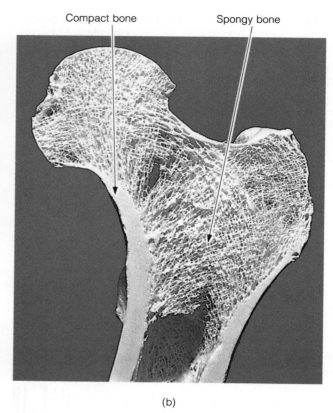

Compact bone | Spongy bone

(b)

FIGURE 6–9 **Bone** (a) Light micrograph of compact bone. (b) Compact and spongy bone shown in section of the humerus. (c) Osteoclast digesting surface of bony spicule.

may weaken or tear, allowing the central part of the disk to bulge or herniate. This condition, called a slipped or herniated disk, can result in a significant amount of pain in the neck, back, or one or both legs, depending on the location of the damaged disk. Pain results when the disk presses against nearby nerves.

Bone. Bone is a specialized form of connective tissue. Contrary to what many might think, bone is a dynamic living tissue. Besides providing internal support and protecting internal organs such as the brain, heart, and lungs, bone plays an important role in maintaining blood calcium levels. Calcium is required for many body functions: muscle contraction, normal nerve functioning, and even blood clotting (for more on the function of bones, see Chapter 14).

Like all connective tissue, bone consists of cells embedded in an abundant extracellular matrix (Figure 6–9a). Bone matrix consists primarily of collagen fibers and needlelike salt crystals containing calcium, phosphate, and hydroxide ions. The collagen fibers and salt crystals combine to give bone its unique characteristics. The strength and resiliency of bone are imparted by the collagen, while its hardness results from the salt crystals. Demineralized bone matter, described earlier in the chapter, is the rubbery collagen skeleton of a bone that remains when the salts are dissolved away. The collagen in a bone can also be selectively removed, producing a brittle mineral replica. Clearly, both components are required for normal functioning.

Two types of bone tissue are found in the body: compact bone and spongy bone. Compact bone is dense and hard. The bone cells are located in concentric rings of bony matrix surrounding the **central canal** (Figure 6–9a). These channels contain blood vessels and nerve fibers. As illustrated in Figure 6–9a, compact bone also contains bone cells or **osteocytes**. Each osteocyte has numerous processes that course through tiny canals in the bone matrix. The canals, called **canaliculi**, provide a route for nutrients to flow to the osteocytes from the central canal. Wastes also flow through the canaliculi from the osteocytes to the veins in the central canal.

Inside most bones of the body is a tissue called **spongy bone**, consisting of an irregular network of calcified collagen spicules (Figure 6–9b). On the surface of the spicules are **osteoblasts**, cells that form bone and become osteocytes when embedded in a bony matrix. Between the spicules are numerous cavities. The small cavities in the spongy bone often communicate with much larger cavities in the center of bones. In some bones the large and small cavities are filled with fat cells and form the **yellow marrow**. In other bones, they are filled with blood cells and cells that give rise to new blood cells, thus forming **red marrow** (Chapter 14).

On the surfaces of some bony spicules in spongy bone tissue are large, multinucleated cells called **osteoclasts** ("bone breakers") (Figure 6–9c). These cells are part of a homeostatic system that ensures proper blood calcium levels. When calcium levels in the blood fall, osteoclasts are activated and digest small portions of the spongy bone, releasing calcium into the bloodstream and restoring normal levels.

Spongy bone is also remodeled when bones are subjected to new stresses. The leg bones of a desk-bound executive from Atlanta who goes on a skiing vacation in Jackson Hole, Wyoming, for example, are remodeled to accommodate the new stresses. During this process, osteoclasts tear down some of the spongy bone, while osteoblasts rebuild new bone in other areas to meet the new stresses. By the end of the two-week ski trip, the executive's bones have been refashioned to produce a stronger bone. Back at the desk, the worker's bones revert to their previous, weaker state.

Blood. Blood is also a specialized form of connective tissue and consists of two components: cells and a large amount of extracellular material, a fluid called **plasma** (Figure 6–10). The cellular component of blood consists of red and white blood cells. **Red blood cells** transport oxygen and carbon dioxide to and from the lungs and body tissues. **White blood cells** are involved in fighting infections. Blood also contains **platelets**, fragments of large cells (megakaryocytes) found in the red bone marrow, the principal site of blood cell formation. Not considered to be cells because they lack nuclei and other organelles, platelets play a key role in blood clotting (Chapter 8).

Muscle Tissue

Muscle, the third in our series of primary tissues, is found in virtually every organ in the body. Muscle gets its name from the Latin word for mouse (mus). Early observers likened the contracting muscle of the biceps to a mouse moving under a carpet.

Muscle is an excitable tissue that, when stimulated, is capable of contracting, thus producing mechanical force. Muscle cells working in large numbers can create enormous forces. Muscles of the jaw, for instance, create a pressure of 200 pounds per square inch, forceful enough to snap off a finger. Muscle also moves body parts, propels food along the digestive tract, and expels the fetus from the uterus during birth. It also contracts the heart to pump blood through the 50,000 miles of blood vessels in the human body. Acting in smaller numbers, muscle cells may be responsible for more intricate movements, such as those required for playing a piano or harp or moving the eyes.

FIGURE 6–10 Blood Blood is about 55% liquid (plasma) and 45% formed elements: red blood cells, white blood cells, and platelets.

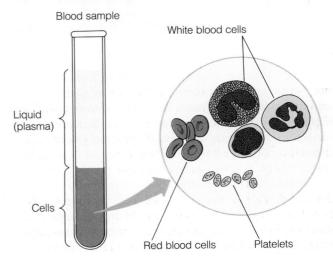

HEALTH NOTE 6–1

Healthy Hearts: Prevention and Transplantation

The human heart is a remarkable organ. Day and night, it pumps blood through the 50,000 miles of vessels in the human circulatory system, flawlessly responding to increases and decreases in demand. Despite its importance, many people abuse their hearts. Stress in our lives and salt in our diets conspire to make the work of the heart more difficult in many people, so difficult, in fact, that the heart may wear out prematurely. Cholesterol also takes its toll; it blocks the vessels supplying blood to the heart, preventing oxygen and nutrients from reaching heart muscle cells and, eventually, killing them.

In the high-tech search for cures to heart disease, many of us seem to have forgotten an old adage: "An ounce of prevention is worth a pound of cure." The risk of heart disease, a major killer of people in the United States, can be greatly reduced by prevention: by diet, exercise, stress management, and quitting smoking. Any one of these factors can help reduce the chances of having a heart attack, but for the greatest risk reduction, health professionals recommend a combined strategy. Consider dietary prevention first.

Cutting back on salt reduces blood pressure, which takes stress off the heart. Many people who cut back gradually report that they don't notice a difference in the taste of their food. In fact, many notice flavors they have been covering up with salt for years. Reducing animal fat (especially red meat) and eggs, both rich in cholesterol, also benefits the heart and other organs. Reducing dietary animal fat reduces cholesterol deposits called plaque that form on arterial walls, cutting off blood supply to vital organs (see Health Note 7–1 for more on cholesterol). If you're a big beef eater, try eating smaller portions and buying leaner meat. Cut the fat off a steak. Eat more fish, turkey, and chicken (without the skin), which are all meats with lower saturated fat content.

Exercise also helps the heart, but start slowly. If you have questions about the strength of your heart, see a physician first. Then start a gradual program. For example, instead of driving to work or school, walk or ride a bike. Instead of taking the elevator to class, try the stairs. Exercise programs need not be intense. Walking or riding a bike or aerobics can burn off extra calories, reduce cholesterol levels, and improve the health of your heart, adding years to your life. Exercise also helps take your mind off your problems, reduces stress, may improve the way you feel, could improve your sleep, and may even help your studies.

What do you do about stress? Stress seems to be a part of our lives. From the moment we wake up to the moment we go to sleep, there always seems to be something that needs tending. Stress can be relieved in many ways, described in Health Note 1–1.

Finally, there's smoking. If you smoke, quit. If you're thinking about taking up the habit, don't. Approximately 1000 Americans die from smoking each day, 365 days a year. About a third of the deaths are from lung cancer, the rest are from heart attacks, strokes, and other forms of cancer. Some smokers escape the effects of cigarette smoke, which is a remarkable tribute to homeostasis, but many will die early. For more on smoking see Health Note 10–1.

Three types of muscle are found in humans: skeletal, cardiac, and smooth. The cells in each type of muscle contain two contractile microfilaments, **actin** and **myosin**. When stimulated, these filaments slide together, shortening the muscle cells and causing contraction (Chapter 14).

Skeletal Muscle.   The majority of the body's muscle is called skeletal muscle—so named because it is frequently attached to the skeleton. When skeletal muscle contracts, it causes body parts to move. Most skeletal muscle in the body is under voluntary control—that is, it is controlled consciously.

As illustrated in Figure 6–11a, skeletal muscle cells are long cylinders that are formed during embryonic development by the fusion of many embryonic muscle cells. Consequently, skeletal muscle cells are referred to as **muscle fibers**. Each muscle fiber contains many nuclei. Because of the dense array of contractile fibers in the cytoplasm of the muscle fiber, the nuclei are generally pressed against the cell membrane. As you might suspect, this highly specialized cell cannot divide; therefore, damaged muscle cells cannot be replaced.

For many people, stress, poor diet, lack of exercise, and smoking have already taken their toll. The arteries to their hearts are filled with cholesterol, making even the slightest exercise painful. These people have no choice except surgery. Clogged heart arteries can be replaced with veins from the leg, but in many cases these bypasses fill up with cholesterol in a few years. Researchers are now experimenting with arterial transplants—using the internal mammary artery to replace clogged coronary arteries—in hopes that these will not clog with cholesterol, or at least not so quickly.

Each year, 400,000 people in the United States develop end-stage heart failure, a serious weakening of the heart. This results from the death of cardiac muscle cells caused by excess stress on the heart and clogged coronary arteries. Nearly all of these people die within a year. Many heart failure patients each year could benefit from heart transplants, but because of a lack of donors, only 1400 heart transplants were performed in 1987. Artificial hearts may provide some hope, but they are costly and plagued with problems.

Skeletal Muscle Wrap Patients with ailing hearts may be aided by skeletal muscle wrapped around the failing organ. Pacemakers can stimulate the muscle to beat in step with the heart muscle, thus increasing cardiac output.

Furthermore, they strap patients to a machine for the rest of their life.

Another promising technique called a skeletal muscle wrap could benefit thousands of patients. As the figure shows, in this procedure, a segment of the latissimus dorsi, a large muscle of the back, is wrapped around the diseased heart. The muscle is stapled in place, forming a contractile basket. Then it is attached to a pacemaker, which senses the heart's natural electrical pulses and delivers bursts of electricity to the skeletal muscle, causing it to contract in step with the heart. This technique leaves the blood supply and nerve supply of the muscle intact so there is little worry over graft survival.

Skeletal muscle wraps improve cardiac function in patients by about 20%, a change that means the difference between the life of an invalid and a fairly normal life. Using a patient's own skeletal muscle has many benefits. First, it is readily available, and, in most cases, patients are more than willing to "donate" tissue for their own survival. Second, because the graft is taken from a person's own body, surgeons do not have to worry about the immune system rejecting the graft. Promising as this technique is, it is no substitute for prevention.

Skeletal muscle fibers appear banded or **striated** when viewed under the light microscope, as illustrated in Figure 6–11a. As discussed in Chapter 14, the striations result from the unique arrangement of actin and myosin filaments inside muscle cells.

Cardiac Muscle.   Like skeletal muscle, cardiac muscle is striated. Unlike skeletal muscle, cardiac muscle is involuntary—that is, it contracts without conscious control. Found only in the walls of the heart, cardiac muscle cells branch and interconnect freely (Figure 6–11b). Individual cells are tightly connected to one another, a feature that helps maintain the structural integrity of the heart. The points of connection also provide pathways for the electrical impulses to travel from cell to cell, allowing the heart muscle to contract uniformly when stimulated.

Smooth Muscle.   The third and final type of muscle is smooth muscle. **Smooth muscle,** so named because it lacks visible striations, is involuntary. Contractile filaments are present but are not organized in the same fashion as those in striated muscle (Figure 6–11c). Smooth muscle cells may occur singly or in

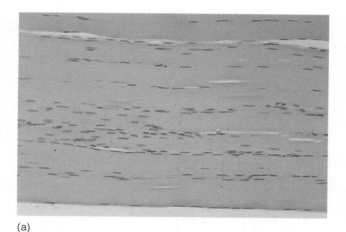

(a)

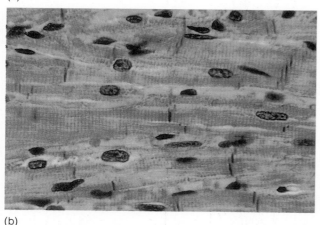

(b)

(c)

FIGURE 6–11   Light Micrograph of the Three Types of Muscle (*a*) Skeletal. (*b*) Cardiac. (*c*) Smooth.

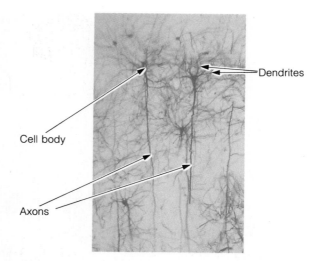

Dendrites

Cell body

Axons

FIGURE 6–12   Multipolar Neuron Attached to the cell body of the multipolar neuron are many highly branched dendrites, which deliver impulses to the cell body. Multipolar neurons have one long, unbranching fiber called the axon, which transmits impulses away from the cell body.

Nervous Tissue

Last but not least of the primary tissues is nervous tissue. **Nervous tissue** consists of two types of cells: conducting cells and supportive cells. The conducting cells, called **neurons**, are specially modified cells capable of responding to various stimuli and producing bioelectric impulses, which they transmit from one region of the body to another. The ability to respond to stimuli is a characteristic of all living things and is called irritability (Chapter 1). The ability to transmit a message is called **conductivity**. The properties of irritability and conductivity in neurons allow humans and other animals to be aware of their environment and to respond to stimuli.

The supportive cells of the nervous system are a kind of nervous system connective tissue. These cells are incapable of conducting impulses, but help transport nutrients from blood vessels to neurons and help guard against toxins by creating a barrier to many potentially harmful substances. Together, the neurons and their supporting cells combine to form the brain, spinal cord, and nerves of the nervous system (Chapter 12).

Several types of neurons are found in the body. All neurons contain two types of fibers—those that conduct impulses to the cell body and those that conduct impulses away from the cell body. One of the most common nerve cells is the **multipolar neuron**, shown in Figure 6–12. It contains a prominent, multiangular

small groups. Small rings of smooth muscle cells, for example, surround tiny vessels in the circulatory system. When these cells contract, they shut off or reduce the supply of blood to tissues. Smooth muscle cells are most often arranged in sheets in the walls of organs, such as the stomach, uterus, and intestines. Smooth muscle cells in the wall of the stomach, for example, churn the food, mixing the stomach contents, and force tiny spurts of liquified food into the small intestine. Smooth muscle contraction propels the food along the intestinal tract.

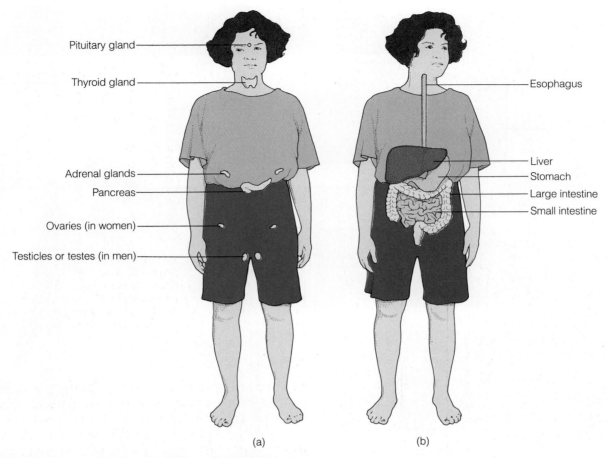

FIGURE 6–13 Comparison of the Endocrine and Digestive Systems
(*a*) The endocrine system. (*b*) The digestive system.

cell body and several short, highly branched fibers, called **dendrites**, which carry impulses to the cell body. Also attached to the cell body is a large, fairly thick process, the **axon**, which transports bioelectric impulses away from the cell body. Like muscle cells, nerve cells are highly differentiated and cannot divide. When a cell is destroyed, it degenerates and cannot be replaced by cell division. A cut nerve axon, however, may partially regenerate, forming a new axon that reestablishes previous connections and restores some degree of sensation or control over muscle.[3] New research, however, may someday provide ways to stimulate regeneration of nerve cells, helping physicians restore nerve function to victims of accidents.

[3]Some evidence in birds suggest that nerve cell regeneration may be possible.

Organs and Organ Systems

The cell contains organelles ("little organs") that carry out many of its functions in isolation from the biochemically active cytoplasm. Compartmentalization is an evolutionary strategy that also occurs at the organismic level. Discrete structures called **organs** evolved to perform specific functions, such as digestion, enzyme production, and hormone synthesis. Most organs, however, do not function alone. They are part of a group of cooperative organs, called an **organ system**. Sometimes components of an organ system are connected—for example, as in the digestive system; and sometimes they are dispersed throughout the body—for example, as in the endocrine system (Figure 6–13). Some organs belong to more than one system. For example, the pancreas produces digestive enzymes that are secreted into the small intestine where they break down food materials. The pancreas

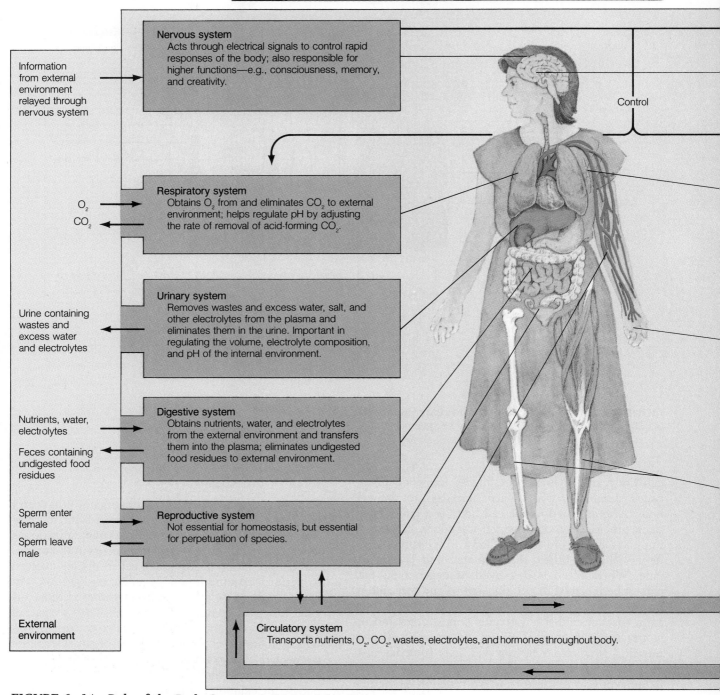

BODY SYSTEMS
Made up of cells organized according to specialization to maintain homeostasis.

Nervous system
Acts through electrical signals to control rapid responses of the body; also responsible for higher functions—e.g., consciousness, memory, and creativity.

Information from external environment relayed through nervous system

Control

Respiratory system
Obtains O_2 from and eliminates CO_2 to external environment; helps regulate pH by adjusting the rate of removal of acid-forming CO_2.

O_2
CO_2

Urinary system
Removes wastes and excess water, salt, and other electrolytes from the plasma and eliminates them in the urine. Important in regulating the volume, electrolyte composition, and pH of the internal environment.

Urine containing wastes and excess water and electrolytes

Digestive system
Obtains nutrients, water, and electrolytes from the external environment and transfers them into the plasma; eliminates undigested food residues to external environment.

Nutrients, water, electrolytes

Feces containing undigested food residues

Reproductive system
Not essential for homeostasis, but essential for perpetuation of species.

Sperm enter female

Sperm leave male

External environment

Circulatory system
Transports nutrients, O_2, CO_2, wastes, electrolytes, and hormones throughout body.

FIGURE 6–14   Role of the Body Systems in Maintaining Homeostasis

is, therefore, part of the digestive system. The pancreas also contains cells that produce insulin and glucagon, two hormones that control blood glucose levels. Consequently, the pancreas also belongs to the endocrine system.

The following nine chapters describe the major organ systems, each of which performs functions necessary for life. Figure 6–14 summarizes the functions

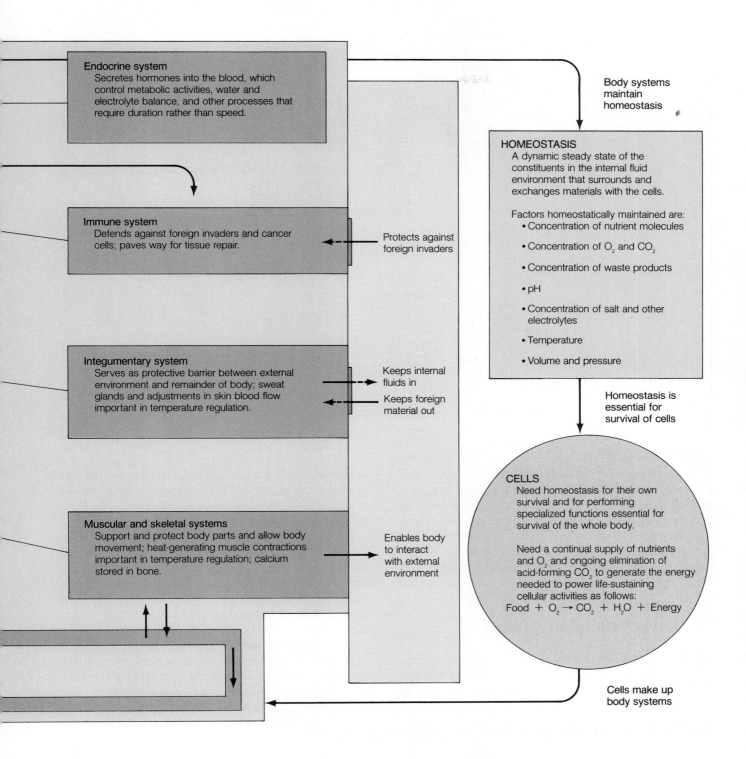

Endocrine system
Secretes hormones into the blood, which control metabolic activities, water and electrolyte balance, and other processes that require duration rather than speed.

Immune system
Defends against foreign invaders and cancer cells; paves way for tissue repair.

Protects against foreign invaders

Integumentary system
Serves as protective barrier between external environment and remainder of body; sweat glands and adjustments in skin blood flow important in temperature regulation.

Keeps internal fluids in

Keeps foreign material out

Muscular and skeletal systems
Support and protect body parts and allow body movement; heat-generating muscle contractions important in temperature regulation; calcium stored in bone.

Enables body to interact with external environment

Body systems maintain homeostasis

HOMEOSTASIS
A dynamic steady state of the constituents in the internal fluid environment that surrounds and exchanges materials with the cells.

Factors homeostatically maintained are:
• Concentration of nutrient molecules
• Concentration of O_2 and CO_2
• Concentration of waste products
• pH
• Concentration of salt and other electrolytes
• Temperature
• Volume and pressure

Homeostasis is essential for survival of cells

CELLS
Need homeostasis for their own survival and for performing specialized functions essential for survival of the whole body.

Need a continual supply of nutrients and O_2 and ongoing elimination of acid-forming CO_2 to generate the energy needed to power life-sustaining cellular activities as follows:
Food + O_2 → CO_2 + H_2O + Energy

Cells make up body systems

of the organ systems. Take a moment to read the descriptions in the boxes. Later you will see how these separate organ systems contribute to the economy of the entire organism in much the same way that

the actions of individuals, businesses, and governments contribute to the U.S. economy.

Organ systems are the functional machinery of the body. In sickness and in health, the functions of the

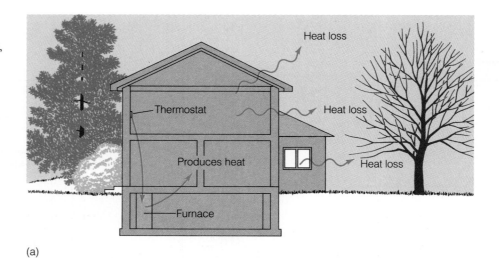

FIGURE 6-15   Homeostasis and the House (a) Heat is maintained in a house by a furnace, which produces heat to balance heat loss. The thermostat monitors the internal temperature and switches the furnace on and off in response to temperature changes. (b) A hypothetical temperature graph showing temperature fluctuation around the operating point.

(a)

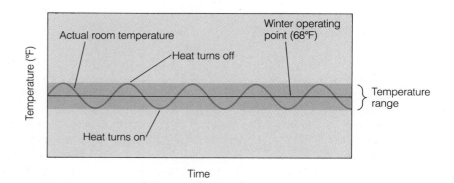

(b)

nine major organ systems remain interconnected and mutually dependent. This interdependence is apparent when one system "breaks down." A breakdown in just one can have catastrophic effects on the others and the organism as a whole. When defective kidneys, for instance, fail to remove toxins from the blood, other cells and other organ systems are poisoned, and death results in a matter of days.

PRINCIPLES OF HOMEOSTASIS

Homeostasis is one of the principal themes of this book. Defined in Chapter 1 as the internal constancy of the body, *homeostasis occurs on a variety of levels—in cells, tissues, organs, organ systems, organisms, and even the environment.* Homeostatic systems at all levels of biological organization have several common features. First and foremost, *homeostatic systems maintain constancy chiefly through negative feedback mechanisms.* Feedback mechanisms were briefly

described in Chapter 3. To understand how negative feedback works, we will first examine a typical house, containing a homeostatic system of its own—the heating system (Figure 6-15a).

On a cold winter night, the heating system (the furnace and thermostat) maintains a constant internal temperature, even though the outside temperature may be falling. Heat that is lost through ceilings, walls, windows, and tiny cracks is replaced by heat generated from the combustion of natural gas or oil in the furnace.

The thermostat detects changes in room temperature. When indoor temperatures decline, the thermostat sends a signal to the furnace, turning it on. Heat from the furnace is distributed through the house and raises room temperature. When the temperature reaches the desired setting, the thermostat shuts the furnace off. Thus, the product of the system (heat) shuts the furnace off.

All homeostatic feedback mechanisms contain at least two components: a sensor (or receptor), which detects

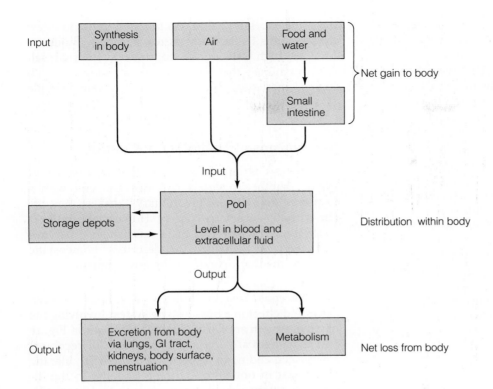

FIGURE 6–16 Generalized View of a Homeostatic System Inputs and outputs are balanced to maintain more or less constant levels of chemical and physical parameters.

changing conditions, and an effector, which responds. In our homes, the thermostat is the sensor and the furnace is the effector.

A graph of a hypothetical house temperature is shown in Figure 6–15b and illustrates another important principle of homeostasis: that *homeostatic systems do not maintain absolute constancy.* Rather, they maintain a variable (such as body temperature) within a certain range.

The human body has its own internal furnace: the catabolism of glucose and other molecules. In many ways, the body's internal furnace is regulated by negative feedback. In Chapter 3, you learned that the cell breaks down glucose to make ATP. During this process, heat is given off. Each cell, then, is a tiny furnace whose heat radiates outward and is distributed throughout the body by the blood.

Maintaining a constant body temperature requires a balance between heat production and heat loss—just as it does in our homes. Suppose you were to walk outdoors on a cold winter day, dressed in a light sweater and blue jeans. Receptors in the skin would sense your body temperature and send signals to the brain, setting into motion a series of physiological changes to keep your body temperature from falling. Body temperature can be maintained by reducing heat loss and by increasing heat production. For example,

the brain sends signals to blood vessels in the skin, causing them to constrict, thus reducing the flow of blood in the skin. The restriction of blood flow through the skin reduces heat loss. If it is cold enough outside, the brain may also send signals to the muscles, causing them to undergo rhythmic contractions—shivering. Shivering burns additional glucose, releasing extra heat. Many voluntary actions may also be "ordered" by the brain to help reduce heat loss or generate more heat. For example, you might put on a hat or turn around and go back inside. *Voluntary action, in fact, is often a crucial component of homeostasis.*

All homeostatic mechanisms in the human body operate similarly, maintaining chemical and physical parameters within a narrow range around the **operating point.** The operating point is like the setting on a thermostat. The body maintains constant levels of most hormones, nutrients, wastes, and ions. It also maintains body temperature, blood pressure, blood flow, and other functions. Each physiological parameter is controlled by negative feedback.

Homeostasis is maintained by balancing inputs and outputs. The generalized diagram in Figure 6–16 illustrates this important concept. First consider the input sides. Ions (e.g., calcium) and other chemical substances (e.g., glucose) are ingested in food or water. Other essential chemicals (e.g., oxygen) enter in the air we breathe. Still others (e.g., some amino acids) are produced in the body. All three routes provide chemical input.

Now consider the output side. The levels of glucose and other substances in the body are reduced by excretion and metabolism. Glucose in the food we eat, for instance, is broken down in cells to make energy. Some glucose may also be excreted in the urine. The level of glucose and other chemicals in the blood and body fluids depends on the balance between input and output. *Internal storage depots (regions where chemicals are stored) also participate in this balance.* As you learned in previous chapters, glucose is stored in the liver and muscle of the body. Glucose levels in the blood depend on the input and output at these two storage depots. Immediately after a meal, blood glucose levels begin to rise. This extra glucose is quickly taken up by the liver and muscle, where glucose is stored as glycogen. Storage depots, therefore, help conserve glucose after a meal. Between meals, glycogen is broken down to provide a continuous supply of energy. The level of glucose in the blood, therefore, is controlled in large part by the movement of glucose to and from storage depots.

The relative importance of the various homeostatic pathways depends on the substance in question. For water, ingestion and excretion are the primary homeostatic pathways. For iron, the rate of absorption by the intestines is the key to determining blood iron levels. When iron levels decrease, say, because of a decrease in iron intake, the body reestablishes the balance by increasing the rate of absorption in the small intestine.

Homeostasis can be upset by a change in a single pathway in the body's complex homeostatic network. Heavy perspiration, for instance, can result in a severe decrease in water volume that may lead to death. Decreased water intake can have the same effect. Diarrhea in children results in a severe depletion of body fluids and kills millions of Third World children each year. *An imbalance in either the input or the output side of the homeostatic equation can have dramatic impacts on human health.* Excess salt intake, for example, can result in hypertension (high blood pressure). If the input and output of salt are balanced by homeostatic mechanisms, you might ask, wouldn't the body merely increase excretion and prevent hypertension?

Under such circumstances, the body does indeed balance input and output, but it usually takes three or four days for the necessary adjustments to occur. Even though input and output are balanced within a few days, the excess salt that was ingested in the first few days remains in the body tissues and body fluids (as long as high salt intake continues). Internal salt concentrations, therefore, remain about 2% or 3% higher, increasing blood pressure in some people (Chapter 8).

Mechanisms of Homeostatic Control

Homeostasis is largely a **reflex**, that is, a series of events that begin with a stimulus and end with a response. Reflexes are involuntary—that is, they occur without conscious control. Homeostatic reflexes in the body involve two main mechanisms: nervous and chemical. Chapter 20 describes homeostatic mechanisms that operate in the environment.

The Nervous System Reflex. The lines between cause and effect in a homeostatic system involving the nervous system are fairly evident. As shown in Figure 6–17, changes are detected by a sensor, or **receptor**, a nerve cell ending or a special structure that, like the thermostat in our homes, detects changes in the internal or external environment (Chapters 12 and 13). The sensor is connected to the brain or spinal cord by a nerve. Thus, the brain and spinal cord receive the input from the body, keeping them attuned to any changes. The brain and spinal cord, in turn, direct an appropriate response to counterbalance the change.

At any one time, the brain and spinal cord receive a great many signals from body receptors, alerting the nervous system of change. The process of making sense of the numerous and varied inputs is called **integration**. The brain is the center of integration. Unlike the furnace, the brain usually cannot make the correction needed; it merely sends a message to some other part of the body to carry out the needed change. For example, when you are chilled, nerve impulses alert the brain to your condition. The brain, in turn, sends signals to the muscles to contract, creating shivering, or signals to the blood vessels in the skin, causing them to constrict and reduce blood flow through the skin. These organs are called **effectors**, for they "effect" (bring about) the brain's command.

Chemical Control. Hormones also participate in reflexes—sometimes involving the brain, and sometimes not. Consider an example free of nervous system intervention. Four tiny glands situated in the neck, called the parathyroid glands, produce a hor-

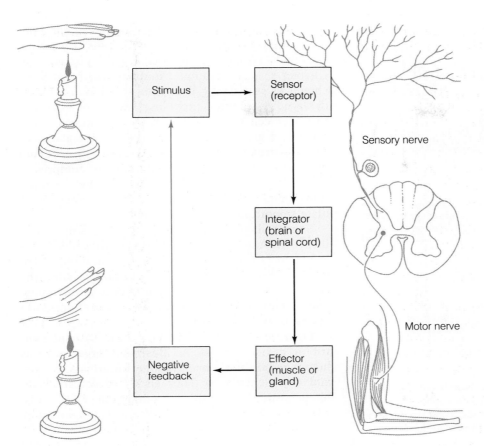

FIGURE 6–17 Nervous System Reflex Reflexes involve some kind of stimulus, a sensor, an integrator, and an effector. The sensor detects a change and sends a signal to the integrator, the brain or spinal cord, which elicits a response in the effector organs. Negative feedback from the effector eliminates the stimulus.

Stimulus

Sensor (receptor)

Sensory nerve

Integrator (brain or spinal cord)

Motor nerve

Negative feedback

Effector (muscle or gland)

mone called parathyroid hormone or PTH.[4] PTH is released from the cells of the glands when calcium levels in the blood fall. PTH travels in the blood to the bone, where it stimulates the release of calcium, correcting the deficiency. This reflex involves a stimulus and response. Distinct receptors, integrators, and effectors are not found in this reflex. Instead, the cells of the parathyroid gland take on several functions: that of receptor, which senses the drop in blood calcium; integration center, which determines a response; and effector, which produces a response (Figure 6–18).

Some endocrine glands work through the nervous system, making the lines of cause and effect a little more difficult to follow. Consider one example. The thyroid gland produces a hormone called thyroxine (Figure 6–19). It increases the metabolic rate, increasing heat production. Thyroxine levels are monitored by cells in the brain (the receptors). When thyroxine levels fall, these cells release a chemical substance that travels from one region of the brain

[4]The parathyroid glands are located within the thyroid gland.

(the hypothalamus) to the pituitary gland, located just beneath the brain. The pituitary responds by releasing another hormone, thyroid-stimulating hormone, or TSH, which travels in the blood to the thyroid gland. Here, TSH steps up the production of thyroxine, correcting the deficiency. This system, like all of the reflexes, involves a stimulus and response. The line between the two is just a little longer and a little more involved.

Local Chemical Control. Nervous and endocrine mechanisms generally occur over considerable distances. In the nervous system, for example, nerves carry impulses from the brain and spinal cord to distant parts of the body, often several feet away. In the endocrine system, the bloodstream carries the messages from endocrine glands to distant effectors. But not all systems require messages to be transported over long distances. In some cases, chemical control is exerted through the extracellular fluid only a cell away. Chemicals that elicit effects in the local regions are called **paracrines**. Paracrines are produced by in-

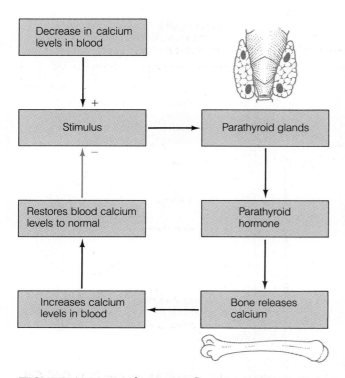

FIGURE 6–18 Endocrine Reflex Not Involving the Nervous System This reflex operates through the bloodstream. A decrease in calcium in the blood stimulates a series of reactions that helps restore normal blood calcium levels.

dividual cells and released into the extracellular fluid, where they diffuse to neighboring cells. Epidermal growth factor produced by skin cells, discussed in Chapter 4, is one example. You may recall that these chemicals stimulate cell division when skin cells are damaged or lost, thus providing a degree of local control over cell growth. One of the best-known paracrines is a group of chemicals called **prostaglandins**. These unusual fatty acids are produced by cells and migrate to nearby cells. Prostaglandins comprise a rather large group of chemicals with diverse functions. Some function in blood clotting. Others stimulate smooth muscle contraction.

Some cells produce chemical substances that affect their own function. These chemicals are called **autocrines** and are part of the simplest chemical reflexes in the body. Some prostaglandins, for example, affect the cells that produce them, as well as cells in the immediate vicinity. These chemicals are, therefore, autocrines as well as paracrines.

Figure 6–14 shows how the organ systems contribute to homeostasis. As illustrated, organ systems help maintain the concentration of nutrients, wastes, and other chemicals. They also help control body temperature, keeping it at a comfortable 37° Celsius, and they control blood pressure and blood volume.

FIGURE 6–19 A Neuroendocrine Reflex This reflex involves both the endocrine and nervous systems. Sensors are located in the brain.

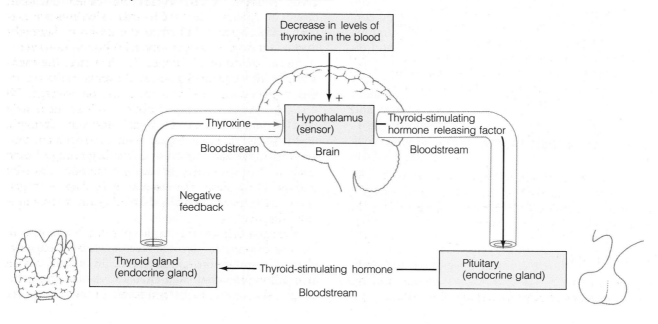

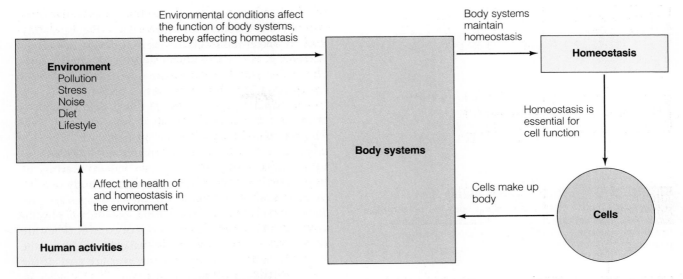

FIGURE 6–20   Health, Homeostasis, and the Environment   Human health is dependent on maintaining homeostasis. Homeostasis, however, is affected by the condition of our environment. Stress, pollution, noise, and other environmental factors upset the function of body systems, thus upsetting homeostasis and human health. As a result, human health is dependent on a healthy, balanced (homeostatic) environment.

Health, Homeostasis, and the Environment

Human health is dependent on homeostasis—maintaining internal constancy. Human health is also dependent on an environment that is healthy socially, as well as healthy physically and chemically (Figure 6–20). Upsets in the environment in which we live—for example, pollution, stress, and noise—affect our health by interrupting the functioning of organ systems that participate in human health. Thus, a balanced and healthy environment helps create healthy humans. Numerous examples in this book will illustrate how environment impinges on our health. You will see how stress, diet, pollution, and other "environmental factors" affect our bodies, principally by altering internal balance.

BIOLOGICAL RHYTHMS

The previous discussion of homeostasis gives the impression that homeostasis establishes an unwavering condition of stability that remains more or less the same day after day, year after year.

Natural Rhythms

In truth, many physiological measurements undergo rhythmic change. Body temperature, for instance, varies during a 24-hour period by as much as 0.5° Celsius. Blood pressure may change by as much as 20%, and the number of white blood cells, which fight infection, varies by 50% over a 24-hour period. Alertness also varies considerably during the day. About 1:00 P.M. each day, for instance, most people go through a slump. For most people, activity and alertness peak early in the evening, making it an excellent time to study. These cycles are called circadian rhythms ("about a day"). **Circadian rhythms** are natural body rhythms linked to the 24-hour day/night cycle. Not all cycles occur over 24 hours, however.

Some can be much longer. For instance, the **menstrual cycle**, a recurring series of events in the reproductive functions of women, lasts, on average, 28 days. During the menstrual cycle, levels of the female sex steroid hormone estrogen undergo dramatic shifts. As illustrated in Figure 6–21, estrogen concentrations in the blood are low at the beginning of each cycle and peak on day 14 when ovulation normally occurs. Throughout the remaining 14 days, estrogen levels are rather high, then drop off again when a new cycle begins.

Estrogen follows this cycle month after month in women of reproductive age. Recent research suggests that estrogen levels may exert profound influences on neural functions in women. One recent study, for example, showed that estrogen levels in the blood alter

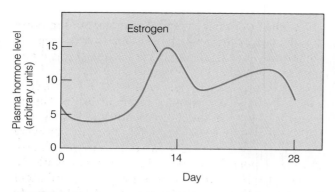

FIGURE 6–21 Estrogen Levels during the Menstrual Cycle Note that over a 28-day period estrogen levels vary considerably. The variation is part of a finely tuned reproductive cycle that prepares the woman for pregnancy.

a woman's ability to perform fine motor functions. Early in the cycle, when estrogen levels are low, women excel at tasks involving spatial relations—for example, solving a three-dimensional puzzle. When levels peak, many women may find the task more difficult.

Motor skills follow a different pattern. When estrogen levels are low, some women are less able to perform complex motor skills, such as the finger movements required to play a musical instrument, than they are during the remainder of the cycle. When estrogen concentrations in the blood are high, motor skills become easier.

Many hormones follow daily rhythmic cycles. The male sex steroid hormone testosterone follows a 24-hour cycle. The highest levels occur in the night, particularly during dream sleep. Dream sleep occurs primarily in the early morning hours—the later the hour, the longer the periods of dream sleep (Figure 6–22). Since testosterone levels are highest during dream episodes and since most dream sleep occurs in the early hours just before waking, many men wake up—how shall I say it?—eager for amorous adventure.

The important point of all of this is that the body is not static. *While many chemical substances are held within a fairly narrow range by homeostatic mechanisms, others fluctuate widely in normal and quite essential cycles.* Just as the seasons change, altering the face of the landscape, so do the internal body seasons change. But over the long run, these changes are predictable. They also occur within prescribed physiological limits; they do not run out of control. They are part of the body's dynamic balance—just as the weather changes throughout the year are part of the dynamic balance of our climate.

The study of biological rhythms is a fascinating field yielding important information. Research on body rhythms has also shown that animals respond to drugs differently at different times of the day and night. By administering drugs at the body's most receptive times, physicians may be able to reduce the doses, thereby reducing toxic side effects. Selecting the optimal time of administration may also enable physicians to fight diseases more effectively.

Biological Clocks: Controlling Internal Body Rhythms

Research has shown that the brain controls many biological cycles. The **suprachiasmatic nucleus**, a clump of nerve cells in the base of the brain in a region called the hypothalamus (Chapter 12), is

FIGURE 6–22 Stages of Sleep Numbers indicate the stages of sleep: the higher the number, the deeper the sleep. Note that around midnight sleep is deepest. As morning approaches, a person's sleep is lighter and REM sleep or dream sleep occurs in longer increments.

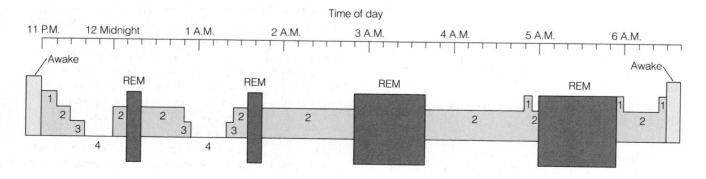

thought to play a major role in coordinating several key rhythms. It also regulates other control centers and is, therefore, often referred to as the "master clock." The suprachiasmatic nucleus has its own inherent rhythm like a clock. When wound, it ticks off the minutes, faithfully imposing its control on the body, turning body functions on and off like the master control in a highly automated factory. Even in complete darkness, the master clock ticks on, directing many circadian rhythms. But under such conditions, the clock seems to impose a 25-hour cycle; that is, most people who are isolated in a dark room with no outside time cues enter into a 25-hour sleep/wake cycle. Some individuals have even longer cycles.

Researchers believe that the natural 25-hour clock is modulated by the 24-hour day/night cycle. Our environment, therefore, imposes its control on the clock and may do so through a gland in the brain called the **pineal gland**. This tiny organ secretes a hormone thought to keep the suprachiasmatic nucleus in sync with the 24-hour day/night cycle.

Jet lag, that drowsy, uncomfortable feeling people get when they travel long distances by jet, results from the body's biological clocks ticking out of synchrony with the day/night cycle. A trip from Los Angeles to New York, for instance, leaves the traveler weary and irritable. At ten o'clock New York time, when New Yorkers are heading to bed, the traveler is still wide awake. That's because his body is still on Los Angeles time—three hours earlier. When the alarm goes off at 6:00 A.M. New York time, the weary traveler's body crawls out of bed totally exhausted. It's only 3:00 A.M. Los Angeles time! To avoid the ill feelings, experts suggest that when in a new time zone, you either abide by your internal clock, following "home time," or reset your biological clock. You can reset your clock by going to bed an hour or two earlier when you will be traveling from west to east. Do just the opposite for east-to-west travel.

 ENVIRONMENT AND HEALTH: TINKERING WITH OUR BIOLOGICAL CLOCKS

Research shows that the hectic pace of modern life, stress, noise, pollution, and the weird work schedules many workers follow can upset internal rhythms, illustrating how a healthy external environment affects our own internal environment—and our health.

Dr. Richard Restak, a neurologist and author, notes that the "usual rhythms of wakefulness and sleep . . . seem to exert a stabilizing effect on our physical and psychological health." The greatest disrupter of our natural circadian rhythms, says Restak, is the variable work schedule, surprisingly common in the United States and other industrialized countries. Today, one out of every four working men and one out of every six working women has a variable work schedule— shifting frequently between day and night work. In many industries, workers are at the job day and night to make optimal use of equipment and buildings. As a result, more restaurants and stores are open 24 hours a day and more health care workers must be on duty at night to care for accident victims.

What American business has forgotten is that people are cyclic creatures. For millions of years humans have slept during the night and been awake during the day. Turn that around and you're asking for trouble. Make a person work at night when he or she normally sleeps and you can expect more accidents and lower productivity.

To spread the burden, many companies that maintain shifts round the clock alter their workers' schedules. One week, employees work the day shift. The next week, they're shifted to the "graveyard shift" from 12:00 midnight to 8:00 A.M. The next week, they work the night shift from 4:00 P.M. to 12:00 midnight. Many shift workers report that they feel rotten most of the time and have trouble staying awake at the job. Their work performance suffers because of fatigue and general malaise. When workers arrive home for bed, they're exhausted but can't sleep, because they're trying to doze off at a time when the body is trying to wake them up. Unfortunately, weekly changes in schedule never permit workers' internal alarm clocks to fully adjust. Most people, in fact, require 4 to 14 days to adjust to a new schedule.

Workers on alternating shifts suffer more ulcers, insomnia, irritability, depression, and tension than workers on unchanging shifts. Their lives are never the same. To make matters worse, tired, irritable workers whose judgment is impaired by fatigue pose a threat to society as a whole. Consider an example.

At 4:00 A.M. in the control room of the Three Mile Island Nuclear Reactor in Pennsylvania, three operators made the first mistake in a series of errors that led to the worst nuclear accident in U.S. history. The operators did not notice warning lights and failed to observe that a crucial valve had remained open. When the morning shift operators entered the control room the next day, they quickly discovered the problems, but it was too late. Pipes in the system had burst, sending radioactive steam and water into the air and into two buildings. John Gofman and Arthur Tamplin, two radiation health experts, estimate that the radiation released from the accident will cause at least 300 and possibly as many as 900 additional fatal cases of cancer in the residents living near the troubled reactor, although other "experts" (especially in the nuclear industry) contest these projections, saying the accident will have an unnoticeable effect. Whatever the outcome, the 1979 accident at Three Mile Island will cost several billion dollars to clean up.

Late in April of 1986, another nuclear power plant went amok. This one, in the Soviet Union, was far more severe (Figure 6–23). In the wee hours of the

FIGURE 6–23   Disaster at Chernobyl Nuclear Power Plant   This costly accident was caused primarily by operator error. What price do we pay to have industries running 24 hours per day to keep production levels high?

morning, two engineers were testing the reactor. They deactivated key safety systems in violation of standard operational protocol. This single error in judgment (possibly due to fatigue) led to the largest and most costly nuclear accident in the history of the world. Steam built up inside the reactor and blew the roof off the containment building. A thick cloud of radiation rose skyward and then spread throughout Europe and the world. While workers battled to cover the molten radioactive core that spewed radiation into the sky, the whole world watched in horror.

The Chernobyl disaster, like the accident at Three Mile Island, may have been the result of workers operating at a time unsuitable for clear thinking. One has to wonder how many plane crashes, auto accidents, and acts of medical malpractice can be traced to judgment errors resulting from our insistence on working against the inherent body rhythms.

Thanks to studies of biological rhythms, researchers are finding ways to reset the biological clocks, which could help lessen the misery and suffering of shift workers and could increase the performance of the graveyard shift workers. For instance, one simple measure is to put shift workers on three-week cycles to give their clocks time to adjust. And instead of shifting workers from daytime to a graveyard shifts, transfer them forward, rather than backward (for example, from a daytime to a nighttime shift). It's a far easier adjustment. Bright lights can also be used to reset the biological clock as well. It's a small price to pay for a healthy workforce and a safer society.

SUMMARY

THE BODY HUMAN: FROM CELLS TO ORGAN SYSTEMS

1. The basic structural unit of the human body is the cell. Cells and extracellular material form tissues, and tissues combine to form organs.
2. Four primary tissues are found in the human body: epithelial tissue, connective tissue, muscle tissue, and nervous tissue. Most of these tissues have two or more subtypes.
3. Each organ contains all four primary tissues in varying proportions.
4. Epithelial tissues consist of two types: membranous epithelia, which form coverings or linings of organs, and glandular epithelia, which form exocrine glands and endocrine glands.
5. Connective tissues bind other tissues together, provide protection, and support body structures. All connective

tissues consist of two basic components: cells and extracellular fibers.

6. Two types of connective tissue are found in the body: connective tissue proper (tendons, ligaments, and loose connective tissue packing material) and specialized connective tissue (bone, cartilage, and blood).

7. Cartilage consists of specialized cells embedded in a matrix of extracellular fibers and extracellular material. The cells are supplied with nutrients from blood vessels in the periphery of the cartilage, which is why damaged cartilage is repaired slowly.

8. Bone is a dynamic tissue that provides internal support, protects organs such as the brain, and helps regulate blood calcium levels.

9. Bone consists of bone cells (osteocytes) and a calcified cartilage matrix. Two types of bone tissue exist: spongy and compact.

10. Osteoclasts play a major role in reshaping bone to meet changing demands of the body and in releasing calcium to help maintain blood calcium levels.

11. Blood is another form of specialized connective tissue that consists of numerous blood cells and platelets and an extracellular fluid, called plasma.

12. Muscle is an excitable tissue that contracts when stimulated. Three types of muscle tissue are found in the human body: cardiac, skeletal, and smooth muscle.

13. Cardiac muscle is found in the heart and is involuntary. Skeletal muscle, for the most part, is under voluntary control and forms the muscles that attach to bones. Smooth muscle is involuntary and forms sheets of varying thickness in the walls of organs and blood vessels.

14. Nervous tissue is the fourth primary tissue. It consists of two types of cells: conducting and nonconducting (supportive). The conducting cells, called neurons, transmit impulses from one region of the body to another. The nonconducting cells are a type of nervous system connective tissue.

15. Tissues combine to form organs in which specialized functions are carried out. Most organs are part of an organ system, a group of organs that cooperate to carry out some complex function.

PRINCIPLES OF HOMEOSTASIS

16. The nervous and endocrine systems coordinate the functions of organ systems, helping the body achieve homeostasis, or internal constancy. Homeostasis also occurs at the level of the cell and the environment.

17. All homeostatic systems maintain constancy by balancing inputs and outputs. Homeostatic systems do not maintain absolute constancy and can be upset by an alteration in input and output. Imbalances can have serious effects on an individual's health.

18. Homeostasis is largely a reflex, a series of events that begin with a stimulus and end with a response. Reflexes are involuntary, occurring without conscious control.

19. Homeostatic reflexes occur at all levels of biological organization and involve two main mechanisms: nervous and chemical.

20. Nervous and endocrine mechanisms generally occur over considerable distances. In some cases, control is exerted locally through the extracellular fluid.

BIOLOGICAL RHYTHMS

21. Many physiological processes undergo definite rhythmic change. These natural rhythms may take place over a 24-hour period, or over much longer or shorter periods.

22. The brain controls many biological cycles. A clump of nerve cells in the base of the brain in the hypothalamus is thought to play a major role in coordinating several key functions and several other control centers. It is, therefore, sometimes called the "master clock."

ENVIRONMENT AND HEALTH: TINKERING WITH OUR BIOLOGICAL CLOCKS

23. The hectic pace of modern life, the weird work schedules many workers follow, and the stressful environments in which we live can upset internal rhythms with sometimes disastrous consequences.

24. The greatest disrupter of our natural circadian rhythms is the variable work schedule, which is surprisingly common in industrialized nations.

25. Workers on alternating shifts may suffer more ulcers, insomnia, irritability, depression, and tension than workers on regular shifts. Making matters worse, tired, irritable workers whose judgment is impaired by fatigue pose a threat to society.

26. Researchers are finding ways to reset the biological clock, which could help lessen the misery and suffering of shift workers and could increase the performance of the graveyard shift workers.

◨ EXERCISING YOUR CRITICAL THINKING SKILLS

Health care costs are skyrocketing and many critics are arguing that our health care dollars are being misused. In particular, they argue that expensive procedures, such as organ transplants, costing $100,000 or more each, are draining dollars that could be invested in preventive medicine, such as prenatal care. Prenatal care is medical care and advice given to pregnant women on nutrition and other matters that are crucial to the development of a healthy fetus. What is your opinion on this matter? Should Americans spend more on prevention and less on dramatic life-saving measures, such as organ transplants? After you have given your opinion, put yourself in the position of a person needing a new kidney. How does this affect your position?

TEST OF TERMS

1. The first three cell types to emerge during embryonic development are _____ , _____ , and _____ .
2. All tissues are made of cells and a variable amount of _____ _____ .
3. The four primary tissues are _____ , _____ , _____ , and _____ .
4. Glands that secrete their products into ducts that, in turn, carry the products to some other part of the body are called _____ glands.
5. The outer layer of the skin is called the _____ ; it is a type of _____ tissue.
6. Tendons are a form of _____ tissue.
7. The cell that produces collagen in tendons and other similar primary tissues is the _____ _____ .
8. The soft tissue forming most of the nose is made of _____ _____ .
9. The intervertebral disk contains a ring of _____ that holds the soft, cushiony center of the disk in place.
10. The blood vessels of bone are found in tunnels called _____ canals. Nutrients flow out of these canals through even smaller tunnels in the bony matrix called _____ , providing bone cells, or _____ , with nourishment.
11. The dense outer layer of most bones is called _____ bone.
12. The fluid component of blood is also known as _____ .
13. The blood contains two components that do not have nuclei or organelles; they are the _____ and the _____ _____ .
14. The involuntary muscle found in the lining of the small intestine and stomach is called _____ muscle. It contains contractile proteins, chiefly, _____ and _____ , which are also present in other muscle types.
15. A nerve cell or _____ is a specialized cell that generates and transmits bioelectric impulses from one part of the body to the other.
16. A group of organs that cooperate to perform some key function is called an _____ _____ .
17. All homeostatic systems maintain chemical and physical parameters within a narrow range, called the _____ .
18. Endocrine and neural _____ are a series of events that begin with a stimulus and end with a response and are crucial for maintaining homeostasis.
19. The process of making sense of various chemical and nervous inputs is called _____ .
20. Organs and glands that carry out the instructions of the brain and spinal cord are called _____ .
21. A chemical released into the bloodstream that travels to distant sites where it elicits some response is called a(n) _____ ; a chemical that acts on cells close to its point of origin is called a(n) _____ .
22. Body temperature and some hormone levels vary over a 24-hour period; these and other similar cycles are called _____ rhythms.
23. A clump of nerve cells called the _____ nucleus in the hypothalamus of the brain is thought to be the master clock, controlling several key internal body rhythms and several other clocks.
24. Travellers suffer from _____ _____ because their biological clocks are out of synchrony with the 24-hour day/night cycle.

Answers to the Test of Terms are located in Appendix B.

TEST OF CONCEPTS

1. Define the following terms: tissues, extracellular material, organs, and organ systems.
2. List the four primary tissues and their subtypes.
3. Describe the similarities and differences in the embryonic origin of endocrine and exocrine glands. Explain why the two secrete differently.
4. From your study of biology, discuss some examples where structure reflects function.
5. Describe the two types of connective tissue and how they function.
6. Why do cartilage injuries repair so slowly? Bone repairs much more easily than cartilage. Why would you think this is true?
7. Why is bone part of a homeostatic system?

8. Describe the chief differences between cardiac, skeletal, and smooth muscle.
9. What is an organ system? List some examples.
10. Define homeostasis and describe the major principles of homeostasis presented in this chapter. Use an example to illustrate your points.
11. How do storage depots enter into the input/output concept of homeostasis?
12. Homeostatic mechanisms are largely reflexes, involving chemical or nervous impulses. Describe each type of reflex, giving an example. You may find it helpful to include drawings of the systems to help explain these reflexes.
13. Are biological rhythms an exception to the principle of homeostasis?
14. Describe the biological clock and explain how it is synchronized with the 24-hour day/night cycle.
15. How does shift work upset the biological clock? How can these problems be mitigated?

SUGGESTED READINGS

Coleman, R. 1986. *Wide awake at 3:00 AM.* New York: W. H. Freeman. An account of sleep/wake cycles.

Dworetzky, J. P. 1988. *Psychology.* 3d Ed. St. Paul, Minn.: West. Chapter 5 presents a brief, but well-written account of circadian rhythms and the problems arising from shift work.

Long, P. 1989. Fat chance. *Hippocrates* 3(5): 38–47. This article helps you identify if you are overweight and discusses ways of losing weight.

Montoye, H. J., J. L. Christian, F. J. Nagle, and S. M. Levin. 1988. *Living fit.* Menlo Park, Calif.: Benjamin/Cummings. Chapters 7 and 12 deal with stress and cardiovascular disease and are excellent reading for students wanting to further their knowledge.

Restak, Richard. 1984. *The brain.* New York: Bantam. See Chapter 3 for a more detailed account of the research on natural body rhythms.

7 Nutrition and Digestion

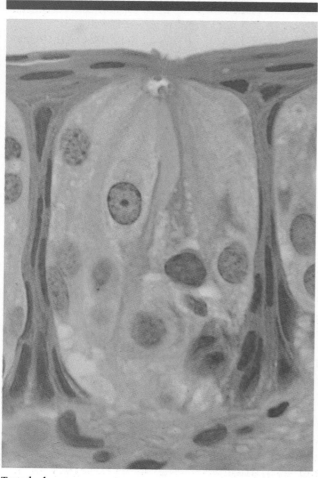

Taste bud.

I t must be a law of human nature. Ask almost any couple and they will tell you: she lies shivering under the covers on a cold winter night while he bakes. Out on a hike in winter, he wears a light jacket and is warm, while she bundles up in stocking cap, down coat, and gloves to keep warm. What causes this marked difference in many men and women?

New research suggests that part of the answer lies in iron—not pumping iron, but dietary iron. Many American women do not consume enough iron to offset losses that occur during **menstruation**, the monthly discharge of blood and tissue from the lining of the uterus (Chapter 16). Iron deficiencies in women may reduce internal heat production.[1]

John Beard, a researcher at Pennsylvania State University, recently reported on a study that supports this conclusion. In his research, Beard compared two groups of women, one with low levels of iron in the blood and the other with normal iron levels. Beard found that body temperature drops more quickly in iron-deficient women exposed to cold than those with normal iron levels. He also found that iron-deficient women generated 13% less heat. Furthermore, iron-deficient women warm themselves less efficiently than women with normal levels. For reasons not yet well understood, iron-deficient women rely more on

[1]Critical thinking suggests that the difference in body temperature between men and women may also result from other factors, such as body mass and surface area.

162

glucose than fat for energy production. Ounce for ounce, glucose provides half the energy of fats. After 12 weeks of iron supplements, though, women who were formerly iron-deficient responded normally to cold.

At least two hypotheses can explain these results. The first hypothesis is that iron deficiencies may reduce the amount of oxygen available for metabolism, thus lowering heat production. Iron is a vital component of the hemoglobin molecule (a protein) in the red blood cells (RBCs). RBCs transport oxygen from the lungs to body tissues. Iron in the hemoglobin molecule binds to oxygen, which is required for cellular respiration (Chapter 3). Cellular respiration, in turn, produces ATP and heat. Reduced oxygen transport could result in reduced body heat.

The second hypothesis is based on the role of iron in energy production. Iron is also a vital component of the electron transport proteins found in the mitochondria. As noted in Chapter 3, these proteins are part of the electron transport system, which produces most of the ATP in the cell. If iron levels are low, electron transport and energy production could be impaired. ATP production releases heat. Iron deficiencies, therefore, could reduce internal heat.

The research on iron may help explain not only why many women experience cold feet in bed at night, but also why many women who take iron supplements during pregnancy report an improvement in heat production. This example also illustrates how important a healthy, balanced diet is to normal physiological function. This chapter introduces other examples of the dependence of our health on our diet as it outlines the major nutritional requirements of humans and describes the process of digestion.

A PRIMER ON HUMAN NUTRITION

Some people eat to live and a great many others seem to live just to eat. No matter what your orientation, food probably occupies a central part of your day-to-day activities. Without food, you could survive six to eight weeks, but much of that time you would be too weak to move.

In our study of nutrition, however, we will be concerned not just with survival, but with living and eating well—thriving. The Greek philosopher Socrates put it best, "Not life, but a good life, is to be chiefly valued." Despite the increased emphasis on nutrition, most Americans still pay little attention to their diet. To perform our very best, though, we must eat well,

TABLE 7–1	Macronutrients and Micronutrients
NUTRIENTS	FOUND IN
Macronutrients	
Water	All drinks and many foods
Amino acids and proteins	Cheese, meats, eggs
Lipids	Cheeses, meats, eggs
Carbohydrates	Breads, pastas, cereals
Micronutrients	
Vitamins	Many vegetables
Minerals	Many vegetables

acquiring a full complement of the nutrients required by our cells, tissues, and organs each day.

Humans require two basic types of nutrients: macronutrients and micronutrients (Table 7–1). This section describes these nutrients and notes their importance in human physiology.

Macronutrients

Macronutrients are required in relatively large quantities and include four substances: water, proteins, lipids, and carbohydrates.

Water. Water is one of the most important of all nutrients. Without it, a person can only survive one or two days. Despite its importance, water does not usually show up on the nutrition charts. In part, that is because water is supplied in so many different ways. For example, water is in the liquids we drink and in virtually all of the solid foods we eat; water is even produced internally during cellular metabolism (Chapter 3).

Evolution has provided humans (and other organisms) with a variety of mechanisms to control the amount of water in the body (Chapter 11). Adequate water intake is important for several reasons. First, water participates in many chemical reactions in body cells. Thus, a decrease in water levels may impair metabolism, including energy production. According to some studies, athletic performance may drop significantly when water levels fall even slightly. Maintaining an adequate water volume is also important because water helps maintain normal body temperature. A decline in body water decreases the volume of the blood and the volume of the extracellular fluid. A decrease in body fluids can cause body temperature to rise, because the heat normally produced by the cells is being absorbed by a smaller volume of water. A rise in body temperature may impair cellular function and can lead to death. Maintaining an adequate water intake also helps maintain normal concentrations of nutrients and toxic waste products in the blood and extracellular fluid. Urine becomes more concentrated when a person fails to drink enough liquid; the increased concentration of wastes in the urine increases the likelihood of developing kidney stones. Kidney stones are deposits of calcium and other materials in the kidney that can block the flow of urine, causing extensive damage to this organ (Chapter 11).

Sources of Energy. The body also requires a continuous supply of energy. Energy is provided by three macronutrients: carbohydrates, lipids, and, to a lesser degree, protein. Cells require energy to carry out thousands of functions needed to maintain homeostasis, to grow, and to divide. All cells need energy to transport molecules across their membranes. Surpris-

FIGURE 7–1 Glucose Balance Glucose levels in the blood result from a balance of input and output.

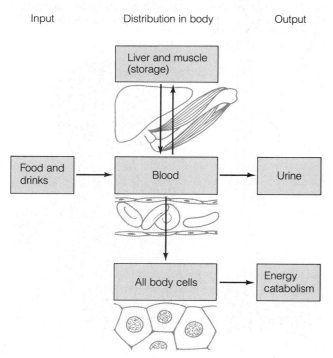

ingly, 70 to 80% of the total energy required by humans goes to perform basic functions—metabolism, food digestion, absorption, and so on. The remaining energy is used to power body movements, such as walking, talking, and running. For more active people, though, these percentages shift considerably. A basketball player, for example, uses proportionately more energy for vigorous muscular activity than a sedentary office worker.

At rest, the body relies on nearly equal amounts of fat and carbohydrate to supply basic body needs. Contrary to popular myth, proteins provide energy only under certain circumstances, described later in the chapter. As you may recall from Chapter 3, carbohydrates are catabolized during cellular respiration. They provide about 4 calories per gram. Lipids (fats) are also catabolized in cells to provide energy. The by-products of fat catabolism enter the biochemical pathways of cellular respiration, yielding—gram-for-gram—twice as much ATP as carbohydrates.

Carbohydrates come in many forms and from many different foods (Table 7–1). Starch from plants (grains and vegetables), for instance, contributes about 45% of the total energy requirements of humans. Much of the rest of our energy comes from glycogen (a polysaccharide) in meat, simple sugars in fruits, and lactose (a disaccharide) in milk.

Glycogen and starch are broken down into glucose molecules in the digestive system and are absorbed into the bloodstream for distribution to body cells. Most of this glucose is stored as glycogen in muscle and the liver for later use. Some is catabolized immediately by the cells to produce energy; and a small amount is excreted in the urine.

Figure 7–1 illustrates the homeostatic balance of body carbohydrates, showing the inputs and outputs that determine the level of blood glucose. Since most of us eat only a few times a day, body stores of glucose play a crucial role in maintaining homeostasis. Between meals and during exercise, glycogen is broken down to provide glucose. When muscular activity is prolonged, however, as in the case of a long-distance runner, glycogen stored in the liver and muscle may be depleted. Eighteen to twenty miles into a marathon, runners become exhausted and weak, a sensation they describe as "hitting the wall," as a result of muscle glycogen depletion.

A knowledge of energy metabolism is more than an academic exercise. Consider an example. Determined to lose a few extra pounds, two friends of mine spent a year working out in a health club. They exercised hard and often—running, swimming, and bicycling three or four nights a week. At the end of the year, though, they both weighed about the same as they had on the day they began their program. Disappointed, they turned to an exercise physiologist for help. She explained an often-overlooked fact about energy metabolism: heavy exercise depletes muscle oxygen levels, and when muscles become anaerobic, they use carbohydrates—mostly glucose—to generate energy. Body fat stores are relatively untouched. Try as you may, you generally do not lose weight if your exercise program is so vigorous that it depletes muscle oxygen levels.

The exercise physiologist advised my friends to tone down their program—to work out often and long, but not quite so hard. The object? To avoid depleting muscle oxygen during exercise. This is called **aerobic exercise**. It helps the heart and cardiovascular system and also helps people "burn off" weight because it relies primarily on the body's fat reserves. Soon after the couple shifted to aerobic workouts, the weight started disappearing.

Amino Acids and Protein. *Although many people think of protein as a source of energy, it is really only important under two conditions: either when dietary intake of carbohydrates and fats is severely restricted or when protein intake far exceeds demands.* Dietary deficiencies occur in millions of children throughout the world. The arms and legs of children deprived of protein are often thin and emaciated, because their muscle cells break down protein in an effort to provide energy (Figure 7–2). Their abdomens are bloated because of fluid accumulation, caused by the increased levels of protein in their blood.

On the other end of the spectrum are the well-fed populations of the world. Many Americans eat far more food than they need. On average, U.S. protein intake is twice the daily requirement. Amino acids released from the surplus dietary protein are broken down in the body to generate energy and to make fat.

Protein in the food we eat is used to synthesize enzymes and structural proteins. Dietary proteins cannot be used directly by cells, but must first be broken down into their constituent amino acids in the digestive tract. Amino acids liberated during this process are absorbed into the bloodstream and distributed throughout the body.

In cells, some amino acids are used to make body protein directly. Others are converted into different amino acids, then used in protein synthesis. Proteins are synthesized from 20 different amino acids. Humans can make 11 of the 20 different types of amino

acids needed to produce protein. The remaining nine amino acids must be provided by the diet (that is, they are not synthesized from other amino acids) and are called **essential amino acids**. A deficiency of even one of the essential amino acids can cause severe physiological problems. Nutritionists, therefore, recommend a diet containing many different protein sources. In this way, individuals can be assured they are getting all of the amino acids they need. Milk and eggs provide the best assortment of essential amino acids. These proteins are, therefore, said to be **complete**. Unfortunately, many American adults lack the enzyme (lactase) needed to digest milk sugar (lactose), and many people avoid eggs to reduce cholesterol levels in their blood (Health Note 7–1). These individuals get their proteins from other sources. The next best sources of proteins are meats, fish, poultry, cheese, and soybeans. Each of these has low levels of several (but not all) essential amino acids. To avoid deficiencies, nutritionists recommend combining these protein sources. The lowest quality proteins are in legumes (peas, beans), nuts, seeds, grains, and vegetables. Each of these foods contains low levels of only one or two essential amino acids. Care must be exercised to prevent deficiencies. Vegetarians, for instance, can acquire all of the amino acids they need by

FIGURE 7–2 A Protein-Deficient Child This child suffers from severe protein deficiency. His arms and legs are emaciated because muscle protein has been broken down to supply energy. His belly is swollen because of a buildup of fluid in the abdomen.

FIGURE 7–3 Complementary Protein Sources By combining protein sources, a vegetarian can be assured of getting all of the amino acids needed. Legumes can be combined with foods made from grains or nuts and seeds.

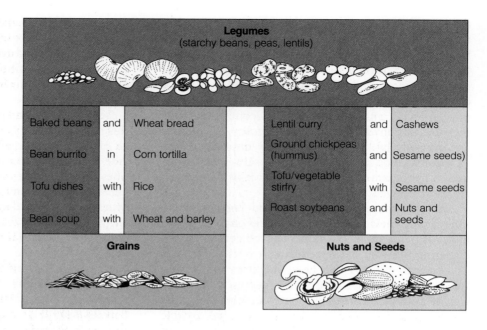

Legumes (starchy beans, peas, lentils)					
Baked beans	and	Wheat bread	Lentil curry	and	Cashews
Bean burrito	in	Corn tortilla	Ground chickpeas (hummus)	and	Sesame seeds)
Tofu dishes	with	Rice	Tofu/vegetable stirfry	with	Sesame seeds
Bean soup	with	Wheat and barley	Roast soybeans	and	Nuts and seeds
Grains			**Nuts and Seeds**		

combining protein sources. Figure 7–3 shows how this is done.

Lipids. Although lipids provide cellular energy, they serve other functions as well. The layer of fat beneath the skin in humans, for instance, helps insulate the body from heat loss. Certain lipids are used to synthesize hormones, and lipids are a principal component of the plasma membrane. Lipids in the intestine increase the intestinal uptake of the fat-soluble vitamins (A, D, E, and K).

Lipids are consumed in excess by most Americans. On average, fats provide about 45% of the dietary caloric intake, but to lower the risk of heart attack, fat intake should only be about 30%. (For more on the effects of lipids on the circulatory system, see Health Note 7–1.)

Carbohydrates. Carbohydrates, like lipids, are a principal source of energy. However, carbohydrates are important for other reasons as well. Dietary **fiber**, for example, is nondigestible polysaccharide (such as cellulose) found in fruits, vegetables, and grains. As noted in Chapter 2, cellulose cannot be digested because the body lacks the enzymes needed to break the covalent bond joining the monosaccharide units (glucose) in the molecule. In humans, fiber passes through the intestine largely unaffected by enzymes or stomach acidity.

Fiber exists in two basic forms: water soluble and water insoluble. Water-soluble fibers are gummy polysaccharides in fruits, vegetables, and some grains, including apples, bananas, carrots, barley, and oats. In contrast, water-insoluble fibers are rigid cellulose molecules in celery, cereals, wheat products, and brown rice. Some foods, such as green beans and green peas, contain a mixture of both types.

Water-insoluble fiber increases the water content of the **feces**, the semisolid waste produced by the large intestine. This makes the feces softer and facilitates their transport through the large intestine. Increasing the water content also reduces constipation and helps prevent pressure from building up in the large intestine. In some cases, pressure from constipation causes small pouches, or diverticulae, to form in the wall of the large intestine, resulting in a condition known as **diverticulitis**. The pouches in the intestinal wall often become inflamed. Bacteria may enter the bloodstream, causing high fever. Occasionally, the diverticulae burst, releasing feces into the abdominal cavity. Since feces contain billions of bacteria, their release into the abdominal cavity results in a massive infection that is difficult to treat and may result in death.

Diverticulitis, although not common, has increased substantially in America since the introduction of white flour. Flour is made from wheat, a grain grown in many states. When wheat grain is ground, it produces whole wheat flour, a powdery mixture containing the entire products of the wheat grain. White flour, however, is produced from wheat whose shells have been removed. Removing the shells eliminates most, if not all, of the fiber. With the widespread use of white flour, several problems, such as diverticulitis, arose. Not surprisingly, diverticulitis is rare in areas, such as Africa, where fiber intake is higher.

Water-insoluble fiber is thought to reduce colon cancer, which afflicts about 3% of the U.S. population. In rural Africa, colon cancer is practically unheard of. Although critical thinking skills suggest that other differences may also be responsible for the difference, this and other evidence led researchers to recommend an increase in the amount of fiber we eat. How does fiber protect us from colon cancer?

A recent study showed that some bacteria in the large intestine produce a potent chemical mutagen that may cause cancer. Researchers suggest that, by accelerating the transport of wastes through the intestine, fiber may decrease colon cancer by reducing the formation of the mutagen or by reducing the time in which the intestinal cells are exposed to it.

Water-soluble fiber is also beneficial to our health; several studies suggest that it helps lower blood cholesterol. Some water-soluble fibers act as sponges, absorbing cholesterol inside the digestive tract and preventing it from being absorbed into the bloodstream. Others change the pH of the intestine, making cholesterol insoluble and difficult to absorb. (The chemical action of water-soluble fibers is discussed in more detail in the section on the liver.)

Micronutrients

Micronutrients are substances needed in small quantity and include two broad groups: vitamins and minerals.

Vitamins. Vitamins are a diverse group of organic compounds that occur in foods in small quantities and play an important role in many metabolic reactions. Vitamins serve as **enzyme cofactors**, chemicals required by a number of enzymes for proper functioning. Since *vitamins are recycled many times during these reactions, they are only needed in small amounts.* One gram of vitamin B-12, for example, would supply over 300,000 people for a single day.

Vitamins are absorbed in the digestive tract without being broken down and generally cannot be synthesized in the body cells in sufficient amounts to satisfy cellular demands, making dietary intake essential. Vitamin D, for instance, is manufactured by the skin when exposed to sunlight. However, most Americans spend so much time indoors that dietary input is essential to good health. Table 7–2 lists vitamins and their functions.

Vitamins are needed in almost all cells of the body. As a result, a dietary deficiency in just one vitamin can cause wide-ranging effects in the body. Vitamins also interact with other nutrients in the dynamic balance of the body. Vitamin C, for example, increases the absorption of iron in the small intestine. Large doses of the vitamin, however, decrease copper utilization by the cells. *Maintaining good health, therefore, requires the proper balance of vitamins and other nutrients.*

Vitamins fall into two broad categories: water soluble and fat soluble. Water-soluble vitamins include vitamin C and 11 different forms of vitamin B. Water-soluble vitamins are transported in the blood plasma and are eliminated by the kidneys. Because they are water soluble, they are not stored in the body in any appreciable amount.

Water-soluble vitamins generally work in conjunction with enzymes, promoting the cellular reactions that supply energy or synthesize cellular materials. Contrary to common myth, the vitamins themselves do not provide energy.

Experts long believed that megadoses of water-soluble vitamins were harmless because these vitamins were excreted in the urine and did not accumulate in the body. New research, however, shows that this is not entirely true. *Some water-soluble vitamins, such as vitamin C, when ingested in excess, can be toxic* (see Table 7–2 for some examples).

The fat-soluble vitamins are vitamins A, D, E, and K. They perform many different functions. Vitamin A, for example, is converted to a light-sensitive pigment present in receptor cells of the retina, the light-sensitive layer of the eye (Chapter 13). These pigments play an important role in vision. Like other vitamins, vitamin A exists in several forms, serving many different functions. One member of the vitamin A group, for example, removes harmful oxidants from the body. Retin-A, a derivative of vitamin A, is used by dermatologists to treat acne and reduce wrinkling that occurs with aging and exposure to sunlight (Chapter 17).

Unlike water-soluble vitamins, the fat-soluble vitamins are stored in body fat and accumulate in the fat reserves. The accumulation of fat-soluble vitamins

Lowering Your Cholesterol

Let there be no question about it: diseases of the heart and arteries are leading causes of death in the United States.

Atherosclerosis, the buildup of plaque on artery walls, and the problems it creates are responsible for nearly two of every five deaths in the United States each year. Thanks to improvements in medical care and diet, the death rate from atherosclerosis has been falling steadily in recent years, but it is still a major concern. New research, in fact, shows that atherosclerotic plaques are present even in children. Researchers believe atherosclerotic plaques begin to form after minor injuries to the lining of blood vessels. High blood pressure, they think, may damage the lining, causing platelets and cholesterol in the blood to stick to the injured site. The blood vessel responds by producing cells that cover the fatty deposit. This thickens the wall of the artery, reducing blood flow. Additional cholesterol is then deposited in the thickened wall, forming a larger and larger obstruction to blood flow. Cholesterol deposits impair the flow of blood in the heart and other organs, cutting off oxygen to vital tissues. Blood clots may form in the restricted sections of arteries, further obstructing blood flow. When the oxygen supply to the heart is disrupted, cardiac muscle cells can die, resulting in heart attacks and death. Blood clots originating in other parts of the body may also lodge in the diseased vessels, obstructing blood flow.

Oxygen deprivation can weaken the heart, impairing the heart's ability to pump blood. When the oxygen supply to the heart is restricted, the result is a type of heart attack known as a **myocardial infarction**. Victims of a myocardial infarction feel pain in the center of the chest and down the left arm. If the deprived area is extensive, the heart may cease functioning altogether.

Atherosclerotic plaque also impairs the flow of blood to the brain, which can cause cell death. Victims often lose the ability to speak or to move limbs. If the damage is severe enough, they may die. Thanks to advances in medical treatment, however, many victims can be saved. Over time, they can recover many lost functions as other parts of the brain take over the functions of destroyed regions. Atherosclerosis also affects other organs, such as the kidneys.

Atherosclerosis and cardiovascular disease are the result of nearly 40 factors, some more important than others. Several of these risk factors, such as old age and sex (being male), cannot be changed. Other factors are controllable. These include high blood pressure, high blood cholesterol, smoking, inactivity, and excessive food intake. Of all the risk factors, three emerge as the primary contributors to cardiovascular disease: elevated blood cholesterol, smoking, and high blood pressure.

Consider cholesterol. It may surprise you, but for most of us the cholesterol in our blood is produced by the liver. The liver synthesizes and releases about 700 milligrams of cholesterol per day. Only about 225 milligrams of cholesterol are derived from the food we eat each day. Normally, the concentration of cholesterol in the blood is constant. If dietary input falls, the liver increases its output. If the amount of cholesterol in the diet rises, the liver reduces production. So what's all the fuss about cholesterol in a person's diet?

Even though the liver regulates cholesterol levels, it cannot work fast enough. It may simply be unable to absorb, use, and dispose of cholesterol quickly enough. Consequently, excess cholesterol circulates in the blood after a meal and is deposited in arteries.

Cholesterol is carried in the bloodstream bound to protein. These complexes of protein and lipid fall into two groups: **high-density lipoproteins (HDLs)** and **low-density lipoproteins (LDLs)**. LDLs and HDLs function very differently. LDLs, for example, transport cholesterol from the liver to body tissues. In contrast, HDLs are scavengers, picking up excess cholesterol and transporting it to the liver where it is removed from the blood and excreted in the bile. Research shows that the ratio of HDL to LDL is an accurate predictor of cardiovascular disease. The higher the ratio, the lower the risk of cardiovascular disease.

High cholesterol level (or hypercholesterolemia) tends to run in families. Thus, if a parent has died of a heart attack or suffers from this genetic disease, his or her offspring are more likely to have high cholesterol levels than others.

For many years, physicians have advised cutting back on high-cholesterol foods, especially eggs and red meats, to reduce cholesterol levels. However, a reduction in dietary cholesterol in one individual

Testing Blood Cholesterol Level Blood cholesterol tests can be easily performed in a doctor's office, with results available almost immediately.

may result in very little decline in total blood cholesterol, but in another a reduction may result in a much larger drop. The difference in response can be attributed to exercise, genetics, initial cholesterol levels, and age.

Despite the differences in response, the American Heart Association recommends (1) limiting dietary fat to less than 30% of the total caloric intake, (2) limiting dietary cholesterol to 300 milligrams/day, and (3) acquiring 50% or more of one's calories from carbohydrates, especially polysaccharides. Reductions in saturated fatty acids (animal fats) can also help lower cholesterol levels for reasons not fully understood. You can cut back on saturated

fat by using margarine instead of butter, reducing your consumption of red meat, and trimming the fat off all meats before cooking. You can also increase your consumption of fruits, vegetables, and grains, letting these low-fat foods displace some of the fatty foods you might otherwise eat.

New and still controversial research also indicates that a diet rich in fish oils can help reduce blood cholesterol. Fish oils contain polyunsaturated fatty acids called omega-3 fatty acids. These fatty acids stimulate the release of prostaglandin, which increases the flexibility of the red blood cells and reduces their stickiness, which is essential for blood clotting.

Research on mice shows that a diet rich in omega-3 fatty acids doubles life span. One of the conclusions of the study, though, is that omega-3 fatty acids, which are extremely susceptible to oxidation, are only effective if oxidation can be prevented. Mincing fish prior to cooking increases oxidation and lowers the levels of omega-3 fatty acids.

Cholesterol levels can be lowered with drugs, diet, and exercise. Research spanning several decades shows that lower cholesterol levels translate into a decline in cardiovascular disease. Unfortunately, experts disagree on several key issues. One is exactly who will benefit from a reduction in cholesterol. Some researchers say that only high-risk people with cholesterol levels over 250 milligrams per 100 milliliters of blood should take steps to cut back. Since two-thirds of the American adult population have blood cholesterol levels over 200 mg/100 ml, some experts believe that the entire adult population should take steps to reduce cholesterol.

High cholesterol is also surprisingly common in children, leading many health experts to believe that steps should be taken to prevent problems later in life. In children under the age of two, however, diets should not be restricted. A diet that is too restrictive may actually impair physical growth and development. What children need is a well-balanced diet, low in fats, with sufficient calories from other sources. If nothing else, it could help create the good eating habits necessary for good health throughout adult life.

TABLE 7-2 Important Information on Vitamins

VITAMIN	MAJOR DIETARY SOURCES	MAJOR FUNCTIONS	SIGNS OF SEVERE, PROLONGED DEFICIENCY	SIGNS OF EXTREME EXCESS
Fat soluble				
A	Fat-containing and fortified dairy products; liver; provitamin carotene in orange and deep green produce	Component of rhodopsin; still under intense study	Keratinization of epithelial tissues including the cornea of the eye (xerophthalmia); night blindness; dry, scaling skin; poor immune response	From preformed vitamin A: damage to liver, kidney, bone; headache, irritability, vomiting, hair loss, blurred vision. From carotene: yellowed skin.
D	Fortified and full-fat dairy products; egg yolk	Promotes absorption and use of calcium and phosphorus	Rickets (bone deformities) in children; osteomalacia (bone softening) in adults	Gastrointestinal upset; cerebral, CV, kidney damage; lethargy
E	Vegetable oils and their products; nuts, seeds; present at low levels in other foods	Antioxidant to prevent plasma membrane damage; still under intense study	Possible anemia	Debatable; perhaps fatal in premature infants given intravenous infusion
K	Green vegetables; tea, meats	Aids in formation of certain proteins, especially those for blood clotting	Severe bleeding on injury: internal hemorrhage	Liver damage and anemia from high doses of the synthetic form menadione
Water soluble				
Thiamin (B-1)	Pork, legumes, peanuts, enriched or whole-grain products	Coenzyme used in energy metabolism	Beriberi (nerve changes, sometimes edema, heart failure)	?
Riboflavin (B-2)	Dairy products, meats, eggs, enriched grain products, green leafy vegetables	Coenzyme used in energy metabolism	Skin lesions	?

can have many adverse effects (Table 7-2). An excess of vitamin D, for example, can cause hair loss; nausea; joint, bone, and muscle pain; and even diarrhea. Large doses taken during pregnancy can cause birth defects.

Vitamin excess is a condition largely encountered in the developed countries. Each year, approximately 4000 Americans are treated for vitamin supplement poisoning. To avoid problems from excess vitamins, nutritionists recommend eating a balanced diet that provides all of the vitamins the body needs, rather than taking vitamin pills, especially megadoses.

Dietary deficiencies of vitamins, like dietary ex-

TABLE 7–2 Important Information on Vitamins (continued)

VITAMIN	MAJOR DIETARY SOURCES	MAJOR FUNCTIONS	SIGNS OF SEVERE, PROLONGED DEFICIENCY	SIGNS OF EXTREME EXCESS
Niacin	Nuts, meats; provitamin tryptophan in most proteins	Coenzyme used in energy metabolism	Pellagra (which may be multiple vitamin deficiencies)	Flushing of face, neck, hands; liver damage
B-6	High protein foods in general, bananas, some vegetables	Coenzyme used in amino acid metabolism	Nervous and muscular disorders	Unstable gait, numb feet, poor hand coordination, abnormal brain function
Folacin	Green vegetables, orange juice, nuts, legumes, grain products	Coenzyme used in DNA and RNA metabolism; single carbon utilization	Megaloblastic anemia (large, immature red blood cells); gastrointestinal disturbances	Masks vitamin B-12 deficiency
B-12	Animal products	Coenzyme used in DNA and RNA metabolism; single carbon utilization	Megaloblastic anemia; pernicious anemia when due to inadequate intrinsic factor; nervous system damage	?
Pantothenic acid	Widely distributed in foods	Coenzyme used in energy metabolism	Fatigue, sleep disturbances, nausea, poor coordination	?
Biotin	Widely distributed in foods	Coenzyme used in energy metabolism	Dermatitis, depression, muscular pain	?
C	Fruits and vegetables, especially broccoli, cabbage, cantaloupe, cauliflower, citrus fruits, green pepper, strawberries	Maintains collagen; is an antioxidant; aids in detoxification; still under intense study	Scurvy (skin spots, bleeding gums, weakness); delayed wound healing; impaired immune response	Gastrointestinal upsets, confounds certain lab tests, poorer immune response

Source: From J. L. Christian and L. L. Greger, *Nutrition for Living*, 2d ed., copyright © 1988 by the Benjamin/Cummings Publishing Company. Used with permission.

cesses, can lead to problems. Vitamin deficiencies, for example, can reduce immunity, making people more susceptible to infectious disease. Most people afflicted by vitamin deficiencies are those who do not get enough to eat. One of every five people living in the nonindustrialized nations of the world goes to bed hungry; most of these people suffer from multiple deficiencies.[2] All told, about 10 million children under the age of 5 suffer from extreme malnutrition in the less-developed nations of the world. Another 90

[2]About 700–800 million people in the less-developed countries do not get enough to eat.

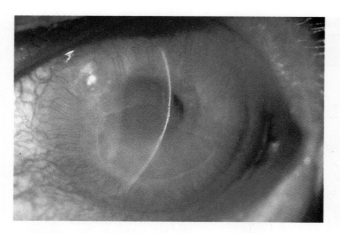

FIGURE 7–4 Vitamin A Deficiency Vitamin A deficiency causes the cornea of the eye to dry and become irritated. If the deficiency is not corrected, ulcers may form and break, causing complete blindness.

million under the age of 5 are moderately malnourished.

Deficiencies of vitamin A afflict over 100,000 children worldwide each year (Figure 7–4). If not corrected, vitamin A deficiency causes the eye to dry. Ulcers may form on the eyeball and can rupture, causing complete blindness. People suffering from vitamin deficiency typically complain of weakness and fatigue. Children with insufficient vitamin intake fail to grow.

Minerals. Humans require many minerals, such as sodium and iron, for normal body functions. These micronutrients are derived from the food we eat and the water we drink. Minerals are divided into two groups: the **major minerals** and the **trace minerals** (Table 7–3). Calcium and phosphorus, for example, are major minerals. They form part of the dense extracellular matrix of bone and are required in a much greater quantity than zinc or copper, two trace minerals that are components of some enzymes. Mineral deficiencies, like excesses, can result in many physiological problems (Table 7–3).

THE HUMAN DIGESTIVE SYSTEM

The food we eat is broken down by physical and chemical processes in the digestive system (Figure 7–5). In the mouth, for example, food is sliced, crushed, and torn by the teeth into smaller particles. In the stomach and intestines, these particles are fur-

ther digested by enzymes. In the small intestine, amino acids, monosaccharides, and other small molecules produced by enzymatic digestion are absorbed into the bloodstream for distribution to body cells.

The Mouth: Physical Breakdown of Food

The mouth is a complex structure in which food is broken down mechanically and, to a much lesser degree, chemically. The jaws and teeth perform the mechanical breakdown. The sharp teeth in front slice and cut our food, while the flatter teeth toward the back of the mouth grind food into a pulpy mass. As the food is pulverized in the mouth, saliva is added by the **salivary glands**, three sets of exocrine glands located around the oral cavity (Figure 7–6). The release of saliva is triggered by the smell, feel, taste, and—sometimes—even the thought of food.

Saliva (1) liquifies the food, making it easier to swallow, (2) kills or neutralizes some bacteria via enzymes and antibodies it contains, (3) dissolves substances so they can be tasted, and (4) begins to break down starch molecules with the aid of the enzyme **amylase**. Saliva also cleanses the teeth, washing away bacteria and food particles. Since the release of saliva is greatly reduced when we sleep, bacteria tend to accumulate on the surface of the teeth. Here they break down microscopic food particles, producing some foul-smelling chemicals that give us "dragon breath" or "morning mouth."

The bacteria that live on the teeth secrete a sticky material called **plaque**. Plaque adheres to the surface of the teeth trapping bacteria. Entrapped in their own secretions, these bacteria release small amounts of a weak acid, which dissolves the hard outer coating of our teeth, the **enamel**. Small pits or **cavities** may form on the surface of the teeth and may deepen as the acids eat into the softer layer beneath the enamel. Brushing helps remove plaque and thus helps reduce the incidence of cavities. Most toothpastes also contain small amounts of fluoride, a chemical that hardens the enamel, reducing the incidence of cavities. Small amounts of fluoride are also added to domestic drinking water. Some recent research suggesting that fluoride may be carcinogenic has forced health officials to reexamine this practice .

Salivary flow during the day is much greater and tends to keep the teeth clean. A recent study showed that chewing gum also increases the release of saliva. Thus, by chewing sugarless gum also within 5 minutes of a meal, and for at least 15 minutes, you can

TABLE 7–3   Important Information on Minerals

MINERAL	MAJOR DIETARY SOURCES	MAJOR FUNCTIONS	SIGNS OF SEVERE, PROLONGED DEFICIENCY	SIGNS OF EXTREME EXCESS
Major minerals				
Calcium	Milk, cheese, dark green vegetables, legumes	Bone and tooth formation; blood clotting; nerve transmission	Stunted growth; maybe bone loss	Depressed absorption of some other minerals
Phosphorus	Milk, cheese, meat, poultry, whole grains	Bone and tooth formation; acid-base balance; component of coenzymes	Weakness; demineralization of bone	Depressed absorption of some minerals
Magnesium	Whole grains, green leafy vegetables	Component of enzymes	Neurological disturbances	Neurological disturbances
Sodium	Salt, soy sauce, cured meats, pickles, canned soups, processed cheese	Body water balance; nerve function	Muscle cramps; reduced appetite	High blood pressure in genetically predisposed individuals
Potassium	Meats, milk, many fruits and vegetables, whole grains	Body water balance; nerve function	Muscular weakness; paralysis	Muscular weakness; cardiac arrest
Chloride	Salt, many processed foods (as for sodium)	Plays a role in acid-base balance; formation of gastric juice	Muscle cramps; reduced appetite; poor growth	Vomiting
Trace minerals				
Iron	Meats, eggs, legumes, whole grains, green leafy vegetables	Component of hemoglobin and enzymes	Iron deficiency anemia, weakness, impaired immune function	Acute: shock, death. Chronic: liver damage, cardiac failure
Iodine	Marine fish and shellfish; dairy products; iodized salt; some breads	Component of thyroid hormones	Goiter (enlarged thyroid)	Iodide goiter
Fluoride	Drinking water, tea, seafood	Maintenance of tooth (and maybe bone) structure	Higher frequency of tooth decay	Mottling of teeth; skeletal deformation

Source: Adapted from J. L. Christian and L. L. Greger, *Nutrition for Living,* 2d ed., copyright © 1988 by the Benjamin/Cummings Publishing Company. Used with permission.

also reduce the incidence of cavities. If you can't brush after every meal, chew sugarless gum.

After food is chewed it must be swallowed. The tongue plays a key role in swallowing by pushing food to the back of the oral cavity into the pharynx. The **pharynx** is a chamber that connects the oral cavity with the **esophagus,** a long muscular tube that leads to the stomach (Figure 7–5).

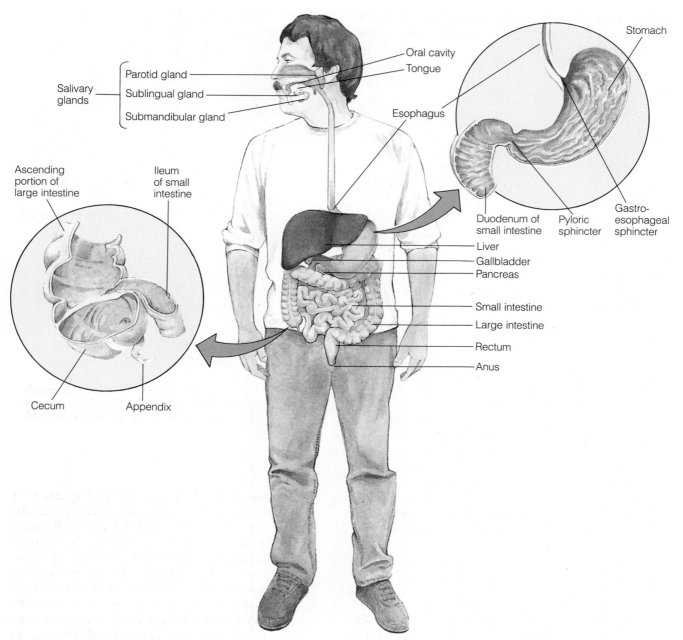

FIGURE 7–5   The Digestive System

The tongue, which also helps us form words, contains taste receptors, the **taste buds**, on its upper surface. Taste buds are stimulated by four basic flavors: sweet, sour, salty, and bitter (Figure 7–7). Various combinations of these (combined with odors we smell) give us the rich assortment of tastes (Chapter 13). Food propelled from the pharynx into the esophagus is prevented from entering the trachea, which

also opens into the pharynx, by the epiglottis. The **epiglottis** is a flap of tissue that acts like a trap door, closing off the trachea during swallowing.

Swallowing begins with a voluntary act—the tongue pushing food into the back of the oral cavity. Once food enters the pharynx, however, the process becomes automatic. Food stimulates receptors in the wall of the pharynx, which trigger the **swallowing**

reflex, an involuntary contraction of the muscles in the wall of the pharynx, which forces the food into the esophagus.

The Esophagus and Peristalsis

Involuntary contractions of the muscular wall of the esophagus propel the food to the stomach. The muscles of the esophagus contract above the swallowed food mass, squeezing it along (Figure 7–8a). This involuntary muscular action is called **peristalsis**. Peristalsis propels food (and waste) along the digestive tract. It is so powerful that you can swallow when hanging upside down.

Esophageal peristalsis sometimes proceeds in the opposite direction; this is known as **reverse peristalsis** or, more commonly, vomiting. Vomiting is a reflex action that occurs when irritants are present in the stomach. Vomiting is, therefore, a protective measure that allows the body to rid the stomach of bad food, viruses, or bacteria.

The Stomach: Liquification, Storage, and Release of Food

The stomach is an expandable, muscular organ that performs several key functions (Figure 7–9). First, the stomach stores and liquifies food. Second, it begins the chemical breakdown of some substances, and, third, it releases the highly liquified and partially digested food in timed pulses into the small intestine. The stomach, shown in Figures 7–5, lies on the left side of the abdominal cavity, partly under the protection of the rib cage.

Food enters the stomach from the esophagus. The opening to the stomach, however, is usually constricted to prevent stomach acid and food from percolating upward and irritating the esophagus, causing "heartburn." The opening to the stomach is closed off by the **gastroesophageal sphincter**, a functional valve produced by a slight thickening in the muscular wall of the stomach where the esophagus joins it.

As food enters the lower esophagus, the gastroesophageal sphincter opens to allow it to enter the stomach, then promptly closes, preventing food and acid from escaping (Figure 7–8b). Inside the stomach, food is liquified by acidic secretions of glands in the stomach wall, the **gastric glands**. The food is churned by peristalsis and converted into a liquid, called **chyme**.

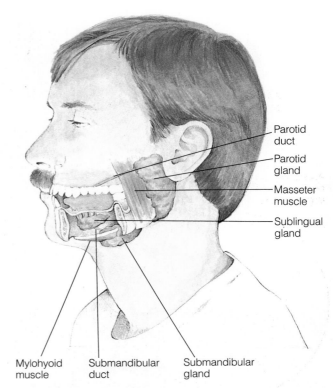

FIGURE 7–6 Salivary Glands Three salivary glands (parotid, submandibular, and sublingual) are located in and around the oral cavity and empty into the mouth via small ducts.

The stomach holds 2 to 4 liters (2 quarts to 1 gallon) of chyme and releases it gradually into the small intestine, at a rate suitable for proper digestion and absorption. Glands in the wall of the stomach secrete hydrochloric acid (HCl) and a watery liquid that turns our meals into a paste. The churning action of the stomach's muscular walls breaks down large pieces of food. In some animals, such as dogs, the stomach has a considerable effect on food, allowing dogs to swallow their food with very little chewing. In humans, however, food must be chewed more thoroughly to ensure proper digestion.

Very little enzymatic digestion occurs in the stomach. Instead, the stomach's role is merely to prepare most food for enzymatic digestion that will occur in the small intestine. There are some exceptions, however. Protein, for example, is denatured by hydrochloric acid (Chapter 2). Denaturation, in turn, allows **pepsin**, an enzyme produced by the gastric glands, to begin breaking proteins into large peptide fragments.

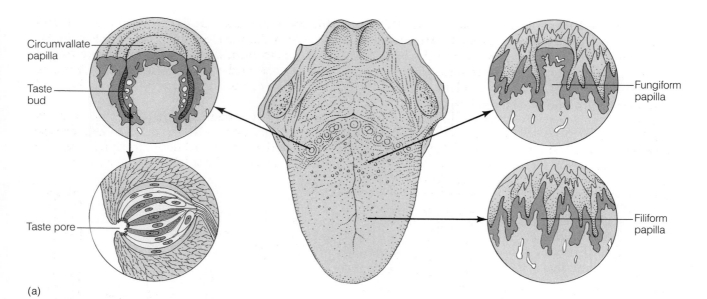

Circumvallate papilla

Taste bud

Taste pore

Fungiform papilla

Filiform papilla

(a)

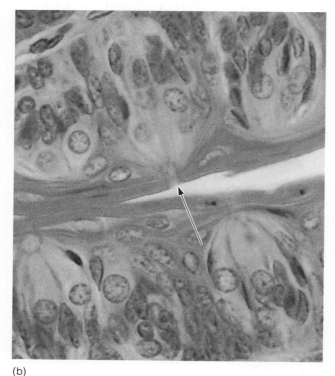

(b)

FIGURE 7–7 The Tongue and Taste Buds (*a*) The tongue is a muscular organ that aids in swallowing and phonation (producing sounds). Its upper surface is dotted with protrusions called papilla. Three types are present: the fungiform, filiform, and circumvallate. Taste buds are located on the fungiform and circumvallate papillae shown in this figure. Taste buds have specialized cells that detect salt, bitter, sweet, and sour flavors. Foods dissolved in saliva enter the taste pore and stimulate these cells, which in turn trigger nerve impulses to the brain. (*b*) A photomicrograph of taste buds; the arrow indicates a taste pore.

Lipids are also partially digested in the stomach with the aid of lipase, an enzyme produced by the salivary glands and transported to the stomach with the food.

Contrary to popular belief, the stomach does not absorb foodstuffs. Only a few substances, such as alcohol and aspirin, can actually penetrate the lining of the stomach and enter the bloodstream. Alcohol consumed on an empty stomach passes quickly through the stomach lining into the bloodstream, resulting in a rather immediate effect. The presence of food in the stomach, however, retards alcohol absorption, giving credence to the advice never to drink on an empty stomach.

Aspirin is also absorbed through the stomach lining. Excess aspirin can irritate the lining and cause bleeding. Large doses taken for pain over considerable periods may result in ulcers.

Hydrochloric acid creates an acidic environment in the stomach that is both useful and potentially harm-

These fragments are further broken down in the small intestine, the next stop in the digestive process. Pepsin is secreted in an inactive form called **pepsinogen**, which is activated by HCl. Once some pepsin molecules are formed, they begin activating other pepsinogen molecules.

ful. Consider the benefits first. As noted above, HCl denatures protein, rendering it digestible. Moreover, HCl kills most bacteria, helping to protect the body. HCl also activates pepsinogen molecules. However, the dangerous mix of acid and proteolytic (protein-digesting) enzymes can also destroy the delicate stomach lining, forming sores or **ulcers**. Normally, the stomach lining is protected from destruction by an alkaline secretion called **mucus**, produced by some of the cells in the lining of the stomach. Mucus coats the stomach lining, protecting it from acid. The tissues beneath the epithelium are protected from acid leakage by the cells of the epithelium, which are tightly joined to one another, forming a leak-proof barrier.

Unfortunately, the stomach's protective mechanisms can break down. A number of factors, such as stress, coffee consumption, excess aspirin, and alcohol—or combinations of them—can increase acid levels in the stomach, overwhelming the mucous layer. Hydrochloric acid and pepsin come in contact with the epithelial cells and may digest parts of the wall of the stomach, forming painful ulcers. When detected early, most ulcers can be treated by reducing stress and changing one's diet—eating bland foods and reducing coffee, aspirin, and alcohol (Health Note 1–1). When ulcers are not detected early and when damage is severe, parts of the stomach may have to be removed surgically.

Chyme in the stomach is ejected into the small intestine by peristaltic muscle contractions. A peri-staltic wave travels across the stomach every 20 seconds. When the wave of contraction reaches the far end of the stomach, the **pyloric sphincter** (a ring of smooth muscle at the juncture of the small intestine and stomach) opens, and chyme squirts into the small intestine.

The stomach contents are emptied in two to six hours, depending on the size of the meal and the type of food. Peristaltic contractions continue after the stomach is empty and are felt as hunger pangs.

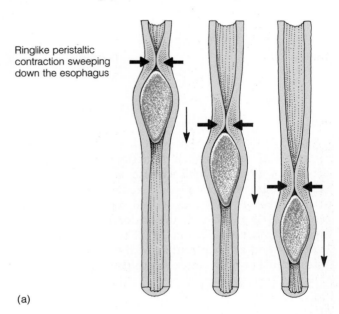

Ringlike peristaltic contraction sweeping down the esophagus

(a)

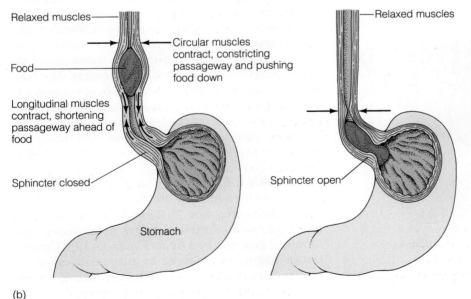

Relaxed muscles

Food

Circular muscles contract, constricting passageway and pushing food down

Longitudinal muscles contract, shortening passageway ahead of food

Sphincter closed

Stomach

Relaxed muscles

Sphincter open

(b)

FIGURE 7–8 Peristalsis
(a) The involuntary contraction of the muscular wall of the esophagus forces food to the stomach and propels chyme and waste throughout the remainder of the digestive tract as well. (b) When food reaches the stomach, the gastroesophageal sphincter opens, allowing food to enter.

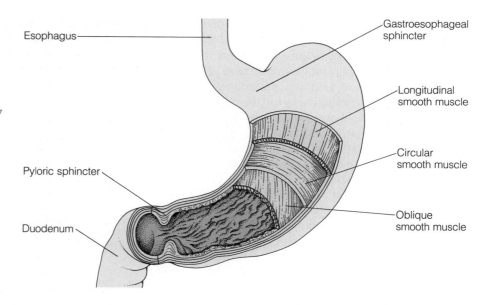

FIGURE 7–9 The Stomach
The stomach lies in the abdominal cavity. In its wall are three layers of smooth muscle that help mix the food and force it into the small intestine where most digestion occurs. The gastroesophageal and pyloric sphincters control the inflow and outflow of food.

Esophagus

Gastroesophageal sphincter

Longitudinal smooth muscle

Circular smooth muscle

Oblique smooth muscle

Pyloric sphincter

Duodenum

Regulating the Stomach Function.   The stomach does not produce hydrochloric acid continuously. Continuous production would endanger the stomach lining. Instead, HCl is produced on demand. Its secretion is controlled by the endocrine and nervous systems (Chapters 12 and 15). The sight, smell, and taste of food activate centers of the brain, which, in turn, transmit nerve impulses to the stomach via the

vagus nerve. The vagus nerve terminates in the stomach wall and activates HCl production by cells in the gastric glands. Nerve impulses also stimulate the synthesis and release of a stomach hormone called gastrin. **Gastrin** increases the output of HCl. Amino acids and peptides in the stomach also activate acid production.

The Small Intestine and Associated Glands: Digestion and Absorption of Food

The small intestine is a coiled tube in the abdominal cavity about 6 meters (20 feet) long in adults (Figure 7–5). So named because of its small diameter, the small intestine digests macromolecules enzymatically, forming smaller molecules that are transported into the bloodstream and the lymphatic system. The lymphatic system, discussed in Chapter 8, is a network of vessels that carries extracellular fluid from the tissues of the body to the circulatory system. It also transports fats absorbed by the intestine into the bloodstream.

The Intestinal Epithelium.   In the small intestine, macromolecules in food are first broken into large fragments by enzymes. These enzymes are produced by the pancreas, described below, and released into the small intestine by a duct (Figure 7–10). The molecular fragments produced by enzyme digestion are further broken down by enzymes produced by the small intestine (Table 7–4). The intestine's enzymes

FIGURE 7–10 Organs of Digestion The liver, gallbladder, and pancreas all play key roles in digestion and empty by the common bile duct into the small intestine in which digestion takes place.

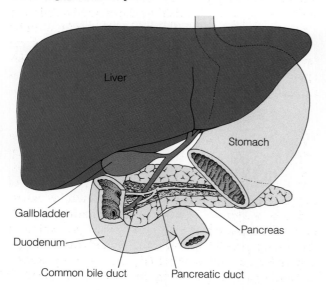

Liver

Stomach

Gallbladder

Pancreas

Duodenum

Common bile duct

Pancreatic duct

TABLE 7-4   Digestive Enzymes

SITE OF PRODUCTION	ENZYME	ACTION
Salivary glands	Lipase	Works in the stomach to digest fats
	Amylase	Helps digest polysaccharides in the oral cavity
Stomach	Pepsin	Works in the stomach to break proteins into peptides
Pancreas	Trypsin	Cleaves peptide bonds of peptides and proteins
	Chymotrypsin	Same as trypsin
	Carboxypeptidase	Cleaves peptide bonds on carboxy end of peptides
	Amylase	Breaks starches into smaller units
	Phospholipase	Cleaves fatty acids from phosphoglycerides
	Lipase	Cleaves two fatty acids from triglycerides
	Ribonuclease	Breaks RNA into smaller nucleotide chains
	Deoxyribonuclease	Breaks DNA into smaller nucleotide chains
Epithelium of small intestine	Maltase	Breaks maltose into glucose subunits
	Sucrase	Breaks sucrose into glucose and fructose subunits
	Lactase	Breaks lactose in glucose and galactose subunits
	Aminopeptidase	Breaks down peptides

are embedded in the membranes of the cells lining the small intestine and are produced by the cells themselves. As a result, the final phase of digestion occurs just before the nutrient is absorbed into the cell.

Absorption in the Small Intestine.   Absorption is facilitated by three structural modifications of the small intestine. Figure 7–11a shows, for instance, that the lining of the small intestine is thrown into circular folds, which increase the overall surface area. The circular folds are comprised of many fingerlike projections called **villi** (Figures 7–11b and 7–11c). Like the circular folds, the villi also increase the surface area available for absorption. The intestinal surface area is further increased by **microvilli**, projections on the plasma membranes of the epithelial cells (Figure 7–11d). Each cell contains approximately 3000 microvilli. Two hundred million microvilli occupy a single square millimeter of intestinal lining. Thus, the surface area of the lining of the small intestine is 600 times greater than it would be without the circular folds, the villi, and the microvilli.

Although numerous mechanisms are involved in absorption, most molecules enter the intestinal epithelium by active transport, then move out of the intestinal cells and into the bloodstream or lymphatic vessels by diffusion. Each villus of the small intestine has a rich supply of blood and lymph capillaries that carry off the nutrients (Figure 7–11b). Most of the nutrients diffuse into the capillaries, minute vessels of the circulatory system. Fatty acids and monoglycerides, however, are an exception. They first diffuse into the cells lining the villi where the triglycerides are reassembled. Triglycerides combine with cholesterol and phospholipids absorbed by the epithelial cells and are released from the cells into the interstitial fluid by exocytosis. In the extracellular fluid, the lipids form small globules. Blood capillaries, however, are relatively impermeable to fat globules. Most of the lipid globules, therefore, enter the more porous lymph capillaries.

The Liver.   Three organs work in concert with the small intestine, providing digestive secretions. They are the liver, the pancreas, and the intestinal glands. The liver is the largest organ in the body and one of the most versatile, performing many functions essential to homeostasis. Situated in the upper right quadrant of the abdominal cavity, under the protection of the rib cage, the liver is one of the body's main storage depots for glucose and fats (Figure 7–5). By storing fats and lipids and releasing them as they are needed,

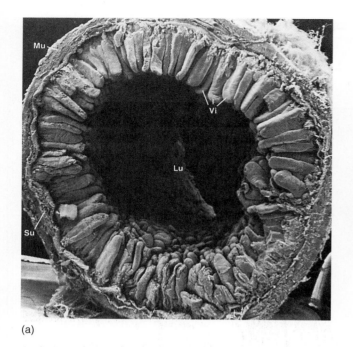

(a)

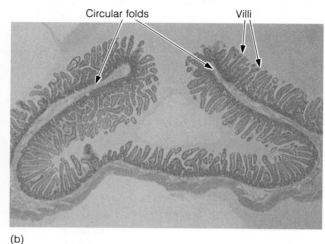

Circular folds   Villi

(b)

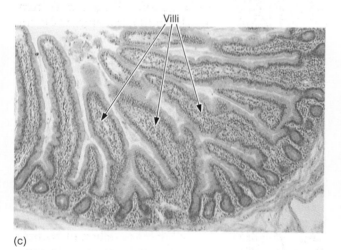

Villi

(c)

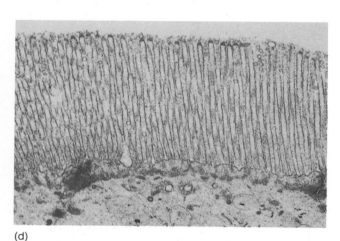

(d)

FIGURE 7–11 The Small Intestine The small intestine is uniquely "designed" to increase absorption. (*a*) A cross section showing the folds. Vi are villi; Su is submucosa layer; Mu is muscularis layer; and Lu is lumen. (*b*) A light micrograph of folds and villi. (*c*) Higher magnification of villi. (*d*) An electron micrograph of the apical surface of the absorptive cells showing the microvilli.

Source: Part (a) from *Tissues and Organs: A Text-Atlas of Scanning Electron Microscopy* by Richard G. Kessel and Randy H. Kardon. Copyright © 1979 by W. H. Freeman and Company. Reprinted with permission.

the liver helps to ensure a constant supply of energy-rich molecules to body cells. The liver synthesizes some key blood proteins involved in clotting and stores iron and certain vitamins, releasing them as needed. The liver is an efficient detoxifier of poten-tially harmful chemicals, such as nicotine, barbiturates, and alcohol.

Last but not least, the liver plays an important role in the digestion of fats. The liver produces a fluid called **bile**, which contains water, ions, and chemicals, such as cholesterol, fatty acids, and bile salts. **Bile salts** are steroids that are produced by liver cells and are required for lipid digestion. Bile is transported to the **gallbladder**, a sac attached to the underside of the liver (Figure 7–5). The gallbladder removes water from the bile, thus concentrating the bile. Upon demand, bile is released into the small intestine through the common bile duct (Figure 7–10).

Bile salts are **emulsifying agents**—that is, chemical substances that break up fat globules into much smaller globules. This reduction in the size of the globules is essential for lipid digestion because the small intestine does not digest large globules efficiently.

Bile flow to the small intestine may be blocked by **gallstones**, deposits of cholesterol and other materials that form in the gallbladder. Gallstones may lodge in the ducts draining the organ, thus reducing—even completely blocking—the transport of bile to the small intestine. In the absence of bile salts, lipid digestion is greatly reduced. Because lipid digestion is reduced, fat globules remain in the undigested food mass, or feces. Fats are decomposed by bacteria in the large intestine, but not absorbed. The decomposition of these fats, therefore, gives the feces a foul odor. The higher percentage of fat also makes the feces quite buoyant and difficult to flush.

Approximately one of every ten American adults has gallstones, although many (30–50%) of these people exhibit no symptoms whatsoever. Gallstones occur more frequently in older individuals, and the incidence in the elderly is about one in every five. When they cause problems, gallstones are usually removed surgically. This procedure requires that the entire gallbladder be removed. Scientists are now testing a new drug that dissolves gallstones in many patients, thus eliminating the need for surgery.

The Pancreas.   The pancreas lies in the abdominal cavity under the stomach, nestled in a loop formed by the first portion of the small intestine, the **duodenum** (Figure 7–10). The pancreas is a dual-purpose organ: it produces enzymes needed to digest foodstuffs in the small intestine and also produces hormones that regulate blood glucose levels.

The digestive enzymes of the pancreas are produced in small glandular units and are carried away by ducts that converge to form the large pancreatic duct. The pancreatic duct joins with a duct from the gallbladder and drains into the duodenum. Each day approximately 1200–1500 milliliters (1.0 to 1.5

quarts) of pancreatic juice are produced and released into the small intestine. This liquid is composed of water, sodium bicarbonate, and several important digestive enzymes (Table 7–4).

Sodium bicarbonate in the pancreatic juice neutralizes acid in the chyme released by the stomach. This helps protect the small intestine from stomach acids and also creates conditions optimal for the function of the pancreatic digestive enzymes.

Pancreatic enzymes, listed in Table 7–4, break down lipids and macromolecules—protein, polysaccharides, and nucleic acids—into smaller molecules. Pancreatic **amylase**, for instance, digests starch molecules, forming smaller polysaccharide chains and the disaccharide maltose. Two other pancreatic enzymes, **chymotrypsin** and **trypsin**, break down proteins to form peptides. Pancreatic **lipase** removes some of the fatty acids from the glycerol molecule, forming monoglycerides, while **ribonuclease** and **deoxyribonuclease** break RNA and DNA into shorter nucleotide chains.

Like the stomach, the pancreas secretes its enzymes in an inactive form, helping protect the pancreas from self-destruction. **Trypsinogen**, for example, is the inactive form of the protein-digesting enzyme trypsin. Trypsinogen is activated by a substance on the epithelial lining of the small intestine. Trypsin, in turn, activates other enzymes.

The Large Intestine: Water Resorption

Undigested materials pass from the small intestine into the large intestine, so named because of its larger diameter (Figure 7–5). The large intestine is about 1.5 meters (5 feet) long and receives a mixture of water and undigested foods—fats, protein, and carbohydrate (mostly fiber)—from the small intestine. The large intestine supports a huge population of bacteria that feed on the unabsorbed nutrients. These bacteria synthesize several key vitamins: B-12, thiamine, riboflavin, and, most importantly, vitamin K, which is often deficient in the human diet. The large intestine absorbs vitamins produced by its bacteria, as well as sodium ions, potassium ions, and much (90%) of the water remaining in the feces.

The feces contain indigestible wastes and dead bacteria. Bacteria, in fact, account for about one-third of the dry weight of the feces.[3] The feces are propelled

[3]The feces are about 30% dead bacteria, 10–20% fat, 10–20% inorganic matter, 2–3% protein, and about 30% undigested roughage (cellulose).

by peristaltic contractions until they reach the rectum. When the rectum distends, it stimulates the defecation reflex. Fortunately, this reflex can be consciously overridden beginning early in life.

Controlling Digestion

Digestion is a complex and varied process that is controlled largely by nerves and hormones. Consider some key events. Digestion begins in the oral cavity and is largely under the control of the nervous system. As noted earlier, the sight, smell, taste, and sometimes even the thought of food stimulate the release of saliva. Chewing has a similar effect. Besides activating salivary production, these stimuli also cause the brain to send nerve impulses along the vagus nerve to the stomach, initiating the secretion of HCl and the hormone **gastrin**. Gastrin stimulates additional HCl secretion.

The concentration of HCl in the stomach is regulated by a negative feedback mechanism (Chapter 6). When the acid content rises too high, it inhibits gastrin secretion, thus shutting off acid production. Protein in the stomach tends to reduce the concentration of HCl by accepting free H^+. This reduces the acidity, which stimulates gastrin production, thus increasing acid secretion.

Chyme next enters the small intestine where its acids stimulate the release of another hormone, secretin. **Secretin** is produced by the cells of the duodenum and travels in the bloodstream to the pancreas where it stimulates the release of sodium bicarbonate. Sodium bicarbonate, in turn, is secreted into the small intestine where it neutralizes the acidic chyme and creates an environment optimal for pancreatic enzymes.

The release of pancreatic enzymes is triggered by another hormone, **cholecystokinin (CCK)**, produced by cells of the duodenum when chyme is present. CCK also stimulates the gallbladder to contract, releasing bile into the small intestine.

Recent evidence links abnormally low secretion of CCK to **bulimia**, an eating disorder characterized by recurrent binge eating, followed by vomiting. CCK has been found in the brain's hypothalamus with other hormones and may be involved in a range of behaviors, including bulimia. Approximately 4% of America's young adult women, and a far smaller fraction of men, are bulimic. Bulimia is thought to have both biological and psychological roots, but researchers have failed to identify a biochemical cause until

now. No single chemical is likely to control a complex behavior like appetite, but it appears that CCK plays an important role.

The small intestine produces one additional hormone, **gastric inhibitory peptide (GIP)**, in response to fatty acids and sugars in chyme. GIP inhibits acid production and peristalsis in the stomach, slowing down the rate at which chyme is released into the small intestine and providing additional time for digestion and absorption to occur.

ENVIRONMENT AND HEALTH: EATING RIGHT/LIVING RIGHT

In few places is the delicate balance between homeostasis and human health as evident as in human nutrition. Homeostasis requires an adequate supply of nutrients. Many studies suggest that a healthy, balanced diet can decrease the risk of cancer, heart disease, hypertension, and other diseases. All of these diseases may be caused by imbalance.

Consider a few examples. Magnesium is one of the major minerals, but is routinely ingested in insufficient amounts. New research suggests that such deficiencies may underlie a number of medical conditions including diabetes, high blood pressure, pregnancy problems, and cardiovascular disease.

Research shows that adding magnesium to the drinking water of rats with hypertension can eliminate high blood pressure. Studies in rabbits show that magnesium reduces lipid levels in the blood and also reduces plaque formation in blood vessels. Rabbits on a high-cholesterol, low-magnesium diet, for example, have 80 to 90% more atherosclerosis than rabbits on a high-cholesterol, high-magnesium diet.

Researchers have found that in humans magnesium deficiencies during pregnancy result in migraines, high blood pressure, miscarriages, stillbirths, and babies with low birth weight. Magnesium supplements greatly reduce the incidence of these problems. Research suggests that magnesium deficiency causes spasms in blood vessels of the placenta, reducing blood flow to the fetus.

Researchers believe that 80 to 90% of the American public may be magnesium deficient. One reason the American diet may be deficient in magnesium is that phosphates in many carbonated soft drinks bind magnesium in the intestine, preventing it from being absorbed into the blood. Magnesium deficiencies can be reversed by eating more green leafy vegetables, seafoods, and whole grain cereals. Mineral supplements could help as well, but they should be used with caution.

Zinc is a trace mineral that has also been implicated in a wide range of health problems. Rats fed diets severely deficient in zinc, for example, have more birth defects, are often stunted, and reach sexual maturation later than their counterparts fed normal diets. Concerned that less-severe zinc deficiencies may cause problems in humans, researchers followed 10 monkeys from birth through adolescence. One group was fed a diet low in zinc. The other received far more than they required. Monkeys fed the zinc-deficient diet showed several curious symptoms. Their immune function was suppressed 20 to 30%, making them more susceptible to disease. Significant learning impairments were also observed. The monkey studies substantiate studies in rodents and suggest concern for people in less-developed countries who often subsist on low-zinc diets consisting primarily of cereals.[4]

Over the years, numerous dietary recommendations have been issued to help Americans live healthier lives and reduce the risk of cancer and heart attack. Nutritionists recommend that we daily (1) consume fruit and vegetables, especially cabbage and greens, (2) consume high-fiber foods, such as whole wheat bread and celery, (3) and that we consume foods high in vitamins A and C. A healthy diet also minimizes the consumption of animal fat, red meat, and salt-cured, nitrate-cured, smoked or pickled foods including bacon and lunch meat.

Has the American public taken these recommendations to heart? The results of one recent study suggest that the answer is no. A survey of the eating habits of nearly 12,000 Americans conducted from 1976 to 1980 showed that the diets of both black and white Americans typically were deficient in the very foods that nutritionists recommended.[5] When asked to recall everything they had eaten in the previous 24-hour period, fewer than one in five people in the study reported eating any of these foods. In sharp contrast, many of the people surveyed had eaten red meat, bacon, and lunch meat.

---

[4]To determine if you are receiving an adequate supply of micronutrients, you can undergo a blood test or an analysis of your diet at a nutritional clinic. If there are problems, a trained nutritionist will be able to make recommendations to correct the problem.

[5]Critical thinking suggests caution in interpreting these results. Studies of the dietary habits of people in 1990 would be more informative.

A healthy diet results largely from habit and circumstance or environment. How does our environment affect our nutrition and health? In the hustle and bustle of modern society, many of us ignore proper nutrition, grabbing snack foods when we are hungry because we haven't the time to sit down to a nutritionally balanced meal. The fast-paced world we live in—that places a high premium on speed—often ignores the importance of eating right.

SUMMARY

1. Studies suggest that iron levels in the body may affect heat production in women, an example of the importance of diet to physiological processes.

A PRIMER ON HUMAN NUTRITION

2. Humans require two types of nutrients: macronutrients, substances needed in large quantity, and micronutrients, substances required in much lower quantities.
3. The four major macronutrients are water, carbohydrates, lipids, and proteins.
4. Water is in the liquids we drink and the foods we eat. Maintaining adequate water intake is important because water is involved in many chemical reactions in the body. Water helps maintain body temperature and a constant level of nutrients and wastes in body fluids.
5. Carbohydrates and lipids are major sources of cellular energy; 70 to 80% of all energy required by the body goes for basic functions.
6. Contrary to popular myth, protein does not supply much energy, except when lipid and carbohydrate intake is low or if protein intake exceeds daily requirements. Dietary protein is chiefly a source of amino acids for building proteins.
7. Amino acids produced by protein digestion can be used to build new protein or may be chemically modified to produce other amino acids. Amino acids that cannot be synthesized in the body and must be supplied in the food we eat are essential amino acids.
8. To ensure an adequate supply of all amino acids, individuals should eat complete proteins, such as milk or eggs, or combine protein sources.
9. Lipids provide energy during rest and aerobic activity. Lipids serve many other functions in the body, such as insulation.
10. Besides providing energy, carbohydrates serve other important functions. Dietary fiber, for example, increases the liquid content of the feces, reducing constipation, the incidence of diverticulitis, and the risk of colon cancer.

11. Micronutrients are needed in much smaller quantities and include two groups: vitamins and minerals.
12. Vitamins are a diverse group of organic compounds that act as enzyme cofactors. Vitamins are required in relatively small quantities.
13. Human vitamins fit into two categories: water soluble and fat soluble. The water-soluble vitamins include vitamin C and the B-complex vitamins. The fat-soluble vitamins include vitamins A, D, E, and K. Vitamin deficiencies and vitamin excesses can result in health problems.
14. Minerals fit into one of two groups: trace minerals, those required in very small quantity, and major minerals, those required in greater quantity. Deficiencies and excesses of both types of minerals can lead to serious health problems.

THE HUMAN DIGESTIVE SYSTEM

15. Food is chemically and physically broken down in the digestive system. Small molecules produced in this process are transported from the digestive system to the bloodstream or lymphatic system.
16. Food digestion begins in the mouth. The teeth mechanically break down the food. Saliva liquifies the food making it easier to swallow. Salivary amylase begins to digest starch molecules.
17. Food is pushed to the pharynx by the tongue where the food triggers the swallowing reflex. Peristaltic contractions propel the food down the esophagus to the stomach.
18. The stomach is an expandable organ that stores and liquifies the food. The churning action of the stomach, brought about by peristaltic contractions, mixes the food, turning it into a paste called chyme. The stomach releases food into the small intestine in timed pulses, ensuring efficient digestion and absorption. Very limited chemical digestion and absorption occur in the stomach.
19. The stomach produces hydrochloric acid, which denatures protein, allowing it to be acted on by enzymes. The stomach also produces a proteolytic enzyme called pepsin, which breaks proteins into peptides. The lining of the stomach is protected from acid by mucus. When the mucous protection fails, however, the lining may be eroded by acids, creating an ulcer.
20. The functions of the stomach are regulated by neural and hormonal mechanisms.
21. The small intestine is a long, coiled tubule in which most of the enzymatic digestion of food and absorption takes place. Enzymes for digestion come from the pancreas and the lining of the intestine itself.

22. The liver also plays an important role in digestion. It produces a liquid called bile that contains, among other chemicals, bile salts. Bile is stored in the gallbladder and released into the small intestine when food is present. Bile salts emulsify fats, breaking them into small globules that can be acted on by enzymes.
23. The pancreas produces sodium bicarbonate, which neutralizes the acid entering the small intestine with the chyme and also produces enzymes that act on macromolecules, breaking them into smaller ones.
24. Undigested food molecules pass from the small intestine into the large intestine, which carries the waste, or feces, to the outside of the body. The large intestine absorbs water, sodium, potassium, and vitamins produced by intestinal bacteria.
25. The digestive process is controlled by the nervous and endocrine systems.
26. Digestion begins in the mouth with enzymes released by the salivary glands. The release of saliva is stimulated by the sight, smell, taste, and sometimes even the thought of food. These stimuli also cause the brain to send nerve impulses to the gastric glands of the stomach, initiating the secretion of HCl and gastrin, a hormone that also stimulates HCl secretion.
27. Chyme entering the small intestine stimulates the release of two hormones, secretin and cholecystokinin. Secretin travels in the bloodstream to the pancreas where it stimulates the release of sodium bicarbonate. Cholecystokinin stimulates the release of pancreatic enzymes and also stimulates the gallbladder to contract, releasing bile into the small intestine.

ENVIRONMENT AND HEALTH: EATING RIGHT/LIVING RIGHT

28. Human health is dependent on good nutrition. Numerous studies suggest that a healthy, balanced diet can decrease the risk of cancer, heart disease, hypertension, and other diseases.
29. Nutritionists recommend the daily consumption of (a) fruit and vegetables, especially cabbage and greens, (b) high-fiber foods, such whole wheat bread and celery, and (c) foods high in vitamins A and C.
30. In addition, they recommend reducing consumption of animal fat, red meat, and salt-cured, nitrate-cured, smoked or pickled foods including bacon and lunch meat.
31. Studies suggest, however, that Americans have not taken these recommendations to heart. Many of us ignore proper nutrition, because we haven't the time to sit down to a nutritionally balanced meal. The fast-paced world we live in often ignores the importance of eating right.

Corn is a staple for 200 million people worldwide, including nearly half of the world's chronically malnourished people. Because corn is such an important source of calories and protein throughout the world, researchers have developed a new strain of corn called Quality-Protein Maize (QPM).

QPM has about the same amount of protein as common maize, but has twice the usable protein because its protein is more complete—that is, it supplies more of the essential amino acids. Normal maize has about 40% of the biological value of milk protein. QPM approaches that of milk, a common standard of nutritional excellence.

In countries where maize is a staple, QPM could double the efficiency of subsistence farming. A study by the National Research Council reports that QPM could allow families to combat malnutrition without outside help. QPM could prove helpful in Mexico, Central America, and Africa where hunger and starvation are common.

Using your critical thinking skills, examine this optimistic new finding. You might want to begin with the big picture. For example, will QPM be affordable to peasants in the Third World? Will it be more susceptible to insect pests than currently used native strains? Will it require costly pesticides? Will it require additional irrigation? Will it deplete the soil? How does it compare economically to other measures? What other questions can you think of that should be answered?

You may want to refer to an environmental science book (see Suggested Readings in Chapter 21) to find out more on world hunger. What other ways are there to solve the problem? How have efforts to improve crop yield worked in the past? Should QPM be supplemented by other strategies?

TEST OF TERMS

1. Water, carbohydrates, lipids, and proteins are all members of a class of nutrients called _____ .

2. The two main sources of energy in a well-balanced diet are _____ and _____ .

3. _____ exercise strives to maintain oxygen levels in muscles and relies primarily on _____ for energy.

4. Excess protein in the diet can be used to generate energy or be used to make _____ .

5. The amino acids the body cannot produce are called _____ amino acids. Proteins that supply all of these amino acids are said to be _____ .

6. Indigestible carbohydrates in fruits, vegetables, and grains are called _____ and are thought to reduce the incidence of _____ cancer.

7. _____ are a diverse group of organic compounds that act as cofactors for enzymes. They fit into two broad groups: _____ and _____ .

8. Minerals required in minute quantities such as zinc and copper are called _____ _____ .

9. The salivary glands produce two enzymes, _____ and _____ .

10. The taste receptors on the tongue are called _____ _____ .

11. The _____ connects the oral cavity with the esophagus.

12. The involuntary muscle contractions that propel food along the digestive tract are called _____ .

13. At the juncture of the esophagus and stomach is a ring of muscle called the _____ sphincter.

14. Food is converted to a liquified mass called _____ in the stomach.

15. The proteolytic enzyme, _____ , produced by the stomach, is released in an inactive form, _____ .

16. The ring of muscle at the junction of the stomach and small intestine that controls the passage of food into the small intestine is called the _____ sphincter.

17. The _____ produces bile, a fluid that is stored in the gallbladder and later released into the small intestine where its chief chemical component, _____ _____ emulsify fats.

18. The pancreas produces two major products: _____ and _____ _____ .

19. Three major structural modifications increase the surface area for absorption in the small intestine; they are _____ _____ , _____ , and _____ .

20. Fats absorbed by the small intestine are carried away by _____ capillaries.

Answers to the Test of Terms are located in Appendix B.

TEST OF CONCEPTS

1. The body requires proper nutrient input to maintain homeostasis. Give an example and explain how the nutrient affects homeostasis.
2. You are an exercise physiologist in charge of a weight-control clinic. A client comes to you and asks your advice on a weight-loss program. Should he start lifting weights or join an aerobic class? Why?
3. Describe the conditions in which protein is used to provide cellular energy.
4. Using your knowledge of homeostasis and energy metabolism, describe the homeostatic control of blood glucose. Be sure to indicate the input, storage depots, and output.
5. If you were considering becoming a vegetarian, how would you be assured of getting all of the amino acids your body needs?
6. Describe how fiber may help reduce colon cancer.
7. What are vitamins and why are they needed in such small quantities?
8. A dietary deficiency of one vitamin can cause wide-ranging effects. Why?
9. What organs physically break food down and what organs participate in the chemical breakdown of food?
10. Describe the process of swallowing.
11. Describe the function of hydrochloric acid and pepsin in the stomach. How does the stomach protect itself from these substances?
12. How do ulcers form and how can they be treated?
13. Describe the endocrine and nervous system control of the stomach function.
14. The small intestine is the chief site of digestion. Where do the enzymes needed for this process come from, and how is release of these enzymes stimulated? What other chemicals are needed for proper digestion?
15. Describe the functions of the large intestine.

SUGGESTED READINGS

Boskin, W., G. Graf, and V. Kreisworth. 1990. *Health dynamics: Attitudes and behaviors.* St. Paul, Minn.: West. Excellent information on health and nutrition.

Christian, J. L., and L. L. Greger. 1988. *Nutrition for living,* 2d ed. Menlo Park, Calif.: Benjamin/Cummings. Well-written and informative guide to nutrition with lots of practical advice and important biochemical information.

Guyton, A. C. 1986. *Textbook of medical physiology,* 7th ed. New York: Saunders. Excellent, in-depth coverage of digestion.

Long, P. 1989. Fat chance. *Hippocrates* 3(5): 38–47. This article helps you identify if you are overweight and discusses ways of losing weight.

McKenzie, A. 1989. A tangle of fibers. *Science News* 136(22): 344–45. Good discussion of the various types of dietary fiber.

Sizer, F. S., and E. N. Whitney. 1988. *Life choices: Health concepts and strategies.* St. Paul, Minn.: West. Contains excellent information on nutrition and health.

Vander, A., J. Sherman, and D. Luciano. 1985. *Human physiology: The mechanisms of body function.* 4th ed. New York: McGraw-Hill. General coverage of human physiology.

Whitney, E. N., E. M. N. Hamilton, and S. R. Rolfes. 1990. *Understanding nutrition,* 5th ed. St. Paul, Minn.: West. Superb coverage of nutrition for students interested in learning more.

8 Circulation and Blood

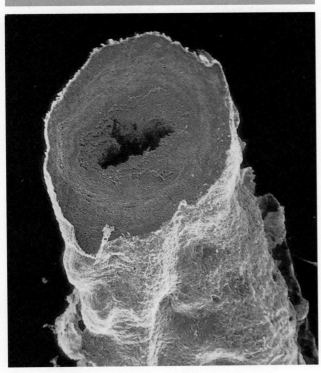

Colorized scanning electron micrograph of a sectioned varicose saphenous vein.

David Adams, a 48-year-old New York attorney, collapsed in his office one morning. Dazed, barely conscious, and in great pain, he was rushed to the hospital. There a team of cardiologists discovered a blood clot lodged in a narrowed section of Adams's right coronary artery, which supplies blood and oxygen to much of the heart. The physicians began immediate action to prevent further damage to the oxygen-starved heart muscle. Through a small incision in the man's groin, they inserted a small plastic catheter and snaked it through the femoral artery, the large artery that brings blood to the leg. They then threaded the catheter up through the arterial system to the heart (Figure 8–1). The physicians then guided the catheter into the clogged coronary artery. When they reached the clot, they injected an enzyme, called streptokinase, through the catheter. Streptokinase dissolves blood clots, restoring the blood flow to heart muscle.

David Adams survived. However, many others are not so lucky. They either arrive at the hospital too late or do not receive blood-clot–dissolving agents or other treatments in time to avoid widespread damage to heart muscle cells.

Heart attacks strike thousands of Americans each year. They are one of a handful of diseases afflicting

187

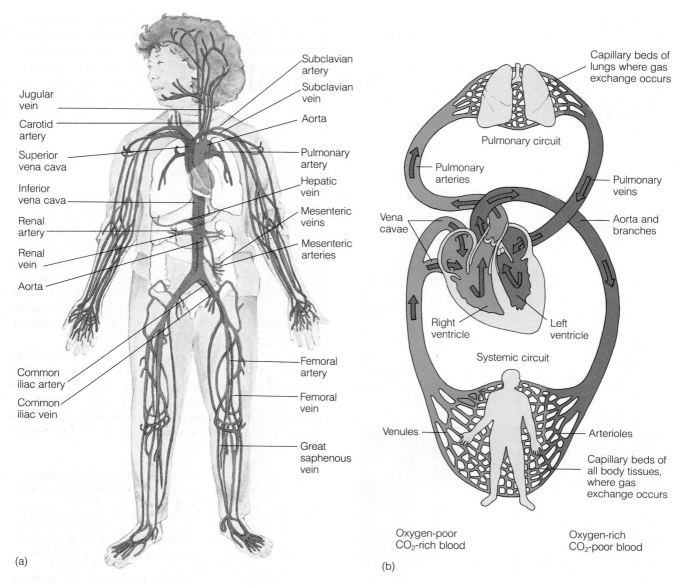

Labels on figure (a):

Jugular vein
Carotid artery
Superior vena cava
Inferior vena cava
Renal artery
Renal vein
Aorta
Common iliac artery
Common iliac vein

Subclavian artery
Subclavian vein
Aorta
Pulmonary artery
Hepatic vein
Mesenteric veins
Mesenteric arteries
Femoral artery
Femoral vein
Great saphenous vein

(a)

Labels on figure (b):

Capillary beds of lungs where gas exchange occurs
Pulmonary circuit
Pulmonary arteries
Pulmonary veins
Aorta and branches
Vena cavae
Right ventricle
Left ventricle
Systemic circuit
Venules
Arterioles
Capillary beds of all body tissues, where gas exchange occurs

Oxygen-poor CO$_2$-rich blood

Oxygen-rich CO$_2$-poor blood

(b)

FIGURE 8−1 The Circulatory System   (a) The circulatory system consists of a series of vessels that transport blood to and from the heart, the pump. (b) The circulatory system consists of two major circuits, the pulmonary circuit, which delivers blood to the lungs, and the systemic circuit, which delivers blood to the rest of the body.

the circulatory system, which are caused, in large part, by the stressful conditions of modern society and the habits of modern people. This chapter examines the cardiovascular system, the heart, blood vessels, and the blood. It describes the structure and function of the parts of the cardiovascular system and notes how our lifestyles and eating habits can be modified to reduce the likelihood of cardiovascular disease.

THE CIRCULATORY SYSTEM'S FUNCTION IN HOMEOSTASIS

The circulatory system is one of the chief homeostatic systems of the body. It consists of a muscular pump and two circulatory loops—one that delivers blood to the lungs and one that delivers blood to the body (Figure 8−1).

TABLE 8–1   Functions of the Blood

Transports oxygen to body cells

Transports nutrients from the digestive system to body cells

Transports hormones to body cells

Transports wastes from body cells to excretory organs

Distributes body heat

Helps maintain constant pH in tissue fluids

Prevents infections

The cardiovascular system operates tirelessly day and night (Table 8–1). The blood pumped through the cardiovascular system transports oxygen from the lungs to the body cells, where it is used in cellular energy production. The circulatory system also distributes nutrients (absorbed by the digestive tract) and hormones (produced by the endocrine glands) to body tissues and cells. The blood also transports the waste products of cellular metabolism to excretory organs, such as the kidneys. The excretory organs rid the body of potentially harmful chemical substances; the circulatory system, therefore, helps maintain low levels of carbon dioxide and other wastes in the body. The circulatory system also helps regulate and distribute body heat and helps protect the body against bacteria and viruses (through white blood cells) and against blood loss (through clotting). Each of these functions—and a great many more—helps to maintain homeostasis.

THE HEART

The heart is a muscular pump in the thoracic (chest) cavity. The heart is the workhorse of the cardiovascular system, propelling blood through the 50,000 to 60,000 miles of blood vessels. Each day, the heart beats approximately 100,000 times, adjusting its rate to meet the changing needs of the body. If you had a dollar for every heartbeat, you would be a millionaire in 10 days. Over a 70-year lifetime, you would collect $2.5 billion for your heart's work.

The heart, shown in Figure 8–1, is a fist-sized organ whose walls are composed of three layers, the pericardium, the myocardium, and the endocardium (Figure 8–2). The **pericardium** is a thin, closed sac surrounding the heart and the bases of large vessels that enter and leave the heart. The sac is filled with a

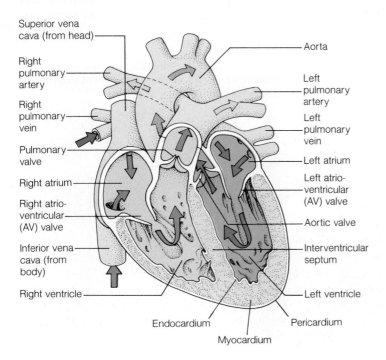

FIGURE 8–2 Blood Flow through the Heart
Deoxygenated (carbon dioxide–enriched) blood (blue) flows into the right atrium from the systemic circulation, then is pumped into the right ventricle. The right ventricle, in turn, pumps the blood into the pulmonary artery, which delivers it to the lungs where carbon dioxide is released and oxygen is picked up. Reoxygenated blood (red) is returned to the left atrium, then flows into the left ventricle, which pumps it to the rest of the body through the systemic circuit.

clear, slippery fluid that reduces friction resulting from continual heart contraction. The middle layer, the **myocardium**, is the thickest part of the wall and is composed chiefly of cardiac muscle cells. The inner layer, the **endocardium**, is the endothelial layer, which forms the lining of the heart chambers.

Pulmonary and Systemic Circulation

As illustrated in Figure 8–2, the heart consists of four chambers—two on the right side of the heart and two on the left. Each side of the heart is really a separate pump. The right side of the heart, for example, pumps blood to the lungs and is part of the **pulmonary circulation**, the relatively short loop between the heart and lungs (Figure 8–1b). The left side of the heart pumps blood to the rest of the body through the **systemic circulation**.

The systemic and pulmonary loops of the circulatory system may be anatomically separate, but they are functionally interdependent. As shown in Figure 8–1b, the pulmonary circuit delivers blood to the lungs where it loses most of its carbon dioxide and replenishes its supply of oxygen. The oxygenated blood is then returned to the heart and distributed to body tissues via the systemic circulation. Within the outlying body tissues, the blood loses much of its oxygen and is recharged with carbon dioxide, released by cells during energy production. The blood is then returned to the pulmonary circulation where much of

its carbon dioxide is lost and oxygen supplies are replenished.

Figure 8–2 illustrates the course that blood takes through the heart. Drawn in blue, blood low in oxygen (and rich in carbon dioxide) enters the right side of the heart through the **superior** and **inferior vena cavae**. These large veins empty directly into the **right atrium**, the uppermost chamber on the right side of the heart. The blood is pumped from here into the **right ventricle**, the lower chamber on the right side. When the right ventricle is full, the muscles in its wall contract, forcing blood into the pulmonary arteries, which lead to the lungs.

Blood whose oxygen supply has been restored, drawn in red in Figure 8–2, flows back to the heart via the **pulmonary veins**. The pulmonary veins, in turn, empty directly into the **left atrium**, the upper chamber on the left side of the heart. From here, the oxygen-rich blood is pumped to the **left ventricle**. When the left ventricle is full, its thick, muscular walls contract, propelling the blood into the aorta. The **aorta** is the largest artery in the body. It carries the oxygenated blood away from the heart, delivering it to the cells and tissues of the body via many smaller branches (discussed below).

Figure 8–3 shows that both atria fill simultaneously, then contract, delivering blood to the ventricles. The right and left ventricles also fill simultaneously; when both ventricles are full, they contract, pumping the blood into the systemic and pulmonary

FIGURE 8–3 Blood Flow through the Heart
(*a*) Blood enters both atria simultaneously from the systemic and pulmonary circuits. When full, the atria pump their blood into the ventricles. (*b*) When the atria are full, they contract simultaneously, (*c*) delivering the blood to the pulmonary and systemic circuits.

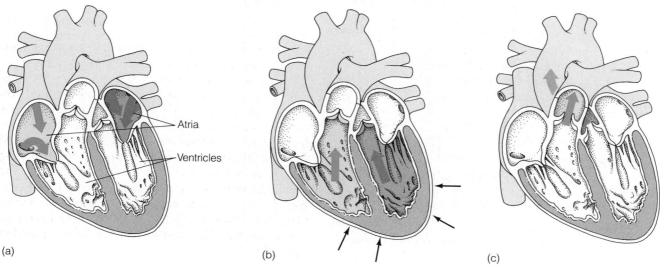

(a)   (b)   (c)

Atria

Ventricles

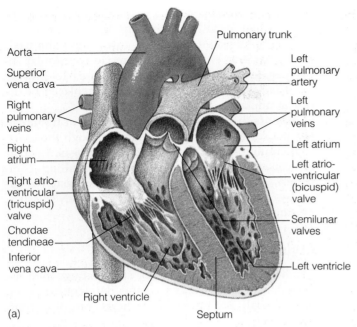

Aorta

Superior vena cava

Right pulmonary veins

Right atrium

Right atrio-ventricular (tricuspid) valve

Chordae tendineae

Inferior vena cava

Right ventricle

Septum

Pulmonary trunk

Left pulmonary artery

Left pulmonary veins

Left atrium

Left atrio-ventricular (bicuspid) valve

Semilunar valves

Left ventricle

(a)

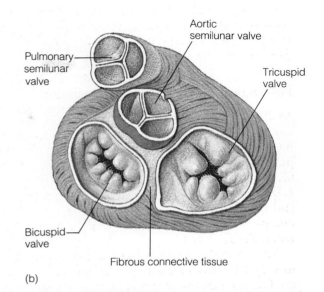

Aortic semilunar valve

Pulmonary semilunar valve

Tricuspid valve

Bicuspid valve

Fibrous connective tissue

(b)

FIGURE 8–4 Heart Valves (a) A cross section of the heart showing the four chambers and the location of the major vessels and valves. (b) A view of the heart from above, with the major vessels removed to show the valves. (c) Pulmonary semilunar valve.

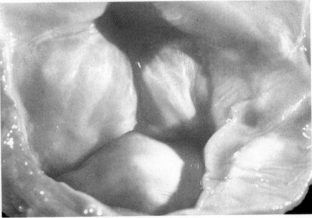

(c)

circulations. The coordinated contraction of heart muscle is brought about by an internal timing device or pacemaker, described later.

Heart Valves

The heart contains four valves that help control the direction of blood flow (Figure 8–4). The valves between the atria and ventricles are called **atrioventricular valves**. Each valve consists of two or three flaps of tissue that are anchored to the inner walls of the ventricles by slender tendinous cords, the **chordae tendineae**, resembling the strings of a parachute. The right atrioventricular valve between the right and left ventricles is called the **tricuspid valve**, because it contains three flaps. The left atrioventricular valve is the **bicuspid valve**.[1] To remember the valves, imagine you are wearing a jersey with the number 32 on the front. This reminds you that the tricuspid valve is on the right side and the bicuspid valve is on the left.

Between the right and left ventricles and the arteries they empty into (pulmonary artery and aorta, re-

[1]The bicuspid valve is also called the mitral valve because it resembles a miter, a hat worn by the pope and bishops.

spectively) are the **semilunar valves**. The semilunar valves consist of three semicircular flaps of tissue.

The atrioventricular and semilunar valves are one-way valves, opening when blood pressure builds on one side and closing when blood pressure increases on the other, much like the purge valves in scuba diving masks, which allow divers to force water out of their masks but also prevent the ocean from rushing in. When the ventricles contract, blood forces the semilunar valves to open. Blood flows out of the ventricles into the large arteries. The semilunar valves then close, preventing blood from flowing back into the ventricles. The atrioventricular valves function similarly.

Heart Sounds

By placing an ear or stethoscope to the chest, two distinguishable sounds can be detected. These are called the **heart sounds** and may be described as "lub-DUPP." The first heart sound (lub) results from the closure of the atrioventricular valves. It is longer and louder than the second heart sound (DUPP), produced when the semilunar valves snap shut.

The right and left atrioventricular valves and the two semilunar valves do not close at precisely the same time. Thus, by careful placement of the stethoscope, a physician can listen to each valve individually. For most of us, our heart valves function flawlessly throughout life. However, some diseases can alter the function of the valves. Rheumatic fever, for example, is caused by a bacterial infection and affects many parts of the body, including the heart. Now rare in developed countries, rheumatic fever is still a significant problem in the Third World. Rheumatic fever begins as a sore throat caused by certain types of **streptococcus** bacteria. The sore throat (called strep throat) is usually followed by general illness. Antibodies to the streptococcus bacteria damage the heart valves. Damage to the valves can prevent them from closing completely. Scar tissue results in a condition known as **valvular incompetence** in which defective valves allow blood to leak back into the atria and ventricles after contraction, thus reducing the efficiency of the organ. A defective valve creates a distinct "sloshing" sound called a **heart murmur**.

Valvular incompetence causes the heart to work harder than usual to make up for the inefficient pumping. Increased activity in turn causes the walls of the heart to enlarge. In severe cases, valvular incompetence can result in heart failure. To prevent heart failure, damaged valves can be replaced by artificial implants.

Tumors (benign and malignant) and scar tissue can block a valve, reducing blood flow through the heart. This condition is known as **valvular stenosis**. Valvular stenosis prevents the ventricles from filling completely. As a result, the heart must beat faster to ensure an adequate supply of blood to the body's tissues, putting additional stress on the organ.

Controlling the Heart's Rate

The heart beats day and night, accelerating when tissues demand more oxygen, and decelerating when demands fall. But how does the heart control its beating?

As noted earlier, the heart contains an internal pacemaker, the **sinoatrial (SA) node**, which is located in the wall of the right atrium, as shown in Figure 8–5. The SA node is a small clump of specialized cardiac muscle cells that causes the heart muscle to contract rhythmically. On their own, cardiac muscle cells contract independently. The SA node, however, imposes a single rhythm on the atrial heart muscle cells, bringing about coordinated muscle contraction

FIGURE 8–5 Conduction of Impulses in the Heart The sinoatrial node is the heart's pacemaker. Located in the right atrium, it sends timed impulses into the heart muscle, coordinating muscle contraction. The impulse travels from cell to cell in the atria, then passes to the atrioventricular node and into the ventricles via the atrioventricular bundle and its branches.

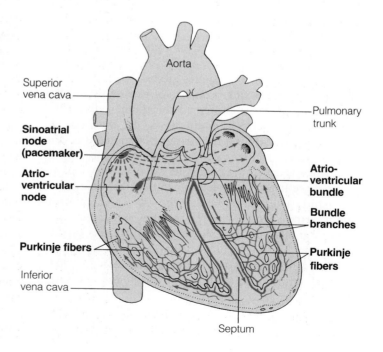

and the orderly movement of blood into the ventricles.

The cells of the SA node contract spontaneously and rhythmically, and each contraction produces a bioelectric impulse akin to those produced in nerve cells (Chapter 12). This impulse spreads rapidly from muscle cell to muscle cell in the atria. Because cardiac muscle cells are tightly joined and because the impulse travels quickly, both atria contract simultaneously and uniformly.

The electrical impulse next passes from the atria to the ventricles. However, its passage is briefly delayed by a barrier of unexcitable tissue that separates the atria from the ventricles. At this point the impulse is channeled through a second mass of specialized muscle cells, the **atrioventricular (AV) node** shown in Figure 8–5. The impulse is delayed approximately one-tenth of a second. This delay gives the atria ample time to empty and also gives the ventricles plenty of time to fill before they are stimulated to contract.

From the AV node, the impulse travels rapidly along a tract of specialized cells. Called the **atrioventricular bundle**, it consists of modified cardiac muscle fibers that conduct the impulse to the walls of the ventricles. The tract divides into two branches that travel on either side of the wall separating the two ventricles (Figure 8–5). The impulse travels down these branches, then onto smaller terminal branches, the **Purkinje fibers**, which terminate on individual heart muscle fibers.

Unlike the muscle of the atria, the cardiac muscle of the two ventricles does not contract in unison, in large part, because the impulse is not transmitted as quickly and as uniformly through the ventricles. Because of the arrangement and distribution of the fibers, contraction begins at the tip of the heart and proceeds upward, squeezing the blood out of the ventricles into the aorta and pulmonary arteries.

Left on its own, the SA node of the human heart produces a steady rhythm of about 100 beats per minute. This, of course, is much too fast for most human activities. Nervous impulses reaching the heart from the brain bring the heart rate into line with body demands at rest or during nonstrenuous activity. These signals dampen the SA node, slowing the heart to about 70 beats per minute. (They dampen much as the brakes reduce a car's speed.) During exercise or stress, when the heart rate must increase to meet body demands, the decelerating impulses are reduced. (In other words, the body lets up on the brakes, permitting the heart rate to increase.) Other nerves also affect heart rate. These nerves carry impulses that ac-

celerate the heart rate even further, allowing the heart to attain rates of 180 beats or more (Chapter 12). (Continuing with the car analogy, these nerves are akin to the accelerator pedal of an automobile, which increases the RPMs of the engine.)

Some hormones also play a role in controlling the heart rate. During stress or exercise, for example, the adrenal glands, located on top of the kidneys, secrete epinephrine (adrenalin). Epinephrine stimulates heart rate, increasing cardiac output.

Fibrillation and Defibrillation. Without control from the SA node, the heart may be converted into an ineffective, quivering mass of muscle tissue. In some heart attack victims, the heart goes into spasms, during which individual cardiac muscle cells contract at their own rate. Such cardiac anarchy is called **fibrillation**. During fibrillation, the heart pumps very ineffectively and may cease pumping altogether, a potentially fatal condition known as **cardiac arrest**.

Physicians treat fibrillation by applying a strong electrical current to the chest. During this procedure, known as **defibrillation**, the electrical current passes through the wall of the chest, often restoring normal electrical activity and heartbeat. A normal heartbeat can also be restored by cardiac massage, a procedure in which the heart is pumped externally by applying pressure to the sternum (breastbone).

The Electrocardiogram

When the electrical impulse that stimulates muscle contraction in the heart reaches a cardiac muscle cell, it causes the cell to contract. Normally, the outside surface of the cardiac muscle cell is slightly positive. The inside surface is slightly negative. When the impulse arrives, it causes a rapid change in the permeability of the cardiac muscle plasma membrane to sodium ions. Sodium ions flow inward, changing the polarity of the plasma membrane and temporarily making the inside of the cell more positive than the outside. This change in polarity causes the cell to contract (Chapter 14). The shift in cardiac muscle cell polarity, or depolarization, can be detected by surface electrodes, small metal plates connected to wires and a voltage meter (Figure 8–6a). The electrodes are placed on a person's chest where they detect the flow of current caused by cardiac muscle cell depolarization. The resulting reading on a voltage meter is called an **electrocardiogram** or **ECG** (Figure 8–6b).

The tracing produced on the voltage meter has three distinct waves (Figure 8–6b). The first wave,

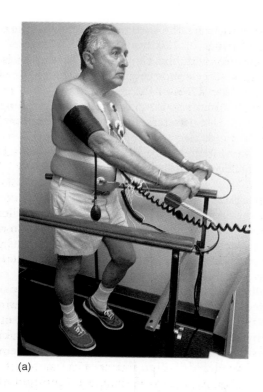

(a)

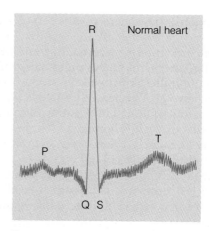

R   Normal heart

P   T

Q S

P = Atrial depolarization, which triggers atrial contraction.

T = Repolarization of ventricles

QRS = Depolarization of AV node and conduction of electrical impulse through ventricles; ventricular contraction begins at R.

P to R interval = Time required for impulses to travel from SA node to ventricles.

(b)

FIGURE 8–6   The Electrocardiogram   (*a*) This patient is wired to a meter that detects electrical activity produced by the heart. (*b*) An electrocardiogram.

the **P wave**, represents the electrical changes occurring in the atria of the heart. The second wave, the **QRS wave**, is a record of the electrical activity taking place during ventricular contraction, and the third wave, the **T wave**, is a recording of electrical activity occurring as the ventricles relax.

Diseases of the heart may disrupt one or more waves of the ECG. As a result, an ECG is often a valuable diagnostic tool for cardiologists. The ECG detects only those defects that alter the heart's electrical activity, however.

Cardiac Output

The amount of blood pumped by the ventricles each minute is called the **cardiac output**. In an individual at rest, the heart pumps about 5 liters (about 5 quarts) of blood per minute.

Cardiac output is a function of two variables: **heart rate**, the number of contractions the heart undergoes per minute, and **stroke volume**, the amount of blood pumped by each ventricle during each contraction. At rest, the heart beats approximately 72 times per minute, and the stroke volume is about 70 milliliters, yielding a cardiac output of 5000 milliliters or 5 liters.

Cardiac output can increase dramatically. The heart of a trained athlete, for example, can pump 35 liters per minute (seven times the cardiac output at rest). Most nonathletes, however, can only increase the cardiac output to about 20 liters per minute.

Heart Attacks: Causes, Cures, and Treatment

Heart attacks come in several varieties. The most common heart attack is called a **myocardial infarction** (described below) and is caused by **thrombosis**, blockage of one or more of the coronary arteries by a blood clot. A blood clot lodged in a coronary artery restricts the flow of blood to the heart muscle, cutting off the supply of oxygen and nutrients. Depriving the heart muscle of oxygen can damage and even kill the cells. A damaged region is called an **infarct**, hence the name myocardial infarction.

Heart attacks generally occur only if the coronary arteries are already narrowed by atherosclerotic plaque. As you learned in Health Note 7–1, atherosclerotic plaque results from a combination of stress, poor diet, lack of exercise, smoking, and genetic factors. Under normal circumstances, narrowing caused by plaque does not result in a heart attack, but when a clot lodges in the vessel, trouble begins.

If the size of the infarct is small and if the change in electrical activity of the heart is transient or minor, a heart attack generally will not be fatal. If the damage is great or electrical activity is severely disrupted, however, myocardial infarctions can prove fatal.

The major symptom of a heart attack is often a crushing pain in the center of the chest. Pain may also appear in the neck, jaw, arms, and abdomen. Heart attacks can occur quite suddenly, without warning, or may be proceeded by several weeks of **angina**, pain that is felt when the supply of oxygen to the myocardium is reduced. Anginal pain appears in the center of the chest and can spread to a person's throat, upper jaw, back, and arms (usually just the left one). Angina is a dull, heavy, constricting pain, which appears when an individual is active, then disappears when he or she ceases the activity.

Angina attacks may also be caused by stress and exposure to carbon monoxide, a pollutant that reduces the oxygen-carrying capacity of the blood, described in the last section of this chapter. Angina usually occurs after the age of 30 in men and is nearly always caused by coronary artery disease. In women, angina tends to occur at a much later age.

Medical Treatment for Heart Attacks. Proper diet and exercise can help reduce the risk of heart problems later in life (Health Notes 6–1 and 7–1). Research also shows that a daily dose of aspirin taken over long periods can help cut an individual's chances of a heart attack. Aspirin apparently reduces heart attacks because it impairs blood clotting. A recent study of 22,000 male physicians in the United States, in fact, showed that aspirin in small doses reduced the risk of first heart attacks by nearly half.[2]

Prevention should be the first line of attack against heart disease. It could save Americans hundreds of millions of dollars each year in medical bills, lost work time, and decreased productivity. But given human nature and the fast pace of modern life, heart disease will probably be around for a long time. To reduce the death rate, physicians therefore are also looking for ways to treat patients after they have had a heart attack.

One promising development is the use of blood-clot–dissolving agents, similar to streptokinase, introduced at the beginning of the chapter. When administered within a few hours of the onset of a heart attack, streptokinase can reduce the damage to heart muscle and accelerate a patient's recovery.

Streptokinase is an enzyme ironically derived from the bacterium that causes rheumatic fever. Thus, streptokinase is a foreign substance that evokes an immune reaction. In some people, the reaction is so severe that it causes death. This has forced scientists to look for other similar chemicals. One promising chemical is urokinase, an enzyme produced by human cells. Urokinase dissolves blood clots and does not trigger an immune reaction.

Scientists are also testing another naturally occurring clot dissolver, called TPA (tissue plasminogen activator). Tests in humans suggest that TPA may be free of the dangerous side effects of streptokinase. Consequently, TPA has been approved for use in humans since 1987. Nevertheless, TPA use is not without its problems. Two of the most significant are (1) the recurrence of blood clots in many patients and (2) the high cost of the drug.

In cases where the coronary arteries are completely blocked by atherosclerotic plaque, physicians must also reestablish full blood flow to the heart muscle. To restore blood flow, physicians can surgically implant segments of the leg veins to the heart, which carry blood around the clogged coronary arteries (Figure 8–7). This technique, called **coronary bypass surgery**, was once hoped to be a long-term solution to obstructed coronary arteries. Studies show that bypass surgery patients have a significantly higher rate of survival in the five years following surgery than patients who just received drugs. In the next seven years, however, studies show that long-term survival from coronary bypass surgery is about the same as the survival of patients treated with diet and medications. Thus, in the long run, bypass surgery is only slightly more effective than nonsurgical medical treatments.

The problem with coronary bypasses is that the venous grafts fill fairly quickly with cholesterol plaque and the heart becomes starved for oxygen once again, especially when forced to work harder (for example, during exercise). The recurrence of plaque in venous grafts has led researchers to explore the use of marginally important arteries, such as the internal mammary artery, for coronary bypass surgery. Re-

[2]You should consult your physician if you are thinking about taking aspirin as a preventive measure.

Atherosclerosis, Hypertension, and Aneurysms: Causes and Cures

The stresses of modern life and the unhealthy diets of many people take their toll on the heart and blood vessels. One of the most common problems of modern times is atherosclerosis, the buildup of cholesterol plaque in artery walls. Arteries clogged with cholesterol force the heart to work harder. Blocked coronary arteries reduce blood supply to the myocardium with devastating consequences.

Hypertension, or high blood pressure, is another problem of our times. In some individuals, hypertension is hereditary, passed from parent to offspring. Hypertension is also thought to result from stress and diet, especially high salt intake. In other instances, it is caused by disorders in other organs, such as kidney failure or hormonal imbalances. Pregnancy can lead to hypertension, and so can oral contraceptives.

A few facts about hypertension are clear, though. People who are overweight when they are young are more likely to have the disease when they are adults. An adult who is hypertensive and overweight can often control the disease by losing weight and by reducing salt intake.

If untreated, blood pressure rises steadily over the years. For many people, hypertension is nearly a symptomless disease. People feel fine for years, despite the fact that they have high blood pressure, and may not exhibit any physical ailments, until the disease has progressed to the dangerous stage. That is why it is important for people over the age of 40 to have their blood pressure checked each year.

Hypertension is more common in men than women and is more common in African Americans than in Caucasians. The disease is dangerous because the increased pressure in the circulatory system forces the heart to work harder. Elevated blood pressure may also damage the lining of arteries, creating a site for atherosclerotic plaque to form. Atherosclerosis increases the risk of heart attack. A hypertensive person is six times more likely to have a heart attack than an individual with normal blood pressure. Hypertension also increases the chances for an occlusion in the arteries supplying the brain, which can result in a stroke (Chapter 12).

Hypertension, certain infectious diseases (such as syphilis), and atherosclerotic plaque weaken arteries. A degeneration of the tunica media causes a ballooning of the arterial wall, as illustrated in the figure. A bulge in an arterial wall is called an **aneurysm**. Like a worn spot on a tire, it can blow out when pressure builds inside or when the wall becomes too thin.

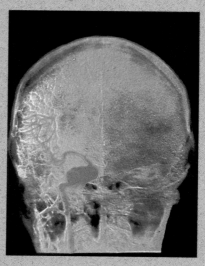

Aneurysm This X-ray shows a ballooning of one of the arteries in the brain. If untreated, an aneurysm can break, causing a stroke.

searchers hope that these arteries will prove more resistant to plaque buildup than veins.

Physicians can clean clogged blood vessels using a small catheter with a tiny balloon attached to its tip in conjunction with the clot-dissolving agents. After the chemical clot dissolvers are given to the patient, the balloon is inflated. This procedure, called balloon angioplasty, forces the artery open and apparently loosens the plaque from the wall, allowing it to be washed away by the blood.

Scientists are also experimenting with lasers to burn away plaque in artery walls. So far, clinical trials on human subjects have proved encouraging. Catheters containing fine glass fibers can be inserted into the blocked arteries during open-heart surgery. They can also be inserted through the artery in the thigh and snaked through the arterial system until they reach the clot. Laser beams transmitted through the glass fibers burn away the plaque. Unfortunately, as in other techniques, cholesterol builds up again in the

When an aneurysm breaks, blood pours out of the circulatory system. Because it happens so quickly, most aneurysms lead to death. An estimated 30,000 Americans die each year from ruptured aneurysms in the brain, and nearly 3000 die from ruptured aortic aneurysms.

As in most diseases, the first line of defense against aneurysms is prevention. By reducing or eliminating the two main causes—atherosclerosis and high blood pressure—an individual can greatly lower his or her risk. Physicians recommend a number of steps to cut your chances of these diseases: stop smoking; lose weight, if you are overweight; reduce salt intake; reduce stress at work and at home; drink alcohol in moderation, if at all; and learn to relax. Reducing cholesterol intake is also helpful. Eat high-cholesterol foods, such as eggs, in moderation. Certain types of fiber can also help reduce cholesterol. The water-soluble fiber found in apples, bananas, citrus fruits, carrots, barley, and oats, for example, reduces cholesterol uptake by the intestine, as explained in Chapter 7. Drug treatment is also possible.

The second line of defense against cardiovascular disease is early detection and treatment before trouble sets in. Hypertension can be detected by regular blood pressure readings. Atherosclerosis can be discovered by blood tests. Aneurysms can be detected by X-ray. Pain alerts patients and physicians that something is wrong. Once any of these diseases is detected, physicians have many options.

In the event of an aneurysm, for example, surgeons can remove the weakened section of the artery and replace it with a section of a vein. In other instances, where venous grafts are more difficult (for example, in the brain), surgeons can clamp or tie off the artery just before the bulge, preventing blood flow through the damaged section. This works only when other arteries provide adequate blood flow to the area served by the damaged artery. In larger arteries, pieces of dacron or other synthetic materials can be sewn into the wall of the artery, protecting it from breaking.

Researchers have also been studying the use of an alloy of nickel and titanium, called nitinol. Nitinol is a "metal with a memory." When a fine nitinol wire is wrapped around a cylinder the size of the interior of an artery and heated, it forms a tightly coiled spring. When the spring is cooled, it reverts to a straight wire, but when reheated the metal returns to a coil.

Scientists are testing ways to reinforce weak arterial walls with nitinol coils. After creating a wire coil that corresponds to the internal diameter of damaged arteries, they cool the wire, causing it to revert to the straight form. Then, they push the wire through a catheter inserted into a damaged artery. As the wire emerges from the cooled catheter inside the damaged artery, body heat causes it to coil once again. In place, the coil adds strength to the wall of the artery, preventing rupture. Experiments with dogs show that the endothelial cells of the tunica intima soon grow over the implant, making it a permanent part of the artery's wall.

This procedure could help save hundreds, perhaps thousands, of lives each year, but it is no substitute for a healthy diet and a healthy environment. Reducing stress and reducing cholesterol in our diets require a lifetime of commitment, but pay huge dividends in the long run.

walls of arteries within a few months. (See Health Note 8–1 for a discussion of other aspects of cardiovascular disease.)

THE BLOOD VESSELS

The circulatory system can be divided into four functional parts. The first is the heart, which pumps the blood throughout the body. The second is the arteries, a delivery system that carries blood to the body tissues. The third is the capillaries, which form an extensive exchange system. The fourth is the veins, which constitute a return system, carrying blood back to the heart after it has given off its oxygen and picked up wastes.

Arteries transport blood away from the heart and, as they course through the body, branch many times, forming smaller and smaller blood vessels. The smallest of all arteries is the **arteriole**. As illustrated in

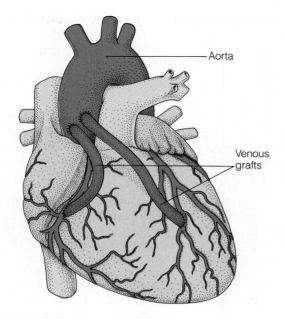

FIGURE 8–7 Coronary Bypass Surgery Venous grafts bypass coronary arteries occluded by atherosclerotic plaque.

Figure 8–8, arterioles empty into **capillaries**, thin-walled vessels through the walls of which nutrients and wastes pass. Capillaries, therefore, provide an avenue for exchange between the blood and the tissue fluid surrounding the cells of the body (Figure 8–8).

Blood flows from the capillaries into the smallest of all veins, the **venules**. Venules, in turn, converge to form small veins, which unite with other small veins, in much the same way that small streams unite, forming larger and larger streams.

Figure 8–9 shows a cross section of an artery and a vein. These two vessels are quite different. Veins, for example, tend to be smaller, have thinner walls, and are irregularly shaped. Despite their obvious differences, arteries and veins share a common architecture. For example, both consist of three layers: (1) an external layer of connective tissue, the **tunica adventitia**, which binds the vessel to surrounding tissues; (2) a middle layer, the **tunica media**, which is primarily made of smooth muscle, and (3) an internal layer, the **tunica intima**, which is composed of a layer of flattened cells, the **endothelium**, and a thin, nearly indiscernible layer of connective tissue, which binds the endothelium to the tunica media (Figure 8–10).

Arteries and Arterioles: The Delivery System

The largest of all arteries is the aorta. As noted earlier, this massive vessel carries oxygenated blood from the left ventricle of the heart to the rest of the body. The aorta loops over the back of the heart, then descends through the chest and abdomen, giving off large branches along its route. These branches carry blood to the head, the extremities, and major organs, such as the stomach, the intestines, and the kidneys. The very first branches are the coronary arteries, which supply the tissues of the heart with oxygenated blood.

The aorta and many of its chief branches are **elastic arteries**, so named because they contain numerous wavy elastic fibers interspersed among the smooth muscle cells of the tunica media (Figure 8–11a). As blood pulses out of the heart, the elastic arteries expand to accommodate the blood. Like a stretched rubber band, though, the elastic fibers in the tunica media cause the arterial walls to recoil. This provides a backup pump, which helps push the blood along the arterial tree and helps maintain an even flow of blood through the capillaries.

The elastic arteries branch to form smaller vessels, the **muscular arteries** (Figure 8–11b). Muscular arteries contain fewer elastic fibers, but can still expand and contract with the flow of blood. You can feel this expansion and contraction in the arteries lying near the skin's surface in your wrist and neck. It's the pulse health care workers use to measure your heart rate.

The smooth muscle of the tunica media of muscular arteries responds to a number of stimuli, including nerve impulses, hormones, carbon dioxide, and lactic acid. These stimuli cause the blood vessels to open or close to varying degrees. This, in turn, allows the body to adjust blood flow through its tissues to meet increased demands for nutrients and oxygen. Arterioles in muscles, for instance, open when an individual is threatened by danger. This increases blood flow to the muscle, allowing a person to flee or to meet the danger head on. At the same time, vessels in the digestive system are closed down, temporarily shutting down the digestive process and increasing the amount of blood available to the muscles. Regulating the flow of blood to body tissues is required to control body temperature as well.

Blood Pressure. Blood pressure is highest in the aorta and drops considerably as the arteries branch.

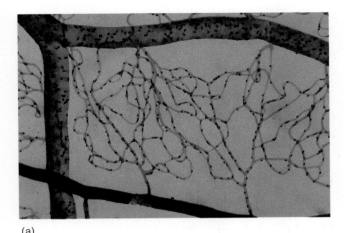

(a)

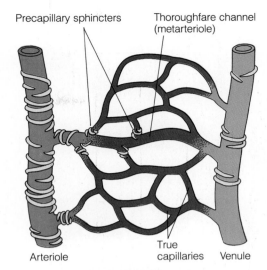

Sphincters open

FIGURE 8–8 Capillaries (*a*) Capillaries form extensive branching networks between the arteriole on the left and the venule on the right. (*b*) Flow into the capillary bed is controlled by the metarterioles and the precapillary sphincters. When the metarterioles are open, they allow blood to flow into the capillary network. Precapillary sphincters must open to permit further influx. When precapillary sphincters close they reduce blood flow through the capillary bed. (*c*) A scanning electron micrograph of an arteriole showing the smooth muscle cells that contract and relax, controlling flow through these vessels.

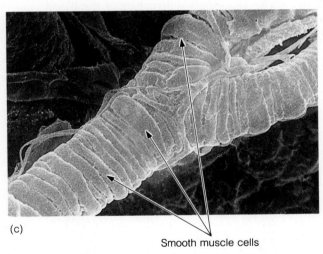

(c)

Smooth muscle cells

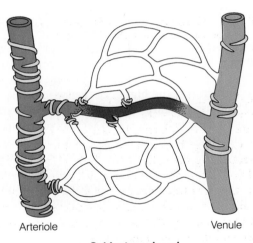

Sphincters closed

(b)

When blood reaches the capillaries, the flow of blood is greatly reduced. The reduction in flow and pressure increases the rate of exchange between the blood and the tissues.

Blood pressure is measured with an inflatable device, which goes by a tongue-twisting name, **sphygmomanometer**, or more commonly a blood pressure cuff (Figure 8–12a). The blood pressure cuff is first wrapped around the upper arm. A stethoscope is positioned over the artery just below the cuff. Air is pumped into the cuff until the pressure stops the flow of blood through the artery (Figure 8–12b). The pressure in the cuff is then gradually reduced as air is released. When the blood pressure in the artery exceeds the external pressure of the cuff, the blood starts flowing through the vessel once again. This

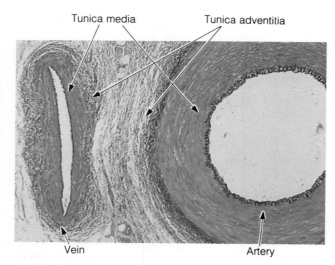

FIGURE 8–9 Artery and Vein A cross section through a vein shows that the muscular layer, the tunica media, is much thinner than in an artery. Veins typically lie along side arteries and have irregular lumens.

point represents the **systolic pressure** and is detected by a thump heard through the stethoscope. Systolic pressure is the peak pressure at the moment the ventricles contract. It is the higher of the two numbers in a blood pressure reading (120/70).[3] The pressure at the moment the heart relaxes to let the ventricles fill again is the **diastolic pressure** and is the lower of the two readings. It is determined by continuing to release air from the cuff until no arterial pulsation is audible. At this point, blood is flowing continuously through the artery. A typical reading would be 120/70.

Capillaries: The Exchange System

The heart, arteries, and veins are an elaborate system that transports blood to and from the capillaries. Capillaries form branching networks, or **capillary beds**, among the cells of body tissues. It is in these extensive capillary beds that wastes and nutrients are exchanged between the cells of the body and the blood. As illustrated in Figure 8–13a and b, the walls of the capillaries consist of flattened endothelial cells that allow dissolved substances to pass through them with great ease—an illustration of the remarkable correlation between structure and function. In some places, such as the kidneys and small intestine, small win-

[3]Blood pressure is measured in millimeters of mercury (mm Hg).

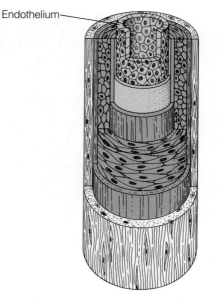

FIGURE 8–10 General Structure of the Blood Vessel The artery shown here consists of three major layers, the tunica intima, tunica media, and tunica adventitia.

dows, or **fenestrae**, are present in the cells of capillary walls (Figure 8–13d). Fenestrae permit even greater movement of molecules to and from the capillary.

The extensive branching of capillaries brings the capillaries in close proximity to body cells, which also facilitates the exchange of nutrients and wastes. If you could remove all of the capillaries from the body and line them up end to end, they would extend over 80,500 kilometers (50,000 miles). That's enough to circle the globe at the equator twice! Extensive capillary branching slows the rate of blood flow through capillary networks and decreases pressure, both of which increase the efficiency of capillary exchange.

A "typical" capillary network is shown in Figure 8–8b. As illustrated, most capillary beds contain **thoroughfare channels** or **metarterioles**, circulatory "short cuts" that connect the arterioles with the venules. The metarterioles give off smaller branches, the **true capillaries**, the site of the exchange of nutrients and wastes.

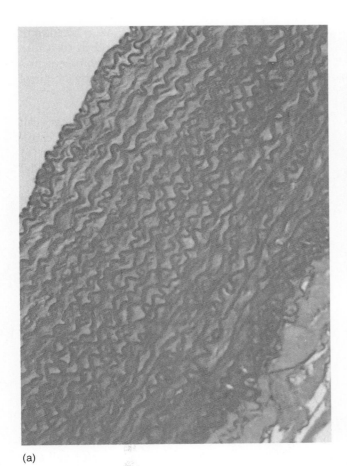

(a)

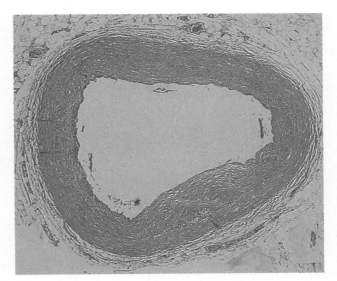

(b)

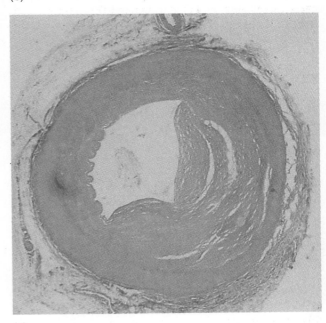

(c)

FIGURE 8–11 Arteries (*a*) A cross section of the wall of the aorta showing numerous wavy elastic fibers common in all elastic arteries. (*b*) The muscular artery has a thick tunica media as well but fewer elastic fibers. (*c*) Muscular artery partly occluded by atherosclerotic plaque.

The flow of blood through the capillary bed, and therefore the supply of nutrients to tissues, is largely controlled by the constriction and dilation of the metarteriole. When the metarteriole is open, blood can flow into the capillary network, but when it constricts, blood passes by, traveling on to other tissues in need of nutrients.

The constriction and dilation of the metarteriole help regulate body temperature. On a cold winter day, the metarterioles of the capillary networks in the skin, for example, close down, restricting blood flow and conserving body heat. Just the reverse happens on a warm day. The flow of blood through the skin increases, sometimes creating a pink flush and releasing body heat.

Blood flow is controlled at another level as well, thus increasing the body's ability to deliver blood on demand to cells that need it or reduce the flow to cells that do not. As Figure 8–8b shows, tiny rings of muscle surround the capillaries as they branch from the metarterioles. These muscle rings are called **precapillary sphincters**. They open and close like floodgates in response to local chemical signals (such as carbon

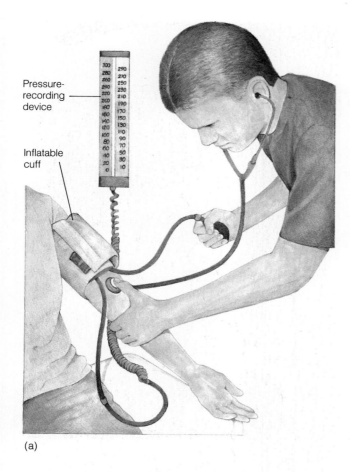

Pressure-recording device

Inflatable cuff

(a)

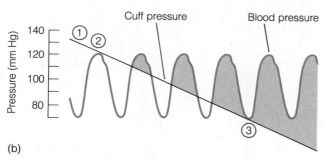

(b)

FIGURE 8–12 Blood Pressure Reading (*a*) A sphygmomanometer (blood pressure cuff) is used to determine blood pressure. (*b*) When the pressure in the cuff exceeds the arterial peak pressure, blood flow stops ①. No sound is heard. Cuff pressure is gradually released. When arterial pressure exceeds the cuff pressure, blood starts flowing. This is the systolic pressure ②. The first sound will be heard. Cuff pressure continues to drop. When cuff pressure is equal to the lowest pressure in the artery, no sound is heard ③. This is the diastolic pressure.

dioxide levels) from nearby tissues. The relaxation of the precapillary sphincters causes blood to rush into the capillary bed. The precapillary sphincters are, therefore, a means of fine-tuning the body's control over blood flow.

Capillary Exchange. Wastes and nutrients can travel (1) between the endothelial cells of the capillary, (2) through the endothelial cells by diffusion, (3) through fenestrae, and (4) through the endothelial cells in minute pinocytotic vesicles (Chapter 3).

As blood flows *into* a capillary bed, nutrients, gases, water, and hormones immediately begin to diffuse out of the tiny vessels. As the blood flows *through* the capillary bed, however, water-dissolved wastes, such as carbon dioxide, begin to flow into the capillaries by diffusion, a movement of materials from areas of higher concentration to areas of lower concentration. The forces that control movement across the capillary wall are explained in Figure 8–14.

Veins and Venules: The Return System

Blood leaves the capillary beds stripped of its nutrients and loaded with cellular wastes. Blood draining from the capillaries first enters the smallest of all veins, the venules. The venules converge with others, forming small veins. **Veins** carry blood toward the heart. Unlike the arteries, the veins start off small and converge with other veins, forming larger and larger vessels. Eventually, all blood returning to the heart from the systemic circulation enters the inferior and superior vena cavae, the two main veins that empty into the right atrium of the heart.

Veins and arteries generally run side by side throughout the body much like opposing lanes on a freeway. The artery takes blood away from the heart and toward body tissues, and the veins return blood to the heart.

Blood pressure in the veins is low, and veins have relatively thin walls with few smooth muscle cells (Figure 8–9). Because the veins' walls are so thin, obstructions can cause them to balloon out, in much the same way that a tree down across a small stream can cause water to pool upstream. Blood pools in the obstructed veins, forming rather ugly bluish bulges, called **varicose veins**.

Some people inherit a tendency to develop varicose veins, but most cases can be attributed to factors that reduce the flow of blood back to the heart. Abdominal

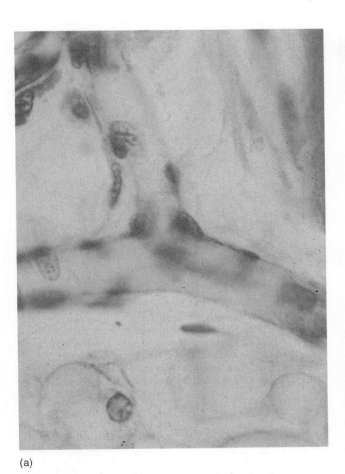

(a)

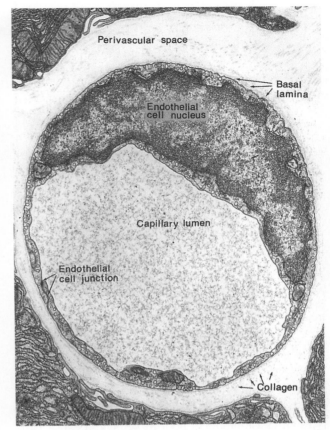

(b)

FIGURE 8-13 The Capillary (*a*) A light micrograph of a capillary showing the endothelial cells that make up the wall of this vessel. (*b*) A cross section of a capillary showing the nucleus of an endothelial cell and capillary lumen. (*c*) A cross section through the wall of a capillary showing how materials flow through via endocytosis. (*d*) A section through the wall of a highly porous capillary showing the fenestrae. Note that each window is spanned by a thin membrane that permits rapid movement of molecules into and out of the capillary.

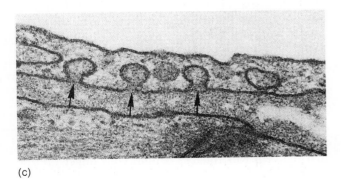

(c)

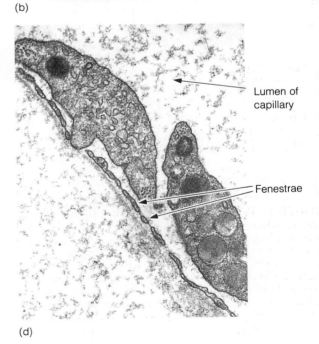

(d)

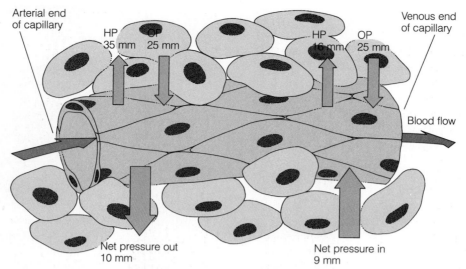

Arterial end
of capillary

HP
35 mm

OP
25 mm

HP
16 mm

OP
25 mm

Venous end
of capillary

Blood flow

Net pressure out
10 mm

Net pressure in
9 mm

FIGURE 8–14 Capillary Exchange Fluid begins to flow out of the capillary as soon as it enters the vessel from the arteriole. A slight hydrostatic pressure (HP), resulting from blood pressure, creates a slight outward force of 35 mm Hg. Osmotic pressure (OP), the tendency for water to flow inward because of a higher internal concentration of dissolved solute, is only 25 mm Hg. As a result, there is a net outward movement on the arterial side of the capillary network of 10 mm Hg. The hydrostatic pressure declines as the blood flows through the capillary bed, and, therefore, the osmotic pressure forcing water inward exceeds the outward pressure. This drives water back into the capillary. Notice that the net movement inward does not equal the net movement out. As a result, excess fluid accumulates in the tissue and must be drained by the lymphatic system.

tumors, pregnancy, obesity, and even sedentary lifestyles all cause blood to pool in veins and cause varicose veins to appear in the lower extremities. The restriction of blood flow may also result in muscle cramps and the buildup of fluid, **edema** (swelling), in the lower extremity, the ankles and legs.

Varicose veins may also form in the wall of the anal canal. When the internal hemorrhoidal veins become swollen, they result in a condition often known as **hemorrhoids**. The internal hemorrhoidal veins are supplied by numerous pain fibers, making this condition extremely painful.

How Do the Veins Work? Figure 8–15 shows that blood pressure declines rapidly as it travels through the arteries and continues to decrease, but at a slower rate, as it flows through the capillaries and veins. The lowest blood pressure readings occur in the superior and inferior vena cavae. With such low pressure, how do the veins return blood to the heart?

For blood in veins above the heart, gravity is the chief means of propulsion. But for veins below the heart, which have very little pressure to force the blood along, the body must rely on the movements of body parts to squeeze the blood upward. As you walk to class, for example, the contraction of muscles in your legs squeezes blood upward, slowly and surely. Even the nervous muscle contractions that occur during study help move the blood back to the heart. But these movements are not enough. Blood flow would probably not occur without valves. **Valves** are flaps of tissue that span the veins and prevent the backflow of blood. The structure of the valves is shown in Figure 8–16. As illustrated, the semilunar flaps of the veins resemble those found in the heart. Just as in the valves of the heart, blood pressure, however slight, pushes the flaps open (Figure 8–16). This allows the blood to move forward. As the blood fills the segment of the vein in front of the valve, it pushes back on the valve flaps, forcing them shut. The valves of the veins resemble the locks of a canal. Designed to move ships uphill, locks open to let water in, then close when the downstream section is full. You can locate the valves by running a finger along one of the superficial veins on your forearm. Start by pressing gently on a vein, then run your finger toward your wrist. You will see that the vein will collapse behind your finger until it crosses a valve.

THE BLOOD

In 1966, Dr. Leland C. Clark of the University of Cincinnati's College of Medicine immersed a live laboratory mouse in a clear fluorocarbon solution saturated with oxygen. To the amazement of the audience, the mouse continued to "breathe." Some time later, Dr. Clark pulled the animal from the solution, and after a brief moment the rodent began to move, apparently unharmed by the ordeal.

The importance of this discovery was not that an animal could "breathe" this fluid into its lungs and survive, but that the solution could hold so much oxygen. The researchers hoped that it could be injected into the bloodstream where it would serve as an oxygen-transport medium. Clark and a colleague had discovered a potential substitute for blood.[4] "Artificial blood," they hoped, would someday be a boon to medical science, helping emergency medical personnel sustain accident victims while en route to the hospital. In rural America alone, artificial blood could save thousands of lives a year.

Human blood is a far cry from Clark's artificial substitute. The **blood** in our circulatory systems, for example, is a watery fluid, not a fluorocarbon, and consists of two basic components: plasma and "formed elements"—cells or fragments of cells. **Plasma** is the liquid portion of the blood, consisting of about 90% water. Many dissolved substances are found in the plasma. The blood contains three different formed elements: (1) white blood cells (or leukocytes), (2) red blood cells (or erythrocytes), and (3) platelets (or thrombocytes).

Blood accounts for about 8% of our total body weight. In an adult male weighing 70 kilograms (150 pounds), the cardiovascular system contains about 5 to 6 liters of blood, or approximately 1.3 to 1.5 gallons. An adult female has about a liter less.

As illustrated in Figure 8–17, the plasma constitutes about 55% of the volume of a person's blood. The remaining portion consists of blood cells. The volume of the blood occupied by blood cells is called the **hematocrit**. To determine the hematocrit, whole blood is placed in a test tube, which is inserted into a centrifuge, a device that spins the samples at high speeds (Figure 8–17). Centrifugation causes the red

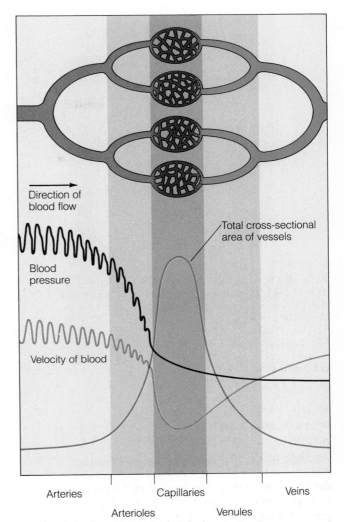

FIGURE 8–15 Blood Pressure in the Circulatory System Blood pressure declines in the circulatory system as the vessels branch. Arterial pressure pulses because of the heartbeat, but pulsation is lost by the time the blood reaches the capillary networks, creating an even flow through body tissues. Blood pressure continues to decline in the venous side of the circulatory system.

blood cells (RBCs), white blood cells, and platelets to separate from the plasma and "settle" to the bottom of the test tube. The average hematocrit in humans is approximately 45%. During centrifugation, the white blood cells and platelets settle out on top of the RBCs and comprise only about 1% of the total blood volume. Therefore, the hematocrit is largely determined by the concentration of RBCs.

---

[4]Artificial blood, of course, is a misnomer. The fluorocarbon solution takes over the role of the plasma and the red blood cells, but not the white blood cells.

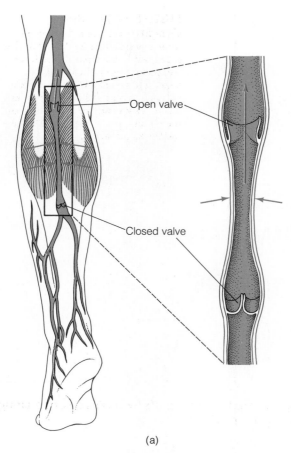

(a)

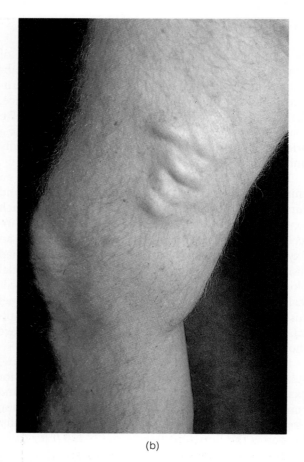

(b)

FIGURE 8–16 Valves in Veins (*a*) The slight hydrostatic pressure in the veins and the contraction of skeletal muscles propel the blood along the veins back toward the heart. The one-way valves stop the blood from flowing backward. (*b*) Any restriction of venous blood flow to the heart causes veins to balloon out, creating bulges commonly known as varicose veins.

The hematocrit varies in individuals living at different altitudes. In a person living in the Mile High City of Denver, Colorado, for example, the hematocrit is about 5% higher than in a person living at sea level. The slight increase in the hematocrit compensates for the slightly lower level of oxygen in the atmosphere at higher altitudes.

Plasma

The plasma is a light yellow (straw-colored) fluid. Dissolved in the plasma are (1) gases, such as nitrogen, carbon dioxide, and oxygen; (2) ions, such as sodium, chloride, and calcium; (3) nutrients, such as glucose and amino acids; (4) hormones; (5) proteins; and (6) various wastes. Lipids are also found in the plasma, but are suspended in tiny globules. Lipid mol-

ecules may also bind to plasma carrier proteins, which transport them through the bloodstream.

The most abundant of all dissolved substances in the plasma are the plasma proteins. Three major proteins are found in the blood plasma: (1) albumins, (2) globulins, (3) fibrinogen. Their functions are summarized in Table 8–2. The albumins and some of the globulins are carrier molecules that bind to hormones, ions, and fatty acids, helping to transport these molecules through the bloodstream. Carrier proteins are large, water-soluble molecules. Lipid molecules hitch a free ride on the proteins and can be transported throughout the bloodstream. The carrier proteins also protect molecules from destruction by the liver.

Some globulins serve as **antibodies**, proteins that neutralize viruses and bacteria or target them for de-

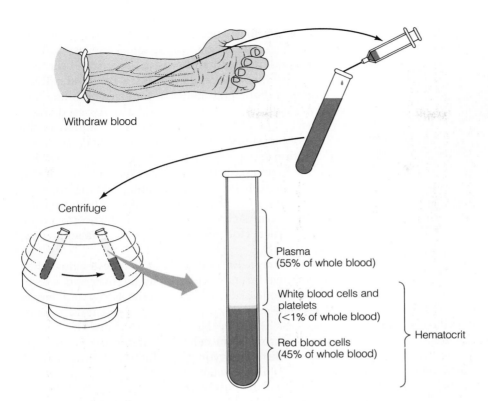

Withdraw blood

Centrifuge

Plasma
(55% of whole blood)

White blood cells and
platelets
(<1% of whole blood)

Red blood cells
(45% of whole blood)

Hematocrit

FIGURE 8–17   Blood Composition Blood removed from a person can be centrifuged to separate plasma from the cellular component. Red blood cells constitute about 45% of the blood volume, except at higher altitudes where they make up about 50% of the volume to compensate for the lower oxygen levels.

struction by macrophages (Chapter 10). **Fibrinogen,** the third major protein in blood, helps form blood clots, and is described below.

All plasma proteins contribute to osmotic pressure. As you may recall from Chapter 3, the osmotic pressure of the blood helps regulate the chemical equilibrium between blood plasma and the extracellular fluid. Capillary exchange, discussed earlier, is largely the result of osmotic concentration gradients.

Chapters 2 and 6 noted that the blood maintains a fairly constant pH, thanks to the buffering action of bicarbonate ions (HCO_3^-), formed when carbon dioxide dissolves in the plasma. Blood proteins also help maintain a constant pH by accepting or donating hydrogen ions (H^+).

Red Blood Cells

The **red blood cell** (**RBC** or **erythrocyte**) is a highly specialized "cell" in humans. Unlike most other cells, RBCs lose their nuclei and the rest of their organelles during cellular differentiation (Figure 8–18). Consequently, fully formed RBCs cannot divide, and new cells must be produced by the bone marrow to replace the billions of RBCs that die each day.

TABLE 8–2 Summary of Plasma Proteins	
PROTEIN	FUNCTION
Albumins	Maintain osmotic pressure and transport smaller molecules, such as hormones and ions
Globulins	Alpha and beta globulins transport hormones and fat-soluble vitamins
	Gamma globulins (antibodies) bind to foreign substances
Fibrinogen	Converted into fibrin network that helps to form blood clots

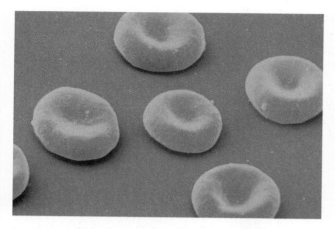

FIGURE 8–18   Red Blood Cells   A scanning electron micrograph showing the biconcave disk shape.

The RBC transports oxygen and, to a lesser degree, carbon dioxide in the blood (Table 8–3). Human RBCs are flexible, biconcave disks (Figure 8–18). The biconcave shape of the RBC increases the surface area for absorption of gases, which facilitates the rapid and efficient exchange of gases between the cell and the plasma.

Twenty drops of blood equal one milliliter and contain approximately 5 billion RBCs.[5] If RBCs were people, a single milliliter of blood would contain the entire world population.

RBCs are swept along in the bloodstream, traveling many times around the circulatory system each day. Just slightly larger than the internal diameter of the capillaries, the highly flexible RBCs bend and twist to get through narrow capillaries. A genetic mutation carried by some African Americans illustrates just how important RBC flexibility is.[6] In some African Americans, a mutated gene that codes for the hemoglobin protein produces a molecule containing an incorrect amino acid. This slight change in the primary structure of the hemoglobin molecule alters the tertiary structure of the hemoglobin molecule and is responsible for the disease called **sickle-cell anemia**. The alteration in protein structure causes the RBCs to transform from biconcave disks to long, sickle-shaped

cells when they encounter low levels of oxygen in capillaries. The sickle-shaped cells are less flexible and unable to bend and twist as they travel through the maze of body capillaries. As a result, the RBCs collect at branching points in capillary beds like logs in a logjam. Here, they may block capillaries, disrupting the flow of blood to tissues. A disruption in the flow of blood reduces oxygen levels in body tissues, a condition known as **anoxia**. Anoxia causes considerable pain and tissue damage. Blockage in lungs, heart, and brain can be life threatening, leading to heart attacks and brain damage. Many people who have the disease die in their late twenties and thirties; some die even earlier (for more on sickle-cell anemia, see Chapter 18).

On average, RBCs live about 120 days before they are removed from circulation by the liver and spleen. The iron released from RBCs in the spleen and liver is recycled—as many nutrients are—and used to produce new RBCs in the red bone marrow. In the red bone marrow, **stem cells**, undifferentiated cells that trace back to embryonic development, give rise to 2 million RBCs per second! The recycling of iron is not 100% efficient, however, and small amounts of iron must be ingested each day in the diet. Loss of blood from an injury or, in women, loss of blood during menstruation increases the body's demand for dietary iron. Without adequate intake, anemia may occur.

In infants and children, almost all of the bone marrow is dedicated to the production of RBCs. As growth slows, though, the red marrow of many bones becomes inactive and gradually fills with fat, forming **yellow marrow**, a fat storage depot. By the time an individual reaches adulthood, only a few bones—such as, the hip bones, sternum (breastbone), ribs, and the bodies of the vertebrae—are engaged in RBC production. In severe, prolonged anemia, yellow marrow can be reactivated—converted back into red marrow to produce RBCs.

The number of RBCs in the blood remains more or less constant over long periods. Maintaining a constant concentration of RBCs is essential to homeostasis and is controlled by the hormone erythropoietin. **Erythropoietin** is secreted by the kidney when blood oxygen levels decline—for example, when a person moves to high altitudes or loses a significant amount of blood. Erythropoietin is carried throughout the bloodstream. In the red bone marrow, it stimulates the stem cells to multiply, increasing RBC production. As the RBC concentration in the blood increases, oxygen supplies increase. When oxygen levels return to normal, erythropoietin levels fall, reducing the rate of

---

[5]Many texts note the concentration per cubic millimeter, which is about 5 million RBCs. There are, of course, 1000 cubic millimeters in a cubic centimeter or one milliliter.

[6]Approximately 1 of every 500 to 1000 African Americans is afflicted with sickle-cell anemia. One of every 12 carries the gene. Carriers are not afflicted with the disease, but can pass it to others. For an explanation of this disease, see Chapter 18.

TABLE 8-3 Summary of Blood Cells

NAME	LIGHT MICROGRAPH	DESCRIPTION	CONCENTRATION (NUMBER CELLS/MM³)	LIFE SPAN	FUNCTION
Red blood cells (RBC)		Biconcave disk; no nucleus	4-6 million	120 days	Transports oxygen and carbon dioxide
White blood cells Neutrophil		Approximately twice the size of RBC; multilobed nucleus; clear-staining cytoplasm	3000-7000	6 hours to a few days	Phagocytizes bacteria
Eosinophil		Approximately same size as neutrophil; large pink-staining granules; bilobed nucleus	100-400	8-12 days	Phagocytizes antigen-antibody complex; attacks parasites
Basophil		Slightly smaller than neutrophil; contains large purple cytoplasmic granules; bilobed nucleus	20-50	Few hours to a few days	Releases histamine during inflammation
Monocyte		Larger than neutrophil; cytoplasm grayish-blue; no cytoplasmic granules; U- or kidney-shaped nucleus	100-700	Lasts many months	Phagocytizes bacteria, dead cells, and cellular debris
Lymphocyte		Slightly smaller than neutrophil; large relatively round nucleus that fills the cell	1500-3000	Can persist many years	Involved in immune protection, either attacking cells directly or producing antibodies
Platelets		Fragments of megakaryocytes; appear as small darkstaining granules	250,000	5-10 days	Play several key roles in blood clotting

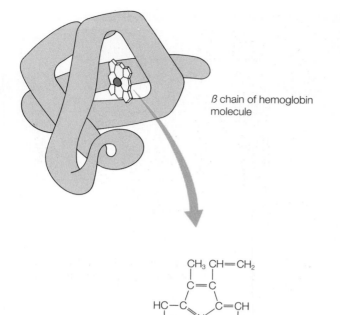

β chain of hemoglobin molecule

FIGURE 8–19 Porphyrin Ring The porphyrin ring of the hemoglobin molecule holds an iron ion that binds to oxygen and carbon monoxide.

RBC formation in a classical negative feedback mechanism common in homeostatic mechanisms.

Hemoglobin. Hemoglobin is a large protein molecule composed of four protein subunits. Found exclusively in the RBCs, it accounts for about a third of the RBC's weight. As shown in Figure 8–19, each hemoglobin subunit contains a heme group. Each **heme group** consists of a large, organic ring structure, called a **porphyrin ring**. In the center of the ring is an iron ion.[7]

[7]The iron is Fe^{+2}, the ferrous ion. To be effective, iron supplements should contain ferrous iron.

When blood flows through the capillary beds of the lungs, oxygen diffuses into the blood, then into the RBCs. Inside the RBC, oxygen binds to the iron ions in hemoglobin for transport through the circulatory system. The importance of this binding is underscored by the fact that 98% of the oxygen in the blood is transported bound to iron of hemoglobin. The remaining 2% is dissolved in the blood plasma.

Carbon dioxide also binds to hemoglobin, but to a much lesser degree. Most carbon dioxide, in fact, is carried as bicarbonate ions (HCO_3^-) dissolved in the plasma.

Diseases Involving Red Blood Cells. Homeostasis requires the normal operation of the heart and blood vessels. Homeostasis also requires the blood to absorb a sufficient amount of oxygen as it passes through the lungs. Unfortunately, these requirements are not always met. Blood disorders and nutritional deficiencies, for example, can reduce the oxygen-carrying capacity of the blood, resulting in anemia. Anemia is a blood disorder that may result from (1) a decrease in the number of circulating RBCs, (2) a reduction in the hemoglobin content of the RBCs, and (3) the presence of abnormal hemoglobin. The number of RBCs in the blood may decline because of excessive bleeding, tumors in the bone marrow that reduce RBC production, and several infectious diseases (such as malaria). The amount of hemoglobin in RBCs may be caused by iron deficiency, described in the nutrition chapter (Chapter 7) or a deficiency in vitamin B-12. Abnormal hemoglobin is produced in sickle-cell anemia and other genetic disorders.

Anemia generally results in weakness and fatigue. Individuals are often pale and tend to faint easily or become breathless. The heart often beats faster to compensate for a reduction in the oxygen-carrying capacity of the blood. Anemia is very unlikely to lead to death, but does weaken one's resistance to the effects of other diseases or injury. It also limits a person's productivity and energy. No matter what the cause, anemia should be treated quickly (Chapter 7).

White Blood Cells

White blood cells (WBCs) are nucleated cells that are part of the body's protective mechanism to combat microorganisms, such as bacteria and viruses (Chapter 10). White blood cells are produced in the bone marrow and circulate in the bloodstream; they constitute less than 1% of the blood volume.

White cells do most of their work outside of the bloodstream, in the tissues. The bloodstream is, therefore, a vehicle of transportation, delivering them to sites of infection. When WBCs arrive at the scene, they squeeze between the endothelial cells, a process called **diapedesis** (Figure 8–20).

Table 8–3 lists and describes the five WBCs found in the blood. The three most numerous, which are discussed here, are neutrophils, monocytes, and lymphocytes.

Neutrophils are the most abundant WBC. Approximately twice the size of the RBC, these cells are distinguished by their multilobed nuclei. So named because their cytoplasm has a low affinity for stains, neutrophils patrol the blood like a cellular police force awaiting microbial invasion. Attracted by chemicals released from infected tissue, neutrophils escape, then migrate to the site of infection by ameboid movement.

Neutrophils are usually the first WBC to arrive on the scene. Here, they phagocytize microorganisms, helping prevent the spread of bacteria and other organisms. When a neutrophil's lysosomes are used up, however, the cell dies and becomes part of the yellowish liquid, or **pus**, which exudes from wounds. Pus is a mixture of dead or dying neutrophils, cellular debris, and bacteria, both living and dead.

Monocytes are also phagocytic cells. Approximately the size of neutrophils, monocytes contain distinctive U-shaped or kidney-shaped nuclei. Like neutrophils, monocytes leave the bloodstream to do their "work," migrating through tissues via ameboid motion. Once on the scene, they begin phagocytizing microorganisms, dead cells, and dead neutrophils. Thus, while neutrophils are the "first-line" troops, the monocytes are something of a mop-up crew.

Monocytes take up residence in connective tissues of the body where they are referred to as **macrophages**. These cells remain more or less stationary, like watchful soldiers.

The second most abundant WBC is the **lymphocyte**. Most lymphocytes exist outside the circulatory system in **lymphoid organs**, such as the spleen, thymus, and lymph nodes, and **lymphoid tissue**, aggregations of lymphocytes beneath the lining of the intestinal and respiratory tracts. It is in these locations that they attack microbial intruders.

Two types of lymphocytes are found in the body, both of which play a vital role in immune protection (Chapter 10). The first type, the **T-lymphocyte** or **T cell**, attacks foreign cells, such as fungi, parasites, and

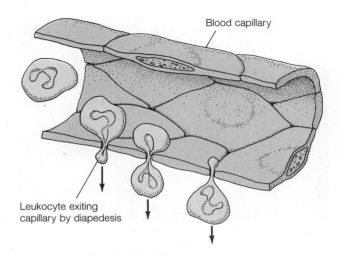

FIGURE 8–20 Diapedesis White blood cells (leukocytes) escape from capillaries by squeezing between endothelial cells.

tumor cells.[8] T-lymphocytes are thus said to provide "cellular immunity." The second type is called the **B-lymphocyte** or **B cell**. It transforms into a **plasma cell**, which synthesizes and releases proteins called antibodies. **Antibodies** circulate in the blood and are capable of binding to foreign substances, neutralizing and targeting microorganisms and tumor cells for destruction by macrophages (Chapter 10).

The WBCs, like the other formed elements of blood, are part of the body's homeostatic mechanism. Their numbers can increase greatly during a microbial infection and other diseases. An increase in the number of WBCs, called **leukocytosis**, is a normal homeostatic response to foreign intruders. It ends when the invaders have been destroyed. Increases and decreases in various types of blood cells can be used to diagnose many medical disorders. Thus, a blood test is a standard procedure for patients undergoing diagnostic testing.

Diseases Involving White Blood Cells. By protecting the body, WBCs are part of one of the body's most important homeostatic mechanisms. The function of WBCs, however, can run amok. White blood cells, for

[8]T-lymphocytes usually attack large eukaryotic cells, such as fungi and parasites. Most bacteria are controlled by antibodies. Only the few bacteria that are intracellular parasites, such as *M. tuberculosis*, are attacked by T cells, but even then, the lymphocyte attacks the host cell, not the bacterium directly.

FIGURE 8–21 Rosy Periwinkle Many tropical plants provide extraordinary medical benefits. The rosy periwinkle helps physicians treat leukemia. Unfortunately, the tropical rainforests are being cut at an alarming rate, reducing our chances of finding other cures.

example, can become cancerous, dividing uncontrollably in the bone marrow, then entering the bloodstream. A cancer of WBCs is called **leukemia** (literally, "white blood"). The most serious type of leukemia is **acute leukemia**, so named because it kills victims quickly. Children are the primary victims of this disease.

In acute leukemia, WBCs fill the bone marrow and crowd out the marrow cells that produce RBCs and platelets. A decline in the production of RBCs leads to anemia. A reduction in platelet production reduces clotting, increasing internal bleeding. Making matters worse, the cancerous WBCs produced in leukemia are often incapable of fighting infection. Because of this and because the production of platelets is reduced, victims of acute leukemia typically succumb to infections and internal bleeding.

Leukemia can be treated by irradiating the bone marrow and by administering a drug that stops mitosis. Twenty years ago, only one of every four children with leukemia survived. Today, thanks to this drug, the odds have changed dramatically: three of every four children with the disease survives! The drug, like thousands of others in use today, was derived from a plant, called the rosy periwinkle, found in the tropical rain forests (Figure 8–21).

Another common disorder of the WBCs is **infectious mononucleosis**, commonly called "mono" or "kissing disease." Mono is caused by a virus transmitted through saliva and may be spread by kissing; by sharing silverware, plates, and glasses; and possibly even through drinking fountains. The virus spreads through the body, affecting many organs. Even though the virus only infects lymphocytes, the number of monocytes and lymphocytes in the blood increases rapidly during an infection. Individuals suffering from mono complain of fatigue, aches, sore throats, and low-grade fever. Physicians recommend that victims get plenty of rest and drink lots of liquids while the immune system eliminates the virus. Within a few weeks, symptoms generally disappear, although weakness may persist for two more weeks.

Platelets and Blood Clotting

The circulatory system is a delicate apparatus. Even minor bumps and scrapes can cause it to leak. Leakage is normally prevented by **blood clotting**, one of the most complex homeostatic systems encountered in the human body. A simplified version is discussed here.

Blood clotting requires numerous participants. Among the most important agents of blood clotting are the platelets, tiny cell fragments produced in the bone marrow by a large cell called the **megakaryocyte** (Figure 8–22). Platelets are not true cells and, like RBCs, they cannot divide.

Platelets are carried passively in the bloodstream and are coated by a layer of a sticky material, which causes them to adhere to irregular surfaces, such as

FIGURE 8–22 Megakaryocyte A light micrograph of a megakaryocyte, a large, multinucleated cell found in bone marrow, which fragments, giving rise to platelets.

Megakaryocyte   Platelets

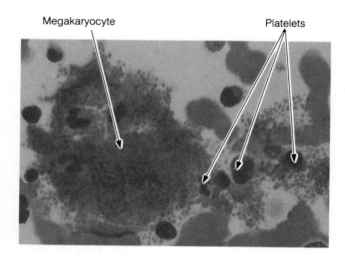

tears in blood vessels. This stickiness is extremely important in the clotting mechanism.

The process of blood clotting is summarized in Figure 8–23. As illustrated, cells in the damaged tissue release a substance into the bloodstream called **thromboplastin**. Thromboplastin is a lipoprotein. It converts an inactive plasma enzyme, **prothrombin**, produced by the liver, into its active form, **thrombin**. Thrombin, in turn, acts on another blood protein, **fibrinogen**, also produced by the liver. Fibrinogen is

converted into **fibrin**, long, branching fibers, that create a weblike network in the wall of the damaged blood vessel (Figure 8–24). The fibrin web traps RBCs and platelets, forming a plug that cuts off the flow of blood to the tissue. Platelets captured by the fibrin web release platelet thromboplastin, which causes more fibrin to be laid down. Blood clotting occurs fairly quickly. In most cases, a damaged blood vessel is sealed by a clot within 3 to 6 minutes of an injury; 30 to 60 minutes later, the platelets in the clot

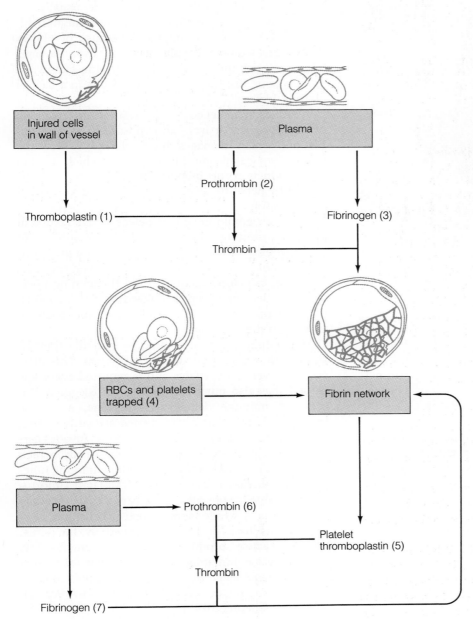

FIGURE 8–23   Blood Clotting Simplified   Injured cells in the wall of blood vessels release the chemical thromboplastin (1). Thromboplastin stimulates the conversion of prothrombin, found in the plasma, into thrombin (2). Thrombin, in turn, stimulates the conversion of the plasma protein, fibrinogen, into fibrin (3). The fibrin network captures RBCs and platelets (4). Platelets in the blood clot release platelet thromboplastin (5), which converts additional plasma prothrombin into thrombin (6). Thrombin, in turn, stimulates the production of additional fibrin (7).

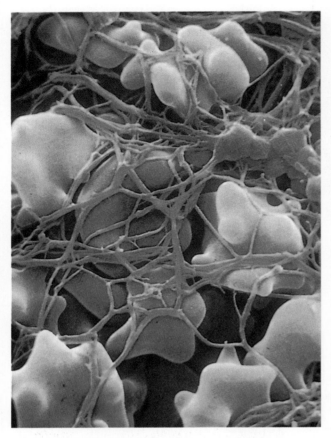

FIGURE 8–24 Blood Clot An electron micrograph of a fibrin clot that has already trapped platelets and RBCs, helping to plug up a leak in a vessel. RBCs are yellow. Fibrin network is red.

begin to stitch the wound together. How? Platelets contain contractile proteins like those in muscle cells. Contraction of the protein fibers draws the fibrin network inward, pulling the edges of the cut or damaged blood vessel together. Like stitches, this closes the wound and accelerates healing.

Blood clots cannot stay in place indefinitely. If they did, the circulatory system would eventually become clogged and blood flow would come to a halt. Clots are dissolved by an enzyme called **plasmin**, produced from an inactive form, **plasminogen**, found in the blood. Plasminogen is incorporated in the clot as it forms and then is gradually converted to plasmin by an activating factor secreted by the endothelial cells of the blood vessel. Plasmin dissolves the clot after the damage has been repaired. TPA, discussed earlier in the chapter, is one such activator used to break up clots in heart attack victims.

As important as blood clots are in homeostasis, they can also cause problems. For example, blood clots can obstruct arteries narrowed by atherosclerosis, further restricting blood flow and causing heart attacks. Clots form near sutures in the walls of small blood vessels and can dislodge, traveling to narrower vessels in the heart, brain, and other vital organs where they block the flow of blood and oxygen.

Clotting Disorders.   Tiny breaks occur in blood vessels with surprising regularity, but are usually repaired without our ever knowing it. In some individuals, blood clotting is impaired because of (1) an insufficient number of platelets; (2) liver damage, which impairs the production of clotting factors; or (3) a genetic disorder, which also results in a lack of clotting factors.

A reduced platelet count may result from leukemia, as noted earlier, or from an exposure to excess irradiation, which damages bone marrow where the megakaryocytes reside. Liver damage, which impairs the production of blood-clotting factors, can be caused by hepatitis, liver cancer, or excessive alcohol consumption.

The most common genetic defect, which results in a lack of clotting factors, is called **hemophilia**. In these people, certain clotting factors are not produced at all. Problems set in early in life, for even tiny cuts or bruises can bleed uncontrollably, threatening a person's life. Because of repeated bleeding into the joints, victims suffer great pain and often become disabled; they often die at a young age.

Hemophiliacs can be treated by periodic transfusions of the missing blood-clotting factors. This therapy, however, is expensive and is required every few days. It has also put hemophiliacs at risk for **AIDS**, acquired immune deficiency syndrome, a disease caused by a virus that is transmitted primarily by sexual contact (Chapter 10). The AIDS virus invades certain white blood cells, called T helper cells, which results in a gradual deterioration of the immune function. The disease is considered 100% fatal. Unfortunately, testing for the AIDS virus began late, and many transfusions of whole blood, blood plasma, and clotting factors have been contaminated by the virus. Clotting factors produced by genetic engineering, however, are eliminating the need for transfusions of clotting agents taken from whole blood, reducing the risk of hemophiliacs contracting AIDS. (For more on AIDS, see Chapter 10.)

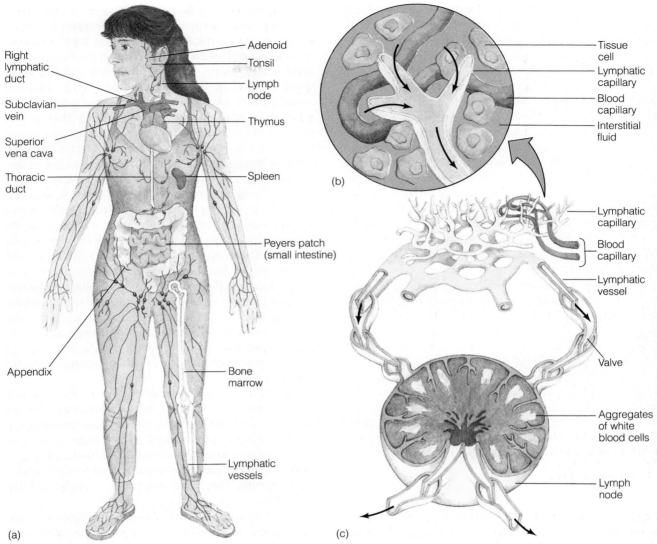

FIGURE 8–25   The Lymphatic System   (a) The lymphatic system consists of a series of vessels that transport lymph, excess tissue fluid, back to the circulatory system.

(b) Like the veins, the lymphatic vessels contain valves that prohibit backflow. (c) Lymph nodes are interspersed along the vessels and serve to filter the lymph.

THE LYMPHATIC SYSTEM

The lymphatic system is functionally related to two systems: the circulatory system and the immune system. This section examines the role of the lymphatic system in circulation.

You may recall from Chapter 6 that the cells of the body are bathed in a liquid called **interstitial fluid.** The interstitial fluid is in equilibrium with the blood plasma and provides a medium through which nutri-

ents, gases, and wastes can diffuse between the capillaries and the cells.

Tissue fluid is replenished by water and dissolved chemicals that enter by diffusion from the capillaries. The flow of water out of the capillaries, however, normally exceeds the return flow by about three liters per day. The "excess" water is picked up by the **lymphatic system,** a network of vessels that transports it to the large veins in the base of the neck where it joins the blood (Figure 8–25).

The lymphatic system also consists of several lymphatic organs: the lymph nodes, the spleen, the thymus, and the tonsils. The lymphatic organs function primarily in immune protection and are discussed in more detail in Chapter 10.

The vessels of the lymphatic system form a one-way avenue from the body tissues to the circulatory system. Interstitial fluid is picked up by small **lymph capillaries** in tissues. Like the capillaries of the circulatory system, these vessels have thin, highly permeable walls through which water and other substances pass with ease. As illustrated in Figure 8–26, the cells in the walls of the lymph capillaries overlap, creating a number of one-way valves, which resemble swinging doors. The accumulation of interstitial fluid in the tissues forces the doors to open, causing the fluid to flow into the lymphatic capillaries. Once inside the fluid is called **lymph**.

Fluid drains from the capillaries into larger ducts, the collecting vessels. These vessels, in turn, merge with others, creating larger and larger ducts and eventually forming the **thoracic duct** and the **right lymphatic duct**, which empty into the large veins at the base of the neck (Figure 8–25).

Lymph moves through the vessels of the lymphatic system much the same way that blood is transported in veins. In the upper parts of the body, lymph flows by gravity. In regions below the heart, lymph is propelled largely by muscle contraction. Breathing and walking, for example, pump the lymph out of the extremities. Lymphatic flow is also assisted by valves nearly identical to those found in the veins.

Along the lymphatic highway are small nodular organs called **lymph nodes** (Figure 8–25). Varying in size and shape, the lymph nodes are found in small clusters in the armpits, groin, neck, and numerous other locations. A lymph node consists of a network of fibers and irregular channels that reduce the flow of lymph. Lining the channels are numerous **macrophages**, which phagocytize bacteria, viruses, cellular debris, and other particulate matter transported by the lymph. Consequently, lymph nodes are filters. The lymph node also plays a key role in the body's immune response (Chapter 10). Lymphocytes in the node multiply when viruses and bacteria are present, destroying the intruders. The swollen glands in your neck when you have got a sore throat or cold are lymph nodes that have become engorged with lymphocytes and dead cells. The swelling indicates the presence of a nearby infection. T-lymphocytes and antibodies produced by plasma cells in the lymph node exit with the outgoing lymph and eventually reach the circulatory system to fight microbial invaders elsewhere, thus providing systemic protection.

Under normal circumstances, lymph is removed from tissues at a rate equal to its production. In some instances, however, lymph production exceeds the ability of the lymphatic capillaries. A burn, for example, may cause extensive damage to blood capillaries, increasing their leakiness and overwhelming the lymphatic system. This causes edema, the buildup of fluid in tissues.

Lymphatic vessels may also become blocked. One of the most common causes of blockage worldwide results from an infection by tiny parasitic worms that are transmitted by mosquitoes in the tropics. The worm larvae (an immature form) enter lymph vessels and take up residence in the lymph nodes. An inflammatory reaction causes the buildup of scar tissue in the nodes, which blocks the flow of lymph. After several years, the lymphatic drainage of certain parts of the body may become almost completely obstructed. A leg may swell so much, in fact, that it weighs as much as the rest of the body (Figure 8–27). This condition, known as **elephantiasis**, often affects the scrotum, the sac of skin that holds the testes, the male gonads. Edema in the scrotum causes it to enlarge so much in some cases that a man must carry his scrotum in a wheelbarrow in order to move about.

FIGURE 8–26 Lymphatic Capillaries These thin-walled vessels absorb excess tissue fluid. The cells of the wall push inward to allow the fluid to enter the vessel.

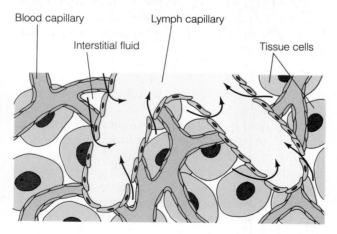

Blood capillary

Lymph capillary

Interstitial fluid

Tissue cells

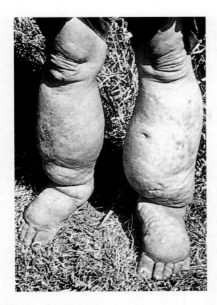

FIGURE 8–27 Elephantiasis A parasitic worm that invades the body blocks the flow of lymph through the lymph nodes, causing tissue fluid to build up. This condition is known as elephantiasis.

ENVIRONMENT AND HEALTH: CARBON MONOXIDE

Modern society is powered by fossil fuels—coal, oil, and natural gas. When burned, they produce energy and a wide variety of pollutants. One of the most dangerous to human health is carbon monoxide. This colorless, odorless gas results from incomplete combustion of organic fuels.[9]

Carbon monoxide (CO) emanates from stoves and furnaces in our homes and spews out of the tailpipes of our automobiles, causing increased levels along highways, in parking garages, and in tunnels. It spews out of power plants and factory smokestacks, hovering around our cities. It is even a major pollutant in tobacco smoke. According to estimates by the Environmental Protection Agency, over 40 million Americans, or one of every six people in this country, are exposed to levels of CO in the outside air thought to be harmful to health.

What makes CO so dangerous? Carbon monoxide, like oxygen, binds to hemoglobin. However, hemoglobin has a much greater affinity for CO (about 200 times greater attraction) than for oxygen. Thus, CO outcompetes oxygen for the binding site on the hemoglobin molecule and reduces the blood's ability to carry oxygen. For healthy people, CO does not create much of a problem, as long as levels are low. The body simply produces more RBCs and increases the heart rate to augment the flow of blood to tissues. It does create a problem, however, when CO levels are so high that the body cannot compensate adequately. At very high levels, CO becomes a deadly killer.

New research suggests that CO levels currently deemed acceptable by federal standards can trigger chest pain caused by a lack of oxygen. Levels such as those thought safe at work or levels in the blood encountered after one hour in heavy traffic cause angina and abnormal ECGs in moderately active adults with coronary artery disease, according to a study published in the *New England Journal of Medicine*.

In levels commonly found in and around cities, CO can be a problem for the elderly or for people suffering from coronary artery and lung diseases. Carbon monoxide places additional strain on their hearts, making the already weakened organ work harder. Thus, the elderly and infirm are advised to stay inside on high-pollution days.

Carbon monoxide is just one of many pollutants we breathe. No one knows what, if any, long-term effects it will have on our health. In the short term, however, it is clear that CO upsets homeostasis in the elderly and infirm. It has become another risk factor in an increasingly risky society, but it is not an insolvable problem. More efficient automobiles, mass transit, overall reductions in driving, and alternative automobile fuels (such as hydrogen) can all reduce the level of CO in and around the cities where most Americans live and work. More efficient factories, home furnaces, and power plants can also help reduce CO levels in our cities.

SUMMARY

THE CIRCULATORY SYSTEM'S FUNCTION IN HOMEOSTASIS

1. The circulatory system is one of the body's chief homeostatic systems. It helps maintain constant levels of nutrients and wastes, helps regulate body temperature, distributes body heat, protects against microorganisms, and, through clotting, protects against blood loss.

---

[9]When gasoline, coal, and other organic fuels burn completely, they produce carbon dioxide and water.

2. The circulatory system consists of a pump, the heart, and two circuits, the pulmonary circuit, which transports blood to and from the lungs, and the systemic circuit, which delivers blood to the body and returns the blood to the heart.

THE HEART

3. The heart consists of four chambers: two atria and two ventricles. The right atrium receives blood from the superior and inferior vena cavae. The blood returning from body tissues is low in oxygen and rich in carbon dioxide.

4. From the right atrium, blood is pumped into the right ventricle, then to the lungs, where it is resupplied with oxygen and stripped of carbon dioxide. Blood returns to the heart via the pulmonary veins, which empty into the left atrium.

5. Blood is pumped from the left atrium to the left ventricle, then out the aorta to the body where it supplies cells with oxygen and picks up cellular wastes.

6. Heart valves help control the direction of blood flow. The atrioventricular valves, between the atria and ventricles, prevent blood from flowing from the ventricles to the atria when the ventricles contract. The semilunar valves in the base of the aorta and pulmonary artery prevent blood from flowing back into the ventricles.

7. The closing of the valves produces distinct sounds, heart sounds, which can be detected through the chest using a stethoscope. Irregularities in heart sounds indicate the presence of diseased or damaged valves.

8. Cardiac muscle cells contract rhythmically and independently; contraction is coordinated by the heart's pacemaker, the sinoatrial (SA) node. The SA node is located in the upper wall of the right atrium.

9. The cells of the SA node discharge periodically, sending impulses to all cardiac muscle cells in the atria, which cause them to contract in unison.

10. The impulse next travels to the ventricles, but its passage is delayed, providing time for the ventricles to fill before contracting.

11. The impulse travels from the atrioventricular node down a specialized bundle of cardiac muscle cells, the atrioventricular bundle, into the myocardium of the ventricles.

12. Contraction begins at the tip of the heart and proceeds upward, squeezing the blood out of the ventricles.

13. Left on its own, the SA node would produce 100 contractions per minute. Nerve impulses, however, reduce the heart rate to about 70 beats per minute. During exercise or stress, the heart rate increases to meet body demands.

14. Depolarization of the heart muscle produces weak electrical currents that can be detected by surface electrodes. The change in electrical activity is detected by a voltage meter.

15. The tracing produced on the voltage meter has three distinct waves. Diseases of the heart may disrupt one or more waves of the ECG, making the ECG a valuable diagnostic tool for heart disease.

16. The amount of blood pumped by the ventricles each minute is called the cardiac output and is about 5 liters (about 5 quarts) per minute in a person at rest.

17. The most common type of heart attack is a myocardial infarction, caused by a blood clot that lodges in an already narrowed coronary artery. The obstruction decreases the flow of blood and oxygen to heart muscle, sometimes killing the cells.

18. Heart attacks can occur quite suddenly without warning, or may be preceded by several weeks of angina, pain felt when blood flow to heart muscle is reduced. Angina appears when an individual is active, then disappears when he or she ceases the activity. Angina attacks result from the partial blockage of the coronary arteries by atherosclerotic plaque, stress, or exposure to carbon monoxide.

19. The risk of heart attack can be reduced by proper diet, exercise, and a low dose of aspirin taken daily. Numerous treatments, including clot dissolvers, laser surgery, balloon angioplasty, and coronary bypass are also available to treat patients who have had a heart attack.

THE BLOOD VESSELS

20. The heart pumps blood into the arteries, which distribute the blood to the body tissues. In the body tissues, nutrient and waste exchange occurs in capillary beds. Blood is returned to the heart in the veins.

21. The largest of all arteries is the aorta. It carries oxygenated blood from the left ventricle of the heart to the rest of the body. It loops over the back of the heart, then descends through the chest and abdomen, giving off large branches that carry blood to the head, extremities, and major organs. The first branches are the coronary arteries, which supply the heart.

22. The aorta and many of its chief branches are elastic arteries, so named because they contain numerous wavy elastic fibers in the tunica media. As blood pulses out of the heart, the elastic arteries expand to accommodate the blood, then contract, helping to pump the blood and ensuring a steady flow of blood through the capillaries.

23. The elastic arteries branch, forming smaller vessels, the muscular arteries, which also expand and contract with the flow of blood.

24. The smooth muscle in the walls of muscular arteries responds to a variety of stimuli. These stimuli cause the blood vessels to open or close to varying degrees, controlling the flow of blood through body tissues.

25. Blood pressure is highest in the aorta and drops considerably as the arteries branch. By the time the blood reaches the capillaries, the flow of blood and blood

pressure are greatly reduced. This increases the rate of exchange between the blood and the tissues.

26. Capillaries are thin-walled vessels that form branching networks, or capillary beds, among the cells of body tissues. Cellular wastes and nutrients are exchanged between the cells of body tissues and the blood in capillary networks.

27. Blood flow through a capillary network is regulated by constriction and relaxation of the metarterioles, circulatory short cuts, which connect the arterioles on one side with the venules on the other. Blood flow is also controlled by the precapillary sphincters, tiny rings of smooth muscle that surround the capillaries that branch from the metarterioles.

28. Blood draining from capillary beds enters venules, which join to form veins. Veins return blood to the heart and generally run alongside the arteries.

29. Since the walls of veins have very little smooth muscle, they are easily affected by obstructions, which cause the walls to balloon out, forming rather ugly bluish bulges, called varicose veins.

30. In the veins above the heart, blood drains by gravity. Veins below the heart, however, rely on the movement of body parts to squeeze the blood upward and also on valves, flaps of tissue that span the veins and prevent the backflow of blood.

THE BLOOD

31. Blood is a watery tissue containing dissolved nutrients, proteins, gases, wastes, red blood cells, white blood cells, and platelets. The functions of the platelets and blood cells are summarized in Table 8–3.

32. The liquid portion of the blood, plasma, contains water and many dissolved substances. Plasma constitutes about 55% of the volume of a person's blood, and the blood cells and platelets make up the remainder. The volume occupied by the blood cells and platelets is called the hematocrit.

33. Red blood cells (RBCs) are highly specialized cells that lack organelles and are produced by the red bone marrow. The RBCs transport oxygen in the blood. RBCs are flexible, biconcave disks that are passively swept along in the bloodstream.

34. The concentration of RBCs in the blood is maintained by the hormone erythropoietin, produced by the kidneys when oxygen levels decline.

35. White blood cells (WBCs) are nucleated cells and are part of the body's protective mechanism to combat microorganisms. WBCs are produced in the bone marrow and circulate in the bloodstream, but do most of their work outside it, in the body tissues.

36. The most abundant WBCs are the neutrophils, which are attracted by chemicals released from infected tissue. Neutrophils leave the bloodstream and migrate to the site of infection by ameboid movement.

37. Neutrophils are the first WBCs to arrive at an infection where they phagocytize microorganisms, helping prevent the spread of bacteria and other organisms.

38. The second cell to arrive is the monocyte, which phagocytizes microorganisms, dead cells, cellular debris, and dead neutrophils.

39. Lymphocytes are WBCs that play a vital role in immune protection.

40. Platelets are fragments of large bone marrow cells called megakaryocytes. Platelets are involved in blood clotting, a process that requires a number of blood constituents.

41. Platelets are coated by a layer of a sticky material, which causes them to adhere to irregular surfaces, such as tears in blood vessels. The process of blood clotting is summarized in Figure 8–23.

THE LYMPHATIC SYSTEM

42. The lymphatic system is a network of vessels that drains interstitial fluid from body tissues and transports it to the blood.

43. The lymphatic system also consists of several lymphatic organs, such as the lymph nodes, the spleen, the thymus, and the tonsils, which function primarily in immune protection.

44. Along the system of lymphatic vessels are small nodular organs called lymph nodes, which filter the lymph.

45. Under normal circumstances, lymph is removed from tissues at a rate equal to its production, keeping tissues from swelling. In some cases, however, lymph production exceeds the ability of the lymphatic capillaries, and swelling results.

ENVIRONMENT AND HEALTH: CARBON MONOXIDE

46. Carbon monoxide is produced by the incomplete combustion of organic fuels in our homes, automobiles, factories, and power plants.

47. Carbon monoxide binds to hemoglobin, outcompeting oxygen, and reducing the oxygen-carrying capacity of the blood.

48. At high concentrations, carbon dioxide can be harmful to human health. At lower concentrations, it is harmful to people with cardiovascular disease and lung disease. For individuals with heart disease, carbon monoxide puts additional strain on the heart.

In 1989, the Environmental Protection Agency (EPA) announced new guidelines for lead levels in drinking water. Lead is a neurotoxin, which, even in low doses, has been shown to affect mental development and coordination. The EPA's new guideline for drinking water allows lead levels no higher than 5 parts per million, compared to an old standard of 50 parts per million. Some newspapers carried the report, hailing the move as a major accomplishment.

Would it affect your view of this announcement, if you knew that the measurements for the old standard were taken at people's faucets and the measurement for the new standard is to be taken at the water treatment plants, before the water is released into the pipes that distribute it to customers? Would it affect your view of the announcement if you knew that most of the lead in a water system came from lead pipes in old homes and lead solder used in copper pipes? What questions need to be answered before you can determine the importance of the EPA's announcement? What do these facts suggest about analyzing news reports?

TEST OF TERMS

1. The vessels that supply blood to the lungs are part of the _____ circuit. The vessels that supply blood to the rest of the body are part of the _____ circuit.
2. The superior and inferior vena cavae empty into the _____ _____ of the heart.
3. The large artery carrying blood from the left ventricle of the heart is called the _____ . It is an _____ artery.
4. The _____ valves are found in the larger arteries leading from the heart.
5. The _____ valves are anchored to the inside wall of the heart via the chordae tendineae.
6. The heart's internal pacemaker is called the _____ node and is located in the wall of the _____ _____ .
7. The impulse generated by the pacemaker travels down the septum between the ventricles along the _____ .
8. Arteries and veins contain three layers; starting from the inside, they are the _____ _____ , _____ _____ , and _____ _____ .

9. _____ _____ give the large arteries the ability to recoil when blood is pumped into them.
10. The pressure created by the contraction of the heart is called _____ pressure.
11. The exchange of nutrients and wastes occurs in the _____ , thin-walled vessels that form extensive networks in body tissues. These vessels empty into _____ , the smallest veins.
12. _____ _____ result when the wall of a vein balloons outward because blood flow is impaired.
13. A swelling resulting from the buildup of tissue fluid is called _____ .
14. The backward flow of blood in veins is prevented by _____ .
15. The liquid portion of blood is called _____ .
16. The proteins in the blood that destroy or inactivate viruses and bacteria are _____ .
17. The _____ is a biconcave disk that contains the protein _____ , which binds to oxygen.
18. Blood cells are produced in

_____ _____ marrow.
19. The hormone, _____ , produced by the kidney, stimulates the synthesis of _____ when oxygen levels in the blood fall.
20. The heme group consists of a _____ ring and an atom of iron.
21. _____ is a reduction in the oxygen-carrying capacity of the blood.
22. The two white blood cells that phagocytize bacteria and viruses are _____ and _____ .
23. The _____ is the white blood cell involved in immune protection.
24. _____ is a cancer of the white blood cells.
25. The _____ is a cell fragment produced from megakaryocytes.
26. _____ , an inactive protein in the blood, is activated by thrombin and converted into long branching fibers that form a web-like network in the wall of damaged blood vessels.
27. _____ is a digestive enzyme in the blood that dissolves blood clots.

28. The lymphatic system picks up excess _____ _____ and transports it back to the circulatory system.

29. _____ _____ filter the lymph traveling through the lymphatic vessels.

30. _____ _____ is a colorless, odorless gas that binds to the hemoglobin molecule, reducing the oxygen-carrying capacity of the blood.

Answers to the Test of Terms are located in Appendix B.

TEST OF CONCEPTS

1. List and describe several ways in which the circulatory system and blood cooperate in homeostasis.
2. Describe how the heart coordinates muscular contraction through its pacemaker.
3. Trace the path of blood through the heart and the pulmonary and systemic circuits.
4. Based on what you know about the heart and blood, what would happen if the septum separating the two ventricles failed to form completely during embryonic development?
5. Describe how the valves help control the direction of blood flow in the heart.
6. How does atherosclerosis affect the heart?
7. Describe the general structure of the arteries and veins. How are they different? How are they similar?
8. How do the elastic fibers in the major arteries help ensure a continuous blood flow through the capillaries?
9. Capillaries illustrate the remarkable correlation between structure and function. Do you agree with this statement? Why or why not?
10. Describe how the body controls the movement of blood through capillary beds. Why is this homeostatic mechanism so important? Give some examples.
11. Explain how the blood is returned via the veins although blood pressure in this part of the circulatory system is so low.
12. Describe the structure and function of each of the following: red blood cells, platelets, lymphocytes, monocytes, and neutrophils.
13. Define each of the following terms: leukemia, anemia, and infectious mononucleosis.
14. Explain how a blood clot forms and how it helps prevent bleeding.
15. Explain the effects of a blockage of the lymphatic vessels draining the leg.

SUGGESTED READINGS

Fackelmann, K. A. 1989a. Hidden heart hazards. *Science News* 136(12): 184–86. Looks at some controversial findings regarding the relationship between high blood insulin levels and heart disease.

_____. 1989b. Japanese stroke clues. *Science News* 135(16): 250, 253. Examines several studies of cholesterol risks.

Greg, H. M., A. Sette, and S. Buus. 1989. How T cells see antigen. *Scientific American* 261(5): 68–74. Explains how T-lymphocytes function.

Lerner, R.A., and A. Tramontano. 1988. Catalytic antibodies. *Scientific American* 258(3): 58–60, 65–70. Looks at a new class of molecules that combine the diversity of antibodies with the catalytic power of enzymes.

Loupe, D. E. 1989. Breaking the sickle cycle: Potential treatments emerge for sickle cell anemia. *Science News* 136(23): 360–62. Discusses some possible treatments for a disease long thought untreatable.

Marieb, E. N. 1989. *Human anatomy and physiology*. Redwood City, Calif.: Benjamin/Cummings. Succinct and detailed coverage of human anatomy and physiology with excellent illustrations.

Plotkin, M. J. 1990. The healing forest: The search for new jungle medicines. *The Futurist* 24(1): 9–14. Discusses potential medicines from tropical plants and the rapid decline of tropical forests.

Raloff, J. 1989. Do you know your HDL? *Science News* 136(11): 171–73. Examines how the "good cholesterol," high-density lipoprotein, may prove to be a strong predictor of risk for coronary heart disease.

Vander, A., J. Sherman, and D. Luciano. 1985. *Human physiology: The mechanism of body function*, 4th ed. New York: McGraw-Hill. Good coverage of cardiovascular physiology.

Weiss, R. 1989. Postponing red-cell retirement: Can aging blood cells get a new lease on life? *Science News* 136(26): 424–25. Interesting look at the aging process of RBCs and potential ways of slowing it.

9 The Vital Exchange: Respiration

Mammal respiratory system.

George F. Eaton was a robust and handsome Irishman who grew up in eastern Massachusetts where he married and raised three children. To support his family, Eaton worked as a firefighter for 30 years. Fire fighting is dangerous work in large part because it exposes men and women to smoke containing numerous potentially harmful air pollutants.

Eaton retired from the fire department and moved to Cape Cod where he fished the ocean. Unfortunately, his life was cut short by emphysema, a debilitating respiratory disease. **Emphysema** results from the breakdown of the air sacs in the lungs where oxygen and carbon dioxide are exchanged between the air and the blood. As the walls of the air sacs break down, the surface area for the diffusion of oxygen and carbon dioxide decreases. Emphysema is an incurable disease that worsens by the month. Victims complain of shortness of breath; eventually, even mild exertion becomes trying.

The degeneration of the lungs in victims of emphysema creates a domino effect. One of the dominoes is the heart. Since oxygen absorption in the lungs declines substantially, the heart of an emphysemic patient must work harder. This additional strain can lead to heart failure.

George Eaton died a slow and painful death as his lungs grew increasingly more inefficient. Relatives mourned his death and blamed it on the pollution to which he was exposed while working for the fire department. Fire fighting, however, was probably only part of the cause, for George Eaton was also a smoker, who had smoked most of his adult life. Cigarette smoking is the leading cause of emphysema.

This chapter describes the respiratory system and diseases like emphysema, many of which can be prevented.

THE STRUCTURE OF THE HUMAN RESPIRATORY SYSTEM

The respiratory system supplies oxygen to the body and gets rid of carbon dioxide. Oxygen is used by cells to produce energy; carbon dioxide is a waste product of cellular energy production. In its role as a provider of oxygen and a disposer of carbon dioxide waste, the respiratory system helps to maintain a constant internal environment necessary for normal cellular metabolism. Thus, like many other systems, it plays an important role in homeostasis. The respiratory system functions automatically, drawing air into the lungs, then letting it out in a cycle that repeats at least 23,000 times per day or 8 million times per year!

Figure 9–1 shows the structure of the human respiratory system. As illustrated, the respiratory system consists of two basic parts: an air-conducting portion and a gas-exchange portion (Table 9–1). The air-conducting portion is an elaborate set of passageways that transports air from the oral cavity and nose to and

FIGURE 9–1   The Human Respiratory System
A cutaway showing the air-conducting portion and the gas-exchange portion of the human respiratory system. The insert shows a higher magnification of the alveoli where oxygen and carbon dioxide exchange occurs.

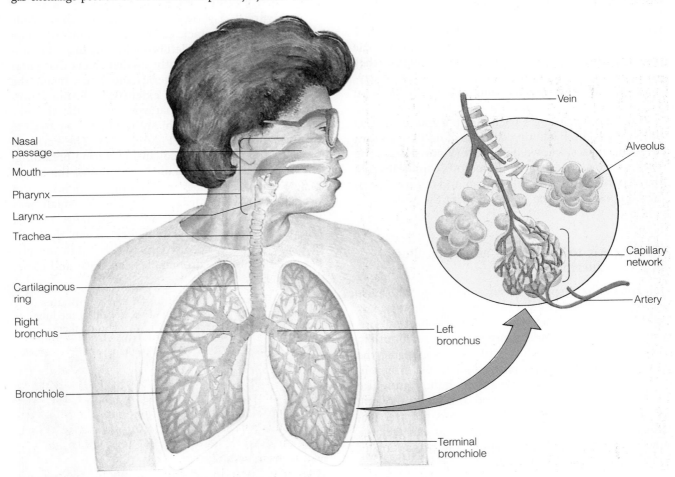

TABLE 9–1   Summary of the Respiratory System

ORGAN	FUNCTION
Air conducting	
Nasal cavity	Filters, warms, and moistens air; also transports air into pharynx
Oral cavity	Transports air to pharynx; warms and moistens air; helps produce sounds
Pharynx	Transports air to larynx
Epiglottis	Covers the opening to the trachea during swallowing
Larynx	Produces sounds; transports air into trachea; helps filter incoming air; warms and moistens incoming air
Trachea and bronchi	Warm and moisten air; transport air to lungs; filter incoming air
Bronchioles	Control air flow in the lungs; transport air to alveoli
Gas exchange	
Alveoli	Site of gaseous exchange

from the **lungs**, two large saclike organs in the thoracic cavity (Figure 9–1). Like the arteries of the body, the passageways start out large, then become progressively smaller and more numerous as they penetrate the lung tissue where the exchange of oxygen and carbon dioxide between the air and the blood takes place.

The lungs are the gas-exchange portion of the respiratory system. Each lung consists of millions of tiny, thin-walled sacs, known as **alveoli** (Figure 9–2). The walls of the alveoli contain numerous capillaries that absorb oxygen from the inhaled air and release carbon dioxide, which will be removed from the body.

Getting Air to the Lungs and Back Out: The Conducting Portion

Air enters the respiratory system through the nose and mouth, then is drawn backward into the pharynx (Figure 9–3). The **pharynx** is the portion of the respiratory system that opens into the nose and mouth in the front and joins below with the esophagus, the muscular tube that transports food to the stomach, and the larynx. The **larynx** is a rigid, hollow structure that houses the vocal cords and aids in swallowing (Figure 9–3). The larynx opens into the **trachea**, which transports air to the lungs. You can feel the trachea below your Adam's apple, the protrusion of cartilage on your neck. Occasionally food accidentally enters the trachea; Health Note 9–1 explains what to do when this happens.

As Figure 9–1 shows, the trachea is a short, wide duct that, in the thoracic cavity, divides into two main branches, the right and left bronchi. The **bronchi** (singular, bronchus) are slightly smaller ducts that enter the lungs alongside the arteries and veins. Inside the lungs, the bronchi branch extensively, forming progressively smaller tubes that carry air to the alveoli.

The trachea and bronchi are reinforced by hyaline cartilage in their walls (Figure 9–4). The cartilage prevents the ducts from collapsing during breathing, ensuring a steady flow of air in and out of the lungs.

The smallest bronchi in the lungs branch to form **bronchioles**, which lead to the **alveoli**, small sacs where oxygen and carbon dioxide exchange occurs. Like the arterioles of the circulatory system, the walls of the bronchioles consist largely of smooth muscle. Because of this, the bronchioles can open and close,

FIGURE 9–2   The Alveoli and Their Capillaries
A scanning electron micrograph of the alveoli of the lung showing the rich capillary network surrounding them.

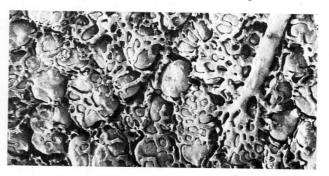

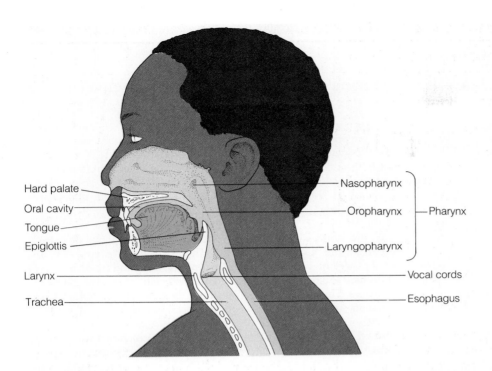

Hard palate

Oral cavity

Tongue

Epiglottis

Larynx

Trachea

Nasopharynx

Oropharynx — Pharynx

Laryngopharynx

Vocal cords

Esophagus

providing a means to control air flow in the lung. During exercise or during times of stress, the smooth muscle in the bronchioles relaxes, opening the tubes and increasing the flow of air into the lungs. This helps meet the body's need for more oxygen, in much the same way that the arterioles of capillary beds dilate to let more blood into body tissues.

The conducting portion is a highway to the lungs, but also serves as a filter that removes many impuri-

ties from the air we breath, especially airborne particles, such as dust and bacteria. Particles in the air exist in many sizes. Some are small and can penetrate deeply into the lung. Larger particles are deposited in the nose, trachea, and bronchi. Particles containing toxic metals, such as mercury, can cause cancer.

Particles are deposited as the inhaled air travels through the maze of passageways leading to the lungs. Particles are also filtered in the nose. The convoluted

FIGURE 9–4 Cross Section of the Trachea and Bronchus (*a*) The trachea contains hyaline cartilage ribs, C-shaped segments of cartilage that give the organ internal support. (*b*) An intrapulmonary bronchus showing hyaline cartilage.

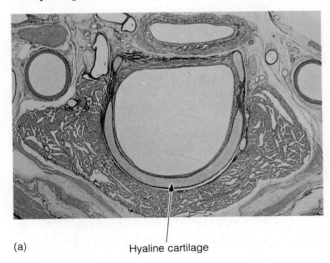

(a)

Hyaline cartilage

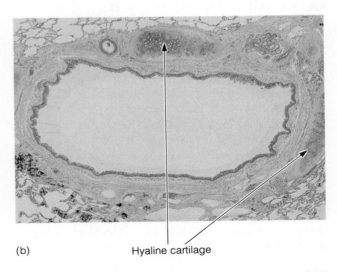

(b)

Hyaline cartilage

First Aid That May Save Someone's Life: The Heimlich Maneuver

During swallowing, food is normally excluded from the trachea by the epiglottis (Chapter 7). Each year, however, approximately 3000 people—about 8 people each day—will choke to death on food that gets stuck in the larynx or trachea. Many of these deaths could be prevented. Here's what should be done if you encounter a person who is choking. First, stand behind the victim (who may be either standing or sitting), positioning yourself slightly to one side, as shown in the figure (part a). Place an arm across the victim's chest for support and lean the person forward. With the heel of your other hand, give four hard thumps between the shoulder blades. This should clear the trachea. If it doesn't, don't give up. Try the Heimlich maneuver.

The **Heimlich maneuver** or abdominal thrust is shown in the figure (part b). The Heimlich maneuver is best performed on an individual who is standing. First, pull the person to his feet while standing behind him. Next, wrap your arms around his waist. Grasp your fist with the other hand, pressing it against the abdomen with the thumb pointing inward just above the person's navel. Now, give your fist a sharp pull inward and upward.

The Heimlich maneuver should dislodge the food. If it doesn't, try it three more times. If that doesn't work, repeat the back blows. If the victim loses consciousness, mouth-to-mouth resuscitation will be needed.[1]

Children are often victims of choking, and many parents make the mistake of trying to dislodge food by sticking their fingers down their child's throat to extract the food. Unfortunately, this may force the food deeper into the larynx or trachea. Part c of the figure shows how to treat babies or young children who are choking on food or other objects. For a toddler, sit down and put the child across your knees with his head down. Give several thumps on her back between the shoulder blades with the heel of your hand (but softer than you would for an adult). Babies can be held face down with one arm supporting the body. Several light raps on the back should dislodge the food. Babies can also be held by the ankles while you rap on the back. It is a good idea to practice these techniques, so that when the time comes you will be prepared. When you practice the Heimlich maneuver, however, be gentle and be sure your arms are around the person's waist, not the chest.

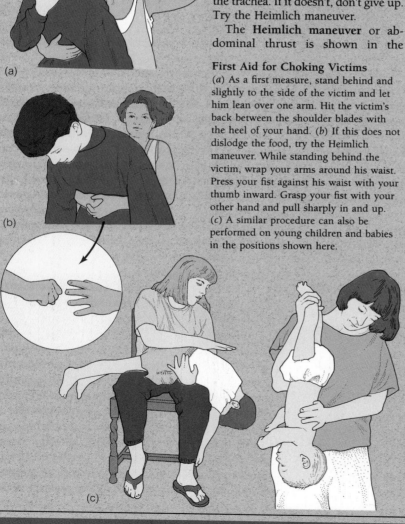

First Aid for Choking Victims
(*a*) As a first measure, stand behind and slightly to the side of the victim and let him lean over one arm. Hit the victim's back between the shoulder blades with the heel of your hand. (*b*) If this does not dislodge the food, try the Heimlich maneuver. While standing behind the victim, wrap your arms around his waist. Press your fist against his waist with your thumb inward. Grasp your fist with your other hand and pull sharply in and up. (*c*) A similar procedure can also be performed on young children and babies in the positions shown here.

[1]Students should consider learning mouth-to-mouth resuscitation and other important first aid procedures.

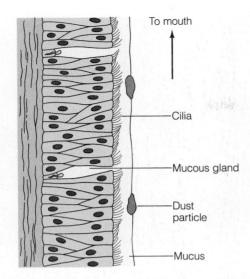

To mouth

Cilia

Mucous gland

Dust particle

Mucus

FIGURE 9–5 The Mucous Trap Mucus produced by the mucous cells in the lining of much of the respiratory system traps bacteria, viruses, and other particulates in the air. The cilia located on other cells transport the mucus toward the mouth.

interior of the nose, shown in Figure 9–3, slows the flow of air. This causes the larger particles to drop out, in much the same way that sediment falls to the bottom in a slow-moving section of a stream. Hairs in the nasal cavity may also trap particles.

Particles that precipitate from the air in the nose and trachea are trapped in a layer of **mucus**, a thick, slimy secretion deposited on inside of much of the respiratory tract (Figure 9–5). Mucus is produced by glands along the respiratory tract and by cells in the epithelium, called **mucous cells**.[1] The epithelium of the trachea also contains numerous ciliated cells. These cilia beat upward toward the mouth, slowly transporting mucus and its cargo of bacteria and dust particles. The cilia of the trachea operate day and night, sweeping mucus toward the oral cavity, where it can be swallowed or expectorated (spit out). This cleansing mechanism protects the respiratory tract and lungs from bacteria and other particles.

Like all homeostatic mechanisms, however, the respiratory mucous trap is not invulnerable. Bacteria do occasionally penetrate the lining, where they may proliferate, causing respiratory infections. Making matters worse, sulfur dioxide, a pollutant in cigarette smoke and urban air pollution, temporarily paralyzes

[1]You will notice that the spelling of the word *mucus* varies. When used as an adjective (e.g., *mucous* cell), it is spelled *mucous*. When used as a noun, it is spelled *mucus*.

cilia. This reduces the respiratory system's natural cleansing mechanism. Bacteria and toxic particulates deposit on the respiratory tract and may enter the lungs. Sulfur dioxide gas in the smoke of a single cigarette, for instance, can paralyze the cilia for an hour or more; ironically, the cilia of a smoker are paralyzed when they're needed the most! It should come as no surprise, then, that smokers suffer more frequent respiratory infections than nonsmokers. Alcohol can also paralyze cilia. As a result, alcoholics are much more prone to certain types of respiratory infections.

The conducting portion of the respiratory system also moistens and warms the incoming air. Beneath the epithelium of the respiratory tract is a rich network of capillaries that releases moisture and heat. Moisture protects the lungs from drying out, and heat protects the lungs from cold temperatures. Except in extremely cold weather, by the time inhaled air reaches the lungs, it is nearly saturated with water and is warmed to body temperature.

The Site of Oxygen and Carbon Dioxide Exchange

The air we breath consists principally of nitrogen and oxygen with small amounts of carbon dioxide (Table 9–2). Oxygen must be supplied in relatively constant amounts to maintain cellular metabolism.

The conducting portion of the respiratory system delivers warmed, moistened, filtered air to the bronchioles, which deliver the air to the alveoli. Each alveolus is surrounded by a rich capillary network. When viewed through a microscope, lung tissue resembles a spongy angel food cake (Figure 9–6a and b).

TABLE 9–2 Composition of Air	
GAS	PERCENTAGE COMPOSITION
Nitrogen (N)	78
Oxygen (O)	21
Argon (Ar)	0.9
Carbon dioxide (CO_2)	0.03
Water vapor (H_2O)	Variable (0–4)
Pollutants	Variable

Alveoli

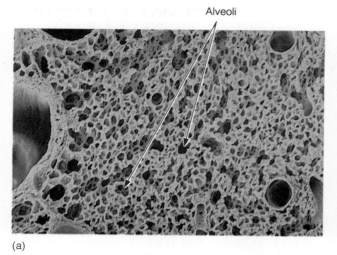

(a)

Alveoli

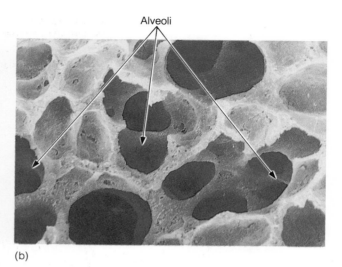

(b)

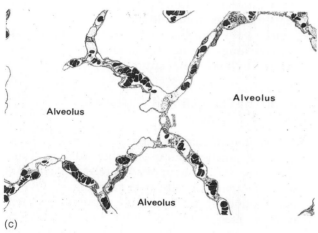

Alveolus

Alveolus

Alveolus

(c)

FIGURE 9–6 The Alveoli (*a*) A scanning electron micrograph of the lung showing many alveoli. The smallest openings are capillaries in the alveolar walls. (*b*) A scanning electron micrograph of lung tissue showing alveoli. Arrows indicate the direction of air flow. (*c*) An electron micrograph showing several alveoli and the close relationship of the capillaries.

Each lung contains an estimated 150 million alveoli. Both lungs provide a surface area of about 60 to 80 square meters or 760 square feet, an area approximately the size of a tennis court. Alveoli are lined by a single layer of flattened cells, called **Type I alveolar cells**, shown in Figure 9–7a. The large surface area of the lungs and the relatively thin barrier between the blood and the alveolar air are responsible for the rapid and efficient diffusion of gases across the alveolar wall.

Dust and other particulates that reach the lungs are removed by **alveolar macrophages**, or **dust cells**, so named because they phagocytize particulate matter (dust) that penetrates the lungs (Figure 9–7a). Dust cells wander through the alveoli, engulfing foreign material and helping rid the lungs of potentially harmful particulates.

Helping to keep the alveoli from collapsing are the **Type II alveolar cells**, large, round cells that produce

a phospholipid called surfactant (Figure 9–7a). **Surfactant** is a detergentlike substance that dissolves in the thin layer of water that lines each alveolus. The water covering the alveolar lining produces surface tension. **Surface tension** is a compaction of water molecules at the surface of a liquid containing water, which results from the hydrogen bonds of water molecules (Chapter 2). Hydrogen bonds draw water molecules together more tightly at the surface of a liquid than in the interior, which is why a drop of water beads up on your car windshield. In the alveoli, surface tension creates an elastic watery layer that draws the walls of the alveoli inward. Surfactant reduces this surface tension, decreasing forces that might otherwise cause the alveoli to collapse.

At birth, in some very small, premature infants, the surfactant-producing cells fail to produce enough surfactant. Consequently, surface tension causes the larger alveoli to collapse. This reduces the surface area for absorption. In these infants, alveoli begin to collapse within a few hours of birth. This condition, known as **respiratory distress syndrome** or **hyaline membrane disease**, is characterized by labored and rapid breathing, which can lead to exhaustion. If untreated, the lungs may collapse, killing the infant.

New treatments are now available. In one technique, physicians blow a surfactantlike chemical into

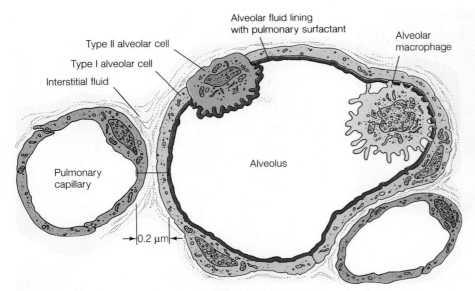

Type II alveolar cell

Type I alveolar cell

Interstitial fluid

Alveolar fluid lining
with pulmonary surfactant

Alveolar
macrophage

Pulmonary
capillary

Alveolus

0.2 μm

(a)

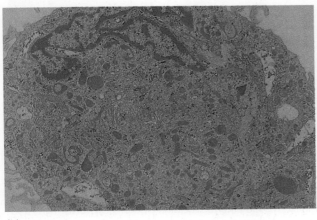

(b)

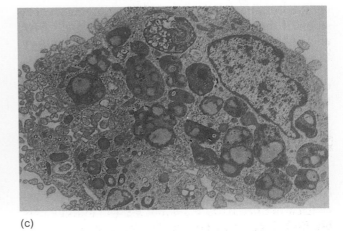

(c)

FIGURE 9–7   The Alveolar Macrophage or Dust Cell (*a*) Drawing of the alveolus showing Type I and Type II alveolar cells and dust cells. (*b*) An electron micrograph of a dust cell from the lung of a nonsmoker. (*c*) Compare this to a dust cell from the lung of a smoker that is filled with carbon particles that have been engulfed by the cell.

the lungs of affected babies soon after birth. Although fairly successful, this procedure has not yet been approved by the Food and Drug Administration because of some unanswered medical and ethical questions.

THE FUNCTIONS OF THE RESPIRATORY SYSTEM

The chief functions of the respiratory system are to conduct air to and from the lungs, to replenish the blood's oxygen supply, and to rid the blood of excess carbon dioxide. The respiratory system serves other functions as well. The vocal cords, described below, produce sounds allowing people to communicate through speech and song. The **olfactory membrane**,

the epithelium in the roof of the nasal cavity, allows us to perceive odors (Chapter 13). The respiratory system also helps maintain pH balance by its influence on carbon dioxide levels (Chapters 2 and 8).

Phonation: Producing Sounds

Phonation, the production of sounds, is critical to many members of the animal kingdom. The eerie cry of the coyote is a signal to the pack of a member's whereabouts (Figure 9–8). It helps the coyote pack stay in touch. When cornered, the coyote's fierce growl communicates the self-protective intentions of the animal.

Humans exhibit a wider range of sounds during communication. These sounds are produced in large

FIGURE 9–8 A Coyote Howls

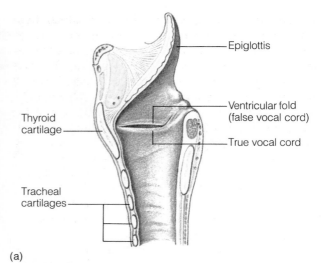

(a)

(b)

FIGURE 9–9 Vocal Cords (*a*) Longitudinal section of the larynx showing the location of the vocal cords. (*b*) View into the larynx of a patient also showing the vocal cords.

part by the **vocal cords**, elastic ligaments inside the larynx. The vocal cords vibrate as air is expelled from the lungs (Figure 9–9). The sounds generated by the vocal cords are further modified by changing the position of the tongue and the shape of the oral cavity, allowing guttural utterances to be transformed into an amazing variety of sounds.

The vocal cords vary in thickness from one person to the next. They also vary between men and women. Most males have thicker vocal cords than women and, therefore, have deeper voices. Testosterone, a male sex hormone produced by the testes (the male gonads), is responsible for the thickening.

Certain muscles in the larynx also affect sounds. The muscles attach to the vocal cords, for example, changing the tension on the cords. Relaxing the muscles lowers the tension on the cords, dropping the tone. Tightening the vocal cords has the opposite effect.

The larynx is a rather vulnerable organ. Bacterial and viral infections of the larynx, known as laryngitis, are rather common. **Laryngitis** is characterized by inflammation of the lining of the larynx and the vocal cords. Laryngitis may also be caused by irritation of the larynx from tobacco smoke, alcohol, excessive talking, shouting, or singing. In young children, inflammation results in a swelling of the lining that may impede the flow of air and impede breathing. This condition is commonly called croup.

The Vital Exchange

Blood entering the lungs arrives from the right ventricle of the heart. This blood, you may recall from Chapter 8, is laden with carbon dioxide that was picked up by the blood as it traveled through the body tissues, giving off oxygen. In the lungs, carbon dioxide is released and oxygen is added.

Oxygen and carbon dioxide readily diffuse across the capillary and alveolar walls. The driving force for diffusion is the concentration difference.[2] Figure 9–10 illustrates the process. As shown, oxygen in the

[2]Physiologists actually speak of differences in partial pressure. The partial pressure of a gas is caused by the collision of moving gas molecules with a surface. The partial pressure of oxygen is proportional to the force of impaction of all the oxygen molecules striking the alveolar wall. Thus, the total pressure is directly proportional to the concentration of the gas molecules.

alveolar air first diffuses into the extracellular fluid surrounding the capillary bed. The difference in oxygen concentration between the alveolus and the extracellular fluid is responsible for this rapid movement. Oxygen next diffuses through the capillary wall and into the blood plasma (Figure 9–11). From here, oxygen molecules cross the plasma membrane of the red blood cells (RBCs) and bind to hemoglobin molecules in their cytoplasm (Figure 9–11). About 98% of the oxygen in the blood is carried in the RBCs bound to hemoglobin. The rest is dissolved in the plasma and the cytoplasm of the RBCs.

Carbon dioxide travels in the opposite direction— from the capillaries into the alveoli. The movement of carbon dioxide molecules into the alveoli is also driven by concentration differences. The concentration of carbon dioxide inside the capillary is slightly higher than concentrations in the alveolar air (Figure 9–10).

Carbon dioxide is carried in the blood in three ways. A small percentage (about 15 to 25%) is bound to hemoglobin. An even smaller amount (7 to 8%) is

dissolved in the plasma. The bulk of the carbon dioxide is transported in the blood as bicarbonate ions (60 to 70%).

As shown in Figure 9-12, carbon dioxide enters the RBCs and is chemically converted to carbonic acid, H_2CO_3, in a reaction catalyzed by the enzyme **carbonic anhydrase**. Carbonic acid molecules readily dissociate, forming bicarbonate ions and hydrogen ions. The bicarbonate ions diffuse out of the RBCs into the plasma.

When blood rich in carbon dioxide reaches the lungs, bicarbonate ions combine with hydrogen ions, reforming carbonic acid (Figure 9–13). Carbonic acid dissociates, forming carbon dioxide, which then diffuses out of the blood into the alveoli. Carbon dioxide is then expelled from the lungs during exhalation.

FIGURE 9–11   Oxygen Diffusion   Oxygen travels from the alveoli into the blood plasma, then into the RBCs where much of it binds to hemoglobin. When the oxygenated blood reaches the tissues, oxygen is released from the RBCs and diffuses into interstitial fluid and then into body cells.

FIGURE 9–10   Close-Up of the Alveolus   Oxygen diffuses out of the alveolus into the capillary. Carbon dioxide diffuses in the opposite direction, entering the alveolar air that is expelled during exhalation.

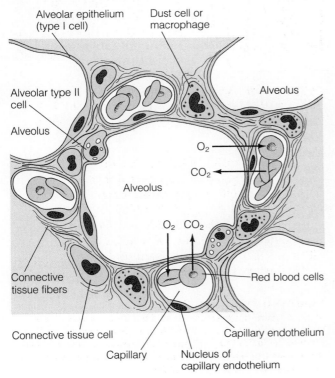

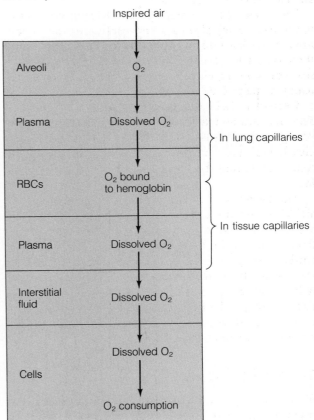

FIGURE 9–12   Bicarbonate Ion Production Carbon dioxide (CO_2) also moves by diffusion. It diffuses out of body cells where it is produced and into the tissue fluid, then into the plasma. Although some carbon dioxide is dissolved in the plasma and the cytoplasm of the RBCs, most is converted to carbonic acid (H_2CO_3) in the RBCs. Carbonic acid dissociates and forms bicarbonate, most of which is transported in the plasma.

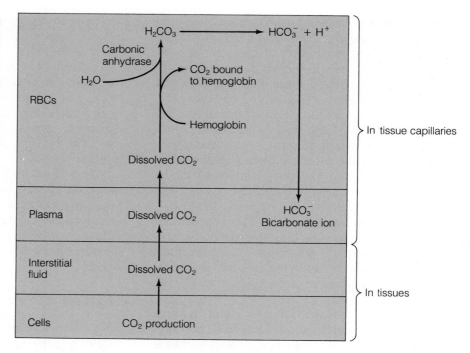

After picking up oxygen and releasing carbon dioxide, the blood flows back to the left atrium of the heart, traveling in the pulmonary veins. The blood then travels into the left ventricle, from which it is pumped to the body tissues via the aorta and its many branches.

FIGURE 9–13   Carbon Dioxide Production from Bicarbonate When the carbon dioxide-laden blood reaches the lungs, bicarbonate ions combine with hydrogen ions to form carbonic acid, which dissociates, forming carbon dioxide gas. CO_2 diffuses into the alveoli.

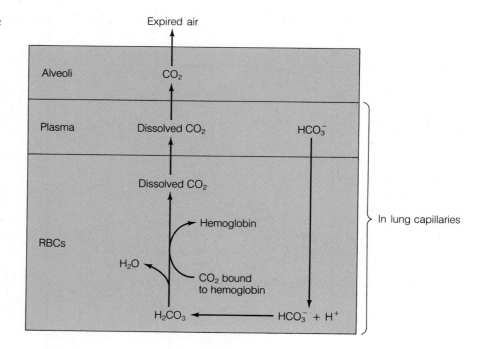

Diseases of the Respiratory System

The respiratory system is one of the main entryways for bacteria and viruses. The mucous layer, the cilia along much of the epithelium, and the phagocytic cells in the lung provide a degree of protection; they are not infallible. Bacteria and viruses may penetrate the epithelium, entering the underlying tissues where they may proliferate. A good many viruses, such as the influenza virus, multiply in the epithelial cells lining the respiratory tract. Once inside the respiratory tract, bacteria, viruses, and other microorganisms can spread to other organ systems. (Meningitis, a bacterial or viral infection of the brain, is a good example.)

Bacterial and viral infections of the respiratory tract can cause considerable discomfort. Some are even fatal. Respiratory tract infections are named by their location. Thus, an infection in the bronchi is called bronchitis, and an infection of the sinuses is known as sinusitis. Bacteria and viral infections are discussed in more detail in the next chapter, but a few of the more common ones are listed in Table 9–3.

One common disease of the respiratory system worth noting is asthma. **Asthma** is a chronic disorder—a disease that persists for many years—and is marked by periodic episodes of wheezing and difficult breathing. Asthma is not an infectious disease. Most cases of asthma, in fact, are caused by allergic reactions (abnormal immune reactions) to common stimulants such as dust, pollen, and skin cells (dander) from pets. Certain foods, such as eggs, milk, chocolate, and food preservatives can also trigger asthma attacks. Still other cases are caused by drugs, vigorous exercise, and physiological stress.

In asthmatics, irritants, such as pollen and dander, cause a rapid increase in the production of mucus by the bronchi and bronchioles. Breathing becomes difficult. Irritants also stimulate the contraction of the smooth muscle cells in the walls of the bronchioles, making it even more difficult for an asthmatic to move air in and out of the lungs.

Asthma is fairly common in school-age children, but often disappears as these children grow older. As a result, only about 2% of the adult population suffers from asthma. Nevertheless, asthma is a serious disease. Asthma attacks can be quite disabling; some may even lead to death. By one estimate, several thousand Americans die each year from severe asthma attacks. Victims are generally elderly individuals who are suffering from other diseases as well.

Although asthma is incurable, the severity of an attack can be greatly lessened by treatment. One of

TABLE 9–3 Common Respiratory Diseases

DISEASE	SYMPTOMS	CAUSE
Emphysema	Breakdown of alveoli; shortness of breath	Smoking and air pollution
Chronic bronchitis	Coughing, shortness of breath	Smoking and air pollution
Acute bronchitis	Inflammation of the bronchi; yellowy mucus coughed up, shortness of breath	Many viruses and bacteria
Sinusitis	Inflammation of the paranasal sinuses; mucus discharge; blockage of nasal passageways; headache	Many viruses and bacteria
Laryngitis	Inflammation of larynx and vocal cords; sore throat; hoarseness; mucus buildup and cough	Many viruses and bacteria
Pneumonia	Inflammation of the lungs ranging from mild to extremely severe; cough and fever; shortness of breath at rest; chills; sweating; chest pains; blood in mucus	Bacteria, viruses, or inhalation of irritating gases
Asthma	Constriction of bronchioles; mucus buildup in bronchioles; periodic wheezing; difficulty breathing	Allergy to pollen, some foods, food additives; dandruff from dogs and cats; exercise

FIGURE 9–14 Asthma Relief Constriction of the bronchioles can be released by epinephrine inhalant spray.

the most common treatments is an oral spray containing the hormone epinephrine, which stimulates the bronchioles to open (Figure 9–14). Screening tests can help a patient find out what substances trigger an asthmatic attack so they can be avoided.

BREATHING AND THE CONTROL OF RESPIRATION

Air moves in and out of the lungs in much the same way that air moves in and out of the bellows blacksmiths use to fan their fires. Breathing, however, is generally an involuntary action, controlled by the nervous system.

The Mechanics of Breathing

During breathing, air must first be drawn into the lungs. This process is known as **inspiration** or **inhalation** (Table 9–4). Following inspiration, air must be expelled, a process called **expiration** or **exhalation**.

Inhalation is stimulated by the brain. Nerve impulses traveling from the brain stimulate the **diaphragm**, a dome-shaped muscle that separates the abdominal and thoracic cavities (Figure 9–15a). These impulses cause the diaphragm to contract. When the diaphragm contracts, it flattens and lowers. Much as pulling out the plunger of a syringe draws in air, the contraction of the diaphragm draws air into the lungs.

Nerve impulses also travel to **intercostal muscles**, short, powerful muscles that lie between the ribs. (They're the meat on barbecued ribs.) When these muscles contract, the rib cage lifts up and out (Figure 9–15a). The contraction of the intercostal muscles and the diaphragm increases the volume of the thoracic cavity by about 500 milliliters (about one-half of a quart). (Figure 9-15b).

The increase in the volume of the thoracic cavity decreases the **intrapulmonary pressure**, the pressure in the alveoli. This creates a small vacuum that draws air in through the mouth or nose downward into the trachea, bronchi, and lungs. At rest, each breath delivers about 500 milliliters of air to the lung. This is

TABLE 9–4 Summary of Inhalation and Exhalation	
Inhalation	• Stimulated by nerve impulses from the breathing center to the muscles of inspiration: the diaphragm and intercostal muscles
	• Contraction of intercostal muscles causes rib cage to move up and out
	• Contraction of diaphragm causes it to flatten
	• Volume of thoracic cavity increases
	• Intrapulmonary pressure decreases
	• Air flows into the lungs through the nose and mouth
Exhalation	• Nerve impulses from breathing center feed back on breathing center, shutting off stimuli to muscles of inspiration
	• Intercostal muscles relax and rib cage falls
	• Diaphragm relaxes and rises
	• Lungs recoil
	• Air is pushed out of the lungs

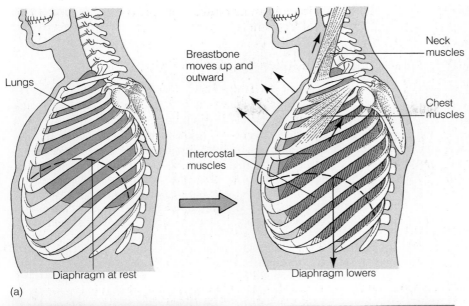

Lungs

Breastbone moves up and outward

Diaphragm at rest

Neck muscles

Chest muscles

Intercostal muscles

Diaphragm lowers

(a)

FIGURE 9–15   The Bellows Effect (*a*) The rising and falling of the chest wall caused by contraction of the intercostal muscles is shown in this diagram, illustrating the bellows effect. Inspiration is assisted by the diaphragm, which lowers, like the plunger on a syringe, drawing air into the lungs. (*b*) X-rays showing the size of the lungs in full exhalation (*left*) and full inspiration (*right*).

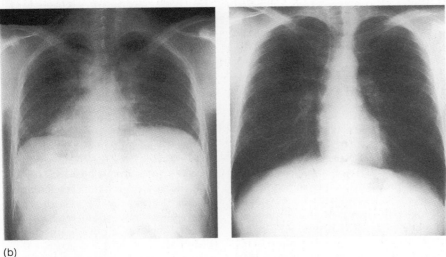

(b)

known as the **tidal volume**, the amount of air inhaled or exhaled with each breath when a person is at rest.

Inhalation is an active process; that is, it requires muscle activity. In contrast, exhalation in an individual at rest is a passive process. When the lungs fill, the diaphragm and intercostal muscles relax. The diaphragm rises, resuming its domed shape, and the chest wall falls slightly inward. Both changes reduce the volume of the thoracic cavity, raising the pressure and forcing the air out.

The lungs also play an important role in exhalation. The lungs contain numerous elastic connective tissue fibers. When full, the lungs are like two filled balloons. When inhalation ceases, the lungs recoil, forcing air out.

Exhalation can be augmented by enlisting additional muscles, making it an active process, known as **forced exhalation**. During forced exhalation, air is expelled from the lungs by contracting the muscles in the walls of the abdomen and chest (try this). Contraction of the abdominal muscles increases the intra-abdominal pressure, forcing the abdominal organs upward against the diaphragm. Contraction of the muscles in the wall of the chest reduces the volume of the chest, forcing air out.

Inhalation can also be consciously augmented by a

Smoking and Health: The Deadly Connection

Urban air pollution worries many Americans—and for good reason. However, the air in many cities is benign compared to the smoke that 65 million Americans voluntarily inhale from cigarettes. Smoking takes a huge toll on citizens of the world. In the United States, for example, an estimated 390,000 people die each year from the adverse health effects of tobacco smoke. That's the equivalent of more than 1000 people a day whose lives are cut short by lung cancer, heart disease, and emphysema.

Smoking costs society a great deal in lost lives, medical bills, and lost productivity. According to the Worldwatch Institute, every pack of cigarettes sold in the United States costs American society between $1.25 to $3.17 in medical costs, lost wages, and reduced productivity—that's about $40 million to $100 million a year! Anyway you look at it, smoking is a costly and dangerous habit.

Smoking is a principal cause of lung cancer throughout the world, claiming the lives of an estimated 127,000 men and women in the United States each year. By various estimates, depending on how many packs they smoke each day, smokers are 11 to 25 times more likely to develop lung cancer than nonsmokers.

Nonsmokers are also affected by the smoke of others. Nonsmokers inhale tobacco smoke in meetings, in restaurants, at work, and at home. As a result, they are often called "passive smokers." Until recently, passive smokers had little to say about their exposure to other people's smoke. Today, as a result of a growing awareness of the dangers of smoking, new regulations are banning smoking in many public places and in the workplace, except in designated areas.

This trend has been spurred, in part, by research showing that passive smokers are more likely to develop lung cancer than people who manage to steer clear of smokers. In a study of Japanese women married to men who smoked, researchers found that the passive smoking wives were as likely to develop lung cancer as people who smoked a half a pack of cigarettes a day! A recent report by the U.S. Environmental Protection Agency estimates that passive smoking annually causes 500 to 5000 cases of lung cancer in the United States. Passive smokers who are exposed to tobacco smoke for long periods also suffer from impaired lung function equal to that seen in light smokers (people who smoke under a pack a day).

Cigarette smoke in closed quarters may also cause angina, chest pains, in individuals who are suffering from atherosclerosis of the coronary arteries (Chapter 8). Carbon monoxide in cigarette smoke is responsible for angina. Research also shows that smokers are more susceptible to colds and respiratory infections. And smoking also affects children. Children who are raised in families in which both parents smoke suffer twice as many upper respiratory infections as their playmates from nonsmoking families. Researchers from the Harvard Medical School reported finding a 7% decrease in lung capacity in children raised by mothers who smoked. The researchers believe that this may lead to other pulmonary problems later in life.

Smoking has been shown to affect fertility in women as well. Research shows, for example, that women who smoke over a pack of cigarettes a day are half as fertile as nonsmokers. Smoking may also affect the outcome of pregnancy. According to the 1985 Surgeon General's Report, women who smoke several packs a day during pregnancy are much more likely to miscarry and give birth to children who are, on average, 200 grams (nearly 0.5 pounds) lighter than children born to non-smoking mothers.[1] Finally, children of women who smoke heavily during pregnancy score, on average, lower on mental aptitude tests during early childhood than children whose mothers do not smoke.

Tobacco smoke contains numerous hazardous substances that damage the delicate tissues of the respiratory system. Nicotine and sulfur dioxide, for example, paralyze the cilia lining the respiratory tract. One cigarette can knock the cilia out of action for an hour or more, destroying the natural cleansing mechanism, described in this chapter.

Tobacco smoke is also laden with microscopic carbon particles. Many toxic chemicals adhere to the carbon particles and are transported into the lungs. A dozen or so of these chemicals are known to cause cancer. Carbon particles penetrate deeply into the lungs where they accumulate in the alveoli and alveolar walls, turning healthy tissue into a blackened mass that often becomes cancerous, as illustrated in the figure. Tobacco smoke may also paralyze the phagocytic cells, making a

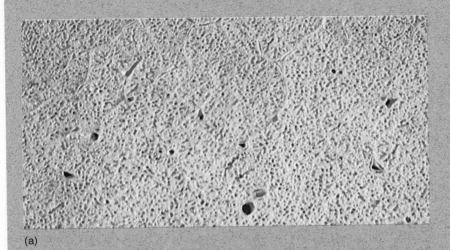

(a)

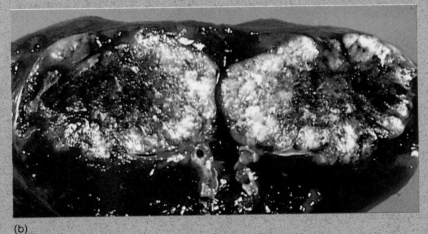

(b)

The Normal and Cancerous Lung (*a*) The normal lung appears spongy. (*b*) The cancerous lung from a smoker is filled with smoke and laden with tumor tissue.

bad situation even worse.

Toxin-carrying particles adhere to the lungs, the larynx, trachea, and bronchi. Virtually any place they stick, they can cause cancer. That is why smokers are five times more likely to develop laryngeal cancer and four times more likely to develop cancer of the oral cavity than nonsmokers.

Nitrogen dioxide and sulfur dioxide in tobacco smoke penetrate deep into the lungs where they dissolve in the watery layer inside the alveoli. Nitrogen dioxide is converted to nitric acid; sulfur dioxide is converted to sulfuric acid. Both acids erode the alveolar walls, leading to emphysema.

If tobacco smoke is so dangerous, why don't we ban smoking or discontinue generous government subsidies to tobacco growers? Surely, an air pollutant or food contaminant that killed hundreds of thousands of Americans each year would be banned. Tobacco use has a long history in the United States, and many people think that because smoking is a voluntary act, individuals should have the right to make their own decision. Furthermore, smoking supports the $30 billion-a-year tobacco industry that employs about 2 million people, including tobacco farmers, advertising people, retailers, and so on. The tobacco industry lobbies diligently to protect the rights of smokers. (For more on the controversy, see the Point/Counterpoint in this chapter.)

The dangers of smoking are becoming well known. As a result of widespread publicity and public pressure, smoking has dropped substantially in the United States. In 1990, for example, only 26% of the American population smoked. That is down from 34% in 1985 and down substantially from the 1950s and 1960s when well over half the adult male population and over one-third of the adult female population engaged in this potentially lethal habit.

Despite the downturn in smoking, an estimated 65 million Americans still smoke. Literally millions of people will continue to die from smoking over the coming decade.

[1]The Surgeon General's office publishes an annual report on smoking that summarizes new findings on the effects of smoking on reproduction and health.

forceful contraction of the muscles of inspiration (try this). By enlisting a more forceful contraction, you can increase the amount of air entering your lungs. Athletes often breath deeply just before an event to increase oxygen levels in their blood. A competitive swimmer, for example, may take several deep breaths before diving into the pool for a race. Deep breathing, while effective, can be dangerous, for reasons explained shortly.

The Control of Breathing

Breathing is controlled principally by the brain and is influenced by certain chemical substances in body fluids.

Neural Control. The **breathing control center** is located in a region of the brain called the brain stem. In some respects, the breathing center is similar to the sinoatrial node of the heart. In the breathing center, certain nerve cells give off periodic impulses that stimulate contraction of the intercostal muscles and the diaphragm, resulting in inhalation. When the lungs fill, the impulses cease.

Several mechanisms are responsible for the termination of these impulses. The first is a negative feedback loop, shown in Figure 9–16. Here's how it works. The breathing center sends nerve impulses to the diaphragm and intercostal muscles; it also sends impulses to a nearby region of the brain stem (Figure 9–16). This region is a kind of relay center, which transmits nerve impulses back to the breathing center,

thus inhibiting the breathing center and allowing the muscles of inspiration to relax.

Changes in the depth and rate (frequency) of breathing are thought to result from neural input arising from still other sections of the brain stem (Figure 9–16). These areas receive nerve impulses from chemical receptors in the brain and certain arteries. These receptors detect the concentration of carbon dioxide and other chemicals in the body. As carbon dioxide levels rise, breathing increases and often becomes deeper.

The third control mechanism consists of sensory nerve fibers in the lung that respond to stretch. When the lung is full, nerve impulses from the stretch receptors are transmitted to the breathing center. These impulses turn off the pacemaker. Stretch receptors, however, probably only function during exercise when large volumes of air are moved in and out of the lungs.

If the breathing center or the nerves that convey impulses to these muscles are destroyed—for example, by the polio virus or a head injury—breathing ceases.

Chemical Control. Breathing is influenced by certain chemicals in the blood. Three of the most important chemicals are carbon dioxide, hydrogen ions, and oxygen. Carbon dioxide levels are one of the most powerful stimulants.

Carbon dioxide receptors are found in certain arteries—notably, the aorta and the carotid arteries, which transport blood to the brain. When carbon di-

FIGURE 9–16 Breathing Center The breathing center controls respiration. It sends periodic impulses along the nerves to the muscles of inspiration, causing them to contract. The center also sends impulses along another route to an additional relay center in the brain stem. Impulses from here feed back on the breathing center, shutting off the impulses that stimulate inspiration. Chemical receptors in the brain and certain arteries and stretch receptors in the lung also alter the activity of the breathing center.

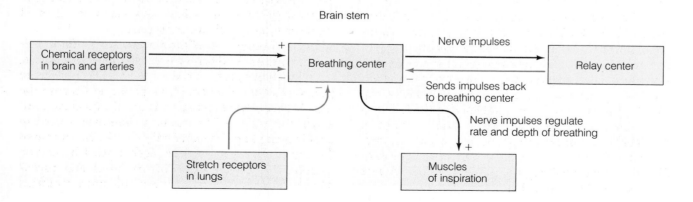

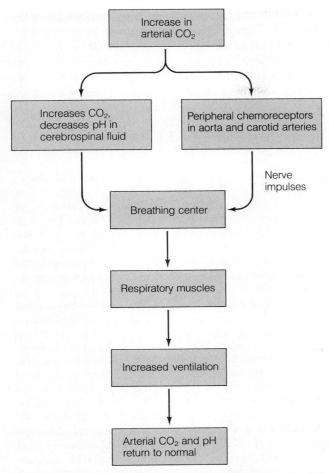

FIGURE 9–17 Chemical Control of Breathing
CO_2 and hydrogen in concentrations are the chief chemical controls of breathing.

oxide levels rise, these receptors generate impulses that are transmitted to the breathing center, increasing the rate of breathing. A decline in carbon dioxide levels has the opposite effect.

In the brain, carbon dioxide diffuses into the **cerebrospinal fluid** (CSF), a clear liquid found within cavities in the brain called the ventricles (Chapter 12). In the CSF, carbon dioxide is converted to carbonic acid. Carbonic acid dissociates, forming bicarbonate and hydrogen ions. A rise in carbon dioxide in the blood, therefore, results in an increase in the H^+ concentration of the CSF. The increase in H^+, in turn, is detected by chemical receptors, or **chemoreceptors**, located in the brain. These receptors send impulses to the breathing center, triggering an increase in the rate and depth of breathing (Figure 9–17).

Chemoreceptors allow the body to align respiration with cellular demands. During exercise, for example, cellular energy production increases to meet body demands. As noted in Chapter 3, energy production requires oxygen and generates carbon dioxide waste. Carbon dioxide produced during exercise increases the depth and rate of breathing, ultimately lowering the concentration of carbon dioxide in the blood and returning levels to normal. Increased ventilation also makes more oxygen available for energy production.

The body also contains a third set of sensors, the oxygen receptors. These receptors are not as sensitive as the H^+ receptors. Consequently, oxygen levels must fall considerably before the oxygen receptors begin generating impulses. This fact can have profound consequences, especially for divers and swimmers. Repeated deep and rapid breathing, or **hyperventilation**, for example, makes it possible for divers to hold their breath longer. Hyperventilation decreases carbon dioxide levels in the blood and H^+ concentrations in the cerebrospinal fluid, reducing the urge to breathe. The problem begins when the diver enters the water. Oxygen levels in the blood of the diver may fall so low that the brain is deprived of oxygen, causing the diver to lose consciousness. Ironically, the decrease in blood oxygen levels is not enough to stimulate breathing; the diver blacks out well before the H^+ concentration in the CSF reaches the level needed to stimulate breathing.

ENVIRONMENT AND HEALTH: AIR POLLUTION

The air of the industrialized world contains a multitude of potentially harmful chemicals, or pollutants. The diseases caused by air pollutants generated in our homes, factories, and cities now claim the lives of thousands of Americans each year.

Air pollution upsets the normal homeostatic balance, affecting millions of Americans in often subtle, but sometimes pronounced ways. But most people are unaware of the dangers of air pollution because the line between cause and effect is not always clear. Consider, for instance, the headache you experienced in traffic going home from school or work. Was it caused by tension or could it have been caused by carbon monoxide emissions from the cars, buses, and trucks? And what about the runny nose and sinus condition

ENDING THE DESTRUCTION

Dr. Louis W. Sullivan

Dr. Louis W. Sullivan has been secretary of Health and Human Services since 1989. Sullivan oversees work on health, food and drug safety, medical research, and income security programs of the U.S. government.

My view, and the view of my department, on smoking is straightforward and simple: *No smoking.* If we are serious about promoting health and preventing disease, then Americans must work to establish a smoke-free society.

The glamorization of smoking must end. The real story must be heard and heeded—the *serious personal* health risk confronting the smoker and those who passively inhale the deadly fumes of smokers and the cumulative and devastating impact on our economy.

It is high time that we also stopped allowing smoking advertisers such a high hand. A cigarette is the only legal product that, when used as intended, causes death. Advertisers who disproportionately target women, minorities, or young Americans have gone too far. They must stop their irresponsibility.

I recently spoke against the plans by R. J. Reynolds Tobacco Company to test market a new cigarette called Uptown to blacks in Philadelphia. Because of the justified public indignation, the company cancelled its plans.

I noticed something about that experience. Some tobacco industry spokespeople, when they couldn't explain away the death rates or adequately tell us why black Americans should be targeted with deadly products, started to talk about the legality of cigarette advertising. Certainly such advertising is allowed under current law, but that is not the point. I'm talking about the health of the user and those who come in contact with the user's smoke. It is *morally wrong* to promote a product which, when used as intended, causes death, trading death for corporate profits. This is a difference I will not let the industry spokespeople obfuscate.

Responsibility is needed, by the advertisers, the media, and the public. Advertisers and the media should question their own participation. I hope to see some of our ad agencies step forward to renounce tobacco ads and join in our antismoking efforts.

Certainly, all of us are concerned that targeted tobacco advertising, including the sponsorship of athletic events, may perpetuate smoking by adults and lure young people into a smoking habit. It is frightening to realize that studies have found that the younger the age at which one begins to smoke, the more likely that a person will become a long-term smoker and develop smoking-related diseases. In fact, 90% of smokers begin a cigarette addiction as children or adolescents.

Knowing all this, I am shocked that advertisers and the tobacco industry continue to paint a misleading picture about smoking. Even though the carnage of smoking is well documented, we still see young, good-looking, seemingly healthy models in their ads. Advertising that paints a *different* picture, especially advertising designed to lure young people, women, minorities, or blue-collar workers, is dishonest, irresponsible, and unconscionable.

I am especially concerned about the dangerous mixed message sent when sporting events are sponsored by the tobacco industry. Sporting events should encourage good health practices for the participants and fans. And yet, when the tobacco industry sponsors an event to push their deadly product, they are trading on the health, the prestige, and the image of the athlete to barter a product that will kill the user. The sponsorship itself uses the vigor and energy of athletes as a subtle, but incorrect and dishonest message, that smoking can be compatible with good health, which it is not. It is ironic—but telling—that most of the superb athletes competing in organized sports do not use tobacco, and could not use it, if they want to be successful and healthy.

We must put a premium on good health. National and local competitive sporting associations and athletes should reject sponsorship by the tobacco industry.

I hope that universities and other institutions will not host events that are sponsored by tobacco companies. I also urge the tobacco industry to voluntarily withdraw from such sponsorship, which would be the responsible and honest course. I call upon advertisers themselves to shun the temptation of this tainted money, stained by addiction, disease, and death. The most courageous, prudent, and morally correct action would be for advertisers to *kick* the tobacco habit.

FREE SPEECH, FREE CHOICE

Daniel L. Jaffe

Daniel L. Jaffe is executive vice president of the Association of National Advertisers (ANA). Mr. Jaffe directs the ANA's government relations programs at the state, local, and federal levels from the ANA's office in Washington, D.C.

The First Amendment to the Constitution is the cornerstone of our nation's democracy. Thus, efforts of opponents of smoking to restrict or ban truthful, nondeceptive advertising of tobacco products raise concerns about the First Amendment protection of free speech and free choice.

If the federal government succeeds in banning or restricting tobacco ads, it is reasonable to assume that the government will then be free to restrict other truthful, nondeceptive advertising.

A clear distinction must be drawn between advertising targeted to illegal audiences and advertising targeted to the public at large. It is both illegal and reprehensible for *any company* to target underage smokers, drinkers, or any other group that is barred from legally using a product. The Federal Trade Commission (FTC), which is empowered by Congress to protect the public from false or deceptive advertising, has both the authority and responsibility to stop targeting to any illegal audience.

But the U.S. Supreme Court has ruled that the protection of children cannot be used as an excuse to censor truthful, nondeceptive advertising to the general public. Louis W. Sullivan, secretary of the Department of Health and Human Services and a number of members of Congress, however, have now declared war on tobacco companies that have designed their sales pitches specifically for blacks, Hispanics, women, and blue-collar workers. This attack, unfortunately, is based on the condescending assumption that major segments of the American public are not capable of making a rational decision when confronted with a barrage of tobacco advertising.

In our society the government rightly considers all adults, be they black, Hispanic, women, or blue-collar workers, to have sufficient judgment to vote and serve on juries. Can we then accept the proposition that these same groups cannot be trusted to receive advertising messages directed to them?

Furthermore, it simply is false to suggest that the public has not been informed of the health consequences of tobacco. Clearly, in the United States, only a "Rip Van Winkle" could be unaware of the heath concerns associated with tobacco. Every pack of cigarettes and every cigarette advertisement contains extremely strong health warnings.

As the U.S. Supreme Court stated in *Virginia Board of Pharmacy v. Virginia Citizens Consumer Council*, "... people will perceive their own best interests if only they are well enough informed, and the best means to that end is to open the channels of communication rather than to close them."

Finally, it has been suggested that tobacco advertising should be treated not as a legal, but as strictly a moral issue. In fact, the tobacco issue raises both legal and moral questions. But, in a democracy, we allow each individual—not the government—to make his or her own moral determinations.

Robust debate in regard to the tobacco issue is essential. But in a society in which tobacco can be legally grown, manufactured, sold, and smoked, we cannot accept that the one thing that will not be allowed is truthful advertising of tobacco products.

We have learned in the United States that censorship is habit forming. Therefore, the American Civil Liberties Union, the Washington Legal Foundation, and numerous other constitutional experts have gone on record in opposition to proposals to ban or severely restrict tobacco advertising. Once we head down this road, we have gone a long way toward restricting everyone's right to publish and receive information about tobacco or any other product the goverment might dislike but is unwilling, or unable, to ban outright.

SHARPENING YOUR CRITICAL THINKING SKILLS

1. Summarize Jaffe's and Sullivan's main point. How does each author support his argument?
2. Give your views on this issue and explain your position.

you experienced last winter? Was that caused by a virus or bacterium or was it caused by pollution?

One classic study of air pollution on the East Coast showed an often overlooked relationship between air pollution and upper respiratory problems. The researchers found that the level of sulfur dioxide, a pollutant produced by automobiles, power plants, and factories increased during the winter months in New York City. Weather conditions trapped the pollutants, raising ground-level concentrations. During episodes of slightly elevated sulfur dioxide concentrations, upper respiratory illness in New York residents skyrocketed (Figure 9–18). Colds, coughs, nasal irritation, and other symptoms increased fivefold in just a few days. When the pollution levels returned to normal, the symptoms rapidly subsided.

FIGURE 9–18 Air Pollution and Health Graph of respiratory illnesses associated with an air pollution episode in New York City in 1962. Sulfur dioxide (SO_2) levels rose on days 16 and 17 from 0.2 ppm to 0.8 to 0.9 ppm.

Source: Redrawn from "Health and the Urban Environment: Health Profiles versus Environmental Pollutants," J. R. McCarroll, E. J. Cassell, W. T. Ingram, and D. Wolter in *American Journal of Public Health* 56(1966): 266–75.

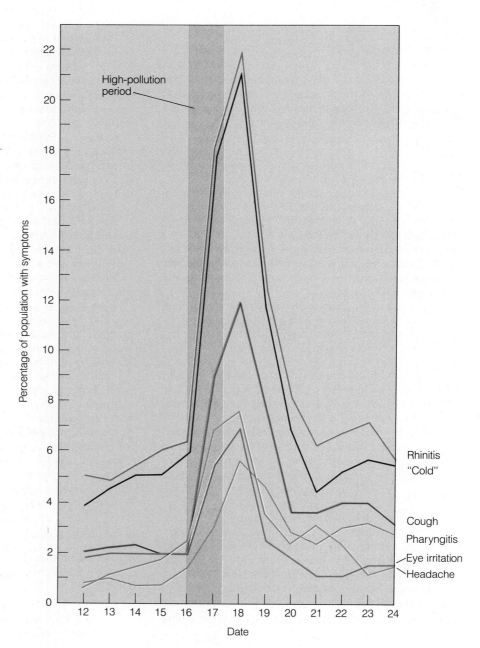

Air pollution is partly responsible for other long-term diseases. One of these is **chronic bronchitis**, a persistent irritation of the bronchi, which causes mucous buildup, coughing, and difficulty breathing. Today, one out of every five American men between the ages of 40 and 60 suffers from chronic bronchitis. The leading cause of chronic bronchitis is cigarette smoking, but studies show that urban air pollution also contributes to this disease. Three pollutants in particular have been linked to chronic bronchitis: sulfur dioxide, nitrogen oxides, and ozone. Each of these irritates the lung and bronchial passages.

A far more troublesome disease is emphysema. Emphysema, discussed earlier in the chapter, is the fastest-growing cause of death in the United States. Caused principally by smoking and air pollution, emphysema afflicts over 1.5 million Americans and is more common in men than women. As noted earlier, emphysema is a progressive, incurable disease. As the disease worsens, lung function deteriorates until the victim requires supplemental oxygen to perform even routine functions, such as walking or speaking.

The leading cause of emphysema is smoking, a habit of 65 million Americans. Emphysema is also caused by urban air pollution. As a result, smokers who live in polluted urban settings have the highest incidence of the disease. Like chronic bronchitis, emphysema is caused by the three lung irritants: ozone, sulfur dioxide, and nitrogen oxides.

Researchers believe that 85% of the nearly 150,000 cases of lung cancer in the United States are caused by smoking (for more on the effects of smoking and the controversy surrounding tobacco advertising, see Health Note 9–2 and the Point/Counterpoint). The remaining 15,000 or so cases are thought to be caused by urban and workplace air pollutants as well as natural pollutants, such as radioactive radon gas. Radon is emitted from radioactive radium, which occurs naturally in the soil in many parts of the country.

No one knows the exact toll of urban air pollution. It is not something that researchers can determine very easily, if at all, because people are exposed to so many different pollutants over their lifetimes. However, a recent report issued by the federal government estimates that approximately 51,000 Americans die each year from lung disease caused by urban air pollution. The authors of the study predict that, by the year 2000, the number of victims will climb to nearly 60,000 per year, illustrating once again that human health is clearly dependent on a clean environment.

SUMMARY

THE STRUCTURE OF THE HUMAN RESPIRATORY SYSTEM

1. The respiratory system consists of an air-conducting portion and a gas-exchange portion.
2. The air-conducting portion consists of the oral and nasal cavities, the pharynx, larynx, trachea, bronchi, and bronchioles. It transports air from outside the body to the alveoli in the lungs, the site of gaseous exchange.
3. The alveoli are tiny, saclike pockets in the lung formed by a single layer of flattened epithelial cells that facilitate diffusion. Surrounding the alveoli are capillary beds that pick up oxygen and expel carbon dioxide.
4. The lining of the alveoli is kept moist by water. Surfactant, a phospholipid produced in the lung, reduces the surface tension inside the alveoli and prevents them from collapsing.

THE FUNCTIONS OF THE RESPIRATORY SYSTEM

5. The respiratory system conducts air to and from the lungs, exchanges gases, and helps produce sounds. Sound is generated as air rushes past the elastic vocal cords, causing them to vibrate. The sounds are modified by movements of the tongue and changes in the shape of the oral cavity.
6. Oxygen and carbon dioxide gases diffuse across the alveolar wall, a process driven by concentration differences between the blood and alveolar air. In the lungs, oxygen diffuses from the alveoli into the blood plasma, then into the RBCs. Most of it binds to hemoglobin. Carbon dioxide diffuses in the opposite direction.
7. Carbon dioxide, a waste product of cellular energy production, is released by the blood flowing through capillaries. Carbon dioxide enters the RBCs in the bloodstream where it is converted into carbonic acid, which then dissociates, forming hydrogen and bicarbonate ions. Bicarbonate is the principal form found in the blood plasma.

BREATHING AND THE CONTROL OF RESPIRATION

8. Breathing is an involuntary action with a conscious override. It is controlled by the breathing center in the brain stem. Nerve cells in the breathing center send bursts of impulses to the diaphragm and intercostal muscles. Their contraction increases the volume of the chest cavity, which draws air into the lungs through the nose or mouth.
9. When the impulses stop, inspiration ends. Air is then expelled passively as the lungs recoil, the chest wall returns to the normal position, and the diaphragm rises.
10. The breathing center is regulated by a negative feedback loop that it generates itself. It is also regulated by outside influences.

11. Expiration can be augmented by enlisting the aid of abdominal and chest muscles. Inspiration can also consciously be made more forceful.
12. The rate of respiration can be increased by rising blood carbon dioxide levels, falling oxygen levels, and an increase in physical exercise.

ENVIRONMENT AND HEALTH: AIR POLLUTION

13. The proper functioning of the respiratory system is essential for health. Respiratory function can be dramatically upset by microorganisms as well as by pollution from factories, automobiles, power plants, and even our own homes.
14. Chronic bronchitis, a persistent irritation of the bronchi, is brought on by sulfur dioxide, ozone, and nitrogen oxides, lung irritants sometimes found in dangerous levels in urban air.
15. Emphysema, a breakdown of the alveoli that gradually destroys the lung's ability to absorb oxygen, is similarly induced. Despite the role of air pollution in causing emphysema and chronic bronchitis, smoking remains the number one cause of these diseases.

⌘ EXERCISING YOUR CRITICAL THINKING SKILLS

An inventor has devised a pollution control device that removes particulates from the smokestack of factories. He claims that the device removes 80% of all particulates and will bring factories into compliance with federal law and will therefore protect nearby residents from harmful pollutants. What is your reaction? What questions might you ask? Would you approve installing the device, given this information? Would your decision be affected by data showing that, while the device does reduce total particulates by 80%, it does not capture finer particulates? These are particles that can be inhaled deeply into the lung because they do not precipitate out of the respiratory tract. Would your decision be affected if you found that the small (respirable) particles in the smokestack contained toxic metals, such as mercury and cadmium? What rule(s) of critical thinking does this example illustrate?

TEST OF TERMS

1. _____ is a disease of the lung characterized by the progressive breakdown of the alveoli.
2. The _____ is that part of the respiratory system that conducts air from the nose and mouth to the larynx.
3. Air travels from the larynx to the _____ , a ribbed duct that leads to the lungs then splits into right and left _____ , which penetrate the lung.
4. The _____ , the muscular ducts that conduct air to the alveoli, can contract and relax like arterioles, thus providing a way to control the flow of air inside the lung.
5. Much of the lining of the conducting portion of the respiratory system contains _____ cells, which produce a secretion that traps dust particles and bacteria.
6. The chemical, _____ , is produced by certain cells in the alveoli and helps reduce surface tension, preventing the alveoli from collapsing.
7. The phagocytic cell that wanders in and out of the alveoli is called a _____ cell.
8. Inside the larynx are two elastic cords that vibrate when air breezes past. These are the _____ .
9. The patch of epithelium in the roof of the nasal cavity that perceives smell is called the _____ .
10. Most oxygen is carried in the blood bound to _____ in the RBCs. Most carbon dioxide is transported as _____ .
11. The region of the _____ that controls inspiration is called the _____ _____ .
12. The active process in which air is drawn into the lungs is called _____ . It is caused by an enlargement of the chest cavity, which results from the contraction of the _____ and the _____ muscles.
13. The passive expulsion of air is called _____ .
14. Persistent irritation of the bronchi leading to mucous buildup and coughing is called _____ _____ and is often caused by pollution in cigarette smoke and urban air.

Answers to the Test of Terms are located in Appendix B.

TEST OF CONCEPTS

1. Trace the flow of air from the mouth and nose to the alveoli and describe what happens to the air as it travels along the various passageways.
2. Draw an alveolus, including all cell types found there. Be sure to show the relationship of the surrounding capillaries. Show the path that oxygen and carbon dioxide must take.
3. Trace the movement of oxygen from alveolar air to the blood in alveolar capillaries. Describe the forces that cause oxygen to move in this direction. Do the same for the reverse flow of carbon dioxide.
4. Why would a breakdown of alveoli in emphysemic patients make it more difficult for them to receive adequate oxygen?
5. Describe how sounds are generated and modified by the respiratory system.
6. A baby is born prematurely and is having difficulty breathing. As the attending physician, explain to the parents of the child what the problem is and how it could be corrected.
7. Smoking irritates the trachea and bronchi, causing mucus to build up and paralyze cilia. How does this affect the lung?
8. Describe inspiration and expiration, being sure to include discussions of what triggers them and the role of muscles in bringing about these actions.
9. How does the breathing center regulate itself to control the frequency of breathing?
10. Exercise increases the rate of breathing. How?
11. Debate the statement: Urban air pollution has very little overall impact on human health.
12. Do you agree or disagree with the following statement: The single most effective way of reducing deaths in the United States would be to ban smoking.

SUGGESTED READINGS

Chiras, D. D. 1991. *Environmental science: Action for a sustainable future*, 3d. ed. Redwood City, Calif.: Benjamin/Cummings. Provides a more detailed look at the effects of air pollution and smoking.

Guyton, A. C. 1986. *Textbook of medical physiology*, 7th ed. Philadelphia: Saunders. Excellent coverage of respiration.

Sherwood, L. 1989. *Human physiology: From cells to systems*. St. Paul, Minn.: West. Detailed coverage for students wishing more information.

Spence, A. P., and E. B. Mason. 1990. *Human anatomy and physiology*, 3d ed. Redwood City, Calif.: Benjamin/Cummings. Excellent coverage of the human respiratory system.

Vander, A. J., J. H. Sherman, and D. S. Luciano. 1985. Human physiology: The mechanisms of body function, 4th ed. New York: McGraw-Hill. Superb information on the respiratory function for students wanting more detail.

10 Immunity: Protecting Homeostasis

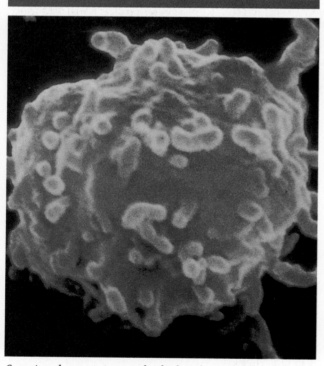

Scanning electron micrograph of a lymphocyte, one of the key players in immunity.

D amion Knight, a bright, young cabinet-maker, began to lose weight and experience bouts of unexplained fever. His lymph nodes became swollen, and he began feeling weak and drowsy. Friends urged him to see a doctor. When he did, he learned the bad news: he had contracted AIDS (acquired immune deficiency syndrome), a deadly disease caused by a virus that attacks and weakens the immune system.

Like thousands of others, Damion died a few years after diagnosis. His doctor could not help him. Medical science has only begun to understand this disease that could kill tens of thousands, perhaps millions, of people the world over in the next decade.

This chapter looks at the body's defense mechanisms, which protect us against disease-causing (pathogenic) microorganisms and cancer. It also discusses AIDS—what causes the disease and how it can be prevented.

VIRUSES AND BACTERIA: A PRIMER ON INFECTIOUS AGENTS

Ralph Waldo Emerson once wrote, "As soon as there is life, there is danger." In Emerson's time, there were

246

many dangers. Viruses and bacteria, for example, often spread through populations, killing large numbers of people. Thus, even a small cut could be life-threatening if an infection developed and spread to the bloodstream.

Microorganisms remain a problem today in the industrialized nations, despite the introduction of vaccines and antibiotics and marked improvements in sanitation, water purification, and living conditions. Although these measures have helped public health officials control infectious diseases, they have not eliminated them. Except perhaps for the smallpox virus, the microorganisms that caused widespread death only a few decades ago are still present. Public health officials urge continued vigilance and recommend vaccination of all school-age children to keep a lid on infectious diseases that once caused so much pain and suffering.

In the United States, an estimated 100,000 Americans die from resistant strains of bacteria to which people are exposed, ironically, while being treated in hospitals. Each year, hundreds of people in this country, mostly young children and the elderly, die of the flu, caused by viruses.

Infectious diseases are also a major problem in the Third World; there the flu alone kills millions of people each year. Starvation and hunger make many people more susceptible to diseases long since brought under control in the United States and other industrialized countries.

Viruses: Cellular Pirates

Viruses confound the study of biology. Viruses are not really living organisms, because they cannot reproduce on their own and cannot respond to stimuli (Chapter 1). Viruses are not cells either, for they lack cytoplasm and organelles. So what are they?

Viruses are tiny packets of nucleic acid, either DNA or RNA, bound by a protein coat (Figure 10–1). Viruses are carried in water and through the air on dust particles or in fine moisture droplets released when someone sneezes. Viruses may also be transmitted on inanimate objects and may be passed from person to person—either by a friendly handshake or by sexual contact, depending on the type of virus. Colds are caused by viruses and are spread chiefly by hand contact. Physicians, therefore, recommend (1) washing your hands frequently during the cold season, especially if you have been around a person with a cold, and (2) keeping your hands away from your eyes, nose, and mouth until you have had a chance to wash them.

Viruses spread from person to person with remarkable speed. So rapid is this spread, in fact, that half the world's population is exposed to a new strain of the influenza virus within two years of its emergence. The rapid transmission of viruses results, in large part, from the fact that many people travel freely and frequently. In addition, many of the world's people live in densely populated settlements, through which infectious diseases disperse rather quickly.

As Figure 10–1 shows, viruses contain a nucleic acid core, consisting of RNA or DNA, which is surrounded by a protein coat, or capsid. The **capsid** protects the nucleic acid core and is formed from 100 to 3000 globular proteins, or **capsomeres**. Some of these proteins bind to plasma membrane receptors, allowing viruses to be phagocytized by cells. Others are enzymes that digest holes in the plasma membranes of cells, allowing viruses to enter cells. Many viruses have an additional protective coat, the **envelope**, which consists of a layer of lipid and protein resembling the plasma membrane of cells.

Viruses come in many shapes and sizes. The simplest is the globular virus (Figure 10–2a). Others are polyhedral (many sided), resembling multifaceted diamonds (Figure 10–2b). Some viruses consist of long cylinders of protein surrounding the nucleic acid core (Figure 10–2c). Other viruses are odd-shaped structures. The bacteriophage T4 virus is a good example (Figure 10–2d). The bacteriophage T4 virus invades bacterial cells (hence the name "bacteria eater"). Resembling a lunar landing module from a science fiction story, the T4 virus consists of a head containing the nucleic acid core, a tail, and tail filaments. This virus "lands" on the surfaces of bacterial cells with its

FIGURE 10–1 General Structure of Viruses
(*a*) Naked virus with no envelope. (*b*) Virus with envelope.

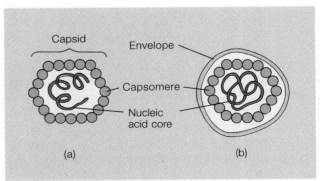

FIGURE 10-2 Viral Architecture Viruses come in many shapes and sizes. Each virus consists of a nucleic acid core surrounded by a capsid made of proteins. Some, like the influenza virus, have an additional coating, an envelope made chiefly of phospholipid molecules. This illustration shows the four main viral shapes: (a) globular, (b) polyhedral, (c) cylindrical, and (d) odd shaped.

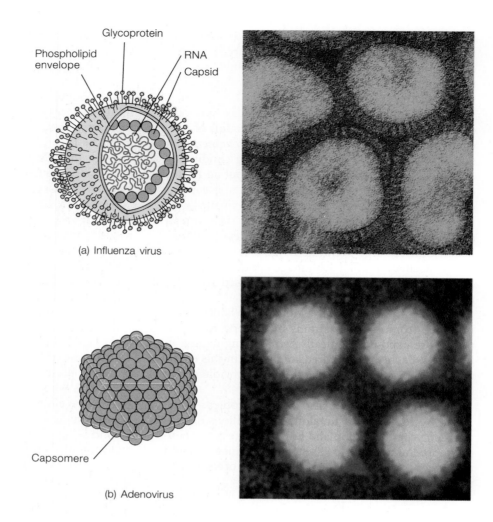

(a) Influenza virus

(b) Adenovirus

tail down. Enzymes in the tail section digest a small hole in the plasma membrane of the host cell, and the virus then injects its nucleic acid into the cell.

Viruses can also enter cells via phagocytosis. In order to be phagocytized, viruses must first bind to protein receptors on the plasma membrane of the host cell (Figure 10–3). Inside the cell, the virus releases its nucleic acid core. The nucleic acid of DNA viruses duplicates, producing complementary strands. The complementary strands so produced also replicate, generating more viral DNA. The genes a DNA virus carries into the cell also code for certain enzymes needed for viral replication and for messenger RNA needed to produce capsomeres. (The action of RNA viruses will be explained shortly.)

The energy and building blocks needed to synthesize viral nucleic acid and proteins are drawn from the cell. Commandeered by viruses, like a ship taken over by a band of pirates, an infected cell stops producing the materials it needs to survive. New viruses form from the proteins and viral nucleic acid that has been produced using the host cell's ribosomes and nutrients. New viruses may be released a few at a time via exocytosis, a process virologists refer to as **budding**. As Figure 10–3b illustrates, budding adds the envelope found in some viruses. Viruses may also be released *en masse* when the host cell dies and breaks down.

A cell infected by a virus can produce as many as 80,000 to 120,000 new viruses. Viruses released by a host cell into the tissue fluid surrounding the cell are transported in the blood or lymph, traveling throughout the body to new sites.

Unfortunately, the spread of viruses cannot be stopped by antibiotics, since antibiotics are specific for bacteria (Chapter 5). Viruses, however, can be

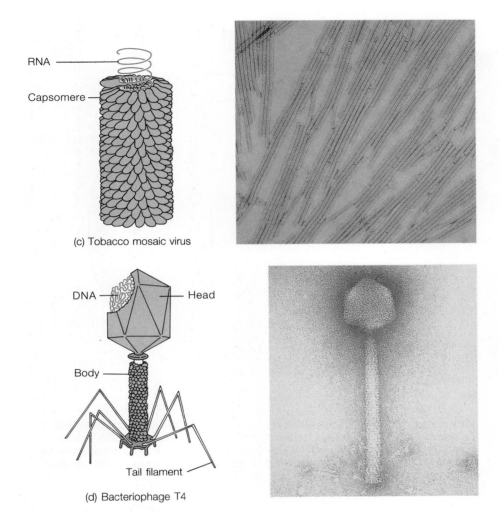

RNA

Capsomere

(c) Tobacco mosaic virus

DNA — Head

Body

Tail filament

(d) Bacteriophage T4

destroyed by the body's immune system. Even though it normally brings viral infections under control, the immune system does not always eliminate every single virus from the body. DNA from some viruses, for example, may become incorporated in the DNA of body cells. The herpes simplex virus is a case in point. This virus consists of several strains. The herpes simplex type II produces genital lesions and afflicts an estimated 20 million American men and women (Figure 10–4).[1] Soon after an individual is infected, the virus produces blisters on the genitals, thighs, and buttocks. The blisters rupture, leaving painful ulcers in the skin that heal within one to three weeks. The virus is not destroyed, however. Instead, it enters a small clump of nerve cells alongside the sacrum, the wedge-shaped bone just above the tailbone. Here, the

[1]Herpes simplex type I causes cold sores.

virus remains, periodically flaring up under certain conditions. Stress, menstruation, sex, and even exposure to sunlight can cause the virus to reemerge. (For more on the herpes virus, see Chapter 16.)

Viral Replication. Figure 10–5 shows how a DNA virus replicates. Inside a cell, the viral DNA produces complementary strands, using DNA polymerase and nucleotides from the host cell. The complementary DNA strands, in turn, code for the production of messenger RNA, using cellular supplies of RNA nucleotides and RNA polymerase. Proteins are produced on the strands of viral mRNA, also using cellular ribosomes, transfer RNA, and amino acids. Inside a cell, viral replication, RNA synthesis, and protein synthesis take precedence over all other activities. The cell forgoes its activities and may even die.

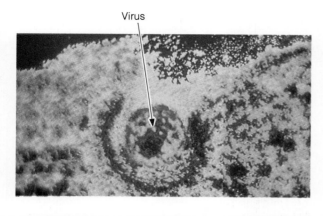

Virus

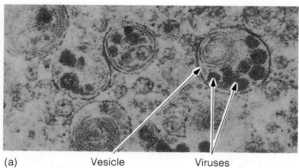

(a) Vesicle   Viruses   (b)

Virus-with envelope

Viruses budding from plasma membrane

Viruses

Viruses being phagocytized by cell

New viruses assemble in cell

Nucleus

Nucleic acid and viral protein formed

Viral DNA "escapes" from phagasome

Viral DNA replicates inside cell

FIGURE 10–3   Viral Infection   (*a*) Viruses enter a cell either by phagocytosis or by fusing with the plasma membrane and injecting their nucleic acid into the cell. Viruses that are phagocytized fuse with the membrane of the phagosome and inject their nucleic acid into the cytoplasm. Once inside the cell, the viral nucleic acid replicates and codes for the proteins needed for the capsid. (*b*) Viruses reassemble inside the cell and are released via exocytosis or when the cell dies and the plasma membrane disintegrates.

RNA viruses replicate inside cells as well. At least two mechanisms for RNA replication are known. The first occurs in a group of RNA viruses that are called **retroviruses**. These viruses contain an enzyme that they carry into the host cell. The enzyme, called **reverse transcriptase**, catalyzes the synthesis of DNA on the virus's RNA, hence the name.[2] The DNA produced in the host cell is then used to produce additional viral RNA. Viral RNA is needed to make new viruses and to synthesize viral capsomeres and enzymes.

Viral infections can cause a variety of problems, depending on the nature of the virus and the tissues they infect. Approximately 200 viruses are known to cause the common cold, a minor illness characterized by a runny nose, sneezing, sore throat, and coughing. Others produce influenza or the flu. Influenza is characterized by chills, fever, sneezing, headache, muscular aches, and sore throat. Still other viruses cause pneumonia, an infection of the lungs. And still other viruses cause cancer. These viruses may insert cancer-causing genes in the host cell genome (Chapter 5).

Viruses infect all organisms—from bacteria to plants to cows. Some viruses even cross species boundaries, although such crossings are rare. The virus that causes distemper in dogs, for example, can be spread to seals. The rabies virus can be transmitted from dogs to humans. The AIDS virus is thought to have been passed to humans from the African green monkey, either through contact or when humans ate viral-contaminated meat. The AIDS virus is now passed from person to person by sexual contact and blood transfusion.

[2]Other retroviruses produce reverse transcriptase after they infect the host cell, using viral RNA as a template for enzyme synthesis.

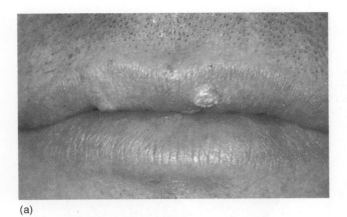

(a)

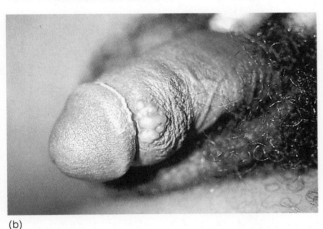

(b)

FIGURE 10–4 Herpes Simplex Virus The herpes simplex virus comes in several forms. Two of the most common cause (a) cold sores and (b) genital lesions.

Bacteria

Unlike viruses, **bacteria** are single-celled organisms that are equipped with all of the metabolic machinery necessary for survival and reproduction. A typical bacterium is delimited by a plasma membrane, which, in turn, is surrounded by a thick, protective cell wall.

Bacteria contain a single circular molecule of DNA. Unlike the chromosomes found in eukaryotic cells, bacterial DNA is not associated with protein and is not surrounded by a nuclear envelope. Bacteria also lack cellular organelles, except ribosomes (Figure 10–6).

Bacterial replication is a rather simple affair. Unlike eukaryotes, bacteria do not have a distinct cell cycle, but divide more or less continuously if an adequate supply of nutrients is available. Bacteria divide by binary fission, splitting into two separate parts, as illustrated in Figure 10–7.

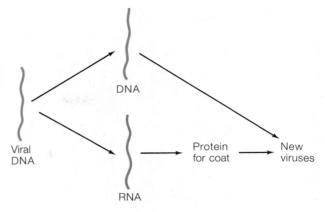

(a) DNA viruses

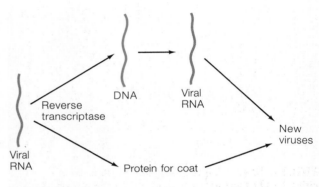

(b) Retroviruses

FIGURE 10–5 Details of Viral Replication (a) DNA viruses produce additional DNA, needed to make more viruses. The DNA is also transcribed, producing the RNA needed to make protein for the capsid. (b) Some RNA viruses carry with them an enzyme, reverse transcriptase, which is used to produce complementary DNA from the RNA. The DNA, in turn, makes viral RNA, allowing the virus to replicate itself.

Like viruses, bacteria can be harmful to cells, tissues, and organs. Most pathogenic (disease-causing) bacteria release some kind of chemical toxin or enzyme that injures cells or disrupts their function. Bacteria cause many common diseases, such as strep throat, some forms of pneumonia, food poisoning, and urinary tract infections.

Not all bacteria are harmful, however. Many bacteria are useful organisms. Bacteria in the stomach of cows and other hoofed animals, for example, allow these animals to digest cellulose, gaining nutrients from grasses and other plant material that cannot be digested by their own digestive enzymes (Chapter 7).

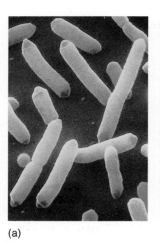

(a)

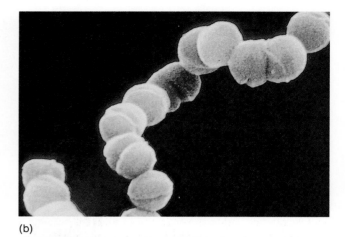

(b)

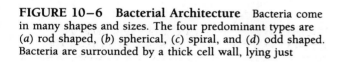

(c)

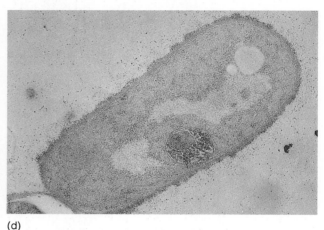

(d)

FIGURE 10–6 Bacterial Architecture Bacteria come in many shapes and sizes. The four predominant types are (*a*) rod shaped, (*b*) spherical, (*c*) spiral, and (*d*) odd shaped. Bacteria are surrounded by a thick cell wall, lying just outside the plasma membrane. The DNA exists in a circular strand in the nucleoid. The only organelle present is the ribosome, which is needed for protein synthesis.

In humans, intestinal bacteria produce several vitamins that are absorbed by the large intestine. Certain bacteria in the soil degrade animal wastes and the remains of dead plants and animals (Chapter 20). The decomposition of these organic materials releases nutrients, which replenish the soil. Other soil bacteria are associated with the roots of certain plants. These bacteria absorb nitrogen gas from the air spaces in the soil and convert it into a chemical form that plants can use to make amino acids and nucleic acids. It is from plants that humans and most other animals ultimately derive their nutrients. Certain bacteria can degrade some of the organic chemicals found in oil, when it is accidentally spilled on land and at sea, thus helping reduce long-term environmental damage.

THE FIRST AND SECOND LINES OF DEFENSE

The human body is a fortress, which repels a host of potentially harmful microscopic organisms and destroys most of those that enter. The human fortress consists of three basic lines of defense.

The First Line of Defense

Like any good fortress, the human body has an outer protective wall, the skin, which repels a great many harmful microorganisms. The skin is the first line of defense. It consists of a relatively thick layer of epidermal cells overlying the rich vascular layer called the dermis. Epidermal cells are produced by cell division in the base of the epidermis. As the basal cells proliferate, they move outward, become flattened, and die. The dead cells form a fairly waterproof protective layer that reduces moisture loss and protects underlying tissues from microorganisms.

The epithelial linings of the respiratory, digestive, and urinary systems are also part of the body's wall of protection, keeping potentially harmful microorganisms from invading the body. A break in these linings can result in the invasion of microorganisms. In Chapter 7, for example, we saw that a break in the digestive tract, caused by diverticulitis, can result in widespread infection in the blood or abdominal cavity.

The first line of defense also includes several protective chemicals. The skin, for example, produces slightly acidic secretions that impair bacterial growth. The stomach lining produces hydrochloric acid, which destroys many bacteria. Tears and saliva contain an enzyme called **lysozyme**, which dissolves the cell wall of bacteria, killing them. Cells in the lining of the trachea and bronchi produce mucus, which traps bacteria. Cilia in the lining of the respiratory system sweep this mucus toward the mouth where it can be expectorated or swallowed (Chapter 9).

The Second Line of Defense

The first line of defense is not impenetrable. Even tiny breaks in the skin or in the lining of the respiratory, digestive, and urinary tracts can admit viruses, bacteria, and other microorganisms, such as single-celled fungi. The second line of defense handles most of the invaders.

The Inflammatory Response. Damage to body tissues triggers a response called inflammation. **Inflammation** is a protective response, characterized by heat, redness, swelling, and pain.

Tissue injury results in the influx of bacteria, which are often destroyed by macrophages and neutrophils. Tissue injury also stimulates the release of several chemical substances (Figure 10–8). Injured cells, for example, release **histamine**, a substance that stimulates the arterioles to dilate, causing the capillary networks to swell with blood.[3] The increase in blood flow creates the heat and redness around a cut

---

[3]Histamine is produced by mast cells, platelets, and basophils, a type of white blood cell. Mast cells are a connective tissue cell described later in the chapter.

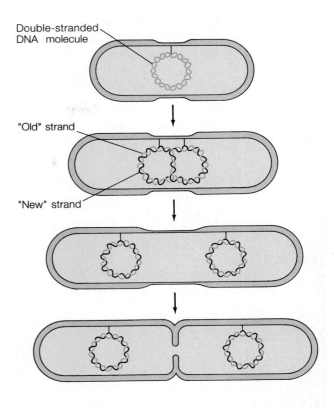

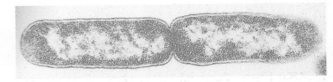

FIGURE 10–7 Binary Fission Most bacteria reproduce by binary fission, a splitting of the cells. The circular DNA molecule reproduces prior to fission, as shown here; then the cell splits, forming two new bacteria.

or abrasion. Increased heat increases the metabolic rate in the injured area, accelerating healing.

Still other chemicals increase the permeability of capillaries, allowing plasma to flow into a wounded area. Plasma carries with it clotting factors, (Chapter 8). Clots wall off injured areas and help contain the infection.

The fluids that accumulate in an injured tissue cause swelling, which stimulates pain receptors in the injured area. Pain receptors send nerve impulses to the brain. Pain also results from chemical toxins released by bacteria and from chemicals, such as prostaglandins, released by injured cells. Aspirin and

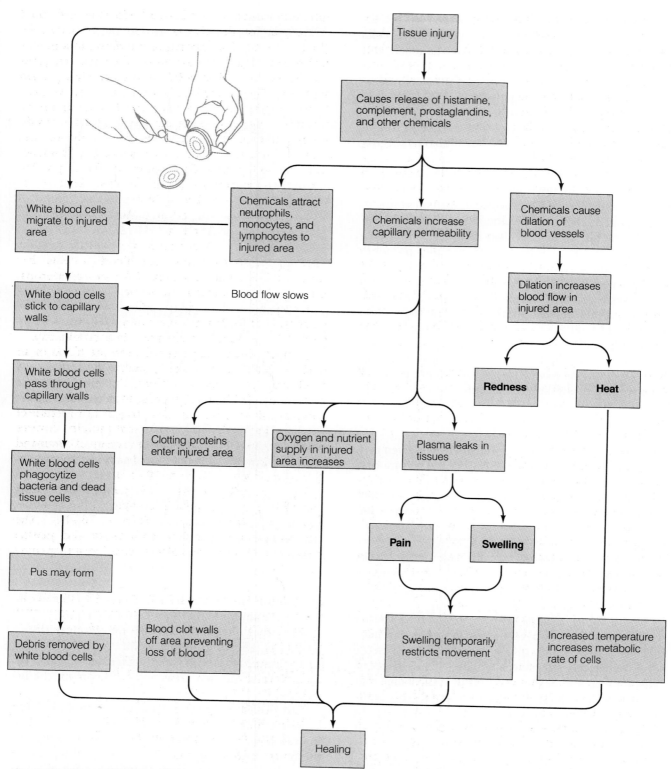

FIGURE 10–8 The Inflammatory Response

other mild painkillers inhibit the synthesis and release of prostaglandins, reducing pain. The flow of fluid into body tissues, though painful, has several biological benefits. For one, it increases the supply of oxygen and cellular nutrients, making them available to cells that are fighting the infection and to cells involved in repair. The fluid also dilutes harmful chemical substances, such as bacterial toxins. Swelling in a joint helps immobilize the joint, like a brace, allowing the tissue to repair itself.

The increase in blood flow, the release of chemical attractants, and the flow of fluids into a wound constitute the **inflammatory response** (Figure 10-8). Inflammation occurs in virtually all tissues invaded by bacteria and viruses. It is the second line of defense and even comes equipped with its own cleanup crew—late-arriving monocytes—that mop up after the battle, disposing of dead cells, cell fragments, dead bacteria, and viruses. As noted in Chapter 8, the yellowish fluid containing dead cells, microorganisms, and cellular debris that accumulate at the site of inflammation is known as pus.

Additional Chemical Protection.   The inflammatory response is a kind of a chemical and biological warfare waged against bacteria, viruses, and other microorganisms. The second line of defense, however, includes three additional substances: pyrogens, interferons, and complement.

Pyrogens are chemicals released primarily from macrophages that have been exposed to bacteria and other foreign substances. Pyrogens travel to a region of the brain called the hypothalamus. In the hypothalamus is a group of nerve cells that control the body's temperature, in much the same way that a thermostat regulates the temperature of a room. Pyrogens turn the thermostat up, increasing body temperature.

Mild or moderate fever causes the spleen and liver to remove iron from the blood, reducing blood levels. Many pathogenic bacteria require iron to reproduce. Thus, fever reduces bacterial replication and helps the body battle infectious agents. Fever also increases metabolism, which facilitates healing, and accelerates cellular defense mechanisms, such as phagocytosis. Fever can also be debilitating and a severe fever (over 105° F) is dangerous.

Another chemical safeguard is a group of small proteins, the interferons. **Interferons** are released from cells infected by viruses. Research suggests that each type of cell produces a slightly different form of interferon. Interferons released by infected cells bind to receptors on the plasma membranes of body cells (Figure 10-9). The binding of interferon to a noninfected cell triggers the synthesis of cellular enzymes that break down viral mRNA and block viral protein synthesis. These enzymes, however, are inactive until a virus infects the cell. A cell primed by interferon is, therefore, equipped to combat viruses. Interferons do nothing to protect infected cells, but simply stop the spread of viruses. Their release is a dying cell's last act to protect other cells of the body. Interferons provide a temporary protection against a wide range of viruses, helping control viral infections until the immune response is fully activated. Additional effects of the interferons are listed in Table 10-1.

Another group of chemical agents used to fight infection consists of numerous blood proteins that form the complement system. Like the interferons, the **complement system** (so named because it complements the action of antibodies) is a nonspecific response to infection. In other words, it is activated by all foreign cells and attacks them indiscriminately.

Complement proteins circulate in the blood in an inactive state, much like the proteins (fibrinogen and prothrombin) involved in blood clotting. When the body is invaded by foreign cells, such as bacteria, the complement system is activated, triggering a cascade of reactions in which one complement protein activates the next in the series, much like the chain of command being awakened when a nation's surveillance system detects an invasion (Figure 10-10).

The five final proteins in the complement system join to form a large protein complex, known as the **membrane-attack complex** (Figure 10-11). The membrane-attack complex embeds in the plasma membrane of microorganisms, creating an opening

---

TABLE 10-1   Functions of Interferons

Protect cells against viruses by destroying viral mRNA and inhibiting protein synthesis

Enhance the phagocytic activity of macrophages

Stimulate the production of antibodies

Stimulate the activity of cytotoxic T cells

Suppress tumor growth

FIGURE 10-9 The Mechanism of Interferon's Action Interferon protects cells from viral infection.

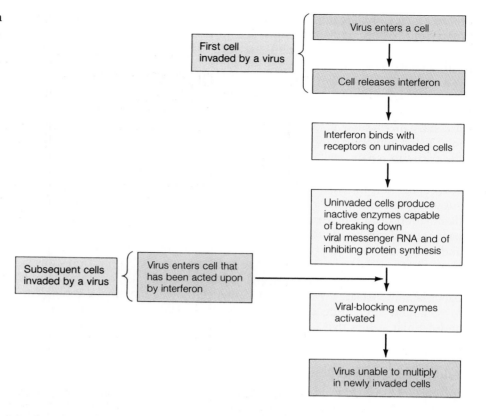

through which water flows, causing the cell to swell and burst. This kills invading microorganisms.

Several of the activated complement proteins also function on their own and are part of the inflammatory response. Some complement proteins, for example, stimulate the dilation of blood vessels in an infected area. The resulting increase in blood flow causes an increase in the supply of nutrients needed for cellular repair and increases the local supply of white blood cells needed to help fight off an infection. Some complement proteins increase the permeability of the blood vessels, allowing white blood cells to pass more readily into an infected zone. Certain proteins may also act as chemical attractants, drawing macrophages, monocytes, and neutrophils to the site of infection where they phagocytize foreign cells. Phagocytosis is enhanced by still another complement protein (called C3b). This protein binds to microorganisms, forming a rough coat on the intruders that enhances their phagocytosis.

The physical barriers, the chemical weapons, and the cellular defense mechanisms discussed so far are nonspecific—they operate indiscriminately. In other words, they do not target a particular foreign material.

The skin, for example, repels most bacteria and fungi. Macrophages and neutrophils devour whatever foreign substances enter the body tissues. Fever battles all dividing bacteria by reducing iron levels. The first and second lines of defense lighten the work load of the third and final line of defense, the immune system.

THE THIRD LINE OF DEFENSE: THE IMMUNE SYSTEM

The **immune system** is not a distinct organ system, such as the digestive or respiratory system. Instead, it is a functional system consisting of many millions of cells (lymphocytes) that circulate in the blood and lymph and take up residence in the lymphoid organs, such as the spleen, thymus, lymph nodes, and tonsils, as well as other body tissues.[4] The cells of the immune system selectively target foreign substances and foreign organisms.

[4]Each individual contains an estimated two trillion lymphocytes.

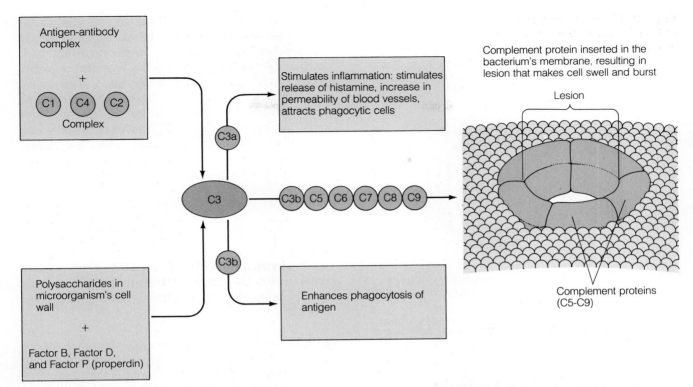

FIGURE 10–10   The Complement System
The complement system is activated by the presence
of antibodies bound to their antigens and by the
polysaccharides in the cell wall of some bacteria and fungi.
These triggers activate C3 protein, which splits forming two
fragments. The C3b fragment binds to bacteria making them
easier to phagocytize. The C3a fragment stimulates the
inflammatory response. C3b enhances phagocytosis and also
activates additional complement proteins, which are inserted
into the membrane of bacteria. They create an opening or
lesion in the membrane, which, if large enough, can kill the
cell.

The immune system is an important homeostatic
mechanism that eliminates bacteria, viruses, and
single-celled fungi that penetrate the outer defenses of
the body. The immune system also helps reduce cellular dissent from within—that is, the emergence of
cancer.

Antigens: Evoking the Immune Response

One of the chief functions of the immune system is to
identify what belongs in the body and what does not.
The ability to recognize foreign materials is a vitally
important, but difficult, task. It is a little like having to
sort through the thousands of items in your family's
home, including every drawer in the kitchen and every jar of nails in the garage, every day to determine if
someone has brought something new into the house.

Once a foreign substance has been located, the immune system mounts an attack to eliminate it. Like all
homeostatic systems, the immune system requires
both detection and action through receptors and effectors. In the immune system, the lymphocytes serve
both of these functions.

*How does the body recognize the foreign microorgan-
isms and materials that enter our bodies?* The immune
system is triggered by large molecules, such as proteins and polysaccharides. Molecules that stimulate
the immune system are called **antigens** (*antibody-
generating substance*). All antigens are foreign substances—materials not normally found in the body.
As a rule, small molecules do not elicit an immune
reaction; of those molecules that do, the larger the
molecule is, the greater its antigenicity. Nevertheless,
small nonantigenic molecules sometimes bind to naturally occurring proteins in the body, forming complexes. The large complex, although partly natural,
may be recognized as a foreign substance and may
evoke an immune response. Two examples include

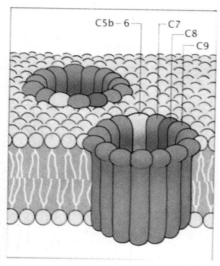

FIGURE 10–11 The Membrane-Attack Complex
Complement proteins embed in a cell's membrane, causing it to leak, swell, and burst.

poison ivy toxin and the antibiotic penicillin.

The immune system also responds to viruses, bacteria, and single-celled fungi in the body. Even parasites, such as the protozoan that causes malaria, can elicit an immune response. Viruses, bacteria, fungi, and parasites elicit a response because they are enclosed by a membrane or coat that contains large molecular-weight proteins or polysaccharides—that is, antigens.

Finally, the immune response is also triggered by cells that are transplanted from one person to another. As Chapter 3 notes, the cells of each individual contain a unique "cellular fingerprint," resulting from the unique array of plasma membrane glycoproteins. The immune system detects the presence of foreign cells and destroys them. Although they arise from a person's own cells, cancer cells present a different chemical fingerprint. They are essentially foreign cells within our bodies, whose antigens elicit an immune reaction.

Antigens stimulate the proliferation and differentiation of two types of lymphocytes: the T-lymphocytes, commonly called T cells, and the B-lymphocytes, also known as B cells (Chapter 8). The **immune reaction**, therefore, is the response of T and B cells. Lymphocytes can respond to a virtually unlimited number of antigens. As you will see in later sections,

however, B and T cells recognize and respond to antigens differently. As a rule of thumb, B cells recognize and react to small, free-living microorganisms, such as bacteria, bacterial toxins, and a few viruses. When activated, B cells produce antibodies to these antigens. In contrast, T cells recognize and respond to body cells that have gone awry, such as cancer cells or cells that have been invaded by viruses. T cells also respond to transplanted tissue cells and larger disease-causing agents, such as single-celled fungi and parasites. Unlike B cells, T cells attack their targets directly.

Immunocompetence: Lymphocytes Come of Age

Lymphocytes are produced in the red bone marrow and released into the bloodstream. These immature cells circulate through the blood and lymph (Figure 10–12).

T Cells. Some of the lymphocytes take up residence in the thymus, a lymphoid organ located above the heart. Inside the thymus, the lymphocytes undergo a process of maturation that takes two to three days. Maturation is called **immunocompetence** because the cells gain the capacity to respond to specific antigens. The cells become fully formed T cells (T for thymus). *Immunocompetence is a kind of cellular specialization. During this process T cells are preprogrammed to respond to specific antigens during fetal development long before they encounter the antigens.* During this process, each cell produces a unique type of membrane receptor that will bind to one—and only one—type of antigen. Over a lifetime, millions of antigens will be encountered. Thus, each of us is equipped with millions of different preprogrammed T cells.

All offspring of a particular T cell form a family of identical cells called a **clone**. Like the original T cell, the members of each clone respond to only one antigen. Many clones will never be used, but nevertheless remain in the body in a kind of cellular reserve. After becoming immunocompetent, many T cells leave the thymus and take up residence in other organs, especially the lymph nodes, spleen, and liver. A great many T cells circulate in the blood. When an antigen enters the body, it activates only the T cell preprogrammed to respond to it.

B Cells. B cells mature and differentiate similarly, but gain immunocompetence in the bone marrow, not the thymus. Like the T cells, B cells become part of the body's vast cellular reserve. B cells circulate in the blood and take up residence in connective tissue and lymphoid organs. All told, several million immunologically different B and T cells are produced in the body early in life. Over a lifetime, only a small fraction of these cells will be called into duty.

Humoral Immunity: Antibodies and B Cells

The immune response consists of two separate but related reactions: humoral immunity, provided by the B cells, and cell-mediated immunity, involving T cells. Let's consider humoral immunity and the B cells first.

When an antigen enters the body, it binds to B cells preprogrammed during their residence in the bone marrow (Figure 10–13).[5] The B cells soon begin to divide, producing additional B cells. Some of the B cells differentiate, forming plasma cells. **Plasma cells** contain a prominent rough endoplasmic reticulum, on which antibodies are produced. Antibodies released from plasma cells circulate in the blood and lymph, where they encounter and bind to the free antigens of the type that triggered the response. Since the blood and lymph were once referred to as body humors, this arm of the protective immune response is called **humoral immunity**.

B Cells and the Primary and Secondary Responses. The first time an antigen enters the body, it elicits an immune response, but the initial reaction—or **primary response**—is relatively slow and weak. During the primary response, antibody levels in the blood do not begin to rise until approximately the beginning of the second week *after* the intruder is first detected (Figure 10–14). This lag occurs because B cells take time to multiply and form a sufficient number of plasma cells. Antibody levels usually peak about the end of the second week, then decline over the next three weeks, partly explaining why it takes most people about a week to 10 days to get over a cold or the flu.

If the same antigen reappears in the body, the immune system acts much more quickly and more forcefully (Figure 10–14). This is called the **secondary response**. As Figure 10–14 illustrates, a few days after the antigen enters the body, antibody levels skyrocket. Consequently, the antigen is quickly destroyed, and a recurrence of the illness is prevented.

[5]As you shall soon see, this process is a bit more complex and involves the macrophage.

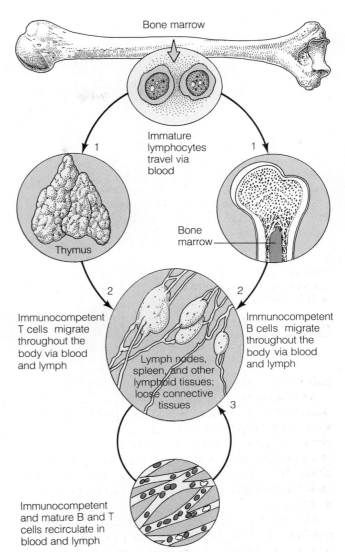

FIGURE 10–12   B- and T-Cell Immunocompetence Immunocompetence, the ability to respond to specific antigens, is conferred in the bone marrow, in the case of B cells, or thymus, in the case of T cells. The cells then migrate in the blood and lymph to lymphoid organs, such as the lymph nodes, spleen, and loose connective tissue underlying many epithelia.

The secondary response is so much more rapid because of the production of memory cells during the primary response. **Memory cells** are formed from B cells. They constitute a relatively large reserve force of antigen-specific B-lymphocytes. When the antigen reappears, these cells proliferate and differentiate to form numerous plasma cells that quickly begin producing antibodies.

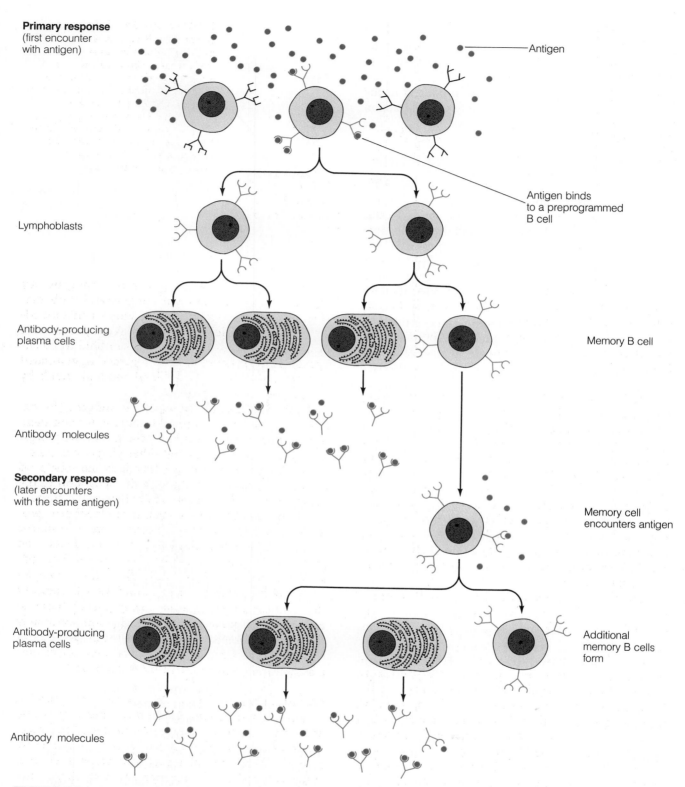

Primary response
(first encounter
with antigen)

Antigen

Antigen binds
to a preprogrammed
B cell

Lymphoblasts

Antibody-producing
plasma cells

Memory B cell

Antibody molecules

Secondary response
(later encounters
with the same antigen)

Memory cell
encounters antigen

Antibody-producing
plasma cells

Additional
memory B cells
form

Antibody molecules

FIGURE 10–13 B-Cell Activation Immunocompetent
B cells are stimulated by the presence of an antigen,
producing an intermediate cell, the lymphoblast. The
lymphoblasts divide, producing plasma cells and some
memory cells. Memory cells respond to subsequent antigen
encroachment, yielding a rapid, secondary response.

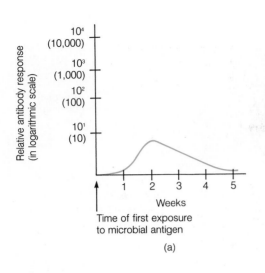

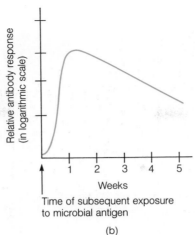

FIGURE 10–14 Primary and Secondary Responses (*a*) The primary (initial) immune response is slow. It takes about 10 days for antibody levels to peak. Almost no antibody is produced during the first week as plasma cells are being formed. (*b*) The secondary response is much more rapid. Antibody levels rise almost immediately after the antigen invades. T cells show a similar response pattern.

Immune protection can last 20 years or longer and is the reason that once you've had a childhood disease, such as the mumps or chicken pox, it is unlikely that you will contract it again. Resistance to disease that is provided by the immune system is known as **immunity**.

Antibodies: Agents of Destruction.   Antibodies belong to a class of blood proteins called the globulins, which were introduced in Chapter 8. Antibodies are specifically called **immunoglobulins**. Each antibody consists of four peptide chains (Figure 10–15). The chains are joined by disulfide bonds. Two small chains intertwine with two larger chains, forming Y-shaped molecules. It is the arms of the Y that bind to antigens and confer specificity, in much the same way that the active sites of enzymes result in enzyme specificity. Immunologists have discovered five different classes of antibodies, each with a slightly different role (Table 10–2).

Antibodies bring about the destruction of antigens chiefly through four mechanisms: neutralization, agglutination, precipitation, and complement activation.

During **neutralization** antibodies bind to viruses, coating them completely and preventing them from binding to plasma membrane receptors on body cells. If a virus cannot bind to a plasma membrane receptor, it cannot get inside the cell. Neutralization also helps destroy bacterial toxins. A toxic protein, for instance, may be so heavily coated with antibody that it is rendered ineffective. Toxins and viruses neutralized by their antibody coating are eventually engulfed by macrophages and other phagocytic cells (Figure 10–16).

Antibodies can also deactivate antigens by **agglutination**—a clumping of antigens and antibodies (Figure 10–16). Agglutination occurs when antibodies bind to more than one antigen. A single molecule of the IgM antibody, for instance, can bind to 10 antigens, causing them to clump together. Agglutinated antigens are removed from blood and body fluids by phagocytic cells.

Antibodies can also cause soluble antigens (for example, a protein) to become insoluble, forcing them to **precipitate** out of solution, where they are phagocytized by macrophages and other phagocytic cells.

The final mechanism by which antibodies help rid the body of bacteria is through the activation of the complement system. As we saw earlier in the chapter, the complement system is a family of bloodborne proteins that are part of the nonspecific immune response to antigen. The complement system is activated by the presence of the antigen-antibody complexes (antibodies bound to antigens). When activated, the complement system produces a membrane-attack complex that embeds in the plasma membrane of bacterial cells, causing them to leak, swell, and eventually burst. Some proteins of this system stimulate the inflammatory process, and some coat microorganisms, facilitating the phagocytosis by macrophages.

Role of the Macrophage.   Macrophages are found in connective tissue, lymphoid tissue, and organs and arise from monocytes, a type of white blood cell. Macrophages play an important role in the immune response. These cells phagocytize bacteria and other antigens at the site of infection and also phagocytize antigen-antibody complexes. However, the macrophage also plays a role in stimulating T- and B-cell dif-

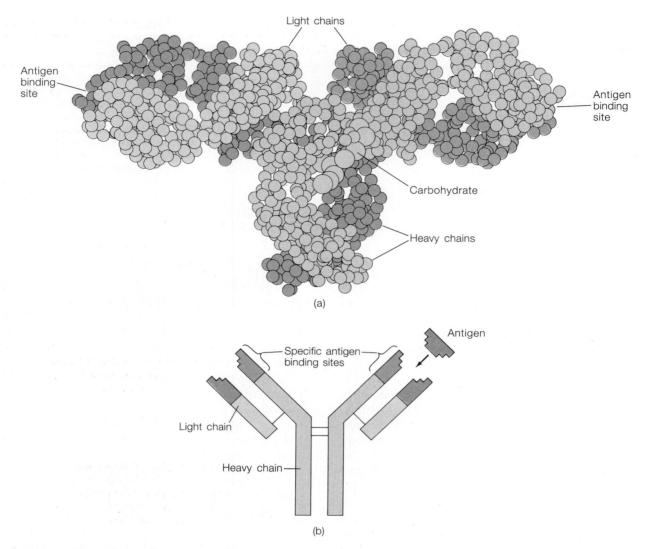

Light chains

Antigen binding site

Antigen binding site

Carbohydrate

Heavy chains

(a)

Specific antigen binding sites

Antigen

Light chain

Heavy chain

(b)

FIGURE 10–15 Antibody Structure (*a*) A three-dimensional model of an antibody showing the four chains. The molecule is T shaped before binding to an antigen. After it binds, it becomes Y shaped, as shown in part b. (*b*) A diagrammatic representation of the structure of an antibody molecule. It shows the four protein chains, two large (heavy chains) and two small (light chains). Note that the antigens bind to the arms of the molecule.

ferentiation. B cells, in fact, generally cannot differentiate into plasma cells and produce antibodies without the participation of macrophages.

Figure 10–17 is a simplified illustration showing how the macrophage operates. As illustrated, macrophages phagocytize foreign materials, such as bacteria. Then the macrophages transfer antigen, which was once part of the surface of the bacterium, to their own plasma membrane. The macrophages then cluster around B cells, presenting the bacterial antigen to them. The B cells, in turn, are activated when they encounter the bacterial antigen to which they are preprogrammed to respond. The B cells begin to divide and differentiate into antibody-producing plasma cells. As illustrated in Figure 10–17, macrophages also secrete a chemical called **interleukin 1**, which enhances the proliferation and differentiation of B cells.

Macrophages also present antigen to certain T cells, called helper T cells (described in more detail

TABLE 10–2 Types and Functions of the Immunoglobulins

CLASS		LOCATION AND FUNCTION
IgD	Monomer	Present on surface of many B cells, but function uncertain; may be a surface receptor for B cells; plays a role in activation of B cell
IgM	Pentamer / J chain	Found on surface of B cell and in plasma; acts as a B-cell surface receptor for antigen; secreted early in primary response; powerful agglutinating agent
IgG	Monomer	Most abundant immunoglobulin in the blood plasma; produced during primary and secondary response; can pass through the placenta, entering fetal bloodstream, thus providing protection to fetus
IgA	J chain	Produced by plasma cells in the digestive, respiratory, and urinary systems where it protects the surface linings by preventing attachment of bacteria to surfaces of epithelial cells; also present in tears and breast milk; protects lining of digestive, respiratory, and urinary systems
IgE		Produced by plasma cells in skin, tonsils, and the digestive and respiratory systems; responsible for allergic reactions, including hay fever and asthma

FIGURE 10–16   Antibody
Functions

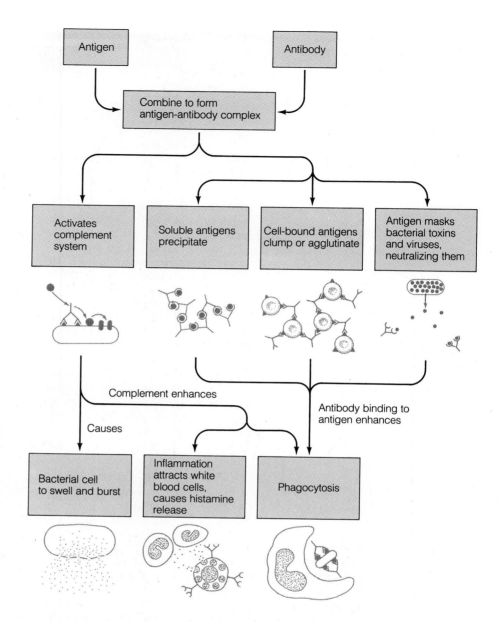

shortly). When activated, the helper T cells produce a chemical substance that has two effects. It enhances B-cell proliferation and differentiation, and it enhances antibody production by plasma cells (Figure 10–17).

Cell-Mediated Immunity: T Cells

T cells provide a much more complex form of protection. Like B cells, they respond by undergoing rapid proliferation. T cells differentiate into at least four cell types: (1) memory cells, (2) cytotoxic T, or killer, cells, (3) helper T cells, and (4) suppressor T cells (Table 10–3). Memory cells play a crucial role in the secondary response.

Cytotoxic T cells perform a variety of functions (Table 10-3). Some cytotoxic T cells attack and kill body cells that have become infected by viruses. When a virus infects a body cell, antigenic proteins in the virus's envelope become incorporated in the plasma membrane of the host cell. Cytotoxic T cells bind to that antigen and destroy the host cell. As noted earlier, cytotoxic T cells also attack and kill bacteria, parasites, single-celled fungi, cancer cells,

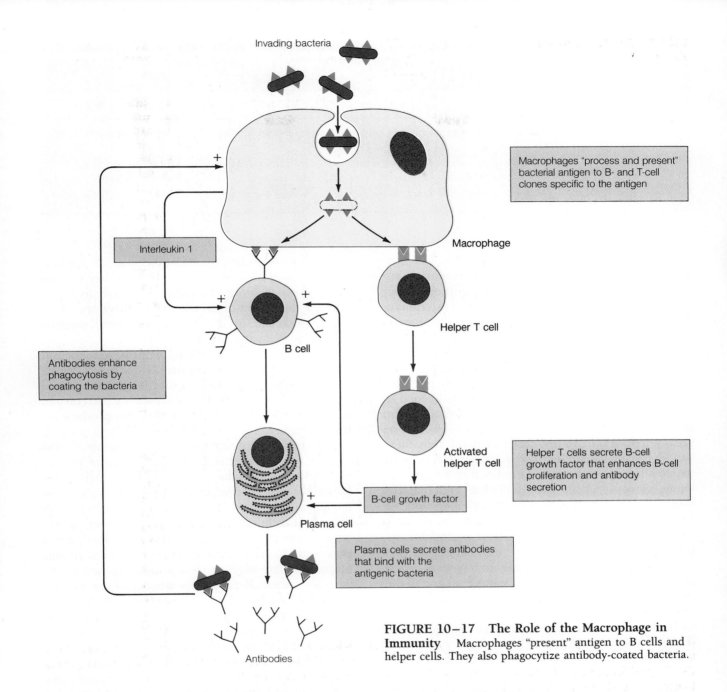

Invading bacteria

Macrophages "process and present" bacterial antigen to B- and T-cell clones specific to the antigen

Macrophage

Interleukin 1

Helper T cell

+

+

B cell

Antibodies enhance phagocytosis by coating the bacteria

Activated helper T cell

Helper T cells secrete B-cell growth factor that enhances B-cell proliferation and antibody secretion

+

B-cell growth factor

Plasma cell

Plasma cells secrete antibodies that bind with the antigenic bacteria

Antibodies

FIGURE 10–17   The Role of the Macrophage in Immunity   Macrophages "present" antigen to B cells and helper cells. They also phagocytize antibody-coated bacteria.

and foreign cells introduced during blood transfusions or tissue or organ transplants. Cytotoxic T cells bind to antigenic molecules in the membranes of these cells and release a chemical called **perforin-1** (Figure 10–18). Perforin-1 molecules become embedded in the plasma membrane of the target cell. The perforin 1 molecules aggregate in the plasma membrane of the target cell and form pores, similar to

those produced by the membrane-attack complex of the complement system. The pores cause the plasma membrane to leak, destroying the target cell within a few hours. After it has delivered its lethal payload, the cytotoxic cell detaches and is free to hunt down other antigens.

Helper T cells enhance cell-mediated immunity. As noted earlier, helper T cells produce a growth factor

TABLE 10–3 Summary of T Cells

CELL TYPE	ACTION
Cytotoxic T cells	Destroy body cells infected by viruses, and attack and kill bacteria, fungi and parasites, and cancer cells
Helper T cells	Produce a growth factor that stimulates B-cell proliferation and differentiation and also stimulates antibody production by plasma cells; enhance activity of cytotoxic T cells
Suppressor T cells	May inhibit immune reaction by decreasing B- and T-cell activity and B- and T-cell division

that stimulates the proliferation of B cells and their differentiation into plasma cells. It also stimulates antibody production by plasma cells. In addition, helper T cells release a chemical substance called **interleukin 2**, which increases the activity of cytotoxic T, suppressor T, and even helper T cells.

Helper T cells are the most abundant of the T cells (comprising about 60–70% of the circulating T cell population). Helper T cells are activated by the presence of antigen. Helper T cells have been likened to the immune system's master switch. Without them, antibody production and T-cell activity would be greatly reduced. In fact, without helper T cells there would be almost no immune response. An antigen would stimulate a few B and T cells, then the process would come to a halt. The AIDS virus preferentially infects helper T cells. Consequently, people suffering from the disease are unable to mount an effective immune response, and many die from bacterial infections and cancer.

The role of **suppressor T cells** is less well understood. Research suggests that they serve to turn off the immune reaction as the antigen disappears. Their activity, therefore, increases as the immune system finishes its job. Suppressor cells release chemicals that reduce B- and T-cell division.

Active and Passive Immunity

One of the major medical advances of the last century has to be the discovery of vaccines. **Vaccines** contain deactivated or weakened viruses or bacteria that, when injected in the body, elicit an immune response. Inactivated bacterial toxins can also be used as vaccines. Many vaccines provide immunity or protection from microorganisms for long periods, sometimes for life. Others, however, give only short-term protection.

Vaccines stimulate the immune reaction because the weakened or deactivated organisms they contain still possess the antigenic proteins or carbohydrates. Because the organisms in the vaccine have been seriously weakened or deactivated, however, vaccines usually do not cause disease.

Vaccination provides a kind of protection immunologists call **active immunity**—so named because the body actively produces T and B memory cells to protect a person against future attacks by bacteria and viruses. Viral or bacterial infections also produce active immunity.

A second type of immunity is also possible. Called **passive immunity**, it is a temporary form of protection, resulting from the injection of immunoglobulins (antibodies to specific antigens). Immunoglobulins remain in the blood for a few weeks, protecting an individual from infection. The liver, however, slowly removes the immunoglobulins from the blood, and a person gradually loses his or her protection. Immunoglobulins are used to treat people exposed to the virus that causes hepatitis (liver infection) or individuals who have been bitten by poisonous snakes.

Passive immunity can also occur naturally. A fetus, for instance, receives antibodies from its mother's bloodstream. These antibodies cross the placenta, the organ that transfers nutrients from the mother's bloodstream to the fetus's blood. Maternal antibodies remain in the blood of an infant for several months, protecting it from bacteria and viruses. The mother also transfers antibodies to her baby in her breast milk. The maternal antibodies attack bacteria and viruses in the intestine of the baby, thus protecting the infant from infection. (For more on this topic, see Health Note 10–1.)

Vaccination Decreases in the United States. Vaccines have helped eliminate many infectious diseases

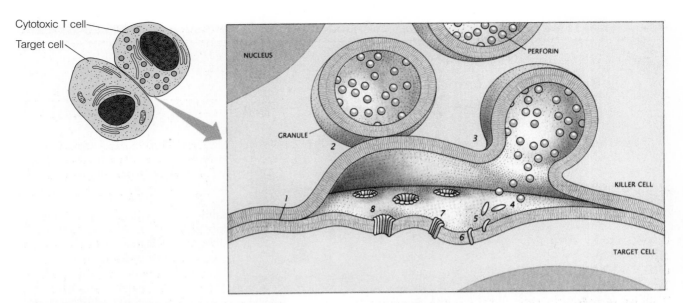

Cytotoxic T cell
Target cell

NUCLEUS
PERFORIN
GRANULE
KILLER CELL
TARGET CELL

FIGURE 10-18 How Cytotoxic T Cells Work
Cytotoxic T cells, containing perforin 1 granules, bind to their target and release perforin 1, then detach in search of other invaders. Perforin 1 molecules congregate in the target plasma membrane, forming a pore that disrupts the plasma membrane, causing the cell to die.

Source: By Dana Burns from "How Killer Cells Kill" by John Ding-E Young and Zanvil A. Kohn. Copyright © January 1988 by Scientific American, Inc. All rights reserved.

in the United States. Vaccines for diphtheria, tetanus, whooping cough, polio, measles, mumps, and congenital rubella (German measles), for example, have reduced the incidence of these often-lethal diseases in the United States by more than 99%. Despite the successes of vaccines, publicity concerning their rare, but sometimes-devastating, side effects has created something of a medical dilemma in the United States, Japan, and Great Britain. In 1976 and 1977, for example, a mass-immunization program in the United States for the swine-flu, a type of influenza that infects people, caused paralysis in a number of people. Because of public concern over this and other incidents, many parents are choosing not to have their children vaccinated. Some observers also believe that the media attention given to the rare but serious complications has harmed efforts to promote vaccination. Pharmaceutical companies are increasingly reluctant to invest the huge sums needed to perfect a vaccine for fear of lawsuits.

Parents may be choosing not to have their children vaccinated for other reasons as well. Researchers believe that another major reason for this trend ironically stems from the success of previous immunization programs, which have greatly reduced the incidence of most infectious diseases. Parents raised in an environment free of such diseases, researchers say, are unaware of the true dangers. Having their children immunized seems unimportant. Public health officials are concerned that the incidence of infectious diseases, such as polio, may increase as a consequence.

Harmful side effects from conventional vaccines may be caused by reactions to certain "nonessential" antigens on the injected microorganism. These antigens often play little or no role in developing immunity. By eliminating these antigens from vaccines, researchers hope to develop safer alternatives to vaccines in use today. These vaccines should be available in the early 1990s.

Practical Applications: Blood Transfusions and Tissue Transplantation

The immune system evolved to protect animals from microorganisms. This natural mechanism, however, presents something of a problem during blood transfusions and tissue transplants.

Blood Transfusions. The surface of the red blood cell (RBC) membrane contains certain inherited antigens. These antigens are determinants of blood type. The major blood types constitute the **ABO system**. In this system, four possible blood types exist: A, B, AB, and O. The letters refer to the type of antigen present on the RBCs of an individual.

HEALTH NOTE 10-1

Bringing Baby Up Right:
The Immunological and Nutritional Benefits of Breast Milk

A baby is born into a dangerous world in which bacteria and viruses abound. Complicating matters, the immune system of a newborn child is poorly developed. Fortunately, newborn babies are protected by passive immunity—antibodies that have traveled from their mothers' blood. Antibodies also travel to the infant in breast milk (see the figure).

Several immunoglobulins are present in breast milk. One of those is called **secretory IgA**. It is present in very high quantities in **colostrum**, a thick fluid produced by the breast immediately after delivery—before the breast begins full-scale milk production. Colostrum is so important, in fact, that some hospitals give "colostrum cocktails" to newborns who will not be breast fed by their mothers. Nurses remove the colostrum from the mother's breast with a breast pump and feed it to the baby in a bottle.

Colostrum, says infant nutritionist Sarah McCamman from the University of Kansas Medical School, coats the lining of the intestines. The

Breast Feeding Mother breast feeding her newborn.

IgA antibodies in colostrum prevent bacteria ingested by the infant from adhering to the epithelium and gaining entrance. Breast milk also contains lysozyme, an enzyme that breaks down the cell walls of bacteria, destroying them.

Unfortunately, not all medical personnel agree on the benefits of breast milk. One problem, they say, is that breast milk has unusually low levels of iron. This fact has led many physicians to recommend iron supplements for newborns. A more careful analysis, required of critical

thinkers, suggests that this recommendation is not valid. Breast-fed infants generally do not suffer from iron deficiency because the percentage of iron absorbed from breast milk is extraordinarily high. Thus, low levels of iron in breast milk are offset by the high absorption.

The wisdom of iron supplements has also been questioned on other grounds. Researchers, for example, have found that iron supplements increase the incidence of harmful bacterial infection in newborns. As noted earlier in the chapter, iron is a limiting factor in many pathogenic bacteria. Low levels of iron in breast milk, therefore, reduce bacterial replication in an infant's intestinal tract, aiding in protecting the newborn.

In general, breast-fed babies are healthier than bottle-fed babies. The incidence of gastroenteritis (inflammation of the intestine), otitis (ear infections), and upper respiratory infections is lower in breast-fed babies than bottle-fed babies. Studies also show that children breast fed for at least six months contract fewer

As illustrated in Table 10–4, individuals with type A blood contain RBCs whose plasma membranes contain the A antigen. RBCs in individuals with type B blood contain type B antigens. People with AB blood, however, have both A and B antigens, and people with type O contain neither antigen.

To perform a successful blood transfusion, physicians must match donors and recipients. Individuals with type A blood, for example, can receive blood from people with the same blood type, but cannot receive blood from people with type B. Individuals with type B blood can generally receive blood from people having type B blood, but not from type A.

Cross-matching the blood is essential to prevent an immune reaction, which results from antibodies in the bloodstream of most people. Table 10–4 notes the antibodies found in each blood type. People with type A blood, for example, naturally contain antibodies to the B antigen. People with type B blood contain antibodies to the A antigen. These antibodies appear in human blood during the first year of life.

Problems arise when incompatible blood types are mixed. For example, imagine that an individual with type A blood is accidentally given type B blood. The recipient's blood contains antibodies to type B blood. These antibodies bind to the transfused RBCs (type B

childhood cancers than their bottle-fed counterparts. The incidence of childhood lymphoma, a cancer of the lymph glands, in bottle-fed babies is nearly double the rate in breast-fed children for reasons not yet understood.

New research also suggests that certain proteins in breast milk may stimulate the development of a newborn's immune system. In laboratory experiments, the still-unidentified proteins speed up the maturation of B cells and prime them for antibody production. These soluble proteins may also activate macrophages, which play a key role in the immune system.

Breast milk is also more digestible and more easily absorbed by infants than formula. Formula is a mixture of cow's milk, proteins, vegetable oils, and carbohydrates. It is only an approximation of mother's milk and is not broken down and absorbed as completely as breast milk.

Because of a growing awareness of the benefits of breast feeding, today virtually every national and international organization involved with maternal and child health supports breast feeding, says McCamman.

Breast feeding can have a major health impact in this country. Unfortunately, the benefits are not as widely known as many people would like, even among health care professionals. Fortunately, more and more health workers, including physicians, are beginning to understand the benefits of breast feeding and are promoting this option. As a result, many middle-class American women are now choosing to breast feed.

Unfortunately, says McCamman, there's "a huge population of low-income, poorly educated...women who choose not to nurse." Ironically, the federal government may be playing an unwitting role in their decision. National legislation aimed at improving child nutrition provides money for programs that distribute free formula to needy women. Free formula may discourage mothers from breast feeding.

Another reason is economic. Low-income women often work at jobs that do not provide maternity leave. Thus, these women must return to work soon after giving birth.

Still another reason for the low rate of breast feeding in low-income women is that women need a lot of support to nurse. "People think nursing is innate, natural, and easy," claims McCamman. That's not always the case, however. In some instances, getting started requires guidance and education. Without that education and support, breast feeding can be a difficult and painful experience that discourages many women.

Given the many benefits of breast feeding, McCamman recommends it to all mothers who can. Physicians can help by educating their patients on the benefits of breast feeding. "Doctors should present the information on breast and bottle feeding," says McCamman, "outlining the pros and cons of both methods. Then, let the woman choose. Too few doctors do that today, so women aren't making informed decisions."

---

antigens), causing them to agglutinate (clump together) or hemolyze (burst). Hemolysis and agglutination constitute the **transfusion reaction**. RBC clumping restricts blood flow through capillaries, reducing oxygen and nutrient flow to cells and tissues. Massive hemolysis results in the release of large amounts of hemoglobin into the blood plasma. Hemoglobin precipitates in the kidney, blocking tiny tubules that produce urine, which often results in acute kidney failure.

Successful transfusions require careful matching of the blood types of the donor and recipient. As Table 10–4 shows, RBCs from individuals with type O blood have neither A nor B antigens. Therefore, type O blood does not cause an immune reaction in a recipient and can be transfused into individuals with all four types: A, B, AB, and O. Type O individuals are said to be **universal donors**.

Type O blood, while free of antigens, contains antibodies to both A and B antigens. Therefore, individuals with type O blood can only receive type O blood. Any other type of blood would cause a transfusion reaction.

As shown in Table 10–4, individuals with type AB blood contain RBCs with both A and B antigens. AB blood, however, contains no antibodies related to the

TABLE 10-4 Summary of Blood Types

BLOOD TYPE	ANTIGENS ON RBCS	ANTIBODIES IN BLOOD	SAFE TO TRANSFUSE To	From
A	A	B	A, AB	A, O
B	B	A	B, AB	B, O
AB	A + B	—	AB	A, B, AB, O
O	—	A + B	A, B, AB, O	

ABO system. These people can, therefore, receive blood from all others and consequently are referred to as **universal recipients**. AB blood can only be safely transfused into individuals with AB blood.

The terms *universal donor* and *universal recipient* are somewhat misleading, however, because RBCs also contain other antigens that can cause transfusion reactions. The most important of these is called the Rh factor. This antigen was first identified in rhesus monkeys, hence the designation, Rh. People whose cells contain the Rh antigen or **Rh factor** are said to be **Rh positive**. Those without it are **Rh negative**.

Unlike the ABO system, in the Rh system antibodies are produced *only* when Rh-positive blood is transfused into the bloodstream of a person with Rh-negative blood. The first transfusion of Rh-positive blood into an Rh-negative person generally does not result in a transfusion reaction, but a second transfusion will. To reduce the likelihood of a transfusion reaction, however, Rh-negative people should only receive Rh-negative blood, and Rh-positive people should only receive Rh-positive blood.

The Rh factor becomes particularly important during pregnancy. Problems can arise if an Rh-negative mother has an Rh-positive baby. Even though the maternal and fetal bloodstreams are separate, small amounts of fetal blood can enter the maternal bloodstream at birth, causing the immune reaction. Rh antibodies will form in the maternal bloodstream, and the woman will be sensitized to the Rh factor.

If the woman is not treated and becomes pregnant again with an Rh-positive baby, antibodies to the Rh factor will cross the placenta. These antibodies will destroy RBCs in the fetal bloodstream, resulting in anemia and hypoxia (lack of oxygen to tissues). Unless the baby receives a blood transfusion (of Rh-

negative blood) *before* birth, and several after birth, it is likely to have brain damage and may even die.

To prevent antibody production in Rh-negative women who have given birth to Rh-positive babies, physicians can administer antibodies to Rh-positive RBCs soon after the women have given birth. These antibodies destroy Rh-positive RBCs from the fetus before a woman's immune system can respond. This way, the woman is not sensitized. B cells are not stimulated, and no memory cells are generated. To be effective, however, the treatment must begin immediately after the baby is born.

Tissue Transplants. Tissue transplantation is a much more complex matter. Only two conditions exist in which a person can receive a transplant and not reject it. One is if the tissue comes from the individual himself or herself. For burn victims, surgeons might use healthy skin from one part of the body to cover a badly damaged region. The new skin can take hold there and cover the wound. The second instance is when a tissue is transplanted between identical twins—individuals formed from a single fertilized ovum that splits and forms two embryos. These individuals are genetically identical and, therefore, have identical cellular antigens.

What about the heart, liver, and kidney transplants that you hear about in the news? How do recipients keep from rejecting their new tissue? Basically, the only way an organ transplant can remain intact and escape rejection by T cells is if the patient is treated with drugs that suppress the immune system. This treatment must be continued throughout the life of the patient. One of the most effective drugs in use today is **cyclosporin**. Unfortunately, most immune suppressors have numerous side effects and often leave the patient vulnerable to infection bacteria and viruses. Research suggests that cyclosporin works by suppressing the formation of interleukin 2 by helper T cells, thus greatly reducing cell-mediated immunity without affecting B cells.

Diseases of the Immune System

The immune system, like all other body systems, can malfunction. The most common malfunctions of the immune system are allergies. An **allergy** is an extreme overreaction to some antigens, such as pollen or foods (Figure 10–19). Those antigens that stimulate an allergic reaction are called **allergens**. Allergens cause the production of IgE antibodies (Table 10–2).[6] In some people, IgE antibodies are produced in excess. As Figure 10-19 shows, the antibodies bind to mast cells. **Mast cells** are cells found in many tissues, especially in the connective tissue surrounding blood vessels. Mast cells contain large granules comprised of the chemical histamine.

Allergens bind to the IgE antibodies that have attached to the mast cells, which triggers the release of histamine (Figure 10–19). Histamine, in turn, causes nearby arterioles to dilate. The increased blood flow increases the production of mucus in the respiratory tract, creating congestion. Histamine released in the lungs causes the bronchioles to constrict, cutting down airflow and making breathing difficult. This condition is called **asthma** (Chapter 9).

The allergic reaction usually occurs in specific body tissues, where it creates local symptoms that, while irritating, are not life-threatening. However, the allergic response can also occur in the bloodstream where it may cause death if not treated quickly. For example, in certain people penicillin or a bee venom circulating in the bloodstream can cause the release of massive amounts of histamine and other chemicals. The sudden release of these chemicals causes massive dilation of blood vessels in the skin and other tissues. This causes the blood pressure to fall precipitously, essentially shutting down the circulatory system. These chemicals also cause severe constriction of the air ducts in the lungs, making breathing difficult. The patient suffers from anaphylactic shock. Death may follow if measures are not taken to reverse the physiological nightmare. An injection of the hormone epinephrine (commonly known as adrenalin) can rapidly reverse the symptoms.

AIDS: FIGHTING A DEADLY VIRUS

In the late 1970s, medical science discovered a new disease called AIDS, acquired immune deficiency syndrome. One researcher likened it to the plague that spread through Europe in the fourteenth and fifteenth centuries. In the 1300s, the plague killed one-quarter of the adult population and numerous children. By 1985, approximately 20,000 Americans had been diagnosed with AIDS. Nearly half of them had died. By September 1990, the figure had risen to 120,000. The death toll hovered around 60,000, but all of the victims are expected to die, and many more will join their ranks in the coming years.

Unraveling the Mystery

Early suspicions that the disease was caused by a virus were confirmed by researchers in France and the United States (Figure 10-20a). They discovered an RNA virus—now called the **HIV virus** (human immunodeficiency virus)—that attacks helper T cells, severely impairing the immune system. AIDS patients grow progressively weaker and fall victim to other infectious agents. Many die from a rare form of pneumonia.

The HIV virus affects more than a person's immune system. AIDS patients, for example, may contract a rare form of skin cancer called Kaposi's sarcoma (Figure 10-20b). At first, researchers thought that Kaposi's sarcoma resulted from immunologic suppression. They hypothesized that the cancer developed because the AIDS virus had incapacitated the T helper cells. Without them, cancer could develop.

Although the AIDS virus does affect the T helper cells, recent evidence suggests that this cancer is caused by something else. The HIV virus carries a regulatory gene that is incorporated into body cells. The gene may cause certain body cells to proliferate uncontrollably, forming a cancer, or may stimulate the production of a chemical substance that causes rapid cell growth in neighboring cells.

AIDS patients also suffer from a number of neurological disorders, such as early memory loss and progressive general mental deterioration. A recent laboratory study may explain how the brain is affected

---

[6]Some allergies involve IgG or IgM, and some apparently do not involve antibodies at all.

Sensitization stage

Antigen (allergen) enters the body

Plasma cells synthesize and release large amounts of IgE antibodies

IgE antibodies bind to mast cells located in many body tissues

Subsequent (secondary) responses

More of same allergen enters body

Allergen combines with IgE on mast cells, triggering release of histamine from mast cell

Histamine stimulates dilation of blood vessels, causing fluid to leak out; stimulates release of copious amounts of mucus; and causes contraction of smooth muscle in bronchioles.

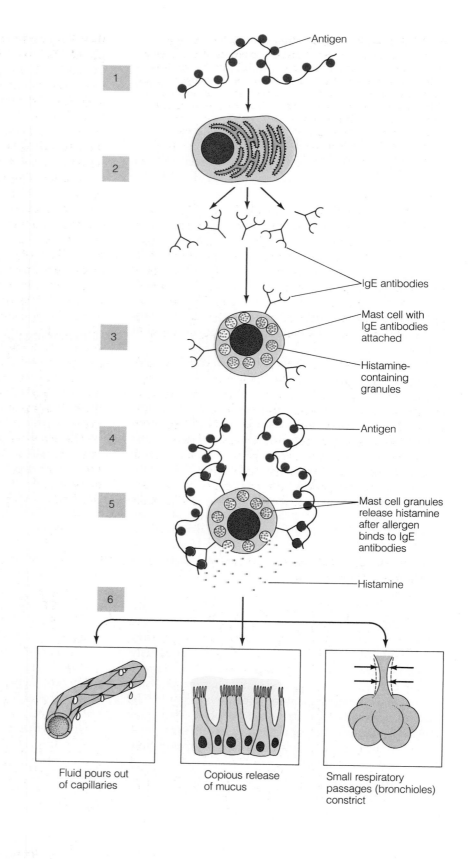

Antigen

IgE antibodies

Mast cell with IgE antibodies attached

Histamine-containing granules

Antigen

Mast cell granules release histamine after allergen binds to IgE antibodies

Histamine

Fluid pours out of capillaries

Copious release of mucus

Small respiratory passages (bronchioles) constrict

FIGURE 10–19   Allergic Reaction (left)   Antigen stimulates the production of massive amounts of IgE, a type of antibody produced by plasma cells. IgE attaches to mast cells. This is the sensitization stage. When the antigen enters again, it binds to the mast cells, triggering a massive release of histamine and other chemicals. Histamine, in turn, causes blood vessels to dilate and become leaky. This triggers the production of mucus in the respiratory tract. In some people, the chemicals released by the mast cells also cause the small air-carrying ducts in the lungs to constrict, making breathing difficult.

by this virus. AIDS viruses inside host white blood cells (helper cells) produce a number of proteins that are incorporated into the viral capsid. One of those proteins is gp120. Researchers recently found that gp120 kills fetal brain cells in culture. Brain cells, like the immune system cells, have membrane receptors that bind to gp120. The researchers believe that gp120 may, in some patients, travel in the blood to the brain where it kills brain cells, causing neurological defects.

Research has shown that the AIDS virus is passed via sexual contact, blood transfusions, and "dirty" needles shared by intravenous drug abusers. Homosexual men, hemophiliacs, and drug addicts are the chief victims. Researchers have also found that the AIDS virus is being transmitted among the heterosexual population and can even be transmitted from a mother to her baby through the placenta.

Victims of the genetic disorder hemophilia, described in the last chapter, are at risk for AIDs because

they frequently receive clotting factors from human plasma. Before 1984, blood donors were not screened for AIDS. Consequently, many of the preparations were contaminated with the AIDS virus. As a result, a majority of the estimated 15,000 hemophiliacs in the United States who received clotting factors between 1975 and 1984 are thought to have HIV antibodies in their blood. Whether or not they will develop AIDS no one knows. Evidence suggests that they will.

Health workers determine the presence of the AIDS virus by using an immunological test, which detects antibodies to the AIDS virus in the blood. Recently, scientists announced the development of a new and sensitive genetic test to determine the presence of the virus. It could help screen blood donors for the AIDS virus more carefully, helping protect the blood supply.

By current estimates, over 1 million Americans carry the AIDS virus. No cure has yet been discovered. A new drug called AZT seems to prolong a victim's life, but does not halt the deadly virus. AZT is costly and is also thought to be carcinogenic. New drugs are currently under investigation. Because of an outcry among the gay community, the Food and Drug Administration, which regulates all drug testing on humans, has relaxed its standards, permitting wider testing of a variety of possible drugs.

The AIDS virus, while lethal, does not spread as readily as the flu virus or cold viruses; individuals can protect themselves by practicing safe sex—using condoms. By avoiding multiple partners, one also reduces the likelihood of contracting AIDS. To prevent the

FIGURE 10–20   AIDS Virus and Kaposi's Sarcoma   (*a*) HIV viruses. (*b*) Kaposi's sarcoma on foot.

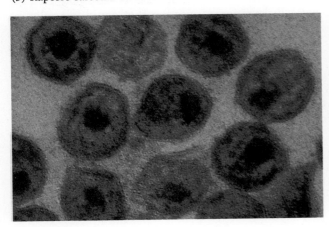

(a)

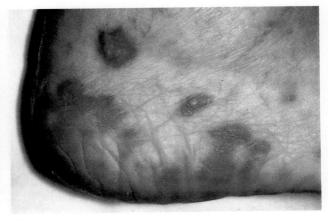

(b)

spread of the HIV virus among intravenous drug users, some countries and some states permit the distribution of clean hypodermic needles.

AIDS will remain a significant public health threat here and abroad for many years. In parts of Africa, the incidence may be as high as 50%. Without better education and other measures to control the spread of the disease, millions of people worldwide could die from AIDS. (For a discussion of some of the social and political implications of controlling the spread of AIDS, see the Point/Counterpoint in this chapter.)

Stopping the Deadly AIDS Virus

Will there be a cure? Can the deadly AIDS virus ever be stopped?

The Bad News about AIDS. The AIDS virus is notorious for its ability to mutate, a feature that will make the task of developing a vaccine extremely difficult.[7] Even within the body, the AIDS virus mutates. In one study, researchers analyzed AIDS viruses isolated from two infected patients over a 16-month period. They found 9 to 17 different varieties, all thought to have been formed from the original virus.

Making matters worse, the AIDS virus may be taking refuge in the body. In 1988, a research team announced that out of 100 homosexual men studied, 4 initially showed antibodies to the AIDS virus, but slowly lost them. This process usually only occurs in the late stages of AIDS, when the immune system is too weak to produce antibodies. These four men, however, had no overt symptoms of AIDS. Had the virus been conquered by the immune system or had it simply entered a latent phase, hiding out much like the herpes virus?

Research suggests that the AIDS virus may take up residence in bone marrow stem cells that give rise to lymphocytes. The virus, therefore, may remain in these cells and be passed on to new white blood cells produced by cell division. Thus, once the virus is in the body, it may remain forever. Completely eliminating the virus from the body may be impossible. If the AIDS virus does indeed go into hiding, researchers fear that AIDS-infected donors may escape detection, even with the new genetic tests. AIDS-infected blood cells could unknowingly be passed to thousands of patients.

[7]The AIDS virus is a retrovirus, and reverse transcription is fairly inaccurate, producing many mutant forms of the virus.

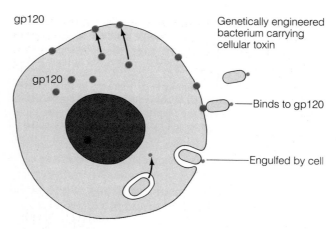

AIDS-infected cell

FIGURE 10–21   Duping the AIDS Virus   The AIDS virus has a protein, gp120, in its capsid. When it infects cells, it produces more gp120 to make new capsids. However, some gp120 ends up in the infected cell's plasma membrane, marking it. By genetically engineering a bacterium that can locate the infected cells by finding gp120, medical researchers may be able to hunt and kill infected cells, stopping the spread of the virus.

The Good News about AIDS.   Despite the bad news, researchers and much of the general public remain determined to find a cure or preventive tool for AIDS. One encouraging development was revealed in 1988. Two California research groups announced that they had successfully transplanted parts of the human immune system into mice. Human immune cells in mice may provide scientists with a tool to study the disease and to test potential cures for it, thus accelerating the pace of research.

On another front, researchers have developed a genetically engineered weapon that could kill cells infected with the AIDS virus, possibly curing the disease once it has developed. As noted above, cells infected with the AIDS virus produce a protein called gp120, which is part of the capsid (Figure 10–21). This protein also ends up in the plasma membrane of infected cells.

Researchers have genetically engineered a bacterium that binds to the gp120 protein. The bacterium carries with it a toxin that kills the HIV-infected cells. Preliminary studies indicate that noninfected cells are unharmed by this treatment. Although initial studies suggest that this treatment will not work, researchers hope that the technique can be improved or modified, making it possible to kill enough infected cells in

AIDS patients to halt the disease. One question that must be answered before this procedure can be tried in people is whether AIDS-infected cells killed by this technique will release active AIDS viruses that could infect other cells.

Researchers are also working on an AIDS vaccine that would be administered as a preventive measure. The rapid rate at which the virus mutates, however, has stymied these efforts.

 ## ENVIRONMENT AND HEALTH: MULTIPLE CHEMICAL SENSITIVITY

Richard Sharp was a physicist who worked for a major aviation firm in California. Today, he is confined to two stripped-down rooms equipped with special air filters that remove all air contaminants.

Sharp is one of countless Americans who has developed a condition known as Multiple Chemical Sensitivity (MCS). He has literally become a prisoner in his own world, unable to venture forth into modern society without suffering extreme discomfort, even debilitation.

MCS is thought to be caused by exposure to a number of common household and industrial chemicals, including formaldehyde, solvents, acrylic resins, mercury compounds, and pesticides. These chemicals result in changes in the immune system that make it hypersensitive—or overreactive. Chemicals such as formaldehyde, which is commonly found in carpeting, plywood, furniture, and many other household products, are thought to bind to naturally occurring proteins in the body. This produces foreign substances against which the immune system mounts an attack. In other words, common household and industrial substances turn the immune system against the body.

Individuals become sensitive to low levels of chemicals over long periods. They also exhibit cross-reactions—that is, they react to other chemically similar substances. Massive exposures to certain chemicals may also elicit a hypersensitivity reaction.

The symptoms of MCS vary, ranging from life-threatening to mild. The most common symptoms are tension, memory loss, fatigue, sleepiness, headaches, confusion, and depression. Many victims of MCS suffer from gastrointestinal problems as well—nausea, indigestion, and cramps. Some exhibit respiratory symptoms as well, including frequent colds, bronchi-

tis, and shortness of breath. Skin rashes are not uncommon. Many people report allergy-like symptoms such as nasal stuffiness and sinus infections.

MCS remains one of the most mysterious of all human diseases. Victims often witness a sudden deterioration of their health. Physicians, family members, friends, and even the victims themselves are baffled by the disease. Victims are often likely to be labeled "psychiatric cases." Victims must lock themselves in their own homes and must often get rid of all cleaning agents, pesticides, perfumes, deodorants, and other common household chemicals.

At first many physicians and scientists dismissed the symptoms, but a growing body of evidence shows that many common chemicals can indeed alter the immune system. Some chemicals result in hypersensitivity—that is, they make the body more sensitive to chemicals. The National Research Council estimates that 15% of the U.S. population experiences chemical hypersensitivity. Studies show that 5% of the workers exposed to an agent used in the manufacture of plastics, called TDI (toluene diisocyanate), develop asthmalike symptoms. TDI apparently binds to proteins in the respiratory tract, creating foreign substances that stimulate an allergic reaction technically known as a hypersensitivity reaction. Individuals who have been hypersensitized have difficulty breathing due to bronchial constriction when exposed to TDI, tobacco smoke, and air pollutants. In Japan, 15% of all cases of asthma in men can be directly attributed to industrial exposure to chemicals.

Other chemicals bind to proteins in the skin, creating foreign substances to which the immune system reacts. Formaldehyde, for example, results in a condition called contact dermatitis. Contact dermatitis is characterized by a skin rash. T cells attack skin cells, destroying them. Even low levels found in newsprint dyes, some cosmetics, and photographic papers are sufficient to induce a rash.

Other chemicals apparently act by suppressing immune function, making individuals more susceptible to infectious agents. Chronic workplace exposure to benzene, for example, results in a number of symptoms, notably a depression in the number of circulating lymphocytes. Benzene probably depresses lymphocyte production in the bone marrow. In rabbits, benzene exposure results in an increased susceptibility to various infectious agents.

Dioxins, PCBs, ozone, certain pesticides, and a variety of other chemicals suppress the immune response in laboratory animals and humans. At this

ANONYMOUS TESTING'S THE ANSWER

Earl F. Thomas

Earl F. Thomas is the vice-president of the National Association of People with AIDS (NAPWA). He is also the executive director of the People with AIDS (PWA) Coalition of Colorado, a member of the board of directors of the Colorado AIDS Project, and a member of the Governor's AIDS Coordinating Council for Colorado. Thomas was diagnosed as having AIDS in 1986.

Confidential testing for the HIV virus (where names and addresses of individuals who test positive are reported to health officials) versus anonymous testing continues to be a highly controversial topic. In today's society, the word AIDS still breeds fear, especially the fear of discrimination. These fears, quite simply, are keeping people from being tested.

Health departments and AIDS organizations stress that persons who have reason to believe that they may have been infected with the HIV virus should be tested for several reasons. First, the earlier a person knows he or she is HIV positive, the earlier treatment can be started to slow the progression of the disease or simply to buy time while researchers explore better treatments. Second, knowledge of one's HIV status is crucial in determining what behavioral changes need to be made.

If the testing system discourages individuals from obtaining knowledge about their HIV status, however, no one benefits. Unfortunately, name reporting tends to create an atmosphere of distrust between health officials and those who wish to be tested. These people already feel disfranchised from society and cannot believe that the public health system is concerned about their personal rights and welfare. They believe that information concerning their HIV status goes beyond the health department to others who may have a need to know and fear that eventually the information will find its way to persons who have no need to know. For instance, numerous links in the information chain (nurses, laboratory workers, therapists, and secretaries) all have access to a person's medical information. Confidential medical information is not so confidential after all. As a result of these concerns, many persons, even in high-risk groups, refuse to be tested at all.

In a survey conducted by the Educational Department of the Colorado AIDS Project in 1989, 32% of 1,112 respondents (gay and bisexual men) cited confidentiality concerns as their reason for not being tested. Despite the health department's insistence that fears of information leaks are poorly founded, they are very real to many individuals. More importantly, these fears prevent individuals who have engaged in high-risk behaviors from being tested.

Further information that supports anonymous testing comes from the state of Oregon. When HIV anonymous testing was offered along with confidential testing, there was a 50% increase in the demand to be tested during the first four months of the program. People from all segments of society sought out the anonymous test sites. Confidential testing sites reported *no* increase during the same period. These data strongly suggest that reporting by name is not the solution to HIV testing.

Fear of testing is justified. There is documented evidence of discrimination against people who are HIV positive. Individuals have been denied housing and, in certain cases, have been evicted from their homes, have lost their jobs, have been denied access to public education, and have been shunned by families, friends, and co-workers.

Partner notification (contact tracing) may have a place when trying to control the spread of AIDS, but at what cost? Some feel that contact tracing is not a viable option due to the monetary cost. In six months in 1988 the state of Colorado spent $450,000 on partner notification. The result of this enormous expenditure was that 52 people were found.

The experiences of other health departments clearly show that partner notification can be carried out with reasonable success when testing is anonymous. If lists are being maintained, people shy away from testing and forfeit the possibility of early intervention treatment and partner notification information. When health officials act as contact tracers, they are perceived as police. In our society, police-state tactics will never work, and no one benefits.

Anonymous testing would diffuse a serious impediment to a powerful collective anti-AIDS effort. And most important, we could get an answer to whether people are truly avoiding testing because of reportability.

NOTIFICATION WORKS

John Potterat

John Potterat is an authority on AIDS and sexually transmitted disease (STD) control. He has published numerous articles in medical journals dealing with STD/AIDS control and is currently director of the STD/AIDS programs in Colorado Springs.

The process of reporting people with the AIDS virus (HIV) by name to public health officers and in turn tracing their contacts should not be controversial. Such procedures have been standard public health practice for serious communicable diseases for nearly a century. Formal notification allows society to define the disease burden (surveillance) and to counterattack (control). You cannot control a communicable disease if you do not know who has it and who might be next to have it; moreover, you need to find those directly affected.

Notifying partners of people infected with sexually transmitted disease (STD) has been an effective control tool for 50 years. The fundamental reason that health officers are involved in this notification process is that STD patients are not good at referring their own sexual partners. Such "self-referral" fails more often than it succeeds: less than a third of STD partners are successfully referred for medical evaluation. Partner referral by HIV patients is even less successful (despite frequent assurances by patients that they "will take care of it!"). Part of this failure is due to the reluctance of HIV patients to face their partners (fear of anger of reprisal); part is due to selective notification (denial that "nice" partners can be infected); and part to failure to convince partners (partner denial). Trusting the notification process to infected people alone is a luxury that society can ill afford.

Those exposed to HIV have a right to know. Important sexual (safer practices) and reproductive (postponing pregnancy) decisions depend on knowledge of exposure and its outcome. Many persons are unaware that their partners have histories of needle exposures or of bisexuality. The duty to warn people has compelling moral, legal, and historical foundations. In free societies, notification is a straightforward, confidential process. Medical workers who detect HIV infection report the case by name and address to the local health officer who then discreetly contacts the patient to counsel him or her and to obtain identifying information on sexual and needle partners. People are persuaded, not coerced, into voluntary cooperation. While counseling is "mandatory," blood testing is optional.

Even if "treatment" for partners were to consist solely of personal counseling to discourage behaviors that facilitate transmission or accelerate disease progression, partner notification would be worthwhile.

A disease control procedure should be acceptable to people. Partner notification by health officers has been well received by the affected populations. The majority (70–80%) of notified partners accept blood testing, and almost all who decline testing accept counseling. Although organized gay advocacy groups have generally opposed both HIV reporting by name and partner notification by health officers, when approached individually and sympathetically, gay men have generally cooperated.

Health officers are responsible for maintaining the physical security of HIV records; such records are also immune from any discovery process. They cannot be subpoenaed or released to potentially adversarial agents like insurance, police, or employer investigators. Whatever discrimination is suffered by infected people, none of it stems from disease notification to or by public health officers.

Notification initiatives are affordable, acceptable to patients, and an effective means of reaching high-risk people. It is well known to health officers that those at highest risk are least inclined to appear for counseling and least likely to use safer practices. While notification is not a panacea, it is one of the most useful measures for containing this tragic epidemic.

SHARPENING YOUR CRITICAL THINKING SKILLS

1. Summarize Thomas' reasons for keeping AIDS testing confidential.
2. Summarize Potterat's views in support of name reporting.
3. Do you agree or disagree with the following statement? Why or why not? Both writers believe that their approach will provide the greatest protection to the public health, but they differ in their approach.
4. Of the two basic approaches, which do you think would be most effective in reducing the spread of AIDS?

writing, however, the overall significance of immune suppression and hypersensitivity in human populations remains unknown. This section illustrates the subtle but potentially far-reaching effect of toxic chemicals and underscores the importance of a healthy environment for maintaining a healthy population.

SUMMARY

VIRUSES AND BACTERIA:
A PRIMER ON INFECTIOUS AGENTS

1. A virus consists of a nucleic acid core bound by a protein coat, the capsid. Viruses invade cells, take over their metabolic machinery, and convert the host cell into a virus factory, often killing the host cell in the process.
2. The immune system kills many viruses, but some may take refuge in certain cells, reemerging under stress or some other influence.
3. Bacteria are independent microorganisms that perform many useful functions. Some bacteria cause sickness and death, however.
4. Bacteria contain cytoplasm enveloped by a membrane. Outside the plasma membrane is a thick, rigid cell wall.
5. Unlike viruses, bacteria reproduce on their own.

THE FIRST AND SECOND LINES OF DEFENSE

6. The first line of defense against viruses, bacteria, and other infectious agents is the skin and the epithelia of the respiratory, digestive, and urinary systems. These epithelia also produce protective chemical substances that ward off invaders.
7. The second line of defense consists of cells and chemicals the body produces against infectious agents that penetrate the epithelia.
8. One of the chief combatants in the second line of defense is the macrophage, a cell derived from the monocyte. Macrophages are found in connective tissue lying beneath the epithelia where they phagocytize infectious agents, preventing their spread. Neutrophils and monocytes also invade infected areas from the bloodstream, destroying bacteria and viruses.
9. Another portion of the second line of defense consists of the chemicals released by damaged tissue, which stimulate arterioles in the infected tissue to dilate. The increase in blood flow raises the temperature of the wound. Heat stimulates macrophage metabolism, accelerating the rate of the destruction of infectious agents. Heat also speeds up the healing process.
10. Still other chemicals increase the permeability of the capillaries, causing plasma to flow into the wound. Clotting walls off the damaged area. Fluids pouring into the wound site increase the supply of nutrients for macrophages and other cells fighting the invader.

11. The increase in blood flow, the release of chemical attractants, and the flow of plasma into the wound constitute the inflammatory response.
12. Pyrogens, chemicals released primarily by macrophages exposed to bacteria, raise body temperature and lower iron availability. Lower levels of iron decrease bacterial replication.
13. Interferons, a group of proteins released by cells infected by viruses, travel to other virus-infected cells where they inhibit viral replication.
14. The blood also contains a group of proteins. Called the complement proteins, they circulate in the blood in an inactive state, becoming activated when the body is invaded by bacteria. Some of the complement proteins stimulate the inflammatory response. Others embed in the plasma membrane of bacteria. There, they combine to form a membrane-attack complex, which creates a hole in the bacterial plasma membrane, killing the invader. Another complement protein binds to the invader, making it more easily phagocytized by macrophages.

THE THIRD LINE OF DEFENSE:
THE IMMUNE SYSTEM

15. The immune system consists of billions of cells that circulate in the blood and lymph and take up residence in the lymphoid organs.
16. The cells of the immune system recognize antigens—foreign cells and foreign molecules, mostly proteins and large-molecular-weight polysaccharides.
17. T and B cells are produced in red bone marrow. The T cells become immunocompetent—able to respond to a particular antigen—in the thymus. B cells gain this ability in the bone marrow. During this process, the T and B cells gain specific membrane receptors that respond to specific antigens.
18. Immunocompetent B cells encounter antigens (often presented to them by macrophages) to which they are preprogrammed to respond, then begin to divide, forming plasma cells and memory cells. The plasma cells produce antibodies.
19. Antibodies are small protein molecules produced by plasma cells. They bind to specific antigens destroying them either directly or indirectly. Some antibodies coat their antigens, neutralizing them or increasing their phagocytosis by macrophages. Other antibodies bind to the antigens, triggering the complement system. Still others cause the antigens to clump together (agglutinate) or to form insoluble complexes that precipitate out of solution.
20. When T and B cells first encounter an antigen, they react slowly in what is called the primary response. Since numerous memory cells are produced during the first assault, a reappearance of the antigen elicits a much faster and more powerful reaction, the secondary response.

21. Resistance created by a response to an antigen is called immunity.

22. When activated by antigen, T cells multiply and differentiate forming memory cells, cytotoxic T cells, helper T cells, and suppressor T cells.

23. Cytotoxic T cells bind to foreign cells and inject a lethal chemical into the membrane. Helper T cells stimulate the proliferation of T and B cells. Suppressor T cells release chemicals that put a halt to T- and B-cell division to shut down the immune response.

24. A solution containing a dead or weakened virus, bacteria, or bacterial toxin, which is injected into people to provide immunity, is called a vaccine. Vaccines produce an active immunity, which may last for years.

25. Passive immunity can be acquired by injecting antibodies into a patient or by the transfer of antibodies from a mother to her baby through the bloodstream or milk. Passive immunity is short-lived, lasting at most only a few months.

26. Blood transfusions require careful matching of donor and recipient blood types. The antigens on the membranes of body cells are more complex. Only cells from the same individual or an identical twin will be accepted. All others are rejected by the T cells, unless the system is suppressed with drugs, which often leave the patient vulnerable to bacterial infections.

27. The most common malfunctions of the immune system are allergies, extreme overreactions to some antigens. Allergies are caused by a class of antibodies called IgE antibodies, produced by plasma cells. IgE antibodies bind to mast cells. The antigen binds to the antibody on mast cells, causing the cells to release histamine and other chemical substances that induce the symptoms of an allergy—mucus production, sneezing, and itching.

AIDS: FIGHTING A DEADLY VIRUS

28. AIDS is a disease of the immune system caused by the HIV virus, which attacks helper T cells, severely impairing a person's immune system.

29. Victims grow progressively weak and develop tumors and bacterial infections because of the diminished immune response.

30. AIDS is spread through body fluids during sexual contact and blood transfusions.

31. Stopping the virus has proved difficult in large part because it mutates so rapidly and now appears to take refuge in the body. Recent accomplishments in research, however, offer some promise in stopping the deadly virus.

ENVIRONMENT AND HEALTH: MULTIPLE CHEMICAL SENSITIVITY

32. Many individuals suffer from multiple chemical sensitivity. Chronic exposure to low levels of a pollutant or short-term exposure to high levels may alter the immune system, causing a wide range of symptoms.

33. Research shows that many toxic chemicals affect the immune system. Some chemicals cause immune hypersensitivity, evoking allergy like symptoms. Others stimulate autoimmune responses. Still others cause immune suppression.

EXERCISING YOUR CRITICAL THINKING SKILLS

You have been selected as a juror for a trial involving a physician who refused to perform surgery on an AIDS patient. The patient had been in an automobile accident and suffered considerable internal organ damage. The patient's spleen, for example, had been ruptured by the accident, causing internal bleeding. The physician refused to perform surgery that could have saved the patient's life because she was afraid of contracting AIDS.

Consider the following facts presented by the attorneys for the plaintiffs (the AIDS victim's family). The plaintiffs argued that the physician violated her code of ethics, which obligates her to treat all patients. They also argued that the physician knowingly allowed a patient to die and that she should be punished by having to pay damages as compensation for the lost life. The AIDS patient had received a transfusion of HIV-contaminated blood several years earlier and had only three to six months to live. Nevertheless, these months were valuable to him and to his family.

The defense attorneys admit that the physician refused treatment and caused the premature death of her patient. They say, however, that the physician acted rightfully. During surgery, sharp instruments frequently pierce the protective gloves of surgeons, cutting physicians and their surgical teams, exposing medical workers to the AIDS virus. Refusing to operate protected not only the physician, but her entire surgical team. That team could, over the course of years, save hundreds of lives. In refusing to operate, the physician was considering the greater good—the benefit of her services to the public. The surgeon also had a husband and two children. By protecting herself, she was also taking into account the good of her family.

How would you decide such a case? Should the physician be forced to pay damages? Why? Or was the physician correct in her decision? Why? What would you have done if you were the physician?

TEST OF TERMS

1. A virus contains a nucleic acid core and a coat of protein called the _____ .

2. The enzyme, _____ _____ , allows RNA viruses to make DNA from their RNA.

3. Bacteria divide by _____ _____ .

4. The redness and swelling after a cut or abrasion are part of the _____ _____ .

5. _____ are chemical substances produced primarily by macrophages, which increase body temperature during a bacterial or viral infection.

6. _____ are a group of chemicals produced by virus-infected cells that impair viral replication in other virus-infected cells.

7. A(n) _____ is a large-molecular-weight carbohydrate or protein that evokes an immune response.

8. The process in which B and T cells develop their ability to respond to specific foreign invaders is called _____ .

9. Antibodies provide _____ immunity, protection against bacteria and viruses occurring primarily in the blood and lymph.

10. The secondary response to an antigen occurs more rapidly than the primary response because of the formation of _____ cells.

11. Antibodies belong to a class of proteins called _____ .

12. Coating a virus or bacterial toxin with antibody is called _____ .

13. Blood cells from an incompatible source clump together, because the recipient's blood contains antibodies to the donor's RBCs. This process is called _____ .

14. T cells differentiate into a cell, the _____ _____ , that travels throughout the body, destroying bacteria and virus-infected cells and tumor cells directly.

15. A _____ contains dead or weakened viruses that elicit an immune response without causing disease. It provides a form of _____ immunity.

Answers to the Test of Terms are located in Appendix B.

TEST OF CONCEPTS

1. Describe the structure of a typical virus, and explain how it gets into cells and how it reproduces inside the cell.

2. In what ways are bacteria similar to human body cells and in what ways are they different?

3. Describe some of the environmental benefits of bacteria.

4. The human body consists of three lines of defense. Describe what they are and how they operate.

5. Describe the inflammatory response, and explain how it helps protect the body.

6. Define each of the following terms, and explain how they help protect the body: pyrogen, interferon, and complement.

7. The first and second lines of defense differ substantially from the third line of defense. Describe the major differences.

8. How does the immune system detect foreign substances?

9. Describe how the B cell operates. Be sure to include the following terms in your discussion: bone marrow, immunocompetence, plasma membrane receptors, lag effect, primary response, plasma cell, antigen, antibody, and secondary response.

10. Describe the mechanisms by which antibodies "destroy" antigens.

11. Describe the events that occur after a T cell encounters its antigen.

12. What is the difference between active and passive immunity?

13. A child is stung by a bee, swells up, and collapses, having great difficulty breathing. What has happened? What can be done to save the child's life?

14. What is AIDS? What are the symptoms? What causes it?

SUGGESTED READINGS

Davidoff, L. L. 1989. Multiple chemical sensitivities. *The Amicus Journal* 11(1): 12–23. Well-documented and well-written piece on MCS.

Gallo, R. C., and L. Montagnier. 1988. AIDS in 1988. *Scientific American* 259(4): 40–48. Discusses how the AIDS virus was discovered and current key areas of research.

Greg, H. M., A. Sette, and S. Buus. 1989. How T cells see antigen. *Scientific American* 261(5): 68–74. Good description of how T cells work.

Norton, C. 1989. Not just for kids. *Hippocrates* 3(5): 74–78. Discusses vaccines.

Oldstone, M. B. A. 1989. Viral alteration of cell function. *Scientific American* 261(2): 42–48. Reviews how some viruses can get into a cell's genome and escape an effective immune response.

Young, J. D., and Z. A. Cohn. 1988. How killer cells kill. *Scientific American* 258(1): 38–44. Examines how cytotoxic T cells work.

11 Ridding the Body of Wastes: The Urinary System

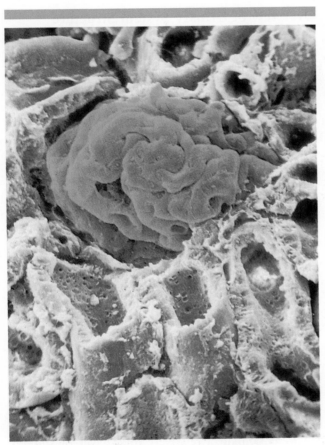

Transmission electron micrograph of the kidney showing the glomerulus and associated tubules.

L eon Markowitz is lowered into a large pool of warm water in a special room in the hospital where he'll spend the next few hours (Figure 11–1). Physicians position a large cylindrical device in the water near his lower back, next to the kidneys, the organs that filter the blood and thus help maintain normal blood concentrations of nutrients and wastes. Over the next few hours, as Leon listens to tapes of the Beatles, ultrasound waves, undetectable by the human ear, will bombard his kidneys, silently smashing a large kidney stone, a mass of calcium phosphate and other chemicals that is obstructing the flow of urine and causing excruciating pain.

A decade earlier, surgeons would have had to cut an incision 15–20 centimeters (6–8 inches) long in Markowitz's side to reach and remove the stone. Leon would have spent a week to 10 days recovering in the hospital and would have had to recuperate at home an additional eight weeks before returning to work. With this new technique, called ultrasound lithotripsy, patients usually return home the same day as the treatment or the day after.

This chapter describes the urinary system. It covers the structure and functions of the system, presenting information that will inevitably be useful some time in your life. This chapter also points out the many ways that the urinary system helps control homeostasis and describes some common diseases—such as kidney stones—that afflict this system.

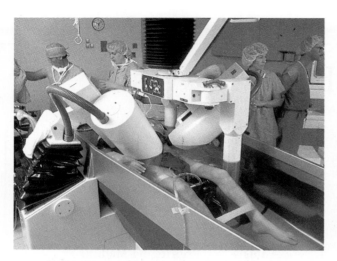

FIGURE 11-1 Lithotripsy Ultrasound waves, undetectable to the human ear, bombard the kidney stones in this man, smashing them into sandlike particles that can be passed relatively painlessly in the urine.

THE URINARY SYSTEM

Human body cells produce enormous amounts of waste, such as carbon dioxide, ammonia, and urea (Table 11-1). Like the toxic wastes produced by factories in an industrial society, these potentially harmful substances must go somewhere. However, just as in an industrial society, they cannot be dumped carelessly. If these hazardous substances were discarded in one of the body's storage depots, they would surely poison our cells and kill us in a few days.

TABLE 11-1 Metabolic Wastes

WASTE	PRODUCED BY	EXCRETED BY
Carbon dioxide	Metabolism in all body cells	Lungs
Ammonia	Breakdown of amino acids	Kidneys
Urea	Ammonia metabolism	Kidneys
Uric acid	Breakdown of nucleotides	Kidneys
Bile salts	Hemoglobin catabolism	Liver

Evolution has provided several ways to get rid of—or excrete—cellular wastes. The lungs, the skin, the liver, the kidneys, and even the large intestine all participate in this process, as described in previous chapters. If the number of organs involved in a body function is an indication of the importance of that process, excretion would have to rate among the most important of all physiological processes. Of these organs, however, the **kidney** ranks as one of the most important, for it rids the body of a variety of dissolved wastes. The kidney also plays a key role in regulating the chemical constancy of the bloodstream.

Kidneys are standard equipment. The paired kidneys are a part of the **urinary system**, shown in Figure 11-2. The urinary system also includes the ureters, the urinary bladder, and the urethra. The functions of these organs are described below and are summarized in Table 11-2.

Overview of Urinary System Structure

Each kidney, which is about the size of a fist, lies on the side of the vertebral column. Surrounded by fat and located high in the posterior abdominal wall beneath the diaphragm, the human kidneys are oval structures, slightly indented on one side like kidney beans (Figure 11-2). Arterial blood flows into the kidneys through the renal arteries, which enter at each indented region or hilus. The **renal arteries** are major branches of the abdominal aorta (Figure 11-2). Inside the kidney, much of the bloodborne wastes is filtered out and eliminated in the **urine**, a watery fluid containing inorganic ions, nitrogenous wastes (such as urea), small amounts of glucose, hormones, ions, and other chemical substances. After the blood has been filtered, it leaves the kidneys via the **renal veins**, which drain into the inferior vena cava.

Dissolved wastes are removed by numerous microscopic filtering units in the kidney, called **nephrons**. The nephrons produce urine, which is drained from the kidney via the **ureters**, muscular tubes that transport urine to the urinary bladder (Figure 11-2). Peristaltic contractions, involuntary smooth muscle contractions like those in the digestive tract, help propel the urine along the ureters. The **urinary bladder** is a temporary receptacle for urine. It lies in the pelvic cavity, behind the pubic bone. The bladder's distensible walls contain a relatively thick layer of smooth muscle that contracts when the bladder is filled. Urine forced out of the bladder then travels to the outside of the body through the urethra.

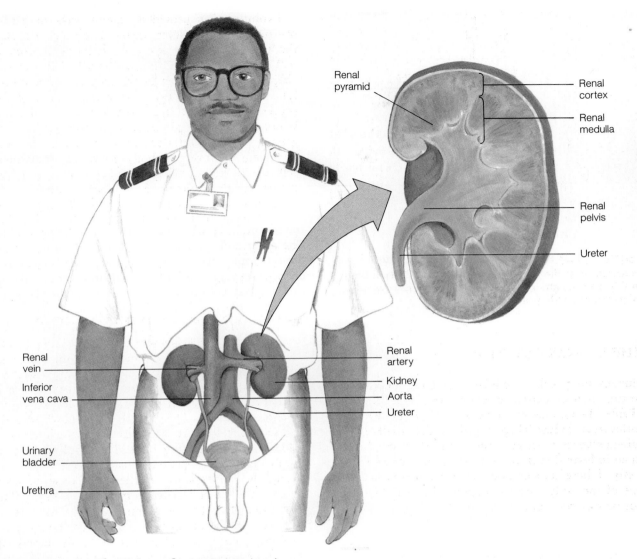

FIGURE 11-2 The Urinary System Anterior view showing the relationship of the kidneys, ureters, urinary bladder, and urethra.

The **urethra** is a narrow tube, measuring approximately 4 centimeters (1.5 inches) in women and approximately 15–20 centimeters (6–8 inches) in men. The additional length in men is largely due to the fact that the urethra travels through the penis (Figure 11–3). The difference in the length of the urethra in men and women has important medical implications. The shorter urethra in women, for example, results in a much higher incidence of bacterial infections of the urinary bladder. Bacteria can travel fairly easily up the urethras of women, invad-

ing the bladder.[1] Bladder infections may result in an itching or burning sensation and an increase in the frequency of urination. They may cause blood to appear in the urine. Urinary tract infections can be treated with antibiotics. If untreated, infections may spread up the ureters to the kidneys, where they can destroy the nephrons and impair kidney function.

[1]Sexual intercourse increases the frequency of urinary bladder infections in many women. To help avoid the problem, physicians advise women to empty their bladders soon after intercourse.

TABLE 11–2 Components of the Urinary System and Their Functions

COMPONENT	FUNCTION
Kidneys	Eliminate wastes from the blood; help regulate body water concentration; help regulate blood pressure; help maintain a constant blood pH
Ureters	Transport urine to the urinary bladder
Urinary bladder	Stores urine; contracts to eliminate stored urine
Urethra	Transports urine to the outside of the body

Human Kidney Structure

Each kidney is surrounded by a delicate connective tissue capsule, the **renal capsule**. The internal structure of the kidney is shown in Figure 11–4. As illustrated, the kidney is divided into two distinct zones. The outer zone is the **renal cortex**. The inner zone, the **renal medulla**, consists of cone-shaped structures called the **renal pyramids** and intervening tissue called **renal columns**. The renal pyramids contain

FIGURE 11–3 The Urinary Bladder and Urethra These drawings show the differences in the urethras of men (*a*) and women (*b*). The smooth muscle at the juncture of the urinary bladder and urethra forms the internal sphincter. The pelvic diaphragm is a flat sheet of muscle covering the lower boundary of the pelvic cavity. It forms the external sphincter and is under voluntary control.

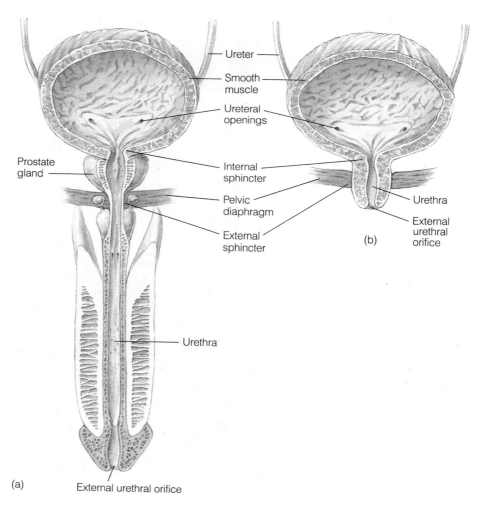

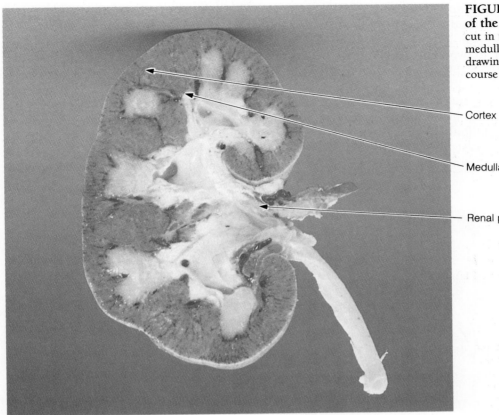

FIGURE 11–4 Cross Section of the Kidney (*a*) Human kidney cut in two showing the cortex, medulla, and renal pelvis. (*b*) A drawing of the kidney showing the course of the arteries and veins.

Cortex

Medulla

Renal pelvis

(a)

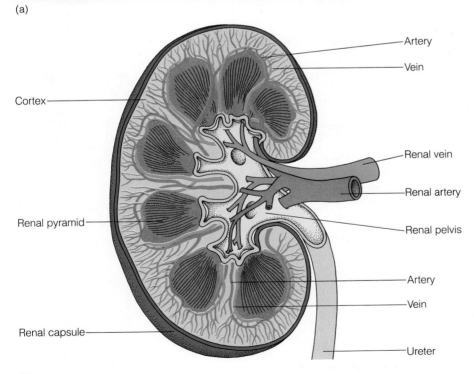

Cortex

Renal pyramid

Renal capsule

Artery

Vein

Renal vein

Renal artery

Renal pelvis

Artery

Vein

Ureter

(b)

small ducts that empty into a central chamber called the **renal pelvis**. Urine produced in the nephrons drains into the renal pelvis, then into the ureters.

Each kidney contains one to two million nephrons, visible only through a microscope (Figure 11–5 a and b; Table 11–3). The nephrons form the bulk of the kidney tissue. Each nephron consists of a tuft of capillaries, the **glomerulus** (Latin for ball of yarn), and a long, twisted tubule, the **renal tubule** (Figure 11–5b). The renal tubule, in turn, consists of four parts: (1) Bowman's capsule, (2) the proximal convoluted tu-

bule, (3) the loop of Henle, and (4) the distal convoluted tubule.

As illustrated in Figure 11–5b, **Bowman's capsule** is a double-walled structure that surrounds the glomerulus. The inner wall of the capsule fits closely over the glomerular capillaries and is separated from the outer wall by a small space, **Bowman's space**. The outer wall is continuous with the second segment of the renal tubule, the proximal convoluted tubule.

The **proximal convoluted tubule** is a sinuous or winding section of the renal tubule. It soon straightens, then descends downward and back up again, forming a thin, U-shaped segment of the renal tubule, known as the **loop of Henle**. The loops of Henle of some nephrons extend into the medulla.

The loop of Henle drains into the fourth and final portion of the renal tubule, the **distal convoluted tubule**, another winding segment. Each distal convoluted tubule drains into a series of rather straight ducts, called **collecting tubules**. These ducts, in turn, merge, forming larger ducts that empty into the renal pelvis.

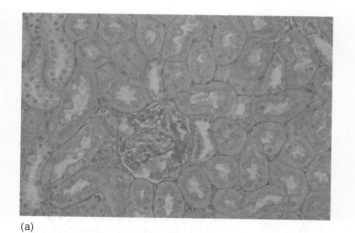

(a)

FIGURE 11–5 The Renal Cortex and Nephron
(*a*) A microscopic view of the cortex of the kidney showing the many tubules packed together and a single glomerulus. (*b*) A drawing of a nephron.

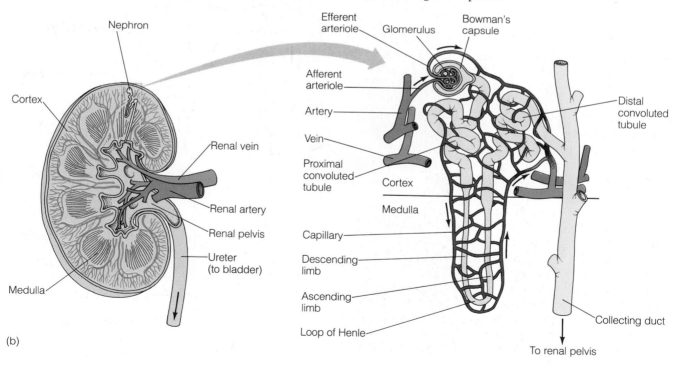

(b)

TABLE 11–3 Components of the Nephron and Their Function

COMPONENT	FUNCTION
Glomerulus	Mechanically filters the blood
Bowman's capsule	Mechanically filters the blood
Proximal convoluted tubule	Reabsorbs 75% of the water, salts, glucose, and amino acids
Loop of Henle	Participates in countercurrent exchange that maintains the concentration gradient
Distal convoluted tubule	Site of tubular secretion of H^+, potassium, and certain drugs

The nephrons filter large amounts of blood and produce from 1 to 3 liters of urine per day. The urinary output depends, in large part, on the intake of fluids. In general, the more fluid you drink, the more urine you produce.

THE FUNCTION OF THE URINARY SYSTEM

With the basic anatomy of the kidney and urinary system in mind, we next turn our attention to the function of the urinary system, examining the function of the nephron and the process of urination.

Physiology of the Nephron

Blood filtration in nephrons involves three processes: (1) glomerular filtration, (2) tubular reabsorption, and (3) tubular secretion (Figure 11–6).

Glomerular Filtration. **Glomerular filtration** is the movement of fluids and dissolved materials out of the glomerulus into Bowman's capsule. To understand this process, however, we need to study the anatomy of the glomerulus and Bowman's capsule in a little more detail.

As noted earlier, blood enters the kidney in the renal artery. Inside the kidney, the renal artery branches, forming smaller arteries that enter the renal cortex. These branches, in turn, form arterioles that supply the glomeruli in each kidney.

As shown in Figure 11–5b, each glomerulus is supplied by an **afferent arteriole**. Blood flows from the afferent arteriole into the glomerulus, a highly branched network of capillaries, which resembles a tangled knot of string (Figure 11–7a). Much of the

plasma minus its proteins is forced out of the capillaries into the space of Bowman's capsule by blood pressure. This resulting liquid is called the **glomerular filtrate**.

Glomerular filtration is a mechanical screening process, akin to the filtration that takes place in a coffee filter. The materials that pass from the blood to Bowman's capsule must be small enough to pass through the holes of the glomerular sieve. The glomerular sieve is composed of two layers of cells: (1)

FIGURE 11–6 Physiology of the Nephron The nephron carries out three processes: glomerular filtration, tubular reabsorption, and tubular secretion. All contribute to the filtering of the blood.

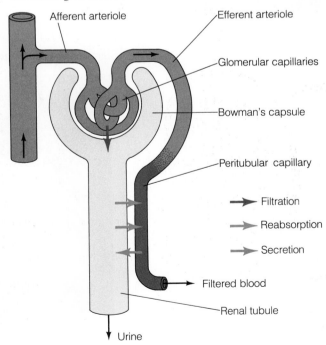

Afferent arteriole

Efferent arteriole

Glomerular capillaries

Bowman's capsule

Peritubular capillary

→ Filtration

→ Reabsorption

→ Secretion

Filtered blood

Renal tubule

Urine

those comprising the endothelium of the capillaries and (2) those of the inner membrane of Bowman's capsule (Figure 11–7a).

The glomerular capillaries contain numerous fenestrae (windows), or openings, that allow relatively large molecules to pass through the capillary wall with ease. Because the glomerular capillaries are so porous, fluids containing many dissolved substances exit freely. Blood cells, platelets, and large blood proteins, however, cannot pass through the fenestrae.

The cells of the inner membrane of Bowman's capsule also assist in the filtration of fluids passing from the blood to Bowman's space (Figure 11–7b). To understand the relationship between the capillaries and Bowman's capsule, imagine that your fist is a glomerulus. Also imagine that you are holding a partially inflated balloon in your other hand. If you push your fist into the balloon, the layer immediately surrounding your fist would resemble the inner layer of the capsule. The inner layer is separated from the outer layer of Bowman's capsule by Bowman's space. Fluids must pass through the capillary wall and the inner layer of Bowman's capsule.

The inner layer of Bowman's capsule consists of highly branched cells that surround the glomerular capillaries (Figure 11–7b). The cells of the inner layer of Bowman's capsule are called **podocytes** (because they contain many footlike processes). To understand the relationship of the cells of the inner layer of Bowman's capsule to the capillaries, hold a piece of plastic tube in your hands. A vacuum cleaner hose

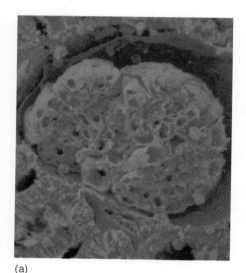

(a)

FIGURE 11–7   The Glomerulus   (a) A scanning electron micrograph of a glomerulus and Bowman's capsule. (b) A scanning electron micrograph of glomerular capillaries.

Source: Part (b) from *Tissues and Organs: A Text-Atlas of Scanning Electron Microscopy* by Richard G. Kessel and Randy H. Kardon. Copyright © 1979 W. H. Freeman and Company. Reprinted with permission.

(b)

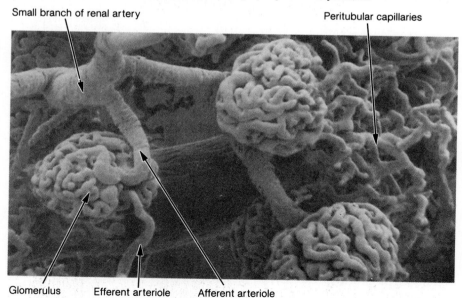

Small branch of renal artery

Peritubular capillaries

Glomerulus   Efferent arteriole   Afferent arteriole

will do. Imagine that this is the capillary. Next, wrap your hands around the tube, interlocking your fingers. Each of your hands now represents a podocyte (Figure 11–8b). The fingers resemble the branching processes of the cells. Looking down on your fingers, you will see small slits between them. These slits form a physical barrier that prevents larger molecules (blood proteins) from entering Bowman's space (Figure 11–8).

The fenestrae of the glomerular capillaries and the filtration slits allow the passage of water, ions, and many small-to-medium-sized molecules, but prevent the passage of blood cells, platelets, and most blood proteins. If an infection spreads to the kidneys, however, the capillary wall and inner membrane of Bowman's capsule of the nephrons may be so severely damaged that they allow blood cells and proteins to pass into the renal tubule where they become part of the urine. If blood appears in your urine, you should see a physician immediately to rule out a kidney in-

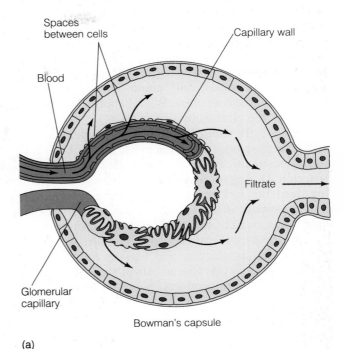

(a)

FIGURE 11–8   The Glomerulus Revisited   (a) A simplified drawing of the glomerulus showing the relationship of the glomerular capillaries to Bowman's capsule. (b) A scanning electron micrograph of the podocytes on the glomerular capillaries.

Source: Part (b) from *Tissues and Organs: A Text-Atlas of Scanning Electron Microscopy* by Richard G. Kessel and Randy H. Kardon. Copyright © 1979 W. H. Freeman and Company. Reprinted with permission.

(b)

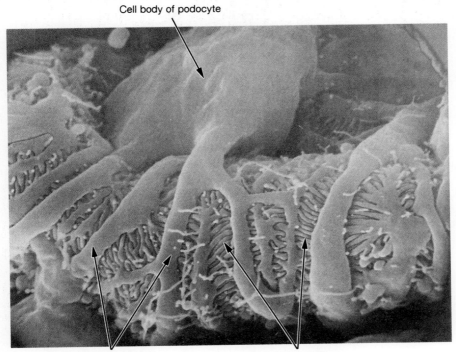

fection. In most cases, blood in the urine is the result of a bladder infection or a kidney stone.

Tubular Reabsorption. Each day, approximately 180 liters or 45 gallons of fluid—or filtrate—pour out of the glomerular capillaries and into the renal tubules. Thus, about one-fifth of the blood plasma entering the glomeruli at one time is filtered by the glomeruli. Despite this massive filtration of blood, the kidneys produce only about 1 to 3 liters of urine each day. In other words, only about 1% of the filtrate leaves the kidneys as urine. What happens to the fluid filtered at the glomerulus?

Most of the fluid filtered by the glomeruli is reabsorbed—passed from the renal tubule back into the bloodstream. This process is called **tubular reabsorption**. During tubular reabsorption, water containing nutrients and ions passes from the renal tubule into a network of capillaries, called the **peritubular capillaries**, which surround each nephron (Figure 11–9). The peritubular capillaries reabsorb water, nutrients, such as glucose, and various ions, such as sodium, which were filtered out of the blood in the glomerulus. Waste products, however, are generally

left alone, passing on to the collecting tubules and being eliminated from the body in the urine. Reabsorption, therefore, helps to maintain normal blood concentrations of water and various dissolved materials. Reabsorption conserves nutrients, water, and ions.

The peritubular capillaries are branches of the arterioles that carry filtered blood out of the glomerulus, the **efferent arterioles**, shown in Figure 11–9. Constriction of the efferent arterioles raises glomerular blood pressure, increasing the rate of filtration. Thus, the kidney's unique anatomical arrangement provides a means by which glomerular blood pressure can be increased.

The blood draining from the peritubular capillaries empties into venules, then into veins. The small veins converge, forming the renal vein, which carries filtered blood out of the kidney and into the inferior vena cava.

Tubular reabsorption occurs in all parts of the renal tubule, but the bulk of it occurs in the proximal convoluted tubule. In the proximal convoluted tubule about 75% of all sodium ions and water are reabsorbed. As the positively charged sodium ions are ac-

FIGURE 11–9 The Nephron and Peritubular Capillaries The peritubular capillaries absorb water, nutrients, and ions that pass out of the renal tubule.

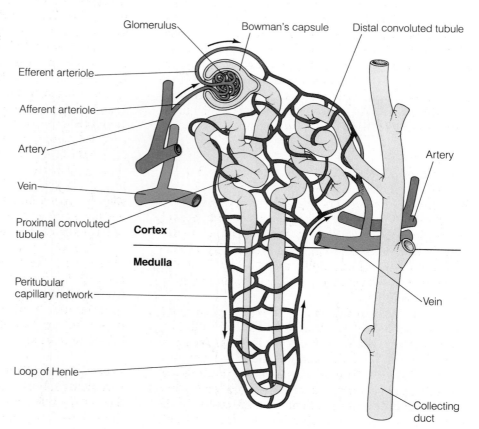

tively transported out, chloride and other negatively charged ions follow. The increase in the concentration of ions in the extracellular fluid causes water molecules to move outward by osmosis. The proximal convoluted tubule is also responsible for the reabsorption of calcium ions, glucose, vitamins, and small plasma proteins that may have entered the filtrate during glomerular filtration.

Tubular Secretion. Waste disposal is supplemented by a third process, known as **tubular secretion**. Here's how it operates. Wastes not removed from the blood during glomerular filtration enter the peritubular capillaries. Some of these wastes are then transported *into* the renal tubule. For example, hydrogen and potassium ions that escaped filtration diffuse out of the blood capillaries of the peritubular capillary network into the surrounding extracellular fluid. From here, the wastes may be actively transported into the renal tubule.

Tubular secretion occurs in both the proximal and distal convoluted tubules, but mostly in the latter. Tubular secretion is particularly important because it helps the body regulate the H^+ concentration of the blood, helping maintain a constant pH. Hydrogen ions are produced during cellular metabolism. Tubular secretion, therefore, supplements buffers in the blood and extracellular fluid, especially bicarbonate ions, which help regulate H^+ concentrations. To prevent the buildup of hydrogen ions in the renal tubule, some tubular cells produce ammonia (NH_3). Ammonia diffuses into the renal tubule where it combines with hydrogen ions forming ammonium ions (NH_4^+).

Concentrating the Urine: The Countercurrent Exchange Mechanism. Urine leaves the distal convoluted tubule and enters the collecting tubules. As noted earlier, the collecting tubules descend through the medulla and converge to form larger ducts that drain into the renal pelvis. As the collecting tubules descend through the medulla, much of the remaining water is lost. Water moves out of the collecting tubules by osmosis.

As illustrated in Figure 11–10, the concentration of sodium chloride (NaCl) in the extracellular fluid of the medulla increases with depth. Thus, as the urine flows down through the medulla, more and more water is reabsorbed. This increasing concentration gradient helps concentrate the urine and conserve water.

The concentration gradient is maintained by the loop of Henle. Understanding how the loop of Henle

operates requires a careful look at its permeability to various substances, notably water and salt. We begin with the ascending loop.

As the urine moves up the ascending loop toward the distal convoluted tubule, sodium ions are actively transported out. Chloride ions follow passively, moving by diffusion (Figure 11–10). Water cannot follow because the ascending loop is rather impermeable to water. The outward movement of sodium chloride is greatest in the lowermost portion of the loop of Henle, as shown in Figure 11–10 by the thick arrows. Thus, the concentration of sodium and chloride ions in the extracellular fluid is greatest at the bottom of the loop. Since the outward movement of sodium and chloride decreases as the urine moves up, the extracellular concentration of sodium chloride decreases. The differential movement of sodium chloride creates a concentration gradient.

As indicated by the solid arrows in Figure 11–10, sodium chloride ions move back into the descending limb of the loop of Henle. These same ions will be pumped back out as the urine ascends once again. This cyclic movement of ions maintains the concentration gradient. An equilibrium is established in which the outward movement of sodium chloride (in the ascending limb) equals the inward movement of sodium chloride (in the descending limb). The extracellular concentration, therefore, remains the same and is always greatest at the bottom of the loop, for the reasons described above.

This mechanism is called **countercurrent exchange**. The term *countercurrent exchange* applies to any system in which fluids, flowing in opposite directions, exchange chemicals or heat. Arteries and veins, for example, contain blood flowing in opposite directions and often lie side by side. In the arm, warm blood traveling in the artery loses its heat to the slightly cooled blood returning from the limbs. This mechanism helps conserve body heat and is the reason your hands are slightly colder than your torso.

In the kidney, countercurrent exchange exists between the urine flowing in the descending and ascending loops. Sodium chloride is exchanged between the two segments of the loop of Henle. The concentration gradient in the medulla is maintained, thus ensuring the outward movement of water in the collecting tubules.

Kidney Stones: Causes and Cures. Urine contains various dissolved wastes. Approximately 90% of the dissolved waste consists of three substances: urea, sodium ions, and chloride ions. Varying amounts of

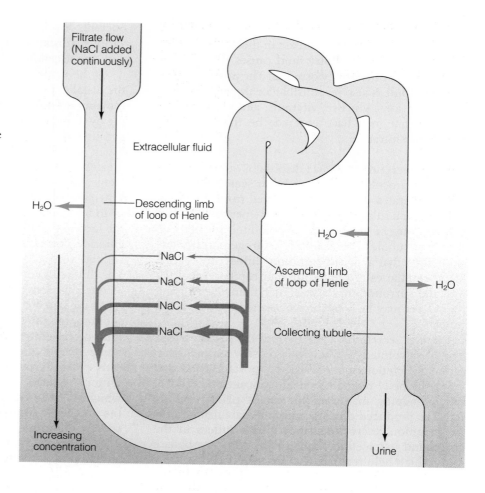

FIGURE 11–10 The Concentration of Urine by the Nephron A countercurrent exchange mechanism ensures the concentration gradient in the extracellular fluid surrounding the nephron, which facilitates the movement of water out of the collecting tubule and helps conserve body water and concentrate urine.

Filtrate flow (NaCl added continuously)

Extracellular fluid

H_2O

Descending limb of loop of Henle

NaCl

NaCl

NaCl

NaCl

H_2O

Ascending limb of loop of Henle

H_2O

Collecting tubule

Increasing concentration

Urine

other materials are also present. Physicians routinely analyze the dissolved components in their patients' urine to check for underlying metabolic problems. Diabetes mellitus, for example, results in an elevated level of urinary glucose.

Excess calcium, magnesium, and uric acid in the urine may crystallize in the renal pelvis of the kidney, forming deposits or kidney stones (Figure 11–11a). These small deposits enlarge by **accretion**—the deposition of materials on the outside of the deposit, which causes it to grow in much the same way that a snowball enlarges as it is rolled through snow (Figure 11–11b).

Many kidney stones enter the ureter and are passed to the bladder. As they are propelled along the ureter, their sharp edges can dig into the wall of the ureter, stimulating pain fibers. Small stones that pass to the urinary bladder are often excreted in the urine. Larger stones, however, may lodge in the ureters or in the renal pelvis where they obstruct the flow of urine to

the bladder. Obstructions increase pressure inside the kidney, which may damage the nephrons.

For years, kidney stones were removed surgically. Today, however, a relatively new medical technique (called ultrasound lithotripsy) can be used. In this technique, physicians shatter kidney stones with ultrasound. The procedure is nearly painless and certainly much safer than surgery. Ultrasound lithotripsy is discussed in Health Note 11–1. Additional urinary system disorders are listed in Table 11–4.

Urination: Controlling a Reflex

The ureters pass through the back wall of the urinary bladder at a slight angle. When the bladder is full, its walls contract to void the urine. The contraction of the smooth muscle forces the ends of the ureters shut, preventing the backflow of urine.

Urine fills the bladder gradually, causing its walls to stretch and become thin (Figure 11–12a). Leakage into the urethra as the bladder fills is prevented by two sphincters—muscular gate valves, shown in Figure 11–3. The first sphincter is called the internal sphincter. The **internal sphincter** is an involuntary gateway—that is, it operates without conscious control. It is formed by a smooth muscle in the neck of the urinary bladder at its junction with the urethra. The second, the **external sphincter**, is a flat band of muscle that forms the floor of the pelvic cavity. The external sphincter is a voluntary gateway controlled by the brain. When both sphincters are relaxed, urine is propelled into the urethra and out of the body.

When the bladder accumulates approximately 200 to 300 milliliters of urine, stretch receptors in the wall of the organ begin sending impulses to the spinal cord via sensory nerves (Figure 11–12b). This is a kind of warning signal that waste is building up inside and must soon be reckoned with. In the spinal cord, incoming nerve impulses stimulate a group of nerve cells that supply the smooth muscle cells of the urinary bladder. Nerve impulses generated in the spinal cord exit, traveling along nerves that terminate on the smooth muscle cells in the wall of the urinary bladder. These impulses stimulate the muscle in the wall of the bladder to contract and also stimulate the muscles in the internal sphincter to relax, releasing urine. Urine does not escape, however, until the external sphincter is opened.

This pathway is one of many reflexes in the body. Because the external sphincter is under voluntary control in most older children and adults, urination can be suppressed until the appropriate time and place. While awaiting the signal, the bladder can expand to hold up to 800 milliliters of urine. At this stage, waiting may become quite painful.

In babies, urination is a purely reflexive—that is, there is no conscious override. When the bladder fills and the stretch receptors are stimulated, urination follows. Not until children grow older (about 2.5 to 3 years) can they begin to control urination.

THE KIDNEYS AS ORGANS OF HOMEOSTASIS

Each drop of blood in your body passes through your kidneys approximately 350 times a day. That means about 400 gallons of blood flow through the kidney each day in an average person. About 20% of the blood plasma is filtered at any one time. Clearly, this results in a very thorough filtering of impurities. This,

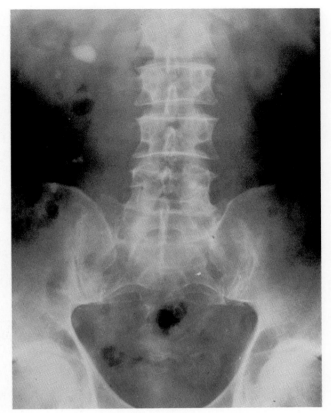

(a)

(b)

FIGURE 11–11 Kidney Stones (*a*) An X-ray of kidney stones. (*b*) Kidney stones removed by surgery.

in turn, helps the body control the composition of the blood to maintain homeostasis.

Maintaining Body Water Levels

Besides ridding the body of wastes, the kidneys also help regulate the water content of the body. Water,

HEALTH NOTE 11–1

New Wave Surgery

Kidney stones are one of the more common and painful ailments of modern times. Stones passing down the ureters cause excruciating pain in the back and side or lower abdomen. The pain is so intense that even powerful painkillers are useless against it.

Kidney stones are four times more common in men than in women. If you're a male, chances are one in ten that you will develop a stone that requires medical attention sometime in your life. Given the size of our population, that means that hundreds of thousands of Americans will be admitted to hospitals each year for treatment. For most of these people, no surgery is required. The stones merely pass into the bladder and out the urethra. While painful, it is not life threatening. Still, as many as 50,000 Americans will require more serious treatment.

Thanks to a new technique, called ultrasound lithotripsy, developed by German researchers and briefly mentioned in the introduction of this chapter, treatment has become as easy as a two-hour hot bath. It has not always been so pleasant, however: even a decade ago, kidney stones had to be removed through a large incision in the patient's side. The trauma was so great that patients required at least a week in the hospital to recover and then could not return to work for another eight weeks.

West German researchers found a way to eliminate this costly and time-consuming operation. They first pioneered a simple surgical technique that allows surgeons to insert a small hollow metal tube through a small incision in the patient's side. Fiber optics, thin tubes that shine light inside the body but also allow surgeons to see where they are, are used to find the kidney stone. If the stone is small, they can suck it out using a vacuum. Larger stones required surgeons to insert a tiny clawlike device with which they could grab the stone and crush it. Fragments could then be sucked out.

For still larger stones, surgeons could use an ultrasound probe. Like a miniature jackhammer, it smashes the stone into tiny bits. The ultrasound probe is pressed against the kidney stone, then switched on. When the hard outer "shell" of the stone cracks, the stone generally falls apart like a clump of sand. Then surgeons aspirate the fragments. This technique, called percutaneous ultrasound lithotripsy, was introduced into the United States in 1981, creating a wave of enthusiasm among kidney specialists. No sooner had it begun to find its way into U.S. hospitals, however, than West German researchers announced a newer and even less traumatic technique.

In this technique, called ultrasound lithotripsy, patients are mildly sedated and given a local anesthetic to reduce the pain around the kidney. Then the patients are placed in a warm tub of water, and ultrasound waves, from a special device placed outside the body near the kidney, bombard the stones, breaking them apart. When the treatment is over, patients may elect to stay in the hospital or may go home to pass the tiny fragments in the urine.

Six U.S. medical centers began to test this technique in 1984. Now most major hospitals are using it. As a result, tens of thousands of Americans are spared the pain and lengthy recovery that once accompanied surgical removal of kidney stones.

---

whose biological significance was discussed in Chapter 2, is passively absorbed by the renal tubules and returned to the bloodstream. The rate of water reabsorption, however, can be increased or decreased. The ability to adjust water reabsorption and, consequently, urine output allows our bodies to get rid of excess water when we have consumed too much and also helps us conserve body water when we have become dehydrated.

Control of Water Reabsorption by ADH. Water reabsorption is controlled in part by **antidiuretic hormone** (ADH). ADH is released by an endocrine gland at the base of the brain called the pituitary gland (Chapter 15).[2] ADH release is regulated by two re-

[2]ADH is manufactured by the hypothalamus and transported to the pituitary via modified nerve cells, called neurosecretory cells, which are described in Chapter 15.

TABLE 11–4 Common Urinary Disorders

DISEASE	SYMPTOMS	CAUSE
Bladder infections	Especially prevalent in women; pain in lower abdomen; frequent urge to urinate; blood in urine; strong smell to urine	Nearly always caused by bacteria
Kidney stones	Large stones lodged in the kidney often create no symptoms at all; pain occurs if stones are being passed to the bladder; pains come in waves a few minutes apart	Deposition of calcium, phosphate, magnesium, and uric acid crystals in the kidney, possibly resulting from inadequate water intake
Kidney failure	Symptoms often occur gradually: more frequent urination, lethargy, and fatigue; should the kidney fail completely, you may develop nausea, headaches, vomiting, diarrhea; water buildup especially in the lungs and skin, and pain in the chest and bones	Immune reaction to some drugs, especially antibiotics; toxic chemicals; kidney infections; sudden decreases in blood flow to the kidney, for example, resulting from trauma
Pyelonephritis	Infection of the kidney's nephrons; sudden, intense pain in the lower back immediately above the waist, high temperature, and chills	Bacterial infection

ceptors (Figure 11–13). The first consists of a group of nerve cells in a region of the brain just above the pituitary gland called the hypothalamus. These cells monitor the osmotic concentration of the blood. ADH release is also controlled by receptors in the heart, which detect changes in blood volume (which reflect water levels).

Consider an example. If you were to play soccer on a hot summer afternoon, you would undoubtedly perspire heavily and lose a significant amount of body water and salts. If you did not replace the water you were losing, two changes would occur. First, your blood volume would fall, and second, the osmotic concentration of your blood would rise (Figure 11–13). The decrease in blood volume results from the loss of water. The osmotic concentration of the blood increases because perspiration carries off salts, but leaves behind the blood's more osmotically active chemicals (blood proteins). The decrease in blood volume and the rise in osmotic concentration trigger the release of ADH (Figure 11–13).

ADH circulates in the blood to the kidney and there increases the permeability of the distal convoluted tubules and the collecting tubules (Figure 11–

13). ADH increases the rate of tubular reabsorption, which reduces urinary output and helps to restore the volume and osmotic concentration of the blood.

Excess water intake has just the opposite effect. Water passes from the small intestine into the blood, increasing the blood volume and decreasing its osmotic concentration. These changes result in a reduction in ADH secretion. As ADH levels in the blood fall, the permeability of the distal convoluted tubules and the collecting tubules decreases. Tubular reabsorption is reduced, and more water is lost in the urine; this helps to decrease blood volume and to restore the osmotic concentration of the blood.

Water in the fluids we drink affects urinary output through ADH secretion, but certain chemicals in common beverages also have a dramatic effect on urine production. Coffee and (nonherbal) tea, for example, both increase urine production because they contain caffeine.[3] Caffeine in teas, coffee, and certain soft drinks is a **diuretic**, a chemical that increases urination. Caffeine has two major effects on the kidney. First, it increases glomerular blood pressure, which,

[3] Part of the increase is due to the fluid contained in these drinks.

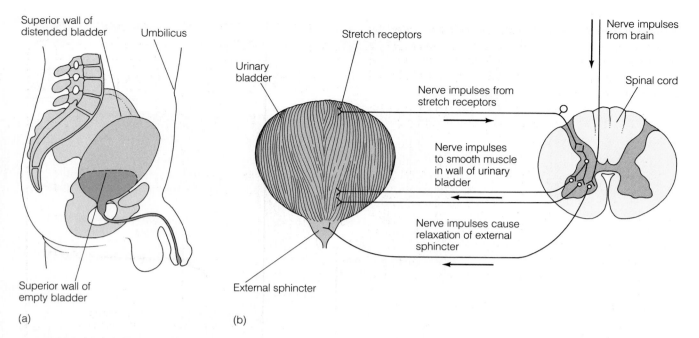

Superior wall of distended bladder

Umbilicus

Superior wall of empty bladder

(a)

Stretch receptors

Urinary bladder

Nerve impulses from stretch receptors

Nerve impulses to smooth muscle in wall of urinary bladder

Nerve impulses cause relaxation of external sphincter

External sphincter

Nerve impulses from brain

Spinal cord

(b)

FIGURE 11–12 Urination (a) The bladder before and after it fills, showing how much this organ can expand to accommodate urine. (b) Stretch receptors signal the distension of the urinary bladder. Nerve impulses travel to the spinal cord as part of a reflex. They, in turn, stimulate nerve cells that send impulses back to the bladder, causing muscle contraction and relaxation of the internal sphincter. Nerve impulses also travel up the spinal cord to the brain, signaling the need to void. In adults, the brain sends signals back to the external sphincter, allowing urine to be voided when the time and place are right.

in turn, increases glomerular filtration. (As a result, more filtrate is formed.) Second, caffeine decreases the tubular reabsorption of sodium ions. As noted earlier, water follows sodium ions out of the renal tubule during tubular reabsorption. Therefore, a decrease in sodium reabsorption results in a decline in the amount of water leaving the renal tubule and an increase in urine output.

Alcoholic beverages also increase urination. Alcohol is also a diuretic, but it works by inhibiting the secretion of ADH. This reduces water reabsorption and increases water loss, giving credence to the quip that you don't buy wine or beer, you rent it!

Severe head injuries can destroy the production of ADH, leading to a disease called diabetes insipidus. This is not to be confused with diabetes mellitus (commonly called sugar diabetes), a disorder involving insulin, described in Chapter 15. **Diabetes insipidus** is characterized by frequent urination (polyuria) and excessive liquid intake (polydipsia). It results from insufficient amounts of ADH. Victims produce up to 20 liters (5 gallons) of colorless, dilute urine per day. Diabetes insipidus gets its name from the fact that the urine is dilute and tasteless (insipid).[4] Sleeping through the night is impossible, for victims are continually awakened by thirst or the urge to urinate.

Diabetes insipidus can be treated in a variety of ways, depending on the severity of the disorder. In patients whose urine output is only slightly elevated, dietary salt restrictions and antidiuretics (drugs that reduce urine output) will work. In severe cases, patients must receive synthetic ADH. ADH can be administered by injections or via nose drops. If the pituitary damage has resulted from a head injury, treatment may only be required for a year or so. If the damage is permanent, however, treatment will be required for the rest of a person's life.

Control of Water Reabsorption by Aldosterone. ADH controls the amount of water in the body, and that, in turn, helps regulate the concentration of dissolved substances. ADH, therefore, is a key component of the body's homeostatic mechanism that main-

[4]Diabetes mellitus is so named because the urine is sweet. The Latin word for honey is *mellifer*.

tains water and chemical balance. ADH is aided by another hormone, called aldosterone. **Aldosterone** is a steroid produced by a pair of glands that sit atop the kidneys like loose-fitting stocking caps, the **adrenal glands** (Figure 11–2).

Aldosterone levels in the blood rise and fall in response to blood pressure, blood volume, and osmotic concentration. A decrease in the blood pressure, for example, triggers the release of this hormone from the outer portion of the adrenal gland, the **adrenal cortex** (Figure 11–14).

As shown in Figure 11–14, aldosterone increases the reabsorption of sodium ions by the distal convoluted tubules and collecting tubules. Sodium is absorbed by the peritubular capillaries, thus elevating sodium levels in the bloodstream. Water follows the salt, moving by osmosis into the bloodstream. The outflow of water, therefore, increases the blood volume and the blood pressure and lowers the osmotic concentration.

As illustrated in Figure 11–14, the release of aldosterone is controlled by a fairly complex sequence of events. A reduction in blood pressure or a reduction in the volume of filtrate in the renal tubule causes certain cells in the kidney to produce an enzyme called **renin**. In the blood, renin cleaves a segment off a large protein, called **angiotensinogen**, which is found in the plasma. The result is a small peptide molecule called **angiotensin I**. Angiotensin I is inactive but is converted into the active form, **angiotensin II**, by further enzymatic action. Angiotensin II stimulates aldosterone secretion.

Kidney Failure

The importance of the kidneys is most obvious when they stop working—a condition called **renal** or **kidney failure**. Renal failure generally results from one of four causes: the presence of certain toxic chemicals in the blood, immune reactions to certain antibiotics, severe kidney infections, and sudden decreases in blood flow (for example, after an injury).

Renal failure may occur suddenly, over a period of a few hours or a few days. This is called **acute renal failure**. Renal function may also deteriorate slowly over many years, resulting in **chronic renal failure** or, more appropriately, **chronic renal impairment** (so named because the kidneys never really quit). Chronic renal failure can lead to a complete or nearly complete shutdown, called **end-stage failure**.

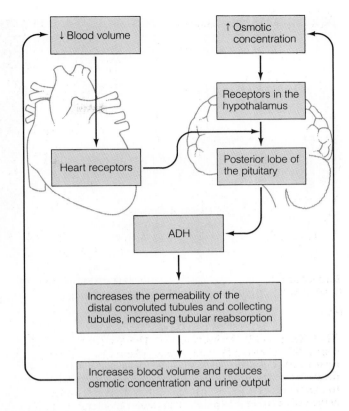

FIGURE 11–13   ADH Secretion   ADH secretion is under the control of the hypothalamus. When the osmotic concentration of the blood rises, receptors in the hypothalamus detect the change and trigger the release of ADH from the posterior lobe of the pituitary. Detectors in the heart respond to changes in blood volume. When it drops, they send signals to the brain, causing the release of ADH.

Kidney failure (whether acute or end stage) is generally fatal unless treated quickly. When the kidneys stop working, water begins to accumulate in the body. Toxic wastes build up in the blood. Homeostasis is disrupted by an imbalance of chemicals that are normally regulated by the kidneys. Death generally occurs in two to three days if a person with kidney failure is not treated. Death usually results from an increase in the concentration of potassium ions in the blood and tissue fluids. Potassium is essential for the normal function of heart muscle. An increase in potassium levels destroys the rhythmic contraction of the heart, causing fibrillation (Chapter 8).

Renal failure, whether it is chronic or acute, requires immediate attention. If the problem is caused by an acute loss of blood, transfusions may be required. Patients whose kidneys have shut down, even

WE NEED TO LEARN TO PRIORITIZE MEDICAL EXPENDITURES

Richard D. Lamm

Richard D. Lamm is the director of the Center for Public Policy and Contemporary Issues at the University of Denver, where his primary research and teaching interests have been in the area of health policy. Before assuming his present position, he served three terms as governor of Colorado, from 1975 until 1987.

Health care in the United States finds itself in (to paraphrase Dickens) "the best of times and the worst of times." It is the epoch of medical miracles, yet at the same time millions go without even basic health care.

The basic dilemma of American medicine is that we have invented more health care than we can afford to deliver. Health care costs are rising at three times the rate of inflation. They are absorbing funds desperately needed elsewhere in our system to educate our kids, rebuild our infrastructure, and revitalize our industries.

Health care costs are rising for many reasons. One major reason is that no one prioritizes the myriad of procedures that health care can deliver. No one asks, "How do we buy the most health for the most people with our limited funds?" Recent policy decisions in Oregon and California, relating to the public funding of transplant operations, illustrate how two states attempted to deal with a crisis in health care.

Oregon recently received adverse publicity for its decision not to publicly fund soft tissue transplant operations. As is often the case, much of the focus has been on the handful of individuals who have been adversely affected by the policy rather than the more numerous, but anonymous, people who will benefit. The Oregon policy sparked a public outcry over society's lack of compassion for individuals who desperately need transplants. Yet, when Oregon policymakers weighed the needs of a few transplant patients against the basic health care needs of the medically indigent, they decided against the former in favor of the latter. They have now set up a process that sets priorities throughout the health care system.

California took a contrary approach. Policymakers in California decided to publicly fund transplant operations.

Then, one week later, they removed 270,000 low-income people from their state's medical assistance program.

Which was the wiser decision? The answer seems clear. Oregon bought more health for more of its citizens for its limited funds. Some say Oregon should have raised its taxes and funded transplants. That is very easy for nonpoliticians to say. Polls show people believe all Americans should have access to quality health care. But the polls also show that most are unwilling to accept even modest tax increases to provide the care they say they favor.

Illinois recently passed legislation giving "universal access to major organ transplants," but appropriated less than one-third of the funds needed. Politically, of course, Illinois played it far safer than Oregon, but, in doing so, failed to confront one of the most pressing social issues of our time. Avoidance may be a politically expedient tactic, but sooner or later our society will be forced to allocate scarce health resources.

Clearly, medical science is inventing faster than the public is willing or able to pay. Realistically, no system of health care can avoid rationing medicine. In fact, rationing already exists. But instead of rationing medicine according to need or a patient's prospects for recovery, we ration by seniority and ability to pay. Those who pretend that we do not ration medicine forget that 31 million Americans do not have full access to health care in America. Most states have staggering numbers of medically indigent, yet they provide a small percentage with a full program of coverage. Oregon is the first state to cover 100% of the people living under the federal poverty line with basic health care.

I suggest this yardstick: Which state best asked itself, "how do we buy the most health for the most people?" Which state tried to maximize limited resources? Which state benefited the greatest number of people and which state harmed the greatest number of people?

Oregon attempted to weigh the basic health needs of the medically indigent against other programs for which the state paid. It recognized that the money the state paid for transplants and other high-cost procedures for a few people could buy more health care elsewhere, providing basic low-technology services to people not covered at that time.

Prioritizing health care does not abandon the poor; it seeks to serve the largest number with the most effective procedures.

MEDICAL PRIORITIZATION IS A BAD IDEA

Arthur L. Caplan

Arthur L. Caplan is director of the Center for Biomedical Ethics at the University of Minnesota.

The American health care system, the experts say, is going bust at a rapid rate. Efforts to contain our burgeoning $500-billion-plus tab have been a total failure.

The dilemma of how to pay for health care is forcing some public officials to think the unthinkable. Alameda County and the state of Oregon recently announced plans to institute rationing policies for health care.

But, before you applaud the realism, consider that these plans would ration access to health care only for the poor. The medically indigent of Alameda County and those eligible for Medicaid in Oregon will be required by law to forgo life-saving medical care.

Officials in both Oregon and Alameda County note that the poor have always had less access to health care than the rich. This is true. But, our society's failure to meet the health care needs of the poor hardly justifies a public policy that asks the poor to bear the burden of rationing as a matter of law.

Who concocted this blatantly unethical scheme? Incredibly, the inspiration for both the California and Oregon plans for pocketbook triage comes in part from those in my line of work—medical ethicists.

A California bioethics consulting firm is being paid by Alameda County and Oregon state officials to provide moral rationales for dropping the poor out of the health care lifeboat. The consultants appear to be approaching their task with gusto.

"You have to draw the line somewhere,"one moralist-for-hire said in a recent newspaper article about Alameda County's decision to begin rationing for the poor. "We'll provide all services to a diminishing segment of the population, and literally we'll throw the rest of the people overboard."

No hint is given of the theoretical position that would justify aiming all rationing efforts at the poor. But it is hard to think of an ethic that holds that when a nation cannot pay its doctor bills, it is the poor and only the poor who should be denied the right to see a doctor.

It is hard to understand how any ethicist could become involved in a scheme so blatantly unfair as that of rationing necessary health care only for the poor. What is worse is that the same ethicists and the officials taking their advice, are not asking whether it is really necessary in 1989 to institute the rationing of necessary medical care for anyone.

Before saying goodby to the indigent, why aren't public officials in Oregon and California thinking about cracking down on practices that add tens of millions of dollars to state-financed health-care costs each year? Before saying no to a bone marrow transplant for a three-year-old whose mother is on Medicaid, couldn't county and state legislators insist that every licensed hospital and physician be required by law to provide a fixed percentage of care for those who cannot pay?

Before creating laws that would send some of the poor to a premature demise, county and state officials ought to require private health insurers to charge subscribers an additional premium that could be used to supplement the pitifully small budgets of Medicaid and public hospitals. And would it not make some sense to insist on a luxury tax, which could be used to help meet the crucial health care needs of the poor, from the rich who avail themselves of psychotherapy, vitamins, cosmetic surgery, diet clinics and stress-management seminars?

It is wrong to make the poor and only the poor bear the burden of rationing. It is unethical to institute rationing of necessary health services for any group of Americans unless we have made every effort to be as efficient as we can be in spending our health care dollars.

At a time when some can indulge their wants by buying a face lift, it seems extraordinarily hard for ethicists or legislators to convincingly argue that they have no other option but to condemn the poor to die for want of money.

SHARPENING YOUR CRITICAL THINKING SKILLS

1. Summarize Lamm's and Caplan's key points. List supporting information given for each position or main point.
2. Do you agree with the views of Lamm or Caplan? Explain why.

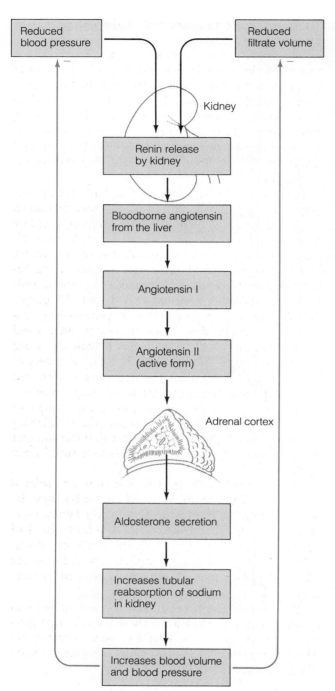

FIGURE 11–14 **Aldosterone Secretion** Aldosterone is released by the adrenal cortex. Its release, however, is stimulated by a chain of events that begin in the kidney.

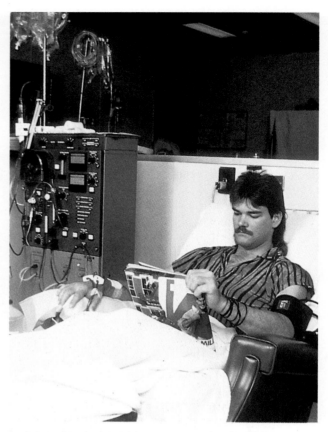

FIGURE 11–15 **Renal Dialysis** When the kidneys fail, the blood must be filtered by a dialysis machine, which runs the blood through filters that remove the impurities.

temporarily, may require renal dialysis. **Renal dialysis** is a procedure in which toxic materials in the patient's blood are filtered using an artificial filter. Patients re-

quiring dialysis are connected to a machine called a **dialysis unit**. Blood is drawn out of a vein and passed through a piece of tubing that transports the blood to the filter. After filtration, the blood is pumped back into the patient's bloodstream (Figure 11–15). Dialysis is a painless procedure, but requires several hours to complete. Complete kidney failure may also be treated by kidney transplants. Transplants are generally most successful when they come from closely related family members (Chapter 10).

For years, the complete or nearly complete destruction of kidney function was almost always fatal. Thanks to renal dialysis and kidney transplantation, today many patients can live normal, fairly healthy lives. These procedures, especially transplants, are costly, however. The Point/Counterpoint discusses some of the issues raised by the growing cost of health care.

ENVIRONMENT AND HEALTH: MERCURY POISONING

The kidneys are elaborate filters that help regulate the levels of many chemical nutrients, such as glucose, amino acids, vitamins, and a number of important ions, such as potassium and chloride. They also help control the level of various wastes, such as urea, a nitrogen-containing organic compound derived during the breakdown of amino acids. The kidneys also eliminate numerous potentially harmful substances—for example, some drugs, food additives, pesticides, and some of the toxic chemicals that enter the bloodstream from cigarette smoke.

The kidneys are wonderfully efficient organs, but they are not a fail-safe protection against toxic chemicals: they cannot rid the body of all toxic chemicals it may encounter. Not only are the kidneys unable to remove some toxic substances, they themselves can be adversely affected by a variety of environmental contaminants. For example, most heavy metals, such as mercury and cadmium, are potent nephrotoxins—toxic chemicals that affect the nephrons directly. Even relatively low doses of heavy metals can be harmful to the kidneys.

Exposure to low levels of various heavy metals increases urine output and increases the levels of amino acids and glucose in the urine—good indicators that tubular reabsorption is not working optimally. At higher levels, heavy metals do considerable damage and can result in renal failure and death.

One of the portions of the nephron most sensitive to heavy metals is the proximal convoluted tubule. The cells of the proximal convoluted tubule are easily destroyed by toxic heavy metals, which disrupts normal kidney function.

The kidney, however, has several protective mechanisms to reduce the impact of heavy metals. Lysosomes inside the cells of the renal tubule, for example, bind to and engulf heavy metals, helping to reduce the cytoplasmic concentration of these harmful substances. As levels increase, however, the protective mechanisms are overwhelmed, and kidney cells begin to die.

The heavy metal mercury is often found in the water we drink and the food, especially the fish, we eat. In high concentrations in the blood, mercury produces acute renal failure. Mercury causes vasoconstriction of the afferent arterioles, reducing the flow of blood into the glomerulus. Numerous signs of cellular deterioration are also observed in the kidney following mercury poisoning.

Mercury is one of the more common water pollutants in the industrialized world. It is a by-product of the production of the plastic polyvinyl chloride (commonly called PVC plastic), which is often used to manufacture children's toys, beach balls, car seats, and other products. Mercury is also emitted into the waterways by a variety of chemical manufacturers and is released by coal-fired power plants and garbage incinerators.

In the 1950s, Japan announced an outbreak of mercury poisoning in residents who consumed fish taken from Minimata Bay, which had been contaminated by a nearby plastics factory. Over 100 people were affected. Victims developed numbness of the limbs, lips, and tongue. Many lost muscular control and became clumsy. Most suffered from blurred vision, deafness, and mental derangement. All told, 17 people died in the tragedy, and 23 were permanently disabled, in large part due to nervous system effects and the inability of the kidney to cope with the toxic metal. The tragedy was deepened by the discovery of birth defects in 19 babies born to women who had eaten contaminated fish. Many of their mothers showed no signs of mercury poisoning, an example of how low levels of a toxin can impact the developing embryo. As a rule, toxicologists note that the younger an organism, the more sensitive it is to harmful toxic substances.

The Minimata Bay tragedy was not an isolated event. A similar tragedy occurred on the Japanese island of Honshu. In Sweden in the early 1960s, mercury poisoning killed large numbers of birds that had been feeding on seeds treated with a mercury fungicide, a coating that retards mildew. Swedes who ate pheasant and other birds were also poisoned by mercury.

These incidents illustrate the importance of a clean environment to our health and have helped many governments recognize the need for tighter controls on hazardous materials produced by factories. Despite this realization, problems still remain.

SUMMARY

THE URINARY SYSTEM

1. Human body cells produce enormous amounts of waste, which are excreted by several organs. One of the most important is the kidney.

2. Kidneys remove impurities from the blood and help regulate the water levels and concentrations of nutrients and ions.

3. Blood enters the kidneys in the renal arteries and is delivered to the millions of filters, the nephrons, found in the cortex and (to a lesser extent) the medulla.

4. The nephrons produce urine, which drains from the kidneys into the ureters, slender muscular tubes that lead to the urinary bladder. Urine is stored in the bladder, then voided through the urethra.

5. The kidneys contain two distinct regions, the outer cortex and the inner medulla. Urine is drained into the hollow renal pelvis before entering the ureters.

6. Each nephron consists of a glomerulus, a tuft of highly porous capillaries, and a renal tubule, a long tubule where urine is produced.

7. The renal tubule consists of four parts: (a) Bowman's capsule, (b) the proximal convoluted tubule, (c) the loop of Henle, and (d) the distal convoluted tubule. The distal convoluted tubules of nephrons drain into collecting tubule ducts, which converge and empty urine into the renal pelvis.

THE FUNCTION OF THE URINARY SYSTEM

8. Blood filtration is accomplished by three processes: glomerular filtration, tubular reabsorption, and tubular secretion.

9. Glomerular filtration occurs in the glomerulus, a knot of capillaries that filters about 20% of the blood's plasma into Bowman's space. The resulting liquid is called the filtrate.

10. The filtrate is processed as it flows along the renal tubule. Water, ions, and nutrients are largely reabsorbed as they travel along the length of the tubule. This process is called tubular reabsorption. What is left is a concentrated liquid, the urine, containing mostly waste products. Water and reabsorbed nutrients and ions pass into a network of capillaries, the peritubular capillaries, surrounding each nephron.

11. Not all wastes leave the blood in the glomerulus. Some must pass into the nephron from the peritubular capillaries in the process of tubular secretion. Hydrogen and potassium ions are secreted into the renal tubule.

12. Calcium, magnesium, and other materials can precipitate out of the urine in the renal pelvis, forming kidney stones, which can block the outflow of urine. Smaller stones may be passed along the ureters to the bladder and are often eliminated during urination.

13. Urination is a reflex in babies and very young children.

When the bladder fills with urine, stretch receptors begin to fire. Nerve impulses from the stretch receptors stimulate nerve cells in the spinal cord to fire; these cells, in turn, send impulses to the smooth muscle cells in the wall of the bladder. Nerve impulses to the bladder cause the walls to contract and also cause the internal sphincter to relax, allowing urine to flow out of the bladder.

14. In older children and adults, the urination reflex still operates, but is overridden by a conscious control mechanism. Nerve impulses from the brain must open the external sphincter to allow urine to flow out of the bladder.

THE KIDNEYS AS ORGANS OF HOMEOSTASIS

15. Each drop of blood in your body flows through the kidneys many times in a single day. This allows for a thorough filtering of the blood and also helps the body control the chemical composition of the blood.

16. Water and the levels of dissolved chemicals in the blood are controlled by two hormones: ADH and aldosterone.

17. ADH is secreted by the posterior lobe of the pituitary gland. ADH is released when the osmotic concentration of the blood increases or when blood volume decreases.

18. ADH increases the permeability of the distal convoluted tubules and the collecting tubules to water. When ADH is present, water reabsorption increases. A lack of ADH causes diabetes insipidus, an overexcretion of urine.

19. Aldosterone is produced by the adrenal cortex. This hormone is regulated by blood pressure, blood volume, and osmotic concentration. When blood pressure falls, aldosterone is released.

20. Aldosterone stimulates the reabsorption of sodium ions by the nephron. Water follows the sodium out of the renal tubule, increasing blood pressure and blood volume.

ENVIRONMENT AND HEALTH: MERCURY POISONING

21. The kidneys help regulate the levels of harmful toxins produced by the body and help eliminate toxins taken into the body from air, water, and food. However, they are not immune to the many potentially harmful substances. Heavy metals, for example, destroy some of the cells of the renal tubule and can restrict blood flow to the glomeruli. As a result, they can damage the kidney, impairing the function of this important homeostatic organ.

After reading the following hypothetical scenario, you will be asked to make a decision. After you have made your decision, you will be asked some questions that may help you begin to clarify your values. Clarifying your values will help you understand your own biases, which should help in future decisions.

Here's the scenario: You are a state legislator considering legislation on prioritizing medical expenditures. Your sub-committee will make a recommendation to the legislature to adopt or reject a plan that would shift state funding from an organ transplant program to a prenatal care program. The state currently pays for organ transplants for needy families, spending over $5 million per year on this program. A group of legislators, however, is proposing that this money be used to fund a prenatal care program, which would offer free checkups for pregnant women as well as advice on drug and alcohol use during pregnancy. The program would also offer advice on maternal and infant nutrition.

The proponents of medical care prioritization say that the money currently required for one organ transplant could fund prenatal care for about a thousand mothers. By spending the money on prenatal care, the state could reach thousands of pregnant women who are too poor to see a doctor. Proponents also estimate that 20% of all newborns in your state are born addicted to cocaine, which affects

mental and physical development.

Opponents of the bill point out that if needy families are denied money for organ transplants, dozens of children will die. They present the case of Jason Lowry to illustrate what will happen. Jason Lowry is 12 years old. His family lives on welfare. Jason needs a liver transplant, which will cost $200,000. Without it, the boy is certain to die. If the state chooses to fund prenatal care instead of organ transplants, dozens like him will die each year.

You have a choice. Would you recommend the bill that transfers funding to prenatal care or continue funding organ transplants? Why? Make a list of reasons why you supported or opposed the bill.

Now take a moment to ponder your reasons. Was your decision based on economics? Was your decision based on relative benefits—that is, the benefits of a few versus the benefits of many? Was your decision based on benefits to future generations? Or were you mostly concerned with immediate effects—for example, saving a few lives now?

Imagine, if you will, that Jason Lowry was your son. How would this affect your decision? Does the issue take on a different meaning? What general observations could you make about your objectivity? Did it change as the issue came "closer to home"? To what extent do personal interests affect your decisions about other issues, such as environmental issues? Give some examples.

TEST OF TERMS

1. The kidneys are the filtering organs of the _____ system. Blood enters each of the kidneys through the _____ arteries.

2. Urine drains into the urinary bladder from the kidneys through two slender muscular tubes, the _____ .

3. Urine drains from the urinary bladder through the _____ to the outside of the body.

4. The outermost region of the kidney is called the _____ .

5. Urine produced in the kidney empties into the _____ _____ , the hollow chamber inside the kidney.

6. The nephron is the filtering unit of the kidney. It consists of two parts, the _____ , a tuft of capillaries, and the _____ _____ , a long twisted tubule.

7. The nephron's tuft of capillaries is supplied by the _____ arteriole. These capillaries are highly porous and allow much of the water and dissolved substances in the blood to enter _____ _____ , the cup-shaped portion of the nephron.

8. The highly branched cells surrounding the capillaries of the nephron are called _____ .

9. The capillary network surrounding much of the nephron, into which water and dissolved substances are

reabsorbed, is called the _____ _____ . The movement of materials from these capillaries into the nephron is called _____ _____ .

10. The band of smooth muscle at the neck of the urinary bladder that is under reflex control is called the _____ _____ .

11. Periodic filtering of the blood by an artificial filter is called _____ .

12. Alcohol inhibits the release of a hormone, _____ _____ , from the posterior lobe of the pituitary. This hormone _____ the permeability of the distal convoluted tubules and collecting tubules.

13. The adrenal glands produce another hormone called _____ ,

which affects the reabsorption of sodium ions and therefore helps control ionic balance and water levels in the body.

14. Mercury is a heavy metal that in high levels can destroy cells of the nephron. It is, therefore, called a _____ .

Answers to the Test of Terms are located in Appendix B.

TEST OF CONCEPTS

1. Draw the various parts of the urinary system and describe what each one does.
2. Trace the flow of blood into and out of the kidney. Be sure to include details of the pathway once it reaches the afferent arteriole.
3. Describe the three ways in which the kidney filters the blood.
4. A drug inhibits the uptake of water by the distal convoluted tubules and collecting tubules. What effect would this have on urine output, urine concentration, blood pressure, blood volume, and the concentration of the blood?
5. Describe how ADH controls blood pressure and the water balance in the body. Describe the hormonal and physiological changes in the nephron that take place when excess liquid is ingested. Do the same for dehydration.
6. Urination in adults is a reflex with a conscious override. Explain what this statement means. Also describe the changes that occur in urination as a child grows older.
7. Aldosterone helps regulate blood pressure and water content. In what ways is this hormone different from ADH?
8. You have just finished your family medicine residency. A patient comes to your office complaining that he drinks water all day long and spends much of the rest of the day in the bathroom urinating. What tests would you order? What diagnosis would you suspect?

SUGGESTED READINGS

Sherwood, L. 1989. *Human physiology: From cells to systems*. St. Paul, Minn.: West. Detailed coverage of the urinary system.

Spence, A. P. 1986. *Basic human anatomy*, 2d ed. Menlo Park, Calif.: Benjamin/Cummings. Detailed coverage of the structure of the human urinary system.

Vander, A., J. Sherman, and D. Luciano. 1985. *Human physiology: The mechanisms of body function*, 4th ed. New York: McGraw-Hill. Excellent reading for students wanting to delve deeper into renal physiology.

Integration, Coordination, and Control: The Nervous System

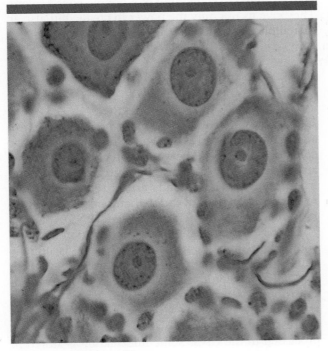

Neurons in spinal ganglia.

Ernest Hemingway grew up in Oak Park, Illinois, a suburb of Chicago. Hemingway never went to college, but instead worked at an odd assortment of jobs. At one time, he even served as a sparring partner for boxers. Hemingway eventually landed a job as a reporter for the Kansas City *Star,* but when World War I broke out, he left to work in an American ambulance unit.

Hemingway remained in Europe after the war and began his writing career, which eventually would earn him the Nobel Prize in literature. Despite his success and widespread popularity, Hemingway committed suicide. What caused him to take such a drastic measure no one knows, but some believe that Hemingway was suffering from a rare, but painful nervous system disorder, called **trigeminal neuralgia**. This disease results in periodic and unexplained flashes of intense pain along the course of the **trigeminal nerve**, which supplies the face (Figure 12–1). A slight breeze or the pressure of a razor can set off the pain, which lasts a minute or more. In some patients, the pain will recur for no apparent reason every few minutes for weeks on end. Some observers believe that the pain

FIGURE 12–1 Ernest Hemingway This master of the written word may have suffered from trigeminal neuralgia, a painful disorder of the nervous system.

Ernest Hemingway felt became unbearable. Combined with personal conflict, it caused the writer to take his life.

Today, physicians can treat this rare disease with drugs. In extreme cases, surgeons must cut the trigeminal nerve as it leaves the brain. This procedure, while effective, leaves half of the victim's face and half of the tongue and oral cavity numb—a little like the feeling you get with novocaine.

This chapter describes the anatomy and physiology of the nervous tissue and examines the ways in which the nervous system helps to regulate homeostasis.

AN OVERVIEW OF THE NERVOUS SYSTEM

The nervous system is a governing body. It exerts control over our muscles, glands, and organs. It controls the heartbeat, breathing, digestion, and urination. The nervous system also regulates blood flow and helps control the osmotic concentration of the blood. In its capacity as a governor of body functions, the nervous system, like any governing body, receives input from a large number of sources. Input from the body helps the nervous system govern in much the same way that citizen input (letters and phone calls) helps elected officials make important decisions.

The human nervous system provides a great many functions not seen in other species. For example, the human nervous system is the place where ideas arise; thus, it allows us to think about and plan for the future. The nervous system is the site of dreams and imagination, and it is the site of reasoning, allowing us to judge right from wrong, logical from illogical. The nervous system is also a repository of information.

This marvelous network of cells allows us to manipulate our environment for our own purposes. Like few other species alive today, we are reshaping the Earth. We topple forests to make pastures, drain swamps for homes and farms, split atoms to generate energy, and catapult men and women into outer space. Joan McIntyre, an author and critic, once wrote, "The ability of our minds to imagine, coupled with the ability of our hands to devise our images, brings us a power almost beyond control." Today, however, humankind has begun to realize that the profound alterations we have made in the planet threaten our long-term future (Chapter 21). Our brains, then, are something of a double-edged sword. They give us not only the power to create, but also an incredible power to destroy.

The Central and Peripheral Nervous Systems

The nervous system consists of the brain, spinal cord, and nerves (Figure 12–2). The brain and spinal cord constitute the **central nervous system (CNS)**. **Nerves** are bundles of fibers that transport messages to and from the CNS. Nerves and various receptors, which respond to a variety of internal and external stimuli, form the **peripheral nervous system (PNS)**.

The CNS receives all sensory information from the body. Right now, for instance, your CNS is literally being bombarded with sensory impulses from receptors. These receptors alert you to the room temperature, the presence of wind, traffic sounds, and the touch of the page. They transmit visual images of figures on the pages and words. This information is processed in the CNS. Some information is stored in memory. Some may be ignored or blocked. Some stimuli elicit physiological responses. A particularly

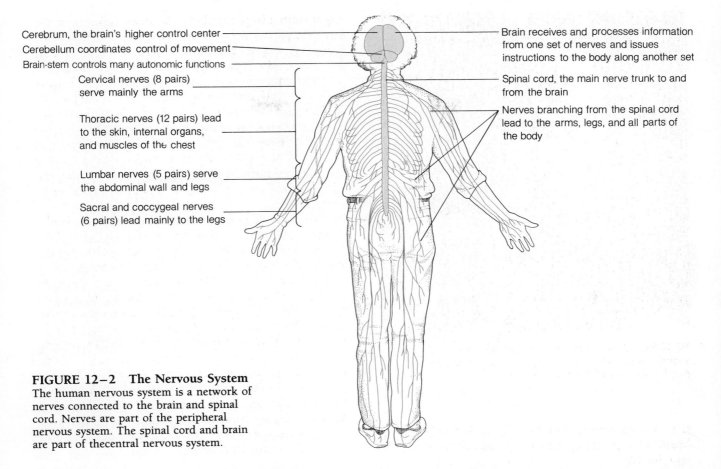

Cerebrum, the brain's higher control center

Cerebellum coordinates control of movement

Brain-stem controls many autonomic functions

Cervical nerves (8 pairs) serve mainly the arms

Thoracic nerves (12 pairs) lead to the skin, internal organs, and muscles of the chest

Lumbar nerves (5 pairs) serve the abdominal wall and legs

Sacral and coccygeal nerves (6 pairs) lead mainly to the legs

Brain receives and processes information from one set of nerves and issues instructions to the body along another set

Spinal cord, the main nerve trunk to and from the brain

Nerves branching from the spinal cord lead to the arms, legs, and all parts of the body

FIGURE 12–2 The Nervous System
The human nervous system is a network of nerves connected to the brain and spinal cord. Nerves are part of the peripheral nervous system. The spinal cord and brain are part of thecentral nervous system.

exciting section you read, for example, might accelerate your heart rate. A frightening thought might cause you to cringe.

The brain responds to the stimuli by sending nerve impulses to effectors (muscles and glands) along your nerves. Nerves, therefore, carry two kinds of information: (1) sensory impulses that travel *to* the CNS from sensory receptors in the body, and (2) motor impulses that travel *away from* the CNS to glands, organs, and muscles.

Sensory information pouring into the CNS is integrated with information stored in memory. Thus, a new fact may trigger memories of previous knowledge, causing you to think about a problem in a new way. Memory also influences the way we respond to stimuli. A pet cat brushing against your leg, for example, elicits a smile. The sensation is not startling because your memory reminds you of the cat's presence.

In summary, then, the CNS receives all sensory information, which it integrates and to which it often responds. The PNS carries sensory information to and from the CNS.

Functional Divisions of the Nervous System

The CNS and PNS are the two main divisions of the nervous system. The peripheral nervous system can also be divided into two parts: the somatic nervous system and the autonomic nervous system. The **somatic nervous system** is that portion of the PNS that controls voluntary functions, such as muscle contractions that lead to the movement of the limbs. That part of the PNS that controls involuntary function, such as heart rate, is called the **autonomic nervous system** (Figure 12–3). Breathing and digestion are also under the control of the autonomic nervous system.[1] Many other body functions are under autonomic control and participate in feedback loops.

[1]Note that breathing can be controlled voluntarily, but for the most part breathing is an involuntary action. It occurs without conscious control.

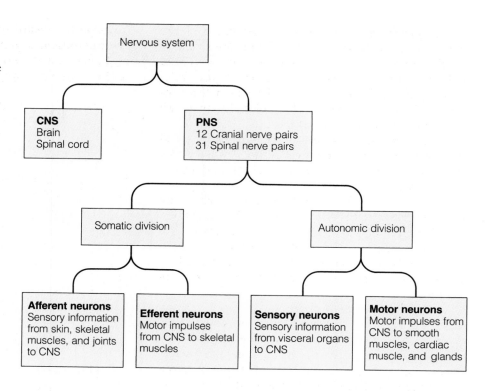

FIGURE 12-3 Subdivisions of the Nervous System The activities of the autonomic and somatic divisions often overlap. The nervous system is divided into two parts, the CNS and PNS. The PNS consists of autonomic and somatic divisions.

THE STRUCTURE AND FUNCTION OF THE NEURON

The fundamental unit of the nervous system is the **neuron**, a highly specialized cell that generates and transmits bioelectric impulses from one part of the body to another. Neurons come in a variety of shapes and sizes. Despite these differences, neurons share several common characteristics.

The Anatomy of the Neuron

All neurons, for example, consist of a more or less spherical central portion, called the cell body (Figure 12-4). The **cell body** houses the nucleus, most of the cell's cytoplasm, and numerous organelles. Metabolic activities in the cell body sustain the cell, providing energy and synthesizing materials necessary for proper cell function. Two organelles in the cell body of particular interest are microtubules and microfilaments; they form the cytoskeleton of the neuron and are responsible for the cell's characteristic shape.

All neurons contain two types of cellular processes that extend from their cell bodies: dendrites and axons. **Dendrites** vary in length, depending on the type of neuron, and transmit impulses to the cell body, as

indicated by the arrows in Figure 12-4b. Also attached to the cell body of the neuron is the axon. **Axons** carry impulses away from the cell body of the neuron. Axons occasionally also give off side branches, **axon collaterals** (Figure 12-4b). When an axon reaches its destination, it often branches profusely, giving off many small fibers. These fibers terminate in tiny swellings called **terminal boutons** (end buttons) or **terminal bulbs**. Nerve impulses reaching the terminal ends of axons may be transmitted to other neurons, muscle fibers, or glands.

The axons of many neurons are coated with a protective layer called the **myelin sheath** (Figure 12-5a). The myelin sheath is formed by a nonconducting cell found in the nervous system. These cells, called **glial cells** (glue), play a supporting role in the nervous system. In the PNS, the cells that form the myelin sheath are called **Schwann cells**, after the early German cytologist, Theodore Schwann. During embryonic development, Schwann cells attach to the growing axon, then begin to encircle it (Figure 12-5d). As they do, they leave behind a trail of plasma membrane, which wraps around the axon, forming many concentric layers.

As Figure 12-5 shows, many Schwann cells align themselves along the length of the axon, forming the myelin sheath. Since the plasma membrane of the Schwann cell is about 80% lipid, the myelin sheath is mostly fat and appears glistening white when viewed with the naked eye. As shown in the figure, each segment of the sheath is separated by a small indentation, the **node of Ranvier**. The nodes of Ranvier are tiny, unmyelinated patches (Figure 12–5b).

The myelin sheath permits nerve impulses to travel quickly down an axon (Figure 12–5a). Unmyelinated axons are also found in the body. They conduct impulses much more slowly.

As Figure 12–6 illustrates, unmyelinated axons are also associated with Schwann cells. The Schwann cells, however, do not wrap around the axons to produce concentric layers of myelin. As shown in Figure 12–6b, the unmyelinated axons are merely embedded in the Schwann cell, which provides support and protection.

Inside all axons are bundles of microtubules, which act like rails, transporting materials produced in the cell body down the axon. The microtubules also play a key role in the development of axons. During embryonic development, nerve cells start out as fairly unremarkable round cells. Housed in the developing brain and spinal cord, these cells undergo a remarkable transformation. Axons form as microtu-

bules push outward against the plasma membrane. As the microtubules grow, the axons migrate from the brain and spinal cord to the muscles and glands they will eventually supply. Axons inside the central ner-

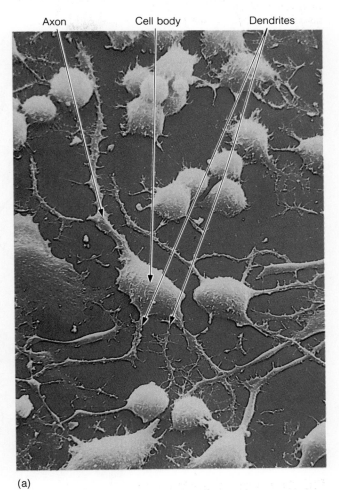

(a)

FIGURE 12–4 A Neuron (a) An electron micrograph of the cell body and dendrites of a multipolar neuron. The multipolar neuron resides within the central nervous system. Its multiangular cell body has several highly branched dendrites and one long axon. (b) Collateral branches may occur along the length of the axon. When the axon terminates, it branches many times, ending on individual muscle fibers.

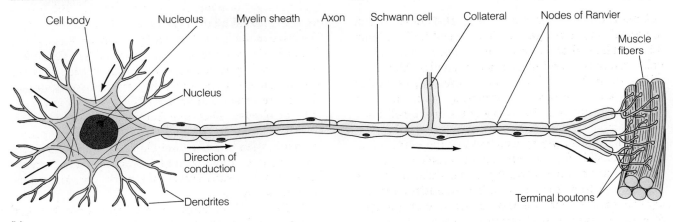

(b)

FIGURE 12–5 The Myelin Sheath and Saltatory Conduction (*a*) The myelin sheath allows impulses "to jump" from node to node, greatly accelerating the rate of transmission. (*b*) A drawing showing the arrangement of Schwann cell membrane in the myelin sheath. (*c*) A transmission electron micrograph of the myelin sheath. (*d*) A drawing showing how the myelin sheath is formed.

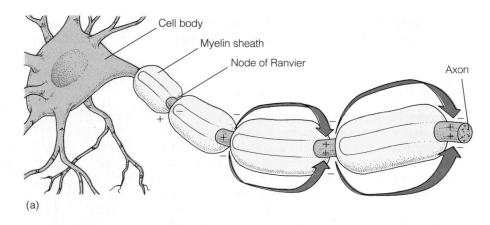

Cell body
Myelin sheath
Node of Ranvier
Axon

(a)

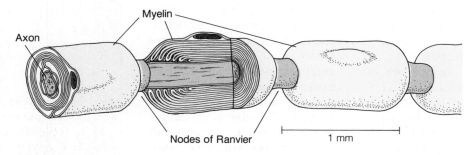

Myelin
Axon
Nodes of Ranvier
1 mm

(b)

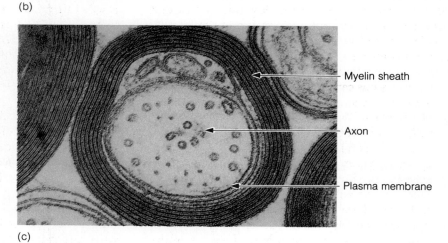

Myelin sheath
Axon
Plasma membrane

(c)

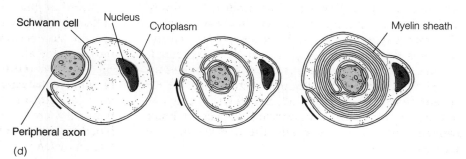

Schwann cell
Nucleus
Cytoplasm
Myelin sheath
Peripheral axon

(d)

vous system (brain and spinal cord) also migrate to other regions within the CNS, connecting its parts.

Special Characteristics of Neurons

Like many cells that have diverged anatomically from their embryonic origins during cellular differentiation, nerve cells lose the ability to divide. Therefore, a nerve cell that dies cannot be replaced by cell division of other nerve cells. As a result, damage to nerve cells often results in permanent impairment. A severed nerve in the arm ends the transmission of sensory information to the brain and spinal cord if the nerve cell dies. A person is left with a numb feeling in the skin the nerve once innervated (supplied). Severing a nerve can also eliminate muscle control if the nerve contained motor fibers from the CNS.

In humans, some regeneration of axons is possible. As Figure 12-7 illustrates, an axon that is severed degenerates from the point of injury to the muscle or gland it supplied. The segment of the axon still attached to the cell body may regenerate, however, as long as the cell does not die. The remaining axon elongates, traveling along the hollow tunnel provided by the myelin sheath. An axon may grow to its previous length and reestablish connections with muscles it previously supplied, providing partial or even nearly complete control.

Neurosurgeons can facilitate neuronal regeneration by microsurgery, surgery performed under a dissecting microscope—like the ones you may have used to dissect frogs in your high school biology class. Surgeons sew the severed ends of nerves together and line up the empty myelin sheaths with the regenerating nerve fibers, thus facilitating axonal regrowth.

Besides being unable to divide, nerve cells have an extraordinarily high metabolic rate. Like teenagers, they require considerable input of nutrients. The brain also requires a constant supply of oxygen. If the amount of oxygen flowing to the brain is drastically reduced, neurons begin to die within a few minutes. To prevent brain damage from occurring in someone who has drowned or been accidentally electrocuted, rescuers must start resuscitating the victim within three to five minutes. Although victims may be revived after this crucial period, the lack of oxygen in the brain often results in varying degrees of brain damage. Generally, the longer the deprivation, the greater the damage.

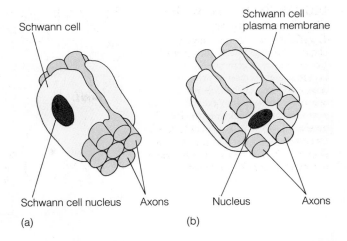

FIGURE 12-6 Unmyelinated Axons (*a*) Schwann cells encompass groups of axons, but do not produce myelin sheaths. (*b*) Axons may be embedded individually as well.

There is one exception to the five-minute rule, however. If a person drowns in icy water, resuscitation may be successful if begun within the hour. A victim will often recover without brain damage. In Fargo, North Dakota, a young boy was under water for five hours before he was recovered and resuscitated. The boy survived the incident with apparently no ill effects. Cold water slows metabolism in the brain, thus reducing oxygen demand considerably. As a result, brain cells are preserved, and brain damage is minimized or avoided.

Nerve cells are also highly dependent on glucose for energy production. Unlike other cells, which can use fatty acids to generate energy, the nerve cell relies exclusively on glucose. When blood glucose levels fall, nerve cells are first to suffer. A decline in blood glucose is normally prevented by homeostatic mechanisms as described in Chapter 7. In diabetics, however, blood glucose levels can fall dangerously low—when a diabetic takes too much insulin, for example. Deprived of glucose, brain cells begin to falter. Individuals may become dizzy and weak. Vision may blur. Speech may become awkward. The person may be mistaken for a drunk. Low blood glucose (hypoglycemia) often results in headaches. People can even become uncharacteristically aggressive. In some cases, low blood sugar triggers convulsions and death. Other body cells are not adversely affected by a decline in blood glucose because they switch to alternative fuels, fats and even proteins.

FIGURE 12–7 Axonal Regeneration (*a*) A severed axon can regenerate in the peripheral nervous system. The segment from the cut to the effector organ degenerates. (*b*) The myelin sheath remains, providing a tunnel (*c*) through which the axonal stub can regrow, often reestablishing previous contacts and restoring motor function.

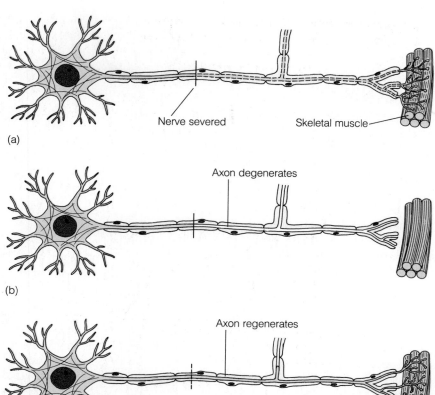

(a) Nerve severed — Skeletal muscle

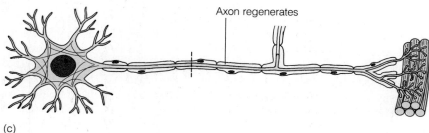

(b) Axon degenerates

(c) Axon regenerates

The Bioelectric Impulse: How Nerve Cells Work

Nerve cells transmit bioelectric impulses from one part of the body to another. These impulses keep us aware of our internal and external environments and help us make the many thousands of adjustments needed to survive in an ever-changing world.

Nerve impulses are not like the electric impulses sent along a wire. The electrical current running your computer, for example, is formed by the flow of electrons inside the wire. In contrast, nerve impulses are small ionic changes in the membrane of the neuron that move along the nerve cell like ocean waves racing toward the shore.

To understand the nerve cell impulse or **bioelectric impulse**, so named to distinguish it from electricity, we begin by looking at the plasma membrane of a neuron. If you placed a tiny electrode on the outside of the plasma membrane of a neuron and inserted another on the inside of the membrane and hooked them up to a voltmeter, a device that measures voltage, you would be able to measure a small voltage, much like that measured in a battery. In simplest terms, **voltage** is a measure of the tendency of charged particles to flow from one pole of the battery to the other. The higher the voltage, the greater the tendency for electrons to flow when a wire is connected between the poles.

In the nerve cell, however, electrons do not flow from one side of the membrane to another, ions do. Sodium ions are among the most important ones. The potential difference, therefore, is a measure of the force that will drive sodium ions from one side of the membrane to the other. For now, however, it is important just to remember that a small voltage exists across the plasma membrane of the neuron. It is so small in fact that it is measured in millivolts. A millivolt is $1/1000$ of a volt.

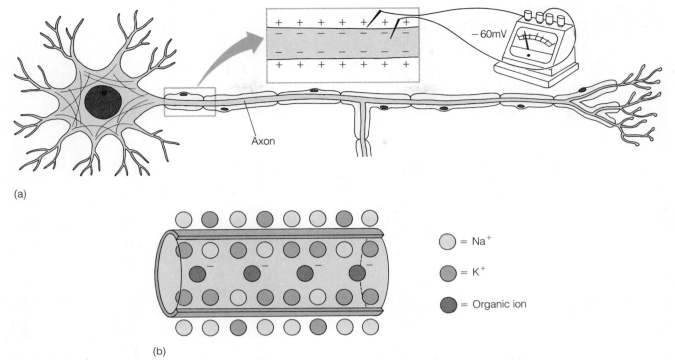

(a)

(b)

FIGURE 12–8 Resting Potential (*a*) Electrodes placed on either side of the plasma membrane of a neuron measure a tiny potential difference, roughly −60 millivolts (mV). This is the resting potential. (*b*) It results from a chemical disequilibrium caused by the active transport of sodium (Na) ions out of the neuron and potassium (K) ions into the cytoplasm.

As shown in Figure 12–8a, the potential difference in a neuron is −60 millivolts. It is called the **membrane** or **resting potential**, for it is the membrane potential of a cell at rest. The minus sign is added because the plasma membrane is positively charged on the outside and negatively charged on the inside (Figure 12–8b). The outside is positive because the concentration of sodium ions is higher on the outside (in the tissue fluid surrounding the neuron) than it is on the inside (in the cytoplasm). The interior of the cell is negative because the concentration of large, negatively charged organic molecules inside the cell is much higher than outside.

Nerve cells expend energy to maintain this concentration imbalance. Energy is used to operate active transport pumps in the membrane that transport sodium ions out of the cytoplasm into the surrounding fluid, thus maintaining the external concentration. Negatively charged ions might be expected to follow the sodium ions to maintain electrical neutrality—to balance the positively and negatively charged ions—but many negatively charged ions are large molecules, such as proteins, that cannot escape. Because sodium can only leak back very slowly and the negatively charged ions cannot leak out, a charge difference is maintained.

The plasma membrane of a nerve cell is like a loaded gun. It has a built-up charge of sodium ions on the outside. When the nerve is stimulated, the membrane undergoes a rapid change to discharge the load. The first change occurring in the membrane is a rapid increase in its permeability to sodium ions. Neurophysiologists believe that stimuli open protein pores through which sodium enters, letting positively charged sodium ions flow into the cell.

Electrodes implanted in a nerve cell would detect the sudden influx of sodium ions and would register a shift in the resting potential from −60 millivolts to +40 millivolts. The change in voltage occurs at the site of stimulation and is called **depolarization**.[2] Im-

[2]Depolarization means that the membrane loses its previous polarization.

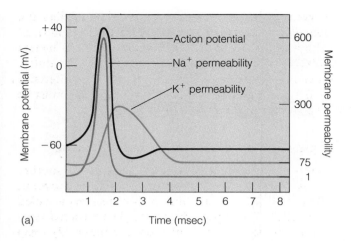

(a)

Time (msec)

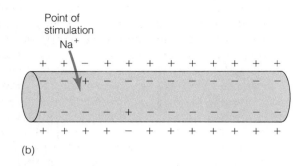

(b)

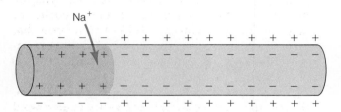

Depolarization and generation of the action potential

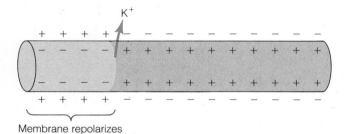

(d)

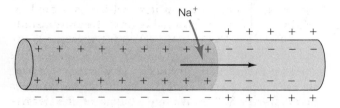

Propagation of the action potential

(c)

FIGURE 12–9   Action Potential   (*a*) Stimulating the neuron creates a bioelectric impulse, which is recorded as an action potential. The resting potential shifts from -60 millivolts to +40 millivolts. The membrane is said to be depolarized. This graph shows the shift in potential and the change in sodium (Na$^+$) and potassium (K$^+$) ion permeability, which is largely responsible for the action potential. (*b*) This drawing shows the influx of sodium ions and the depolarization that occur at the point of stimulation. (*c*) The impulse travels along the membrane as a wave of depolarization. (*d*) The efflux of potassium ions restores the resting potential, allowing the neuron to transmit additional impulses almost immediately.

mediately after depolarization, the membrane returns to its previous state, a process called **repolarization.**

A graphical tracing of the electrical change across the plasma membrane is shown in Figure 12–9a and is called an **action potential.** This graph shows (1) the brief upswing, depolarization, as the voltage goes from −60 millivolts to +40 millivolts, and (2) the rapid downswing, repolarization, which is the return to the resting potential.

The action potential occurs so rapidly and the membrane returns to the resting state so quickly (about ³⁄₁₀₀₀ of a second) that the nerve can be stimulated in rapid succession. Such brisk recovery allows us to respond swiftly and forcefully to danger and to perform rapid muscle movements. A nerve cell can also transmit many impulses in sequence because only a small number of sodium ions are exchanged with each impulse.

During repolarization, the membrane shifts from +40 millivolts to −60 millivolts. Repolarization results from (1) a sudden decrease in the membrane's permeability to sodium ions, which stops the influx of sodium ions, and (2) a rapid efflux of positively charged potassium ions (Figure 12–9d).[3] Both of these changes help reestablish the resting potential.

After repolarization, the neuron reestablishes sodium and potassium ion concentrations; that is, it pumps sodium out of the axon and pumps potassium ions back in—all the while maintaining the resting potential. The plasma membranes of nerve cells contain numerous sodium-potassium pumps that actively transport sodium out of and potassium into the cell. The sodium-potassium active transport pumps, therefore, help neurons reestablish the chemical disequilibrium necessary for normal nerve cell function.

Conducting the Nerve Impulse along the Neuron

Stimulating a nerve cell artificially at one point creates a bioelectric impulse caused by an inward rush of sodium ions, which dramatically shifts the resting potential. The influx of sodium ions is followed by an outward rush of potassium ions that returns the membrane to normal. How does the impulse travel along the nerve cell?

Many studies of the nerve cell have been made using the giant unmyelinated axons of the squid. These studies show that a change in membrane permeability in the stimulated region, which results in depolarization, causes a change in the sodium and potassium permeability of neighboring regions. In other words, depolarization in one region of the plasma membrane of an axon stimulates depolarization in adjacent regions, a process that continues down the length of the axon.

In unmyelinated fibers, nerve impulses travel like waves from one region to the next. In a myelinated fiber, however, the depolarization appears to "jump" from one node of Ranvier (the section between adjacent Schwann cells) to another. This movement is called **saltatory conduction** (from the Latin word *saltare,* meaning to jump) and is shown in Figure 12–5a. The "skipping" of the impulse from node to node greatly increases the rate of transmission. A

nerve impulse travels along an unmyelinated fiber at a rate of 0.5 meters per second (1.5 feet per second). In a myelinated nerve, the impulse travels 200 meters (600 feet) per second—400 times faster. The difference in the rate of transmission is due to a difference in the total amount of axonal membrane that must be depolarized and repolarized.

Transmitting the Impulse: The Synapse

Nerve impulses travel from one neuron to another across a small space that separates them, as shown in Figure 12–10. The juncture of two neurons is called a synapse. A **synapse** consists of (1) a terminal bouton (or some other kind of axon terminus), (2) a gap between the adjoining neurons, called the **synaptic cleft**, and (3) the membrane of the dendrite or postsynaptic cell (Figure 12–10b). The neuron that transmits the impulse to another is called the **presynaptic neuron**; the one that receives the impulse is called the **postsynaptic neuron**.

When a nerve impulse reaches a terminal bouton, depolarization of the plasma membrane causes a rapid influx of calcium ions into the terminal bouton. Calcium ions cause the release of a chemical substance, known as a **neurotransmitter**, which is stored in small vesicles in the terminal bouton. Neurotransmitters diffuse across the synaptic cleft between adjoining nerve cells and bind to receptors in the plasma membrane of the postsynaptic (receiving) neuron. The binding of neurotransmitter to the postsynaptic membrane stimulates a rapid change in the permeability of the membrane of the postsynaptic cell to sodium ions. Sodium channels open, and sodium ions flow inward, causing a localized depolarization.

Neurotransmitters are produced and packaged in vesicles in the cell body, then transported down the axon along the microtubules to the terminal bouton, where they are stored until needed. When the bioelectric impulse arrives, the vesicles bind to the presynaptic membrane and release the neurotransmitter chemicals (by exocytosis) into the synaptic cleft.

A variety of neurotransmitters are present in the body. One of the best understood is a chemical called acetylcholine. **Acetylcholine** is released by nerves that stimulate muscle contraction. In some synapses, the neurotransmitter stimulates an action potential in the postsynaptic neuron. This synapse is, therefore, called an **excitatory synapse**. In other synapses, however, the binding of neurotransmitter may inhibit sodium influx. This makes the resting potential more negative

[3]The permeability of the membrane to potassium is normally low, but increases when the cell is stimulated; this allows potassium ions to move out quickly to help restore the resting potential.

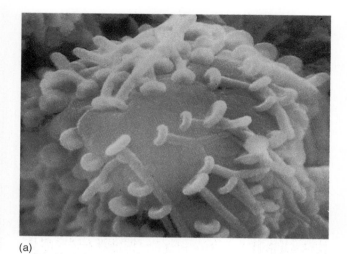

(a)

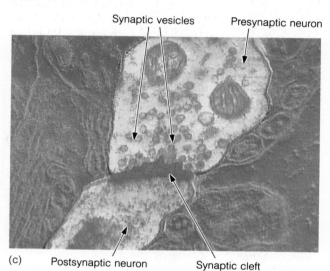

Synaptic vesicles   Presynaptic neuron

(c)   Postsynaptic neuron   Synaptic cleft

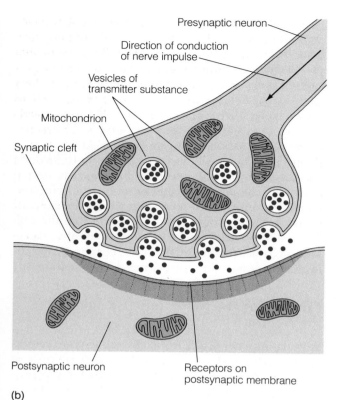

Presynaptic neuron

Direction of conduction of nerve impulse

Vesicles of transmitter substance

Mitochondrion

Synaptic cleft

Postsynaptic neuron

Receptors on postsynaptic membrane

(b)

FIGURE 12–10   The Terminal Bouton and Synaptic Transmission   (*a*) A scanning electron micrograph showing the terminal boutons of an axon ending on the cell body of another neuron. (*b*) The arrival of the impulse stimulates the release of neurotransmitter held in membrane-bound vesicles in the axon terminals. Neurotransmitter diffuses across the synaptic cleft and binds to the postsynaptic membrane where it elicits another action potential that travels down the dendrite to the cell body. (*c*) A transmission electron micrograph showing the details of the synapse.

and less excitable. This kind of synapse is called an **inhibitory synapse.**

Transmission across the synapse is remarkably fast, requiring only about 1/1000 of a second. Synaptic transmission is also a transitory event. A short burst of neurotransmitter is released each time an impulse reaches the terminal bouton. What prevents the neurotransmitter from remaining in the synaptic cleft where it might restimulate the postsynaptic membrane? At least three mechanisms remove neurotransmitter from the synaptic cleft, ensuring that there is no lingering aftereffect: (1) enzymes, (2) reabsorption by the terminal bouton, and (3) diffusion. In some synapses, enzymes found in the synaptic cleft rapidly

destroy neurotransmitter substance. In others, the terminal boutons reabsorb neurotransmitter to be used again. Neurotransmitter substance may also diffuse out of the synaptic cleft.

Practical Applications: Insecticides, Anesthetics, and Other Chemicals.   If neurotransmitters remain in the synaptic cleft, the postsynaptic membrane may be restimulated. Many commonly used insecticides, for example, block the action of **acetylcholinesterase,** an enzyme found in the synapses of insects and people.[4] Acetylcholinesterase enzyme destroys acetyl-

[4]Organophosphates, such as malathion and parathion, are two examples.

choline after it has bound to the receptors of postsynaptic membrane and stimulated a bioelectric impulse. Insecticides that block the enzyme kill insects by disrupting nerve transmission. Unfortunately, nerve poisons, such as these, are also harmful to farm workers and pesticide applicators, who are often exposed to high levels of the pesticides at work. Blocking acetylcholinesterase results in the accumulation of acetylcholine in synapses, causing the postsynaptic nerve to be stimulated repeatedly. Muscles go into spasms. At low levels, these insecticides cause blurred vision, headaches, rapid pulse, and profuse sweating. At higher doses, victims may begin to writhe uncontrollably and can even die.

Each year, an estimated 100,000 to 300,000 Americans (mostly farm workers) are poisoned by pesticides. By various estimates 200 to 1000 of them die. Worldwide approximately 500,000 people are poisoned and 5,000 to 14,000 people die each year from pesticides. These problems and the widespread contamination of the environment have led some farmers to reduce pesticide use and to rely on other, nonpolluting methods of pest control. Crop rotation, insect-resistant crops, and a variety of alternatives are practical and cost-effective options.

A variety of other chemicals also affect nerves and nerve cell transmission. Some anesthetics, for example, which are vital to modern surgery, may impair synaptic transmission. Others may affect the protein pores in the plasma membrane of neurons. These pores regulate the flow of sodium ions into and out of the nerve cells. By blocking the flow of sodium, these anesthetics paralyze sensory nerves carrying pain messages to the brain.

Caffeine and cocaine also affect nerve cell function. Caffeine increases synaptic transmission, thus increasing neuronal activity. It is no wonder that coffee makes some of us so jittery. Cocaine affects neurotransmitters. In the brain, for example, cocaine blocks the uptake of neurotransmitters by terminal boutons, causing an increase in neural activity. Increased neural transmission in the brain results in a heightened state of increased alertness and euphoria (sense of well-being), commonly known as a "high," which lasts for about 20 to 40 minutes. Euphoria, however, is followed by a period of depression and anxiety, which causes many people to seek another high. Excessive cocaine use can result in serious mental derangement—in particular, delusions that others are out to get you. In this state, heavy users may become violent.

Nerve Cell Types: A Functional Classification

Nerve cells can be categorized by structure or function. For our purposes, a functional classification is most useful. According to this system, nerve cells fall into three distinct groupings: (1) sensory neurons, (2) interneurons, and (3) motor neurons.

Sensory neurons carry impulses from body parts to the central nervous system, transmitting impulses from **sensory receptors** located in the body. Sensory receptors come in many shapes and sizes and respond to a variety of stimuli, such as pressure, pain, heat, and movement (Chapter 13).

Motor neurons carry impulses from the brain and spinal cord to effectors, the muscles and glands of the body. Sensory information entering the brain and spinal cord via sensory neurons often stimulates a response. A response is brought about by impulses transmitted via motor neurons to muscles and glands of the body. In some cases, intervening neurons—called **interneurons** or **association neurons**—are present. Interneurons transmit impulses from the sensory neurons directly to motor neurons and may also transmit impulses to other parts of the CNS. The importance of interneurons is underscored by the fact that 99% of the neurons in the human central nervous system are interneurons. As noted below, these neurons play an important role in coordinating complex activities. They are the neural communication network that sends impulses from one part of the CNS to another to help bring about coordinated efforts.

THE SPINAL CORD AND NERVES

With this understanding of the neuron, we can now turn our attention to the spinal cord and nerves.

The Spinal Cord

The spinal cord is a long, cordlike structure about the diameter of your little finger. The spinal cord connects to the brain above and courses downward through the vertebral canal formed by the vertebrae (Figure 12–11). The spinal cord gives off nerves along its course, which innervate the skin, muscles, bones, and joints of the body. These nerves transmit sensory information to the spinal cord and transmit motor information back. The spinal cord extends into the lower back (about the second lumbar vertebrae)

FIGURE 12–11 The Spinal Cord The spinal cord extends from the brain to the upper lumbar region.

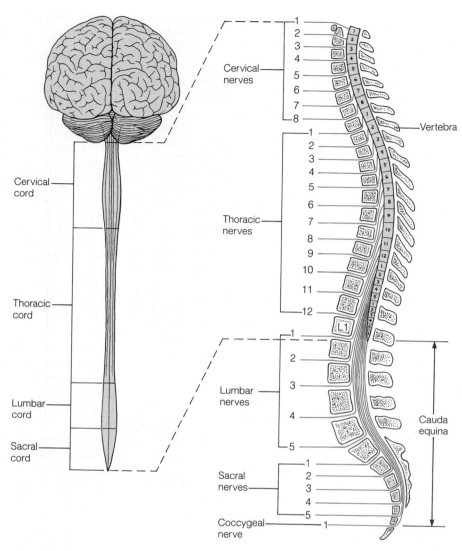

Cervical cord

Thoracic cord

Lumbar cord

Sacral cord

Cervical nerves

Thoracic nerves

Lumbar nerves

Sacral nerves

Coccygeal nerve

Vertebra

Cauda equina

where it ends in a series of nerves called the **cauda equina** (horse's tail), which supply the lower sections of the body.

As shown in Figure 12–12, the central portion of the spinal cord is an H-shaped zone, or gray matter. **Gray matter** consists of nerve cell bodies of interneurons and motor neurons and appears gray to the naked eye. Surrounding the central zone are fiber tracts—axons and dendrites that travel up and down the spinal cord, delivering information to and carrying information from the brain. The fiber tracts form the **white matter.**

The Nerves

As noted earlier, nerves are part of the peripheral nervous system. They carry sensory information to the spinal cord and brain and motor information out. Some nerves are strictly motor and some are strictly sensory, but many are mixed, having both motor and sensory fibers.

Nerves arising from the brain are called **cranial nerves** (Figure 12–13). The trigeminal nerve mentioned in the introduction is one of the 12 cranial nerves. The cranial nerves supply structures of the

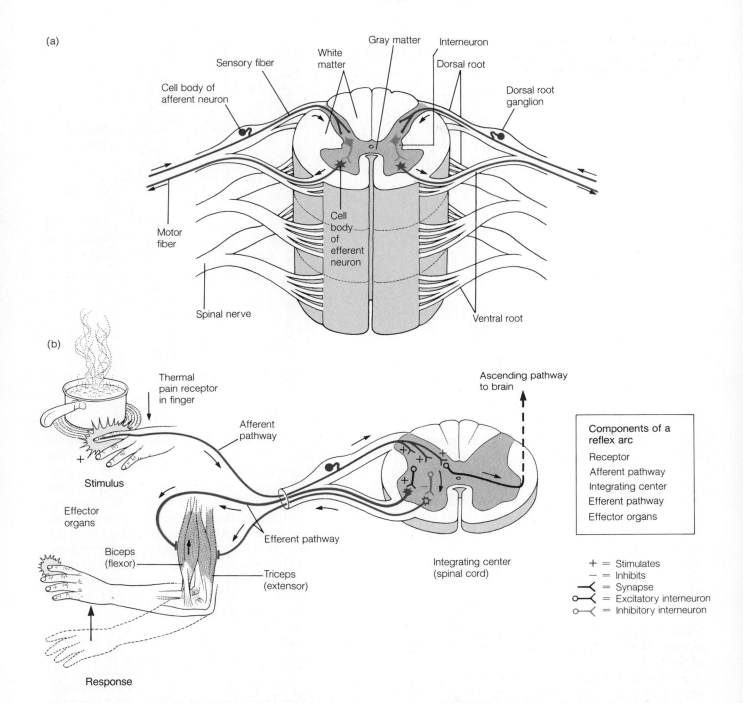

(a)

Cell body of afferent neuron

Sensory fiber

White matter

Gray matter

Interneuron

Dorsal root

Dorsal root ganglion

Motor fiber

Cell body of efferent neuron

Spinal nerve

Ventral root

(b)

Thermal pain receptor in finger

Afferent pathway

Ascending pathway to brain

Stimulus

Effector organs

Biceps (flexor)

Triceps (extensor)

Efferent pathway

Integrating center (spinal cord)

Response

Components of a reflex arc

Receptor
Afferent pathway
Integrating center
Efferent pathway
Effector organs

+ = Stimulates
− = Inhibits
= Synapse
= Excitatory interneuron
= Inhibitory interneuron

FIGURE 12–12 The Spinal Cord and Dorsal Root Ganglia (*a*) Spinal nerves are attached to the spinal cord by two roots, the dorsal and ventral roots. The dorsal root carries sensory information into the spinal cord. The ventral root carries motor information out of the spinal cord. The spinal nerve often contains both sensory and motor fibers. (*b*) When you accidentally touch a hot pan on the stove, you withdraw your hand before the brain even knows what's happening. This occurs because of a reflex arc. Sensory fibers send impulses to the spinal cord. The sensory impulses stimulate motor neurons in the spinal cord. This causes muscle contraction in the flexor muscles (+) and inhibits muscle contraction in the extensor muscles (−), allowing you to withdraw your hand. Nerve impulses also ascend to the brain to let it know what is happening.

FIGURE 12–13 Cranial Nerves The 12 cranial nerves arise from the underside of the brain and brain stem.

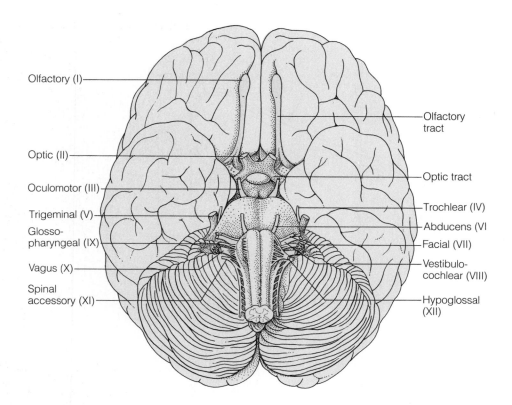

Olfactory (I)

Optic (II)

Oculomotor (III)

Trigeminal (V)

Glosso-
pharyngeal (IX)

Vagus (X)

Spinal
accessory (XI)

Olfactory
tract

Optic tract

Trochlear (IV)

Abducens (VI

Facial (VII)

Vestibulo-
cochlear (VIII)

Hypoglossal
(XII)

head and several key body parts, such as the heart and diaphragm.

Nerves arising from the spinal cord are called **spinal nerves** (Figure 12–12a). Each spinal nerve has two roots, the dorsal and ventral roots, which attach to the spinal cord. The **dorsal root** is the inlet for sensory information. On each dorsal root is a small aggregation of nerve cell bodies, which is called the dorsal root ganglion. The **dorsal root ganglia** house the cell bodies of sensory neurons. As Figure 12–12a shows, the dendrites of these nerves conduct impulses from the body to the dorsal root ganglia; the axons, in turn, carry the impulse into the spinal cord, along the dorsal root.

Sensory fibers entering the spinal cord often end on interneurons (Figure 12–12a). Interneurons receive input from many sensory neurons and process this information, much like a receptionist in a busy corporate office. Interneurons transmit the impulses to nearby motor neurons, whose axons escape via the ventral root of the spinal nerve, carrying impulses to muscles and glands. This anatomical arrangement of neurons allows information to enter and exit the spinal cord quickly and forms the basis of the **reflex arc** (Figure 12–12b). Some reflex arcs contain interneurons and some do not.

When a physician taps her rubber hammer on the tendon just below your kneecap (patellar tendon), she is testing one of your body's many reflexes. The tapping of the hammer on the tendon stretches it; this stimulates stretch receptors in the tendon. These receptors, in turn, generate nerve impulses that travel to the spinal cord via a sensory neuron. In this reflex, the sensory neuron ends directly on a motor neuron, which supplies the muscles of the thigh. Thus, a quick tap on the tendon results in a motor impulse sent to the anterior thigh muscles, causing them to contract and the knee to jerk.

Reflexes are mechanisms that often protect the body from harm. Touching a hot stove, for example, elicits the withdrawl reflex. Your hand is pulled from the stove before you are aware of what is happening.

Babies come equipped with a number of important reflexes. Rub your finger on the cheek of newborn babies, and they immediately turn their head toward your finger. This reflex helps babies find the mother's nipple. Crying is also a reflex. When a baby is hungry, thirsty, wet, or uncomfortable, it cries, a reflex sure to get attention.

The spinal cord is involved in many reflexes. The spinal cord, however, also transmits sensory information to the brain along special tracts lying outside the

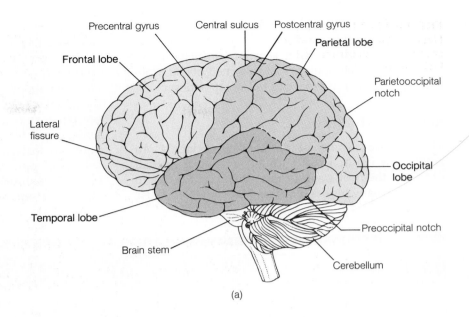

(a)

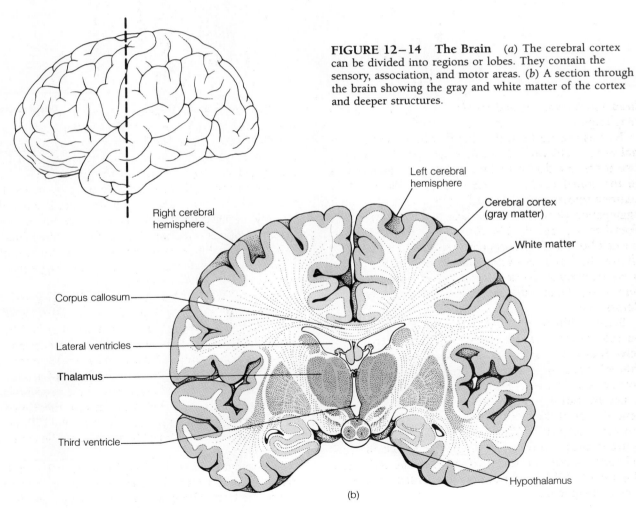

FIGURE 12–14 **The Brain** (*a*) The cerebral cortex can be divided into regions or lobes. They contain the sensory, association, and motor areas. (*b*) A section through the brain showing the gray and white matter of the cortex and deeper structures.

(b)

central H-shaped zone. Therefore, sensory information pouring into the spinal cord can reach vital brain centers. Although incoming sensory information may elicit a reflex, the brain is still informed of the problem, allowing for appropriate follow-up action.

Spinal Cord Injury

The spinal cord may be damaged in accidents. An automobile crash, a bullet wound, or even a bad fall can sever the fiber tracts of the spinal cord. Injury to the cord usually results in permanent damage because neurons of the central nervous system do not, as a rule, regenerate.

The amount of damage depends on the location and severity of the injury. Sensory fibers traveling to the brain and motor fibers traveling from the brain to the spinal cord run in separate tracts. If both tracts are severed, for example, by a severe vertebral fracture, all sensory and motor functions below the level of the injury are lost. As a result, muscles supplied by nerves below the injury become paralyzed and are unable to contract voluntarily. Segments of the body below the injury lose sensation.

TABLE 12–1   Summary of Brain Structures and Functions

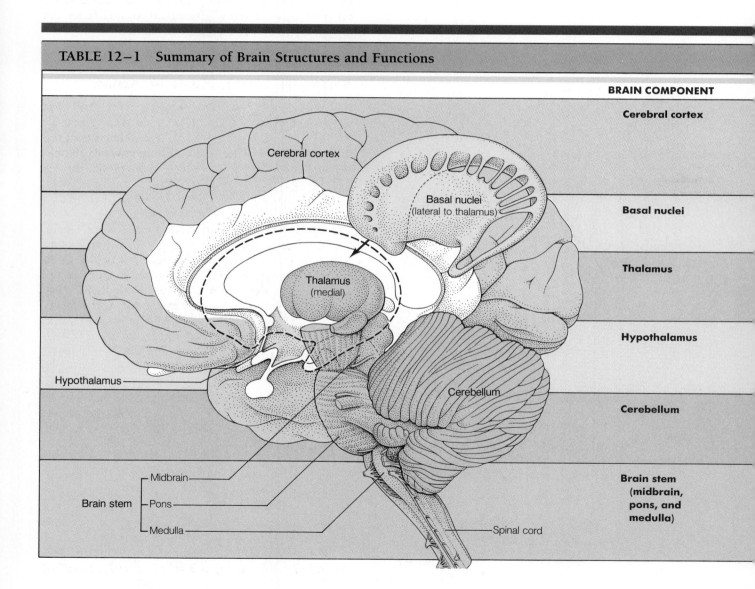

If the spinal cord injury occurs high in the neck (above the fifth cervical vertebra), it severs the nerve fibers traveling to the muscles that control breathing. These muscles become paralyzed, and the person dies fairly quickly. This is how a hangman's noose works.

Damage to the cord just below the fifth cervical vertebra (the fifth vertebra in the neck) does not affect breathing, but paralyzes the legs and arms. The condition is known as **quadriplegia**. If the spinal cord injury occurs below the nerves that supply the arms, the result is **paraplegia**, paralysis of the legs.

Research is now currently underway to find ways to stimulate the regeneration of axons in the brain and spinal cord. Some early results are promising, suggesting that one day physicians may be able to prevent or limit spinal cord damage.

THE BRAIN

About the size of a cantaloupe, the human brain is an extraordinary organ. It is responsible for producing art and music and for feats of engineering and abstract reasoning. It allows us to ponder right and wrong and remember a vast amount of information. Table 12–1 provides an overview of the structures and functions of the major components of the brain.

The Cerebral Hemispheres: Integration, Sensory Reception, and Motor Action

The brain is housed in the skull, a bony shell that protects it from damage. Figure 12–14a (see page 321) shows some of the externally visible parts of the brain. The largest and most conspicuous are the **cerebral hemispheres**. These large, convoluted masses of nervous tissue overlie numerous deeper structures.

The outer layer of the cerebral hemispheres is called the **cerebral cortex**. The cerebral cortex consists of many neurons and nerve cell fibers and can be divided into two sections: the gray matter and white matter (Figure 12–14b). The nerve cell bodies lie in the gray matter, the outer region, which appears gray when viewed with the naked eye. Beneath the gray matter in the cerebral cortex run most of the axons and dendrites of the neurons in the gray matter. The axons and dendrites carry information to and from the gray matter. Because these fibers are myelinated and because myelin is white, this tissue is called white matter.

As shown in Figure 12–14a, the cerebral cortex is thrown into numerous folds, called **gyri** (gyrus, singular). The gyri are interspersed by numerous valleys, called **sulci** (sulcus, singular).

Functional Subdivision of the Cortex. The hemispheres of the cortex can be divided into four major regions or **lobes**, shown in Figure 12–15. The four major lobes are the frontal lobe, the parietal lobe, the occipital lobe, and the temporal lobe. Within each of these lobes are specific subregions that perform special functions. These regions may be broadly classified

MAJOR FUNCTIONS
1. Sensory perception 2. Voluntary control of movement 3. Language 4. Personality traits 5. Sophisticated mental events, such as thinking, memory, decision making, creativity, and self-consciousness
1. Inhibition of muscle tone 2. Coordination of slow, sustained movements 3. Suppression of useless patterns of movement
1. Relay station for all synaptic input 2. Crude awareness of sensation 3. Some degree of consciousness 4. Role in motor control
1. Regulation of many homeostatic functions, such as temperature control, thirst, urine output, and food intake 2. Important link between nervous and endocrine systems 3. Extensive involvement with emotion and basic behavioral patterns
1. Maintenance of balance 2. Enhancement of muscle tone 3. Coordination and planning of skilled voluntary muscle activity
1. Origin of majority of peripheral cranial nerves 2. Cardiovascular, respiratory, and digestive control centers 3. Regulation of muscle reflexes involved with equilibrium and posture 4. Reception and integration of all synaptic input from spinal cord; arousal and activation of cerebral cortex 5. Sleep centers

into three groupings: motor cortex, sensory cortex, and association cortex. Motor cortex stimulates muscle activity. Sensory cortex receives sensory stimuli, and association cortex integrates incoming information, bringing about coordinated responses.

Figure 12–15 shows the major cortical regions and lists some of their functions. Take a moment to study it. As you can see, the cortex contains areas for understanding and generating speech, areas that receive input from the eyes, and regions that receive sensory information from the body. It also contains regions that control muscle movement and areas that allow for planning.

This section discusses some of the major areas to broaden your understanding of how the brain works. We begin with a section of motor cortex called the primary motor cortex. The **primary motor cortex** occupies a single gyrus (ridge) on each hemisphere, just in front of the central sulcus (Figure 12–15). The primary motor cortex controls voluntary motor activity—for example, the muscles in your hand that are turning the pages.

The neurons in the primary motor cortex are arranged according to the part of the body they control. As Figure 12–16a shows, the neurons that control the muscles of the knee are located in the uppermost region of the primary motor cortex. Hip muscle control occurs below that. Muscles of the hand are controlled by neurons located even lower down.

To bring about a voluntary movement, a nerve impulse is first generated in the primary motor area. The impulse then travels from the brain down the spinal cord to the motor neurons in the spinal cord that control the muscles.

FIGURE 12–15 Functional Regions of the Cortex
The cerebral cortex has three principal functions: receiving sensory information, association (integrating incoming and outgoing information), and motor output. Special sensory areas handle vision, smell, taste, and hearing.

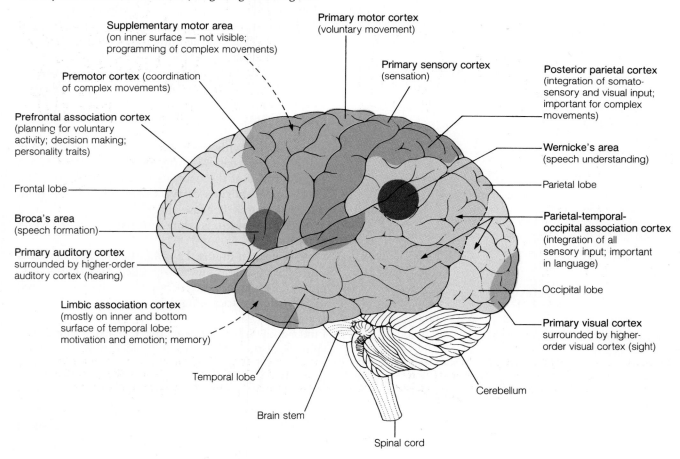

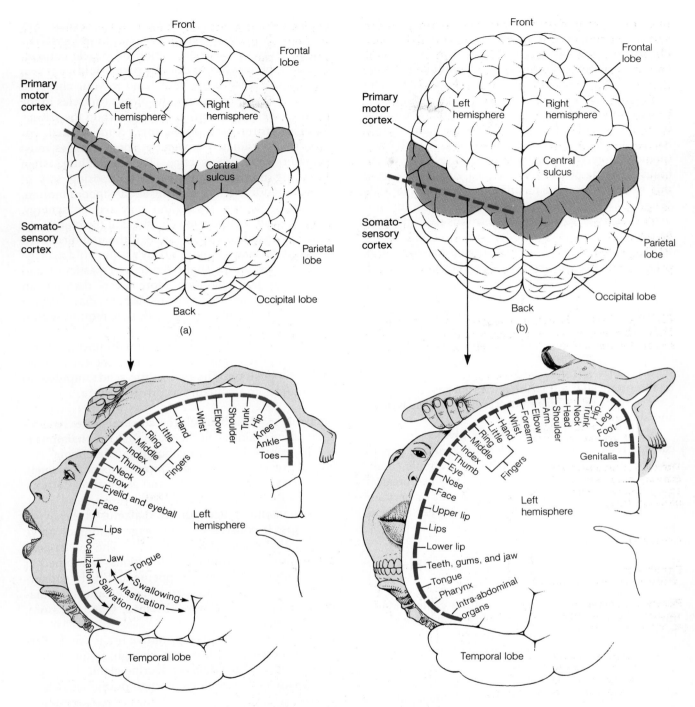

FIGURE 12–16 Primary Motor and Sensory Areas
(a) Top view of the brain showing the primary motor cortex. Below is a map of the location of motor functions within the primary motor cortex. (b) The primary sensory cortex (somatosensory cortex). Below is a map of the location of motor functions within the primary sensory cortex.

In front of the primary motor area is the **premotor cortex**. The premotor area is also involved in controlling muscle contraction. However, the movements the premotor cortex controls are less voluntary—for example, typing or the fingering required to play a musical instrument.

Just behind the central sulcus is another long ridge of tissue running parallel to the primary motor area. Known as the **primary sensory cortex**, it is the destination of many sensory impulses traveling to the brain. Like the primary motor area, different parts of this ridge correspond to different parts of the body (Figure 12–16b). Electrodes inserted into the primary sensory area will elicit sensations that appear to be coming from specific body parts.

In between the regions of motor and sensory cortex lies the **association cortex**, large expanses of brain tissue where integration occurs. In the prefrontal lobe is a region of association cortex that houses complex intellectual activities, such as planning and ideation (creating new ideas). This region also modifies behavior, conforming human actions with social norms. Another important association area lies posterior to the sensory cortex. It interprets sensory information sent to the brain and stores memories of past sensation. Other association areas interpret language in written and spoken form.

Although the cerebral cortex is divided into specialized regions, functions can shift from one part to another in the case of damage. For example, brain cells that control muscle movement may be damaged by a stroke. Undamaged cells may take over their function, however, allowing varying degrees of recuperation.

Unconscious Functions: The Cerebellum, Hypothalamus, and Brain Stem

Consciousness resides in the cerebral cortex. Many functions occur at an unconscious level. Heartbeat, breathing, swallowing, and many homeostatic functions, for example, all take place without conscious control.[5] One region of unconscious control is the cerebellum. The **cerebellum** is a large, visible structure—the second largest structure of the brain. As Figure 12–15 shows, the cerebellum sits below the cortex on the brain stem.

[5]Conscious control is possible for many automatic functions, but for the most part breathing and swallowing are controlled automatically in lower brain centers.

Cerebellum. The cerebellum plays several key roles. One of them is synergy. Neurophysiologists define **synergy** as the coordination of muscle contraction and the movement of body parts. Hold your arm straight out, then bring your hand to your chest. To perform this simple action, the biceps (muscles in the front of the upper arm) contracted while the triceps (muscles in the back of the upper arm) relaxed. For smooth, coordinated movement some muscles must contract while others relax. The cerebellum, when operating normally, ensures that opposing sets of muscles work together to bring about smooth motion.

Damage to the cerebellum may reduce synergy. Damage may occur during childbirth if the blood supply is interrupted—for example, if the umbilical cord accidentally wraps around a baby's neck. Mild damage generally results in slight rigidness (spasticity) and moderately jerky motions. More severe damage can result in serious impairment. Motions may become extremely jerky, and simple tasks may require several attempts.

The cerebellum also helps maintain posture. It receives impulses from sense organs in the ear that detect body position. The cerebellum sends impulses to the muscles to maintain or correct posture.

Thalamus. Just beneath the cerebral cortex is a region of the brain called the thalamus. The **thalamus** is a relay center, like a switchboard in a telephone system. It receives all sensory input, except for smell, then relays it to the sensory and association cortex.

Hypothalamus. Beneath the thalamus is the hypothalamus. It consists of many aggregations of nerve cells, called **nuclei** (Figure 12–17). The nuclei control a variety of autonomic functions—all aimed at maintaining homeostasis. Appetite and body temperature, for example, are controlled by the hypothalamic nuclei. So are water balance, blood pressure, and sexual activity.

The hypothalamus is a primitive brain center. Perhaps its best-known function, though, is the control of the pituitary gland, an endocrine gland that regulates many body functions through hormones (Chapter 15).

Limbic System. Among the most fundamental responses organisms exhibit are instincts—for example, the protective urge of a mother, the territorial assertions of a male, and the flight-or-fight response we

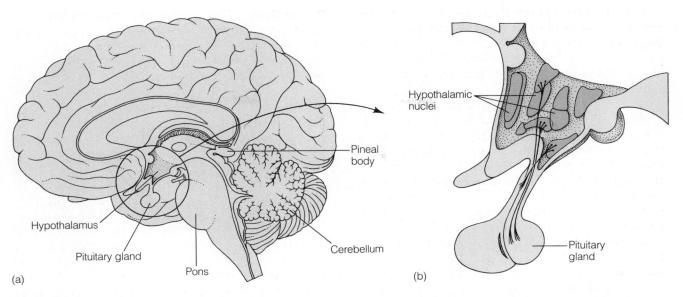

(a)

(b)

FIGURE 12–17 Cross Section of the Brain (*a*) The hypothalamus is clearly visible in a cross section of the brain. (*b*) The hypothalamus is at the base of the brain, just above the pituitary gland. It regulates many autonomic functions and plays a particularly important role in controlling pituitary secretions.

experience in the face of adversity. Instincts and emotions reside in a complex array of structures called the **limbic system**, shown in Figure 12–18. The limbic system works in concert with the hypothalamus. Electrodes placed in some areas of the limbic system may elicit primitive rage. Damage to this area, described later, shows how vital the limbic system is to normal behavior.

Brain Stem. The **brain stem** controls additional primitive functions. As Figure 12–19 shows, the brain stem consists of the medulla and pons. The **medulla** is the anterior continuation of the spinal cord. Nearly all incoming and outgoing information must pass through it. Nerve fibers carrying information to and from the brain send branches into a region of the medulla called the **reticular activating system (RAS)** (Figure 12–19). The RAS monitors incoming and outgoing information, like a security guard at a doorway. The RAS also keeps its "boss" upstairs, the cortex, informed of the informational flow.

The RAS is an arousal system. Nerve fibers from the RAS travel to the cortex, where they activate nerve cells, thus maintaining wakefulness or alertness. When we are asleep, the flow of information through the RAS is greatly reduced, and the cortex sleeps. A biting insect, however, stimulates sensory nerves in the skin, which send impulses to the brain. The impulses travel to the sensory areas of the brain, awakening the sleeper.

The RAS wakes us up and keeps us alert, but may also prevent us from falling asleep from time to time. Pain from a bad sunburn, for example, prevents many people from falling asleep. Pain impulses traveling to the brain enter the RAS, which generates impulses that travel to the cortex.

The RAS also houses centers involved in heart rate, respiration, blood vessel control, swallowing, coughing, and vomiting. In concert with areas in the hypothalamus, the medulla helps control many of these basic functions.

THE AUTONOMIC NERVOUS SYSTEM

Figure 12–3 illustrates the subdivisions of the nervous system. As noted earlier, the nervous system consists of two parts: the peripheral nervous system (nerves) and the central nervous system (brain and spinal cord). The PNS consists of the somatic division

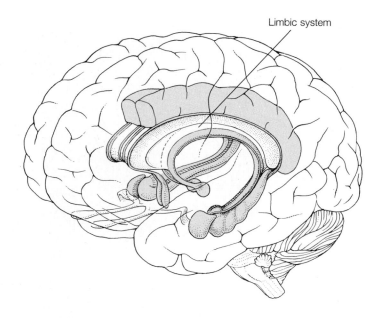

FIGURE 12–18 **The Limbic System** This odd assortment of structures is the seat of emotions and instincts, among other functions.

Limbic system

and the autonomic division, both of which contain motor and sensory neurons.

The somatic division receives sensory information from the skin, muscles, and joints, which it transmits to the CNS. Motor impulses from the CNS, in turn, are sent to skeletal muscles. In contrast, the autonomic nervous system (ANS) transmits sensory information from organs to the CNS, which delivers motor impulses to smooth muscle, cardiac muscle, and glands.

The autonomic nervous system functions automatically, usually at a subconscious level. It innervates all internal organs and has two subdivisions: the sympathetic and the parasympathetic.

The **sympathetic division** of the autonomic nervous system functions in emergencies and is largely responsible for the flight-or-fight response. When an individual is startled, sympathetic nerve impulses cause the heart to accelerate, the pupils to dilate, and the breathing rate to increase. The **parasympathetic division** of the ANS brings about internal responses associated with the relaxed state. It reduces heart rate, contracts the pupils, and promotes digestion.

Most organs are supplied with both parasympathetic and sympathetic fibers. As a general rule, the parasympathetic and sympathetic fibers have antagonistic or opposite effects. One stimulates activity, and the other reduces activity. This provides the body with a means of fine-tuning organ function.

 ENVIRONMENT AND HEALTH: THE VIOLENT BRAIN

Dr. Vernon H. Mark, a neurosurgeon currently with the Harvard Medical School, has a special interest in the violent mind. He has studied and treated a large number of violent patients in his years of medical practice. One of the most memorable was a woman named Julie. She came to Dr. Mark at age 21, suffering from epileptic seizures and sudden, unpredictable outbursts of violence.

Her problems began early in life when a mumps infection spread to her brain, causing severe inflammation. At the age of 10, Julie began suffering from epileptic seizures. After many of her seizures, the girl would burst into fits of anger, attacking people. Depression and remorse followed her outbursts.

At age 18, Julie and her father, a physician, went to a movie together. While sitting in the theater, Julie felt a wave of terror overcoming her. She went to the restroom and stared at herself in the mirror, clutching a dinner knife that she carried to protect herself. Another woman in the restroom accidentally bumped into Julie, triggering her rage. Julie attacked her, plunging the knife into her chest. Her father heard the scream and raced in. He was able to administer first aid to save the woman's life.

Three years later, Julie was placed under the care of Dr. Mark. To find out where the problems arose in

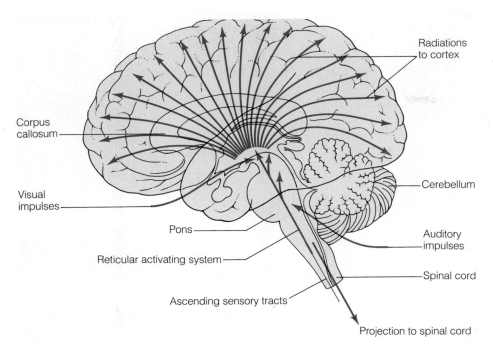

FIGURE 12–19 The Brain Stem and Reticular Activating System The brain stem is composed of the pons and medulla (not labeled). The reticular activating system resides in the brain stem where it receives input from incoming and outgoing neurons.

Julie's brain, Dr. Mark implanted a number of electrodes in her brain that would allow him to study the electrical activity of her brain during seizures. Dr. Mark was also able to stimulate the electrodes by remote control to study the behavioral effects of activating various parts of her brain. Stimulating most of the electrodes had little effect on her behavior. However, stimulating an electrode in a region of the limbic system called the amygdala had a dramatic effect on the young woman. Julie became unresponsive, staring vacantly, although her face turned angry. While this was occurring, the brain cells in her amygdala and hippocampus fired erratically. Dr. Mark had stimulated a full-fledged epileptic seizure. All of a sudden, Julie exploded violently.

Dr. Mark repeated the experiment at a later time and got a similar response, confirming that he had indeed found the focal point of her seizures—and rage. Over the next several months, he set out to destroy the tissue, sending radiowaves along the electrode. This raised the temperature of the electrode and destroyed the cells little by little. Today, Julie's seizures and violent outbursts have ended. She lives a normal life.

This true story is a remarkable testimony to the dramatic effects of brain trauma. Damage in crucial areas, no matter how slight, can upset the normal operations of the brain, resulting in erratic behavior or throwing patients into violent rage. Dr. Richard Restak, a neurologist and author of numerous popular books on the brain, notes that alcohol and drugs can also upset the delicate balance of the brain, creating violent outbursts. Drugs and alcohol can dismantle (at least temporarily) the brain's intricate system of checks and balances, eliciting unusual, and even violent, behavior. "It is humbling and frightening," says Restak, "to consider that our rationality is dependent on the normal function of tissue within our skulls." It is even more frightening to realize that the balance can be so easily upset. "We have the capacity, if everything is operating correctly within our brains, of composing a Bill of Rights or the Constitution," says Restak. "But in the presence of a barely measurable electrical impulse within the limbic system, our much vaunted rationality can be replaced by savage attacks and seemingly inexplicable violence."

Alcohol dampens the inhibitions that hold aggression in check in some people. Many drugs have a similar effect (for more on drugs and alcohol and the brain, see Health Notes 12–1 and 12–2). Barbiturates, for example, increase aggression. Marijuana apparently reduces aggression. Marijuana, however, is sometimes laced with PCP (phencyclidine), commonly known as angel dust. This street drug un-

Drugs and Alcohol: The Causes and Cures of Addiction

Marleen Whitehead lives a lie. Each morning she kisses her husband good-bye, then drives off to work. On her way to the office, she pulls off the highway for a moment, opens a flask she keeps under the seat, and takes a drink. Throughout the day, Marleen laces her coffee with scotch, so no one will suspect that she is drinking.

Like millions of other Americans, her life is on the road to disaster. Her boss is aware of her drinking, and her job is at risk. What makes Marleen and millions of others like her who cannot get through a day without a drink or a fix from a needle so dependent on their drugs?

Many biologists think that genetics plays a key role in alcoholism. Behavioralists, however, argue that a person's upbringing and psychology create alcohol dependency. Most likely, biological and behavioral roots are both involved in alcoholism and other addictive behavior.

Let's review some of the facts. A recent study in mice showed that addictive drugs—alcohol and cocaine, among others—stimulate the brain's pleasure center. The pleasure center is a region of the brain that, when stimulated, brings great pleasure. Stimulation of the pleasure center by drugs, researchers believe, may underlie all forms of drug addiction.

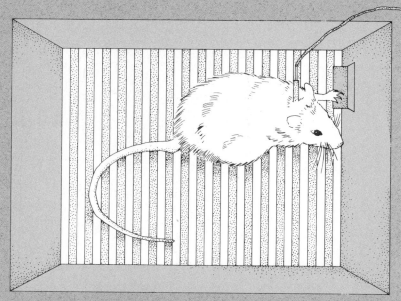

The Pleasure Center An electrode implanted in the pleasure center of a rat's brain can be activated when the rat presses a bar. To obtain this stimulation, the rat will forsake sex, food, and water. Addictive drugs may stimulate the pleasure center.

Experiments in which researchers implanted electrodes in the pleasure centers of rats showed that the rodents can be trained to press a bar to stimulate the center (see the figure). Some rats, in fact, press the bar hundreds of times per hour, ignoring food, water, and sex. After 15 to 20 hours of continual pressing, the rats collapse from exhaustion. But when they awake, they begin pressing again.

Experiments with rats and mice suggest that the pleasure center (the nucleus accumbens) is activated by cocaine and amphetamines. These drugs stimulate the production of large amounts of a neurotransmitter

leashes extremely violent behavior.

PCP acts through the limbic system, shutting off the inhibiting effects of the cortex on primitive rage centers. Episodes of sudden, unprovoked violence due to PCP have become common in many of our major cities, says Restak.

Perhaps one of the most shocking stories of the impacts of chemicals on the brain is that of David

Garabedian, a man described by his family and friends as mild-mannered, passive, even docile. David Garabedian worked for a lawn-care company in Massachusetts. On March 29, 1983, Garabedian arrived at the home of Eileen Muldoon to estimate the cost of treating her lawn with chemical insecticides. He knocked on the door, but found no one at home. Garabedian decided to make an estimate and leave it for her.

known as dopamine. In a recent study, researchers implanted small tubes into rats' pleasure centers and areas involved in muscle movement. The researchers administered various doses of addictive and nonaddictive drugs while the rats moved freely in their cages. The researchers then extracted fluid from the tubes to measure dopamine levels.

Drugs that are addictive to humans and rewarding to rats (amphetamine, cocaine, morphine, methadone, ethanol, and nicotine) increased dopamine concentrations in both brain areas. Levels, however, were much higher in the pleasure center. Although these results cannot be extrapolated to humans, some researchers believe that they help build a case for the theory that dopamine release in the pleasure center is a common denominator in all forms of drug addiction. Other researchers think that other neurotransmitters may also be involved in addiction.

This research helps explain the neural and biochemical basis of addiction, but what about the underlying cause? Why do some people become addicted to drugs while others can take them or leave them?

Research into the genetics of alcoholism has built a case for the assertion that alcoholism is largely the result of defective genes. Donald W. Goodwind of the University of Kansas Medical Center in Kansas City found that children of alcoholics have an increased risk of becoming alcoholic even when reared by adoptive (nonalcoholic) parents. This and a number of other similar studies suggest that the environment is less important than a person's genetic heritage. If your biological parent is an alcoholic, say researchers, you are much more likely to be one yourself.

Nevertheless, some researchers think that the scientific community has accepted the genetic findings too readily and uncritically. They say that the genetic theory of alcoholism may be a simplified view of the causes of the disease. Alcoholism probably results from a complex interaction of environmental factors and genetics.

New research also shows that alcoholism results from physical, personal, and social characteristics that predispose a person to drink excessively. Herbert Fingarette, a professor at the University of California at Santa Barbara who has studied alcohol addiction for years, claims that alcoholism has psychological not biological roots. Fingarette's arguments reflect the findings of many psychological studies. This research indicates that alcoholism and other addictions are more habits than diseases. Addictive behavior, the studies suggest, typically revolves around immediate gratification.

Alcohol enhances social and physical pleasure, increases sexual responsiveness and assertiveness, and reduces tension—up to a point. Unfortunately, the initial physical stimulation, brought on by low doses of alcohol, can lead some people into an addictive cycle. The expectation of improved feelings drives people to drink. But higher doses of alcohol dampen arousal, sap energy, and cause hangovers. This, in turn, say psychologists, leads to a craving for alcohol's stimulating effects—a craving to feel good again. The repetitive cycle of pleasure and displeasure is addiction.

The controversy over the roots of alcoholism and other addictive behaviors will undoubtedly continue for years, pitting biologists against psychologists. Although it is impossible to predict the outcome of future research, it is possible that an intermediate position may hold sway: addiction is caused by genetics and psychology.

When he had completed his work, however, he experienced a sudden urge to urinate. He went to the backyard and urinated near the house. Just then, Mrs. Muldoon appeared, yelling angrily at him. Garabedian became confused, apologized, and tried to explain his plight. She turned away from him, refusing to listen. Garabedian tapped her on the shoulder to say he was sorry, but the woman turned on him and clawed his face with her fingernails. Garabedian exploded. He grabbed the woman by the neck and strangled her. As she lay motionless on the ground, Garabedian hurled large rocks at her head, smashing her skull.

A month before the murder, Garabedian had undergone a remarkable personality change. Usually amicable, he became easily angered and abusive toward his family. He complained of tension, nervous-

HEALTH NOTE 12–2
New Treatments for Alcoholism

Psychological research suggests that conventional treatment strategies for alcoholism must be refurbished. Treatment, says Herbert Fingarette of the University of California at Santa Barbara, must focus not just on the drinking problem, but on developing a satisfying lifestyle that relies on other pleasures—a life that does not revolve around heavy drinking.

Total abstinence is the goal of many medical treatment centers. It is also the central goal of Alcoholics Anonymous (see the figure). Fingarette, however, says that it is an unrealistic goal for many heavy drinkers.

Many psychologists believe that treatment should focus on ways to handle stress and bring pleasure to one's life without drinking.

Professor Allan Marlatt of the University of Washington in Seattle and his co-workers are now developing an alcohol skills training program for college students. They find that students who consume large amounts of alcohol every week cut down considerably after completing

Alcoholics Anonymous   Adults at an Alcoholics Anonymous meeting.

the eight-week session. Says Marlatt, "Children of alcoholics show some of the best responses to the program and are highly motivated to learn how to drink in moderation."

Psychologists strive to teach students how to set drinking limits and how to cope with peer pressure at parties and other social events. Re-

alistic expectations about alcohol's mood-enhancing powers are also developed, and participants learn alternative methods of reducing stress, including meditation and exercise.

A similar approach is now being used to help adult alcoholics. This program was developed by W. Miles Cox of the Veterans' Administration

---

ness, impatience, and bad nightmares. He also complained of numerous physical symptoms—such as nausea, diarrhea, headaches, and frequent urination. These symptoms suggested to some that the young man was suffering from pesticide poisoning.

Dr. Peter Spencer, a toxicologist at the Albert Einstein College of Medicine in New York, testified at the trial that "David Garabedian was involuntarily intoxicated with a chemical in the lawn products that he was exposed to on a daily basis. . . ." David not only sprayed chemicals on lawns and trees, but also mixed them, pouring large amounts into trucks each night in preparation for the next day's spraying.

One chemical in particular has been singled out as

the possible culprit. It is carbaryl, a powerful inhibitor of acetylcholinesterase. Physicians believe that the carbaryl, and possibly other pesticides David Garabedian routinely handled, inhibited acetylcholinesterase in his brain. Acetylcholine then accumulated in the synaptic clefts in the limbic system, triggering uncontrollable rage.

A psychiatrist who testified at the trial said, "In my view, David did not have the capacity to control his behavior because his brain was poisoned." Despite this testimony, David was found guilty of first-degree murder and is currently serving a life sentence. Still, one has to wonder if the wrong villain was convicted.

in Indianapolis and Eric Klinger of the University of Minnesota in Morris. These researchers believe that a number of biological and social factors influence alcohol abuse. The final decision to drink, they say, is motivated by conscious or unconscious expectations that alcohol will brighten one's emotional state and remove stress. Such expectations, they think, often outweigh fears that drinking will lead to other adverse effects—for example, losing a job or getting divorced.

The researchers have also found that alcoholics as a rule often set unrealistically high standards for themselves. Making matters worse, most cannot forgive themselves for failing to meet their unreasonable expectations.

Cox and Klinger are helping patients find emotionally satisfying activities that do not require alcohol. The researchers have developed a questionnaire to assess an alcoholic's major life goals and concerns and then work with the alcoholic to formulate weekly goals that are realistic. The researchers help patients find ways to reduce the use of alcohol as a way of dealing with frustration.

New studies show that alcohol is incredibly effective in diverting one's attention from stressful thoughts. That is why many people hit the bars after a hard test or a hard day at work. Alcohol's ability to shift attention can lead to addiction, especially in individuals who are extremely self-conscious and highly critical of themselves.

On another front, researchers are looking for cues that set off an alcoholic's craving. Many alcoholics feel helpless and bewildered when the craving strikes. Research shows that both internal and external cues create the irresistible urge to drink. In some cases the cues can be quite specific—for instance, driving past a local bar. In a survey of 150 abstinent alcoholics who were asked to identify the cues that triggered their craving, more than half cited internal tension. External cues varied considerably among the individuals. The researchers hope that by identifying the cues that stimulate the urge to drink, alcoholics can be made more aware of emotions and situations that trigger drinking binges and can find ways to control the powerful urges.

Research into the psychology of addiction is beginning to shed new light on alcoholism and other dependencies. Unfortunately, this research remains largely ignored by the biologically oriented advocates of alcoholism as a disease who hold that the only treatment is total abstinence. The new psychological research, suggesting that some (maybe 15 to 20%) alcoholics can continue to drink, has sparked considerable debate among health care professionals.

No doubt the debate over the causes of alcoholism will continue for many years. An issue as complicated as this cannot be easily solved, especially when proponents of separate views line up in distinct camps, viewing the problem narrowly through their own professional biases.

Drugs, alcohol, pesticides, and industrial chemicals enter our bodies. In small amounts, they may be harmless, but in higher concentrations—and in combination—they may disrupt chemical balance and cellular structure, upsetting homeostasis. Nowhere is the effect more pronounced than in the brain.

SUMMARY

AN OVERVIEW OF THE NERVOUS SYSTEM

1. The nervous system controls a wide range of body functions and, in doing so, also helps regulate homeostasis.

2. Besides controlling body functions, the human nervous system performs many other functions, such as ideation, planning for the future, thinking, and remembering.

3. The nervous system consists of two anatomical subdivisions: the central nervous system (CNS) and the peripheral nervous system (PNS). The CNS consists of the brain and spinal cord. The PNS is composed of spinal and cranial nerves.

4. The CNS receives sensory input from sensory receptors in the skin, skeletal muscles, and joints, integrates the input, and generates appropriate responses. Sensory information arrives via sensory neurons. Motor output leaves the CNS in motor neurons that travel in the PNS.

5. The PNS has two functional divisions: the autonomic nervous system, which controls involuntary actions such as heart rate, and the somatic nervous system, the voluntary nervous system.

THE STRUCTURE AND FUNCTION OF THE NEURON

6. The fundamental unit of the nervous system is the neuron, a highly specialized cell adapted to generate and transmit bioelectric impulses from one part of the body to another.
7. Three types of neurons are found in the body: sensory neurons, interneurons, and motor neurons.
8. Despite differences, each neuron has a more or less spherical cell body, containing the nucleus, cytoplasm, and organelles. Extending from the cell body of the neurons are two types of processes: dendrites, which conduct impulses to the cell body, and axons, which conduct impulses away from the cell body.
9. Many axons are covered by an insulating layer of myelin that increases the rate of impulse transmission. Myelin is laid down by Schwann cells, a type of glial cell, during embryonic development.
10. Axons branch profusely when they terminate, forming numerous terminal fibers that end in small knobs, called terminal boutons.
11. During cellular differentiation, nerve cells lose their ability to divide. Nerve cells also have exceptionally high metabolic rates and rely exclusively on glucose for energy.
12. Because nerve cells cannot divide, neurons that die cannot be replaced. In the PNS, axons may regenerate, but axonal regeneration in the CNS is rare.
13. Nerve cells transmit bioelectric impulses from one part of the body to another. A small electrical potential exists across the membrane of nerve cells. Called the membrane or resting potential, it measures about -60 millivolts.
14. When a nerve cell is stimulated, its plasma membrane increases its permeability to sodium ions. Sodium ions rush in, causing depolarization, a shift in resting potential from -60 millivolts to $+40$ millivolts. The wave of depolarization spreads down the membrane.
15. Depolarization is followed by repolarization, a recovery of the resting potential caused by the outflow of potassium ions. The depolarization and repolarization of the plasma membrane of the neuron constitute a bioelectric impulse. A recording of the event is called an action potential.
16. The nerve impulse travels along the plasma membranes of dendrites and unmyelinated axons. In myelinated axons, however, the impulse "jumps" from node to node.
17. When a bioelectric impulse reaches the terminal bouton, it stimulates the release of neurotransmitter held in membrane-bound vesicles in the terminal bouton.
18. Neurotransmitters are substances that are released into synaptic clefts and diffuse across the cleft to the membrane of the effector where they bind to receptors in the postsynaptic membrane, often triggering an action potential there.
19. To end the signal, neurotransmitters may diffuse out of the synaptic cleft, be reabsorbed by the axon terminal, or be removed by enzymes.
20. Some insecticides inhibit the activity of these enzymes, creating a wide range of nervous system effects.

THE SPINAL CORD AND NERVES

21. The spinal cord descends from the brain, through the vertebral canal to the lower back. It carries information to and from the brain, and its neurons participate in many reflexes.
22. Two types of nerves are found: spinal and cranial.
23. The spinal nerves are attached to the spinal cord via two roots, a dorsal root, which brings sensory information into the cord, and a ventral root, which carries motor information out. Sensory fibers entering the cord often end on interneurons, which often end on motor neurons in the spinal cord, thus forming a reflex arc. Interneurons may also send axons to the brain.

THE BRAIN

24. The brain is housed in the skull. The cerebral hemispheres are the largest part of the brain.
25. The outer layer of each hemisphere is the cortex. It contains gray matter, which houses nerve cell bodies, and underlying white matter, which contains myelinated nerve fibers that deliver nerve impulses to and from the gray matter.
26. The cortex is thrown into folds, or gyri, and intervening grooves, called the sulci.
27. The cerebral cortex consists of many discrete functional regions, including motor areas, sensory areas, and association regions.
28. Consciousness resides in the cerebral cortex, but a great many functions occur at the unconscious level in parts of the brain beneath the cortex.
29. The cerebellum, for example, coordinates muscle movement and controls posture. The hypothalamus regulates many homeostatic functions. The limbic system houses instincts and emotions. The brain stem, like the hypothalamus, helps regulate basic body functions.

THE AUTONOMIC NERVOUS SYSTEM

30. The autonomic nervous system (ANS) is a subdivision of the PNS and transmits sensory information from organs to the CNS, which delivers motor impulses to smooth muscle, cardiac muscle, and glands.

31. The ANS innervates all internal organs and has two subdivisions: the sympathetic and the parasympathetic.
32. The sympathetic division of the ANS functions in emergencies and is responsible in large part for the flight-or-fight response, accelerating heart rate and breathing.
33. The parasympathetic division of the ANS brings about internal responses associated with the relaxed state. It reduces heart rate, contracts the pupils, and promotes digestion.
34. Most organs are supplied by both parasympathetic and sympathetic fibers.

ENVIRONMENT AND HEALTH: THE VIOLENT BRAIN

35. Abnormal brain activity can result in bizarre behavior in humans. This is especially true if damage occurs in parts of the limbic system. Violent rage may result.
36. Damage may result from infections or harmful chemicals, such as pesticides and drugs. Alcohol can also bring on rage by eliminating normal inhibitions.

⌐ EXERCISING YOUR CRITICAL THINKING SKILLS

Find a newspaper article that reports on a new medical or scientific discovery. Read it and summarize the article in a paragraph or two. Using your critical thinking skills, critique the study. Who performed the work? Was the experiment appropriately run? Did it have a sufficient number of subjects? Did it have a control group? Were the conclusions drawn from the results valid? Could bias have tainted the interpretation of the study? Does the newspaper account give you enough information to truly determine the validity of the study?

TEST OF TERMS

1. The brain and spinal cord are part of the _____ nervous system. The nerves belong to the _____ nervous system.
2. Heartbeat and breathing are controlled by the _____ nervous system.
3. Many neurons contain short, branching fibers called _____, which conduct impulses to the cell body, and a long, unbranched fiber, the axons.
4. Many axons have a coating, the _____ _____, which is laid down by Schwann cells and accelerates nerve cell transmission.
5. The swellings at the terminal ends of axons are called _____ _____. They release chemical substances called _____, which stimulate a bioelectric impulse in subsequent nerve cells.
6. The voltage measured across the plasma membrane of an inactive neuron is called the _____ _____ and is about _____ millivolts.
7. Stimulating the plasma membrane of a nerve cell results in a drastic change in the membrane's permeability to _____ _____. The change in the resting potential that occurs when the membrane is stimulated is called a(n) _____ _____.
8. Nerves attached to the spinal cord are called _____ _____. Each of these has two roots. Motor fibers exit via the _____ root.
9. The neuron that transmits the impulse from the sensory neuron to the motor neuron in the spinal cord is called a(n) _____ neuron.
10. The portion of the cortex that controls voluntary muscle movement is the _____ _____ area.
11. The ridges in the cerebral cortex are called _____. The grooves between them are _____.
12. Sensory fibers from all over the body terminate in a part of the cortex called the _____ _____ area located just behind the _____ _____.
13. The regions of the cortex that integrate the incoming information are called _____ cortex.
14. The _____ _____ has many functions. It is the site of the primitive rage response.
15. The _____ is in charge of coordinating muscle activity. Problems in this region of the brain cause spasticity.
16. Just above the pituitary and attached to it is the region of the brain called the _____. It controls eating behavior and monitors the composition of the blood. Aggregations of nerve cell bodies in this region are called _____.

Answers to the Test of Terms are located in Appendix B.

TEST OF CONCEPTS

1. The nervous system performs a great many functions. Describe them. What functions do you think are unique to humans?
2. Describe how the resting and action potentials are generated.
3. Draw a typical synapse, label the parts, and explain how a nerve impulse is transmitted from one nerve cell to another.
4. Draw a cross section through the spinal cord showing the spinal nerves. Label the parts and explain a reflex arc and how nerve impulses entering a spinal nerve also travel to the brain.
5. A physician can stimulate various parts of the brain and get different responses. What effects would you expect if the electrodes were placed in the premotor area, the primary motor cortex, and the sensory motor cortex?
6. The brain is a delicate organ. Slight shifts in electrical activity can create bizarre behavior. Do you agree with this statement? Give examples.

SUGGESTED READINGS

Alkon, D. L. 1989. Memory storage and neural systems. *Scientific American* 261(1): 42–50. When learning takes place, changes occur in the electrical and molecular properties of nerve cells. This article discusses how memory linkages are established.

Bower, B. 1988. The brain in the machine. *Science News* 134(22): 344–45. New computer models may provide insight into how the brain operates.

Kimleberg, H. K., and M. D. Norenberg. 1989. Astrocytes. *Scientific American* 260(4): 66–76. Recent findings indicate that astrocytes, brain cells that were formally thought to have a passive supporting role, may actually have an active role in brain function, development, and disease.

Levoy, G. 1989. Nervous energy. *Health* 21(2): 58–61. Engaging article on the intricacies of the nervous system.

Pennisi, E. 1989. Neurobiology gets computational. *Bioscience* 39(5): 283–87. A new approach to studying the brain through neural-network computers.

Restak, R. M. 1988. *The mind*. New York: Bantam. Superb treatise on the human brain.

Rhodes, M. 1989. Countering the chemistry of fatigue. *Health* 21(2): 48. Discusses different approaches to overcoming fatigue.

Sherwood, L. 1989. *Human physiology: From cells to systems*. St. Paul, Minn.: West. Provides a much more detailed look at the nervous system's function.

Vander, A. J., J. H. Sherman, and D. S. Luciano. 1985. *Human physiology: The mechanisms of body function*. New York: McGraw-Hill. Excellent coverage of the nervous system function. An excellent next step for students wanting to learn more.

Organ of Corti. This remarkable structure found in the cochlea is the sensory organ for hearing.

Angela Cartwright noticed something strange one day as she began to eat dinner. For no apparent reason, she had lost her senses of smell and taste. All in all, her health was fine. She did not have a cold, a sinus infection, or even any allergies that might block her nasal passages and account for the symptoms. Since the symptoms persisted, she went to the student health clinic for an examination. A blood test unearthed the cause of her symptoms: low blood levels of the micronutrient zinc (Chapter 7).

Angela's doctor prescribed a zinc supplement, and in a few days her senses of smell and taste returned. This story illustrates the importance of a nutritionally balanced diet and the impact of a dietary deficiency on two important body functions.[1] This is just another of many examples presented in this book of the way basic body functions and, ultimately, homeostasis can be altered by environmental factors—stress, pollution, and diet.

This chapter focuses on the senses. Virtually all of the senses participate in homeostasis. Receptors in sensory systems detect disturbances in our internal and external environments, helping our bodies cor-

[1]Zinc deficiencies are quite rare, and dietary supplements containing zinc are generally not needed if you are eating a well-balanced diet.

rect imbalances and keeping us healthy. This chapter breaks the discussion into two broad categories of senses: the general senses and the special senses.

THE GENERAL SENSES

Sit back in your chair for a moment, close your eyes, and concentrate on the stimuli you feel. A cold draft may be stirring at your feet. You may feel the pressure of the chair on your buttocks and warmth emanating from a nearby reading lamp. You may feel your cat brushing against the hairs on your arm. You may feel some mild discomfort as well: a slight pain from a bruise or gas pressure in your intestines from the burritos you ate for lunch. Now move your arm and feel it moving.

Pain, light touch, pressure, temperature, and **proprioception**, a sense of body and limb position, constitute the general body senses, or **general senses** (Table 13–1). The body is also equipped with a number of **special senses**: taste, smell, vision, hearing, and balance. This section focuses on the general senses.

Pain, temperature, light touch, and other stimuli activate the general sense receptors in the skin, internal organs, joints, and muscles. As Chapter 12 noted, receptors are nerve endings or specialized structures activated by various stimuli.

Sense receptors fall into four broad groups: (1) mechanoreceptors, (2) chemoreceptors, (3) thermoreceptors, and (4) photoreceptors. **Mechanoreceptors** are activated by mechanical stimulation—for example, touch or pressure. **Chemoreceptors** are activated by chemicals in the food we eat or the air we breathe. **Thermoreceptors** are activated by heat and cold. **Photoreceptors** are sensitive to light.

Stimuli that activate general sense receptors arise either internally or externally. Receptors relay messages to the spinal cord and brain via sensory nerves (Chapter 12). Sensory input to the central nervous system may elicit no response at all or may cause a muscle reflex. In some cases, sensory stimuli cause unconscious shifts in body function.

General sense receptors fall into two broad groups, based on their structure: naked nerve endings and encapsulated nerve endings, shown in Figure 13–1.

Naked Nerve Endings

The **naked nerve endings** consist of dendrites of the sensory neurons. These receptors are responsible for at least three sensations: pain, temperature, and light touch.

TABLE 13–1 Summary of General and Special Senses

SENSE	STIMULUS	RECEPTOR
General senses	Pain	Naked nerve ending
	Light touch	Merkel disc; naked nerve endings around hair follicles; Meissner's corpuscles
	Pressure	Pacinian corpuscle
	Temperature	Ruffini's corpuscle*; Krause's end-bulb*; naked nerve ending
	Proprioception	Golgi tendon organs; muscle spindles; receptors similar to Meissner's corpuscles in joints
Special senses	Taste	Taste buds
	Smell	Olfactory epithelium
	Sight	Retina
	Hearing	Organ of Corti
	Balance	Crista ampularis in the semicircular canals; maculae in saccule and utricle

*These may both be alternate forms of the Meissner's corpuscle and may be stimulated by light touch.

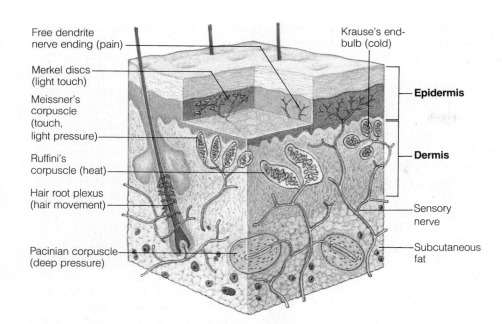

Free dendrite nerve ending (pain)

Merkel discs (light touch)

Meissner's corpuscle (touch, light pressure)

Ruffini's corpuscle (heat)

Hair root plexus (hair movement)

Pacinian corpuscle (deep pressure)

Krause's end-bulb (cold)

Epidermis

Dermis

Sensory nerve

Subcutaneous fat

FIGURE 13–1 General Sense Receptors The skin houses many of the receptors for general senses. Receptors fall into two categories: naked nerve endings and encapsulated receptors, shown here.

Pain. Pain receptors are naked dendritic fibers in the skin, bones, joints, and internal organs. Two types of pain are experienced: somatic pain and visceral pain.

Somatic pain results when pain receptors in the skin, joints, muscles, and tendons are stimulated. Somatic pain fibers respond to several types of mechanical and chemical stimuli. Some pain receptors in the skin, for example, respond to prostaglandins released by injured tissues (Chapter 10). These receptors are, therefore, a kind of chemoreceptor.

Visceral pain results from the stimulation of pain fibers in body organs (the viscera). Here, pain is stimulated by expansion and by **anoxia**, a lack of oxygen. Intestinal pain you feel when you have a bad case of gas, for example, results from the stretching of naked nerve fibers in the wall of the intestine. These receptors are mechanoreceptors. The pain felt during a heart attack results from a lack of oxygen flowing to the heart muscle (Chapter 8).

Visceral pain and somatic pain result from quite different stimuli and are perceived very differently as well. Somatic pain is easily identified whereas visceral pain is vague and often difficult to localize. Visceral pain is often felt on the body surface at a site some distance from its origination. Anginal pain, caused by a lack of oxygen to heart muscle, for example, appears along the inside of the left arm (Figure 13–2). Pain from the lung and diaphragm appears in the neck.

Pain that appears on the body surface away from the location of the pain is called **referred pain**. Phys-

iologists do not know the cause of this phenomenon, but many think that it results from the fact that pain fibers from internal organs enter the spinal cord at the same location the sensory fibers from the skin enter. The brain, therefore, misinterprets impulses from pain fibers supplying the organs as pain from a somatic source (see Health Note 13–1 for some techniques to relieve pain).

Light Touch. Light touch is perceived by two anatomically distinct mechanoreceptors. The first receptor is located at the base of the hairs in our skin. As shown in Figure 13–1, naked nerve endings (dendrites) wrap around the base of the hair follicles. When a hair is moved—for example, by a light breeze or a gentle touch—these nerve fibers are stimulated. The second light touch mechanoreceptor is the **Merkel disc.** Shown in Figure 13–1, Merkel discs consist of small cup-shaped cells on which naked nerve endings terminate. Located in the outer layer of the skin (epidermis), these receptors are activated by gentle pressure on the skin.

Encapsulated Receptors

The **encapsulated receptors** contain a naked nerve ending surrounded by one or more layers of cells, the capsule. Shown in Figure 13–1, the largest encapsulated nerve ending is the **Pacinian corpuscle.** It consists of a free nerve ending surrounded by numerous concentric cell layers. The Pacinian corpuscle resem-

HEALTH NOTE 13–1
Old and New Treatments for Pain

Millions of people suffer from chronic or persistent pain. In the United States, for example, an estimated 70 million people are tormented by back pain. Another 20 million suffer from migraine headaches.

Despite its prevalence, pain is one of the least understood medical problems in the world. The study of pain, in fact, is so poorly funded and so widely ignored by the medical community that some have called it an orphan science. Making matters worse, physicians are often poorly trained in the matter of pain.

For many decades, pain was treated with painkillers, morphine and codeine. Their addictive nature led researchers to look for other techniques. One of the more promising was acupuncture, a technique used by the Chinese for thousands of years. Acupuncture relies on thread-thin needles inserted in the skin near nerves and rotated very quickly (see the figure).

Neurologists are not certain how acupuncture works, but many think that it blocks pain by overloading the neuronal circuitry. Here's the theory: Neurologists believe that two types of nerve fibers transmit sensory information from the body

Patient Undergoing Acupuncture
Acupuncture is generally used to alleviate pain, but has other applications. This patient is being treated with acupuncture in an experimental program to combat drug addiction.

to the central nervous system. Small-diameter fibers carry pain messages. Larger-diameter fibers carry many other forms of sensory information from receptors in the skin. For example, these fibers are thought to carry pressure and light

touch. The dendrites of both small- and large-diameter sensory nerve cells often terminate in the same location in the spinal cord. From here they send impulses to the brain, signaling pain or some other sense.

Nerve impulses traveling from the pain receptors, however, can be blocked by simultaneously stimulating the larger-diameter fibers. During acupuncture, needles inserted in key spots can stimulate the larger nerve fibers, blocking pain. Physicians who are trained to use drugs and surgery to solve most pain remain skeptical as to the usefulness of acupuncture in relieving pain. In the past 10 years, however, a small but steady stream of research has confirmed the painkilling effect of this treatment.

Joseph Helms, a physician with the American Academy of Acupuncture in Berkeley, California, for example, performed acupuncture on 40 women with menstrual pain. Some women received real acupuncture treatment. Others received placebo treatments (shallow needle treatments that did not reach the acupuncture points). In the group of women receiving acupuncture, 10 out of 11 showed a marked decrease in pain. Patients reported an approx-

bles a cocktail onion pierced by a thin wire. Pacinian corpuscles are located in the deeper layers of the skin, in the loose connective tissue of the body, and elsewhere; they are thought to be stimulated by pressure— for example, the pressure you feel sitting in your chair.

Another common encapsulated sensory receptor is the Meissner's corpuscle. **Meissner's corpuscles** are smaller, oval receptors that contain two or three spi-

raling dendritic ends surrounded by a thin cellular capsule (Figure 13–1). Meissner's corpuscles are thought to be mechanoreceptors that respond to light touch and are located just beneath the epithelium in the outermost layer of the dermis. Meissner's corpuscles are most abundant in the sensitive parts of the body, such as the lips and the tips of the fingers.

Two of the more controversial receptors are Krause's end-bulbs and Ruffini's corpuscles (Figure

imately 50% decrease in pain. In the placebo group only 4 out of 11 reported a lessening of pain. Only 1 of 10 people given no treatment showed improvement. Acupuncture also reduced the need for painkilling drugs during treatment by over half. Remember, however, that the Critical Thinking section in Chapter 1 suggested that experiments using small numbers of subjects are themselves subject to question. Further studies are needed to confirm these results.

While acupuncture is slowly earning a respected place in the treatment of pain, researchers are also experimenting with a technique to relieve pain called transcutaneous electrical nerve stimulation, or TENS. Patients are fitted with electrodes that attach to the skin. A small battery supplies energy that stimulates the electrodes placed on the skin over the nerves that transmit pain to the central nervous system. When the pain begins, patients send a tiny current to the electrode. The current is conducted through the skin and blocks the pain impulses.

TENS can be used to reduce pain after surgery. One study showed that this technique reduced the amount of painkillers doctors had to administer by two-thirds and cut hospital stays by one or two days. Someday dentists may use TENS instead of the novocaine. TENS may also be used to reduce the pain of childbirth and could help treat the pain athletes suffer.

TENS works like acupuncture. Part of its success may be psychological, however. In a study of 93 patients with chronic pain, for example, researchers found that over one-third (36%) of them reported no pain or greatly reduced pain when they thought they were being stimulated but really weren't. How effective was it when the battery really worked? About half of the patients reported no pain or greatly reduced pain. The slight improvement suggests to some physicians that TENS may be overrated. To those suffering from chronic pain, it can be a godsend.

Severe pain can also be treated by surgery. Doctors may cut nerves or destroy small parts of the brain to get rid of chronic pain. Unfortunately, pain recurs in 9 of 10 patients who have undergone surgery, usually within a year or so. Recurring pain results from partial regrowth of axons. Even after another surgery, the pain frequently returns, often with much greater intensity. Consequently, physicians are now looking for alternative measures to eliminate chronic pain.

One promising measure is deep brain stimulation. Electrodes can be implanted in parts of the brain and stimulated to block pain impulses before they reach the sensory part of the cortex where pain is perceived. The electrodes are connected to a portable battery worn on the belt or implanted under the skin. When the pain begins, the patient turns on the electrical current, thus blocking the pain impulses.

Research shows that deep brain stimulation is an effective blocker of even the most powerful pain stimuli. Yet it does not upset other brain functions. Unfortunately, this technique requires surgery. Implanting electrodes may cause hemorrhaging in some patients, which may result in permanent paralysis or a loss of feeling in parts of the body. Infections may also develop where the electrodes enter the skull. The 75% success rate and the reduced suffering make most patients more than willing to accept the risks.

13–1). Although many scientists have long thought that Krause's end-bulbs were cold receptors and that Ruffini's corpuscles were heat receptors, not all researchers agree. Some think that these receptors are actually structural variations of the Meissner's corpuscle and are stimulated by light touch. Hot and cold receptors, they say, are probably naked nerve endings.

Proprioception (position sense) is provided by encapsulated receptors located in the joints of the body. Resembling Meissner's corpuscles, these receptors inform us where our limbs are and alert us to body movement. Proprioception is also served by two other encapsulated mechanoreceptors: the muscle spindle and the Golgi tendon organ (Figure 13–3). **Muscle spindles** or **neuromuscular spindles** are found in the skeletal muscles of the body. A muscle spindle consists of several modified muscle fibers with sensory nerve endings wrapped around them; a thin capsule

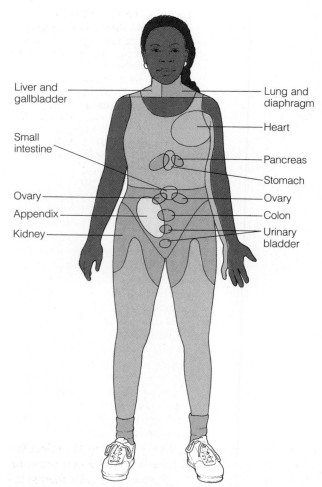

FIGURE 13–2 Referred Pain Visceral pain is often felt on the body surface.

Liver and gallbladder

Small intestine

Ovary

Appendix

Kidney

Lung and diaphragm

Heart

Pancreas

Stomach

Ovary

Colon

Urinary bladder

surrounds the entire structure. When a muscle is extended or stretched, spindle fibers are stimulated. They generate nerve impulses that are transmitted to the spinal cord via sensory nerves. Impulses reaching the spinal cord may ascend to the cerebellum and cortex, alerting the brain to the stretch of body muscles and helping to keep us aware of body position. The nerve impulses generated by muscle spindles may also stimulate motor neurons in the spinal cord, eliciting a reflex contraction of the muscle that is being stretched. This counteracts the stretching.

Another mechanoreceptor, which is functionally similar to the muscle spindle, is found in **tendons**—the connective tissue structures that join muscles to bones. Called **Golgi tendon organs** or **neurotendinous organs**, these receptors consist of connective tissue fibers that are surrounded by dendrites and en-

cased in a capsule (Figure 13–3). When a muscle contracts, the tendon stretches and stimulates the receptor. Like the muscle spindle, this receptor alerts the brain to movement and body position. Impulses from the Golgi tendon organ can also stimulate reflex contraction of muscles. The knee-jerk reflex described in the last chapter is a good example.

Habituation

Pain, temperature, and pressure receptors are all subject to a phenomenon known as habituation. **Habituation** occurs when sensory receptors stop generating impulses even though the stimulus is still present. You have probably witnessed the phenomenon in your lifetime dozens of times. Recall, for example, the first time you wore a ring or contact lenses. At first the sensation may have nearly driven you mad, but after a few days—or perhaps a few hours—the discomfort waned, and the stimulus seemed to have disappeared. Pressure receptors that were originally alerting the brain stopped generating impulses, relieving you of what would have otherwise been a lifetime of discomfort.

Receptors and Homeostasis. The general sense receptors discussed so far are supplemented by a long list of receptors that also play an important role in homeostasis. Chapters 8 and 11, for example, described mechanoreceptors that detect changes in blood pressure and chemoreceptors that respond to the concentration of sodium ions in the blood. These receptors help regulate blood volume and blood concentration through the kidney. In Chapter 9, we examined chemoreceptors that detect carbon dioxide and hydrogen ion levels in the blood and cerebrospinal fluid. These receptors play important roles in respiration, a function vital to normal body function. In Chapter 15, we will study an additional group of chemoreceptors that detect levels of various hormones, nutrients, and ions in the blood and body fluids. These detectors may stimulate hormonal responses that correct potentially disruptive chemical imbalances.

TASTE AND SMELL: THE CHEMICAL SENSES

The special senses include taste, smell, vision, hearing, and balance. The following sections discuss the special senses, beginning with taste.

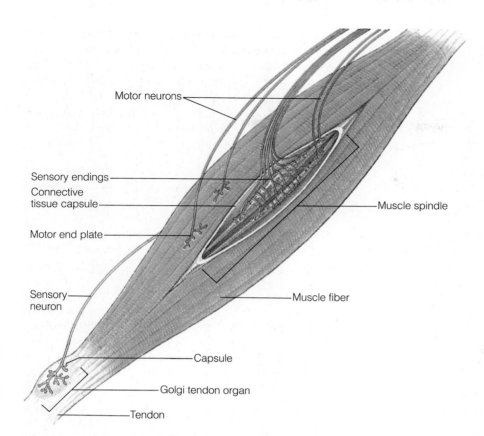

Motor neurons

Sensory endings

Connective
tissue capsule

Motor end plate

Sensory
neuron

Muscle spindle

Muscle fiber

Capsule

Golgi tendon organ

Tendon

FIGURE 13–3 Stretch Receptors The muscle spindle and Golgi tendon organ both detect muscle stretch. The dendrites of the sensory neurons from these receptors send impulses to the spinal cord. These impulses may trigger motor neurons to fire; returning signals cause the muscle to contract. Sensory signals reaching the spinal cord also ascend to the brain, alerting it to the muscle movement.

Taste

In humans and other mammals, the tongue contains receptors for taste or **taste buds**, small onion-shaped structures located in the surface epithelium and on small protrusions on the upper surface of the tongue called **papillae** (Figure 13–4a). Taste buds are also found on the roof of the oral cavity (on the hard and soft palates), the pharynx, and the larynx, but in smaller numbers.

Taste buds are chemoreceptors stimulated by chemicals in the food we eat. These chemicals dissolve in the saliva and enter the **taste pores**, small openings leading to the interior of the taste bud. As Figure 13–4c illustrates, taste buds consist of two cell types, receptor cells and supporting cells (which may be the same cell type, just in a different phase of their life cycle). The **supporting cells** lie on the outside of the taste bud and resemble the staves of a wooden barrel. The **receptor cells** lie inside, like so many pickles jammed into the barrel. As Figure 13–4c shows, the ends of the receptor cells possess large microvilli called gustatory or **taste hairs**, which project into the taste pore. The membrane of the taste

hairs contains receptors that bind to food molecules. The binding of food molecules to the receptors stimulates the receptor cells. The receptor cells then stimulate the dendrites of the sensory nerves wrapped around them.[2] Impulses from the taste buds are then transmitted to the taste centers in the cerebral cortex.

In the opening paragraph of this chapter, you were introduced to an unfortunate student who lost her sense of taste because of a dietary zinc deficiency. Research has shown that zinc, normally found in low concentrations in the saliva, stimulates growth of the cells in the taste buds. A zinc deficiency, therefore, reduces cell division. The cells of the taste buds that are lost from normal wear and tear are not replaced, and the taste buds cease operation.

The taste buds respond to four basic flavors: sweet, sour, bitter, and salty. Sweet flavors result from sugars and some amino acids. Sour flavors result from acidic substances, while salty tastes result from metal ions

[2]Note: unlike the receptors in the previous section, the receptor cells of taste buds are not part of nerve cells, but are independent cells.

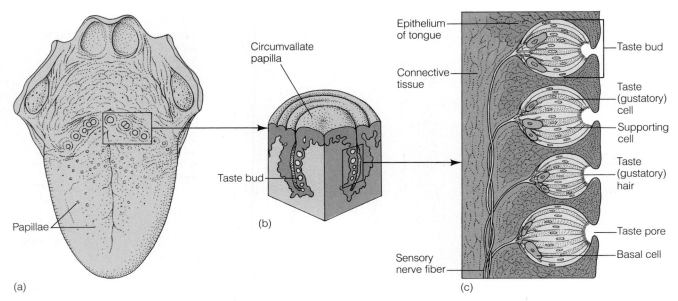

FIGURE 13–4 Taste Receptors (a) Taste buds are located on the dorsal surface of the tongue and are concentrated on the (b) papillae. (c) The structure of the taste bud.

(like sodium in table salt). Bitter flavors result from alkaloids (caffeine) and nonalkaloids (aspirin).

A simple experiment in which drops of these chemicals are placed on the tongue shows that the tip of the tongue is most sensitive to sweet flavors (Figure 13–5). The sides of the tongue are most sensitive to sour flavors, and the back of the tongue is most sensitive to bitter flavors. Salty taste is more evenly distributed, with slightly increased sensitivity on the sides of the tongue near the front. Although there are regions of greater and lesser sensitivity, most taste buds respond to two or more taste sensations.

Food contains many different flavors. What we taste, therefore, depends on the relative proportion of the four basic flavors. For example, grapefruit juice tastes sour because of the predominance of acidic substances. Adding sugar to grapefruit juice, however, gives it a sweet-sour flavor. If you add enough sugar, you can mask the sour taste almost entirely.

FIGURE 13–5 The Distribution of Taste on the Dorsal Surface of the Tongue

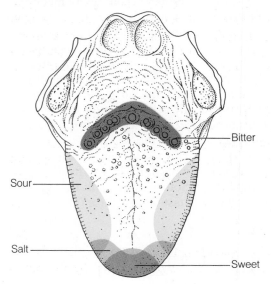

Smell

Smell is also a chemical sense. The receptors for smell are located in the roof of each nasal cavity in a patch of cells called the **olfactory epithelium** (Figure 13–6). The olfactory epithelium contains receptor cells and supporting cells. **Receptor cells** are neurons whose cell bodies lie in the olfactory membrane. The dendrites of these neurons extend to the surface of the olfactory membrane, terminating in six to eight long projections, **olfactory hairs** or **olfactory cilia** (Figure 13–6). The membranes of the olfactory hairs contain receptors for molecules. When airborne molecules bind to these receptors, they activate the neurons,

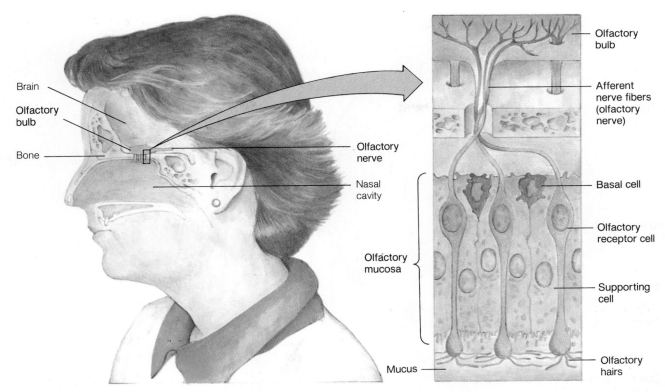

FIGURE 13–6 Location and Structure of the Olfactory Epithelium Olfactory receptors are located in the olfactory epithelium in the roof of the nasal cavity. Chemicals in the air bind to receptors on the plasma membranes of the olfactory hairs. The olfactory receptors terminate in the olfactory bulb. From here, nerve fibers travel to the brain.

causing them to generate impulses that are sent to the brain via the **olfactory nerve. Supporting cells** are interspersed among receptor cells.

Humans can perceive thousands of individual smells. The olfactory receptors are so sensitive that even a single molecule binding to the olfactory hairs can produce a bioelectric impulse.

Receptors for smell habituate within a short period—only about a minute. If you have ever worked on a dairy farm or lived near one, or if you have been around a baby in diapers on a long car trip, you understand (and are grateful for) olfactory habituation.

Hold a piece of hot apple pie to your nose and take a deep breath. It smells so good you can almost taste it. In fact, you are tasting it. Molecules given off by the pie enter the mouth and dissolve in the saliva where they stimulate taste receptors. Just as odors stimulate taste receptors, food in our mouths also stimulates olfactory receptors. Chemicals released by food enter the nasal cavities, stimulating the olfactory epithelia.

The complementary nature of taste and smell is abundantly evident when a person suffers from nasal congestion. As you may have noticed, a viral infection in the nose often makes tasty food seem bland. People complain that they can't taste their food. This phenomenon results from the buildup of mucus, which coats the olfactory epithelium, reducing access to the olfactory hairs. Food loses its "taste" when you have a stuffy nose because your sense of smell is impaired.

THE VISUAL SENSE: THE EYE

The eye is one of the most extraordinary products of evolution. It contains a patch of photoreceptors that detect a remarkably diverse and colorful environment.

Structure of the Eye

The eyes are roughly spherical organs located in the eye sockets or **orbits**, cavities formed by the bones

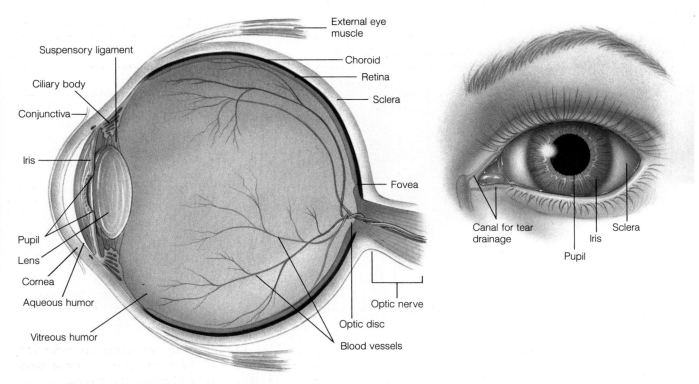

FIGURE 13-7 Cross Section of the Human Eye

of the skull. The eye is attached to the orbit by six muscles, the **extrinsic eye muscles**, which control eye movement. Small tendons connect these muscles to the outermost layer of the eye.

The Sclera and Cornea. As shown in Figure 13-7, the wall of the eye consists of three layers (Table 13-2). The outermost is a durable fibrous layer, which consists of the **sclera**, the white of the eye, and the **cornea**, the clear part in front, which lets light into the interior of the eye (Figure 13-7). Tendons of the extrinsic eye muscles attach to the sclera.

The Choroid. The middle layer is a heavily pigmented and highly vascularized region. As Figure 13-7 shows, the middle layer consists of three parts: the choroid, the ciliary body, and the iris. The **choroid** is the largest portion of the middle layer. Its pigments absorb stray light like the black interior of a camera. The blood vessels of the choroid supply nutrients to the eye. Anteriorly, the choroid forms the ciliary body. The **ciliary body** contains smooth muscle fibers, which control the shape of the lens, permitting us to focus incoming light given off by objects. The third

part of the middle layer is the iris. The **iris** is the colored portion of the eye visible through the cornea. Looking in a mirror, you can see a dark opening in the iris called the pupil. The **pupil** allows light to penetrate the eye. The blackness you see is the choroid layer and the pigmented section of the retina, discussed shortly.

Like the ciliary body, the iris contains smooth muscle cells. The smooth muscle of the iris regulates the diameter of the pupil. Opening the pupil lets more light in, and narrowing it reduces the amount of light that can enter. The pupils open and close reflexively in response to light intensity, thus protecting the light-sensitive inner layer, the retina. The pupils also constrict when the eye focuses on a near object, a process discussed later.

The Retina. The innermost layer of the eye is the **retina**. The retina consists of an outer pigmented layer that complements the light-absorbing function of the choroid layer and an inner layer, the neural layer, consisting of photoreceptors and associated nerve cells. The retina is weakly attached to the choroid and can become separated—for example, in a traumatic

TABLE 13–2   Structures and Functions of the Eye

LAYER	STRUCTURE	FUNCTION
Outer	Sclera	Insertion for extrinsic eye muscles
	Cornea	Allows light to enter; bends incoming light
Middle	Choroid	Absorbs stray light; provides nutrients to eye structures
	Ciliary body	Regulates lens, allowing it to focus images
	Iris	Regulates amount of light entering the eye
Inner	Retina	Responds to light
Accessory structures and components		
	Lens	Focuses images on the retina
	Vitreous humor	Holds retina and lens in place
	Aqueous humor	Supplies nutrients to structures in contact with the anterior cavity of the eye
	Optic nerve	Transmits impulses from the retina to the brain

accident. A detached retina can lead to blindness if not repaired by surgery.

The **photoreceptors** are modified nerve cells located in the outermost portion of the neural layer of retina, adjacent to the pigmented layer. Two additional layers of neurons are described below.

Two types of photoreceptors are present in the retina: rods and cones (Table 13–3). The **rods**, so named because of their rodlike shape, are sensitive to low light (Figures 13-8b and 13-8c). Thus, the rods function on a moonlit evening, yielding grayish, somewhat vague images. The **cones**, also named because of their shape, are photoreceptors that sense colors and operate only in brighter light.

As Figure 13–8b shows, the rods and cones synapse with the bipolar neurons, which, in turn, synapse with ganglion cells. The bipolar neurons and ganglion cells lie in front of the photoreceptors. Light enters the eye through the cornea, passes through the lens, and is projected onto the retina. In order for light to stimulate the photoreceptors, it must pass through the ganglion and bipolar cell layers.

Nerve impulses from the retina travel from the bipolar neurons to the ganglion cells. The axons of the ganglion cells course along the inner surface of the retina and unite at the back of the eye to form the **optic nerve**. The optic nerve exits at the **optic disc** or **blind spot**, so named because it contains no photoreceptors and is, therefore, insensitive to light. The blood vessels that enter and leave the eye do so with the optic nerve. These blood vessels and their branches can readily be seen by an ophthalmologist by shining a light through the pupil onto the posterior wall of the eye (Figure 13–9).

Rods and cones are found throughout the retina, but the cones are most abundant in a tiny region of each eye lateral to the optic disc. This spot is called the **macula lutea**—the yellow spot. In the center of

TABLE 13–3   Summary of Rods and Cones

PHOTORECEPTOR	DAY OR NIGHT	COLOR VISION	LOCATION
Rods	Night vision	No	Highest concentration in the periphery of the retina.
Cones	Day vision	Yes	Highest concentration in the macula and fovea.

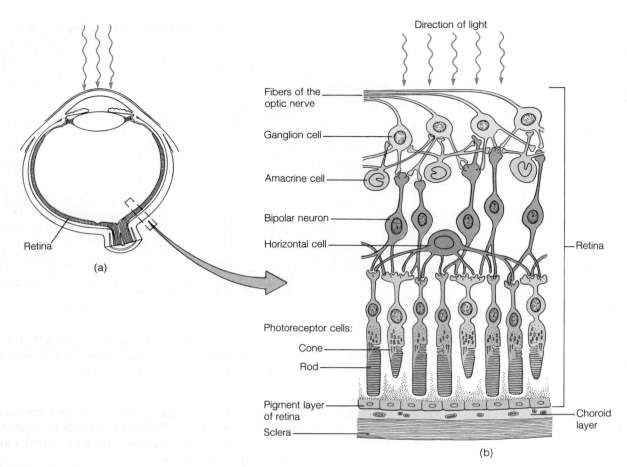

Direction of light

Fibers of the
optic nerve

Ganglion cell

Amacrine cell

Bipolar neuron

Horizontal cell

Retina

Photoreceptor cells:

Cone

Rod

Pigment layer
of retina

Sclera

Choroid
layer

(b)

Retina

(a)

FIGURE 13–8 The Retina (*a*) Cross section through the wall of the eye, showing (*b*) the arrangement of the cellular components of the retina. (*c*) The structure of the rods and cones.

the macula is a minute depression, about the size of a pin head, known as the **fovea centralis** (central depression). The fovea contains only cones. The number of rods in the retina increases progressively from this point outward and is greatest in the periphery of the retina. While the number of rods increases from the fovea outward, the number of cones decreases.

Images from our visual field are cast onto the retina, and impulses are transmitted to the visual cortex of the occipital lobe where they are interpreted. When focusing on an object, the image is projected onto the fovea. The sharpest vision occurs here because of the high concentration of cones and because the bipolar neurons and ganglion cells do not cover the cones in this region as they do throughout the rest of the retina.

The Lens. The **lens** is a transparent structure that lies behind the iris and focuses light on the retina (Figure 13–10). This flexible, clear structure is attached to the ciliary body by thin fibers, **zonular fibers** or **suspensory ligament**. This connection allows the smooth muscle of the ciliary body to alter the shape of the lens, an action necessary for focusing the eye.

In older individuals, the lens may develop cloudy spots or **cataracts**. The loss of transparency is especially prevalent in people who have been exposed to excessive sunlight or excessive ultraviolet light at work or from other sources. Victims of this disease complain of cloudy vision. Looking out on the world to them is a little like looking at the world through frosted glass.

Ophthalmologists once treated cataracts surgically by removing the afflicted lens. Patients were once fitted with a thick pair of glasses or a pair of contact lenses, which compensated for the missing lens. Today, however, lenses are routinely replaced by artifi-

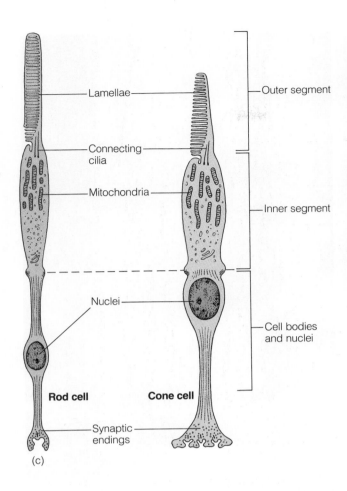

- Lamellae
- Connecting cilia
- Mitochondria
- Nuclei
- Rod cell
- Cone cell
- Synaptic endings
- Outer segment
- Inner segment
- Cell bodies and nuclei

(c)

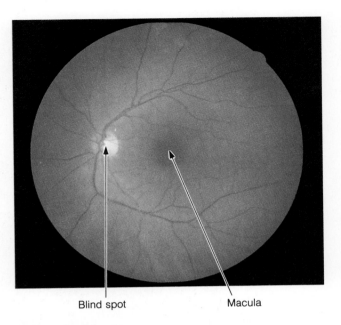

Blind spot   Macula

FIGURE 13–9 View of the Inside Back Wall of the Eye Seen through an Ophthalmoscope The optic disk (blind spot) and macula are both indicated.

cial plastic lenses. The implant is attached to the iris and will restore nearly normal vision.

Research has shown that cataracts may result when a lens protein called **crystallin** denatures. Researchers are hoping that new drugs may be able to reverse the process, restoring vision without surgery.

The lens separates the interior of the eye into two cavities of unequal sizes. Everything in front of the lens is the anterior cavity; everything behind it is the posterior cavity. The posterior cavity is filled with a clear, gelatinous material, the **vitreous humor** (glassy liquid). Formed during embryonic development, the vitreous humor remains throughout life, holding the lens and retina in place. Being clear, the vitreous humor transmits light faithfully to the retina.

The anterior cavity is further divided into two parts: the **anterior chamber** and the **posterior chamber** (Figure 13–10). The anterior chamber lies between the iris and the cornea; the posterior chamber

lies between the iris and the lens. A thin watery liquid chemically similar to blood plasma fills the anterior and posterior chambers of the eye and is called **aqueous humor** (watery fluid).

Unlike the vitreous humor, the aqueous humor is constantly replaced. New fluid is produced by capillaries in the ciliary body. The fluid enters the posterior chamber, then flows forward into the anterior chamber where it drains into a venous sinus, called the **canal of Schlemm**, located at the junction of the sclera and cornea. From here, the plasmalike fluid flows into the bloodstream.

Aqueous humor provides nutrients to the cornea and lens and draws away cellular wastes. In normal, healthy individuals, aqueous humor production is balanced by absorption. If the outflow is blocked, however, aqueous humor builds up inside the anterior chamber, creating internal pressure. This disease, called **glaucoma**, occurs gradually and imperceptibly. If untreated, the intraocular pressure can damage the retina and optic nerve, causing blindness. Because the incidence of glaucoma increases after age 40, doctors recommend an annual eye examination for men and women over 40. If diagnosed early, glaucoma can be treated with eye drops that increase the rate of drainage, thus reducing intraocular pressure.

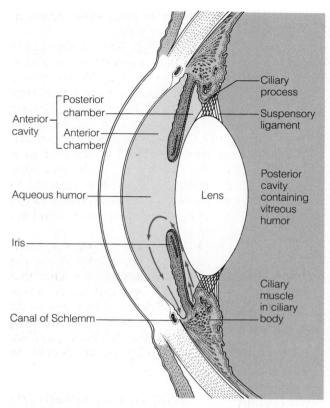

FIGURE 13–10   Detailed Cross Section of the Anterior Cavity Showing the Flow of Aqueous Humor

FIGURE 13–11   **Refraction**   The pencil in the glass of water appears to bend. What is happening, though, is that the light rays coming to our eyes are bent because they must pass through the water and glass. When they do, they bend, making the pencil appear bent.

Physiology of the Eye

To understand the eye's function, you have to understand a little bit about light.

Refraction.   Visible light travels in waves. Light waves travel at a constant rate in any given medium such as air or water. When light passes from one medium to another, however, its velocity changes—either speeding up or slowing down. When light passes from a less dense medium such as air to a denser medium such as the cornea, it slows down. Anytime light changes speed in passing from one substance to another, it bends. The bending of light is called **refraction** (Figure 13–11).

Focusing the Image.   Light traveling through a camera lens is bent. The lens of the camera is designed to bend light enough to focus the image on the film. The lens of the eye also bends incoming light rays, focus-

ing images on the photoreceptors of the retina. Lying in front of the lens is the cornea; it also bends incoming light rays (Figure 13-12). Although we usually think of the lens as the structure that allows us to focus, most of the bending of incoming rays takes place in the cornea. However, the cornea is a fixed structure. Like the lens on a fixed-focus camera, the cornea cannot be adjusted to focus on nearby objects. Without the adjustable lens, the eye would be unable to focus on objects close at hand.

The lens is resilient like a rubber ball. Its shape is controlled by the muscles in the ciliary body, which are attached via the zonular fibers. When the muscles of the ciliary body are relaxed, the zonular fibers are

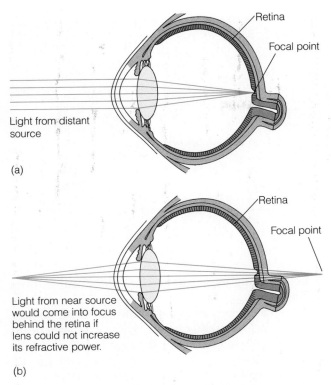

(a)

Light from distant source

Retina

Focal point

(b)

Retina

Focal point

Light from near source would come into focus behind the retina if lens could not increase its refractive power.

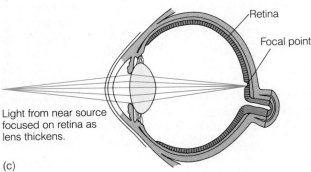

(c)

Retina

Focal point

Light from near source focused on retina as lens thickens.

FIGURE 13–12   Refraction of Light by the Cornea and Lens (*a*) Light rays from distant objects are parallel when they strike the eye. The refractive power of the cornea and resting lens are sufficient to bring them into focus on the retina. (*b*) Light rays from nearby objects are divergent. The cornea has a fixed refractive power and cannot change. The rays would focus behind the retina if the lens could not alter its refractive power. (*c*) To focus the image, then, the ciliary muscles contract. This lessens pressure on the zonular fibers, which allows the lens to thicken and shorten, becoming more curved and more refractive.

taut and the lens is somewhat flattened. As Figure 13–12a shows, light from distant objects comes to the eye as nearly parallel rays. The fixed refractive power of the cornea and the refractive power of the lens in its

relaxed state are sufficient to bend these beams to bring them into focus on the retina.

As Figure 13–13b shows, light from nearby objects is divergent. To focus on nearby objects, the lens must change shape (become more curved). When the eye focuses on nearby objects, the ciliary smooth muscle contracts, causing the zonular fibers to relax (Figure 13–12b). The lens thickens and shortens, becoming more curved, in much the same way that a rubber ball flattened between your hands will return to normal when you reduce the pressure on it. The automatic adjustment in the curvature of the lens required to focus on a nearby object is called **accommodation** (Figure 13–13). Accommodation increases the refractive power of the lens.

Accommodation is enhanced by pupillary constriction. As the eyes focus on a nearby object, the pupils constrict. Pupillary constriction is a reflex that eliminates divergent rays of light that would otherwise strike the periphery of the lens. The lens would be unable to bend these rays sufficiently to bring them into focus on the retina. Without pupillary constriction, images of nearby objects would be blurred.

Convergence.   The human eyes are movable. Six muscles located outside the eye, the extrinsic eye muscles, are responsible for movement (Figure 13-14). These muscles attach to the orbit and to the sclera and allow for a wide range of movements.

The eyes generally move in unison like a pair of synchronized swimmers. Synchronized movements ensure that the image is focused on the foveas of both eyes at the same time. To test the synchronized movement, close one of your eyes. Place an index finger gently over the closed lid. Then hold the other hand in front of your face and move it back and forth, then up and down, following it with your opened eye. You should feel your closed eye moving in sync, even though the lid is shut.

When a nearby object is viewed, the eyes turn inward or converge. Convergence ensures that the image is focused on each fovea. Convergence occurs during all near-point work—reading, writing, and so on—and puts strain on the extrinsic eye muscles, contributing to eye strain.

Common Visual Problems

In the normal eye, objects farther than 6 meters (20 feet) away fall into perfect focus on the back of the

FIGURE 13–13

Accommodation (*a*) The lens is flattened when the ciliary muscles are relaxed. (*b*) The lens shortens and thickens when the ciliary muscles contract, reducing tension on the zonular fibers (also called suspensory ligaments).

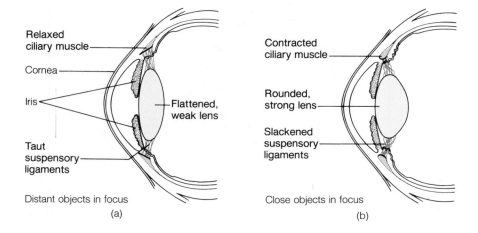

Relaxed ciliary muscle

Cornea

Iris

Flattened, weak lens

Taut suspensory ligaments

Distant objects in focus

(a)

Contracted ciliary muscle

Rounded, strong lens

Slackened suspensory ligaments

Close objects in focus

(b)

relaxed eye (Figure 13–15a). Many individuals, however, have imperfectly shaped eyeballs or defective lenses. These imperfections result in two visual problems: nearsightedness or farsightedness.

Myopia. **Nearsightedness** or **myopia** results when the eyeball is slightly elongated (Figure 13–15b). Without corrective lenses, the parallel light rays arising from distant images would fall into focus in front of the retina, as Figure 13–15b illustrates. Nearby images, with much more divergent light rays, however, tend to be in focus in the uncorrected eye. People with myopia, therefore, can see near objects without corrective lenses—hence the name, nearsightedness. Myopia may also result from a lens that is too concave. A defective lens bends the light too

FIGURE 13–14 Extrinsic Eye Muscles These muscles move the eye in all directions. They attach to the bony orbit and the sclera.

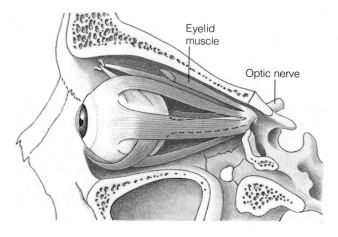

Eyelid muscle

Optic nerve

much, causing the image to come into focus in front of the retina.

Myopia is quite common in the American population. Approximately one of every five Americans needs glasses to correct for myopia. Myopia tends to run in families; it generally appears around age 12, often worsening until a person reaches 20.

Myopia can be corrected by contact lenses or prescription glasses that diverge incoming light rays (Figure 13–15). Contact lenses fit on the surface of the cornea and bend the incoming light rays outward, compensating for the shape of the eye or a defective ocular lens. Surgeons have also developed a method to decrease the refractive power of the cornea, called **radial keratotomy**. Numerous, small superficial incisions are made in the cornea, radiating from the center like the spokes of a bicycle wheel. This procedure flattens the cornea and thus reduces its refractive power, causing the rays to diverge and focusing them on the back of the retina. The long-term effectiveness and safety of radial keratotomy are still under study.

Hyperopia. **Hyperopia** or **farsightedness** is the opposite of myopia. Without corrective lenses, farsighted individuals see distant objects well without correction, but nearby objects are fuzzy. In some people, hyperopia results from an eyeball that is too short; in others, the condition results from a lens that is too weak (Figure 13–15c). In hyperopia, light rays from nearby objects are cast onto the retina out of focus. The ideal focal point lies behind the retina, as Figure 13–15c illustrates. Glasses or contact lenses that bend the light inward help bring objects into sharp focus on the retina.

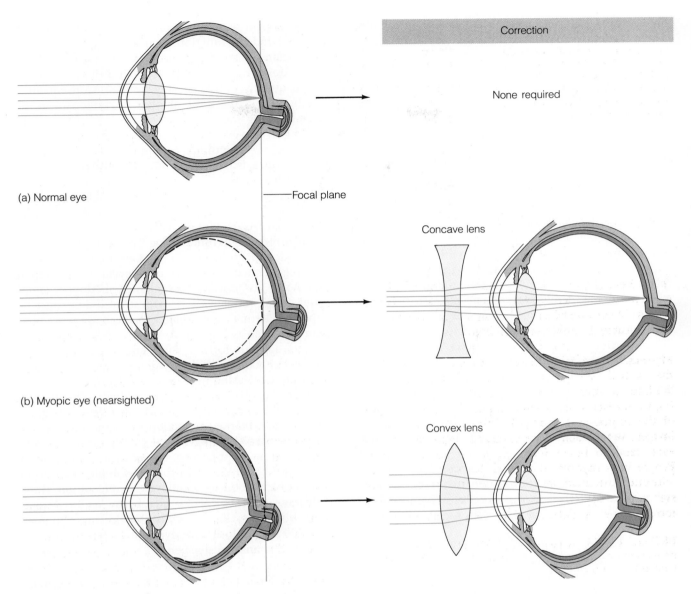

None required

(a) Normal eye

Focal plane

Concave lens

(b) Myopic eye (nearsighted)

Convex lens

(c) Hyperopic eye (farsighted)

FIGURE 13–15 Common Visual Problems (a) The normal eye, (b) the myopic (nearsighted) eye, and (c) the (farsighted) hyperopic eye.

Hyperopia is generally present from birth and is usually diagnosed during childhood. Like myopia, it tends to run in families.

Astigmatism. In the normal eye, the cornea and lens have uniformly curved surfaces. Either of these structures may be slightly disfigured, however. The surface of the cornea, for example, may have a slightly different curvature in the vertical plane than it does in the horizontal plane. This unequal curvature is called an **astigmatism**. It creates fuzzy images because light rays are bent differently by the different parts of the lens or cornea. Astigmatism is usually present from birth and does not grow worse with age. Like other conditions, it can be corrected with specially ground glasses and contact lenses.

Eye Strain. The human eye is best suited for distance vision. Near-point vision—focusing on nearby objects—strains the eyes and can result in a progressive deterioration of eyesight. Frequent readers often become more nearsighted as they age. No one knows why, but research suggests that the eye may elongate as a result of constant near-point use.

To reduce eye strain and deterioration of your eyesight, ophthalmologists advise that you look away from your computer screen or your reading material regularly, letting your eyes focus on distant objects. This relaxes the ciliary muscles, reducing eye strain. Ophthalmologists suggest that computer users "Blink at every period and look up after every paragraph."

Presbyopia. Old age often brings with it many joys: hair loss, hearing loss, and arthritis. Aging also results in a loss in the resiliency of the lens. Thus, when the ciliary muscles contract to allow you to focus on a nearby object, the lens responds slowly or only partially, making it difficult to focus on nearby objects. This condition, known as **presbyopia**, usually begins around the age of 40.[3]

The Function of the Rods and Cones

With this understanding of how the eye focuses light on the retina, we now turn our attention to the retina itself. As we have seen, the cornea and lens focus images on the retina. The image cast on the rods and cones of the retina is transmitted to the brain. The brain, in turn, translates this flood of nerve impulses into a coherent image of our environment.

Rods. Each eye contains about 100 million rods. Rods are sensitive to low light and contain a pigment called **rhodopsin**. When light strikes the rods, rhodopsin molecules in the cytoplasm of the photoreceptors split into two component molecules, retinal and opsin. **Retinal** is a derivative of vitamin A. **Opsin** is an enzyme. In the dark, the rods release a steady stream of neurotransmitter. The neurotransmitter released by the rods inhibits bipolar neurons from firing. When light stimulates the rods, however, the breakdown of rhodopsin inhibits the release of neurotransmitter. This removes the inhibition on the bipolar neurons, allowing them to send impulses to the brain.

[3]You may have seen your parents or grandparents holding a phone number at arm's length to read it. They may have argued that their eyes weren't going bad, their arms were just too short!

Rhodopsin is very sensitive to light, thus permitting the rods to function even in dim light. Because rhodopsin is so easily dissociated, the molecules break down rapidly during daylight hours, dissociating as quickly as they are regenerated. Consequently, bright light reduces the amount of rhodopsin in the rods, making them ineffective in daylight hours. In dim light, however, rhodopsin is regenerated and the rods become functional. Stepping into a dark movie theater on a bright day, you probably have noticed that, at first, your vision was greatly impaired. That is because your rods, which had been bleached by bright outdoor light, require some time to become fully functional. As rhodopsin molecules are regenerated and the rods begin functioning, your eyesight returns.

Rhodopsin molecules in the rods are broken down and regenerated in a continuous cycle. While retinal and opsin are recycled to replenish rhodopsin molecules, some retinal is lost or destroyed and must be replaced. Retinal is produced in the body from vitamin A, which is found in a variety of foods, such as carrots, spinach, fortified milk, and peaches. Rhodopsin concentrations decline when vitamin A intake is reduced, decreasing the sensitivity of the rods so that they are unable to respond to dim light. Therefore, a dietary deficiency of vitamin A often leads to **night blindness**, a marked reduction in night vision.

Besides replenishing retinal supplies, vitamin A is important in maintaining the cornea. As noted in Chapter 7, a deficiency of this fat-soluble vitamin results in corneal dryness, followed by ulceration. If the deficiency persists, corneal ulcers may lead to permanent blindness. In the Third World, an estimated 100,000 people lose their eyesight each year because of severe vitamin A deficiency.

Cones. Each eye contains about 3 million cones. Like the rods, the cones contain photosensitive pigments. However, the cones operate under bright light and are also sensitive to different colors of light, thus making color vision possible.

Three types of cones are found in the human retina: blue cones, green cones, and red cones—so named because of their sensitivity to a particular color of light. Each type contains a unique type of pigment, which responds optimally to one particular color of light (Figure 13–16). The pigments inside the cones dissociate when struck by colored light. As in the rods, the dissociation of the photopigments in the cones decreases the release of neurotransmitter, "unleashing" bipolar neurons. To understand how color vision works, we must first look at the properties of visible light and examine why an object is a certain color.

The sun, light bulbs, and neon signs all emit visible light. White light from the sun and other sources is a blend of all of the colors of the rainbow (Figure 13–17). Shining white light through a prism illustrates this fact.

Light from the sun and other sources strikes objects in the environment. The color of an object, however, is determined by the kinds of pigments it contains. Pigments absorb some wavelengths of white light and reflect others. For example, the dye in a blue flannel shirt absorbs all of the colored light striking it, except blue, which it reflects. The unabsorbed wavelengths, therefore, give an object its characteristic color. Thus, a leaf appears green because the pigments in the leaf absorb all of the colored light striking it, except green, which they reflect.

Color vision occurs because each type of cone responds optimally to one color. Thus, a purely blue object stimulates "blue" cones (Figure 13–16). Red light reflected from an object stimulates "red" cones. Green light stimulates "green" cones. How do we see so many in-between shades, such as yellow? Yellow

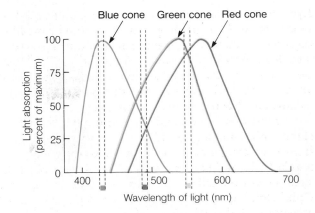

Color perceived	Percent of maximum stimulation		
	Red cones	Green cones	Blue cones
	0	0	100
	31	67	36
	83	83	0

FIGURE 13–16 Sensitivity of the Three Types of Cones to Different Colors of Light The color of light perceived is determined by the type or ratio of cones stimulated. The ratios of stimulation of the three cone types are shown for three sample colors.

FIGURE 13–17 The Electromagnetic Spectrum The sun produces a wide range of electromagnetic radiation, a small portion of which is visible to the eye. Called visible light, it consists of a variety of colors.

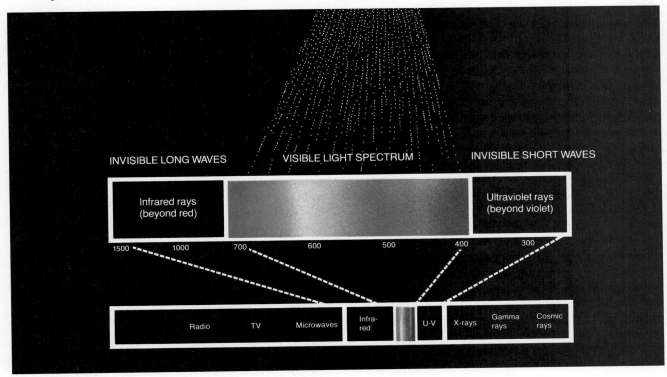

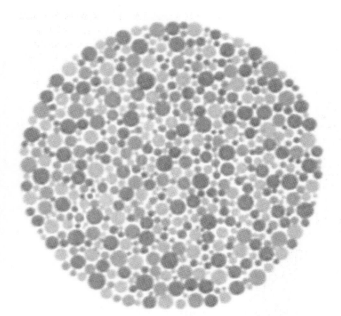

FIGURE 13–18 Color Blindness Chart People with red-green color blindness cannot detect the number 29 in this chart.

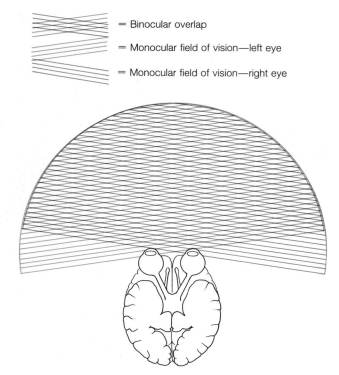

= Binocular overlap

= Monocular field of vision—left eye

= Monocular field of vision—right eye

FIGURE 13–19 Overlapping Visual Fields The overlap helps us perceive depth—three-dimensional relationships.

light has a wavelength between red and green. Blue cones are not stimulated at all. Yellow light, therefore, stimulates both red and green cones. The brain interprets the signal it receives from them as yellow light. Thus, the relative proportion of different cones stimulated determines the color we perceive.

Color Blindness. About 5% of the human population suffers from color blindness. **Color blindness** is a hereditary disorder more prevalent in men than women. Characterized by a deficiency in color perception, color blindness ranges from an inability to distinguish certain shades to a complete inability to perceive color. The most common form of this disorder is red-green color blindness. In individuals with red-green color blindness, either the red or green cones may be missing altogether or may be present in reduced number. If the red cones are missing, red and green appear the same. If the green cones are missing, however, green objects appear red.

Color blindness can be detected by simple tests (Figure 13–18). Many color-blind people are unaware of their condition or untroubled by it. They rely on a variety of cues, such as differences in intensity, to distinguish red and green objects. They also rely on position cues. For example, in traffic lights, the red light is always at the top of the signal. Green is on the

bottom. Although the colors may appear more or less the same, the position of the light helps color-blind drivers determine whether to hit the brakes or step on the gas.

Depth Perception

Human eyes are located in front of the skull, looking forward. Each eye has a visual field of about 170 degrees. Nevertheless, humans do not see in a complete circle because the visual fields overlap considerably (Figure 13–19). The overlapping of visual fields gives us the ability to judge the relative position of objects in our visual field—that is, it gives us **depth perception**.

HEARING AND BALANCE: THE COCHLEA AND MIDDLE EAR

The ear is also an organ of special sense. It serves two important functions: it detects sound, and it detects body position, helping us maintain balance.

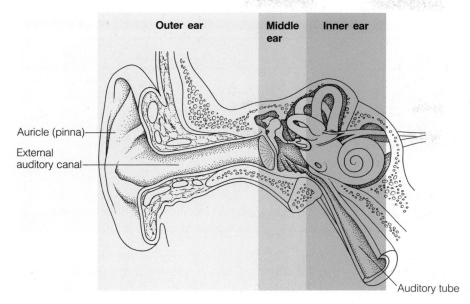

Outer ear | Middle ear | Inner ear

Auricle (pinna)

External auditory canal

Auditory tube

FIGURE 13–20 Cross Section Showing the Structure of the Ear Notice the structures of the outer, middle, and inner ears. Note also that the receptors for balance and sound are located in the inner ear.

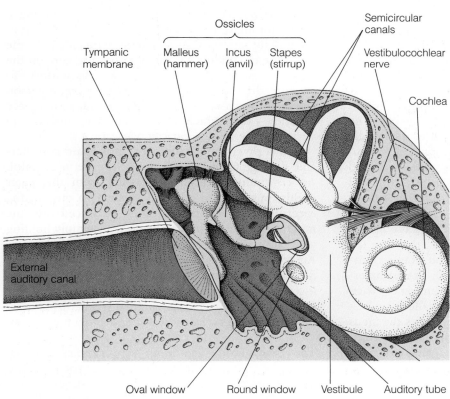

Ossicles

Tympanic membrane

Malleus (hammer)

Incus (anvil)

Stapes (stirrup)

Semicircular canals

Vestibulocochlear nerve

Cochlea

External auditory canal

Oval window | Round window | Vestibule | Auditory tube

The Structure of the Ear

The ear consists of three anatomically separate portions: the outer, middle, and inner ears (Figure 13–20; Table 13–4).

The Outer Ear.   The **outer ear** consists of an irregularly shaped piece of cartilage covered by skin. Called the **auricle** or **pinna**, this flap of tissue attached to the side of the head acts like a radar dish, picking

TABLE 13–4 Structures and Functions of the Ear

PART	STRUCTURE	FUNCTION
Outer ear	Pinna or auricle	Captures sound waves
	External auditory canal	Directs sound waves to the eardrum
Middle ear	Tympanic membrane or eardrum	Vibrates when struck by sound waves
	Ossicles	Transmit sound to the cochlea in the inner ear
Inner ear	Cochlea	Converts fluid waves to nerve impulses
	Semicircular canals	Detects head movement
	Saccule and utricle	Detects head movement and linear acceleration

up sound waves. The outer ear also consists of a short tube, the **external auditory canal**, which directs sound waves from the auricle to the middle ear (Figure 13–20). The external auditory canal is lined by skin containing modified sweat glands that produce earwax or **cerumen**. Earwax traps foreign particles, such as bacteria. It also contains a natural antibiotic substance, which may help reduce ear infections.

The Middle Ear. The **middle ear** lies entirely within the temporal bone of the skull (Figure 13–20). The eardrum or **tympanic membrane** separates the middle ear cavity from the external auditory canal. The tympanic membrane oscillates when struck by sound waves much the same way that a guitar string will vibrate when a note is sounded by another nearby instrument.

Inside the middle ear are three minuscule bones, the **ossicles**. Starting from the outside, they are the **malleus** (hammer), **incus** (anvil), and **stapes** (stirrup). As illustrated in Figure 13–20, the hammer-shaped malleus abuts the tympanic membrane. When the membrane vibrates, it causes the malleus to rock back and forth. The malleus, in turn, causes the incus to vibrate. The incus causes the stapes, the stirrup-shaped bone, to move in and out against the **oval window**, an opening to the inner ear covered with a membrane like the skin on a drum. Thus, the vibrations created in the eardrum are transmitted to the inner ear, which houses the sound receptors. As Figure 13–20 illustrates, the middle ear cavity opens to the pharynx via the **auditory (eustachian) tube**. The auditory tube extends downward at an angle and opens into the nasopharynx. It serves as a pressure valve. Normally, the auditory tube is closed. Yawning

and swallowing, however, cause it to open; this allows air to flow into or out of the middle ear cavity, equalizing the internal and external pressure.

When taking off in an airplane or ascending a mountain highway in your car, pressure builds up in the middle ear cavity. As air pressure outside falls, the air inside the middle ear cavity exerts pressure on the tympanic membrane, creating slight discomfort. To equalize the internal and external pressure, a person can swallow, chew a piece of gum, or suck on a piece of candy—all of which open the auditory tube, letting air inside the middle ear cavity escape.

Scuba divers and swimmers can sustain considerable damage to their eardrum if they are not careful. As a swimmer descends, pressure from the water builds, pushing the eardrum inward. To prevent the eardrum from tearing, air must be forced into the middle ear cavity. This can be done by simply holding your nose, clamping your mouth shut, and blowing. Air is forced through the auditory tube into the middle ear cavity, equalizing the pressure. When a diver ascends, just the opposite happens: air pressure increases inside the middle ear cavity. The diver must release the pressure or else suffer a broken eardrum.

The Inner Ear. The inner ear occupies a much larger cavity in the temporal bone. It contains two sensory organs—the cochlea and the vestibular apparatus. The **cochlea** is shaped like a snail shell and houses the receptor for hearing. The **vestibular apparatus** consists of two parts: the semicircular canals and the vestibule (Figure 13–20). The **semicircular canals** are three ringlike structures set at angles to one another that house the receptors for body position and movement. The vestibule is a bony chamber lying

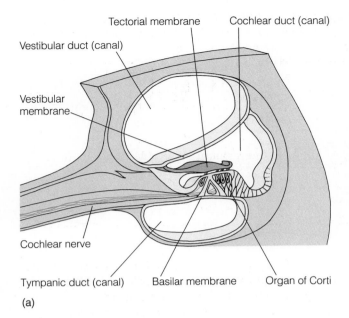

Vestibular duct (canal)

Tectorial membrane

Cochlear duct (canal)

Vestibular membrane

Cochlear nerve

Tympanic duct (canal)

Basilar membrane

Organ of Corti

(a)

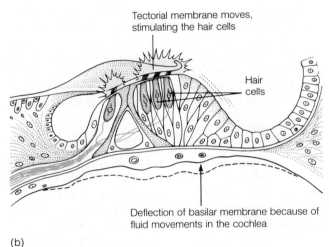

Tectorial membrane moves, stimulating the hair cells

Hair cells

Deflection of basilar membrane because of fluid movements in the cochlea

(b)

FIGURE 13–21 Cross Section through the Cochlea
(*a*) Notice the three fluid-filled canals and the central position of the organ of Corti. (*b*) Hair cells of the organ of Corti are embedded in the overlying tectorial membrane. When the basilar membrane vibrates, the hair cells are stimulated.

between the cochlea and semicircular canals. It houses two receptors that respond to body position and movement.

Hearing

The detection and perception of sound require several structures. The ears, for example, direct sound waves into the external auditory canal, where they strike the tympanic membrane or eardrum. The eardrum vibrates back and forth, causing the ossicles of the middle ear cavity to vibrate, transmitting sound waves to the cochlea, which houses the receptor for sound.

Structure and Function of the Cochlea. As Figure 13–20 shows, the cochlea forms a bony spiral structure. A cross section through the cochlea reveals three fluid-filled canals (Figure 13–21a). Separating the middle canal from the lowermost one is a flexible membrane called the **basilar membrane**. It supports the **organ of Corti**, the receptor for sound.

The organ of Corti contains two rows of receptor cells, called **hair cells** (Figure 13–21b). Cellular projections from the hair cells are embedded in the overlying **tectorial membrane**. When sound waves are transmitted from the middle ear to the inner ear, they create waves of compression and decompression in the fluid in the uppermost canal of the cochlea, the **vestibular canal**. As shown in Figure 13–22a, in which the cochlea is unwound to simplify matters, the stirrup transmits vibrations to the drumlike **oval window**, the opening in the bony cochlea. Fluid pressure waves are created in the vestibular canal; they then travel through the thin vestibular membrane into the middle canal, the **cochlear duct (canal)**. From here, they travel through the basilar membrane into the lowermost canal, the **tympanic canal**. Pressure is relieved by the outward bulging of the **round window**, an opening in the bony cochlea, which, like the oval window, is spanned by a flexible membrane.

In their course through the cochlea, the pressure waves cause the basilar membrane to vibrate up and down. The vibration of the basilar membrane stimulates the hair cells. They respond by releasing a neurotransmitter, which stimulates the dendrites that wrap around the bases of the hair cells. Thus, sound waves are converted to pressure waves and then to nerve impulses. The nerve impulses exit the cochlea via the **vestibulocochlear nerve**, one of the 12 cranial nerves.

Distinguishing Pitch and Intensity. With these facts in mind, we turn our attention to the way humans detect sounds of different pitch (or frequency) and different intensity. Let's start with pitch.

A pitch pipe helps a singer find the note to begin a song. On the pipe are a range of notes from high to low. The ear can distinguish between these various pitches or frequencies in large part because of the

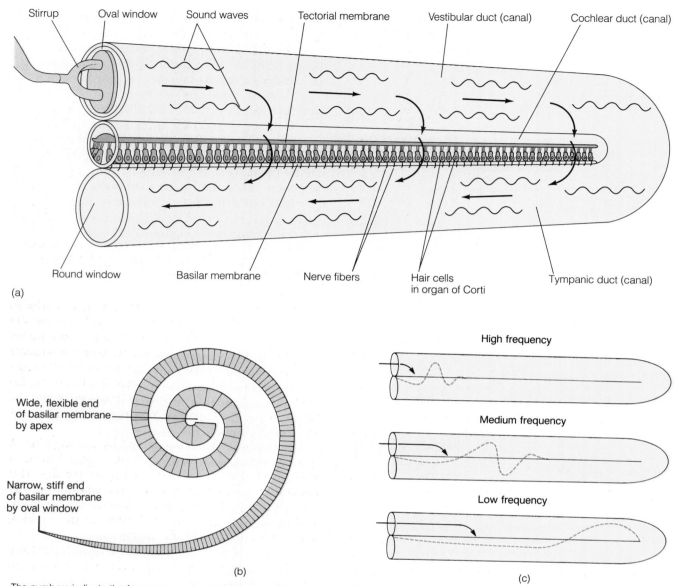

(a)

Stirrup · Oval window · Sound waves · Tectorial membrane · Vestibular duct (canal) · Cochlear duct (canal)

Round window · Basilar membrane · Nerve fibers · Hair cells in organ of Corti · Tympanic duct (canal)

Wide, flexible end of basilar membrane by apex

Narrow, stiff end of basilar membrane by oval window

(b)

High frequency

Medium frequency

Low frequency

(c)

The numbers indicate the frequencies with which different regions of the basilar membrane maximally vibrate.

FIGURE 13–22   The Transmission of Sound Waves through the Cochlea   The cochlea is unwound here to simplify matters. (*a*) Vibrations are transmitted from the stirrup to the oval window. Fluid pressure waves are established in the vestibular canal and pass to the tympanic canal, causing the basilar membrane to vibrate. (*b*) A representation of the basilar membrane, showing the points along its length where the various wavelengths of sound are perceived. Notice that the basilar membrane is narrowest at the base of the cochlea at the oval window end and widest at the apex. (*c*) High-frequency sounds set the basilar membrane near the base of the cochlea into motion. Hair cells send impulses to the brain, which interprets the signal as a high-pitch sound. Low-frequency sounds stimulate the basilar membrane where it is widest and most flexible.

structure of the basilar membrane. The basilar membrane underlying the organ of Corti is stiff and narrow at the oval window where fluid pressure waves are first established inside the cochlea (Figure 13–22b).

As the basilar membrane proceeds to the apex of the spiral, however, it becomes wider and more flexible. The change in width and stiffness results in marked differences in its ability to vibrate. The narrow, stiff end, for example, vibrates maximally when pressure waves from high-frequency sounds (high notes) are present (Figure 13–22c). The far end of the membrane vibrates maximally with low-frequency sounds (low notes). In between the membrane responds to a wide range of intermediate frequencies.

Pressure waves caused by any given sound stimulate one specific region of the organ of Corti. The hair cells stimulated in that region send impulses to the brain, which it interprets as a specific pitch. Each region of the organ of Corti sends impulses to a specific region of the auditory cortex in the temporal lobe of the brain. Thus, the auditory cortex can be mapped according to tone in much the same way the motor and sensory cortex can be mapped.

The intensity of a sound, or its loudness, depends on the amplitude of the vibration in the basilar membrane. The louder the sound, the more vigorous the vibration of the eardrum (although it vibrates at the same frequency). The more vigorous the vibration of the eardrum, the greater the deflection of the basilar membrane in the area of peak responsiveness. The greater the deflection of the basilar membrane, the more hair cells are stimulated.

The auditory system is remarkably sensitive—so sensitive in fact that it can detect sounds that deflect the membrane only a fraction of the diameter of a hydrogen atom. It is no wonder then that loud noises can cause so much damage to hearing. Loud rock music or sirens, for example, cause extreme vibrations in the basilar membrane, destroying hair cells and causing partial deafness.

Hearing Loss. As people grow older, many lose their hearing. Hearing loss usually occurs so slowly that most people are unaware of it. In some cases, though, people lose their hearing suddenly. A loud explosion, for example, can damage the hair cells or even break the ossicles.

Hearing losses may be temporary or permanent, partial or complete. Basically, though, hearing loss falls into two categories, depending on the part of the system that is affected. The first is **conduction deafness**, which occurs when the conduction of sound waves to the inner ear is impaired. Excessive earwax in the auditory canal, a rupture of the eardrum, an infection in the middle ear, damage to the ossicles, and even a rupture of the oval window are all causes of conduction deafness.

Conduction deafness usually results from infections in the middle ear. Bacterial infections can result in the buildup of scar tissue in the middle ear, causing

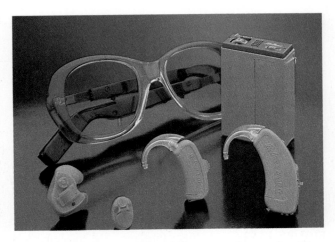

FIGURE 13–23 Hearing Aids Worn by people with conduction deafness, hearing aids send sound impulses through the bone of the skull to the cochlea.

the ossicles to fuse together and lose their ability to transmit sound. Infections of the middle ear usually enter through the auditory tube. Thus, a sore throat or a cold can easily spread to the ear, where it requires prompt treatment. Ear infections are especially common in babies and young children; if undetected, inner ear infections may slow down the development of speech. Untreated, such infections may lead to permanent deafness.

Conduction deafness is treated by hearing aids. A **hearing aid** usually fits in the ear, or just behind it (Figure 13–23). These devices bypass the defective sound conduction system by transmitting sound waves through the bone of the skull to the inner ear. Fluid pressure waves are established in the cochlea, which stimulate the basilar membrane.

The second type of hearing loss is neurological and is called **nerve** or **sensineural deafness**. Sensineural deafness results from physical damage to the hair cells, the vestibulocochlear nerve, and the auditory cortex. Explosions and extremely loud noises, for example, damage hair cells, creating partial or even complete deafness. The auditory nerve, which conducts impulses from the organ of Corti to the cortex, may degenerate, thus ending the flow of information to the cortex. Tumors in the brain or strokes may destroy the cells of the auditory cortex.

Although the ear is quite vulnerable to noise and other problems, it contains a built-in protective mechanism to protect itself from damage. This mechanism consists of two small skeletal muscles (the smallest in the body) located in the middle ear cavity. One of these muscles inserts on the malleus and the other

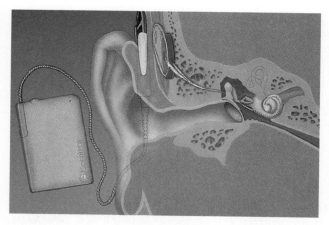

FIGURE 13–24 Cochlear Implant The cochlear implant can correct for nerve deafness. Electrodes convey electrical impulses from a small microphone mounted in the ear to the auditory nerve.

attaches to the stapes. As noted earlier, the malleus attaches to the eardrum and the stapes attaches to the oval window. Loud noises stimulate a reflex contraction of the middle ear muscles, pulling them away from their membrane contacts. Whenever a person is exposed to loud noise, this reflex reduces the conduction of the noise to the inner ear. Unfortunately, the reflex requires about 40 milliseconds, so it cannot protect the ear from explosions.

Correcting Profound Deafness. More than 2 million Americans are profoundly deaf, a condition that has until recently been considered virtually untreatable. The profoundly deaf hear nothing, not even the sound of a siren.

Children who are born deaf, or are deafened before they begin to speak, often fail to mature emotionally. Even reading comprehension may be impaired. Some profoundly deaf children, in fact, never advance beyond third or fourth grade reading levels.

Hearing aids usually cannot help individuals who are born deaf or individuals who suffer from nerve damage. Researchers, however, have developed a new device called a **cochlear implant**, which simulates the function of the inner ear (Figure 13–24).

Dr. William House of the House Ear Institute in Los Angeles has pioneered the cochlear implant. This device picks up sound and transmits it to a receiver implanted inside the skull. The signal then travels to an electrode implanted in the vestibulocochlear nerve in the cochlea. Electrical impulses in the electrode

stimulate the nerve, creating impulses that travel to the auditory cortex.

Cochlear implants provide the deaf with a rudimentary hearing capacity. Today, hundreds of adults and children are equipped with cochlear implants. The single-electrode model, however, is quickly becoming obsolete thanks to the advent of newer multiple-electrode models. These models detect and transmit a wider range of sounds. Recipients of the new models can perceive many distinct words, not just the sound of a telephone or an automobile horn.

Although it is doubtful that "normal" hearing will ever be fully restored by such devices, people who have lived in the silent world note that any sound is better than no sound at all. For them, the cochlear implant provides valuable outside stimuli, such as sirens and horns, the importance of which many of us take for granted. The cochlear implant also helps the deaf monitor and regulate their own voices. Rudimentary hearing also makes lip reading easier. For deaf children, a cochlear implant could mean the difference between learning to speak or a lifetime of silence.

Balance

The cochlea lies alongside the vestibular apparatus. The **vestibular apparatus** contains receptors that detect body movement and head position; it consists of two structures: the semicircular canals and vestibule, which consists of the utricle and the saccule (Figure 13–25).

The Semicircular Canals. Three semicircular canals are found in the inner ear. Each of these bony canals joins the **vestibule**, a bony chamber housing the utricle and saccule, shown in Figure 13–25. The semicircular canals are arranged at right angles to one another. Each semicircular canal is lined by a membrane and filled with a fluid called **perilymph**. The linings of the canals expand at the base of each semicircular canal to form the **ampulla**, which houses the receptor cells (Figure 13–25).

The receptor cells in the semicircular canals are found in the ampullae and are part of a structure called the **crista ampularis** or simply **crista**, shown in Figure 13–25. The cristae are ridges of tissue housing the receptor cells. The receptor cells contain numerous microvilli and a single cilium, which are embedded in a cap of gelatinous material, called the **cupula**. The cupula extends into the cavity of the ampulla and is bathed in endolymph. Dendrites of sensory nerves

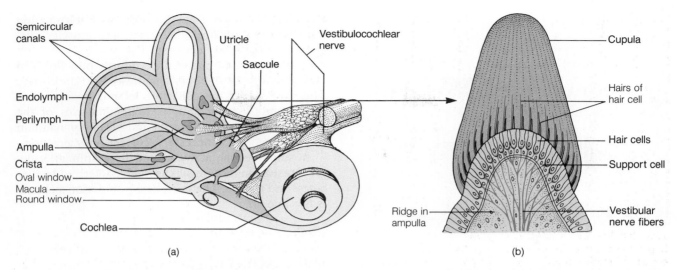

(a)   (b)

FIGURE 13–25   Location and Structure of the Cristae   (a) This illustration shows the membranous canal inside the semicircular canals and the location of the cristae in the ampullae. The semicircular canals are filled with endolymph. (b) When the head spins, the endolymph is set into motion, deflecting the gelatinous cupula of the crista, thus stimulating the receptor cells.

wrap around the base of the receptor cells in a pattern that is rather familiar to us by now.

Rotation of the head causes the endolymph in at least one of the semicircular canals to move. The movement of the fluid deflects the cupula, which stimulates the hair cells. The hair cells release a neurotransmitter that excites the sensory neurons, which send impulses to the brain, alerting the brain to the rotational movement of the head and body.

Since the semicircular canals are set in all three planes of space, movement in any direction can be detected. The semicircular canals contribute to our sense of balance. In particular, they alert the brain to rotation and movement. As a result, they are part of our **dynamic balance**—moving balance.

The Utricle and Saccule.   Balance also requires information during times when the body is not moving and during linear acceleration, acceleration in a straight line. This information is provided by two receptors, the utricle and saccule, housed inside the vestibule. They are part of our **static balance**.

Figure 13–25 shows the location of two membranous sacs, the **saccule** and **utricle**, inside the vestibule. In each of these structures are small receptor organs, called **maculae**. Each macula consists of a patch of receptor cells, structurally similar to those in the cristae. The cilia and microvilli of the receptor cells in the maculae are embedded in a gelatinous cap.

Numerous small crystals of calcium carbonate, called **otoliths** (literally, ear rocks), are embedded in the gelatinous material. The otoliths make the gelatin heavier than the surrounding fluid (Figure 13–26a).

Although the receptors in the utricle and saccule are similar, they are oriented in different directions. For example, when a person is standing, the hair cells of the utricle are oriented vertically. Bending the head forward, as illustrated in Figure 13–26b, causes the otoliths and gelatinous cap to droop forward. Pulled downward by gravity, the gelatinous cap stimulates the receptor cells. Nerve impulses are transmitted to the brain, alerting it to the head movement. Moving forward—say, by walking or running—causes the gelatinous mass to slide backward, as shown in Figure 13–26c. As a result, the utricle also transmits information to the brain on linear acceleration.

In the saccule, the hair cells are oriented horizontally when a person is standing or sitting. The receptor cells are stimulated when the head is tilted back—for example, when you lie down. The saccule also responds to acceleration and deceleration but in a vertical direction—for example, when you ride on an elevator or bounce on a trampoline.

The information provided by the semicircular canals and the maculae is sent to a cluster of nerve cell bodies in the brain stem called the **vestibular nuclei**. Here all of the information the receptors generate on position and movement is integrated. This also in-

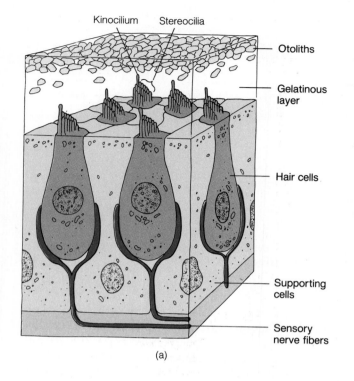

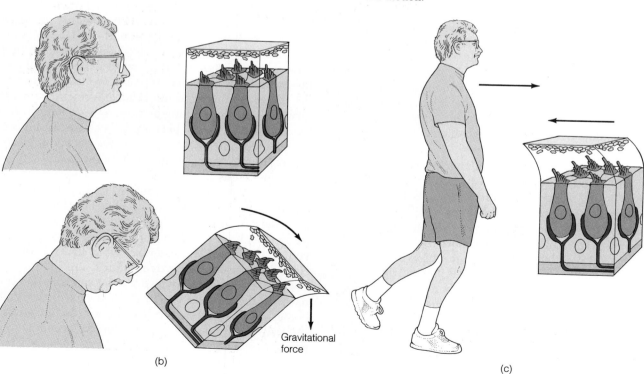

Kinocilium   Stereocilia

Otoliths

Gelatinous
layer

Hair cells

Supporting
cells

Sensory
nerve fibers

(a)

cludes information from the eyes and information from receptors in the skin, joints, and muscles, discussed earlier in the chapter.

From the vestibular center, information flows in many directions. One major pathway is to the cortex. This path makes us conscious of our position and movement. Another path is to the muscles of the limbs and torso. Signals to the muscles help maintain our balance and, if necessary, correct body position.

In some people, activation of the vestibular apparatus results in motion sickness, characterized by dizziness and nausea. The exact cause of motion sickness is not known.

ENVIRONMENT AND HEALTH: NOISE POLLUTION

Noise may be becoming one of the most widespread environmental pollutants in modern, industrial soci-

FIGURE 13–26 The Macula (*a*) Receptor cells in the saccule and utricle are surrounded by supporting cells. Otoliths embedded in the gelatinous cap make the layer heavier than the surrounding fluid. (*b*) Position of the macula of the utricle in an upright position and when head is tilted forward. (*c*) Deflection of the otoliths during forward motion.

Gravitational
force

(b)

(c)

TABLE 13–5   The Decibel Scale

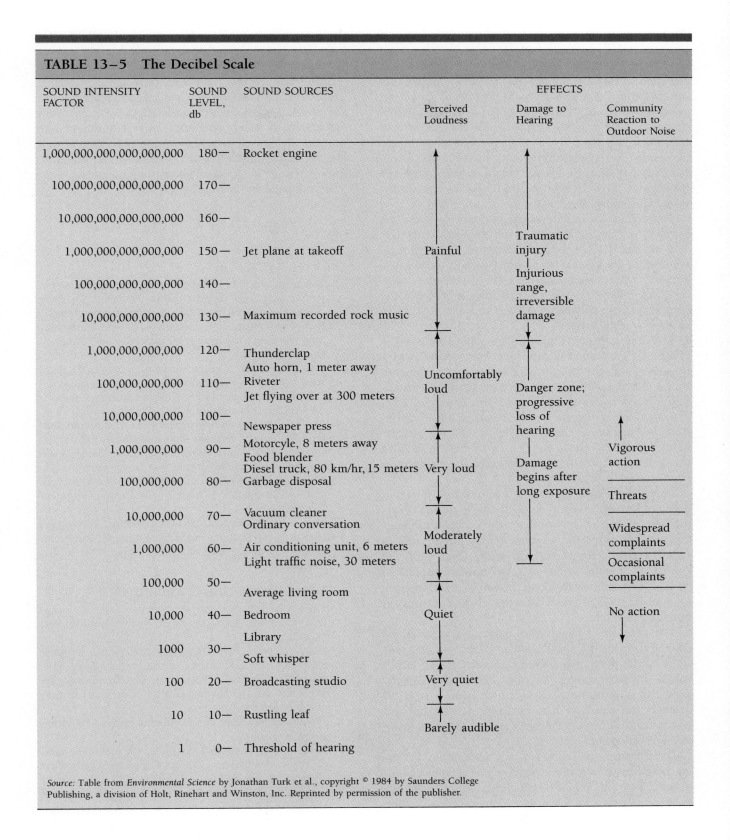

SOUND INTENSITY FACTOR	SOUND LEVEL, db	SOUND SOURCES	EFFECTS		
			Perceived Loudness	Damage to Hearing	Community Reaction to Outdoor Noise
1,000,000,000,000,000,000	180—	Rocket engine	Painful	Traumatic injury	
100,000,000,000,000,000	170—			Injurious range, irreversible damage	
10,000,000,000,000,000	160—				
1,000,000,000,000,000	150—	Jet plane at takeoff			
100,000,000,000,000	140—				
10,000,000,000,000	130—	Maximum recorded rock music	Uncomfortably loud	Danger zone; progressive loss of hearing	
1,000,000,000,000	120—	Thunderclap / Auto horn, 1 meter away			
100,000,000,000	110—	Riveter / Jet flying over at 300 meters			
10,000,000,000	100—	Newspaper press			Vigorous action
1,000,000,000	90—	Motorcycle, 8 meters away / Food blender / Diesel truck, 80 km/hr, 15 meters	Very loud	Damage begins after long exposure	
100,000,000	80—	Garbage disposal			Threats
10,000,000	70—	Vacuum cleaner / Ordinary conversation	Moderately loud		Widespread complaints
1,000,000	60—	Air conditioning unit, 6 meters / Light traffic noise, 30 meters			Occasional complaints
100,000	50—	Average living room	Quiet		No action
10,000	40—	Bedroom			
1000	30—	Library / Soft whisper			
100	20—	Broadcasting studio	Very quiet		
10	10—	Rustling leaf	Barely audible		
1	0—	Threshold of hearing			

Source: Table from *Environmental Science* by Jonathan Turk et al., copyright © 1984 by Saunders College Publishing, a division of Holt, Rinehart and Winston, Inc. Reprinted by permission of the publisher.

eties. Noise is turning our nation into a country of hearing impaired. Traffic noise, airport noise, loud music, crowd noise, gunfire, and other loud noises common in modern society may be slowly destroying our hearing. By the time many New York City residents reach age 20, their hearing has been so impaired that they hear only as well as a 70-year-old African Bushman, who has lived a life free of city noises. Noise also disturbs communications, rest, and sleep. It raises our level of stress, which, in turn, cuts our lives short. Given the profound impact it has on our well-being, noise is truly a type of pollution.

Perhaps the most important impact of noise on humans is hearing loss. Hearing declines with age. The natural decline through aging has long been attributed to middle ear infections, certain antibiotics, and the natural deterioration of the hair cells. New research, however, suggests that noise is probably the principal cause of what was once thought to be age-related hearing loss.

Like any pollutant, the damage caused by noise is related to two factors: exposure level—how loud a noise is—and the duration of exposure. In general, the louder the noise, the more damaging it is. In addition, the longer you are exposed to a damaging level of noise, the more hearing loss you will suffer. Extremely loud noise can result in immediate, permanent damage. An explosion, for instance, can rupture the eardrum or fracture the bony ossicles resulting in conduction deafness.

The noise people are exposed to in factories and at construction sites is sufficient to cause gradual hearing loss. A worker may notice a dulled sense of hearing after working in a noisy environment; this is called a **temporary threshold shift**. Over time the continued assault leads to a **permanent threshold shift**, complete hearing loss. In most cases, hearing loss occurs so gradually that workers do not notice it until it is too late.

The intensity of sound is measured in **decibels (db)**. Table 13–5 (p. 365) shows the scale and lists some common sounds. Surprising new research published by the Environmental Protection Agency shows that continuous exposure to sounds over 55 decibels can result in hearing loss. Light traffic and an air conditioner operating 6 meters (18 feet) from you are 60-decibel sounds.

Good hearing is important to a healthy life. Learning language and communication are heavily dependent on hearing. When hearing is impaired, communication falters and tensions often rise. People who are losing their hearing complain that they feel inadequate in social situations because they have difficulty hearing.

Hearing loss is not inevitable. You can take steps to protect your hearing. Keep your stereo at a reasonable level. Avoid noisy events. Cover your ears when an ambulance or fire engine approaches. Wear ear plugs or ear guards when operating noisy equipment like vacuum cleaners, chain saws, or construction equipment. Get treatment if you develop an ear infection. Prevention is the best medicine, because once you lose your hearing, it is gone forever.

SUMMARY

THE GENERAL SENSES

1. The general senses include pain, light touch, pressure, temperature, and proprioception. Receptors for these senses may be naked or encapsulated nerve endings.
2. Receptors fit into four categories: (a) mechanoreceptors, (b) thermoreceptors, (c) photoreceptors, and (d) chemoreceptors.
3. Sensory stimuli may be internal or external. Some stimuli elicit no response. Others stimulate reflex actions. Still others may stimulate conscious awareness or physiological changes.
4. The naked nerve ending receptors in the body include those stimulated by pain, hair movement, light touch (Merkel disc), and possibly temperature. The encapsulated receptors include those that detect pressure (Pacinian corpuscles), light touch (Meissner's corpuscles), and muscle extension (muscle spindles and Golgi tendon organs).
5. Many sensory receptors cease responding to prolonged stimuli, a phenomenon called habituation. Pain, temperature, pressure, and olfactory receptors can all habituate.

TASTE AND SMELL: THE CHEMICAL SENSES

6. The special senses are served by more elaborate receptors, providing for taste, smell, vision, hearing, and balance.
7. Taste receptors, called taste buds, are located principally on the dorsal surface of the tongue. Taste buds contain sensory cells. Food molecules dissolve in the saliva and bind to the membranes of the hairlike processes of the receptor cells.
8. Taste buds respond to four different flavors: salt, bitter, sweet, and sour.
9. The receptors for smell are located in the roof of the nasal cavities in the olfactory epithelia. The receptor cells in the olfactory membrane respond to thousands of individual chemicals, which bind to receptors on the membrane of olfactory hairs, stimulating nerve impulses that are transmitted to the brain in the olfactory nerve.

THE VISUAL SENSE: THE EYE

10. The eye is the receptor for visual stimuli and is located in the orbit, a bony socket.

11. The wall of the eye contains three coats. The outermost coat is the fibrous layer and consists of the sclera (the white of the eye) and the cornea (the clear anterior structure that lets light shine in).

12. The middle layer consists of the choroid (the pigmented and vascularized section), the ciliary body (whose muscles control the lens), and the iris (which controls the amount of light entering the eye).

13. The innermost layer is the retina, which contains the photoreceptors.

14. Two types of photoreceptors are in the retina: rods and cones. Rods are sensitive to dim light and function in dim light. They provide black-and-white vision.

15. Cones operate in bright light and provide color vision. The highest concentration of cones is found in the fovea centralis.

16. Light is focused on the retina principally by the cornea and lens. The cornea has a fixed refractive power. The lens can be adjusted to bend light according to need. The muscles of the ciliary body play an important role in this process. Focusing on near objects is aided by pupillary constriction.

17. The lens may become cloudy with old age or because of exposure to excess ultraviolet light. This condition, called cataracts, can be corrected by surgically removing the lens and replacing it with a plastic one.

18. The eye is divided into two cavities. The posterior cavity lies behind the lens and is filled with a gelatinous material, the vitreous humor. The anterior cavity is filled with a plasmalike fluid called aqueous humor, which nourishes the lens and other eye structures in the vicinity. If the rate of absorption decreases, however, pressure can build inside the anterior cavity, a condition called glaucoma.

19. As people age, the lens becomes less resilient and less able to focus on nearby objects. This condition is called presbyopia.

20. The eyes move in unison thanks to the extrinsic eye muscles. When they focus on a nearby object, however, the eyes converge—that is, the pupils both turn inward—so that the image can be cast on both foveas.

21. Three common eye problems are myopia, hyperopia, and astigmatism.

22. Myopia or nearsightedness results from a lens that is too strong or an elongated eyeball. In the uncorrected eye, the image comes in focus in front of the retina.

23. Hyperopia results from a weak lens or a shortened eyeball. In the uncorrected eye, the image would come into focus behind the retina.

24. Astigmatism is an irregularly curved lens or cornea that creates fuzzy images.

25. Myopia, hyperopia, and astigmatism can all be corrected with contact lenses or glasses.

26. Three types of cones are present in the eye: red cones, green cones, and blue cones. Each type responds maximally to one specific color of light. Intermediate colors activate two or more types of cone.

27. Color blindness is a genetic disorder, more common in men than women. It results from a deficiency or the complete absence of one or more types of cones. Red-green color blindness is the most common type.

HEARING AND BALANCE: THE COCHLEA AND MIDDLE EAR

28. The ear consists of three portions: the outer, middle, and inner ears.

29. The outer ear consists of the auricle and external auditory canal, which both serve to direct sound to the eardrum.

30. The middle ear consists of the eardrum and ossicles, which send vibrations to the inner ear.

31. The inner ear consists of the cochlea, which houses the organ of Corti, the receptor for sound. The inner ear also houses receptors for movement and head position: the semicircular canals, utricle, and saccule.

32. The auditory or eustachian tube helps equilibrate the pressure inside the middle ear cavity.

33. The semicircular canals are three hollow rings lined internally by a membrane and filled with a fluid called endolymph. The receptors for head movement are located in an enlarged portion at the base of each canal, the ampulla.

34. Fluid movement inside the semicircular canals deflects the gelatinous cap lying over the receptor cells, stimulating them and alerting the brain to head movements.

35. The semicircular canals are set in all three planes of space, so movement in any direction can be detected.

36. Two membranous sacs in the vestibule contain receptors called maculae, which respond to linear acceleration and tilting of the head.

37. The cochlea houses the receptor for sound. This spiral-shaped bony structure contains three fluid-filled canals. Separating the middle canal from the lower one is a flexible membrane called the basilar membrane, which supports the organ of Corti. Hair cells in the organ of Corti are embedded in the relatively rigid tectorial membrane.

38. Sound waves create vibrations in the eardrum and ossicles, which are transmitted to fluid in the cochlea. Pressure waves in the cochlea cause the basilar membrane to vibrate, which, in turn, stimulates the hair cells.

39. Pressure waves caused by any given sound cause one part of the membrane to vibrate maximally. The hair cells stimulated in that region send signals to the brain, which it interprets as a specific pitch. The louder the sound, the greater the vibration of the membrane in any one spot.

40. Hearing loss may occur as a result of damage or block-

age to the conducting portion—the external auditory canal, the eardrum, and the ossicles. Damage to the hair cells, the auditory nerve, or the auditory cortex are forms of nerve or sensineural deafness.

41. Nerve deafness is difficult to cure, although efforts are now underway to perfect cochlear implants that could help people who are profoundly deaf gain some degree of hearing. Far easier to correct is conduction deafness. Hearing aids transmit sound waves through the bone of the skull to the inner ear.

ENVIRONMENT AND HEALTH: NOISE POLLUTION

42. Noise damages the ears. Extremely loud noises can rupture the eardrum or break the ossicles. Less intense noises, however, generally destroy hearing gradually by damaging hair cells.

43. In most people, hearing loss occurs so gradually as to be undetected. Individuals can take steps to avoid hearing loss.

EXERCISING YOUR CRITICAL THINKING SKILLS

In an experiment to determine how chemicals affect vision, two researchers exposed rats to varying levels of a toxin, chemical A. They found that at low doses, the chemical had no effect, but higher doses resulted in severe vision loss and blindness. The researchers immediately asked the Food and Drug Administration (FDA) to ban the chemical from pro- duction for fear it might similarly affect humans. You are head of the FDA. Using your critical thinking skills, how would go about considering the request? What factors would help you determine whether the request was valid? What studies might you want to see done?

TEST OF TERMS

1. The general senses include light touch, pressure, temperature, _____, and proprioception, or position sense.

2. _____ are nerve endings that respond to various stimuli. They fall into two broad groups: naked nerve endings and _____ _____

3. Light touch is perceived by two sensors: naked nerve endings surrounding hair follicles and _____ _____.

4. Pressure is perceived by _____ corpuscles located in the deeper layers of the skin and around various organs.

5. The receptor situated in the superficial layer of the dermis that responds to touch is the _____ _____.

6. The _____ _____ is a stretch receptor found in skeletal muscles.

7. Stretch receptors in tendons are called _____ _____ _____.

8. _____ occurs when a sensory receptor stops sending impulses, even though the stimulus is still present.

9. Taste, hearing, and vision are three of the _____ _____.

10. The taste receptors in the oral cavity and the dorsal surface of the tongue are called _____ _____. They are especially abundant on the _____, small protrusions on the surface of the tongue.

11. Receptors for the sense of smell are located in the _____ epithelium found in the roof of each nasal cavity. Receptor cells in the epithelium are modified _____ neurons.

12. The eye consists of three layers. The outermost, fibrous layer consists of the _____, the white of the eye, and the _____, the clear anterior portion that allows light to enter.

13. The middle layer of the eye is heavily _____ and vascu-

larized. It consists of the _____, the _____ _____, and the iris.

14. The inner layer of the eye is the light-sensitive portion of the eye and is called the _____. It contains two types of receptors, the _____, which confer color vision, and the rods, which operate best at _____.

15. Nerve impulses leave the retina via the optic nerve, which is formed from the axons of the _____ cells.

16. Sharpest vision occurs when an object is cast on the _____ _____, a small depression in the retina lying lateral to the blind spot or _____ _____.

17. The _____ is a flexible structure used to focus light coming from nearby objects on the retina. It is attached to the _____ _____ by the zonular fibers. It may become cloudy with age, a condition known as _____.

18. The _____ _____ is a gelatinous mass that occupies the posterior cavity of the eye.

19. _____ results from the excess buildup of _____ _____ in the anterior cavity of the eye.

20. The bending of light is called _____. It occurs anytime light waves _____ _____.

21. Eye movements are caused by the _____ _____ muscles.

22. Myopia is also called _____. It results from a(n) _____ _____ or a(n) _____ _____.

23. The surgical technique that corrects for myopia is called _____.

24. An irregularly curved lens or cornea results in a condition known as _____.

25. _____ is the pigment of the rods. It breaks down into two molecules when struck by light.

26. Color blindness is a _____-_____ trait.

27. Sound waves are directed into the _____ _____ canal to the _____, which separates the outer ear from the middle ear.

28. Extremely loud noises may damage the _____, the bones in the middle ear, which transmit sound to the _____ of the cochlea.

29. The _____ _____ leads from the middle ear cavity to the nasopharynx and serves as a pressure release valve.

30. Movement of the head is detected by sensors in the _____ of the _____ canals in the inner ear. These canals are filled with a fluid called _____.

31. Static balance is provided in part by two receptor patches, the _____, located in the utricle and saccule.

32. The _____ _____ _____ is the receptor for sound in the cochlea. It consists of three canals: the _____ canal, the cochlear duct, and the tympanic canal.

33. A rupture of the eardrum or a fusion of the bones in the middle ear results in _____ deafness.

34. A transitory loss of hearing is called a _____ _____ shift.

Answers to the Test of Terms are located in Appendix B.

TEST OF CONCEPTS

1. Define the terms *general senses* and *special senses*.

2. Using your knowledge of the senses and of other organ systems gained from previous chapters, describe the role that sensory receptors play in homeostasis. Give specific examples to illustrate your main points.

3. Make a list of both the encapsulated and the nonencapsulated general sense receptors. Note where each is located and what it does.

4. Define the term *habituation*. What advantages does it confer? Can you think of any disadvantages?

5. Describe the receptors for taste, explaining where they are located, what they look like, and how they operate.

6. Taste buds detect four basic flavors. What are they? How do you account for the thousands of different flavors that you can detect?

7. Describe the olfactory epithelium and the structure of the receptor cells. How do these cells operate? In what ways are taste receptors and olfactory receptors similar? In what ways are they different?

8. Explain the following statement: Taste and smell are complementary.

9. Draw a cross section of the human eye and label its parts.

10. Define the following terms: retina, rods, cones, fovea centralis, optic disc, ganglion cells, bipolar neurons, and optic nerve.

11. Compare and contrast the rods and cones.

12 When focusing on a nearby object, your eyes go through several changes. Describe those changes and what they accomplish.

13. Define the following terms: myopia, hyperopia, presbyopia, and astigmatism.

14. You walk into a dark movie theatre and find you can barely make out the aisle. After a while your vision recovers. Explain both phenomena.

15. What is color blindness? What is the most common type? Explain what your world would look like if you were color blind and how you would accommodate for this condition.

16. Describe the anatomy of the ear and the role of the outer, middle, and inner ear in hearing.

17. How do the semicircular canals operate? How do the utricle and saccule operate?

SUGGESTED READINGS

Borg, E., and S. A. Counter. 1989. The middle ear muscles. *Scientific American* 261(2): 74–80. Discusses how the tiny muscles behind the eardrum help prevent sensory overload and enhance our ability to hear.

Koritz, J. F., and G. H. Handleman. 1988. How the human eye focuses. *Scientific American* 259(1): 92–99. Looks at the eye's geometry and biochemistry to examine why the ability to focus on nearby objects declines as people age.

Marieb, E. N. 1989. *Human anatomy and physiology*. Redwood City, Calif.: Benjamin/Cummings. Beautifully illustrated and detailed treatment of the general and special senses.

Nathans, J. 1989. The genes for color vision. *Scientific American* 260(2): 42–49. New insights into the evolution of color vision are gained by examining the genes that encode the color-detecting proteins of the human eye.

Sherwood, L. 1989. *Human physiology: From cells to systems*. St. Paul, Minn.: West. See Chapter 6 for a more detailed discussion of the physiology of the special senses.

Spence, A. P. 1986. *Basic human anatomy*, 2d ed. Redwood City, Calif.: Benjamin/Cummings. Clear, concise coverage of the anatomy of general and special senses.

Vaughan, C. 1988. A new view of vision. *Science News* 134(4) 58–60. Looks at an exciting new theory about how the brain processes visual signals via multiplexed, encoded signals.

14 The Muscles and Skeleton

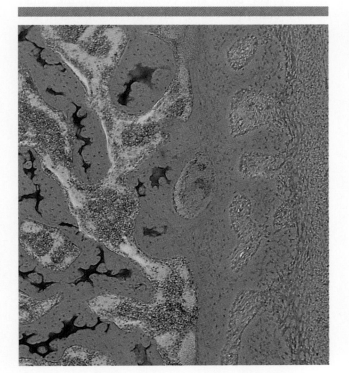

Histological section of the developing femur. Newly formed bone spicules appear red.

T he bones and muscles constitute the musculoskeletal system and comprise about 50 to 60% of the body weight of an adult. Bones and muscles perform several vital functions. For many of us, the first function to come to mind is body movement. Like other organ systems, the musculoskeletal system is also involved in homeostasis. Bone, for instance, helps maintain constant blood calcium levels, necessary for muscle contraction. Muscles contract rhythmically when we are cold, producing a phenomenon called shivering, which helps to maintain body heat.

This chapter examines the human skeleton and the muscles that attach to it, the skeletal muscles.

THE HUMAN SKELETON

The word **skeleton** is derived from a Greek word that means "dried-up body." Many people's first impression of bone isn't much different. To them bones are dry, dead structures. **Bone**, however, is a living, metabolically active tissue. Bone tissue contains numerous cells, known as **osteocytes**. These cells are embedded in a calcified extracellular material, or matrix, which

371

gives the bones of the body their characteristic hardness, strength, and flexibility (Chapter 6).

As Chapter 6 noted, the extracellular material of bone consists of (1) an organic component, collagen, a protein that imparts flexibility, and (2) an inorganic component, chiefly calcium phosphate crystals that are deposited on the collagen fibers, which impart strength.

The Functions of Bones

The human skeleton consists of 206 bones, discrete structures made of bone tissue (Figure 14–1). Bones provide internal structural support, giving shape to our bodies and helping us maintain upright posture. Some bones help protect internal body parts. The rib cage, for instance, protects the lungs and heart, and the skull forms a protective shell for the brain. Bones serve as the site of attachment for many skeletal muscles whose contraction results in purposeful movements. Bones are also home to cells that give rise to red blood cells (RBCs), white blood cells (WBCs), and platelets, all essential to homeostatic function. Red blood cells help maintain homeostasis by supplying oxygen to body tissues and by carrying off carbon dioxide; white blood cells are involved in immune protection, protecting the body from foreign organisms (Chapter 10). Platelets play a key role in blood clotting. Bones are also a storage depot for fat, needed for cellular energy production at work and at rest. Finally, bones are a reservoir of calcium, releasing and absorbing calcium to help maintain normal blood calcium levels. Calcium is essential to muscle contraction, and disturbances in blood calcium levels can impair muscle contraction.

The human skeleton consists of two parts, the axial skeleton and the appendicular skeleton (Figure 14–1). The **axial skeleton** forms the long axis of the body. It consists of the skull, the vertebral column, and the rib cage. The **appendicular skeleton** consists of the bones of the arms and legs and the bones of the shoulders and pelvis, by which the upper and lower extremities are attached to the axial skeleton.

Figure 14–2 illustrates the anatomy of the humerus, the bone found in the upper arm. As shown, the humerus, like other long bones of the body, consists of a long narrow shaft, the **diaphysis**, and expanded ends, the **epiphyses**. The ends of the epiphyses are coated with a thin layer of hyaline cartilage, which reduces friction in joints. The protrusions on the bone mark the site of muscle attachment.

Some Basics of Bone Structure

Take a moment to study the skeleton in Figure 14–1. As you examine it, you may notice that bones come in a variety of shapes and sizes. Some are long, some are short, and some are flat and irregularly shaped. Nevertheless, all bones share some common characteristics. For example, all bones consist of an outer shell of dense bony material called **compact bone**, surrounding a spongy bone interior. **Spongy bone**, so named because of its resemblance to a sponge, is much less dense than compact bone. As shown in Figure 14–2c, spongy bone consists of a latticework of bony material. On the outer surface of the compact bone is a layer of connective tissue, the **periosteum** (around the bone). The outer layer of the periosteum is composed of dense, irregularly arranged connective tissue fibers and serves as the site of attachment for many skeletal muscles. The inner layer of the periosteum contains osteogenic (bone-forming) cells that participate in the production of new bone during remodeling or repair. The periosteum is richly supplied with blood vessels, which enter the bone at numerous locations. Blood vessels travel through small canals in the compact bone and course through the inner spongy bone, providing nutrients and oxygen, and carrying off cellular wastes, such as carbon dioxide. The periosteum is also richly supplied with nerve fibers. The majority of the pain felt after a person bruises or fractures a bone results from pain fibers innervating the periosteum. The spongy bone found in the interior of all bones contains numerous small, adjoining cavities. Together they form the **marrow cavity**. The marrow cavities in most of the bones of the fetus contain **red marrow**. As Chapter 8 noted, red marrow is a blood-cell factory, producing RBCs, WBCs, and platelets to replace those routinely lost each day. As an individual ages, most red marrow is slowly "retired" and becomes filled with fat, becoming **yellow marrow**. Yellow marrow begins to form during adolescence and, by adulthood, is present in all but a few bones. Under certain circumstances, however, yellow marrow may be activated, once again producing RBCs, WBCs, and platelets to meet the body's needs.

The Joints

A gymnast races across the mat and leaps into space, twirling effortlessly before landing on her feet. No sooner has she landed than she takes off again across

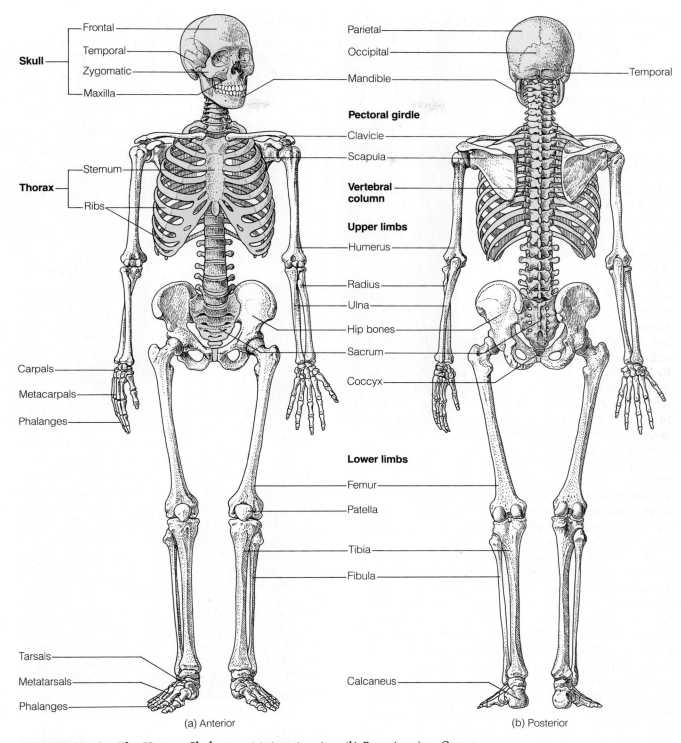

Skull
— Frontal
— Temporal
— Zygomatic
— Maxilla

Thorax
— Sternum
— Ribs

Carpals—
Metacarpals—
Phalanges—

Tarsals—
Metatarsals—
Phalanges—

(a) Anterior

Parietal
Occipital
Mandible

Pectoral girdle
— Clavicle
— Scapula

Vertebral column

Upper limbs
— Humerus

— Radius
— Ulna
— Hip bones
— Sacrum

Coccyx

Lower limbs
— Femur
— Patella

— Tibia
— Fibula

Calcaneus

Temporal

(b) Posterior

FIGURE 14–1 The Human Skeleton (*a*) Anterior view. (*b*) Posterior view. Over 200 bones of all shapes and sizes make up the skeleton. The shaded region shows the axial skeleton. The unshaded region is the appendicular skeleton.

FIGURE 14-2 Anatomy of Long Bones (a) Photograph and (b) drawing of the humerus. Notice the long shaft and dilated ends. (c) Longitudinal section of the humerus showing the position of the compact bone, spongy bone, and marrow.

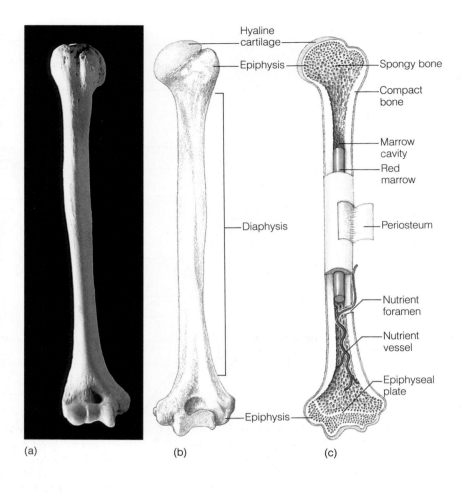

(a)

(b)

(c)

Hyaline cartilage
Epiphysis
Spongy bone
Compact bone
Marrow cavity
Red marrow
Diaphysis
Periosteum
Nutrient foramen
Nutrient vessel
Epiphyseal plate
Epiphysis

the mat in a series of three back handsprings. These delightful movements are the result of long hours of practice and exercise and are made possible by the joints. Wherever two or more bones meet, a joint is formed. **Joints**, therefore, are the structures that connect the bones of the skeleton.

Joints can be classified by the degree of movement they permit. Those that permit no movement are called immovable joints. Those that permit some movement are known as slightly movable joints, and those that permit free movement are called freely movable joints. Consider some examples.

The bones of the skull cap, shown in Figure 14-3a, are held together by immovable joints. As illustrated, opposing bones in the skull cap interdigitate (interlock). The bones are also held together by fibrous connective tissue spanning the small space between the two bones.

The pubic symphysis is the joint formed by the two pubic bones (Figure 14-3b). It is normally quite im-

movable; however, near the end of pregnancy, hormones loosen the fibrocartilage of the pubic symphysis. This allows the pelvic outlet to widen, facilitating the delivery of the baby.

The bodies of the vertebrae are united by slightly movable joints (Figure 14-4). As Chapter 6 noted, each vertebra is separated from its nearest neighbor by an intervertebral disc. The inner portion of the disc acts as a cushion, softening the impact of walking and running. The outer fibrous portion holds the disc in place and joins one vertebra to its nearest neighbor. The joints between the vertebrae offer some degree of movement, resulting in a fair amount of flexibility. If they didn't, we would be unable to bend over to tie our shoes or unable to curl up on the couch for an afternoon snooze.

The most common type of joint is the freely movable **synovial joint**. The synovial joints are more complex than other types and permit varying degrees of movement. Although synovial joints differ consider-

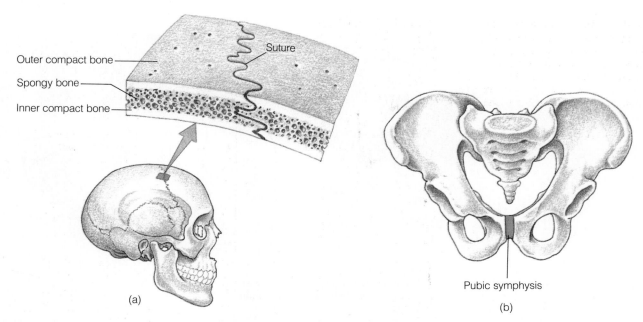

(a)

(b)

Pubic symphysis

FIGURE 14–3   Two Immovable Joints   (a) Many of the bones of the skull are held in place by joints called sutures. The bones are linked by fibrous tissue, and the joints are immovable. (b) The pubic symphysis is another immovable joint.

ably in architecture, they share several common features. The first commonality is the hyaline cartilage located on the articular surfaces of the bones involved in joints (Figure 14–5a). This thin cap of hyaline cartilage reduces friction and thus facilitates movement.

The second commonality of synovial joints is the joint capsule. The **joint capsule** is a double-layered structure that stretches from one bone to another in the joint (Figure 14–5a). The outer layer of the capsule consists of dense connective tissue attached to the periosteum of adjoining bones. Parallel bundles of dense connective tissue fibers in the outer layer of the capsule form **ligaments**, which run from bone to bone, giving support to the joint. As a rule, the ligaments are fairly inflexible, stretching only slightly. However, some individuals have remarkably flexible joints. Some people, for example, can extend their thumbs well beyond the 90 degrees possible for most of us. And some can extend their fingers so much that they can touch the back of their hand. These people are said to be "double jointed."[1]

[1]The term is misleading in that it implies that they have two joints where most of us have one. This is not true. The flexibility is due only to the more stretchable ligaments and tendons.

The inner layer of the joint capsule is called the **synovial membrane**, and it consists of loose connective tissue with a generous supply of capillaries. The synovial membrane produces a fairly thick, slippery substance, called **synovial fluid**, which provides nutrients to the articular cartilage, the hyaline cartilage

FIGURE 14–4   A Slightly Movable Joint   The intervertebral discs allow for some movement, giving the vertebral column some flexibility.

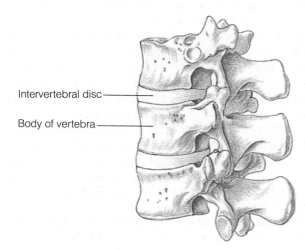

Intervertebral disc

Body of vertebra

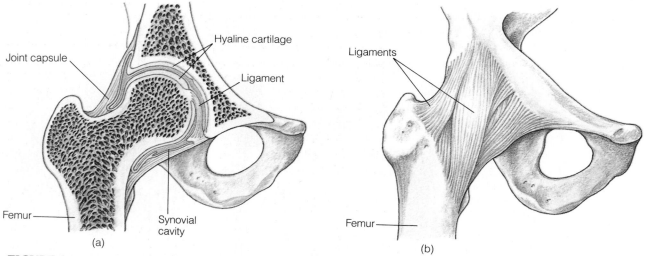

FIGURE 14–5 A Synovial Joint (*a*) A cross section through the hip joint showing the structures of the synovial joint. (*b*) Ligaments help support the joint.

on the articular surfaces of the bone, and also acts as a lubricant, facilitating bone movement. Normally, the synovial membrane produces only enough fluid to create a thin liquid film on the articular cartilage. Injuries to a joint, however, may result in a dramatic increase in synovial fluid production, causing swelling and pain in joints.

Tendons and muscles provide an additional means of support in some joints and are common to synovial joints. In the shoulder joint, for example, the muscles of the shoulder help hold the head of the humerus in the socket formed by the scapula. Muscles in the hip also help hold the head of the femur in place. Because muscles fortify the joint, individuals who are in poor physical shape are much more likely to suffer a dislocation on a ski trip or during exercise than someone who is in good shape. **Dislocation** is an injury in which a bone is displaced from its proper position in a joint due to a fall or some other unusual body movement. In some cases, bones will slip out of place, then back in without assistance. In others, the bone can only be put back in place by a physician.

Synovial joints come in many shapes and sizes and can be classified on the basis of structure. Two of the most common are the hinge joint and the ball-and-socket joint. The knee joint is a **hinge joint**. Hinge joints open and close like hinges on a door and, therefore, provide for movement in only one axis. The joints in the fingers are also hinge joints. The hip joint and shoulder joints are **ball-and-socket joints**. They

provide a wider range of motion. (Compare the movements permitted by the shoulder and hip joints to those permitted by the knee joint.)

Repairing Injured Joints. The acrobatics performed by modern dancers or Olympic gymnasts illustrate the wide range of movement of the human body permitted by the joints. The joints, however, are a biological compromise. They must allow movement, but not too much. The structures of the joint also provide some degree of stability, helping hold bones in place. Full mobility would compromise strength; and absolute strength would compromise mobility.

Given the need for compromise, it is not surprising that joint injuries are common among physically active people, especially athletes (Table 14–1). A hard blow to the knee of a football player, for instance, can tear the ligaments, disabling the joint. A rough fall can dislocate a skier's shoulder. Improperly lifting an object can strain the ligaments that join the vertebrae of the back, resulting in considerable pain.

Torn ligaments, tendons, and cartilage in joints heal very slowly because they are not well endowed with blood vessels.[2] For years, repairing a joint injury required major surgery. Surgeons cut open the knee to access damaged ligaments and cartilage. These surgical procedures were traumatic and put a person out of commission for several months. Today, new surgi-

[2]Because of its poor blood supply, cartilage may not heal at all.

TABLE 14-1 Common Injuries of the Joints

INJURY	DESCRIPTION	COMMON SITE
Sprain	Partially or completely torn ligament; heals slowly; must be repaired surgically if the ligament is completely torn	Ankle, knee, lower back, and finger joints
Dislocation	Occurs when bones are forced out of a joint; often accompanied by sprains, inflammation, and joint immobilization; bones must be returned to normal positions	Shoulder, knee, and finger joints
Cartilage tears	Cartilage may tear when joints are twisted or when pressure is applied to them; does not repair well because of poor vascularization; torn cartilage is generally removed surgically; this operation makes the joint less stable	Cartilage in the knee is the most common

cal techniques allow physicians to repair joints with a minimum of trauma. Through small incisions in the skin over the joint, surgeons insert a device called an **arthroscope**, which allows them to view the damage inside the joint cavity (Figure 14–6). The arthroscope is a hollow instrument through which special instruments can be inserted. A surgeon can, therefore, repair damaged cartilage without opening the joint. This reduces damage caused by large incisions, allowing athletes to be back on their feet and on the playing field in a matter of weeks. New surgical techniques are also used to rebuild torn ligaments, thus returning joints to nearly their original state.

Arthritis. Virtually every time you move, you use one or more of your joints. Problems in joints, therefore, are often quite noticeable. One of the most common problems is called **degenerative joint disease** or **osteoarthritis**. Although its cause is not known, degenerative joint disease may simply result from wear and tear on a joint. Over time, excess wear and tear may cause the articular cartilage on the ends of bones

FIGURE 14–6 Arthroscopic Surgery (*a*) A physician performing arthroscopic surgery. (*b*) Inside view of the knee joint through an arthroscope.

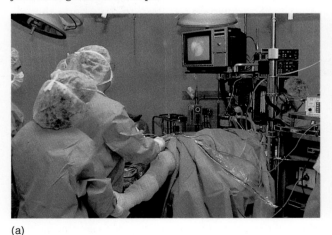

(a)

(b)

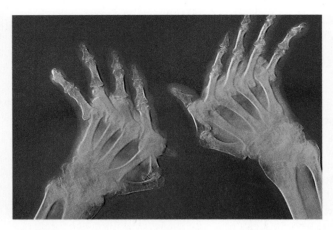

FIGURE 14–7 X-Ray of Hand Disfigured by Rheumatoid Arthritis X-ray has been color-enhanced by computer.

to flake and crack. As the cartilage degenerates, the bones come in contact with one another, grinding against each other during movement and causing considerable pain and discomfort. Swelling usually accompanies these changes, and swelling in a joint tends to reduce mobility.

Osteoarthritis occurs most often in the weight-bearing joints—the knee, hip, and spine—which receive the most wear and tear over time. Osteoarthritis may also develop in the finger joints. The amount of swelling varies considerably. Some patients may experience virtually no swelling; in others, the joints may become enlarged and disfigured.

Osteoarthritis is extremely common. X-ray studies of people over 40 years of age show that most people have some degree of degeneration in one or more joints. Fortunately, many people do not even notice the problem, and the disease rarely becomes a serious medical problem.

Wear and tear on joints is worsened by obesity. The extra pressure on the joints apparently wears the cartilage away more quickly; thus, weight control can help reduce the rate of degeneration in people already suffering from the disease.[3] Painkillers, such as aspirin, and anti-inflammatory drugs can be used to reduce the symptoms (pain and swelling) that accompany osteoarthritis. Injections of steroids may also help reduce inflammation, although repeated injections often damage the joint.

---

[3]Weight control is also a preventive measure that helps people avoid the problem in the first place.

Another common disorder of the synovial joint is rheumatoid arthritis. **Rheumatoid arthritis** is the most painful and crippling form of arthritis and is caused by an inflammation and swelling of the synovial membrane. Inflammation may spread to the articular cartilages, damaging them. If the condition persists, rheumatoid arthritis causes degeneration of the bones. The thickening of the synovial membrane and degeneration of the bone often lead to disfigurement of the joints, loss of mobility, and considerable pain (Figure 14–7). In some cases, afflicted joints may be completely immobilized. In severe cases, the bones may become dislocated, causing the joints to collapse.

Rheumatoid arthritis generally strikes the joints of the wrist, fingers, and feet. It can also affect the hips, knees, ankles, and neck. In many cases, inflammation also occurs in the heart, lungs, and blood vessels. Research suggests that rheumatoid arthritis results from an **autoimmune reaction**—that is, an immune response to the cells of one's own synovial membrane.

Rheumatoid arthritis occurs in people of all ages, but most commonly occurs in individuals between the ages of 20 and 40. Rheumatoid arthritis is usually a permanent condition, although the degree of severity varies widely. Patients suffering from it can be treated with physical therapy, painkillers, anti-inflammatory drugs, and even surgery.

Diseased joints can also be replaced by artificial ones—**prostheses**—thus restoring mobility and reducing pain. This procedure, however, is recommended only in severe cases and generally only if other treatment has failed. Plastic joints can be used to replace the finger joints. These prostheses improve the appearance of the hands, eliminating the gnarled, swollen joints. Moreover, patients regain the use of previously crippled fingers. Day-to-day chores that often required assistance (buttoning a shirt) become noticeably easier. Tasks that had once been impossible because of arthritis—for example, opening screw-top jars and picking up coins—soon become possible. Severely damaged knee and hip joints are replaced with steel or teflon substitutes, which, if fitted properly, may last for 10 to 15 years (Figure 14–8). To put a new hip or knee joint in place, the degenerating bone is first cut away. Then the prosthesis is inserted into the shaft of the bone.

Embryonic Development and Bone Growth

Most of the bones of the human skeleton develop from hyaline cartilage (Chapter 6). This process,

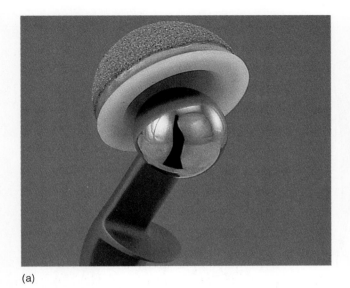

(a)

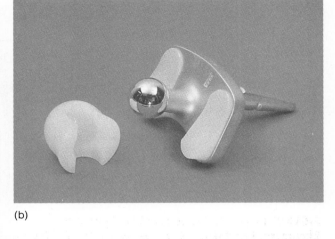

(b)

FIGURE 14–8 **Artificial Joints** (*a*) Photograph of an artificial hip joint. (*b*) Photograph of an artificial knee joint.

called **endochondral ossification**, is illustrated in Figure 14–9. As shown, cells of the hyaline cartilage in the interior of the cartilage mass undergo considerable enlargement. The extracellular material between the cells is compressed, and calcium crystals are deposited on the extracellular material of the cartilage. This region constitutes the **primary center of ossification**.

As these changes are taking place, a thin layer of bone is deposited around the periphery of the cartilage. Peripheral bone is deposited by cells of the **perichondrium**, the connective tissue layer surrounding the cartilage, which becomes the periosteum. The thin shell of bone laid down on the periphery of the cartilage eventually becomes a layer of compact bone.

Blood vessels invade the interior of the slightly calcified cartilage, carrying with them cells that will later give rise to RBCs, WBCs, and platelets. Some cells that enter the interior of the forming bone give rise to the bone-forming cells, **osteoblasts**. The osteoblasts proliferate on the spicules of calcified cartilage in the interior of the embryonic bone and begin to secrete collagen fibers, which are deposited on the slightly calcified spicules. Calcium is deposited on the collagen fibers, thus producing a mass of spongy bone at the primary center of ossification. Ossification centers also form in the ends of the bone, but later. These are called the **secondary centers of ossification**.

As the bone develops, much of the calcified material in the shaft of the bone is destroyed by bone-digesting cells or **osteoclasts**. Osteoclasts hollow out the center of the bone, thus forming the marrow cavity, which soon fills with stem cells that give rise to RBCs, WBCs, and platelets.

During bone formation, most of the hyaline cartilage is converted to bone. When the bone is completely formed, all that remains of the cartilage are two narrow bands located between the shaft of the bone and its two ends (Figure 14–10). Called the **epiphyseal plates**, these bands of cartilage contain actively dividing cells that permit bone to elongate. The process of bone elongation is beyond the scope of this book, but basically results from the proliferation of cartilage on one side of the plate and the ossification on the other. The epiphyseal plates remain active in children and adolescents until their long bones stop growing.[4] Increasing levels of sex steroids cause the plates to be converted to bone (Chapter 15).

Remodeling Bones

Bones are dynamic structures that undergo considerable remodeling in response to changes in our lives. In a newborn baby, for example, the bones of the leg (the tibia and fibula) are quite bowed. Cramped inside the mother's uterus, the bones do not grow very straight. However, during the first two years of life, as the child begins to walk and run, the leg bones generally straighten (Figure 14–11). The bones are literally remodeled to meet the markedly different stresses placed on them by upright posture.

[4]Boys may continue growing until they are 20 or 21. Girls generally stop much earlier, by age 17.

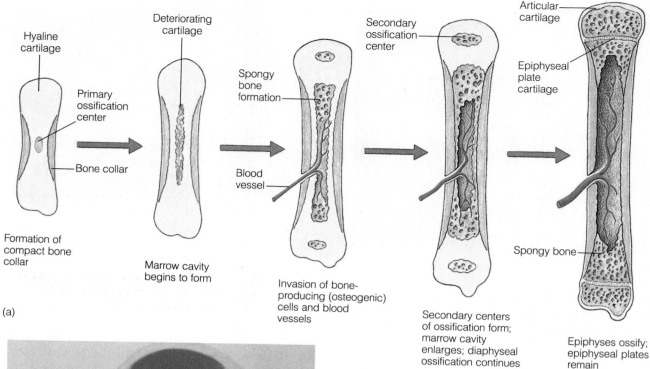

Formation of
compact bone
collar

Marrow cavity
begins to form

Invasion of bone-
producing (osteogenic)
cells and blood
vessels

Secondary centers
of ossification form;
marrow cavity
enlarges; diaphyseal
ossification continues

Epiphyses ossify;
epiphyseal plates
remain

(a)

FIGURE 14–9 Endochondral Ossification   (*a*) Stages of bone formation. (*b*) Eighteen-week-old human embryo showing bones forming by endochondral ossification.

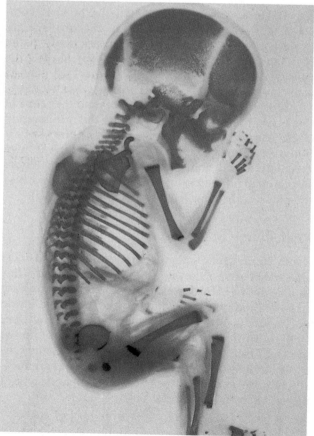

(b)

Remodeling occurs throughout adult life as well. During sedentary periods, for example, the compact bone decreases in thickness. Activity causes the compact bone to thicken, helping to withstand the stress of walking, running, or standing. Spongy bone also undergoes considerable remodeling. Thus, the internal architecture of a bone changes to meet new stresses.

Two cells are responsible for bone remodeling: osteoclasts and osteoblasts. Osteoclasts are multinucleated cells that actively absorb the extracellular material of bone (Figure 14–12). In contrast, osteoblasts lay down new bone during the remodeling phase. When the osteoblasts become surrounded by calcified extracellular material, they are referred to as **osteocytes**.

Maintaining the Balance: Bone as a Homeostatic Organ

In addition to their role in remodeling bone, osteoblasts and osteoclasts help control blood calcium lev-

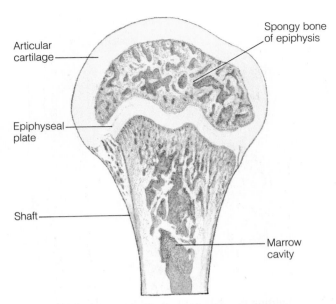

FIGURE 14–10 Epiphyseal Plate The epiphyseal plate allows for bone elongation. Cartilage is added at the epiphyseal end, while new bone is formed at the diaphyseal end, thus elongating the bone.

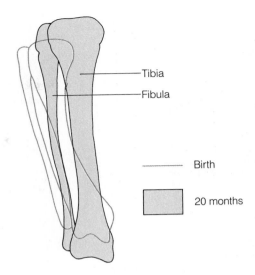

FIGURE 14–11 Remodeling of a Baby's Leg Bones Notice how dramatically the bone changes to meet the changing needs of the toddler.

els. These cells are, therefore, part of a homeostatic mechanism controlled in large part by two hormones: parathormone and calcitonin. When blood calcium levels fall, the parathyroid glands (small glands behind the thyroid glands in the neck) release **parathormone** (PTH). PTH travels throughout the body in the bloodstream. When it reaches the bone, PTH stimulates the osteoclasts. The calcium released by the activity of the osteoclasts replenishes blood levels.

When calcium levels rise—for example, after a meal—the thyroid releases a hormone called **calcitonin**. Calcitonin inhibits osteoclasts, stopping bone destruction. It also stimulates osteoblasts, causing them to deposit new bone. The new bone that is formed helps lower blood calcium levels, returning them to normal.

Bone Repair

Bone fractures result from falls or impact injuries and vary considerably in their severity. Some may involve thin hairline cracks, which mend fairly quickly. Others involve considerably more damage and take longer to repair. In order for repair to occur, the broken ends must often be aligned and immobilized by a cast. Severe fractures may require surgeons to insert steel pins to hold the bones together.

Like other connective tissues, bones are capable of self-repair. As Figure 14–13 illustrates, blood from broken blood vessels in the periosteum and marrow cavity pours into the fracture, forming a blood clot.

FIGURE 14–12 The Bone-Destroying Osteoclast Osteoclasts eat away at bone, releasing calcium to restore blood levels and helping to remodel bone to meet changing needs.

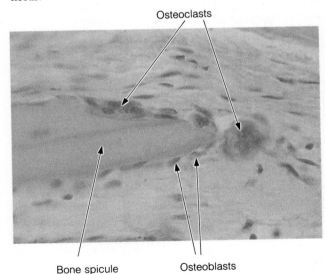

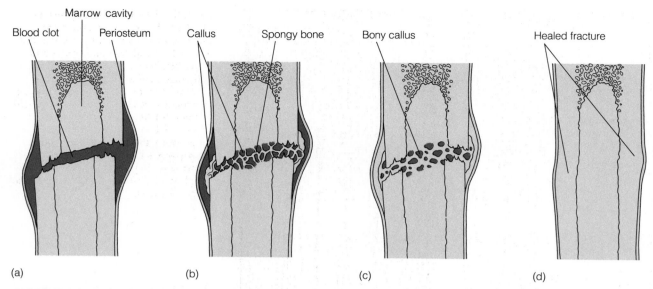

FIGURE 14–13 Fracture Repair (*a*) A blood clot forms. (*b*) The blood clot is invaded by fibroblasts and other cells, forming the callus. (*c*) Calcium is deposited in the callus, knitting the ends together. (*d*) The fracture is repaired.

Within a few days, the blood clot is invaded by fibroblasts, connective tissue cells from the periosteum. Fibroblasts produce and secrete collagen fibers, thus forming a mass of cells and fibers, the **callus**, that bridges the broken ends. The callus protrudes at the surface of the bone.

The callus is next invaded by osteoblasts from the periosteum. Osteoblasts convert the callus to bone, knitting the two ends together. As Figure 14–13 shows, the callus is initially much larger than the bone itself. Excess bone, however, is gradually removed by osteoclasts, and the bone returns to normal.

Osteoporosis

Approximately 20 million Americans suffer from osteoporosis ("porous bone"). **Osteoporosis** is a disease that results in the deterioration of bone (Figure 14–14). Victims of osteoporosis lose more bone than they replace. As a result, their bones become porous and brittle—so much so that even normal activities such as getting out of bed in the morning or dancing may cause fractures. Osteoporosis occurs most often in postmenopausal women. Menopause occurs between 45 and 55 years of age and results from a shutdown of ovarian estrogen production. Estrogen is a reproductive hormone, described in Chapter 16, but like many other hormones in the body, it performs several functions. In women, for example, it reduces bone loss.

Osteoporosis also occurs in people who are immobilized for long periods. Hospital patients who are restricted to bed for two or three months, for example, show signs of osteoporosis. Osteoporosis may also result from environmental factors. In the 1960s, for instance, women living along the Jintsu River in Japan developed a painful bone disease known as itai-itai (which literally means "ouch, ouch"). The women lived downstream from zinc and lead mines that released large amounts of the heavy metal cadmium into the river. River water was used by locals for drinking water and for irrigating rice paddies. Even though men, young women, and children were exposed to cadmium, 95% of the cases occurred in postmenopausal women.

Researchers at the Argonne National Laboratory believe that cadmium may have accelerated bone loss in the postmenopausal Japanese women. To study the connection, they fed mice diets containing various levels of cadmium chloride. The group receiving the highest dose showed significant reductions in bone calcium when their ovaries had been removed.

These findings may also help explain why older women who smoke have an increased risk of osteoporosis. Cadmium is present in cigarette smoke.

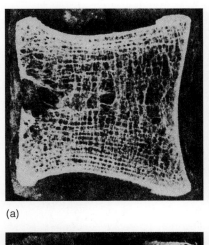

(a)

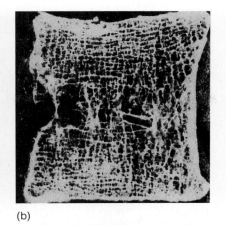

(b)

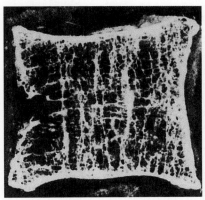

(c)

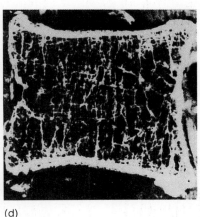

(d)

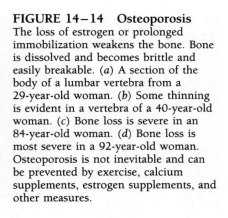

FIGURE 14–14 Osteoporosis
The loss of estrogen or prolonged immobilization weakens the bone. Bone is dissolved and becomes brittle and easily breakable. (*a*) A section of the body of a lumbar vertebra from a 29-year-old woman. (*b*) Some thinning is evident in a vertebra of a 40-year-old woman. (*c*) Bone loss is severe in an 84-year-old woman. (*d*) Bone loss is most severe in a 92-year-old woman. Osteoporosis is not inevitable and can be prevented by exercise, calcium supplements, estrogen supplements, and other measures.

Fetal bone tissue grown in culture dishes in a medium containing cadmium levels similar to those found in a smoker's blood exhibited a 70% reduction in the calcium content of the bone compared to a 25% loss for control samples.

The new findings also help provide a plausible explanation for the fact that female smokers experience more bone fractures and tooth loss than nonsmokers. Smoking, however, may also decrease estrogen levels directly, thus complicating the lines of cause and effect. Whatever the outcome of this research, it suggests the importance of a healthy environment for a healthy body. For a discussion of ways to prevent osteoporosis, see Health Note 14–1.

THE SKELETAL MUSCLES

Purposeful movement is one of the most distinctive features of animal life. Movement is permitted by skeletal muscles acting on bones. Skeletal muscles are under the control of the nervous system.

Figure 14–15 shows the skeletal muscles of the body. Most skeletal muscles cross one or more joints. Therefore, when they contract, they often produce movement. Muscles generally work in groups, rather than alone, to bring about various body movements. Groups of muscles are often arranged in such a way that one set causes one movement and another set on the opposite side of the joint causes an opposing movement. The biceps is a muscle in the upper arm. When it and other members of its group contract, they cause the arm to bend, a movement called flexion. On the back side of the upper arm is another muscle, the triceps. When it contracts, the triceps causes the arm to straighten, or extend.

Individual or groups of opposing muscles are called **antagonists**. Generally, when one muscle contracts to produce a movement, the antagonists relax. Antagonists are under the control of the cerebellum.

Not all skeletal muscles are arranged so that they can make bones move. Some muscles are **synergists**, muscles that steady a joint, allowing other muscles to act. Muscles in the face are another example. An-

Preventing Osteoporosis: A Prescription for Healthy Bones

Twenty million Americans, mostly women, suffer from a painful, debilitating disease called osteoporosis. Caused by a gradual deterioration of the bone, osteoporosis results in nagging pain and discomfort. Bones fracture easily.

If current trends continue, one of every two American women will develop postmenopausal osteoporosis. Each year, nearly 60,000 Americans—mostly women—will die from complications resulting from osteoporosis. Hemorrhage, fat embolisms, and shock are the three most common causes of death. Most women, however, will not realize they have the disease until it has progressed quite far (see the figure).

Recent research shows that osteoporosis begins much earlier than researchers once thought—by the time a woman reaches her mid-twenties. Bone demineralization occurs very rapidly. In fact, by age 30 many women have lost one-third of their bone calcium! Between the ages of 30 and 50, many women's bones continue to deteriorate, becoming extremely brittle.

Calcium loss begins so early in American women because many women in their mid-twenties avoid fatty foods like milk, cheese, and ice cream to help control weight. Although they are fatty, these foods are also a major source of calcium. Milk products are also avoided by many adults because they develop an intolerance to lactose (a sugar) in milk and other dairy products. Lactose intolerance results from a sharp reduction in the production and secretion of the enzyme lactase as one ages, making it difficult for many of us to digest lactose. Because of these and other factors, adult women often ingest only about one-half of the 1000

Osteoporosis An elderly woman suffering from osteoporosis. Notice the hunched back due to collapse of the vertebrae.

to 1500 milligrams of calcium they need every day.

Osteoporosis can be prevented and even reversed by eating calcium-rich foods, such as spinach, milk, cheese, shrimp, oysters, and tofu (soybean curd). Calcium supplements can also halt bone deterioration and restore calcium levels. Vitamin D supplements can also help because vitamin D increases the absorption of calcium in the intestines.

Studies also suggest that osteoporosis can be prevented by exercise. Aerobics, jogging, walking, and tennis—along with the dietary changes noted above—all help prevent the disease.

Research also shows that osteoporosis can be reversed by exercise, even after the disease has reached the dangerous stage. Forty-five minutes of moderate exercise (walking) three days a week, for example, greatly decreases the rate of

calcium loss in older individuals. This exercise regime also stimulates the rebuilding process, replacing calcium lost in previous years. Continued exercise increases bone calcium levels and decrease the rate of bone fractures and fatal complications noted earlier.

Another effective treatment for postmenopausal women is estrogen. Low doses of estrogen halt bone demineralization and promote bone formation. Since women who are given estrogen suffer an increased risk of endometrial cancer, physicians often prescribe a mixed dose of estrogen and progesterone. Progesterone lessens the likelihood of cancer.

Studies have also shown that high doses of fluoride and calcium stimulate bone development. Calcium fluoride treatment increases bone mass approximately 3 to 6% per year and decreases bone fractures. The average patient in one study experienced one fracture every eight months before treatment. After treatment, they only had one fracture every 4.5 years.

Unfortunately, this treatment results in a number of adverse effects as well. Large doses of fluoride, for example, may erode the stomach lining, causing internal bleeding. High doses of fluoride can also stimulate abnormal bone development and cause pain and swelling in joints. To offset these problems, researchers have developed a pill that releases the fluoride gradually. This, they hope, will minimize stomach problems.

For millions of young women, early detection and sound preventive measures, including exercise, vitamin D, dietary improvements, and fluoride treatments, can prevent osteoporosis.

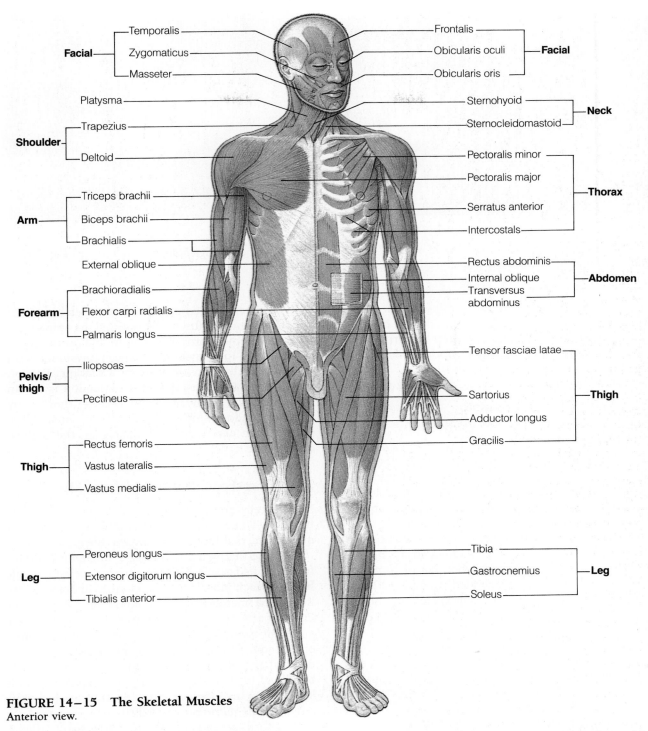

Facial — Temporalis
Facial — Zygomaticus
Facial — Masseter

Platysma

Shoulder — Trapezius
Shoulder — Deltoid

Arm — Triceps brachii
Arm — Biceps brachii
Arm — Brachialis

External oblique

Forearm — Brachioradialis
Forearm — Flexor carpi radialis
Forearm — Palmaris longus

Pelvis/thigh — Iliopsoas
Pelvis/thigh — Pectineus

Thigh — Rectus femoris
Thigh — Vastus lateralis
Thigh — Vastus medialis

Leg — Peroneus longus
Leg — Extensor digitorum longus
Leg — Tibialis anterior

Frontalis — **Facial**
Obicularis oculi — **Facial**
Obicularis oris — **Facial**

Sternohyoid — **Neck**
Sternocleidomastoid — **Neck**

Pectoralis minor — **Thorax**
Pectoralis major — **Thorax**
Serratus anterior — **Thorax**
Intercostals — **Thorax**

Rectus abdominis — **Abdomen**
Internal oblique — **Abdomen**
Transversus abdominus — **Abdomen**

Tensor fasciae latae — **Thigh**
Sartorius — **Thigh**
Adductor longus — **Thigh**
Gracilis — **Thigh**

Tibia — **Leg**
Gastrocnemius — **Leg**
Soleus — **Leg**

FIGURE 14–15   The Skeletal Muscles
Anterior view.

chored to the bones of the skull and to the skin of the face, these muscles allow us to wrinkle our skin, close our eyes, and close our lips.

Muscles help us move about in the environment. They also help us maintain posture. Although few of us are aware of it, our muscles are constantly working to maintain our posture—helping us stand or sit upright despite the never-tiring pull of gravity.

The muscles also produce enormous amounts of heat as a by-product of normal metabolism. When

(a)

(b)

FIGURE 14–16 Light Micrographs of Skeletal Muscle (a) Notice the banding pattern on these muscle fibers. (b) Higher magnification showing nuclei and banding pattern.

working, the muscles produce additional heat—so much that you can cross-country ski in freezing weather wearing only a light sweater. Shivering on a cold day is a reflex reaction that produces body heat as well.

The Anatomy of Movement: The Structure of Skeletal Muscles

Skeletal muscles consist of long, unbranched cells, called **muscle fibers** (Figure 14–16). These fibers are multinucleated cells formed during embryonic development by the fusion of many smaller cells. Viewed with the light microscope, the muscle fibers appear striated.

Like nerve cells, the muscle fiber is an excitable cell. When the membrane is stimulated by neurotransmitter from the terminal bouton of a motor nerve, a nerve impulse is generated (Chapters 12 and 13). The impulse travels along the membrane of the muscle fiber in the same way a nerve impulse travels along an unmyelinated axon or dendrite.

Muscle fibers are contractile. When stimulated, the contractile proteins inside the muscle fibers cause them to shorten. Muscle fibers are also elastic, capable of returning to normal length after a contraction has ended.

Each muscle fiber in a skeletal muscle is surrounded by a delicate layer of connective tissue called the **endomysium** (Figure 14–17). Individual fibers are joined in groups or bundles called **fascicles**. Fascicles are also held together by a connective tissue known as the **perimysium**. Numerous bundles are held together by a connective tissue sheath that surrounds the entire muscle and is called the **epimysium**. This arrangement provides support and protection for muscle cells. In many muscles, the epimysium fuses at the ends of the muscle to form a tendon. Tendons often attach to bones. Because the tendon is continuous with the epimysium and because the perimysium and endomysium are attached to it, muscle contraction can exert a powerful force on the point of attachment.

The Microscopic Anatomy of the Muscles

Muscle cells are uniquely adapted to perform their function. Understanding how a muscle contracts requires a careful look at the muscle fiber.

Imagine that you could tease a single muscle fiber free from its fascicle. Under a microscope, you would find that each muscle fiber is a long cylinder wrapped in plasma membrane and containing many nuclei. Each muscle fiber is characterized by a series of dark and light bands. Inside each muscle fiber are numerous bundles of threadlike filaments, mostly actin and myosin, the contractile proteins (Chapter 6). Each bundle of filaments in the muscle fiber (myofiber) is known as a **myofibril** (Figure 14–18b); each muscle cell contains numerous myofibrils.

Myofibrils are striated. The dark bands are called **A bands**; the narrower, light bands are called **I bands**.[5] The dark bands are arranged in a uniform pattern,

[5]The words *dark* and *light* may help you remember which is which. Dark contains an *a* and light contains an *i*.

which is responsible for the striated appearance of skeletal muscle cells. As illustrated in Figure 14–18c, a fine line runs down the center of each I band. This line is jagged, appearing like many letter Zs stacked on one another, and is appropriately called the **Z line**.

The region between two adjacent Z lines constitutes a **sarcomere**, a subunit of the myofibril. The sarcomere is the functional unit of the muscle cell. As shown in Figure 14–18c, the sarcomere contains thick and thin filaments. The thick filaments are made of myosin and lie in the middle of the sarcomere. The thin filaments are made of the protein actin and extend from the Z line toward the center of the sarcomere but do not join in the middle.

As Figure 14–18c shows, the A band in each sarcomere consists of overlapping actin and myosin fibers. The I bands consist of actin fibers alone. You will also notice that since the actin fibers do not meet in the middle, a thin zone is formed in the middle of the A band. This is the H zone.

How Muscles Contract

Actin and myosin filaments are surprisingly delicate yet are responsible for all muscle contraction. When a muscle contracts, each sarcomere shortens. During contraction, the actin filaments slide toward the center of the sarcomere, touching in the middle. To visualize the process, hold your hands out in front of you with the palms facing toward you. Open your fingers. Imagine that your fingers are the actin filaments. In between them imagine the myosin filaments. When the muscle contracts, the actin filaments from each side of the sarcomere slide toward one another until they touch in the center.

What causes the actin filaments to slide toward the center of the sarcomere? The actin filaments are pulled inward by the myosin molecules. As Figure 14–19 shows, myosin filaments consist of numerous myosin molecules, golf-club–shaped molecules that are arranged with their "club ends" projecting toward the actin filaments. During muscle contraction, the club ends or **cross bridges** attach to actin filaments, then contract, pulling the filaments inward and causing the sarcomere to shorten. During contraction, the I bands narrow and may disappear altogether.

How do the myosin cross bridges function? To answer this question, we must first take a look at Figure 14–19. This illustration shows the molecular makeup of the actin filaments. Each actin filament consists of numerous actin molecules, globular proteins strung together like the beads of a necklace. Figure 14–19

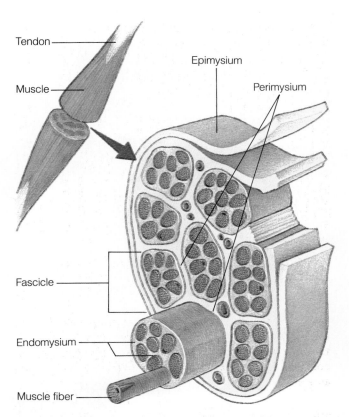

FIGURE 14–17 Connective Tissue Layers Investing Skeletal Muscle Individual muscle fibers are surrounded by the endomysium. Muscle fibers are bundled together to form a fascicle by the perimysium. Fascicles are held together by the epimysium, which also forms the tendons.

shows that each actin molecule contains a binding site to which the myosin cross bridges attach. When the cross bridges attach to the binding site, they undergo a change in shape that causes them to pull the actin filament toward the center of the sarcomere.

In the resting state, the binding sites on the actin molecules of the actin filament are covered by a long stringlike protein molecule called **tropomyosin**. Tropomyosin guards the binding sites, preventing cross bridges from binding to the actin filaments when a muscle is at rest. The tropomyosin molecules are held in place by another protein, **troponin**, shown in the bottom figure in Figure 14–19b.

Muscle contraction is stimulated when this protective guard is removed. In order to see how the tropomyosin "guard" is removed, let's examine the sequence of events that occur when a nerve impulse arrives at the muscle cell. Figure 14–20 illustrates the **neuromuscular junction**, the synapse between the terminal end of a motor neuron and a muscle fiber.

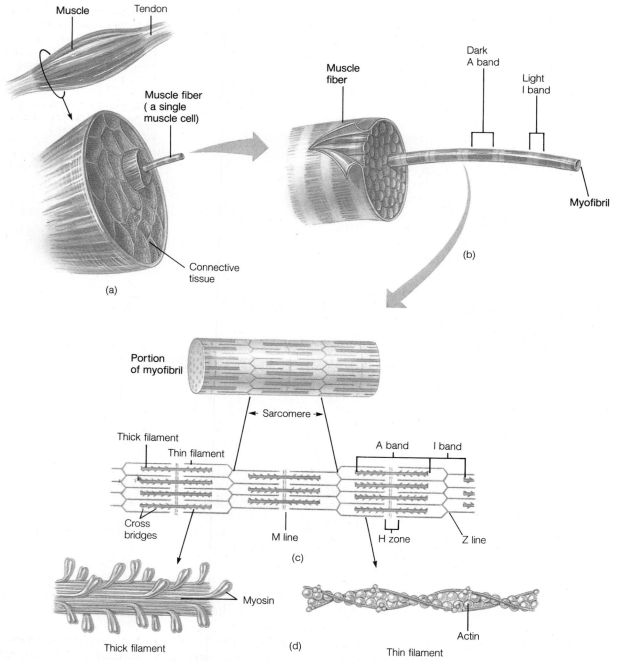

FIGURE 14–18 Structure of the Muscle Fiber, Myofibril, and Sarcomere (*a*) A single muscle fiber teased out of the muscle. (*b*) Each muscle fiber consists of many myofibrils. Note the banded pattern of the myofibril.

(*c*) Sarcomeres consist of thick and thin filaments, as shown here. (*d*) Molecular structure of the thick (myosin) and thin (actin) filaments.

When the nerve impulse arrives at the terminal end of the axon, it triggers the release of acetylcholine. Acetylcholine is a neurotransmitter held in the terminal bouton of the motor nerve axon (Chapter 12). Ace-

tylcholine diffuses into the synaptic cleft, the space between the terminal bouton and the plasma membrane of the muscle fiber, then binds to receptors in the plasma membrane of the muscle fiber.

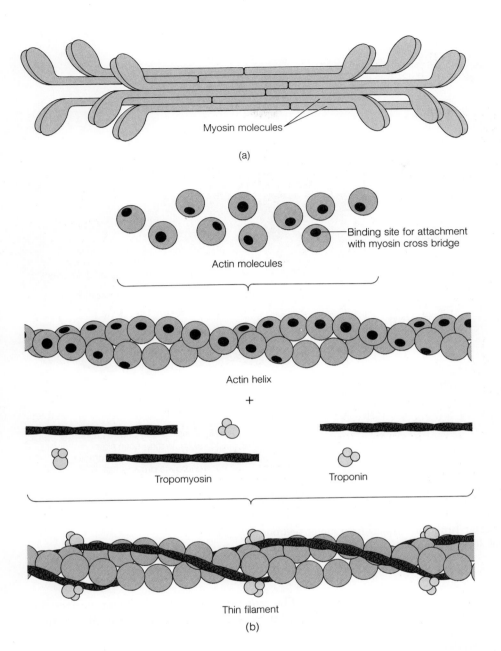

FIGURE 14–19 Structure of Myosin and Actin Filaments (a) Myosin molecules join together to form a myosin filament. Note the presence and orientation of the cross bridges. (b) Globular actin molecules form intertwining strands. Note the presence of the myosin binding sites. Tropomyosin attaches to the binding sites, covering them. Tropomyosin molecules are held in place by the troponin.

Myosin molecules

(a)

Binding site for attachment with myosin cross bridge

Actin molecules

Actin helix

+

Tropomyosin

Troponin

Thin filament

(b)

The binding of acetylcholine to the receptors stimulates changes in the membrane permeability of the muscle fiber, resulting in membrane depolarization. A wave of depolarization travels along the plasma membrane of the muscle fiber. As illustrated in Figure 14–20, the plasma membrane periodically "dips" into the muscle fiber. These deep invaginations, called **T tubules** (transverse tubules), conduct the impulse to the interior of the cell. As it travels inward, the impulse stimulates the release of calcium ions stored inside the muscle cell held in the smooth endoplasmic re-

ticulum (SER), which in muscle fibers is called the **sarcoplasmic reticulum** (Chapter 3). The sarcoplasmic reticulum lies close to the T tubules. Calcium ions released from these saccules diffuse outward into the myofibril and attach to the troponin molecules. Troponin, as noted earlier, holds the tropomyosin molecules in place over the binding sites on the actin filaments. When calcium binds to the troponin, the troponin molecules are released, and the tropomyosin molecules slide off the binding sites on the actin filaments. This allows the cross bridges of the myosin

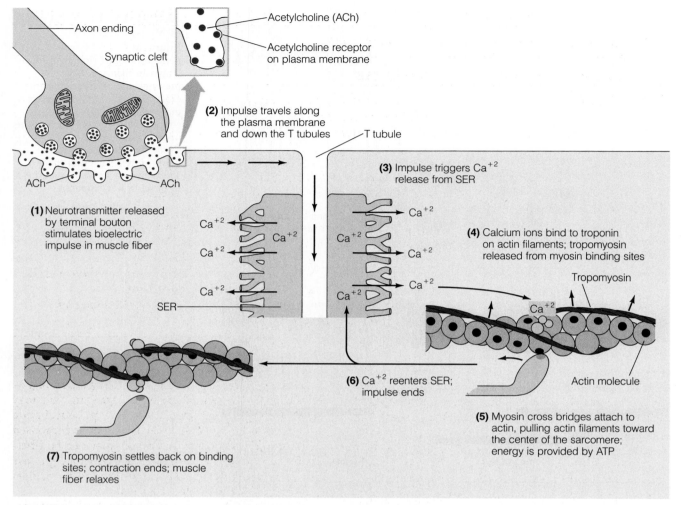

Acetylcholine (ACh)

Acetylcholine receptor on plasma membrane

Axon ending

Synaptic cleft

(2) Impulse travels along the plasma membrane and down the T tubules

T tubule

(3) Impulse triggers Ca^{+2} release from SER

ACh — ACh

(1) Neurotransmitter released by terminal bouton stimulates bioelectric impulse in muscle fiber

Ca^{+2}

Ca^{+2}

Ca^{+2}

Ca^{+2}

Ca^{+2}

Ca^{+2}

Ca^{+2}

Ca^{+2}

Ca^{+2}

Ca^{+2}

SER

(4) Calcium ions bind to troponin on actin filaments; tropomyosin released from myosin binding sites

Tropomyosin

Ca^{+2}

Actin molecule

(6) Ca^{+2} reenters SER; impulse ends

(5) Myosin cross bridges attach to actin, pulling actin filaments toward the center of the sarcomere; energy is provided by ATP

(7) Tropomyosin settles back on binding sites; contraction ends; muscle fiber relaxes

FIGURE 14–20   The Neuromuscular Junction   Axons from motor neurons terminate on the surface of the muscle fiber, forming the neuromuscular junction.

filaments to attach to the actin. The cross bridges contract and pull the actin filaments inward, causing the muscle cell to contract. Myosin cross bridges give a brief tug on the actin filaments, then detach, becoming available to bind again, in a cycle that repeats itself 50 to 100 times during each muscle contraction. The actin filaments are pulled inward, in much the same way that you would pull a boat tied to a rope toward shore.

Contraction ends when the calcium ions are actively transported back into the sarcoplasmic reticulum. When calcium levels fall, tropomyosin slips back into place over the binding sites on the actin filament. This marvelous mechanism is called the

sliding filament theory. Table 14–2 summarizes the roles played by the various components of the muscle cell.

Energy of Contraction.   In this elaborate discussion of the molecular events responsible for muscle contraction, one factor is missing—energy. Energy needed for muscle contraction comes from ATP, the principal form of cellular energy (Chapter 3). ATP binds to the cross bridges of the myosin filaments. The cross bridges contain an enzyme that splits ATP, forming ADP and inorganic phosphate. This reaction releases energy. When ATP is converted to ADP, the energy released is captured and stored in the cross

TABLE 14-2 Components of Muscle Contraction

MUSCLE FIBER COMPONENT	FUNCTIONAL ROLE
Plasma membrane	Conducts impulse from terminal of motor neuron
T tubule	Conducts impulse into the interior of the muscle fiber
Sarcoplasmic reticulum	Releases stored calcium, which stimulates contraction; absorbs calcium to end contraction
Troponin	Holds tropomyosin in place on actin filament, blocking contraction; binds to calcium, releasing tropomyosin to permit contraction
Tropomyosin	Blocks receptors on actin filament, preventing contraction
Actin filaments	Slide toward center of sarcomere during contraction
Myosin filaments	Pull the actin filaments toward center of sarcomere during contraction
Cross bridges	Bind to actin and pull actin filaments. Contain binding site for ATP; contain myosin ATPase, which catalyzes the breakdown of ATP
Calcium ions	Released from the sarcoplasmic reticulum, bind to troponin, causing it to release tropomyosin from binding sites on the actin filaments
ATP	Binds to cross bridges of myosin filaments; broken down by ATPase in cross bridges, providing energy for muscle contraction

bridge momentarily—like energy stored in a spring or in the hammer of a cocked six-shooter. It is released when the cross bridge binds to actin. At this point, the cross bridges draw the actin filaments toward the center of the sarcomere. ATP is quickly regenerated by another high-energy molecule stored in muscle, called **creatine phosphate**. Creatine phosphate has a high energy bond, indicated by the squiggly line. Stored in muscle in high concentrations, it reacts with ADP:

$$\text{creatine} \sim P + ADP \rightarrow ATP + \text{creatine}$$

When you start exercising, creatine phosphate is used immediately to replenish ATP stores. ATP is also generated by glycolysis and aerobic metabolism (the citric acid cycle and the electron transport system), discussed in Chapter 3. The electron transport process requires oxygen. If the muscle's contraction is vigorous, however, oxygen supplies inside muscle cells fall. The circulatory system cannot keep up with oxygen demand. The lack of oxygen shuts off the citric acid cycle and the electron transport system. To generate ATP, the cell must turn to anaerobic metabolism (Chapter 3). Besides being inefficient, this process results in the buildup of lactic acid. The shortage of ATP and the buildup of lactic acid result in **muscle fatigue**.

Muscle fatigue also results from a depletion of glycogen stores in skeletal muscle. Glycogen, you may recall from your earlier studies, is a polysaccharide found in muscle and liver cells. Glycogen is composed of thousands of glucose molecules that can be liberated when needed to help maintain adequate glucose levels inside body cells. Glycogen broken down during exercise is replaced during rest.

Oxygen lost during exercise must also be replaced. This replacement occurs rather quickly. The muscle oxygen deficiency, called the **oxygen debt**, is often largely replaced right after you exercise—that's why you keep breathing hard after you stop exercising.

Increasing the Strength of Contraction: Graded Contractions

Individual muscle fibers obey the **all-or-none law**; that is, when activated by an action potential, muscle fibers contract fully. A single contraction, followed by relaxation, is called a **twitch**. The force of that contraction is shown in Figure 14-21. You will notice that there is a brief lag period after the impulse is generated in the muscle fiber and contraction begins. The latent period results from at least three factors: (1) the time required for the action potential to travel

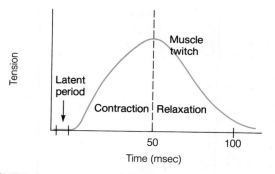

FIGURE 14–21 The Muscle Twitch A graph of the muscle contraction, showing the tension generated and the time required to reach maximum tension and relaxation.

into the T tubules, (2) the time required for calcium to diffuse out of the sarcoplasmic reticulum and to bind to troponin, and (3) the time required for the filaments to begin sliding.

Peak tension in a single muscle cell occurs some time after the impulse arrives. Relaxation requires an equal amount of time as calcium ions are pumped back into the sarcoplasmic reticulum.

Even though individual muscle fibers contract fully when stimulated, whole skeletal muscles are capable of producing contractions of varying strength—or **graded contractions**. Graded contractions result from two different processes. The first is called recruitment. The second is called wave summation.

Recruiting Motor Units. **Recruitment** is the active engagement of many muscle fibers during contraction. To generate the kinds of force needed to move arms and legs and even eyelids requires the action of more than one muscle cell—or recruitment. Each skeletal muscle is served by at least one motor nerve (Chapter 12). However, each motor nerve contains hundreds of axons of motor neurons found in the spinal cord. As we saw in Chapter 12, axons form many branches upon reaching the muscles they supply. As a result, a single motor neuron may innervate (supply) many muscle fibers.

If you could tease out one single motor axon and trace it to the muscle fibers it supplies, you would find that it terminates on many muscle fibers (Figure 14–22). A motor neuron and the fibers it supplies con-

FIGURE 14–22 The Motor Unit Each axon branches at its termination, supplying a few dozen to many thousand muscle fibers. A single axon and its muscle fibers constitute a motor unit.

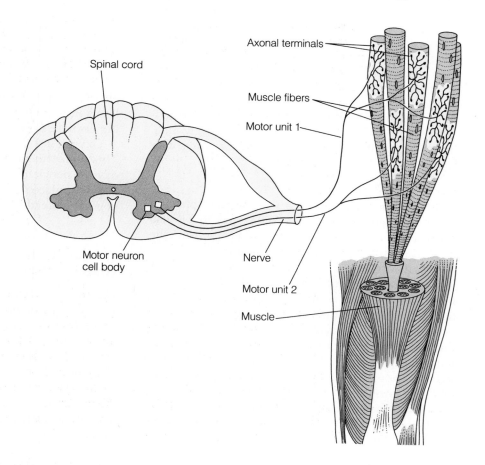

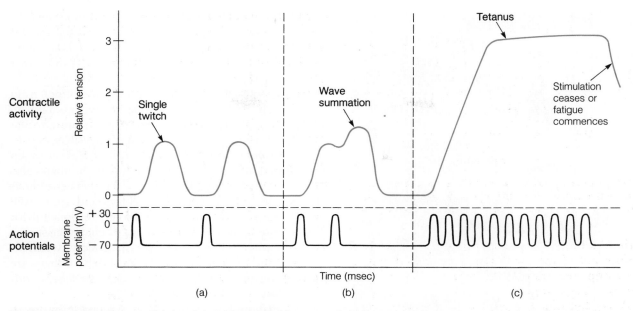

The duration of the action potentials is not drawn to scale but is exaggerated.

FIGURE 14–23 Myograms of Muscle Contraction
(a) The force generated by two separate stimuli. (b) The force generated by two closely timed stimuli. (c) The force generated by many closely spaced stimuli.

stitute a **motor unit**. The finer the control, the fewer muscle fibers in a motor unit. Thus, each neuron that supplies the strong muscles of the leg, which produce crude but strong propulsive force, may end on as many as 2000 muscle fibers. Each time an impulse comes down the axon, it stimulates all 2000 muscle fibers. In contrast, a muscle involved in fine motor movements, such the extrinsic muscles of the eye or the muscles of the hand, contains "smaller" motor units with a few dozen muscle fibers per axon. This provides a greater degree of control.

To increase the force of contraction in all muscles, the central nervous system "recruits" additional motor neurons. The more motor neurons that are stimulated, the stronger the force of contraction.

Wave Summation and Tetanus. Graded contractions also result by increasing the contractile force or tension each fiber generates. Each time a muscle fiber is stimulated, it contracts fully. Figure 14–23a shows the strength of contractions resulting from nerve impulses arriving after muscle fibers contract and relax. If, however, nerve impulses reach the muscle fiber before it has relaxed—that is, while it is still somewhat contracted—additional tension is created (Figure 14–23b). In other words, the muscle fibers contract more forcefully. In a sense, then, the second

contraction "piggybacks" on the first; this process is called **wave summation**.

If the nerve impulses arrive frequently, a smooth sustained contraction will occur in muscle fibers (Figure 14–23c). When you carry a bag of groceries in your arms, for example, your arm muscles contract to support the weight and remain contracted throughout the activity. A sustained contraction at maximal strength is called **tetanus**—not to be confused with the serious, often fatal bacterial infection of the same name. Tetanic contractions eventually cause muscle fatigue. The muscle stops contracting, even though the neural stimuli may continue.

Muscle Tone

Touch one of your muscles. Even if you are not in peak physical condition, you will notice that the muscle is firm—or slightly tense. This firmness is called **muscle tone**. Muscle tone is essential for maintaining posture. Without it, you would literally fall into a heap on the floor when you stood up.

Muscle tone results from the contraction of muscle fibers during periods of inactivity. But not all fibers contract, just enough to keep the muscles slightly tense.

incredible endurance (Figure 14–24). Other athletes (sprinters, for example) seem specialized for sudden bursts of movement. Their muscles contain a higher proportion of **fast-twitch muscle fibers**—fibers that contract swiftly.

Slow-twitch fibers are anatomically distinct from fast-twitch fibers. They are smaller and contain an abundance of **myoglobin**, a cytoplasmic protein that binds to oxygen like its counterpart in RBCs, hemoglobin. Myoglobin in the muscle cells releases oxygen as needed during exercise. Slow-twitch fibers also contain a slow-acting **myosin ATPase**. This enzyme is located in the myosin cross bridges and splits ATP during muscle contraction; it is largely responsible for the slow-twitch fiber's physiological characteristics.

Fast-twitch fibers are larger than slow-twitch fibers. Some of them fatigue quite easily. Others are fatigue resistant. All fast-twitch fatigable fibers contain a fast-acting myosin ATPase.

Skeletal muscles generally contain a mixture of slow- and fast-twitch fibers, giving each muscle a wide range of performance abilities. However, a muscle that performs one type of function more often than another may have a disproportionately higher number of fibers corresponding to the type of activity it performs. In the muscles of the back, for example, slow-twitch fibers are more prevalent. These muscles operate throughout the waking hours to help maintain posture. They do not need to contract quickly, but must be resistant to fatigue. In contrast, the muscles of the arm are used for many "quick actions"—waving, playing tennis, grasping falling objects. Fast-twitch fatigable fibers are more common in the muscles of the arm. Although they allow us to perform quick movements, they tend to tire rather easily.

New research suggests that one of the reasons some people excel in certain sports while others do not may lie in the relative proportion of fast- and slow-twitch fibers in their skeletal muscles. A study of the skeletal muscles of world class long-distance runners suggests that their physical endurance is primarily due to a high proportion of slow-twitch fibers, a trait that may be genetically determined. Biochemical studies also show that the muscle cells in endurance athletes have a higher level of ATP, both at rest and during exercise, thus providing more energy for muscle contraction. Endurance athletes start out with a larger storehouse of energy and maintain a larger supply throughout exercise. (For more on muscles and endurance, see Health Note 14–2.)

FIGURE 14–24   A Long-Distance Runner   An abundance of slow-twitch fibers give the Olympic long-distance runner a better chance than you or I.

Muscle tone is maintained, in part, by the muscle spindle, a receptor that monitors muscle stretching (Chapter 13). The spindle alerts the brain and spinal cord to the degree of stretching. When muscles relax, signals travel to the spinal cord, then back out motor axons, which stimulate a low level of muscle contraction, maintaining muscle tone.

Slow- and Fast-Twitch Muscle Fibers

Physiological studies have revealed the presence of two types of skeletal muscle fibers: fast-twitch and slow-twitch fibers. Endurance athletes (long-distance runners) who can perform for long periods without tiring have a higher proportion of **slow-twitch muscle fibers**—fibers that contract relatively slowly, but have

HEALTH NOTE 14–2
Exercise: Building Muscles and Increasing Endurance

Milo of Croton was a champion wrestler in ancient Greece who stumbled across a revolutionary way to build muscle. His recipe was simple. Milo secured a newborn calf, which be carried around every day. Day after day, Milo faithfully followed this routine, and by the time the calf had become a bull, so had he. Milo's program worked, for he went on to win six Olympic championships in wrestling.

The Greek wrestler discovered a principle of muscle physiology familiar to weight lifters today: when muscles are made to work hard, they respond by becoming larger and stronger. The increase in size and strength results from an increase in the amount of contractile protein inside muscle cells.

Muscle protein is quickly made and quickly destroyed. In fact, half of the muscle you gain in a weight-lifting program is broken down within two weeks after you stop exercising. To build muscle, the rate of formation must exceed the rate of destruction. If you don't keep up with your exercise program, your newly developed biceps will disappear quickly.

High-intensity exercise, such as weight lifting, builds muscle. It takes surprisingly little exercise to have an effect. Working out every other day for only a few minutes will result in noticeable changes in muscle mass. According to some sources, 18 contractions of a muscle (in three sets of six contractions each) are enough to increase muscle mass if the contractions force your muscles to exert over 75% of their maximum capacity.

A Progressive Resistance Machine
Each PRM is designed to exercise one set of muscles.

Low-intensity exercise, such as aerobics and swimming, tends to burn calories, but not build muscle bulk. Such exercise helps individuals increase endurance—that is, increase their ability to sustain muscular effort. Stamina or endurance results from numerous physiological changes. One of the most important changes occurs in the heart.

The heart responds to exercise like any other muscle—it grows stronger and sometimes even enlarges. A well-exercised heart beats more slowly but pumps more blood with each beat. The net result, then, is that the heart works more efficiently, delivering more oxygen to skeletal muscles.

Increased endurance also results from improvements in the function of the respiratory system. For example, exercise increases the strength of the muscles involved in breathing. These muscles become stronger and can operate longer without tiring. Breathing during exercise becomes more efficient.

Increased endurance may also be attributed to an increase in the amount of blood in the body. An increase in blood volume, in turn, results in an increase in the number of RBCs in the body, thus increasing the amount of oxygen available to body cells. This improvement, combined with others, allows an individual to work out longer without growing tired.

When you set out on an exercise program, it is important to establish your goals. If you are interested in increasing your endurance, you should pursue an exercise regime that works the heart and muscles at a lower intensity over longer periods. If you are after bulk, then the answer lies in high-intensity exercises that result in increased muscle mass.

Many health clubs and university gyms offer programs or advice on ways to achieve your goals. And many offer a variety of machines to help you build muscles or tone up flabby flesh (see the figure). Most of the exercise machines work one particular muscle group—for example, the muscles of the upper arm or the muscles of the chest. A half-dozen or more machines, each designed for a specific muscle group, are placed in a line. You simply go down the line, working one set of muscles, then an-

(continued on next page)

other, until you have exercised your entire body.

Exercise machines are popular for several reasons. First of all, they are safer than free weights—barbells and dumbbells. Progressive resistance machines eliminate the chances of your dropping a weight on your toes—or someone else's. Moreover, it is almost impossible to wrench your back if you make a mistake using one. Lifting free weights, on the other hand, requires care and

training as well as brawn.

These machines also reduce the amount of time a person needs to exercise by about one-half. Why? They require work when flexing and extending a joint. The biceps machine, for example, requires you to pull the weights up, then let them return slowly. In both directions, your muscles are being forced to work. The machines also allow you to lift heavier weights. The reason for this is simple: With free weights,

you can only lift a weight that can safely be moved through the part of the exercise where your muscle is the weakest. Any more and you can tear a muscle or drop a weight. One popular brand, the Nautilus machines, alters the resistance automatically as you perform an exercise. In the weakest phase of the exercise, the machine automatically reduces the resistance to prevent damage. Throughout the rest of the exercise, however, the resistance is full.

ENVIRONMENT AND HEALTH: ATHLETES AND STEROIDS

In recent years, many Americans have become increasingly troubled by the use of drugs. Even athletes have come under scrutiny—some for using cocaine and other addictive drugs and others for their use of anabolic steroids.

Anabolic steroids are synthetic hormones, which, when taken in large doses, stimulate muscle formation (anabolism). Taken by weight lifters and other athletes, anabolic steroids increase muscle size and strength by stimulating protein synthesis in muscle cells. Steroids also reduce inflammation that frequently results from heavy exercise, allowing athletes to work out harder and longer.

When it comes to building muscle, steroids and exercise are an unbeatable combination. Some users think that they may even increase aggression, which may be helpful to football players and other competitive athletes on the playing field. Despite steroids' apparent benefits, physicians are concerned about troublesome health effects.

Steroids, for example, may result in psychiatric (mental) and behavioral problems. In interviews with 41 athletes who used steroids, researchers found that one-third of the athletes developed severe psychiatric complications. These athletes routinely took steroids in doses 10 to 100 times greater than those used in

medical studies of the drugs. The athletes also reported using as many as five or six steroids simultaneously in cycles lasting 4–12 weeks. This practice known as "stacking" is quite common and may be responsible for the psychiatric effects.

Athletes in the study reported episodes of severe depression during—and especially after—steroid use. Some athletes reported feelings of invincibility. One man, in fact, deliberately drove into a tree at 40 MPH, while a friend videotaped him. Some subjects reported psychotic symptoms in association with steroid use, including auditory hallucinations. Withdrawal from steroids results not only in depression but also in suicidal tendencies.

This study, based on a relatively small sample, suggests that steroid use may not be worth the benefits. Making matters worse, steroids may also damage the heart and kidney and frequently reduce testicular size in men. Men taking steroids often do not have to shave. In women, steroids deepen the voice and may, studies suggest, cause enlargement of the clitoris. Steroids also cause severe acne and possibly liver cancer. Unfortunately, there are no scientific studies on the long-term health effects of steroids.

Despite the fact that anabolic steroids are banned by the National Football League, the Olympic Committee, and college athletic programs, athletes continue to use them. Most steroids are imported illegally from Mexico and Europe. Because of a federal crackdown on the importation of these illegal drugs, some experts believe that the inflow may be slowing. Others are not so optimistic, noting that the $100-million-a-year black market will not be easily thwarted.

A recent survey of 46 public and private high schools across the nation involving over 3000 teenagers suggests that steroid use is especially prevalent in high school seniors. About 1 of every 15 male high school seniors reported taking anabolic steroids.

The study also suggests that the use of anabolic steroids begins early—in junior high school. Two-thirds of the students surveyed said that they had started using anabolic steroids by age 16. Nearly half the steroid users said they took the drugs to boost athletic performance. Twenty-seven percent said their primary motive was to improve their appearance. Researchers say that adolescents who use steroids may be putting themselves at risk of stunted growth, infertility, and psychological problems.

Steroid use in the United States illustrates our dependence on quick fixes. It also illustrates the almost obsessive focus on performance and achievement in our highly competitive society, a social environment that may be endangering the health of our children and our athletes.

SUMMARY

THE HUMAN SKELETON

1. The human skeleton consists of 206 bones. Bones provide internal support, allow for movement, help protect internal body parts, produce blood cells and platelets, store fat, and help regulate blood calcium levels.
2. Most bones have an outer layer of compact bone and an inner layer of spongy bone. Inside the bone is the marrow cavity filled with either fat cells (yellow marrow) or blood cells and blood-producing cells (red marrow) or combinations of the two.
3. The joints unite bones. Some joints are immovable. Others offer only slight movement. Still others provide relatively free movement.
4. The synovial joint offers the greatest degree of movement. Synovial joints share anatomical similarities: articular surfaces covered by hyaline cartilage and support structures (joint capsules, ligaments, muscles, and tendons).
5. The joint capsule consists of an outer layer of fibrous tissue and an inner layer called the synovial membrane. The synovial membrane has a rich blood supply and produces synovial fluid, which lubricates the articular surfaces of the bones.
6. The movable joints allow for flexion, extension, and several other important movements.

7. The joints are subject to injury and disease. Torn ligaments and ripped cartilage, for example, are common injuries among athletes. Wear and tear on some joints may cause the articular cartilage to crack and flake off, resulting in degenerative joint disease or osteoarthritis.
8. Rheumatoid arthritis results from an autoimmune reaction that produces inflammation and thickening of the synovial membrane, disfiguring and stiffening joints and causing pain.
9. Diseased and worn joints can be replaced by artificial substitutes, or prostheses.
10. Most of the bones form from hyaline cartilage. Bone formation begins in the primary center of ossification.
11. In the primary center of ossification, cartilage cells enlarge. Calcium is then laid down on the cartilage matrix. A thin shaft of compact bone is deposited on the periphery of the hyaline cartilage mass.
12. Cells from the periosteum invade the interior of the cartilage. Some give rise to spongy bone. Others will give rise to blood cells. Secondary centers of ossification form in the epiphyses.
13. At the end of bone development, all that remains of the cartilage is a band of cells, the epiphyseal plate, which allows the bone to elongate.
14. Bone is constantly remodeled after birth to accommodate changing stresses. During bone remodeling, osteoclasts destroy bone. Osteoclasts are stimulated by the parathormone, a hormone produced by the parathyroid glands.
15. Osteoclasts also participate in the homeostatic control of blood calcium levels. When activated, these cells free calcium from the bony matrix, helping to raise blood calcium levels.
16. Osteoblasts are bone-forming cells stimulated by calcitonin, a hormone from the thyroid gland. Calcitonin secretion helps decrease blood calcium levels.
17. Osteoporosis is a disease of the bone caused by progressive decalcification. The bones become brittle and easily broken. Osteoporosis is most common in postmenopausal women where it results from the loss of the ovarian hormone estrogen.
18. Osteoporosis also occurs in people who are immobilized for long periods.
19. Exercise, calcium and fluoride supplements, calcium-rich foods, vitamin D, and estrogen supplements can all help prevent and reverse this potentially fatal disease.

THE SKELETAL MUSCLES

20. Skeletal muscles are involved in body movements, help maintain our posture, and produce body heat both at rest and while working or exercising.
21. Skeletal muscles consist of long, unbranched cells called muscle fibers.
22. Muscle fibers are excitable, contractile, and elastic.
23. Inside each muscle fiber are numerous myofibrils, bundles of contractile filaments. The myofibril is composed of repeating units, called sarcomeres. Each sarcomere

consists of a dark central band, the A band, and two narrow lighter bands, the I bands.

24. Contraction occurs as the actin filaments are pulled toward the center of the sarcomere by the myosin filaments. Cross bridges on the myosin filaments attach to binding sites on the actin filaments, then pull the actin filaments inward.

25. Muscle contraction is stimulated by nerve impulses from motor neurons. Acetylcholine released by motor neurons generates action potentials in the muscle fiber. The impulse travels along the membrane and penetrates the interior of the muscle fiber via the T tubules, stimulating muscle contraction.

26. The energy for muscle contraction comes from ATP. ATP is replenished by creatine phosphate, anaerobic glycolysis, and aerobic metabolism.

27. Muscle fatigue occurs when glycogen stores are depleted and when lactic acid builds up in the muscle fiber.

28. Individual muscle fibers obey the all-or-none law. When activated by an action potential, the muscle fiber contracts fully.

29. Contractions of varying strength (graded contractions) can be generated by recruitment and by wave summation.

30. Recruitment results from the engagement of many motor units. A motor unit is a motor axon and all of the muscle cells it innervates.

31. Wave summation is a piggybacking of muscle fiber contractions caused by stimuli arriving before the fiber can relax, which produces a stronger contraction.

32. Muscle tone is the rigidity of resting muscle caused by low-level contraction of some muscle fibers. It is maintained in part by the muscle spindle, which monitors muscle stretching.

33. The body contains two types of skeletal muscle fibers: slow-twitch and fast-twitch fibers.

34. Skeletal muscles generally contain a mixture of both types, although muscles that perform one type of function over another have a greater percentage of the appropriate muscle fiber.

35. Research suggests that individuals also differ genetically in the relative proportion of fast- and slow-twitch fibers in their muscles.

36. Muscle mass can be increased by exercise. An increase in muscle mass results from an increase in the amount of contractile protein in muscle fibers.

37. To build mass, you generally must use more weight and do fewer repetitions. Building endurance requires less weight and more repetitions.

38. Endurance is a function of three factors: the condition of the heart, the condition of the muscles of the respiratory system, and the blood volume. Improvement in all three factors increases the efficiency of oxygen supply to muscles.

ENVIRONMENT AND HEALTH: ATHLETES AND STEROIDS

39. Many athletes are using synthetic anabolic steroids to improve performance and build muscle.

40. Unfortunately, massive doses of steroids have many harmful effects. They can increase aggression, cause psychiatric imbalance, such as severe depression, and result in damage to the heart and kidneys.

EXERCISING YOUR CRITICAL THINKING SKILLS

A friend is selling an all-natural supplement guaranteed to give you more energy and help you sleep better at night. She cites a number of examples of users who have taken the product and found they feel energized and now can sleep better at night as well. She also notes that the developer of the product has a Ph.D. in nutrition. She wants you to try the product, which costs $40 a bottle. Using the critical thinking rules you learned in Chapter 1, describe what you might do in this situation. How would you evaluate the evidence your friend has given supporting the use of this supplement? What additional information would you want?

TEST OF TERMS

1. The skull, ribs, and _____ form the _____ skeleton.

2. The _____ skeleton consists of the bones of the legs and arms and the bones of the shoulders and pelvis to which they attach.

3. The humerus and femur are examples of _____ bones. The shaft or _____ of this type of bone consists of a layer of _____ bone surrounding a hollow cavity, the _____ cavity.

4. Inside the epiphyses of the hu-

merus is a network of bony spicules forming _____ bone.

5. The knee and shoulder joints are examples of _____ joints. They contain a fluid, called _____ fluid, that allows for easy movement of the bones. These joints are supported by four structures: _____ _____, ligaments, muscles, and _____.

6. Closing a joint is called _____; opening it is called _____.

7. Knee surgery can be performed using an _____, a device that reduces the trauma and allows surgeons to see what they are doing.

8. Degenerative joint disease or _____ results from wear and tear on a joint. It occurs most often in the weight-bearing joints.

9. _____ arthritis is believed to be an autoimmune disease. It results in a thickening of the _____ membrane and degeneration of the joint.

10. An artificial joint or body part is called a(n) _____.

11. Most bone in the body is formed during embryonic development from _____ _____. The first location of bone formation in the diaphysis is called the _____ _____ of ossification.

12. Bone is broken down by large, multinucleated cells called _____; it is laid down by _____.

13. The zone of cartilage lying between the epiphysis and diaphysis of a growing bone is called the _____ _____.

14. Bone deposition is stimulated by the hormone _____ secreted by cells in the _____ gland. Bone dissolution is stimulated by _____.

15. Prolonged bone degeneration, which results in brittle bones and occurs chiefly in postmenopausal women, is called _____.

16. A muscle fiber is formed during embryonic development by the fusion of numerous embryonic muscle cells. Muscle fibers are long, unbranched cells containing many _____. Viewed with the light microscope, they appear banded or _____.

17. Each muscle fiber is surrounded by a thin layer of connective tissue, the _____, which holds it in place.

18. The entire muscle is surrounded by a layer of connective tissue called _____, which forms the tendons that hold the muscle in place.

19. Inside the muscle fiber are numerous bundles of contractile filaments. Each bundle is called a(n) _____.

20. The _____ is the functional unit of the muscle fiber. It extends from one Z line to the next and contains a dark central band, the _____ band, and two lighter bands on either end, the _____ bands.

21. Contraction results when _____ ions are released from the _____ reticulum. These ions bind to _____ molecules, thus releasing tropomyosin from the binding sites on the _____ filament.

22. The _____ _____ of the myosin filaments pull the actin filaments inward, causing contraction.

23. The neurotransmitter _____ is released at the neuromuscular junction. It causes an action potential to be set up in the muscle membrane. The action potential penetrates the interior of the muscle fiber via the _____ _____.

24. _____ _____ occurs when muscle depletes its supplies of glycogen and ATP, and when _____ _____ levels increase.

25. The contraction of a single muscle cell is called a muscle _____. It is an _____ response to stimulation, resulting in a complete contraction.

26. The strength of contraction can be increased by _____ summation, the piggybacking of contractions, and by recruiting more _____ units.

27. A world class long-distance runner would probably have an abundance of _____ muscle fibers in his or her leg muscles. These fibers have an abundance of _____, which holds oxygen and releases it during activity.

28. A sprinter might have an abundance of _____ fibers. They contain a fast-acting _____ _____, which splits ATP, providing energy for muscle contraction.

29. Synthetic hormones called _____ _____ can help an athlete increase muscle mass but have many deleterious side effects.

Answers to the Test of Terms are located in Appendix B.

TEST OF CONCEPTS

1. Describe the functions of bone. In what ways does bone participate in homeostasis?

2. The synovial joints move relatively freely. What structures support the joint, helping to keep the bones in place?

3. A young patient comes to your office with swollen joints and complains about pain and stiffness in the joints. Friends have suggested that the boy has arthritis, but his parents argue that he is too young. Only old people get arthritis, they say. How would you answer them?

4. Describe the process of bone formation. Be sure to define the following terms: primary and secondary centers of ossification, osteoclasts, perichondrium, osteoblasts, and osteocytes.

5. Bone is constantly remodeled from infancy through adulthood. Explain when bone remodeling occurs and what cells and hormones participate in the process.

6. Using what you know about bone, explain why an office worker who exercises very little is more likely to break a bone on a skiing trip than a counterpart who works out every night after work.

7. A 30-year-old friend of yours who smokes, exercises very little, and avoids milk products because of her diet says, "Why should I worry about osteoporosis? That's a disease of old women." Based on what you know about bone, how would you respond to her?

8. Describe the major functions of skeletal muscle.

9. Describe the detailed structure of a skeletal muscle fiber.

10. Describe the molecular events leading to muscle contraction.

11. Using your knowledge of muscle physiology, explain different ways that muscle contraction might be blocked at the cellular and molecular levels.

12. Define the term *graded contraction*. How are graded contractions achieved?

13. Describe the two types of skeletal muscle fibers and explain how they differ.

14. A friend comes to you complaining that he can't seem to lose weight. He works out on barbells three times a week for an hour or so each time. What advice would you offer?

15. A friend is thinking about using anabolic steroids to improve her performance in soccer. She argues that she's young and will only be taking the drug through soccer season for a year or so. She points out that other young women are using steroids and they really help build muscle and endurance. What advice would you give her?

SUGGESTED READINGS

Beil, L. 1988. Of joints and juveniles. *Science News* 134(12): 190–91. Looks at the point at which excessive exercise harms rather than strengthens bone material.

Cooper, K. H. 1989. The basics of bone. *Health* 21(4): 80–82. Discussion of what keeps bones healthy and strong.

Eron, C. 1988. Young hearts. *Science News* 134(15): 234–36. Looks at new ways to prevent heart attacks.

Marieb, E. N. 1989. *Human anatomy and physiology.* Redwood City, Calif.: Benjamin/Cummings. Excellent coverage of the skeletal system and muscles.

Montoye, H. J., J. L. Christian, F. J. Nagle, and S. M. Levin. 1988. *Living fit.* Menlo Park, Calif.: Benjamin/Cummings. See Chapter 5 for a discussion of exercise programs.

Rosato, F. D. 1990. *Fitness and wellness: The physical connection,* 2d ed. St. Paul, Minn.: West. Well-documented treatise on health and fitness.

Sherwood, L. 1989. *Human physiology: From cells to systems.* St. Paul, Minn.: West. Excellent detailed coverage of muscle physiology.

15

The Endocrine System

Histologic section of the thyroid gland, an endocrine organ in the neck.

T he human body is a marvel of design and operation. Thousands of physiological mechanisms operate at any one moment, many of them aimed at maintaining homeostasis in the ever-changing environment. Maintaining this balance requires communication— communication between cells as well as between organs and organ systems. Communication is provided principally by two body systems: the nervous system and the endocrine system. The nervous system carries messages quickly over great distances along axons (Chapter 12). It controls muscles, glands, and organs. The endocrine system relies on chemical messages carried relatively slowly through the bloodstream. These chemical messages and the ways they affect cells, organs, and systems are the principal focus of this chapter.

PRINCIPLES OF ENDOCRINOLOGY

The **endocrine system** consists of numerous small glands scattered throughout the body (Figure 15–1). These glands produce and secrete substances called hormones. A **hormone** is a chemical produced in cells or groups of cells, the **endocrine** (or **ductless**) **glands**,

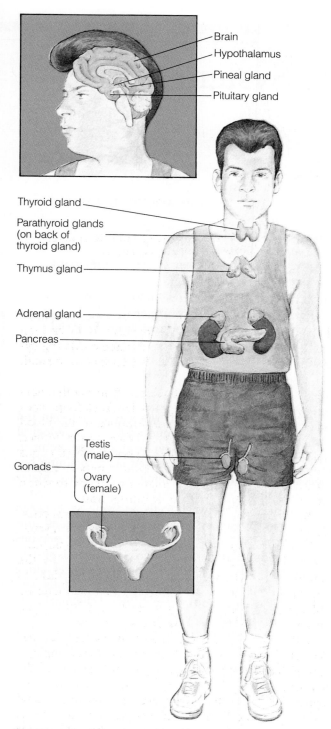

FIGURE 15-1 **The Endocrine System** The endocrine system consists of a scattered group of glands that produce hormones, which help regulate growth and development, homeostasis, reproduction, energy metabolism, and behavior.

which release their products into the bloodstream. Hormones travel in the blood to distant sites where they elicit some response. The hormone insulin (produced by the pancreas), for example, is transported in the blood to skeletal muscle and the liver. The cells a hormone affects are called its **target cells**. Insulin stimulates target cells in the liver and skeletal muscle, causing them to absorb glucose from the blood and produce glycogen.

Human hormones function in five principal areas: (1) homeostasis, (2) growth and development, (3) reproduction, (4) energy production, storage, and use, and (5) behavior. At any one moment, the blood carries dozens of hormones. The cells of the body are, therefore, exposed to many different chemical stimuli. How does a cell keep from responding to all the signals?

Cells respond only to specific hormones. The "selection process" depends on cell **receptors**, proteins in the plasma membrane and in the cytoplasm. Receptors bind specifically to one type of hormone, in the same way that enzymes bind to only one substrate (Chapter 3). Thus, cells containing receptors for a specific hormone are affected by that hormone; cells without the receptors are unaffected.

Types of Hormones

The word hormone comes from a Greek word meaning "I arouse." Hormones fit into two broad categories. The first are the **trophic hormones**, which stimulate the production and secretion of another hormone by another endocrine gland. (The word *trophic* means to nourish.) An example is **thyroid-stimulating hormone** or **TSH**. Produced by the pituitary gland, TSH travels in the blood to the thyroid glands located in the neck on either side of the larynx. Here, it stimulates the production and release of another hormone, thyroxin. Thyroxin is a nontropic hormone. **Nontropic hormones** are hormones that exert their effect principally on nonendocrine tissues. They do not stimulate the synthesis of other hormones. Thyroxin's chief function is to increase the metabolic rate of body cells.

Hormones can also be classified according to their chemical composition into three groups: (1) steroids, (2) proteins and polypeptides, and (3) amines.

Steroid hormones are derivatives of cholesterol (Chapter 2). Figure 15-2 illustrates two common steroid hormones. Very small differences in the chemical structure of these molecules result in profound functional differences.

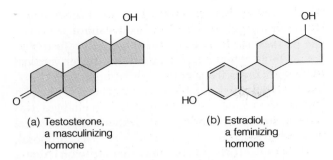

FIGURE 15–2 Two Common Steroids (*a*) Testosterone, a male sex hormone, and (*b*) estradiol, a female sex steroid.

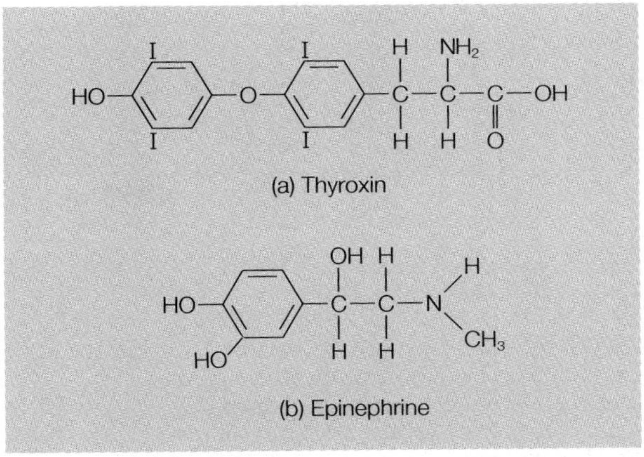

(a) Testosterone, a masculinizing hormone

(b) Estradiol, a feminizing hormone

(a) Thyroxin

(b) Epinephrine

FIGURE 15–3 Representative Amine Hormones (*a*) Thyroxin and (*b*) adrenalin (epinephrine).

Protein and **polypeptide hormones** constitute the largest class of hormones. As Chapter 2 noted, proteins and polypeptides are polymers of amino acids joined by peptide bonds. For convenience, protein and polypeptide hormones are often referred to as polypeptides. Growth hormone and insulin are two examples.

Amines are derivatives of the amino acid **tyrosine**. Figure 15–3 shows the structure of two of the four amine hormones produced in the body.

Hormone Secretion

Feedback Control.   Hormones control homeostasis. Their production and release, however, are generally controlled by negative feedback loops (Chapter 6). Consider the hormone glucagon.

Glucagon is a hormone released by cells in the pancreas (Figure 15–1). When glucose concentrations fall between meals, glucagon is released into the bloodstream. Glucagon stimulates the breakdown of glycogen in liver and muscle, causing the release of glucose molecules into the blood. This helps to restore blood glucose levels. When normal glucose levels are achieved, the glucagon-producing cells in the pancreas end their secretion. Not all feedback loops are as simple as this one, however; some involve intermediary compounds. Nevertheless, all operate on the same basic principle.

Positive feedback loops are also encountered in the endocrine system. A positive feedback loop results when the product of a cell or organ stimulates additional production. Positive feedback loops are rare and have evolved to perform some specialized function. Ovulation, the release of the ovum from the ovary, for example, is stimulated by a positive feedback loop (Chapter 16). Each positive feedback loop has a built-in mechanism that ends the escalating cycle, preventing the response from getting out of hand.

Biorhythms or Biological Cycles.   The fact that hormones are controlled by negative feedback loops does not mean that hormone concentrations in the blood are constant 24 hours a day, 365 days a year. In fact, virtually all hormones undergo periodic fluctuations in their release, causing many body functions also to fluctuate. These natural fluctuations in body function are called **biological cycles** or **biorhythms**.

Biological cycles vary in length. Some hormones, for example, are released in hourly pulses or cycles. Others are released in daily cycles. For example, cortisol is a steroid hormone that is produced by the adrenal cortex in a 24-hour or **circadian rhythm**, as shown in Figure 15–4. Cortisol secretion increases during the night, reaching a peak just before you wake up, then falls sharply during the day.

Other hormones are released in monthly cycles. The 28-day menstrual cycle in women, for example, is controlled by hormones (Chapter 16). Levels of the reproductive hormones vary considerably during the cycle. Still other hormones are released in seasonal cycles. Thyroid hormone in people, for instance, is released in greater amounts during the cold winter months than during the summer. Hormonal cycles are controlled by the biological clocks, regions of the brain that control or regulate biological cycles (Chapter 6).

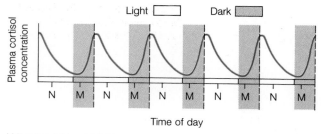

Light ☐   Dark ▨

N is noon; M is midnight.

FIGURE 15–4 Cortisol Secretion Cortisol secretion follows a diurnal (daily) rhythm with highest levels occurring in the night just before waking.

Source: Adapted with permission from George A. Hedge, Howard D. Colby, and Robert L. Goodman, *Clinical Endocrine Physiology* (Philadelphia: W.B. Saunders Company, 1987), Figure 4–4, p. 80.

How Hormones Act on Cells

The chemical nature of a hormone determines how it is transported in the blood and how it acts on cells. Protein and polypeptide hormones, for example, are water soluble (hydrophilic). Consequently, they dissolve in the plasma of the blood, which transports them to their target cells. In contrast, steroid hormones are lipids and do not dissolve in water. To be transported, they must be bound to much larger plasma proteins, like albumin (Chapter 8).

Second Messenger Theory. Protein molecules may travel freely in the blood, but they cannot penetrate the plasma membrane of their target cells. To trigger intracellular changes, proteins must trigger internal cellular changes from outside their target cells.

An hypothesis explaining how protein and polypeptide hormones trigger intracellular changes was first presented by the late Earl Sutherland, a medical researcher at Vanderbilt University. His proposal, called the **second messenger theory**, won him a Nobel Prize. The theory states that protein and polypeptide hormones (first messengers) stimulate intracellular changes through a second messenger, an intermediary synthesized inside the cell. Proteins and polypeptides, said Sutherland, stimulate the formation of second messenger inside the cell. This substance, in turn, triggers physiological changes. Here's a step-by-step explanation of the second messenger theory:

1. Polypeptide hormones bind to receptors of the target cell (Figure 15–5).[1]
2. These receptors are linked to an enzyme, called **adenylate cyclase**, which is bound to the inner surface of the plasma membrane.

[1]The hormones adrenalin and noradrenalin from the adrenal medulla also bring about their effects by the second messenger mechanism.

FIGURE 15–5 The Second Messenger Theory Protein and polypeptide hormones bind to a membrane-bound receptor, thus stimulating the production of intracellular cyclic AMP by adenylate cyclase. Cyclic AMP activates protein kinase, which adds phosphates to intracellular enzymes, activating some and deactivating others.

Source: Adapted with permission from George A. Hedge, Howard D. Colby, and Robert L. Goodman, *Clinical Endocrine Physiology* (Philadelphia: W.B. Saunders Company, 1987), Figure 1–8, p. 18.

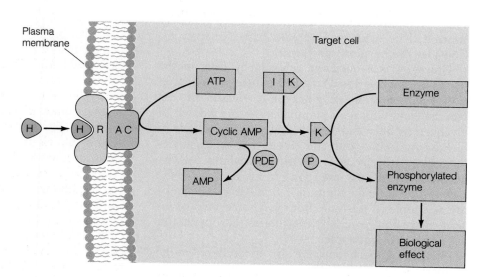

H = Free hydrophilic hormone
R = Surface receptor
AC = Adenylate cyclase
ATP = Adenosine triphosphate
AMP = Adenosine monophosphate

PDE = Phosphodiesterase
IK = Inactive protein kinase
K = Active protein kinase
P = Phosphate

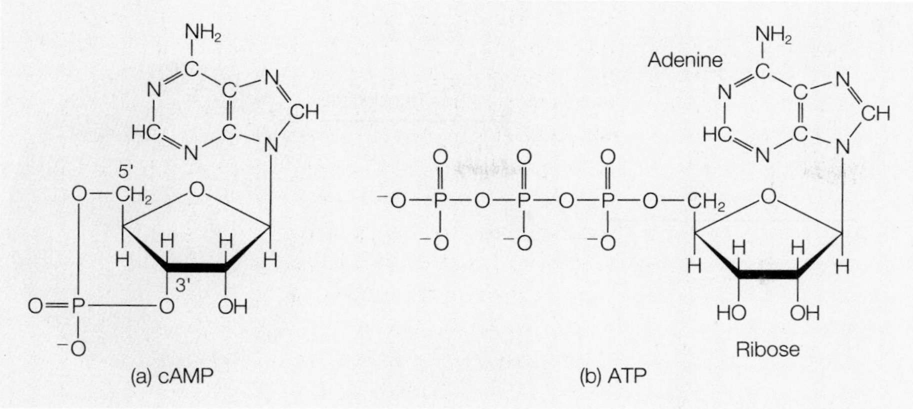

FIGURE 15–6 The Second Messenger, Cyclic AMP (a) Cyclic AMP compared to (b) ATP.

3. Binding of the polypeptide hormone to the receptor activates adenylate cyclase.
4. The activated enzyme catalyzes the conversion of intracellular ATP to **cyclic AMP**, a special nucleotide shown in Figure 15–6.
5. Cyclic AMP is the **second messenger**. Cyclic AMP is water soluble and free to move throughout the cytoplasm. In the cytoplasm, it activates another enzyme called **protein kinase**.
6. This enzyme may "turn on" inactive cytoplasmic enzymes or "turn off" enzymes that are already active. Protein kinase catalyzes the addition of phosphate ions to cellular enzymes, which activates some enzymes and deactivates others. The result is a change in cellular physiology.

Cyclic AMP is not the only second messenger in body cells. Calcium ions and cyclic GMP (guanosine monophosphate) are the intracellular messengers that stimulate metabolic activity.

The Two-Step Mechanism. Steroid hormones stimulate intracellular change differently, by acting on the genes. The process is known as the two-step mechanism. Here's how it works:

1. Steroid hormones can easily penetrate the plasma membrane of the cell (Figure 15–7).
2. Inside the cell (possibly in the nucleus itself), steroid hormones bind to protein receptors. Like their counterparts in the plasma membrane, each receptor recognizes and binds to one type of steroid hormone, thus conferring specificity.
3. When a steroid binds to its receptor, it causes the protein to change its shape slightly. This

activates the receptor-hormone complex. Activation enhances its attraction for chromatin (genetic material in the nucleus).
4. The activated complex binds to the chromatin at a specific site of attachment on the DNA known as an **acceptor site**.[2]
5. The binding of the receptor-steroid complex to the chromatin activates certain genes, resulting in the transcription of DNA—that is, the production of mRNA (Chapter 5).
6. Messenger RNA, in turn, provides a template for the production of structural proteins and/or enzymes. Thus, steroid hormones affect the structure and function of target cells.

THE PITUITARY AND HYPOTHALAMUS

Attached to the underside of the brain by a thin stalk is the **pituitary gland** (Figure 15–8). About the size of a pea, the pituitary lies in a depression in the base of the skull, called the **sella turcica** (Turkish saddle). The pituitary gland is divided into two major parts: the anterior pituitary and the posterior pituitary.

The pituitary gland is one of the most complex of all the endocrine organs. It secretes a large number and variety of hormones and affects a great many of the body's functions. The anterior pituitary, for example, produces seven protein and polypeptide hormones (Table 15–1). The release of the hormones from the anterior pituitary is controlled by a region of

[2]Thyroid hormones also effect intracellular change through the two-step mechanism, but thyroid hormones bind directly to DNA.

FIGURE 15–7 The Two-Step Mechanism Steroid hormones bind to specific receptors in the cell. The receptor-steroid complex then binds to the acceptor sites on the nuclear DNA, activating the genes.

Source: Adapted with permission from George A. Hedge, Howard D. Colby, and Robert L. Goodman, *Clinical Endocrine Physiology* (Philadelphia: W. B. Saunders Company, 1987), Figure 1–9, p. 20.

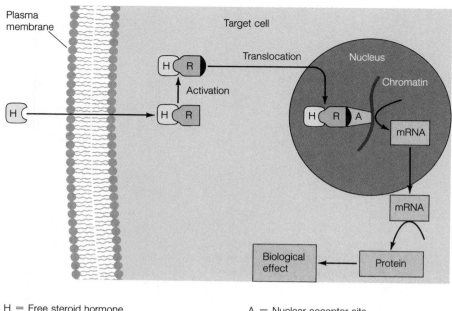

H = Free steroid hormone
R = Cytoplasmic receptor

A = Nuclear acceptor site
mRNA = Messenger RNA

the brain, the **hypothalamus**, lying just above the pituitary gland.

The hypothalamus provides a link between the endocrine and nervous systems. The hypothalamus contains receptors for a variety of chemical substances. These receptors monitor blood levels of hormones, nutrients, and ions. When activated, the receptors stimulate specialized nerve cells within the hypothalamus. These nerve cells are called **neurosecretory neurons**. They synthesize and secrete hormones that act on the anterior pituitary. Some of the hypothalamic hormones stimulate the release of anterior pituitary hormones and are called **releasing hormones**. Others inhibit the release of hormones from the anterior pituitary and are called **inhibiting hormones**.

The releasing and inhibiting hormones travel down the axons of the neurosecretory cells, which terminate in the lower part of the hypothalamus, just above the pituitary gland. The hormones are stored in the axon terminals. When released from the axon terminals, they diffuse into nearby capillaries. These capillaries drain into a series of veins (called the **portal vessels**) in the stalk of the pituitary. The veins, in turn, empty into a capillary network in the anterior pituitary. Thus, the releasing hormones are transported directly to their target cells. This unusual arrangement of blood vessels in which a capillary bed drains to a vein,

which drains into another capillary bed, is called a **portal system**.[3]

Hormones of the Anterior Pituitary

Growth Hormone. People differ considerably in height and body build (Figure 15–9). These differences are largely attributable to **growth hormone** or **GH**, a protein hormone produced by the anterior pituitary. Growth hormone stimulates cellular growth in the body, causing both cellular hypertrophy (enlargement) and hyperplasia (increase in number through division). Growth hormone causes muscle cells to grow by stimulating the uptake of amino acids and protein synthesis. Although growth hormone affects virtually all body cells, its major target organs are bone and muscle. Thus, the more growth hormone produced during the growth phase of an individual, the taller and more massive he or she will be. In men, however, body growth is also stimulated by testosterone, an anabolic steroid produced by the testes. Testosterone stimulates bone and muscle growth, thus explaining why men are generally taller and more massive than women.

[3]Another portal system is found in the liver.

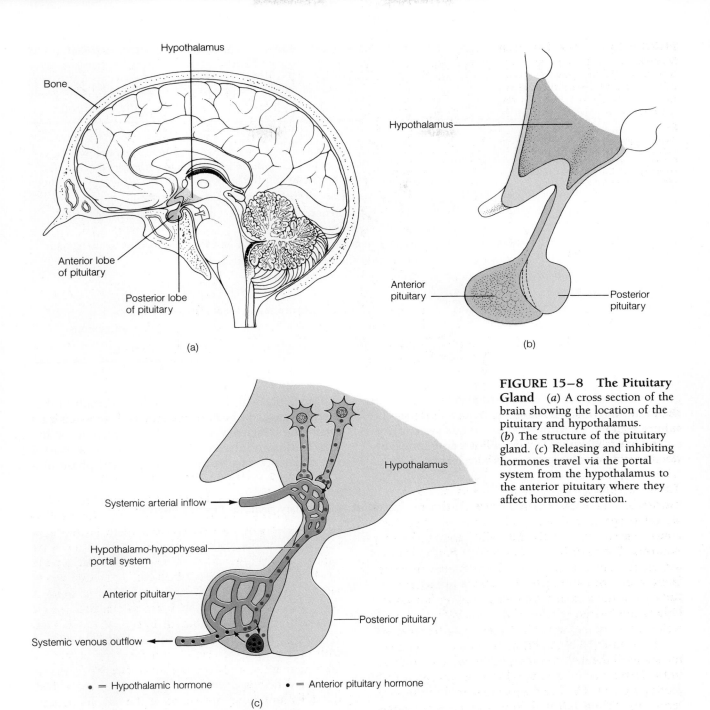

Hypothalamus

Bone

Anterior lobe
of pituitary

Posterior lobe
of pituitary

(a)

Hypothalamus

Anterior
pituitary

Posterior
pituitary

(b)

**FIGURE 15–8 The Pituitary
Gland** (*a*) A cross section of the
brain showing the location of the
pituitary and hypothalamus.
(*b*) The structure of the pituitary
gland. (*c*) Releasing and inhibiting
hormones travel via the portal
system from the hypothalamus to
the anterior pituitary where they
affect hormone secretion.

Hypothalamus

Systemic arterial inflow

Hypothalamo-hypophyseal
portal system

Anterior pituitary

Systemic venous outflow

Posterior pituitary

• = Hypothalamic hormone   • = Anterior pituitary hormone

(c)

Growth hormone secretion undergoes a diurnal
(daily) cycle. Like cortisol, the highest blood levels
are present during sleep. During the day, the level in
the blood declines. It's no wonder, then, that sleep is
so important to a growing child.

Growth hormone secretion declines gradually as
we age. Like many other hormones of the anterior
pituitary, growth hormone secretion is controlled by a
releasing hormone (GH-RH) *and* an inhibiting hormone, both produced by the hypothalamus. Levels of

TABLE 15–1 Hormones Produced by the Pituitary Gland

HORMONE	FUNCTION
Anterior pituitary	
Growth hormone (GH)	Stimulates cell growth; primary targets are muscle and bone where GH stimulates amino acid uptake and protein synthesis; GH also stimulates fat breakdown in the body
Thyroid-stimulating hormone (TSH)	Stimulates release of thyroxin and triiodothyronine
Adrenocorticotropic hormone (ACTH)	Stimulates secretion of hormones by the adrenal cortex, especially glucocorticoids
Gonadotropins (FSH and LH)	Stimulate gamete production and hormone production by the gonads
Prolactin	Stimulates milk production by the breast
Melanocyte-stimulating hormone (MSH)	Function in humans unknown
Posterior pituitary	
Antidiuretic hormone (ADH)	Stimulates water reabsorption by nephrons of the kidney
Oxytocin	Stimulates ejection of milk from breasts and uterine contractions during birth

growth hormone in the blood participate in a negative feedback loop. Growth hormone release can also be stimulated directly through the nervous system (hypothalamus). Stress and moderate exercise, for example, stimulate the hypothalamus to release GH-RH.

Deficiencies in growth hormone can result in dramatic changes in body shape and size, depending on when the deficiency occurs. Undersecretion, or **hyposecretion**, results in stunted growth (dwarfism) if the deficiency begins during the growth phase of a

FIGURE 15–9 Disorders of Growth Hormone Secretion (*a*) Pituitary giant. (*b*) Pituitary dwarves.

(a)

(b)

(a)   (b)

(c)   (d)

FIGURE 15–10   Acromegaly   Hypersecretion of
growth hormone results in a gradual thickening of the bone,
which is especially noticeable in the face, hands, and feet.
(*a*) An individual with acromegaly at age 9; (*b*) at age 16;
(*c*) at age 33; and (*d*) at age 52.

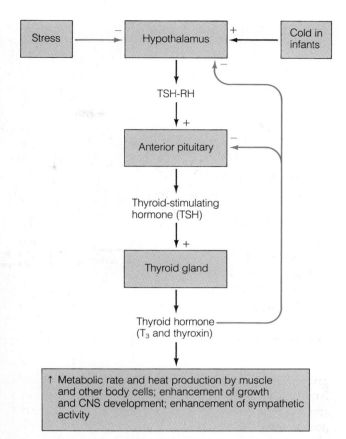

**FIGURE 15–11   Negative Feedback Control of TSH
Secretion** T_3 and thyroxin regulate hypothalamic and
pituitary activity. Other factors, such as stress and cold,
influence the release of TSH via the hypothalamus.

child (Figure 15–9). Oversecretion, or **hypersecretion**, results in giantism, if the excess occurs during
the growth phase. If the pituitary begins producing
excess growth hormone after growth has been completed, the result is a relatively rare disease called
acromegaly. Facial features become coarse and hands
and feet begin to grow (Figure 15–10).

Thyroid-Stimulating Hormone.   Thyroid-stimulating hormone (TSH) is a protein hormone produced by
the anterior pituitary; its release is controlled by
TSH-RH (TSH-releasing hormone) produced by the
hypothalamus. TSH-RH secretion is stimulated by cold
and stress. TSH-RH secretion is also regulated by the
level of thyroid hormone in the blood. Receptors in the

hypothalamus detect the level of circulating thyroid
hormone. When circulating levels of thyroid hormone
are low, the hypothalamus releases TSH-RH. When the
level of thyroid hormone increases, TSH-RH secretion
is inhibited (Figure 15-11). TSH-RH stimulates the
production and release of TSH in the anterior pituitary.
TSH then travels in the blood to the thyroid gland,
where it stimulates the production and release of thyroxin. Thyroxin affects a great many cells. One of its
chief functions is to stimulate the catabolism (breakdown) of glucose by body cells. Since glucose catabolism produces energy and heat, thyroxin raises body
temperature.

Adrenocorticotropic Hormone.   Adrenocorticotro-
pic hormone or **ACTH** is a polypeptide that stimulates cells of the adrenal cortex (another endocrine

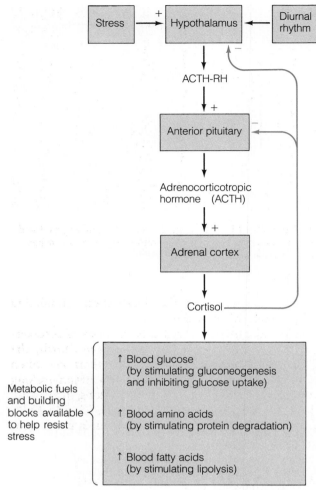

FIGURE 15–12   Feedback Control of ACTH
Cortisol regulates hypothalamic and pituitary activity, but stress and the biological clock also influence the release of ACTH-RH.

ACTH-RH secretion is also controlled by stress, acting through the nervous system. Under stress, the nervous system stimulates the release of ACTH-RH. An increase in ACTH-RH results in an increase in ACTH and an increase in glucocorticoids. Glucocorticoids increase blood glucose levels, thus providing additional energy for cells.

ACTH secretion is also affected by the light-dark cycle. A biological clock in a region of the brain monitors day length and controls ACTH-RH secretion. At night, the clock overrides the negative feedback mechanism involving ACTH-RH and glucocorticoid, causing an increase in ACTH-RH secretion while you sleep. The increased production of ACTH-RH results in elevated levels of ACTH and adrenal steroids. The highest levels of ACTH are present in the blood in the early morning, just before you awake.

To understand how the clock works, consider an analogy. The thermostat in your house operates by a negative feedback mechanism, as described in Chapter 6. When the temperature falls, the furnace turns on. When the temperature rises, the furnace is switched off. Negative feedback loops operate similarly. However, many thermostats can be programmed to maintain a higher temperature for a given period— for example, you can program your thermostat to keep the daytime temperature at 68°F. During the day, the thermostat and furnace maintain a constant temperature via negative feedback. The biological clock is a kind of biological programming mechanism. At night, the clock resets the baseline level, resulting in a marked increase while we sleep.

The Gonadotropins.   The anterior pituitary produces two hormones that affect the male and female gonads. Known as **gonadotropins**, these hormones are discussed in more detail in Chapter 16. One of the gonadotropins is called **follicle-stimulating hormone** or **FSH**; FSH promotes gamete formation in both men and women. The other gonadotropin, **luteinizing hormone** or **LH**, stimulates gonadal hormone production. In men, LH stimulates the production of testosterone, the male sex steroid, by the testes. In women, LH stimulates estrogen and progesterone secretion by the ovary. Both gonadotropins are under the control of a single releasing hormone, **gonadotropin releasing hormone**, produced and secreted by the hypothalamus.

Prolactin.   **Prolactin** is a protein hormone produced by the anterior pituitary. Prolactin performs a great

gland), causing them to synthesize and release hormones. In response to ACTH, the cortex produces a group of steroid hormones, the **glucocorticoids**. Glucocorticoids help control glucose levels, thus supplementing the actions of insulin and glucagon.

ACTH release is under the control of the hypothalamic releasing hormone, ACTH-RH. ACTH-RH secretion is controlled by at least three factors. The first is the level of circulating glucocorticoids, which participate in a negative feedback loop (Figure 15–12). Thus, as the level of glucocorticoid increases, ACTH-RH secretion falls, causing a decline in the release of ACTH by the pituitary.

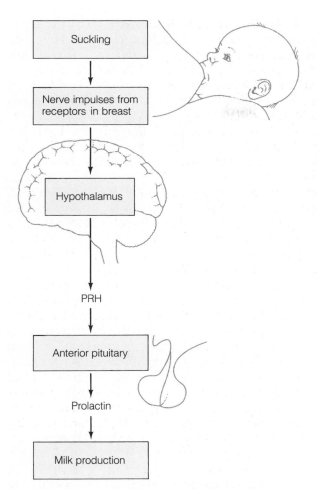

FIGURE 15–13 Neuroendocrine Reflex and Prolactin Secretion Suckling stimulates prolactin release by the anterior pituitary.

FIGURE 15–14 Milking the Cows Milking by hand or machine prolongs milk production long after a calf has been weaned from its mother.

many functions in the animal kingdom. In birds, prolactin stimulates migratory behavior. In fish, it helps regulate electrolyte balance. In women, prolactin is secreted by the anterior pituitary at the end of pregnancy and stimulates milk production by the mammary glands. Suckling prolongs the release of prolactin for at least 3 to 12 months, sometimes much longer.

Prolactin secretion is part of a **neuroendocrine reflex**, a reflex involving both the nervous and endocrine systems. As Figure 15–13 shows, suckling stimulates sensory fibers in the breast. Nerve impulses travel to the hypothalamus via sensory neurons. In the hypothalamus, these impulses stimulate the release of prolactin releasing hormone (PRH), which travels to the anterior pituitary in the bloodstream, stimulating the secretion of prolactin.

The neuroendocrine reflex is the basis of commercial milk production. When a cow gives birth, she produces milk to feed her calf. Calves are often weaned fairly early. Nevertheless, farmers can prolong milk production in their cattle by milking their cows, either manually or by machine (Figure 15–14). Milking machines and hand milking stimulate the neuroendocrine reflex.

Hormones of the Posterior Pituitary

Like the hypothalamus, the **posterior pituitary** is a neuroendocrine gland—that is, a gland made of neural tissue that produces hormones. Derived from brain tissue during embryonic development, the posterior pituitary remains connected to the brain throughout life.

The posterior pituitary consists of neurosecretory cells whose cell bodies are located in the hypothalamus (Figure 15–15). The neurosecretory cells of the posterior pituitary produce two hormones, antidiuretic hormone and oxytocin, each of which consists of nine amino acids. Oxytocin and antidiuretic hormone are synthesized in the cell bodies of the neurosecretory cells and travel down the axons of these cells into the posterior pituitary. The hormones are stored in the axon terminals of the cells and released into the surrounding capillaries.

Antidiuretic Hormone. **Antidiuretic hormone** or **ADH** regulates water balance in humans (Chapter

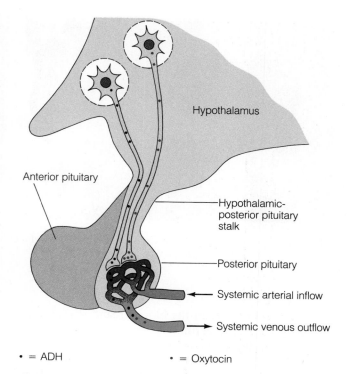

FIGURE 15–15   The Posterior Pituitary
Neurosecretory neurons that produce oxytocin and ADH originate in the hypothalamus and terminate in the posterior pituitary. Hormones are produced in the cell bodies of the neurons and stored and released into the bloodstream in the posterior pituitary.

11). ADH travels in the bloodstream to the kidney where it increases water absorption by the distal convoluted tubules and collecting tubules (Figure 15–16). As a result, water reenters the bloodstream, increasing blood volume and restoring the normal osmotic concentration of the blood.

ADH secretion is controlled by **osmoreceptors** in the hypothalamus. These receptors monitor the concentration of dissolved substances in the blood. When the concentration exceeds the normal level, for example, during dehydration, osmoreceptors stimulate the ADH-producing cells, and ADH is released. Water is reabsorbed by the kidney, diluting the blood. When the concentration approaches homeostatic levels, ADH secretion ceases.

The release of ADH is inhibited by ethanol in alcoholic beverages. The decrease in ADH secretion results in a decrease in water reabsorption in the kidneys and an increase in urine production. The increase in urinary output can lead to dehydration. The dry mouth and intense thirst a person may expe-

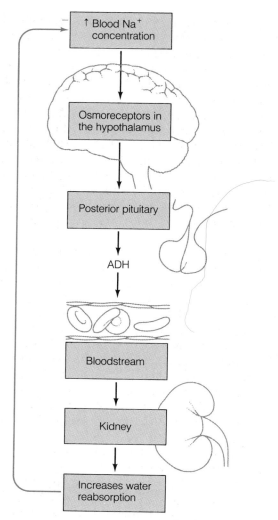

FIGURE 15–16   Role of ADH in Regulating Body Fluid Levels   ADH secretion is stimulated by an increase in the blood concentration of sodium. ADH increases water reabsorption in the kidney, thus eliminating the stimulus for ADH secretion.

rience after drinking are the result of dehydration caused by ethanol.

ADH secretion may also decline as a result of trauma. A blow to the head, for example, can damage the hypothalamus, resulting in a sharp decline in ADH output. In the absence of ADH, the kidneys produce several gallons of urine a day. To keep up with the water loss and to prevent dehydration, an individual must drink enormous quantities of water. This condition is called **diabetes insipidus** (Chapter 11).

ADH is sometimes called **vasopressin** because of its ability to increase arterial blood pressure. Under most conditions, it is doubtful that enough ADH is secreted to increase blood pressure, however. ADH exerts its vasopressive effect only in emergencies. Extreme blood loss—resulting from an automobile accident, for example—stimulates the secretion of large quantities of ADH, causing vasoconstriction. As Chapter 8 explained, the muscle in the walls of severed arteries constricts, helping reduce blood loss. Clotting factors in the blood also help seal off damaged vessel walls. ADH plays an important role in emergencies as well by constricting blood vessels, thus helping to maintain sufficient blood pressure. If blood pressure falls too low, the flow of blood to body cells can decline dramatically, leading to death.

Oxytocin. The posterior pituitary also releases **oxytocin**, a hormone that stimulates the contraction of the smooth muscle of the uterus. Like prolactin, oxytocin release is controlled by a neuroendocrine reflex. During childbirth, the walls of the uterus and cervix stretch. Stretching stimulates nerve fibers. Impulses from the stretch receptors travel to a region of the hypothalamus that controls oxytocin production and release. These impulses cause the neurosecretory cells to release oxytocin into the blood. Oxytocin travels in the blood to the uterus where it stimulates smooth muscle contraction, thus aiding in the expulsion of the baby.

Oxytocin also stimulates contraction of the smooth-muscle–like cells surrounding the glands in the breast. Oxytocin release is stimulated by another neuroendocrine reflex activated by suckling (Figure 15–17). Sensory fibers in the breast conduct impulses to the hypothalamus. These impulses trigger the release of oxytocin. Oxytocin travels in the blood to the breast where it stimulates **milk let-down**—the ejection of milk from the glands—a minute or so after suckling begins.

THE THYROID GLAND

On a 50°F day in the fall, you bundle up in a sweater, but still feel cold. In the spring, however, a day with the same temperature feels warm, even without a sweater. Why the difference?

The answer lies in the thyroid gland and two of its hormones. The **thyroid gland** is a U- or H-shaped gland (it varies from one person to the next) located in the neck just below the larynx (Figure 15–18). The thyroid gland produces three hormones: (1) thyroxin (tetraiodothyronine or T_4), (2) a chemically similar

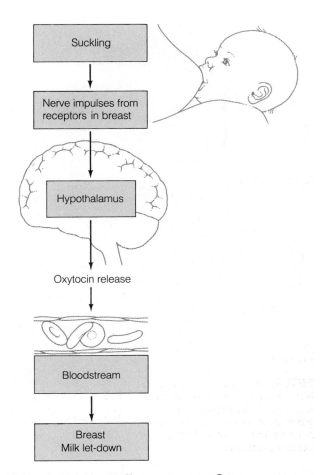

FIGURE 15–17 Milk Let-Down Reflex Suckling also stimulates oxytocin release by the posterior pituitary, causing the expulsion of milk from the glands of the breast.

compound, called triiodothyronine (T_3), and (3) calcitonin. The first two are involved in controlling metabolism and heat production; the last helps regulate blood levels of calcium.

As Figure 15–19 shows, the thyroid gland consists of large spherical structures called **follicles**. Each follicle consists of a central region containing a gel-like material called **thyroglobulin**, which is surrounded by a single layer of cuboidal **follicle cells**. Thyroglobulin is a glycoprotein and is produced by the follicle cells. Thyroglobulin molecules contain certain amino acids to which iodine ions are attached. Two of these iodinated molecules are linked together to form the two thyroid hormones, thyroxin and triiodothyronine.

Thyroglobulin is the storage form of T_3 and T_4. TSH stimulates the follicle cells to engulf a portion of the thyroglobulin reserve. Lysosomes in the cytoplasm of the follicle cells break down the ingested

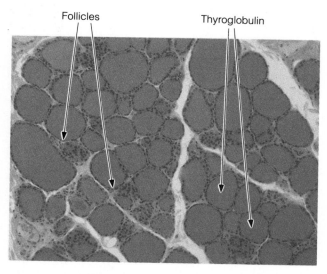

FIGURE 15–19   Histologic Structure of the Thyroid Gland   The thyroid follicles produce thyroxin and triiodothyronine. They are stored in the colloidal material called thyroglobulin.

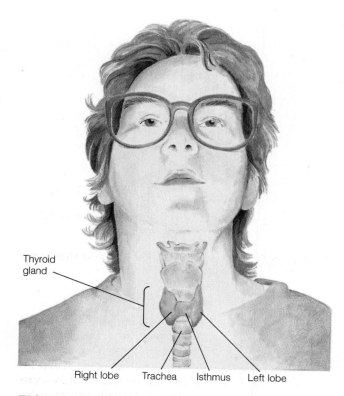

FIGURE 15–18   Location of the Thyroid Gland

material, releasing T_3 and T_4. These hormones then diffuse out of the cell into the bloodstream.

To make T_3 and T_4, the thyroid requires large quantities of iodide, I^-. This ion must be actively transported into the cell where it is added to certain amino acids in the thyroglobulin molecules.

Dietary deficiencies of iodide, if persistent, can result in a condition known as **goiter**. Goiter is characterized by enlargement of the thyroid gland. The shortage of iodide results in a decline in the levels of thyroxin and triiodothyronine in the blood. The hypothalamus responds by producing more TSH. TSH causes the thyroid to increase thyroglobulin production. Thyroglobulin production is dependent on an ample supply of amino acids, not iodide. Without iodide, however, thyroid hormone cannot be produced. The thyroid gland hypertrophies (enlarges), sometimes forming softball-sized enlargements on the neck (Figure 15–20). Goiter was once common in areas in Europe and the United States where iodine was leached out of the soil by rain—for example, near the Great Lakes.

Thyroxin and Triiodothyronine

As Figure 15–21 illustrates, thyroxin and triiodothyronine are virtually identical hormones with nearly identical functions. Both hormones, for example, accelerate the rate of mitochondrial glucose catabolism in most body cells. Although their secretion is controlled principally by a negative feedback mechanism, seasonal differences in temperature also play a role. In cold weather, TSH-RH secretion increases, increasing the release of thyroid hormone.

Thyroid hormones also stimulate cellular growth and development. Bones and muscles are especially dependent on them during the growth phase. Even normal reproduction requires these hormones. A deficiency of T_3 and T_4, for example, delays sexual maturation in both sexes. In children, depressed thyroid output stunts mental as well as physical growth. If the deficiency is not detected and treated, the effects will be irreversible. In adults, reduced thyroid activity, or **hypothyroidism**, is less severe and is fully reversible since growth has been completed. Hypothyroidism, however, decreases the metabolic rate, making a person feel cold much of the time. People suffering from hypothyroidism also feel tired and worn out much of the time. Even simple mental tasks become difficult. Heart rate may slow to 50 beats per minute. Hypothyroidism is treated by pills containing artificially produced thyroid hormone.

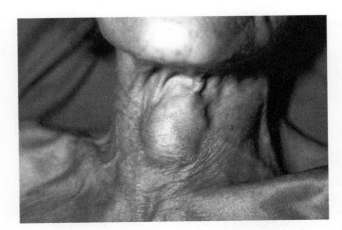

FIGURE 15–20 Goiter Goiter is an enlargement of the thyroid gland, most often resulting from a lack of iodine in the diet.

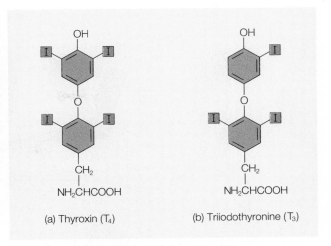

(a) Thyroxin (T$_4$) (b) Triiodothyronine (T$_3$)

FIGURE 15–21 Structure of the Thyroid Hormones (*a*) Thyroxin; (*b*) triiodothyronine.

Excess thyroid activity, **hyperthyroidism**, in adults results in elevated metabolism, excessive sweating (due to overheating), and weight loss, despite increased food intake. The increase in thyroid hormone levels results in increased mental activity, resulting in nervousness and anxiety. People suffering from hyperthyroidism often find it difficult to sleep. Heart rate may accelerate, and individuals may lose their sensitivity to cold. Some people suffering from hyperthyroidism exhibit a condition called **exophthalmos**, or bulging eyes. The eyes may protrude so far that the eyelids cannot close completely. Exophthalmos may cause double vision or blurred vision.

Hyperthyroidism is treated in a number of ways. Patients may be given antithyroid medications, chemicals that antagonize the effects of thyroid hormones. Surgery may be required to remove part or all of the gland if it has become cancerous. The most common treatment, however, consists of taking radioactive iodine. Iodine is concentrated in the thyroid gland. Radioactive iodine damages the overactive cells, reducing their output of thyroid hormone. This procedure, while effective, may lead to other problems (notably cancer) later on in life.

Calcitonin

Large, round cells found in the perimeter of the thyroid follicles produce **calcitonin**, or **thyrocalcitonin**, a polypeptide hormone that lowers blood calcium (Chapter 14). Calcitonin inhibits osteoclasts, bone-resorbing cells, while stimulating bone-forming cells, the osteoblasts. Calcitonin also increases the excre-

tion of calcium (and phosphate) ions by the kidneys. All three effects help lower blood calcium.

Calcitonin is involved in a simple negative feedback loop with calcium ions in the blood. When the calcium ion concentration increases, calcitonin secretion increases. As calcium concentrations fall, calcitonin secretion falls.

THE PARATHYROID GLANDS

The **parathyroid glands** are four small nodules of tissue embedded in the back side of the thyroid gland. These glands, once mistaken by anatomists for undeveloped thyroid tissue, are independent endocrine glands. They produce a polypeptide hormone known as **parathyroid hormone** or **parathormone (PTH)**.

Parathyroid hormone secretion is stimulated when calcium levels in the blood drop. PTH quickly reverses the decline and restores blood calcium levels by three avenues: increasing intestinal absorption of calcium, stimulating bone destruction by osteoclasts, and increasing reabsorption in the kidney.

The most common disorder of the parathyroid gland is **hyperparathyroidism**, excess secretion of parathyroid hormone. Hyperparathyroidism may result from a tumor of the parathyroid glands, which causes the secretion of higher-than-normal amounts of PTH. Excess PTH results in elevated calcium levels in the blood and a loss of calcium from the bones and teeth. It also upsets several metabolic processes re-

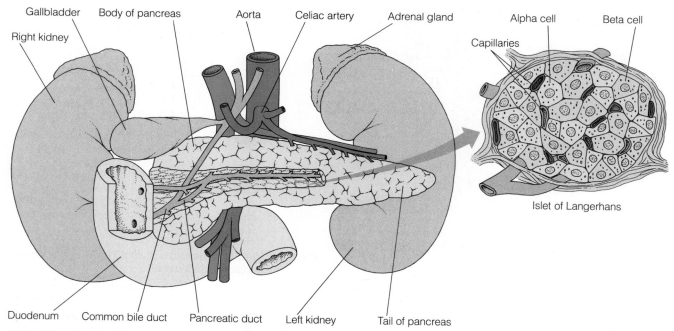

Gallbladder Body of pancreas Aorta Celiac artery Adrenal gland Alpha cell Beta cell

Right kidney Capillaries

Islet of Langerhans

Duodenum Common bile duct Pancreatic duct Left kidney Tail of pancreas

FIGURE 15–22 The Pancreas This dual-purpose organ is located in the abdominal cavity. It produces digestive enzymes, which it releases into the small intestine via the pancreatic duct, and two hormones, which it releases into the bloodstream.

sulting in indigestion and depression. Since the bones contain so much calcium, symptoms of hyperparathyroidism frequently do not appear for two to three years after the onset of the disease. Kidney stones have often formed by the time the condition is discovered. The loss of calcium from the bones leaves them fragile and easily broken. Parathyroid tumors must be removed to prevent further complications, and surgery is usually very effective in reversing the problem.

THE PANCREAS

A teenage girl comes to her doctor's office. She complains of frequent urination, bladder infections, fatigue, and weakness. Tests show that she is suffering from **diabetes mellitus**, a disorder of the endocrine function of the pancreas (Chapter 7). Diabetes mellitus has several causes, but in young people it is generally caused by a lack of insulin output.

Insulin is the glucose-storage hormone—that is, it stimulates the uptake of glucose by body cells. It also stimulates the synthesis of glycogen in liver and muscle cells. Glycogen is the storage form of glucose. Glycogen supplies increase after meals, then decline in the period between feedings as the body taps liver and muscle stores to supply body cells with glucose.

Insulin is produced by the **pancreas**, a dual-purpose organ located in the abdominal cavity (Figure 15–22). The head of the pancreas lies in the curve of the duodenum. Its tail stretches to the left kidney. As noted in Chapter 7, most of the pancreas consists of tiny clumps of cells (acini) that produce digestive hormones. Scattered throughout these enzyme-producing cells are small islands of endocrine cells, the **islets of Langerhans** (Figure 15–23). About 2 million islets are found in the human pancreas; each islet consists of a few hundred cells. The islets contain four different cell types, two of which will be discussed in this chapter, the alpha cell and the beta cell.

Insulin

The **beta cells** produce insulin. This protein hormone is released within minutes after glucose levels in the blood rise. Like many other hormones, insulin plays a major role in homeostasis. Insulin affects a number of cellular processes and a number of different cells. Its principal targets, however, are skeletal muscle, the liver, and fat cells. In muscle cells, insulin increases

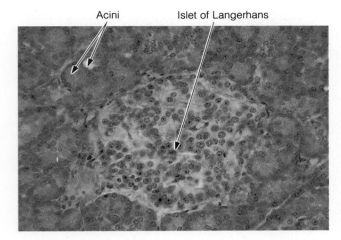

Acini   Islet of Langerhans

FIGURE 15–23   The Islets of Langerhans Scattered among the acini of the pancreas (which produce digestive enzymes) are small islands of cells, the islets of Langerhans. They produce two hormones, glucagon, which increases blood glucose, and insulin, which lowers blood glucose.

the uptake of glucose and stimulates glycogen synthesis (Figure 15–24). Insulin also increases the uptake of amino acids by muscle cells and stimulates protein synthesis in them, thus promoting muscle formation. In the liver, insulin increases glycogen formation, helping store glucose for times of need. In fat cells, insulin increases glucose uptake and stimulates lipid synthesis, again stimulating the storage of foodstuffs for times of need.

Glucagon

The islets of Langerhans also produce an antagonistic hormone, known as **glucagon**. Glucagon is a polypeptide synthesized by the alpha cells. Glucagon increases blood levels of glucose by stimulating the breakdown of glycogen in liver cells (Figure 15–24). This process, called **glycogenolysis**, helps maintain proper glucose levels in the blood between meals. One molecule of glucagon causes 100 million molecules of glucose to be released.

Glucagon also elevates serum glucose levels by stimulating a process known as **gluconeogenesis**, the synthesis of glucose from fatty acids and amino acids. Gluconeogenesis occurs in the liver where amino acids and fatty acids are stored (Chapter 7).

Glucagon secretion is controlled principally by glucose concentrations in the blood. When glucose levels fall, for example, glucagon is released. Glucagon is

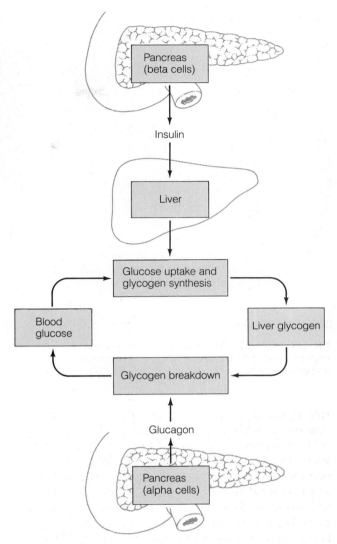

FIGURE 15–24   Control of Blood Glucose Levels Glucagon and insulin are antagonistic hormones that regulate blood glucose levels through different mechanisms.

also stimulated by an increase in the concentration of amino acids in the blood. Thus, amino acids in high-protein meals can be used to produce glucose.

Diabetes Mellitus

Figure 15–25 shows a graph of blood glucose levels in a patient with diabetes. The results were obtained using a procedure known as the **glucose tolerance test**. During this test, physicians give their patients an oral dose of radioactive glucose, then check blood levels at regular intervals afterward. In a normal pa-

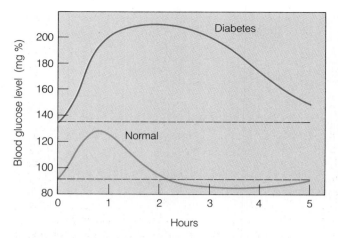

FIGURE 15–25 Glucose Tolerance Test   In a diabetic, blood glucose levels increase rapidly and remain high after ingestion of glucose. In a normal patient, blood glucose levels rise but are quickly reduced by insulin.

tient, blood levels show a slight increase; a much more dramatic increase is seen in patients with diabetes.

The bottom line on the graph in Figure 15–25 represents the blood glucose levels in a normal patient with a healthy pancreas. As illustrated, glucose levels increase slightly after the administration of glucose, but within an hour and a half glucose levels have decreased to normal, thanks to insulin secretion.

The top line shows the response in diabetics—individuals producing insufficient amounts of insulin. In these patients, glucose levels rise considerably after the treatment and remain elevated for several hours. Glucose levels follow a similar pattern in untreated diabetics after they eat a meal. In untreated diabetics, the kidneys excrete excess glucose, thus helping lower blood levels. This wastes a valuable nutrient and also causes the kidneys to lose large quantities of water. Excess urination and constant thirst are the chief symptoms of diabetes. Diabetics may have to urinate every hour on the hour day and night.

Diabetes is really two diseases with similar symptoms but different causes. **Type I diabetes** occurs mainly in young people and results from an insufficient amount of insulin production. Type I diabetes is also called **juvenile diabetes** or **early-onset diabetes**.[4]

---

[4]Type I diabetes also appears at or after maturity; thus, insulin-dependent diabetes is considered the best term to describe it.

Juvenile diabetes is believed to be caused by damage to the insulin-producing cells of the pancreas. One leading theory suggests that viral infections in the pancreas may cause the damage (for more on the causes of juvenile diabetes, see Health Note 15–1). In patients with juvenile diabetes, insulin production varies. In some patients, it is only slightly depressed; in others, it is completely suppressed. The absence of insulin starves the body cells for glucose. To provide energy, body cells must break down fats.

Type II diabetes usually occurs in people over 40 and is sometimes called **late-onset diabetes**. In this disease, the beta cells continue to produce insulin. However, the number of insulin receptors in the target cells declines. Type II diabetes is commonly associated with obesity. Heredity may also be an important contributor. Studies show that about one-third of the patients with type II diabetes have a family history of the disease. Could it be that obesity is inherited rather than the tendency for diabetes?

Although they have different causes, both forms of diabetes exhibit similar symptoms. Excess urination and thirst are obvious signs of trouble. Excess glucose in the urine may result in frequent bacterial infections in the bladder. Patients often feel tired, weak, and apathetic. Weight loss and blurred vision are also common.

Juvenile diabetes is treated with insulin injections and may, therefore, also be called **insulin-dependent diabetes**. Patients receive regular injections. Patients are also required to eat meals and snacks at regular intervals to maintain constant glucose levels in the blood and to ensure that regular insulin injections always act on approximately the same amount of blood glucose.

Insulin injections are tailored to an individual's lifestyle and body demands. To help mimic the body's natural release, medical researchers have developed a device called an **insulin pump**. Health Note 15–2 describes one such device. Medical researchers are also experimenting with ways to transplant healthy beta cells to the pancreas of a diabetic. (Chapter 3 and the Point/Counterpoint in that chapter discuss fetal cell transplantation.)

Type II diabetes is sometimes called **insulin-independent diabetes** because, in most patients, the disease can be controlled by the diet. Physicians restrict carbohydrates and ask patients to eat small meals at regular intervals during the day. Candy, sugar, cakes, and pies are off limits. Glucose must be supplied by complex carbohydrates, such as starches.

Diabetes Mellitus: Advances in Early Detection and Treatment

Diabetes afflicts an estimated 12 million Americans. Five million of these people suffer from juvenile diabetes, which is controlled by insulin injections. A new discovery may help physicians diagnose juvenile diabetes *before* symptoms occur and may even help them find a cure.

As noted in the text, juvenile diabetes is believed to be caused by an autoimmune reaction. In a recent study, researchers from the University of Florida reported that they withdrew blood from 5000 young children who were asymptomatic— that is, showed no symptoms of diabetes. The researchers analyzed the blood for antibodies produced by the body in the autoimmune reaction and also followed the children's health over time to determine which ones developed diabetes. Over the course of the study, 12 of the subjects developed juvenile diabetes.

Blood samples from the 12 who developed juvenile diabetes all contained an antibody that attacks a protein on the surface of the beta cells of the pancreas. The antibody was found in blood samples taken as many as seven years before the onset of diabetes.

At least two other antibodies are also found in the blood of children with diabetes. One of these attacks the insulin molecule itself; the other also attacks the beta cell. These antibodies, however, do not show up consistently in blood samples of children who later develop diabetes and are therefore not predictable markers for the disease. The antibody to the surface protein, however, was present in all samples and, therefore, may provide a marker, a kind of biochemical advance notice of the disease.

This research may help scientists

find ways to prevent diabetes. For example, researchers are now studying the surface protein to which the antibody attaches. Eventually they hope to find ways to attach chemicals to the surface proteins that would neutralize or destroy the antibodies that bind to the proteins, thus preventing the disease from ever developing. If this research proves fruitful, regular blood tests could be used to screen children for the marker. When it is found, treatments could destroy the antibodies before they have time to destroy the islet cells, freeing millions of people from a lifetime of insulin injections and the risk of blindness and other serious medical problems later in life.

These treatments have dramatically changed the prognosis for diabetics. At one time the disease was fatal. Today, patients can live healthy, fairly normal lives. Risks are still present. Type I diabetics, for example, still suffer from diabetic comas, or unconsciousness. Diabetic coma results when patients forget their insulin injection. Without insulin, the body cells become starved for glucose (even though blood levels are high) and begin breaking down fat. Excessive fat catabolism releases toxic chemicals (ketones) that cause the patient to lose consciousness. Both early- and late-onset diabetics also suffer from loss of vision, nerve damage, and kidney failure 15 to 20 years after the onset of the disease, even if they are being treated. These serious complications result from the inevitable elevation of blood glucose levels that occurs over the years.

THE ADRENAL MEDULLA AND ADRENAL CORTEX

You are standing on the banks of a raging river. Your raft is tied to a tree and floating in the calm water of an eddy. As you watch kayakers and rafters head into the rapids, your heart starts to race and your intestines churn in excitement (and fear). Your turn is coming up.

This natural response results from the secretion of two of the many hormones produced by the **adrenal glands**. As Figure 15–26 shows, the adrenal glands perch atop the kidneys, and each adrenal gland consists of two zones. The central region, or **adrenal medulla**, produces the hormones that increase the heart rate and accelerate breathing when a person is excited or frightened. The outer zone, the **adrenal cortex**,

On the Road to an Artificial Pancreas

Diabetics take insulin injections two and three times a day. If they don't, they could easily go into a diabetic coma. Daily injections allow most victims of juvenile diabetes to live a fairly normal life. Occasionally things go awry, however. Too much insulin or too much exercise causes blood glucose levels to fall, making a diabetic feel dizzy and weak. Sometimes the balance is tipped the other way. An excess of blood sugar brings on hyperglycemia and with it depression, fatigue, irritability, and weakness.

Besides worrying over insulin doses and playing a continual balancing act with their blood sugar, most diabetics develop complications 20 to 30 years after the onset of diabetes. Blindness and lethal or disabling diseases of the kidney, nervous system, and cardiovascular system are common. These complications probably result from elevated blood glucose levels, which occur despite good insulin management. High glucose levels damage blood vessels and nerves, causing damage in critical organs.

To prevent complications, physicians try to mimic normal insulin secretion patterns through injections—three times a day just before meals. But mimicking the body's homeostatic system with regular insulin injections is a crude science. The pancreas normally monitors blood glucose levels minute by minute. A conscientious diabetic can only measure blood glucose three or four times a day. He or she may make adjustments for large meals or additional exercise, but such accommodations are primitive in comparison

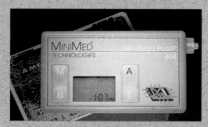

An Insulin Pump The insulin pump delivers preprogrammed doses of insulin to diabetics, helping to mimic the pancreas.

to the body's elaborate system of glucose homeostasis.

Consequently, medical researchers are looking for ways to replace the daily injections. One advance is an insulin infusion pump that delivers tiny amounts of fast-acting insulin to the body day and night via a long plastic tube and needle inserted into the skin in the thigh or abdomen (see the figure). The needle and tubing are usually replaced every three to four days. The insulin pump provides baseline insulin levels needed to maintain proper blood glucose concentrations. Worn outside the body, the pump also delivers a surge of insulin at mealtimes to offset the rise in blood sugar that accompanies a meal. To do this, the diabetic simply presses a button on the pump 30 minutes before eating. The pump delivers a preprogrammed surge of insulin. If the meal is going to be larger than anticipated, a small adjustment can be made to protect against hyperglycemia.

The newer insulin pumps capitalize on computer technology to regulate insulin flow day and night. The

newer models offer great flexibility, allowing an individual to accommodate differing levels of exercise and meals of varying size. They come equipped with memory to store information on the exact doses given over a certain period of time. Using this information, physicians can fine-tune the program to an individual's needs.

The insulin pump is not a popular item among diabetics when it comes to aesthetics and comfort. When it comes to controlling blood sugar, however, the device receives high praise. For many diabetics, the freedom from daily injections outweighs the discomfort. For pregnant diabetic women, the pump may mean the difference between a normal child and no child at all, for even mild hyperglycemia can cause fetal death.

While refinements are being made to the insulin pump, researchers are also testing a biodegradable wafer that can be impregnated with insulin and inserted under the skin of a diabetic. The wafers also contain a sugar-sensitive enzyme. As the blood sugar levels rise, the enzyme is activated. It causes a slight increase in acidity. The acid increases the solubility of the insulin in the wafer, allowing it to be released into the bloodstream. As a result, insulin levels respond very closely to blood glucose levels. Insulin is released only as needed in response to rising blood sugar levels.

This innovative approach could prove even more effective than the insulin pump because it operates on a negative feedback principle like the beta cells in the pancreas.

produces a number of steroid hormones involved in homeostasis.

The Adrenal Medulla

The adrenal medulla produces two hormones: **adrenalin** (epinephrine) and **noradrenalin** (norepinephrine). In humans, about 80% of the adrenal medulla's output is adrenalin. Helping animals to meet the stresses of life, these hormones are instrumental in the flight-or-fight response. Adrenalin and noradrenalin are secreted under all kinds of stress—for example, when an angry dog leaps out at you, when a careless driver cuts in front of you in heavy traffic, or even as you wait outside a lecture hall, anticipating a final exam. Nerve impulses traveling from the brain to the adrenal medulla trigger the release of adrenalin and noradrenalin. The response is part of the autonomic nervous system (Chapter 12).

Consider the physiological changes these hormones cause. Imagine that you are entering Lava Rapid on the Colorado River through the Grand Canyon. Although you are unaware of it, your adrenal glands are actively secreting adrenalin and noradrenalin. These hormones cause blood glucose levels to rise, making more energy available to body cells, particularly skeletal muscle cells. Adrenalin and noradrenalin also increase your breathing rate, providing additional oxygen for skeletal muscles and brain cells. The adrenal medulla's hormones also cause your heart rate to accelerate, thus increasing circulation and ensuring adequate glucose and oxygen for body cells that might be called into action. Blood vessels in your intestinal tract constrict, putting digestion on temporary hold. At the same time, the blood vessels in your skeletal muscles dilate, increasing flow through them. Mental alertness increases.

The Adrenal Cortex

Surrounding the medulla is the adrenal cortex. The adrenal cortex produces three types of steroid hormones with distinctly different functions. The first group, the **glucocorticoids**, affect glucose metabolism and help maintain blood glucose levels. The second group, the **mineralocorticoids**, help to regulate the ionic concentration of the blood and tissue fluids. The final group, the **sex steroids**, are identical to the hormones produced by the ovaries and testes (Chapter 16). In healthy adults, the amount of adrenal sex ste-

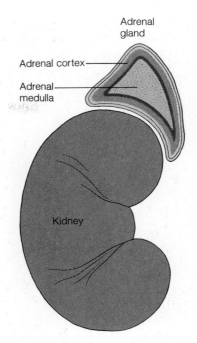

FIGURE 15–26 The Adrenal Gland The adrenal glands sit atop the kidney and consist of an outer zone of cells, the adrenal cortex, which produces a variety of steroid hormones, and an inner zone, the adrenal medulla. It produces adrenalin and noradrenalin. The secretion of these hormones is controlled by the autonomic nervous system.

roids synthesized and released is insignificant compared to the amounts produced by the gonads.

Glucocorticoids. As their name implies, the glucocorticoids are steroids that affect carbohydrate metabolism. Their secretion is governed by ACTH, a hormone of the anterior pituitary, as discussed earlier in the chapter. Several chemically distinct glucocorticoids are secreted, but the most important is **cortisol**. Cortisol increases blood glucose by stimulating gluconeogenesis (the synthesis of glucose from amino acids and fatty acids). This makes more glucose available for energy production in times of stress. Cortisol also stimulates the breakdown of proteins in muscle and bone, freeing amino acids that can then be chemically converted to glucose molecules in the liver.

In **pharmacological doses**, levels beyond those seen in the body, cortisol inhibits inflammation, the body's response to tissue damage (Chapter 10). Pharmacological doses of glucocorticoids also depress the allergic reaction, discussed in Chapter 10. Cortisol

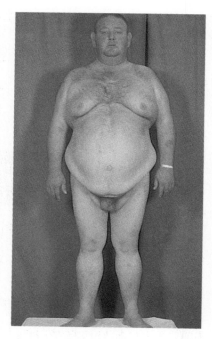

FIGURE 15–27 Cushing's Syndrome This disease results from an excess of glucocorticoid hormone, either cortisol or cortisone. It is most often caused by cortisone treatment for allergies or inflammation. The most common symptoms are a round face due to edema and excess fat deposition.

brings about its effects by inhibiting the movement of white blood cells across capillary walls, thus impeding their migration into damaged tissue. Cortisol also reduces the number of circulating lymphocytes by destroying them at their site of formation.

Because they reduce inflammation, cortisol and other glucocorticoids (particularly cortisone, a synthetic glucocorticoid-like hormone) can be used to treat chronic inflammation resulting from diseases such as rheumatoid arthritis (Chapter 14). However, the benefits must be weighed carefully against the damage that can be caused by upsetting the body's homeostatic balance (discussed below).

Mineralocorticoids. As their name implies, the mineralocorticoids are involved in electrolyte or mineral salt balance. The most important mineralocorticoid is **aldosterone** (Chapter 11). It is the most potent mineralocorticoid and constitutes 95% of the adrenal cortex's hormonal output.

Mineralocorticoids regulate the level of several ions, but principally control sodium and potassium ion concentrations. Their chief target is the kidney. As noted in Chapter 11, aldosterone causes sodium ions to pass out of the nephron into the blood. This process is called tubular reabsorption. Aldosterone also stimulates sodium ion reabsorption in sweat glands and saliva and potassium excretion by the kidney. When sodium is shunted back into the blood in the kidney and the skin, water follows. Aldosterone, therefore, helps conserve body water.

As we saw in Chapter 11, aldosterone secretion is controlled by the sodium ion concentration in a negative feedback loop. As sodium levels fall, aldosterone secretion increases. As sodium levels are restored, aldosterone secretion declines. Aldosterone secretion is also stimulated when potassium levels climb and when blood volume and blood pressure decline.

Diseases of the Adrenal Cortex. Figure 15–27 shows a patient with **Cushing's syndrome.** In most instances, Cushing's syndrome results from pharmacological doses of cortisone, used to treat rheumatoid arthritis or asthma. In a few rare instances, the disease may be caused by a pituitary tumor that produces excess ACTH or a tumor of the adrenal cortex that secretes excess amounts of glucocorticoid.

Patients with Cushing's disease often suffer persistent **hyperglycemia**—high blood sugar levels—because of the excess levels of glucocorticoid. Bone and muscle protein may also decline sharply. Individuals complain of weakness and fatigue. Loss of bone protein increases the ease with which bones fracture.

Water and salt retention are also common in Cushing's patients, resulting in tissue edema or swelling. Cushing's patients have "moon face," a rounded face resulting from facial edema. These symptoms result from the fact that, in high concentrations, glucocorticoids have mineralocorticoid effects.

Since most cases of Cushing's syndrome result from steroid doses, treatment is a simple matter. By gradually reducing the glucocorticoid dose, a physician can eliminate the symptoms. Tumors in the pituitary and the adrenal cortex can be treated with radiation or surgery.

As an individual ages, the adrenal cortex gradually decreases its output of hormone. The decline in hormonal secretion results in **Addison's disease**, which is characterized by a variety of interrelated symptoms. Today, most cases of Addison's disease are the result of autoimmune reactions in which the cells of the adrenal cortex are recognized as foreign and de-

TABLE 15-2 Summary of Some Endocrine Disorders

DISEASE	CAUSE	SYMPTOMS
Gigantism	Hypersecretion of GH starting in infancy or early life	Excessive growth of long bones
Dwarfism	Hyposecretion of GH in infancy or early life	Failure to grow
Acromegaly	Hypersecretion of GH after bone growth has stopped	Facial features become coarse; hands and feet enlarge; skin and tongue thicken
Hyperthyroidism	Overactivity of the thyroid gland	Nervousness; inability to relax; weight loss; excess body heat and sweating; palpitations of the heart
Hypothyroidism	Underactivity of the thyroid gland	Fatigue; reduced heart rate; constipation; weight gain; feel cold; dry skin
Hyperparathyroidism	Excess parathyroid hormone secretion, usually resulting from a benign tumor in the parathyroid gland	Kidney stones; indigestion; depression; loss of calcium from bones
Hypoparathryoidism	Hyposecretion of the parathyroid glands	Spasms in muscles; numbness in hands and feet; dry skin
Diabetes insipidus	Hyposecretion of ADH	Excessive drinking and urination; constipation
Diabetes mellitus	Insufficient insulin production	Excessive urination and thirst; poor wound healing; urinary tract infections; excess glucose in urine; fatigue and apathy
Cushing's syndrome	Hypersecretion of hormones from adrenal cortex or, more commonly, from cortisone treatment	Face and body become fatter; loss of muscle mass; weakness; fatigue; osteoporosis
Addison's disease	Gradual decrease in production of hormones from adrenal gland; most common cause is autoimmune reaction	Loss of appetite and weight; fatigue and weakness; complete adrenal failure

stroyed by the body's own immune system.

Patients with Addison's disease suffer from loss of appetite, weight loss, fatigue, and weakness. Although insulin and glucagon are still present, the absence of cortisol upsets the body's homeostatic mechanism for controlling glucose. The body's reaction to stress is impaired. The lack of aldosterone results in electrolyte imbalance. As noted earlier, aldosterone helps maintain sodium levels and blood pressure. Patients with Addison's disease have low blood sodium levels and low blood pressure. Addison's disease can be treated with steroid tablets that replace the hormones the adrenal cortex no longer produces. Treatment allows patients to lead a fairly normal, healthy life.

ENVIRONMENT AND HEALTH: HERBICIDES AND HORMONES

Hormones orchestrate an incredible number of body functions, creating a dynamic balance that is necessary for good health. When this balance—homeostasis—is altered, our health suffers (Table 15–2). The endocrine system, like others, is sensitive to outside, or environmental, factors. Stress, for example, can lead to an imbalance in adrenal hormones, resulting in high blood pressure and other complications. High blood pressure puts strain on the heart. Hormonal balance can also be altered by toxic pollutants, such as dioxin.

AMERICA'S EPIDEMIC OF CANCER

Lewis G. Regenstein

Lewis G. Regenstein, an Atlanta writer and conservationist, is the author of How to Survive in America the Poisoned *and director of the Interfaith Council for the Protection of Animals and Nature, an affiliate of the Humane Society of the United States.*

America is in the throes of an unprecedented cancer epidemic, caused in large part by the pervasive presence in our environment and food chain of deadly, cancer-causing pesticides and industrial chemicals. Each year, almost a million Americans are diagnosed as having cancer, or almost 3000 people a day! The disease now strikes almost one American in three and kills over a thousand of us every day! This means that of the Americans now alive, some 70 to 80 million people can expect to contract cancer in their lifetimes.

Cancer is the leading cause of death for women between the ages of 30 and 40 and for children aged 1 to 10. Thus the elevated cancer rate is not caused solely by people living longer, as the chemical industry often claims; it has now become a common disease of the young as well as the old.

In 1978, the President's Council on Environmental Quality (CEQ) reported unequivocally that "most researchers agree that 70 to 90 percent of all cancers are caused by environmental influences and are hence theoretically preventable." Only about 30% of the nation's annual cancer deaths (mainly lung cancers) are thought to be caused by cigarette smoking. The remaining 70% are caused by a variety of other factors, with toxic chemicals thought to play a major role. The CEQ stressed the link between the increase in production of such chemicals between 1950 and 1960 and the large increase in the cancer rate showing up 20 to 25 years later, "the lag time one might expect." Evidence continues to mount showing that toxic chemicals contribute heavily to the cancer epidemic. In general, the most polluted areas of the country have the highest cancer rates. In July 1982, the University of Medicine and Dentistry of New Jersey released a study showing a correlation between the presence of toxic waste dumps and elevated cancer death rates

(up to 50% above average) in areas of that highly industrialized state. In February 1984, a report by researchers at the Harvard School of Public Health demonstrated a link between the consumption of chemically contaminated well water near Woburn, Massachusetts, and the incidence of childhood leukemia, stillbirths, birth defects, and other disorders among local residents.

Some scientists and government officials, pointing to the decrease in rates of some cancers, insist that there is no "cancer epidemic" and that toxic chemicals play a small role in causing cancer. But Dr. Samuel Epstein of the University of Illinois Medical Center, perhaps the foremost authority on the subject, points out that apart from AIDS, "cancer is the only major killing disease which is on the increase," with incidence rising by at least 2% a year, and death rates rising at 1% annually over the last decade. He concludes that "the facts show very clearly that we are in a cancer epidemic now," in large part because of carcinogens in our environment.

Today, every American is exposed to a variety of health-destroying chemicals. Dozens of pesticides used on our food are known or thought to cause cancer in animals. By the time restrictions were placed on some of the deadliest chemicals, such as DDT, dieldrin, and PCBs, these carcinogens were being found in the flesh tissues of literally 99% of all Americans tested, as well as in the food chain and even mother's milk. Virtually all Americans carry in their bodies traces of dioxin (TCDD), the most deadly synthetic chemical known. The sources of this dioxin to which we are regularly exposed include food (such as fish and beef), municipal and industrial incinerators (which produce dioxins when plastics are burned), milk from white cardboard cartons, and white paper products (in which the bleaching process creates dioxins).

The response of the U.S. government has been largely weak or nonexistent enforcement of the nation's health and environmental protection laws. For example, with few exceptions, the Environmental Protection Agency has refused to carry out its legal duty to ban or restrict pesticides known to cause cancer. Nor has the government adequately implemented or enforced the laws regulating hazardous waste.

Thus we are even now sowing the seeds for cancer epidemics of the future. Only time will tell what will be the effect on this and future generations of Americans. By the time we know the answers, it may be too late to do anything about the problem.

THE MYTHS OF THE CANCER EPIDEMIC

David L. Eaton

David L. Eaton is an Associate Professor of Environmental Health and Environmental Studies at the University of Washington. His research focuses on natural carcinogens in foods.

There is no debate that cancer is a devastating and deadly disease. However, the contention that we are in the throes of an unprecedented cancer epidemic, and that this "epidemic" is caused in large part by cancer-causing pesticides and industrial chemicals, is not supported by scientific data.

Are cancer rates increasing in epidemic proportions? It is certainly true that both the number of deaths and the fraction of all cancer deaths have increased dramatically in the past 50 years. However, these statistics must be adjusted for changes in the age distribution of our population. A 1988 report from the National Cancer Institute states that "the age-adjusted mortality rates for all (types of) cancers *combined*, except lung cancer, have been declining since 1950 for all age groups except 85 and above."

What proportion of cancers can be related to environmental pollution from synthetic pesticides and industrial chemicals? The often-cited 70 to 90% is frequently incorrectly interpreted as meaning that *chemical pollution* is responsible for 70 to 90% of cancers. Studies in the 1960s suggested that most cancers could not be directly traced to genetic or hereditary factors; therefore, it was concluded that the majority of cancers must have an "environmental" cause. However, the term "environmental" includes not only chemicals, but lifestyle factors such as smoking, dietary factors such as the proportion of fat in the diet, "natural" carcinogens that occur in nearly all foods, cancer-causing viruses, and occupational exposures to substances such as asbestos.

Of the environmental factors other than smoking, dietary factors are now generally thought to represent the largest source of cancer risk, causing perhaps 30 to 40% of all cancers.

Some scientists recently suggested that the largest source of exposure to cancer-causing chemicals may be chemicals that occur naturally in our diet. All plants produce toxic chemicals to protect themselves against insects, fungi, and animal predators; many of these chemicals are potent mutagens and carcinogens. For example, potentially carcinogenic chemicals are found in mushrooms, parsley, and celery, to name a few. The majority of natural chemicals present in foods have never been tested for carcinogenicity. Some scientists believe that due to relatively high exposure to carcinogens from natural sources, complete elimination of synthetic industrial chemicals from our diet, if possible, would have little effect on cancer incidence and mortality.

Occupational exposure to some industrial chemicals can also increase the risk of certain types of cancer. Many sources now estimate, however, that occupational exposures are not likely to account for more than 5% of all cancers, and many of these are a result of extensive exposures to asbestos and a few other industrial carcinogens that were commonplace in the 1950s and 1960s, but have since been greatly reduced.

Finally, recent advances in the understanding of the biology of cancer suggest that "spontaneous" or "background" alterations in DNA may explain much of the cause of cancer. DNA is subject to extensive damage from processes associated with normal cellular metabolism. Within our life span, our cells undergo about 10 million billion cell divisions. Spontaneous errors in this process, which lead to mutations and cancer, accumulate with age. It is not surprising then that cancer seems to be a frequent outcome of old age.

The views that we are in cancer epidemic and that these cancers are largely a result of industrial chemicals are not supported by the majority of cancer researchers throughout the world. Unfortunately, the political arena, influenced greatly by public fears, has not come to grips with the fact that further reduction in public exposure to synthetic chemicals will not by itself solve the cancer problem. The U.S. is currently spending about $80 billion per year on pollution reduction. I believe that much of this is justified to enhance the quality of our environment and ensure the habitability of our planet. However, there is little question that huge sums of money are spent each year to reduce what is in all likelihood a trivial cancer risk. If our society is truly concerned about reducing cancer, more efforts should be focused on eliminating smoking and alcohol abuse, better research and education on dietary risk factors, more research into the biochemical and molecular events that lead to cancer, and continued identification and reduction in occupational exposures to chemicals that pose a significant cancer risk.

SHARPENING YOUR CRITICAL THINKING SKILLS

1. Regenstein asserts that America is in the midst of a cancer epidemic caused in large part by pesticides and industrial chemicals. Eaton argues the opposite. He says that the death rate for all types of cancer except lung cancer is actually declining. Using the critical thinking skills described in Chapter 1, analyze each argument.

Dioxin is a contaminant found in some herbicides and in many paper products. Dioxin was present in the chemical defoliants used in the Vietnam War. Infamous for its carcinogenic properties, dioxin may exert most of its effects through the endocrine system, although more research is needed to be certain.

A number of common herbicides (the thiocarbamates) in use today upset the thyroid's function and may even result in the formation of thyroid tumors. These herbicides are chemically similar to thyroid hormone and therefore block the secretion of TSH, resulting in goiter. At higher levels, these herbicides may cause thyroid cancer.

The impact of toxic chemicals on human health is probably quite small, but not everyone agrees. The Point/Counterpoint in this chapter debates this subject and is worth studying. It is an important debate that will be around for a number of years.

SUMMARY

PRINCIPLES OF ENDOCRINOLOGY

1. The endocrine system consists of a widely dispersed set of ductless glands that produce hormones, chemicals that travel in the bloodstream to their target cells.
2. Hormones function in five areas: (a) homeostasis, (b) growth and development, (c) reproduction, (d) energy production, storage, and use, and (e) behavior.
3. Specificity is conferred by hormone receptors inside cells and in their membranes.
4. The endocrine and nervous systems are similar in several respects. They both send signals to cells and help regulate their function. Both also help coordinate body functions.
5. The endocrine and nervous systems also differ in several key respects. The nervous system elicits rapid, generally short-lived responses. The endocrine system elicits slower, longer-lasting responses.
6. Three types of hormones are produced in the body: steroids, proteins and polypeptides, and amines.
7. Hormone secretion is controlled principally by negative feedback loops. Some of these loops involve several intermediary chemicals. Hormone secretion is also controlled by biological clocks, which create cycles of varying length.
8. Water-soluble hormones (proteins and polypeptides) act on cells via a second messenger. Protein hormones bind to receptors in the plasma membrane. These receptors are linked to a membrane-bound enzyme, adenylate cyclase, which when activated catalyzes the formation of cyclic AMP, a second messenger inside the cell. Cyclic AMP activates another enzyme, protein ki-

nase, which activates or deactivates cellular enzymes, thus eliciting a response.
9. Steroid (lipid-soluble) hormones act via the two-step mechanism. Steroids can penetrate the plasma membrane and bind to a specific receptor molecule inside the cell. This activates the receptor, allowing it to bind to the nuclear DNA. Binding to the DNA activates genes and results in the formation of messenger RNA used to produce enzymes and structural proteins.

THE PITUITARY AND HYPOTHALAMUS

10. The pituitary is a small, pea-sized gland suspended from the hypothalamus by a thin stalk; it consists of two parts: the anterior pituitary and the posterior pituitary.
11. The anterior pituitary produces seven protein and polypeptide hormones. Their release is controlled by releasing and inhibiting hormones produced by the hypothalamus, which are transported to the anterior pituitary via the portal system.
12. The hypothalamic hormones are produced by neurosecretory neurons. Their release is controlled by chemical stimuli and nerve impulses.
13. The hormones of the anterior pituitary and their functions are summarized in Table 15-1.
14. The posterior pituitary consists of neural tissue. Neurosecretory cells whose cell bodies are located in the hypothalamus extend into the posterior pituitary. Hormones are produced in the cell bodies and are transported down the axons of these cells to the posterior pituitary where they are stored and released.
15. The posterior pituitary produces two hormones, antidiuretic hormone (ADH) and oxytocin, whose functions are summarized in Table 15-1.

THE THYROID GLAND

16. The thyroid gland is located in the neck, on either side of the trachea, just below the larynx. The thyroid produces three hormones: thyroxin, triiodothyronine, and calcitonin.
17. Thyroxin and triiodothyronine accelerate the rate of glucose breakdown in most cells, increasing body heat. These hormones also stimulate cellular growth and development.
18. Calcitonin helps control blood calcium levels. When calcium levels in the blood rise, calcitonin is released. Calcitonin inhibits osteoclasts, thus reducing bone destruction. Calcitonin also increases the excretion of calcium in the kidneys. Both responses help return blood calcium levels to normal.

THE PARATHYROID GLANDS

19. The parathyroid glands are located on the back side of the thyroid gland and produce a polypeptide hormone called parathyroid hormone (PTH).

20. PTH increases blood calcium levels by stimulating bone reabsorption by osteoclasts. It also increases intestinal absorption and renal reabsorption.

THE PANCREAS

21. The pancreas produces two hormones, insulin and glucagon, from the islets of Langerhans.
22. Insulin is the glucose-storage hormone. It stimulates the uptake of glucose by body cells and stimulates the synthesis of glycogen in muscle and liver cells. Insulin also increases the uptake of amino acids and stimulates protein synthesis in muscle cells, thus promoting muscle formation.
23. Glucagon is an antagonist to insulin. It raises glucose levels in the blood in the period between meals by stimulating glycogen breakdown (glycogenolysis) and the synthesis of glucose from amino acids and fats (gluconeogenesis).
24. Diabetes mellitus has two principal forms: type I and type II. Type I diabetes, or juvenile diabetes, occurs early in life and is thought to be caused by an autoimmune reaction that destroys insulin or the beta cells of the pancreas. It can be treated by insulin injections.
25. Type II diabetes, or late-onset diabetes, results from insulin production that falls short of body demands. It may be caused by obesity and genetic factors. It can be treated successfully by dietary management.

THE ADRENAL MEDULLA AND ADRENAL CORTEX

26. The adrenal glands lie atop the kidneys and consist of two separate portions: the adrenal medulla, at the center, and the adrenal cortex, a band of tissue surrounding the medulla.

27. The adrenal medulla produces two hormones under stress: adrenalin and noradrenalin. Adrenalin is the chief product. It stimulates heart rate and breathing, elevates blood glucose levels, constricts blood vessels in the intestine, and dilates blood vessels in the muscle.
28. The adrenal cortex produces three classes of hormones: glucocorticoids, mineralocorticoids, and sex steroids.
29. The glucocorticoids affect carbohydrate metabolism and tend to raise blood glucose levels. The principal glucocorticoid is cortisol.
30. In pharmacological doses, cortisol inhibits the immune system and allergic reactions and therefore can be used to treat allergies and swelling in joints. High doses, however, have many adverse impacts on the body.
31. The chief mineralocorticoid is aldosterone. It acts on the kidneys, sweat glands, and salivary glands, causing sodium and water retention and potassium secretion.

ENVIRONMENT AND HEALTH: HERBICIDES AND HORMONES

32. Hormones orchestrate an incredible number of body functions, creating a dynamic balance necessary for good health. When this balance—homeostasis—is altered, our health suffers.
33. The endocrine system, like others, is sensitive to upset from environmental factors, including pollutants.
34. Dioxin, a contaminant in some herbicides and paper products, may exert most of its effects through the endocrine system.
35. A number of common herbicides (the thiocarbamates) in use today upset the thyroid's function and may even cause thyroid tumors.

EXERCISING YOUR CRITICAL THINKING SKILLS

Go to the library and find a copy of the *New England Journal of Medicine*. Find an article on a new discovery in medicine. Read it. How is the article organized? Analyze the article using your critical thinking skills. Was the study performed correctly, using an adequate number of subjects and a control group? Can you tell whether the conclusions supported the data? Can you think of any alternative explanations for the data?

TEST OF TERMS

1. A _____ is a chemical produced in endocrine glands. It travels in the bloodstream to its _____ cells.
2. In the endocrine system, specificity is conferred by the _____
3. As a general rule, _____ hormones are those that stimulate the production and secretion of other hormones.
4. The body produces three types of hormones: _____ , proteins and polypeptides, and _____ .
5. Protein and polypeptide hormones activate cells through a _____

found in the cytoplasm and in the plasma membrane.

_____, usually cyclic AMP. It is produced inside cells from _____ in a reaction catalyzed by a membrane-bound enzyme called _____ _____.
Cyclic AMP activates another enzyme called protein kinase. It in turn adds _____ to other enzymes inside the cell, turning some on and turning others off.

6. Hormones like testosterone activate cells by the _____ mechanism. Steroid hormones activate the _____.

7. The _____ gland located beneath the _____ produces more hormones than any other endocrine gland.

8. The _____ _____ gland is controlled by _____ and inhibiting hormones produced by the _____ cells of the _____.

9. _____ hormone is a protein that stimulates cellular _____ and hyperplasia. In muscle, this hormone stimulates the uptake of _____ _____ and the synthesis of protein.

10. _____ is a condition in which an endocrine gland produces less hormone than needed.

11. _____ is a hormone that stimulates the adrenal cortex to release its hormones.

12. The _____ _____ are hormones that stimulate gamete formation and endocrine production in the gonads in both males and females.

13. Milk production in humans is stimulated by a protein hormone called _____. Milk production begins at the very end of pregnancy and can be continued by _____.

14. Hormone secretion that is stimulated by neural impulses is called a _____ _____ reflex.

15. The posterior pituitary is a _____ _____ gland. It consists of nervous tissue and releases two hormones: _____ and _____.

16. The thyroid gland produces three hormones: _____, triiodothyronine, and _____.

17. The thyroid follicles contain a colloidal material called _____.

18. _____ is a condition that results from a dietary deficiency of iodine.

19. Two hormones control the level of calcium in the blood. The hormone that raises serum calcium levels is _____, and it is produced by the _____ glands. The hormone that reduces serum calcium is _____ and is produced by the _____.

20. _____ is produced by the _____. It stimulates the uptake of _____ by muscle cells and stimulates the synthesis of _____.

21. The synthesis of glucose from amino acids and fatty acids is called _____.

22. Juvenile diabetes is also called _____-_____ diabetes. It results from an _____ reaction.

23. Type II diabetes or _____-_____ diabetes occurs in older people and is often associated with _____. It is controlled by managing _____.

24. The adrenal medulla produces two hormones, _____ and _____, which stimulate heart rate and breathing.

25. Glucocorticoids are produced by the _____ _____; they increase blood glucose by stimulating _____. In high concentrations, they repress the _____ system function.

26. _____ is the principal mineralocorticoid produced by the _____ _____. It affects sodium and potassium ion concentrations.

Answers to the Test of Terms are located in Appendix B.

TEST OF CONCEPTS

1. Define the following terms: endocrine system, hormone, and target cell.

2. Hormones function in four principal areas. What are they? Give some examples of each.

3. Describe the concept of specificity. How is it created in the nervous system? How is it created in the endocrine system?

4. Compare and contrast the functions of the nervous and endocrine systems.

5. The endocrine system controls functions that require duration; the nervous system controls functions that require speed. Do you agree or disagree? Explain your reasons.

6. Name the three types of hormones produced by the endocrine system and give an example of each.

7. Give two examples of negative feedback loops in the endocrine system, a simple feedback mechanism and a more complex one that operates through the nervous system.

8. Define the term *neuroendocrine reflex* and give some examples.

9. Describe the second messenger theory.

10. Describe the two-step mechanism.

11. How are the second messenger theory and the two-step mechanism similar and how are they different?

12. Give several biological reasons for the following observations: (a) the endocrine response tends to be delayed; (b) the endocrine response tends to be prolonged; (c) some

hormones result in several different functions.

13. List the hormone(s) involved in each of the following functions: blood glucose levels, growth, milk production, milk let-down, calcium levels, and metabolic rate.

14. Describe the role of the hypothalamus in controlling anterior pituitary hormone secretion.

15. Explain the biochemical reason for each of the following medical diseases: acromegaly, dwarfism, and giantism. Acromegaly and giantism are both caused by the same problem. Why are these conditions so different?

16. ACTH is controlled by levels of glucocorticoid and by a biological clock. How are the controls different?

17. Describe the neuroendocrine reflex involved in prolactin secretion.

18. Offer some possible explanations for the following experimental observation. Milk production occurs late in pregnancy and is thought to be stimulated principally by prolactin. A nonpregnant rat is injected with prolactin, but does not produce milk.

19. Where is ADH produced? Where is it released? Describe how ADH secretion is controlled. What effects does this hormone have?

20. Where is oxytocin produced? Where is it released? Describe how oxytocin secretion is controlled. What effects does this hormone have?

21. A patient comes into your office. She is thin and wasted and complains of excessive sweating and nervousness. What tests would you run?

22. A patient comes into your office. He is suffering from indigestion, depression, and bone pain. An X-ray of the bone shows some signs of osteoporosis. You think that the disorder might be the result of an endocrine problem. What test would you order?

23. How are the two basic types of diabetes mellitus different? How are they similar? How are they treated and why?

24. Describe the physiological changes that occur under stress. What hormones are responsible for these changes?

25. What is gluconeogenesis? What hormones stimulate the process?

26. Cortisone depresses the allergic response. A patient comes to your office and asks that you treat her allergies with cortisone. What would you tell her?

27. Aldosterone is a mineralocorticoid. Describe its chief functions. How does it help retain body fluid? Under what conditions is it secreted?

SUGGESTED READINGS

Golde, D., and J. C. Gasson. 1988. Hormones that stimulate the growth of blood cells. *Scientific American* 259(1): 62–70. Discusses how scientists have cloned the genes for several hormones that cause certain cell types to differentiate and mature.

Sherwood, L. 1989. *Human Physiology: From cells to systems.* St. Paul, Minn.: West. Detailed coverage of hormones and how they operate.

Uvnas-Moberg, K. 1989. The gastrointestinal tract in growth and reproduction. *Scientific American* 261(1): 78–83. Discusses how the gastrointestinal tract, an endocrine gland, plays an important role in the readjustment of metabolism that accompanies pregnancy as well as in fetal and infant growth.

Wilson, J. D., and D. W. Foster. 1985. *Williams textbook of endocrinology,* 7th ed. Philadelphia: Saunders. Excellent source of detailed information on human endocrinology and endocrinologic disorders.

IV

REPRODUCTION AND DEVELOPMENT

Human Reproduction: New Beginnings

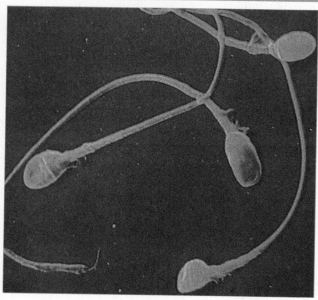

Colorized scanning electron micrograph of human sperm.
Preceding page: Fetal sonogram.

Reproduction is one of the most basic body functions. Like all body functions, reproduction is controlled by the endocrine and nervous systems. This chapter describes the anatomy and physiology of the male and female reproductive systems, laying the groundwork for the discussion of fertilization and development in Chapter 17. We begin with the male.

THE MALE REPRODUCTIVE SYSTEM

The male reproductive system consists of six basic components: (1) the testes, which produce sex steroids and sperm, (2) the **epididymis**, which stores sperm produced in each testes, (3) a pair of ducts that conduct sperm from the epididymis of each testis to the urethra, (4) **sex accessory glands**, which produce secretions that make up the bulk of the ejaculate, (5) the **urethra**, which conducts sperm to the outside, and (6) the **penis**, the organ of copulation (Figure 16–1; Table 16–1).

The Testes and Sex Accessory Glands

The testes are suspended in a pouch known as the **scrotum**. As Figure 16–1 shows, the scrotum is attached to the body below the penis. Although the testes reside in the scrotum throughout most of human life, they do not originate there. During embryological development, the testes form inside the abdominal cavity near the kidneys. Soon after they form, however, the testes descend into the scrotum, passing from inside the body through a small canal, the **inguinal canal**, that links the body cavity with the scrotum. The testes are guided into the scrotum with the aid of a ligament that disappears soon afterward.

By the end of the eighth month of development, the testes generally complete their migration into the scrotum. In some males, however, the testes do not descend during fetal development. This condition is known as **cryptorchidism**. In many of these boys, the testes descend during the first two years of life.

The scrotum provides an environment whose temperature is suitable for sperm development. Inside the

TABLE 16–1	The Male Reproductive System
COMPONENT	FUNCTION
Testes	Produce sperm and male sex steroid
Seminiferous tubules	Produce sperm
Epididymis	Stores sperm
Vas deferens	Conducts sperm to urethra
Urethra	Conducts sperm to outside
Sex accessory glands	Produce seminal fluid that nourishes sperm

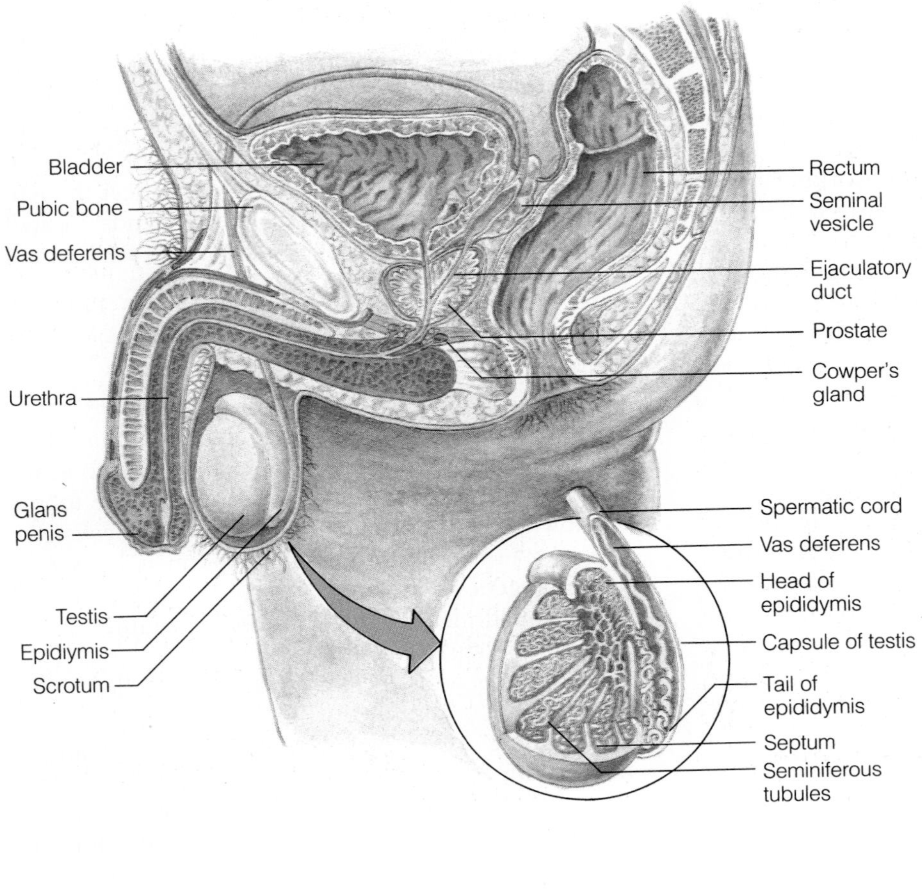

body cavity, however, the temperature is too high for sperm development, and a man whose testes have failed to descend will be sterile. If they have not descended on their own by the time a boy is five years

FIGURE 16–1 The Anatomy of the Male Reproductive System

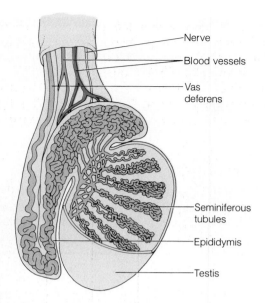

FIGURE 16–2 Interior View of the Testis The lobules and seminiferous tubules are shown.

old, the testes must be lowered into the scrotum surgically to prevent permanent sterility.

The influence of body heat on sperm development is illustrated by the plight of male long-distance runners who jog hundreds of miles a month; elevated body temperature from strenuous workouts often makes long-distance runners temporarily sterile. Even though the testes are suspended in the scrotum, prolonged exercise increases scrotal temperature so much that sperm formation temporarily declines. Tight-fitting pants may have the same effect.

As Figure 16–2 shows, each testis is surrounded by a dense connective tissue layer. This layer is invested with numerous pain fibers, a fact to which most men will attest. The testis is divided into 200 to 400 compartments, or **lobules**, by fibrous tissue that connects to the outer coat (Figure 16–2). Each lobule contains one to four highly convoluted tubules, the **seminiferous tubules**, in which sperm are formed. Stretched end to end, the seminiferous tubules of each testis would extend for about half a mile.

Sperm produced in the seminiferous tubules empty into a network of connecting tubules, the **rete testis**, located in the "back" of the testes. The rete tubules, in turn, empty into the **epididymal duct** where sperm are stored until released during ejaculation, the ejection of sperm. As Figure 16–2 shows, the epididymal duct becomes the **vas deferens**. This muscular duct

courses upward, passing back into the body cavity through the inguinal canal. Inside the pelvic cavity, the vas deferens from each testis joins with the urethra. During ejaculation, the smooth muscles in the walls of the epididymal duct and vas deferens contract, propelling sperm to the urethra. During ejaculation, sperm are joined by fluids produced by the sex accessory glands. The **sex accessory glands** (the seminal vesicles, prostate gland, and Cowper's gland) are located near the neck of the urinary bladder, shown in Figure 16–1. The secretions produced by the sex accessory glands empty into the urethra and the vas deferens and comprise 99% of the volume of the **ejaculate**, the fluid containing sperm and sex accessory gland secretions. The ejaculate nourishes sperm cells and provides a vehicle for sperm transport.

The paired **seminal vesicles**, which empty into the vas deferens, produce the largest portion of the ejaculate. Surrounding the neck of the bladder is the **prostate gland**, which empties directly into the urethra. Routine medical examinations show that in nearly every man over the age of 45, the prostate gland has become enlarged. Small nodules formed by the condensation of prostatic secretions develop inside the gland, but usually cause no trouble. In some men, however, the nodules grow quite large, blocking the flow of urine and making urination painful. In such cases, the prostate must be surgically removed. The prostate is also a common site for cancer in men.

Cowper's glands, a pair of small glands located below the prostate on either side of the urethra, are the smallest of the sex accessory glands.

The Seminiferous Tubules and Spermatogenesis

Sperm are produced in the seminiferous tubules. Figure 16–3a shows a cross section through two seminiferous tubules. The lining of the wall of the tubule is known as the **germinal epithelium**, for its cells give rise to sperm. The formation of sperm is known as **spermatogenesis**; it involves a special type of cell division called meiosis and a process of cellular differentiation (Figure 16–3b).

Spermatogenesis begins with cells called spermatogonia. Located in the periphery of the seminiferous tubule in the germinal epithelium, **spermatogonia** contain 46 single-stranded (nonreplicated) chromosomes. Spermatogonia divide mitotically, ensuring a constant supply of sperm-producing cells. Some of the spermatogonia formed during cellular division, however, enlarge to become **primary spermatocytes**. Two meiotic divisions follow in the formation of sperm.

Meiosis is a special kind of cell division that occurs in the testes and ovaries and nowhere else in the body. The first division in meiosis is called **meiosis I**. During meiosis I, the primary spermatocytes divide to form two **secondary spermatocytes** (Figure 16–3b). During **meiosis II**, the second division, both secondary spermatocytes divide, forming four spermatids. Spermatids undergo a remarkable transformation (differentiation) to become sperm.

As noted in Chapter 3, each gamete (sperm or ovum) formed during meiosis contains 23 single-stranded chromosomes, half the number in a normal somatic cell. When the gametes unite during fertilization, they produce a zygote, a new cell containing 46 single-stranded chromosomes. One-half of its chromosomes come from each parent.

Meiosis reduces the number of chromosomes in germ cells. Each primary spermatocyte, for example, contains 46 double-stranded (replicated) chromosomes; each secondary spermatocyte contains 23 double-stranded chromosomes. Spermatids contain 23 single-stranded chromosomes.

How does this reduction come about? The details of meiosis are shown in Figure 16–4a. This figure also shows the process of mitosis for comparison.

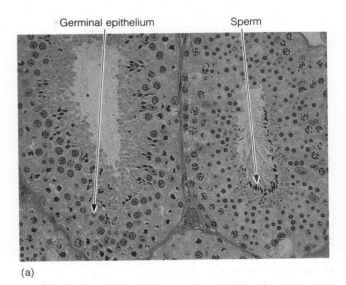

Germinal epithelium   Sperm

(a)

FIGURE 16–3 The Seminiferous Tubules (*a*) A cross section through several seminiferous tubules showing the germinal epithelium and interstitial cells. (*b*) Details of spermatogenesis.

Each contains 46 double-stranded (replicated) chromosomes

Each contains 23 double-stranded chromosomes

Each contains 23 single-stranded chromosomes

Spermatogonium (stem cell)

Primary spermatocyte

Secondary spermatocytes

Spermatids

Spermatids

Spermatozoa

Growth

Mitosis

Enters prophase of meiosis I

Meiosis I completed

Meiosis II

Spermatogenesis (meiosis)

Spermiogenesis

(b)

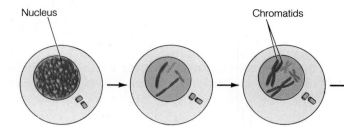

Interphase   DNA replication

(a)

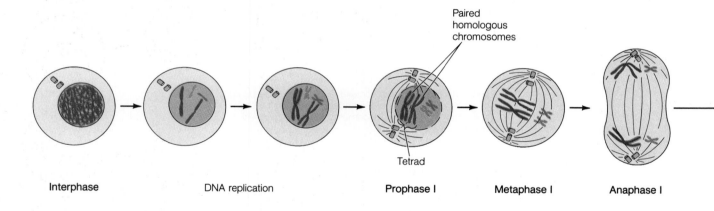

Interphase   DNA replication   Prophase I   Metaphase I   Anaphase I

(b)

FIGURE 16–4   Comparison of Meiosis and Mitosis   (a) Meiosis takes place in two nuclear divisions and results in a reduction in the number of chromosomes. (b) Mitosis occurs in one nuclear division. The number of chromosomes remains the same.

During prophase I of meiosis, the replicated chromosomes condense. The nuclear membrane disappears, the nucleoli vanish, and a spindle forms. The changes taking place during prophase I of meiosis are essentially identical to those occurring during mitosis.

However, there is one major difference: during prophase I, the chromosomes move to the center of the cell and pair up. The primary spermatocyte's 46 chromosomes exist in pairs. Each pair consists of two virtually identical chromosomes—one from the

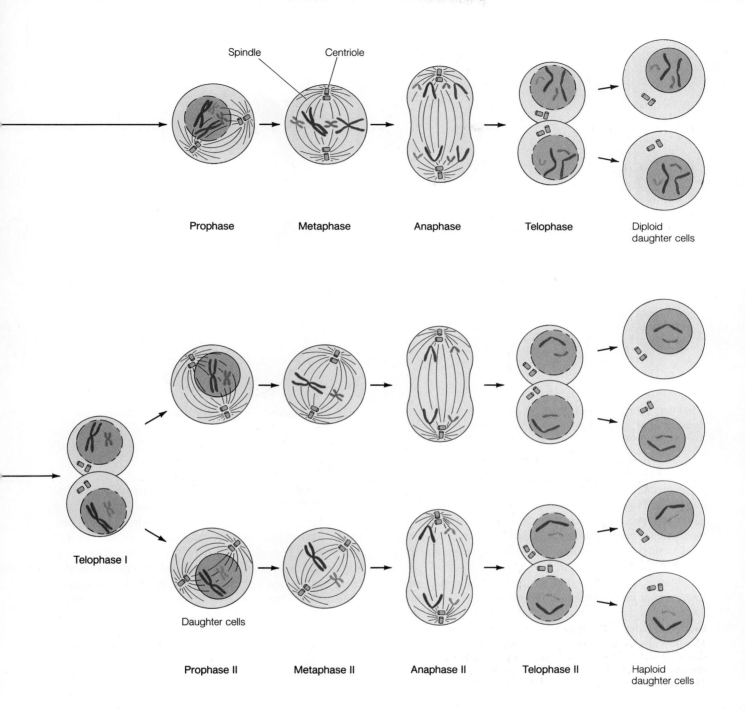

Prophase   Metaphase   Anaphase   Telophase   Diploid daughter cells

Spindle   Centriole

Telophase I

Daughter cells

Prophase II   Metaphase II   Anaphase II   Telophase II   Haploid daughter cells

mother and one from the father. A pair of chromosomes is said to be homologous, because the two members of the pair contain the same genes.

During metaphase I, the homologous pairs become aligned on the equatorial plate (Figure 16–4a). This differs considerably from metaphase in mitosis (Chapter 4). As you may recall, during metaphase in mitosis, the chromosomes line up single file (Figure 16–4b).

In anaphase I of meiosis, the homologous pairs separate. One (double-stranded) member of each pair

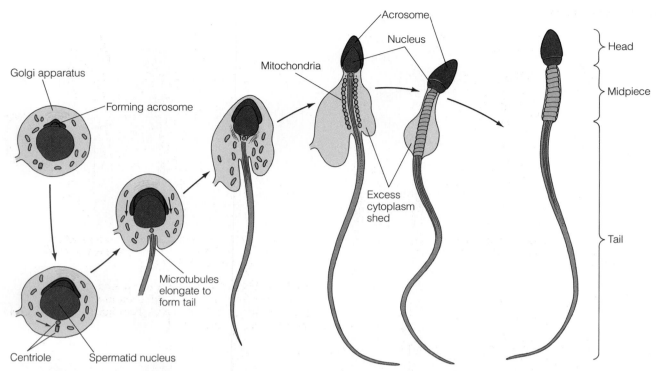

FIGURE 16–5 Sperm Formation Sperm form from spermatids in the germinal epithelium. Note the following changes: nuclear condensation, loss of cytoplasm, tail formation, alignment of the mitochondria, and acrosome formation.

goes to each pole. The cell then divides, producing two offspring, each containing 23 double-stranded chromosomes. Consequently, the first meiotic division is also called a reduction division. Soon after the first division is completed, the second meiotic division, or **meiosis II**, begins.

Meiosis II is equivalent to mitosis in many ways except that each cell contains only 23 chromosomes. During prophase II, the 23 double-stranded chromosomes condense (Figure 16–4a). During metaphase II, they align themselves single file on the equatorial plate. The replicated chromosomes separate at their centromeres during anaphase II, thus forming two single-stranded chromosomes. One single-stranded chromosome goes to each pole.

The two secondary spermatocytes produced during the first meiotic division divide in the second meiotic division, producing four spermatids. Each spermatid contains 23 single-stranded chromosomes and soon gives rise to sperm. During this process, the nuclear material condenses (Figure 16–5). Most of the cyto-plasm is lost as the cell becomes streamlined. The sperm tail forms from the centriole, providing a means for locomotion. The mitochondria of the spermatid congregate around the first part of the tail where they provide energy for propulsion. The Golgi apparatus enlarges and forms an enzyme-filled cap over the condensed nucleus, the head of the sperm. Called the **acrosome**, this cap will help the sperm digest its way through the coatings surrounding the ovum (Chapter 17).

The **spermatozoan** or mature sperm is a marvel of anatomical design. It is rid of excess cytoplasmic baggage and architecturally streamlined for relatively swift movement. Men produce 200 to 300 million sperm every day. Three milliliters of ejaculate contains 240 million or more. Such large numbers are necessary for two reasons. First, many sperm are eliminated as they travel through the female reproductive tract. Second, many sperm may be required to dissolve away the ovum's outer coatings, permitting one sperm cell to reach the ovum and fertilize it.

An acne pimple forms

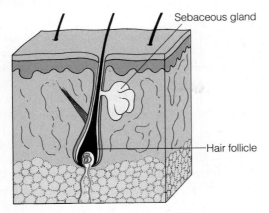

Sebaceous gland

Hair follicle

Sebaceous glands associated with hair follicles secrete sebum, an oily substance that lubricates the skin and hair.

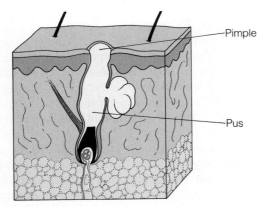

Pimple

Pus

Bacteria present on the skin may infect the sebum, causing inflammation; pus and swelling form an acne pimple.

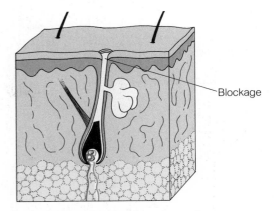

Blockage

A follicle may become blocked by excess sebum and dead skin cells. Unable to escape, the sebum builds up in the hair follicle.

FIGURE 16–6   Formation of an Acne Pimple
Testosterone stimulates oil production in the sebaceous glands. If the outlet is blocked, sebum builds up in the gland and may become infected.

The Interstitial Cells and Testosterone Secretion

The spaces between the seminiferous tubules contain clumps of large cells known as **interstitial cells**, which produce male sex hormones, the **androgens** (Figure 16–3a). The most important androgen is **testosterone**. Testosterone diffuses directly into the seminiferous tubules where it stimulates the stages of meiosis after prophase I. It also stimulates the final maturation of sperm. In the absence of testosterone, the walls of the seminiferous tubules shrink, and sperm production declines, then stops.

Testosterone is also transported in the bloodstream throughout the body where it affects a variety of target cells. Testosterone stimulates cellular growth in bone and muscle and is one reason men are generally more massive and taller than women. Testosterone also promotes facial hair growth and thickening of the vocal cords, generally giving men more facial hair and deeper voices than women. Testosterone stimulates growth of the laryngeal cartilage, producing the prominent bulge called the Adam's apple. Testosterone also stimulates cell growth in the skin, making most men's skin slightly thicker than women's.

Testosterone also stimulates the sebaceous glands of the skin. **Sebaceous glands** secrete oil (**sebum**) onto the skin, helping to moisturize it. During puberty (sexual maturation) in boys, testosterone levels increase dramatically. This causes a marked increase in sebaceous gland activity. Dead skin cells may block the pores in the skin that normally carry the oil to the skin's surface (Figure 16–6). As a result, sebum collects inside the glands. Bacteria on the skin often invade and proliferate in the small pools of oil, resulting in inflammation, pus formation, and swelling. The skin protrudes forming an **acne pimple**.

Mild acne can be treated by washing the skin twice a day with unscented soap. Women should avoid

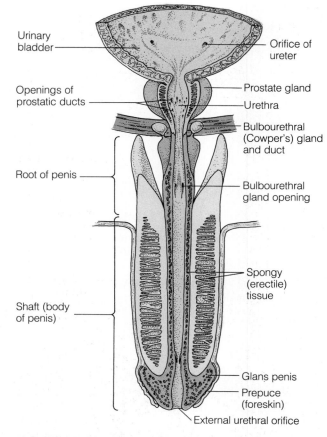

Urinary bladder

Openings of prostatic ducts

Root of penis

Shaft (body of penis)

Orifice of ureter

Prostate gland

Urethra

Bulbourethral (Cowper's) gland and duct

Bulbourethral gland opening

Spongy (erectile) tissue

Glans penis

Prepuce (foreskin)

External urethral orifice

FIGURE 16–7   The Anatomy of the Penis   The penis consists principally of spongy tissue that fills with blood during sexual arousal. The urethra passes through the penis carrying urine or semen.

makeup that has an oily base or use a nonoily type of foundation and wash their faces thoroughly each night. Sunlight also helps clear up acne by drying the oil on the skin and killing skin bacteria. Severe acne can be treated by special ointments and antibiotics.

The Penis, Erection, and Ejaculation

In order for fertilization to occur, many millions of sperm must be deposited in the female reproductive tract. The **penis** serves as the copulatory organ (Figure 16–7). As illustrated, the penis consists of a shaft of varying length and an enlarged tip, the **glans penis.** The glans is covered by a sheath of skin at birth, the **foreskin.** The foreskin gradually becomes separated from the glans in the first two years of life. At puberty, the inner lining of the foreskin begins to produce an oily secretion called smegma. Bacteria can grow in the

protected, nutrient-rich environment created by the foreskin, so special precautions must be made to keep the area clean.[1]

Because of potential health problems and cultural reasons, parents may opt to have the foreskin removed in the first few days of their child's life. The operation, called a **circumcision,** may help reduce penile cancer in men and may also reduce cervical cancer in the wives or sexual partners of circumcised men.

The penis becomes rigid or erect during sexual arousal. Erection and ejaculation are controlled by nerve fibers of the autonomic nervous system. The autonomic nervous system, discussed in Chapter 12, consists of two functionally and anatomically different divisions. The **parasympathetic division** is responsible for erection, among other functions. The **sympathetic division** is responsible for ejaculation.

During sexual arousal, nerve impulses in the parasympathetic division of the autonomic nervous system cause the arterioles in the penis to dilate. Blood floods three bodies of erectile tissue in the shaft of the penis. **Erectile tissue** is a spongy tissue that fills with blood when a male is sexually excited, making the penis harden and become erect. The growing turgidity compresses a large vein on the dorsal surface of the penis, blocking the outflow of blood and further stiffening the organ.

Through the center of the penis runs the urethra, a duct that carries urine from the bladder to the outside of the body during urination and conducts sperm and secretions of the sex accessory glands during ejaculation.

Some men lose their ability to become erect or to sustain an erection. This condition, known as **impotence,** may be caused by psychological, physical, or physiological problems. In most cases, pinpointing the exact cause is difficult. For example, marital conflict, stress, fatigue, and anxiety can all lead to impotence. If the problem is psychological, therapy is often advised. Patients with nerve damage, however, are not so lucky; permanent impotence is likely. Nerve damage may result from diabetes mellitus or from traumatic accidents. For patients with irreversible impotence, urologists can surgically insert an inflatable plastic implant in the penis. The penile implant is attached to a small fluid-filled reservoir in the scrotum. The fluid is pumped into the implant upon de-

[1]Parents must routinely clean the area in children once the foreskin becomes separated from the glans penis.

mand, making the penis erect and permitting sexual intercourse.

Ejaculation is a reflex mechanism. When sexual stimulation becomes intense, sensory nerve impulses traveling to the spinal cord trigger a spinal reflex. Motor neurons in the spinal cord send impulses to the smooth muscle in the walls of the epididymis and vas deferens, causing them to contract. This propels sperm into the urethra. Nerve impulses from the spinal cord also stimulate the smooth muscle in the walls of the sex accessory glands to contract, causing these glands to empty their secretions into the vas deferens and the urethra. The sperm and secretions from the sex accessory glands combine and form a fairly thick fluid called **semen**. Semen is propelled along the urethra by smooth muscle contractions in its wall, which are also caused by nerve impulses from the spinal cord, and sperm is released in spurts.

According to world-renowned sex researchers William Masters and Virginia Johnson, the male sexual response consists of four parts: the excitement phase, the plateau phase, the orgasm, and the resolution phase. Ejaculation occurs during the orgasm or orgasmic phase. The frenzy of muscle contraction that occurs during ejaculation brings with it great pleasure. Sexual arousal is also accompanied by a tensing of the body muscles, rapid breathing, and increased blood pressure. Ejaculation is quickly followed by muscular and psychological relaxation (which is why many men fall asleep after orgasm). Soon after ejaculation, the arterioles in the penis, which were opened to let blood flow in during erection, begin to constrict. This reduces the blood flow, and the penis becomes flaccid once again. In general, another erection is possible in younger men within ten to fifteen minutes. In older men, a repeat performance may take hours or even days.

Hormonal Control of Male Reproduction

The testes produce sex steroid hormones, notably testosterone, which are responsible for **secondary sex characteristics**—that is, male physical features, such as facial hair growth, greater muscle and bone development, and deeper voices. Testosterone also promotes spermatogenesis. Secreted by the interstitial cells, testosterone release is controlled by luteinizing hormone or LH. LH in males is also known as **interstitial cell stimulating hormone (ICSH)** because of its role in stimulating testosterone secretion by the interstitial cells. ICSH secretion, in turn, is controlled by a releasing hormone produced by the hypotha-

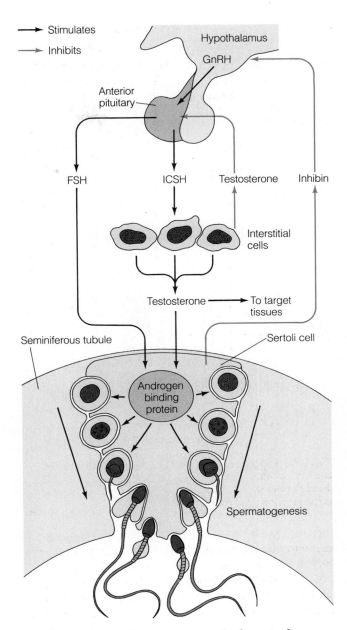

FIGURE 16–8 Hormonal Control of Testicular Function Testosterone, FSH, and ICSH participate in a negative feedback loop. The testes also produce a substance called inhibin, which controls GnRH secretion.

lamus—gonadotropin releasing hormone (GnRH), discussed in Chapter 15. As Figure 16–8 shows, these hormones are participants in a negative feedback loop. A decline in testosterone levels in the blood signals an increase in GnRH secretion, resulting in an increase in ICSH secretion. As testosterone levels climb, GnRH release declines. This feedback loop

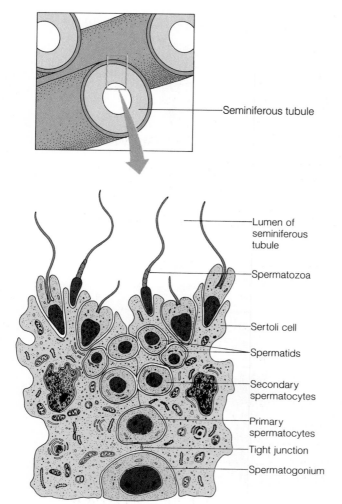

Seminiferous tubule

Lumen of
seminiferous
tubule

Spermatozoa

Sertoli cell

Spermatids

Secondary
spermatocytes

Primary
spermatocytes

Tight junction

Spermatogonium

FIGURE 16–9 Sertoli Cells These cells encompass
the spermatogenic cells as they develop in the germinal
epithelium.

TABLE 16–2	The Female Reproductive System
COMPONENT	FUNCTION
Ovaries	Produce ova and female sex steroid
Uterine tubes	Transport sperm to ova; transport fertilized ova to uterus
Uterus	Nourishes and protects fertilized ova
Vagina	Site of sperm deposition

folds in the plasma membrane of the Sertoli cell, moving slowly to the surface of the germinal epithelium.
The spermatids produced during spermatogenesis remain attached to the Sertoli cells and there differentiate into sperm.

FSH stimulates the Sertoli cells to produce a cytoplasmic receptor protein that binds to androgen, a
male sex steroid. Called **androgen-binding protein**, it
binds to and concentrates testosterone within the Sertoli cell. Testosterone, in turn, stimulates spermatogenesis.

THE FEMALE REPRODUCTIVE SYSTEM

The female reproductive system is shown in Figure
16–10. The female system consists of two basic parts:
the **external genitalia** and the **genital** or **reproductive
tract** (Table 16–2). We begin our discussion with the
reproductive tract.

The Anatomy of the Female Reproductive System

The Reproductive Tract. Shown in Figure 16–10a,
the female reproductive tract consists of four structures: the ovaries, the uterine tubes, the uterus, and
the vagina. The **uterus** is a pear-shaped organ about 7
centimeters (3 inches) long and about 2 centimeters
(less than 1 inch) wide at its broadest point in nonpregnant adult women.[2] The wall of the uterus con-

explains why athletes who use some androgens—
synthetic anabolic steroids—experience a decline in
testicular size (Chapter 14).

The pituitary also produces a gonadotropin known
as follicle-stimulating hormone, or FSH. In males,
FSH works in conjunction with ICSH in stimulating
spermatogenesis; FSH release is also controlled by
GnRH. FSH does not act directly on the spermatogenic cells, however. Instead, it exerts its effects
through another cell in the germinal epithelium of the
seminiferous tubule known as the Sertoli cell. **Sertoli
cells**, shown in Figure 16–9, are large "nurse cells."
The spermatogenic cells (spermatogonia, spermatocytes, and spermatids) divide and differentiate within

[2]The uterus is slightly larger in women who have had children and
grows considerably during pregnancy to accommodate the growing
embryo and fetus.

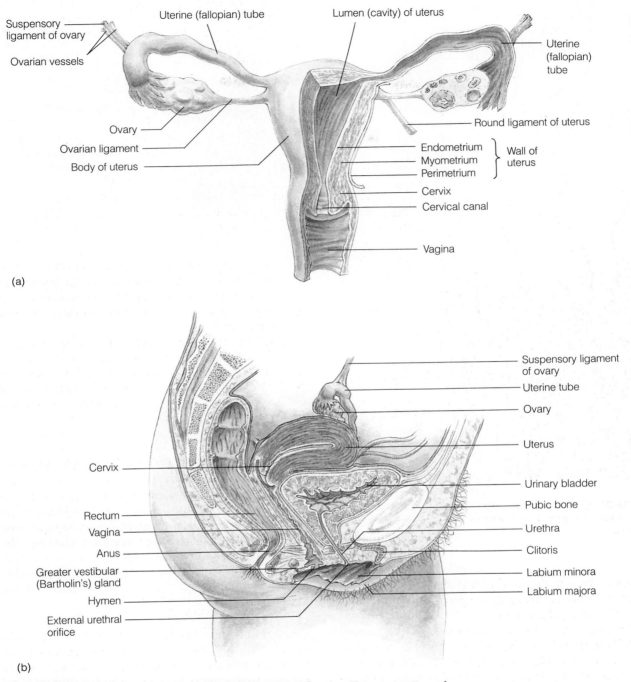

(a)

(b)

FIGURE 16–10   The Anatomy of the Female Reproductive Tract   (*a*) Frontal
view. (*b*) Midsagittal view.

tains a thick layer of smooth muscle cells, known as
the **myometrium.**

The uterus houses and nourishes the developing
embryo and fetus. Attached to the uterus are two hollow tubes, the **Fallopian** or **uterine tubes.** Ova are
produced by the **ovaries,** paired, almond-shaped organs attached to the uterus by the ovarian ligament.
As Figure 16–10a shows, the ends of the uterine

FIGURE 16–11
External Genitalia

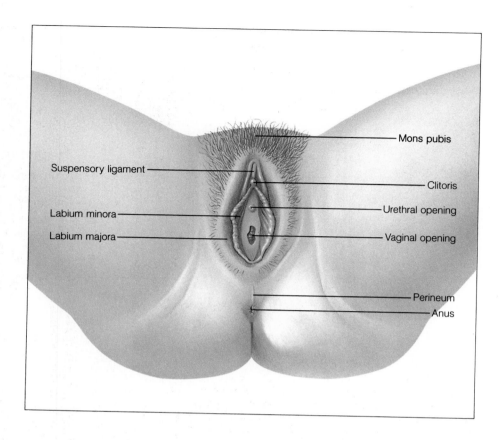

Mons pubis

Suspensory ligament

Clitoris

Urethral opening

Labium minora

Vaginal opening

Labium majora

Perineum

Anus

tubes are widened like a catcher's mitt and fit loosely over the ovaries. Currents created by ciliary beating draw the ova inside and down the uterine tube.

Fertilization occurs in the upper third of the uterine tubes. Fertilized ova are then transported down the uterine tubes to the uterus. In the uterus, the embryo attaches to the lining, the **endometrium**, and embeds itself there, remaining for the duration of pregnancy.

At birth, the fetus is expelled from the uterus through the **cervix**, the lowermost portion of the uterus. As Figure 16–10 shows, the cervix protrudes into the **vagina**, a distensible three-inch tubular organ that leads to the outside of the body.[3] At birth, the cervix stretches open to allow the passage of the baby into the vagina. The vagina also serves as the receptacle for sperm during sexual intercourse. To reach the ovum, sperm must travel through the narrow opening of the cervix and into the lumen (cavity) of the uterus. From here, they move up both the uterine tubes.

The External Genitalia. The external genitalia consist of two flaps of skin on either side of the vaginal opening (Figure 16–11). The outermost folds are the **labia majora**. These large folds of skin are covered with hair on the outer surface and contain numerous sebaceous glands on the inside. The inner flaps are the **labia minora**. Anteriorly, they meet to form a hood over a small knot of tissue called the **clitoris**. The clitoris consists of erectile tissue and is formed from the same embryonic tissue as the penis. Occasionally, a woman will be born with a greatly elongated clitoris.

The Ovaries and Ovulation

During each menstrual cycle, one of the ovaries releases an ovum, the female gamete. The ovum oozes from the ovary and is scooped up by the uterine tube. The release of an ovum is called **ovulation**.[4] Ovulation occurs approximately once a month in women of reproductive age—that is, from puberty (age 11 to 15)

[3]The vagina is often called the birth canal.

[4]The release of the ovum is probably not an explosive event, although many women feel a sharp pain when it occurs.

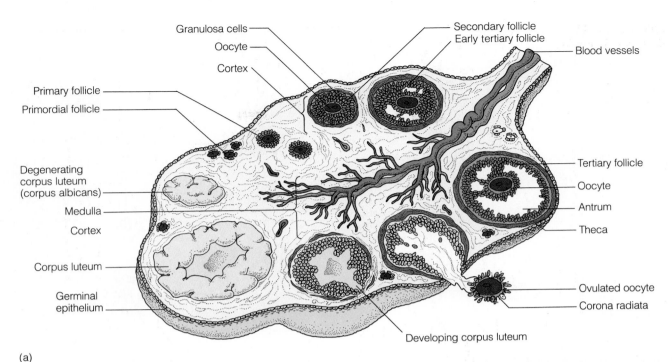

Granulosa cells
Oocyte
Cortex
Secondary follicle
Early tertiary follicle
Blood vessels

Primary follicle
Primordial follicle

Degenerating
corpus luteum
(corpus albicans)
Medulla
Cortex

Corpus luteum

Germinal
epithelium

Tertiary follicle
Oocyte
Antrum
Theca

Ovulated oocyte
Corona radiata

Developing corpus luteum

(a)

FIGURE 16–12   Structure of the Ovary   (*a*) This drawing illustrates the phases of follicular development and also shows the formation and destruction of the corpus luteum (CL). Antral follicles give rise to the CL, and a fully formed CL and antral follicle would not be found in the ovary at the same time. (*b*) Antral follicle.

to menopause (age 45 to 55). However, ovulation is temporarily halted when a woman is pregnant. It may even be temporarily halted under emotional and physical stress.

The structure of an ovary is shown in Figure 16–12. Several ovarian landmarks are immediately visible. One of the most obvious is the germ cell. Germ cells, like those in the seminiferous tubules of the testes, undergo meiotic divisions to produce ova. This process is called **oogenesis.** The first germ cell in oogenesis is the **oogonium.** Containing 46 double-stranded (replicated) chromosomes, the oogonium enlarges and becomes a **primary oocyte.**

The formation of gametes in women occurs by meiosis (Figure 16–13). During prophase I, the homologous chromosomes of the primary oocytes pair up. In metaphase I, the pairs are aligned on the equatorial plate. The members of the pairs separate during anaphase I, migrating toward the poles of the spindle.

Although meiosis I is similar in oogenesis and spermatogenesis, there is a profound difference. During

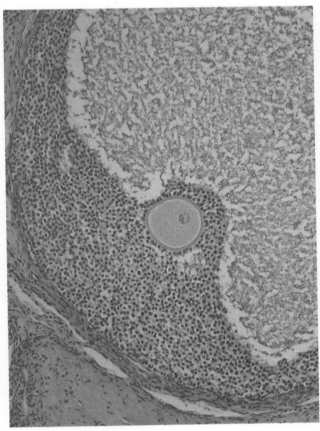

(b)

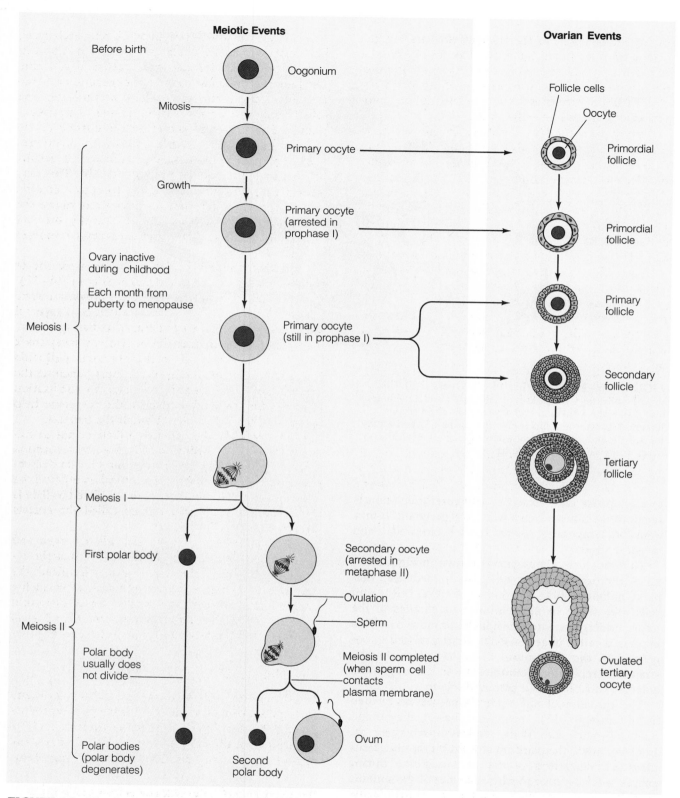

FIGURE 16–13 Oogenesis and Follicle Development

each meiotic division, the nucleus divides, but the cytoplasm does not (Figure 16–13). Therefore, the first meiotic division produces only one cell, the **secondary oocyte** and a small package of discarded nuclear material containing 23 double-stranded chromosomes, called the **first polar body.**

During the second meiotic division, the cytoplasm of the secondary oocyte also fails to divide. This "unequal division" results in the formation of an **ovum,** the female gamete containing 23 single-stranded chromosomes. It also produces another "nuclear discard," the **second polar body.** The second meiotic division occurs only after a sperm penetrates the secondary oocyte. In humans and virtually all other animals, the first polar body usually does not divide.

Germ cells are housed in the ovary in special structures called follicles. A **follicle** consists of a single germ cell surrounded by one or more layers of **follicle cells.** Follicle cells are derived from the loose connective tissue of the ovary. The most abundant of all follicles is the **primordial follicle.** It consists of a primary oocyte surrounded by an incomplete layer of flattened follicle cells. A large number of primordial follicles are found in the periphery of the ovary, and it is from these follicles that the others develop.

Each month, a dozen or so primordial follicles begin to develop. During early development, the oocyte enlarges. The follicle cells divide and enlarge, becoming cuboidal and forming a complete layer around the oocyte. The resulting structure is called a **primary follicle.** As follicular development proceeds, the follicle cells continue to divide, forming two or more layers and giving rise to the **secondary follicle.** The follicle cells and the connective tissue cells surrounding the growing follicle produce a layer of material called the **membrana propria,** which surrounds the entire follicle, walling it off from the outlying tissue.

The cells just outside the membrana propria are compressed by the growing follicle and soon organize into two layers. The inner layer of cells is highly vascularized and produces small amounts of androgen (male hormone) that diffuses into the follicles where it is converted into estrogen by the follicle cells. The outer layer is composed of tightly packed connective tissue cells.

In the largest secondary follicles, a clear liquid begins to accumulate between the follicle cells. The fluid creates small openings among the follicle cells, which enlarge as additional fluid is generated. Eventually, the small cavities coalesce, forming one central cavity. At this point, the follicle is called a **tertiary follicle.**

A dozen or so follicles begin developing each cycle, but only one usually ovulates. The others drop out along the way and, for unknown reasons, degenerate. The follicle or follicles that escape degeneration, however, continue to enlarge. As fluid accumulates, the follicle bulges from the surface of the ovary like a pimple. The pressure exerted on the outside of the ovary causes the ovary's surface to stretch. Blood vessels supplying the tissue may be compressed, resulting in a region of cellular necrosis (death). This may weaken the wall. Enzymes released from ovarian cells in the region are then thought to begin to digest the tissue at the point of weakening. Eventually, the wall of the follicle breaks down, and the secondary oocyte is released.

Around the time of ovulation, the primary oocyte completes the first meiotic division (Figure 16–13). As illustrated in Figure 16–12, the secondary oocyte released from the ovary is surrounded by a layer of follicle cells, the **corona radiata,** or radiating crown. Immediately surrounding the oocyte is a fairly thick layer of gel-like material, called the **zona pellucida** (the clear zone). Incoming sperm must penetrate the corona radiata and the zona pellucida for fertilization to occur. Enzymes released from the acrosome help the sperm "digest" its way through the barriers.

After ovulation, the ovulated follicle begins to collapse. The cells on the inside of the membrana propria enlarge and multiply. The membrana propria disintegrates, and the cells that were once walled off from the follicle invade. The remains of the ovulated follicle is transformed into a cellular mass called the **corpus luteum** or **CL** (yellow body).

The CL produces two sex steroids, estrogen and progesterone, but the fate of the CL ultimately depends on the fate of the oocyte. If it is fertilized, the CL will remain active for several months, producing estrogen and progesterone needed for a successful pregnancy. If fertilization does not occur, the CL soon disappears, and a new menstrual cycle soon begins.

The Menstrual Cycle

The **menstrual cycle** is a series of physiological and anatomical changes occurring in women of reproductive age. The length of the menstrual cycle varies from one woman to the next. In some it lasts 25 days, and in others it may last up to 35 days.[5] On average, how-

---

[5]The length of the menstrual cycle may also vary from month to month in the same woman.

ever, the cycle repeats itself every 28 days. Ovulation usually occurs approximately at the midpoint of the 28-day cycle—about 14 days before the onset of menstruation. The menstrual cycle involves changes in three interdependent cycles (Figure 16–14). The first is a hormonal cycle. Changes in hormones produce cyclic changes in the ovary (the ovarian cycle) and the uterus (the uterine cycle). Understanding the menstrual cycle requires a look at the hormonal and ovarian cycles first.

The Hormonal and Ovarian Cycles.   The first half of the menstrual cycle is known as the **follicular phase**, for it is here that the follicles grow toward ovulation. The second half is called the **luteal phase**—so named because the corpus luteum forms during this time.

As shown in Figure 16–14a, FSH and LH are released from the anterior pituitary during the first half of the cycle, then peak in the middle, just prior to ovulation. FSH stimulates follicular development by promoting mitosis of the follicle cells. LH stimulates estrogen production in the ovary, and estrogen may stimulate mitosis.

Estrogen secreted during the follicular phase of the menstrual cycle controls the release of both FSH and LH by inhibiting the release of GnRH from the hypothalamus. As a result, throughout most of the follicular phase, LH and FSH levels are low and *fairly* constant. Just before ovulation, however, both LH and FSH levels in the blood peak. These surges in LH and FSH secretion are the result of one of the body's rarest events—a positive feedback loop.

During the follicular phase, the amount of estrogen in a woman's blood actually increases very slowly. When estrogen reaches a certain critical level, a positive feedback is triggered. Both the hypothalamus and the anterior pituitary respond with a sudden outpouring of LH and FSH.

The role of the preovulatory surge of FSH, if any, is not known. The LH surge, however, has at least four effects: (1) it causes the primary oocyte to complete its first meiotic division, forming a secondary oocyte; (2) it stimulates the release of the enzymes that break down the ovarian wall, resulting in ovulation; (3) it stimulates estrogen production and release; and (4) it converts the collapsed follicle into a corpus luteum.

During the luteal phase of the cycle, LH secretion gradually declines (Figure 16–14). LH present during the luteal phase stimulates the corpus luteum to produce estrogen and progesterone. However, as LH levels decline, estrogen and progesterone secretion also

decline. If pregnancy does not occur, the CL stops producing hormones and degenerates; the CL regresses (Figure 16–14b). Only a signal from the newly formed embryo can save it. Otherwise, the decline in levels of these steroids permits a new cycle to begin.

The Uterine Cycle.   The uterine lining, or endometrium, also undergoes cyclic changes during the menstrual cycle (Figure 16–14c). These changes result from cyclic changes in ovarian hormones, which, in turn, are controlled by changes in pituitary hormone secretion. As Figure 16–14c shows, the endometrium thickens throughout much of the cycle. The changes that occur in the endometrium prepare the lining for pregnancy. In the absence of fertilization, however, the thickened endometrium is shed or sloughed off, a process called **menstruation**.

Let's begin on day 1 of the average 28-day cycle. Day 1 of the menstrual cycle is the first day of menstruation. During the first 4 or 5 days of the menstrual cycle, the uterine lining is shed. Tissue that formed in the previous menstrual cycle sloughs off from the uterine lining and passes out of the uterus into the vagina along with a considerable amount of blood—on average about 50 to 150 milliliters. The loss of blood during menstruation is the main reason why women are more prone to develop anemia than men and why women should eat iron-rich foods or take iron supplements (Chapters 7 and 8).

As soon as the endometrium has been shed, the lining of the uterus begins to rebuild. Initial regrowth is stimulated by ovarian estrogen. Estrogen stimulates the growth of glands (**uterine glands**) in the uterine lining and stimulates cells in the basal layer (deepest layer) of the endometrium to proliferate. During the regrowth phase, or **proliferative phase**, the uterine glands begin to fill with a nutritive secretion, which will help nourish an embryo should one appear.

The endometrium continues to thicken after ovulation under the influence of both estrogen and progesterone. The uterine glands become distended with a glycogen-rich material. The last half of the uterine cycle is therefore called the **secretory phase**. If fertilization does not occur, approximately four days before the end of the cycle, the uterine lining begins to shrink and the endometrium will be shed, thus starting menstruation.

Menstruation is triggered by a decline in estrogen and progesterone levels. Progesterone acts as a uterine tranquilizer, inhibiting smooth muscle contraction in the myometrium. When progesterone levels fall, the uterus begins to undergo periodic smooth muscle

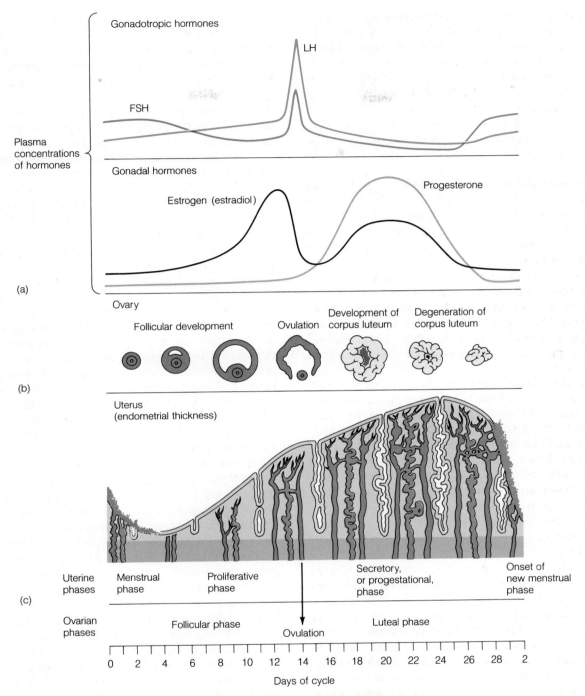

FIGURE 16–14 The Menstrual Cycle (*a*) Hormonal cycles, (*b*) the ovarian cycle, and (*c*) the uterine cycle.

contractions. These contractions propel the sloughed endometrium out of the uterus and are responsible for the cramps that many women experience during menstruation.

If fertilization does occur, however, the newly formed embryo produces a hormone called **HCG (human chorionic gonadotropin)**. HCG is an LH-like hormone that stimulates the corpus luteum, main-

FIGURE 16-15   The Home Pregnancy Test

taining its structure and function. When HCG is present, ovarian estrogen and progesterone continue to be secreted, and the uterine lining remains intact, ready for an embryo. As discussed in the next chapter, the newly formed embryo arrives in the uterus approximately four days after fertilization and embeds in the thickened endometrium from which it derives its nutrients.

HCG maintains the corpus luteum for approximately six months. HCG shows up in detectable levels in a woman's blood and urine within 10 days of fertilization. Pregnancy tests available through a doctor's office or even in your local grocery store detect HCG in a woman's urine. The tests use a commercially prepared antibody to HCG, which binds to the hormone (Figure 16-15). The home pregnancy tests are inexpensive, reliable, and fast.

The Female Orgasm

The sexual response of women is similar to that of men in many respects. During the excitement phase, cervical glands produce a secretion that lubricates the vagina. Erectile tissue beneath the labia fills with blood, and the external genitalia expand. Arousal causes the nipples to become erect. Heart rate and muscle tension also increase, as does blood pressure.

During the next phase, breathing, heart rate, and muscle tension increase. The outer third of the vagina constricts due to vasocongestion, the engorgement of blood vessels in the wall of the vagina. The clitoris also becomes engorged. The clitoris is invested with numerous sensory nerve fibers and, when stimulated, yields considerable pleasure. The nipples become more erect, and the glands near the opening of the

vagina are activated, producing a lubricant that facilitates **intromission**, the insertion of the penis.

Sexual arousal can lead to orgasm. Muscle tension, blood pressure, and breathing all increase in women as they do in men. Rhythmic contractions of the uterus and vagina may occur. These contractions are accompanied by intense physical pleasure. Women often report feelings of warmth throughout their bodies after orgasm.

After orgasm, the body relaxes. Blood drains from the clitoris and labia. Blood pressure, heart rate, and respiration become normal. Unlike men, women generally do not experience a refractory period after orgasm. Thus, a woman can experience multiple orgasms during sexual intercourse.

Additional Effects of Estrogen and Progesterone

As we have seen, ovarian hormones play a key role in regulating the menstrual cycle. Like testosterone in boys, estrogen secretion in girls increases dramatically at puberty. As the levels of estrogen in the blood increase, the hormone begins to stimulate follicle development in the ovaries. Estrogen is also an anabolic hormone—a hormone that stimulates anabolic (synthesis) reactions. In women, estrogen's anabolic effects result in the pubertal growth of the external genitalia and the reproductive tract—the uterus, uterine tubes, and vagina. Estrogen also stimulates rapid bone growth in the early teens. During this time, girls typically grow faster than boys. However, estrogen also stimulates the closure of the epiphyseal plates, described in Chapter 14, ending the growth spurt fairly early. Most girls, in fact, reach their full adult height by the age of 15 to 17. Males experience their most rapid growth later in adolescence and continue until the age of 19 to 21.

Estrogen also stimulates the deposition of fat in women's hips, buttocks, and breasts, giving the female body its characteristic shape. Estrogen also stimulates the growth of ducts in the mammary glands. Progesterone works with estrogen to stimulate breast development; progesterone also promotes endometrial growth.

Premenstrual Syndrome: The Cause and the Cure

For reasons not yet fully understood, many women suffer premenstrual irritability, depression, fatigue, headaches, bloating, swelling and tenderness of the breasts, tension, and even joint pain. Together these symptoms constitute a condition called **PMS** or **premenstrual syndrome**. Premenstrual syndrome is a clinically recognizable condition characterized by one or more of these symptoms. PMS strikes 4 of every 10 women of reproductive age. All told, women report more than 150 different physical and psychological symptoms, which emerge before menstruation begins.

Despite exhaustive studies now underway, it may be years before medical scientists can pinpoint the cause—or causes—of PMS. Nevertheless, dozens of "cures," ranging from massive doses of progesterone to vitamin B-6 to L-tryptophan, have been prescribed by clinics specializing in PMS.[6] Buyers should beware, however, for very little good scientific evidence is available to indicate which, if any, of the "cures" really works. Most of the evidence consists of testimonials—individual accounts. The critical thinking skills you learned in Chapter 1 suggest that anecdotal information such as this is no substitute for controlled studies.

Work is now underway to test various treatments, but the results are not expected for several years. In the meantime, physicians recommend that individuals suffering from PMS see their family doctor to be certain that the symptoms are not in fact caused by some other problem.

Menopause

The menstrual cycle continues throughout the reproductive years, but after a woman reaches 20, the ovaries start to become progressively less responsive to gonadotropins (Figure 16–16). Responsiveness declines slowly. But as it declines, estrogen levels slowly decline. Ovulation and menstruation become erratic.

Menopause is attributed to a reduction in the number of ovarian follicles. At about age 45, most of the primordial follicles have been stimulated to grow and have either degenerated or have ovulated. Consequently, FSH and LH from the pituitary have no follicles to stimulate. The production of ovarian estrogen declines as the number of primordial follicles reaches zero.

---

[6]As noted earlier in the book, L-tryptophan has been implicated in the paralysis of body muscles and death of over a dozen people. At this writing, preliminary studies suggest that the amino acid may not be at fault, but rather that the pills taken by people contained a contaminant that caused the ill effects.

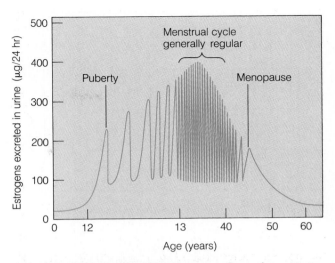

FIGURE 16–16   Ovarian Hormone Secretion   Notice that at about age 20, ovarian estrogen secretion begins a gradual decline.

Source: Adapted with permission from A. C. Guyton, *Textbook of Medical Physiology,* 7th ed. (Philadelphia: W. B. Saunders Company, 1986), Figure 81–8, p. 979.

Between the ages of 45 and 55, ovulation and menstruation cease altogether. The dramatic change in the hormonal climate results in several important physiological changes. The decline in estrogen secretion causes the breasts and reproductive organs, such as the uterus, to begin to atrophy. Vaginal secretions often decline, and in some women sexual intercourse may become painful.

The decline in estrogen levels also results in behavioral disturbances. Many women become more irritable and suffer bouts of depression. Three quarters of all women suffer "hot flashes" and "night sweats" induced by massive vasodilation of vessels in the skin. Fortunately, these symptoms usually pass.

As noted in Chapter 14, declining estrogen levels also accelerate osteoporosis. To counter osteoporosis and other impacts of the decline in ovarian function, physicians sometimes prescribe pills containing small amounts of estrogen and progesterone, as well as a program of exercise and a diet rich in calcium and vitamin D (see Health Note 14–1).

BIRTH CONTROL

Few topics in modern society create as much controversy as birth control. **Birth control** is any method or device that prevents births and includes two broad

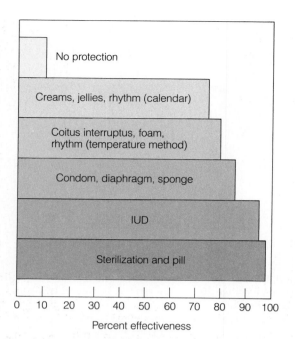

ing means that 95 women out of 100 using a certain method will not become pregnant in a given year.

Sterilization.   Except for complete abstinence, sterilization and the pill are the most effective birth control measures in existence (Figure 16–17). In 1982, sterilization became the leading method of contraception practiced by married women in the United States.

In women, sterilization is performed by cutting and tying off the uterine tubes. This process is called **tubal ligation** and usually involves an overnight stay in the hospital (Figure 16–18a). Surgeons make two small incisions in the abdomen just beneath the navel. An instrument called a **laparoscope** is inserted through each incision. The surgeon locates the uterine tubes through a special viewing lens. An attachment to the laparoscope is then used to cut or seal off the uterine tubes. Physicians may cut the uterine tubes, then tie them off, or may cauterize them—that is, burn them shut using an electrical current. In some cases, surgeons clamp the uterine tubes shut with plastic or metal rings.

Male sterilization requires a far less traumatic surgical procedure called a **vasectomy**, which can be carried out in a physician's office under local anesthesia (Figure 16–18b). To perform a vasectomy, a physician makes a small incision in the scrotum. Each vas deferens is exposed, then snipped, and the free ends are tied off or cauterized.

Vasectomies prevent the sperm from passing into the urethra during ejaculation. They do not impair sex drive and have virtually no effect on ejaculation, since 99% of the volume of the ejaculate is produced by the sex accessory glands.

Vasectomy and tubal ligation are essentially irreversible. However, special surgical methods, called

FIGURE 16–17   Effectiveness of Contraceptive Measures Compared to No Contraceptives   Percent effectiveness is a measure of the number of women in a group of 100 who will not become pregnant in a year.

categories: (1) **contraception**, ways of preventing pregnancy, and (2) **induced abortion**, the deliberate expulsion of a fetus or embryo.

Contraception: Preventing Pregnancy

Figure 16–17 summarizes the effectiveness of the most common means of contraception. Effectiveness is expressed as a percentage. A 95% effectiveness rat-

FIGURE 16–18   Tubal Ligation and Vasectomy (a) In females, the uterine tubes are cut, then tied off and cut or cauterized during a tubal ligation. (b) In males, the vasa deferentia are cut, then tied off during a vasectomy.

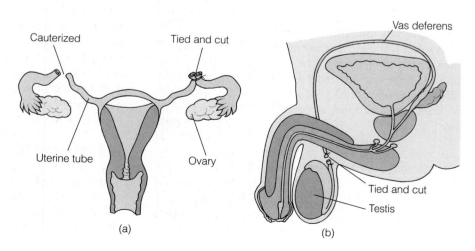

microsurgery, can be used to reconnect the uterine tubes and the vas deferens, but this procedure is costly and not always successful.

The Pill.   The **birth control pill** is the most effective *temporary* means of birth control available (Figure 16–19). Birth control pills come in several varieties, but the most common one in use today is the combined pill. It contains a mixture of synthetic estrogen and progesterone, which act on the pituitary and hypothalamus, preventing the release of LH and FSH. As a result, follicle development and ovulation are inhibited. A minipill is also available. It contains progesterone alone and is less effective than the combined pill, but is more suitable for some women because it results in fewer side effects.

Birth control pills must be taken throughout the menstrual cycle. Skipping a few days releases the pituitary and hypothalamus from the inhibitory influences of estrogen and progesterone, resulting in ovulation and possible pregnancy.

Although effective, birth control pills have some drawbacks or side effects. Overall, the incidence of adverse side effects is rather small, but a woman should carefully consider the possibilities with her physician. The pill is not the best method of birth control for many women.

Table 16–3 compares the risk of death from taking birth control pills (and using other contraceptives) to a number of common risk factors. The risk of a nonsmoker dying from taking birth control pills is 1 in 65,000 in any given year, whereas the risk of dying from pregnancy and childbirth is 1 in 10,000.

The most serious side effects of the birth control pill are heart attacks, stroke, and blood clots. The incidence of these life-threatening side effects is extremely low, especially in nonsmoking women under the age of 30. To reduce the risk even more, pharmaceutical companies have dramatically lowered the estrogen content of the combined pill since estrogen is responsible for most of the adverse side effects.

Early studies of women using birth control pills show a correlation between the use of birth control pills and cancers of the breast and cervix. More recent studies of low-estrogen pills suggest that the new generation of pills is less likely to cause cancer of the breast or cervix. Even with reduced estrogen levels, however, women who take birth control pills are more likely to develop cervical cancer than women who do not. Physicians, therefore, recommend annual pap smears for women who are on the pill. During a **pap smear**, the cervical lining is swabbed. The swab picks

FIGURE 16–19   The Birth Control Pill   One of the most effective means of birth control, the pill consists of a mixture of estrogen and progesterone, which is taken throughout the menstrual cycle to block ovulation. Birth control pills are packaged in numbered containers to help women keep track of them.

up cells sloughed off by the epithelium, which are then examined under a microscope for signs of cancer (Figure 16–20). This procedure helps physicians diagnose cervical cancer early, increasing a woman's chances of survival.

Smoking increases the likelihood of side effects. If you are a smoker and take the pill, for example, you are four times more likely to die from heart attack and stroke than a nonsmoker. The risk of side effects also increases with age. To reduce the chances of developing serious side effects, women over the age of 35 who smoke should use an alternative method of birth control—or should consider giving up smoking. Birth control pills are also not advised for women with a medical history of blood clots, high blood pressure, diabetes, uterine cancer, and cancer of the breast.

Birth control pills have beneficial side effects as well. First of all, they prevent pregnancy. One of every 10,000 women who becomes pregnant and delivers will die from complications, usually during delivery. Thus, even with the risks associated with the pill, using it is six times safer than pregnancy. Birth control pills also reduce the incidence of ovarian cysts, breast lumps, anemia, rheumatoid arthritis, os-

TABLE 16-3 Comparison of Risk of Some Voluntary Activities

RISK	CHANCE OF DEATH IN A YEAR (UNITED STATES)
Smoking	1 in 200
Motorcycling	1 in 1,000
Automobile driving	1 in 6,000
Power boating	1 in 6,000
Rock climbing	1 in 7,500
Playing football	1 in 25,000
Canoeing	1 in 100,000
Using tampons (toxic shock syndrome)	1 in 350,000
Contracting reproductive tract infections through sexual intercourse	1 in 50,000
Preventing pregnancy:	
Oral contraception—nonsmoker	1 in 63,000
Oral contraception—smoker	1 in 16,000
Using intrauterine devices (IUDs)	1 in 100,000
Using barrier methods	None
Using natural methods	None
Undergoing sterilization:	
Laparoscopic tubal ligation	1 in 20,000
Hysterectomy	1 in 1,600
Vasectomy	None
Deciding about pregnancy:	
Continuing pregnancy	1 in 10,000
Terminating pregnancy:	
Nonlegal abortion	1 in 3,000
Legal abortion:	
Before 9 weeks	1 in 400,000
Between 9 and 12 weeks	1 in 100,000
Between 13 and 16 weeks	1 in 25,000
After 16 weeks	1 in 10,000

teoporosis, and pelvic infection. Birth control pills protect a woman from cancer of the ovary and of the uterus, perhaps for life. Nevertheless, there are much safer methods of birth control, such as the condom and diaphragm.

Intrauterine Device. The next most effective means of birth control is the **IUD** or **intrauterine device** (Figure 16–21). The IUD is a small plastic or metal object with a string attached to it. IUDs are inserted into the uterus by a physician, usually during men-

struation because the cervical canal is widest then and because menstrual bleeding indicates that the woman is not pregnant.

No one knows exactly how the IUD works, but there are two major hypotheses. Some researchers think that the IUD increases uterine contractions, making it difficult for an embryo to attach and embed or implant in the wall of the uterus. Others think that the IUD creates a local inflammatory reaction in the uterine lining, resulting in an environment inhospitable to a newly formed embryo. As a result, implan-

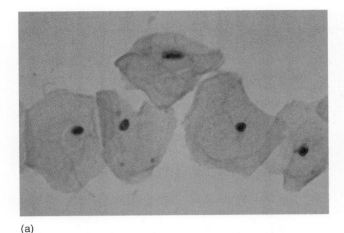

(a)

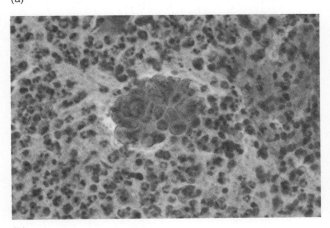

(b)

FIGURE 16–20   The Pap Smear (*a*) A photomicrograph of a normal pap smear. (*b*) A photomicrograph of a cancerous smear.

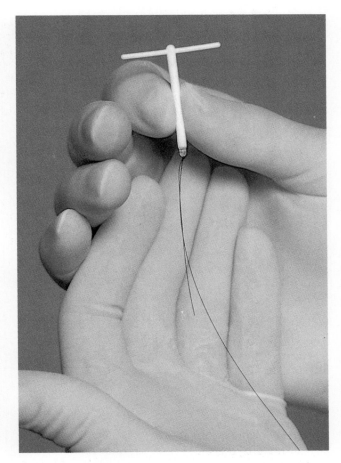

FIGURE 16–21   The IUD IUDs come in a variety of shapes and sizes and are inserted into the uterus where they prevent implantation.

tation is blocked and the embryo dies. It is possible that both mechanisms are operating.

IUDs also have side effects. In some cases, the uterus expels the device, leaving a woman unprotected. Expulsion usually occurs within a month or two of insertion, so couples should check regularly during this period to be certain that the IUD is in place. The IUD may also cause slight pain and may increase menstrual bleeding. These side effects, however, are minor compared to two much rarer but more serious ones: uterine infections and perforation. Women wearing an IUD are more likely to develop uterine infections. If not treated quickly, infections can spread to the uterine tubes, causing scar tissue to develop and blocking the transport of sperm and ova,

causing sterility. Perforation of the uterus is a life-threatening condition requiring surgery to correct.

The Diaphragm, Condom, and Sponge. The next most effective means of birth control are the barrier methods—the diaphragm, condom, and vaginal sponge—all of which prevent the sperm from entering the uterus. The **diaphragm** is a rubber cup that fits over the end of the cervix (Figure 16–22). To increase its effectiveness, a spermicidal (sperm-killing) jelly, foam, or cream is applied to the rim and inside surface of the cup.

Diaphragms must be custom fitted by physicians. To be effective, the diaphragm must be inserted no more than two hours before sexual intercourse and must be worn for at least six hours afterwards. If sexual intercourse is repeated, additional spermicidal cream or jelly should be injected into the vagina onto the diaphragm via a special applicator (Figure 16–22b). If intercourse is desired after six hours, the di-

aphragm should be removed, washed, recoated, and reinserted.

Smaller versions of the diaphragm, called **cervical caps**, are also available. Cervical caps fit over the very tip of the cervix. The cervical cap most often used is held in place by suction. When used with spermicidal jelly or creams, the caps are as effective as diaphragms.

Condoms are thin latex rubber sheaths that are rolled onto the erect penis (Figure 16–23). Sperm released during ejaculation are trapped inside (often in a small reservoir at the tip of the condom) and are, therefore, prevented from entering the vagina. Some condoms are prelubricated with a spermicidal chemical. Besides preventing fertilization, condoms also protect against sexually transmitted diseases, a benefit not offered by any other birth control measure except abstinence. Condoms are widely available and easy to obtain. They do not require a doctor's prescription.

One of the newest methods of birth control is the **vaginal sponge** (Figure 16–24). This small absorbent

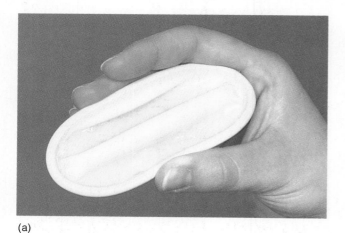

(a)

FIGURE 16–22 The Diaphragm (a) Worn over the cervix, the diaphragm is coated with spermicidal jelly or cream and is an effective barrier to sperm. (b) The insertion of the diaphragm.

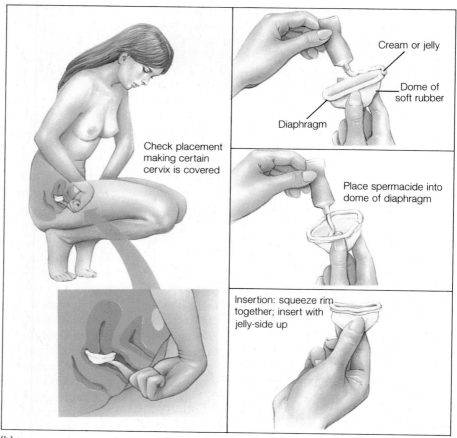

Check placement making certain cervix is covered

Cream or jelly

Dome of soft rubber

Diaphragm

Place spermacide into dome of diaphragm

Insertion: squeeze rim together; insert with jelly-side up

(b)

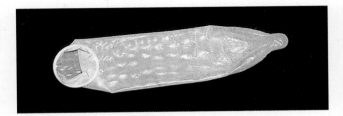

FIGURE 16–23   The Condom   Worn over the penis during sexual intercourse, it prevents sperm from entering the vagina.

sponge is impregnated with spermicidal jelly. Inserted into the vagina, the sponge is positioned over the end of the cervix. The sponge is effective immediately after placement and remains effective for 24 hours. Cervical sponges can be purchased without a doctor's prescription. Like the condom, one size fits all.

Withdrawal and Spermicidal Chemicals.   One of the oldest, but least successful, means of birth control is **withdrawal** or **coitus interruptus**, disengaging before ejaculation. This method requires tremendous willpower and frequently fails for three reasons: because caution is often tossed to the wind in the heat of passion, because the penis is withdrawn too late, or because of pre-ejaculatory leakage—the release of a few drops of semen before ejaculation.

Spermicidal jellies, creams, and foams contain chemical agents that kill sperm, but are harmless to the woman. Spermicidal preparations are most often used in conjunction with diaphragms, condoms, and cervical caps. They can also be used alone, but are only about as effective as withdrawal when so used (Figure 16–25).

The Rhythm Method.   Abstaining from sexual intercourse around the time of ovulation, the **rhythm** or **natural method**, can help couples reduce the likelihood of pregnancy. The natural method relies on the fact that ova remain viable 12 to 24 hours after ovulation, while sperm may remain alive in the female reproductive tract for up to two or three days. If a couple knows the time of ovulation, they can time sexual intercourse to prevent pregnancy.

Simple as it sounds, the rhythm method is the least successful of all birth control measures and one of the more complicated. To practice the natural method successfully, couples must first determine when ovulation occurs and then must abstain from sexual intercourse for a set period before and after ovulation (Figure 16–26).

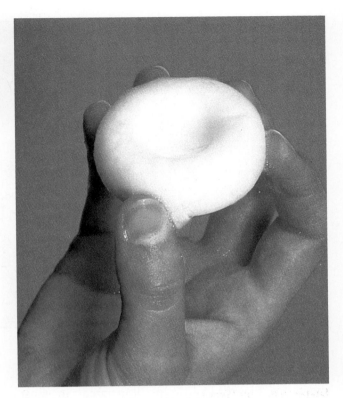

FIGURE 16–24   The Vaginal Sponge   Impregnated with spermicidal chemical, the vaginal sponge is inserted inside the vagina and is effective for up to 24 hours.

Several methods are available to determine when ovulation occurs in a woman's cycle. The most reliable is the **temperature method.** A woman's body

FIGURE 16–25   Spermicidal Jelly   Spermicidal preparations work alone, but are most effective when combined with another form of protection, such as a diaphragm.

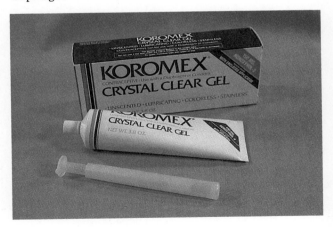

FIGURE 16–26 The Natural Method The yellow shaded area indicates an unsafe period for sexual intercourse, assuming ovulation occurs at the midpoint of the cycle.

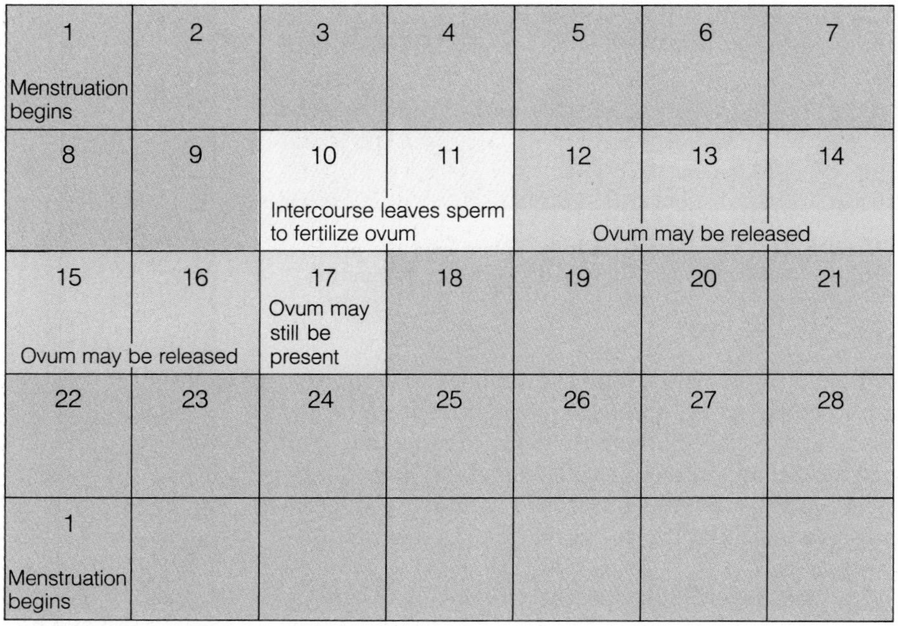

1 Menstruation begins	2	3	4	5	6	7
8	9	10 Intercourse leaves sperm to fertilize ovum	11	12 Ovum may be released	13	14
15 Ovum may be released	16	17 Ovum may still be present	18	19	20	21
22	23	24	25	26	27	28
1 Menstruation begins						

temperature varies throughout the menstrual cycle, as shown in Figure 16–27. As illustrated, in most women body temperature rises slightly after ovulation. By taking her temperature every morning before she gets out of bed, a woman can pinpoint the day she ovulates. By keeping a temperature record over several menstrual cycles, she can determine the length of her cycle and the time of ovulation. Once the length

of the cycle and the time of ovulation are determined, days of abstinence can be determined. Erring on the safe side, some doctors recommend that couples refrain from sexual intercourse from the first day of menstruation until three days after ovulation. Translated, that means no sex for about 17 days of the 28-day cycle.

As most people practice it, the natural method of birth control requires about eight days of abstinence during each menstrual cycle—four days before ovulation and four days after ovulation. This minimizes the chances of a viable sperm reaching a viable ovum. Unfortunately, some women experience the greatest sexual interest around the time of ovulation. Sexual intercourse after a period of abstinence may also advance the time of ovulation. (For a discussion of new methods of birth control, see Health Note 16–1.)

FIGURE 16–27 Body Temperature Measurements during the Menstrual Cycle

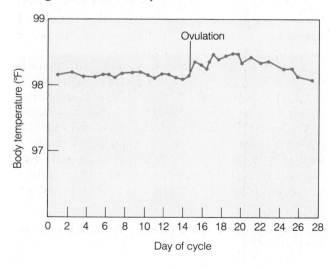

Abortion: The Difficult Choice

Some couples may elect to terminate pregnancy through **abortion**. In the United States, approximately 2 million abortions are performed every year by physicians. Many people in our society view abortion as a legitimate means of family planning, but most people are quick to point out that abortion should not be practiced as a primary means of birth control. Contraception is less costly and less traumatic.

Abortion is not suitable or morally acceptable to many couples. Pro-life factions argue that abortion should be outlawed or severely restricted—that is, allowed only in cases of rape and incest. They advise unmarried women to abstain and if they become pregnant to go through with their pregnancies, keeping their babies or giving them up for adoption.

Pro-choice factions, on the other hand, think that women should have the freedom to choose. Abortion, they say, reduces unwanted pregnancies and untold suffering among unwanted infants, especially in poor families, and it gives women more options than motherhood.

In the first 13 weeks of pregnancy, abortions can be performed surgically in a doctor's office via **vacuum aspiration**. The cervix is dilated by a special instrument, and the contents of the uterus are drawn out through an aspirator tube. The operation is fairly routine and relatively painless. Usually no anesthesia is given. Women bleed for a week or so after the procedure, but have few complications. Between 13 and 16 weeks, a larger aspirator tube is required. The lining of the uterus may have to be scraped with a metallic instrument to ensure complete removal of the fetal tissue. After 16 weeks, abortions are more difficult and more risky. Solutions of salt, urea, or prostaglandins, which stimulate uterine contractions, can be injected into the sac of water surrounding the embryo to induce premature labor. Oxytocin may be administered to the mother with the same effect. Most abortions are performed by the end of the twelfth week of pregnancy.

Abortion is rarely an easy choice for anyone. Therefore, psychological counseling may be advisable before, during, and after the procedure. Contrary to what many think, psychological studies show that the majority of women who choose the abortion route do not suffer lasting emotional harm, especially if they have had counseling.

SEXUALLY TRANSMITTED DISEASES

Several years ago, a group of English professors convened to discuss some pressing questions of language. One of those questions was, "What is the most beautiful word in the English language?" After much deliberation, they settled on a beautiful but unlikely candidate—the word "syphilis." It rolls off the tongue with ease. By most measures, however, syphilis is not a thing of beauty. It is a potentially crippling or deadly disease spread through sexual contact.

Certain viruses and bacteria can be spread from one individual to another during sexual relations. Infections that are transmitted by sexual contact are called **sexually transmitted diseases (STDs)** or **venereal diseases**. Bacteria and viruses transmitted by sexual contact penetrate the lining of the reproductive tracts of men and women and thrive in the moist, warm environment of the body. Most bacteria and viruses that cause STDs are spread by sexual intercourse, but other forms of sexual contact (anal and oral sex) are also responsible for the spread of disease. AIDS, for example, can be transmitted by anal sex as well as vaginal and oral sex (Chapter 10). **Syphilis**, an STD caused by a bacterium, can be spread by oral as well as vaginal sex. Although these diseases are transmitted by sexual contact, the symptoms are not confined to the reproductive tract. In fact, several, including syphilis and AIDS, are primarily systemic diseases—diseases that affect entire body systems.

Some STDs may not result in any obvious symptoms. A good example is gonorrhea. Men and women often show no symptoms whatsoever. As a result, sexually active individuals (who are not monogamous) can transmit gonorrhea without even knowing they have it. In this section, we will examine the most common STDs, except AIDS, covered in Chapter 10.

Gonorrhea

Gonorrhea (referred to as the "clap") is caused by a bacterium that commonly infects the urethra in men and the cervical canal in women. Painful urination and a puslike discharge from the urethra are common complaints in men. Women may experience a cloudy vaginal discharge and lower abdominal pain. Urination may be painful as well if the urethra is infected. Symptoms of gonorrhea usually appear about two to eight days after sexual contact.

If left untreated, gonorrhea in men can spread to the prostate gland and the epididymis. Infections in the urethra lead to the formation of scar tissue, which narrows the urethra, making urination even more difficult. In women, the bacterial infection can spread to the uterus and uterine tubes, causing the buildup of scar tissue. In the uterine tubes, scar tissue may block the passage of sperm and ova, resulting in infertility.[7] Gonorrhea infections can also spread into the abdom-

---

[7]Sexually transmitted diseases are, in fact, a leading cause of infertility.

Advances in Birth Control: Responding to a Global Imperative

Birth control is an issue of world-wide concern. In the wealthy industrialized nations, many parents have chosen to limit their family size for several reasons. First, many parents cannot afford more than one or two children. Raising a single child to college age could cost as much as $85,000. A recent study suggested that another $85,000 will be needed just to send a child born today to a public university. Parents may also limit their family size to provide more personal care for their offspring. And some choose to have fewer children for environmental reasons. Consider some startling statistics about resource demand and pollution:

A child born in the United States today will require 16 pounds of coal, 3.6 gallons of oil, and 240 cubic feet of natural gas *each day* of his or her life.

On average a baby requires 6000 disposable diapers. Altogether, Americans throw away 18 billion disposable diapers per year. Attached end to end, they would extend to the moon and back seven times.

This heavy dependence on resources results in enormous amounts of pollution and environmental damage. Every man, woman, and child, in fact, is responsible for over

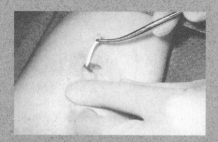

Subcutaneous Progestrone Implant Inserted under the skin, this tiny device releases a steady stream of progesterone, blocking ovulation for months.

one ton of hazardous waste each year.

Birth control is essential in the United States and other developed countries, say many experts, because our impact on the environment is so great. Each American, in fact, uses 25 to 40 times as much of the Earth's resources as a resident of a Third World country. In terms of environmental impact, then, the United States' 248 million population is equal to 6 to 10 billion Third World residents— 1.2 to 2 times the present world population.

Birth control is also important in Third World nations. Each year 89 million new residents are added to the global population—or about 3 new residents per second. Ninety percent of the new residents are born into the Third World where hunger and starvation abound.

Family planning is making headway in some areas, but the task has only just begun. To help in the process, researchers are working on safer, more convenient, and perhaps even more effective methods of birth control.

One line of research has led to the development of nasal sprays that deliver certain drugs. Researchers in California are working on a new generation of nasal sprays containing mild detergents that enhance the solubility of proteins, allowing them to pass through the tissues of the nasal epithelium and into the bloodstream. Tests are now underway to study the effectiveness and safety of nasal spray contraceptives.

Research is also proceeding quickly on transdermal contraceptive patches. The small Band-Aid-like patches are impregnated with a blend of hormones and hormone analogs, including estrogen and progesterone. The patches are worn for a week, then replaced.

In many countries outside the United States, slow-acting, injectable contraceptives are now being used. Women are given a shot of crystalline progesterone under the skin. The crystals dissolve over a period of three months, blocking ovulation. Approval in the United States has been withheld because of animal tests suggesting that this treatment may cause some forms of cancer.

Another novel approach involves

inal cavity through the opening of the uterine tubes. If the infection gets into the bloodstream in men or women, it can spread throughout the body. Fortunately, gonorrhea can be treated by antibiotics, but early diagnosis is essential to limit the damage.

Nonspecific Urethritis

Nonspecific urethritis (NSU) has become the most common sexually transmitted disease known to medical science and is one of several STDs whose inci-

matchstick-sized capsules containing an even more potent progesterone implanted under the skin of a woman's arm; these prevent pregnancy for up to five years (see figure). Contraceptive implants have been approved for use in 12 countries. Population experts hope that they will be widely used in Third World nations because these devices require virtually no effort on the part of the couple. Approval of the Food and Drug Administration for use in the United States is still pending.

Researchers are also experimenting with biodegradable implants. Clinical trials are now underway on a biodegradable material that is impregnated with progesterone and may prevent pregnancy for 18 months or more. The biodegradable material is broken down and gradually disappears.

Experimentation is continuing on a morning-after pill—a pill that can be taken after sexual intercourse to prevent pregnancy. At least two morning-after pills exist. One contains a synthetic estrogen called DES (diethylstilbestrol). DES stimulates muscle contraction in the uterus and uterine tubes, expelling the fertilized ovum from the reproductive tract. DES is sometimes used in cases of incest or rape, but is generally avoided because it causes vomiting and nausea.

A synthetic steroid called RU486 is also in use in certain countries, but not the United States. RU486 binds to progesterone receptors in the uterus, blocking progesterone from binding. Because progesterone inhibits muscular contractions of the uterus, RU486 probably has the same effect as DES, causing an expulsion of the fertilized ovum. Vocal pro-life forces in the United States have taken a strong stand against the use of this method. At this writing, three hospitals in California are seeking permission and funding to test the drug, which is already in use in France.

You may have noticed that when it comes to birth control, there are few options for men. Why not, some people ask, develop a pill for men and shift some of the contraceptive burden to the male of our species?

To be effective, a pill would have to shut down spermatogenesis. Testosterone injections would do the job since testosterone blocks the release of pituitary FSH and LH. FSH is required for spermatogenesis, and its absence would depress sperm production. Unfortunately, testosterone injections might create aggressive behavior. Excess androgen in males is converted to estrogen, causing feminizing side effects. Androgen treatments may depress spermatogenesis, but not enough to lower sperm count to a level where a couple would feel confident.

Another route is selectively inhibiting FSH secretion. The seminiferous tubules produce a substance called **inhibin**, which selectively inhibits the production of FSH by the pituitary gland. If inhibin could be produced and administered to men, it might give men an option to participate in birth control.

Many experts agree that the world's population problem will not be solved through new contraceptive technologies. What is required is a change in the attitudes of millions of men and women throughout the world. Controlling family size must be a conscious decision followed by conscientious action.

In Africa, where population is doubling in some countries every 17 years, contraceptive use is a paltry 10 to 20%. Worldwide only about half the women of reproductive age are using contraceptives. Education is needed to involve more people in a race to stem the swelling tide that, if unchecked, will add another 5 billion people to the world population in the next 40 years. Funds are needed to help pay for contraception and family planning. When one condom costs more than the average annual per capita medical expenditure, we can hardly expect widespread use. Many Third World countries, however, divert enormous amounts of money to pay for weapons and almost nothing on family planning. If the world population is to stabilize, if our children are to inherit a world worth living, many experts agree that contraceptive use must increase.

dence is steadily rising in the United States. Caused by any of several different bacteria, this infection is generally less threatening than gonorrhea or syphilis, although some infections can result in sterility. About half of the reported cases of NSU are caused by a bacterium called **chlamydia**.

Many men and women often exhibit no symptoms whatsoever and, therefore, can spread the disease without knowing it. In men, when symptoms occur, they resemble those of gonorrhea—painful urination and a cloudy mucous discharge from the penis. In women, symptoms resemble those of a urinary tract infection. Urination becomes painful and more frequent. NSU can be treated by antibiotics, but individuals should seek treatment quickly to avoid spread of the disease and more serious complications.

Syphilis

Despite its linguistic appeal, syphilis is a serious sexually transmitted disease caused by a bacterium that penetrates the linings of the oral cavity, vagina, and penile urethra or enters through breaks in the skin. If untreated, syphilis proceeds through three stages. In stage 1, between one and eight weeks of infection, a small, painless red sore develops, usually in the genital area. Easily visible when on the penis, these sores often go unnoticed when they occur in the vagina or cervix of a woman. The sore heals in one to five weeks, leaving a tiny scar.

Approximately six weeks after the sore heals, individuals complain of fever, headache, and loss of appetite. Lymph nodes in the neck, groin, and armpit swell as the bacteria spread throughout the body. This is stage 2.

The symptoms disappear for several years. Then, without warning, the disease flares up. The final stage, stage 3, is an autoimmune reaction that causes paralysis, senility, or even insanity. Individuals may lose their sense of balance and may lose sensation in their legs. The bacterium can weaken the walls of the aorta, causing an aneurysm (Chapter 8).

Fortunately, syphilis can be successfully treated with antibiotics—but only if the treatment begins early. Suspicious sores in the mouth and genitals should be brought to the attention of a physician. In the late stages, antibiotics are useless. Tissue or organ damage is permanent.

Herpes

Herpes is one of the most common sexually transmitted diseases. Approximately 200,000 to 300,000 people contract the disease each year. Herpes is caused by a virus and is essentially incurable—once the virus enters the body, it is there for life. The first sign of infection is pain, tenderness, or an itchy sensation on the penis or external genitalia, which occurs about six days after contact with someone infected by the virus. Soon afterward, painful blisters appear on the penis and female genitalia. The blisters may also form on the thighs and buttocks, in the vagina, and on the cervix.

The blisters break open and become painful ulcers, lasting for one to three weeks, and then disappear. Because the herpes virus is a lifelong resident of the body, however, new outbreaks may occur from time to time, especially when an individual is under stress. Recurrent outbreaks are generally not as severe as the initial one, and in time the outbreaks generally cease altogether.

Herpes can be spread to other individuals during sexual contact, but only when the blisters are present or (as recent research suggests) just beginning to emerge. When the virus is inactive, sexual intercourse can occur without infecting a partner.

Although herpes cannot be cured, doctors may prescribe an antiviral drug called acyclovir. Acyclovir suppresses the virus, reducing the incidence of outbreaks and accelerating healing of herpes blisters.

Women who have herpes run the risk of transferring the virus to their infants at birth. Since the herpes virus can be fatal to newborns, women are often advised to deliver by cesarean section (an incision made just above the pubic bones) if the virus is active at the time of birth.

INFERTILITY

A surprisingly large percentage of American couples cannot produce offspring. The inability to conceive (to become pregnant) is called **infertility**. About 50 to 60% of the time, infertility results from problems occurring in women; the remainder are due to problems in the male.

A couple who have been actively trying for a year or more to conceive, but have not, should see a physician. The physician will first check some of the obvious problems, such as infrequent or poorly timed sex since only intercourse around the time of ovulation will be successful. If timing is not the problem, the physician will test the male's sperm count. A low sperm count is one of the most common causes of male infertility.

Low sperm count may result from a variety of factors—overwork, emotional stress, and fatigue. Excess tobacco and alcohol consumption are also contributors. Tight-fitting clothes and excess exercise, which raise the scrotal temperature, may also reduce the sperm count. But one of the most common causes of low sperm count is an enlargement of the veins draining one or both testes, a condition called a **varicocele**. The testes are also sensitive to a wide range of chemicals and drugs, and some physicians believe that the myriad of chemicals people are exposed to in everyday life may be lowering sperm production in males (see the next section).

If infertility results from a low sperm count, a couple may choose to undergo artificial insemination, using sperm from a sperm bank. These sperm are generally acquired from anonymous donors and are stored frozen. When thawed, the sperm are reactivated, then deposited in the woman's vagina or cervix around the time she ovulates.

If sperm production and ejaculation are normal, a physician checks the woman's reproductive tract. First comes a test of ovulation. A sample of the mucus produced by the cervix and a biopsy of the uterine lining can indicate whether or not a woman is cycling. If ovulation is not occurring, **fertility drugs** may be injected. Several kinds of drugs are available. One of the more common is HCG (human chorionic gonadotropin). As we noted earlier, HCG is an LH-like hormone. Like LH, it can induce ovulation. Unfortunately, fertility drugs often result in superovulation—the ovulation of many fertilizable ova. As a result, couples often end up with four to six babies. Most of the multiple births you hear about on the news are the result of fertility drugs.

If ovulation is occurring, the problem may be an obstructed uterine tube. A previous gonorrheal and chlamydial infection may have spread into the tubes, causing scarring and obstructing the uterine tubes. In that case, a couple may be advised to try adoption or *in vitro* fertilization. During *in vitro* fertilization, ova are surgically removed from the woman, then fertilized by the partner's sperm. The fertilized ovum can be implanted in the uterus of the woman and can grow successfully to term. This procedure is expensive, time-consuming, has a low success rate, and is not currently widely available. It also places heavy emotional demands on the couple.

 ## ENVIRONMENT AND HEALTH: THE SPERM CRISIS

In September 1979, Professor Ralph Dougherty, a chemist at Florida State University, announced findings from a study of 130 healthy male college students. In his test group, Dougherty found extremely low sperm counts—only 20 million sperm per milli-liter of semen compared with an expected value of 60 to 100 million per milliliter. Biochemical analyses of the testes also revealed high levels of four toxic chemicals: DDT, polychlorinated biphenyls (PCBs), pentachlorophenol, and hexachlorobenzene.

An article in the Sierra Club's magazine proclaimed that America is facing a "sperm crisis" caused by toxic chemicals. But not everyone agrees. Health officials, in fact, have challenged these findings. Some officials contend that the low sperm counts Dougherty recorded may have resulted from improved counting techniques. Over the years, advances in technology have allowed scientists to count sperm more accurately. As a result, estimates of the normal sperm concentration have been markedly lowered. Dougherty claims that such improvements are not wholly responsible for the decline.

Other health officials argue that Dougherty's findings are not representative of the American public. Floridians, they say, may be exposed to high levels of pesticides used on farms. These chemicals may be contaminating water supplies of urban and rural residents.

Additional studies in Florida and other states show that sperm counts in American men have been falling since the 1950s. Prior to 1950, the average sperm count was about 110 million per milliliter. By 1980 and 1981, sperm counts had dropped to about 60 million per milliliter. Statistical studies suggest that the decline may be related to growing pesticide use, air pollution, and other factors.

Studies in Hawaii support the belief that certain environmental chemicals may be causing a decline in sperm count. These studies show that Hawaiian men have a considerably higher sperm count than men residing in the continental United States. This observation has been attributed to a generally cleaner environment. There are, say researchers, fewer factories on the Hawaiian islands than on the mainland. People are exposed to fewer agricultural chemicals, and frequent winds probably also keep the air cleaner.

Reductions in sperm count are of concern to many people because a sperm count below 20 million per milliliter is generally insufficient for fertilization. Today, low sperm counts account for about 40 to 50% of all infertility in U.S. couples.

Human reproduction, like other bodily processes, depends on a healthy environment. Research shows that a wide range of factors—from drugs to radiation to industrial chemicals—are toxic, or potentially toxic, to human reproduction (Table 16–4).

TABLE 16–4	Some Agents Potentially Toxic to Male and Female Reproduction

MALES	FEMALES
Natural and synthetic androgens	Natural and synthetic estrogens
Heat	Natural and synthetic progestins
Radiation	
Dioxin	Amphetamine
PCBs	DDT
Vinyl chloride	Parathion (insecticide)
Ethanol	Carbaryl (insecticide)
Benzene	Diethylstilbestrol
Diethylstilbestrol	PCBs
EDB (ethylene dibromide)	
Paraquat (herbicide)	
Carbaryl (insecticide)	
Cadmium	
Mercury	

"There has been an explosion of spermatotoxins in the environment," says Dr. Bruce Rappaport, former director of an infertility clinic in San Francisco. "The problem is environmental pollution." Today, at least 20 common industrial chemicals are known to be reproductive toxins. Ten commonly prescribed antibiotics can reduce sperm count. Even Tagamet, a drug that is used to relieve stress and is now the most prescribed drug in the United States, reduces sperm count by over 40%. By one estimate, at least 40 commonly used drugs depress sperm production, and thousands of other drugs and environmental pollutants have not been tested.

These facts do not necessarily mean that the United States is in a sperm crisis, but they do suggest the need for caution. Further research is needed to determine potential impacts, if any, of the many thousands of chemicals now commonly used or released into the environment. Research may prove that we need to clean up our act, or it may show that the fears are unwarranted. Whatever the outcome, the costs of not acting could be enormous.

SUMMARY

THE MALE REPRODUCTIVE SYSTEM

1. The male reproductive tract consists of six basic parts: (a) the testes, (b) the epididymis, (c) the vas deferens, (d) the sex accessory glands, (e) the urethra, and (f) the penis.
2. The testes lie in the scrotum, which provides a suitable temperature for sperm development.
3. Each testis contains hundreds of sperm-producing seminiferous tubules. Lying between the seminiferous tubules are the interstitial cells, which produce the male sex steroids.
4. Sperm produced in the testes are stored in the epididymis of each testis, from which they are ejaculated. During ejaculation, sperm pass from the epididymis to the vas deferens, then to the urethra. During ejaculation, secretions from the sex accessory glands are added to the sperm, forming semen.
5. Sperm are formed from spermatogonia, located in the periphery of the seminiferous tubules. Spermatogonia divide by mitosis, replenishing their supply. Some spermatogonia enlarge and become primary spermatocytes, each containing 46 double-stranded chromosomes.
6. Primary spermatocytes divide meiotically and form two secondary spermatocytes, each with 23 double-stranded chromosomes. Secondary spermatocytes divide in the second meiotic division, forming spermatids each with 23 single-stranded chromosomes.
7. Spermatids undergo cellular differentiation, forming sperm. In this process, the nuclear material condenses, an acrosome forms from the Golgi apparatus, most of the cytoplasm is lost, a tail is formed, and the mitochondria align themselves along the first section of the tail.
8. Testosterone production by the testes is controlled by a hormone from the anterior pituitary known as interstitial cell stimulating hormone (ICSH). ICSH stimulates the interstitial cells to produce testosterone. ICSH release is regulated by a hypothalamic hormone, gonadotropin releasing hormone.
9. Testosterone stimulates spermatogenesis, facial hair growth, thickening of the vocal cords, laryngeal cartilage growth, sebaceous gland secretion, and bone and muscle development.
10. The penis is the organ of copulation. It contains erectile tissue, which fills with blood during sexual arousal, making the penis turgid.
11. Ejaculation is under reflex control. When sexual stimulation becomes intense, sensory nerve impulses trigger a spinal reflex. Neurons in the spinal cord send impulses to the smooth muscle in the walls of the epididymis, the vas deferens, the sex accessory glands, and the urethra, causing the ejection of sperm and secretions of the sex accessory glands.

THE FEMALE REPRODUCTIVE SYSTEM

12. The female reproductive system consists of two basic components: the reproductive tract and the external genitalia.

13. The reproductive tract consists of (a) a pear-shaped muscular organ, the uterus, (b) two uterine tubes, (c) two ovaries, and (d) the vagina.

14. The external genitalia consist of two flaps of skin on both sides of the vaginal opening: the labia majora and the labia minora. Anteriorly, the labia minor meet to form a hood over the clitoris.

15. Female germ cells are housed in follicles in the ovary. A follicle consists of a germ cell and an investing layer of follicle cells.

16. A dozen or so follicles enlarge during each menstrual cycle, but most follicles degenerate. Usually only one follicle makes it to ovulation.

17. The oocyte and an investing layer of follicle cells are released during ovulation and drawn into the uterine tubes.

18. The menstrual cycle consists of a series of changes occurring in the ovaries, uterus, and endocrine system of women.

19. The first half of the menstrual cycle is called the follicular phase. FSH from the pituitary stimulates follicle growth and development. LH stimulates estrogen production.

20. Estrogen levels rise slowly during the follicular phase, then trigger a positive feedback mechanism that results in a preovulatory surge of FSH and LH, which triggers ovulation.

21. The ovum is expelled from the tertiary follicle at ovulation. The follicle then collapses and is converted into a corpus luteum (CL). The CL produces estrogen and progesterone.

22. In the absence of fertilization, the CL degenerates. If fertilization occurs, however, HCG (human chorionic gonadotropin) from the embryo maintains the CL for approximately six months.

23. Hormones from the pituitary control the activity of the ovary. Ovarian hormones, in turn, stimulate growth of the uterine lining during the menstrual cycle. This growth is necessary for successful implantation. If fertilization does not occur, the uterine lining is sloughed off during menstruation. Menstruation is triggered by a decline in ovarian estrogen and progesterone.

24. Like testosterone in boys, estrogen levels in girls increase at puberty. Estrogen stimulates growth of the external genitalia, the reproductive tract, and bone and stimulates the deposition of fat in women's hips, buttocks, and breasts.

25. Progesterone works with estrogen to stimulate breast development. It also promotes endometrial growth and inhibits uterine contractions.

26. Many women suffer premenstrual irritability, depression, fatigue, headaches, bloating, swelling and tenderness of the breasts, tension, and even joint pain. These are symptoms of premenstrual syndrome and appear before menstruation begins each month.

27. The menstrual cycle continues throughout the reproductive years, but after a woman reaches 20, the ovaries begin to become progressively less responsive to gonadotropins. As a result, estrogen levels slowly decline as a woman ages. Ovulation and menstruation become erratic as a woman approaches 45.

28. Between the ages of 45 and 55, ovulation and menstruation cease altogether. The end of reproductive function in women is known as the menopause.

29. The decline in estrogen levels results in atrophy of the reproductive organs and behavioral disturbances. Many women become more irritable, suffer bouts of depression, and experience hot flashes and night sweats induced by intense vasodilation of vessels in the skin.

BIRTH CONTROL

30. Birth control refers to any method or device that prevents births and includes two general strategies: contraception, ways of preventing pregnancy, and induced abortion, the deliberate expulsion of a fetus or embryo.

31. Figure 16–17 summarizes the effectiveness of the various birth control measures.

32. One of the most effective means of birth control is sterilization. In men, the vasa deferentia are cut and tied off during a procedure called a vasectomy. In women, the uterine tubes are cut and tied during a tubal ligation.

33. The pill is also a highly effective means of birth control. The most common pill in use today is the combined pill, containing a mixture of estrogen and progesterone, which inhibits ovulation. In some women, however, estrogen causes serious side effects.

34. The intrauterine device is a plastic or metal coil that is placed inside the uterus where it prevents the fertilized ovum from implanting.

35. The diaphragm, condom, and vaginal sponge are less effective than the previously described measures. The diaphragm is a rubber cap fitted over the cervix by the woman prior to sex. To be effective, it must be coated with a spermicidal jelly, foam, or cream.

36. The condom is a thin latex rubber sheath worn over the penis during sexual intercourse. It prevents sperm from entering the vagina.

37. The vaginal sponge is a tiny absorbent sponge worn by the woman. It is impregnated with a spermicidal chemical.

38. One of the oldest but least successful methods of birth control is withdrawal—removing the penis before ejaculation. Spermicidal chemicals used alone are about as effective as withdrawal.

39. Abstaining from sexual intercourse around the time of

ovulation, the rhythm or natural method, can help couples prevent pregnancy. Unfortunately, the natural method is the least successful of all birth control measures and one of the more complicated.

40. Some couples may elect to terminate pregnancy through an abortion. Abortion may be an acceptable means of family planning, but contraception is less costly and less traumatic.

41. In the first 13 weeks of pregnancy, abortions can be performed surgically in a doctor's office via vacuum aspiration. Between 13 and 16 weeks, a larger aspirator tube is required. The lining of the uterus may be scraped with a metallic instrument as well to ensure that the operation is complete.

42. After 16 weeks, abortions are more difficult and a little more risky. Solutions of salt, urea, or prostaglandins, which stimulate uterine contractions, can be injected into the fluid surrounding the fetus to induce premature labor. Oxytocin may be administered to the mother with the same effect.

SEXUALLY TRANSMITTED DISEASES

43. Certain viruses and bacteria can be spread from one individual to another during sexual relations. Infections transmitted by sexual contact are called sexually transmitted diseases.

44. Gonorrhea is caused by a bacterium that commonly infects the urethra in men and the cervical canal in women. Overt symptoms of the infection are frequently not present, so people can spread the disease without knowing they have it. If left untreated, gonorrhea can spread to other organs, causing considerable damage.

45. Nonspecific urethritis is the most common sexually transmitted disease. Caused by any of several different bacteria, it is less threatening than gonorrhea or syphilis. Many men and women show no symptoms and can spread the disease without knowing it. Symptoms, when they occur, resemble those of gonorrhea.

46. Syphilis is a serious sexually transmitted disease caused by a bacterium that penetrates the linings of the oral cavity, vagina, and penile urethra. If untreated, syphilis proceeds through three stages. It can be treated with antibiotics during the first two stages, but in stage 3, when damage to the brain and blood vessels is evident, treatment is ineffective.

47. Herpes is also a common sexually transmitted disease. It is caused by a virus and is incurable. Once the virus enters the body, it remains for life. Blisters form on the genitals and sometimes on the thighs and buttocks. The blisters break open and become painful ulcers. At this stage, an individual is highly infectious. New outbreaks of the virus may occur from time to time, especially when an individual is stressed.

INFERTILITY

48. The inability to conceive is called infertility.

49. Infertility may result from a variety of problems in men and women: poorly timed sex, low sperm count, impotence, failure to ovulate, or obstruction in the uterine tubes.

ENVIRONMENT AND HEALTH: THE SPERM CRISIS

50. A variety of drugs and chemical pollutants affect sperm development and may be causing a decline in the sperm count of American men.

EXERCISING YOUR CRITICAL THINKING SKILLS

1. Devise an experiment to test the hypothesis that the United States is in the midst of a sperm crisis—a decline in male sperm production—caused by chemicals in the environment. What type of evidence would support or refute your hypothesis? How can you devise your experiment to avoid bias?

2. You are a journalist for a major urban newspaper. You receive a press release from the local medical school announcing that one of the researchers has discovered that a common household chemical reduces fertility in rats and mice. Large doses were given to both males and females before conception. The results showed a decline in the litter size and several abnormalities. The researcher suggests that the chemical should be banned from use in homes. Using your critical thinking skills, what questions would you ask before writing your article?

TEST OF TERMS

1. The testes lie in the _____, a sac that provides a suitable temperature for the development of sperm. Sperm are produced inside the _____ _____ tubules.

2. Sperm are stored in the _____ _____ and delivered to the urethra during ejaculation via the _____, two muscular ducts.

3. The ejaculate consists of secretions from three glands, the _____ _____ glands.

4. Spermatogenic cells are found in the _____ epithelium. Spermatogenesis begins with _____. They divide mitotically. Some of these cells, however, enlarge to form the _____ _____, which undergo the first meiotic division, forming two _____ _____.

5. The second meiotic division produces four _____, each containing _____ single-stranded chromosomes.

6. The nucleus of a mature sperm is capped by an enzyme-containing structure called the _____, which helps the sperm penetrate the barriers around the ovum.

7. Male sex steroids are produced by the _____ in the testes. One of the chief steroids is _____. Its secretion is controlled by a pituitary hormone called _____ _____.

8. The glans penis is covered by a flap of skin at birth called the _____. Surgical removal of this part of the penis is called a(n) _____.

9. The penis contains _____ tissue, which fills with blood during sexual arousal.

10. _____ is under reflex control. Muscle contractions in the reproductive tract cause sperm to be propelled to the outside. As the sperm pass along the tract, they are mixed with secretions from certain glands along the way. Sperm and the secretions of these glands constitute the _____.

11. The _____ is a pear-shaped organ that houses and nourishes the developing embryo. Ova are produced by the ovaries and picked up by the _____.

12. Sperm are deposited in the _____ and make their way through the _____ canal.

13. The _____ _____ are two flaps of skin covered with hair externally; they are part of the external genitalia in women.

14. The _____ in females is formed from the same embryological tissue that forms the penis in males.

15. Release of an ovum from the ovary is called _____. It occurs at the midpoint of the _____ cycle.

16. A _____ follicle consists of two or more layers of follicle cells and a large central cavity. The primary oocyte is surrounded by a gel-like layer called the _____.

17. During oogenesis, the primary oocyte divides during the _____ meiotic division, producing a secondary oocyte and the _____ body containing _____ (give a number) _____-stranded chromosomes.

18. The _____ _____ is a structure in the ovary produced from a follicle that releases its ovum. It produces two steroid hormones, _____ and _____.

19. The first half of the menstrual cycle is called the _____ phase. LH released from the pituitary at this time stimulates _____ production by the large follicles in the ovary, while pituitary _____ stimulates follicle growth.

20. The lining of the uterus, the _____, thickens during the first half of the menstrual cycle. It continues to grow throughout the cycle, but in the absence of fertilization, it is sloughed off. The loss of blood and tissue from the lining is called _____.

21. In women, depression, irritability, and swelling and tenderness of the breast are symptoms of _____ _____.

22. A cessation of ovarian function occurs in women sometime after age 45. This is called the _____. The decline in _____ from the ovaries at this time results in behavioral changes and atrophy of the breasts and uterus.

23. Male sterilization is a surgical procedure in which the vasa deferentia are cut and is called _____. In women, sterilization is achieved by severing the _____, an operation called a(n) _____.

24. The combined birth control pill contains two synthetic female steroid hormones, _____ and _____.

25. Cervical cancer can be diagnosed by a _____ _____, a swab of the cervical lining.

26. A(n) _____ _____ is a plastic coil inserted in the uterus, which prevents pregnancy.

27. A(n) _____ is a rubber cup that fits over the cervix as it protrudes into the vagina. It is coated with _____ jelly or foam.

28. A thin latex rubber device worn over the penis during sexual intercourse to reduce disease and the chance of pregnancy is called a _____.

29. Timing sexual intercourse to avoid the deposition of sperm in the vagina near the time of ovulation constitutes the _____ method and is one of the least

effective contraceptive techniques available.

30. A woman can determine when she ovulates by taking daily _____ measurements or by examining the consistency of her _____ mucus.

31. Gonorrhea and syphilis are _____ _____ _____ caused by bacteria.

32. _____ _____ is the most common sexually transmitted disease known to medical science and is one of several STDs whose incidence is steadily rising in the United States.

33. _____ is a viral disease transmitted during sexual intercourse and results in small blisters in the genital region, thighs, and buttocks.

Answers to the Test of Terms are located in Appendix B.

TEST OF CONCEPTS

1. You are a fertility specialist. A young woman arrives in your office complaining that she has been trying to get pregnant for two years but to no avail. Describe how you would go about determining where the problem lay with her and her husband.

2. Describe the anatomy of the male reproductive system. Where are sperm produced? Where are they stored? What structures produce the semen?

3. Describe the process of spermatogenesis, noting the cell types and the number of chromosomes in each type.

4. Why is the first meiotic division called a reduction division?

5. Describe the hormonal control of testicular function. What hormones are involved, where are they produced, and what effects do they have on the testes?

6. You are a family doctor. A man comes to your office complaining of impotence. What are the possible causes? How would you go about testing for possible causes?

7. Trace the pathway for a sperm from the seminiferous tubule to the site of fertilization.

8. Describe the process of ovulation and its hormonal control.

9. What is the corpus luteum? How does it form? What does it produce? Why does it degenerate at the end of the menstrual cycle if fertilization does not occur?

10. What is menstruation?

11. Describe the effects of estrogen and progesterone on the reproductive tract and the body.

12. A woman comes to your office. She is 47 years old and complains of irritability and depression. She asks for the name of a reliable psychiatrist who could help her. She says that she wakes up in the middle of the night in a sweat. Would you give her the name of a psychiatrist? Why or why not? If not, what would you do?

13. Describe each of the following birth control measures, explaining what they are and how they work: the pill, IUD, diaphragm, cervical cap, condom, spermicidal jelly, and natural method.

14. Describe ways to prevent the spread of sexually transmitted diseases.

SUGGESTED READINGS

Boskin, W., G. Graf, and V. Kreisworth. 1990. *Health dynamics: Attitudes and behaviors*. St. Paul, Minn.: West. Excellent introductory chapters on reproduction and sexuality.

Davis, L. 1989. The myths of menopause. *Hippocrates* 3(3): 52–59. Interesting discussion of the facts and fallacies of menopause.

Mareib, E. N. 1989. *Human anatomy and physiology*. Redwood City, Calif.: Benjamin/Cummings. Excellent coverage of reproduction.

Sizer, F. S., and E. N. Whitney. 1988. *Life choices: Health concepts and strategies*. St. Paul, Minn.: West. Chapters 12 and 13

offer additional information of practical importance on reproduction and contraception.

Wassarman, P. M. 1988. Fertilization in mammals. *Scientific American* 259(6): 78–84. Looks at the process that allows only one sperm to fertilize an ovum.

17

Human Development and Aging

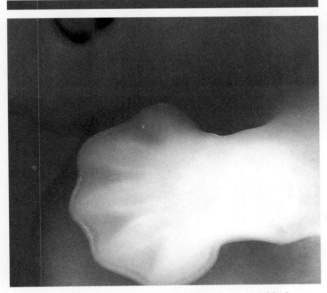

Human embryo at 40 days. The hand begins as a webbed structure. The webbing is eventually destroyed, producing fingers.

Human reproduction is one of the most fascinating of all body functions. That a single sperm and ovum can give rise to a new human being boggles the mind, raising our appreciation of the miracle of life. This chapter examines the process of human development, beginning with fertilization and culminating with a discussion of aging and death.

FERTILIZATION: A NEW LIFE BEGINS

The sperm and the ovum unite in the upper third of the uterine tube. This process, called **fertilization**, results in the formation of a zygote (Figure 17–1).

Oocytes are released from the ovary during ovulation and are drawn into the uterine tube by cilia and by fingerlike projections called **fimbriae**. Like massaging fingers, the fimbriae contract rhythmically, helping to sweep oocytes into the uterine tube.

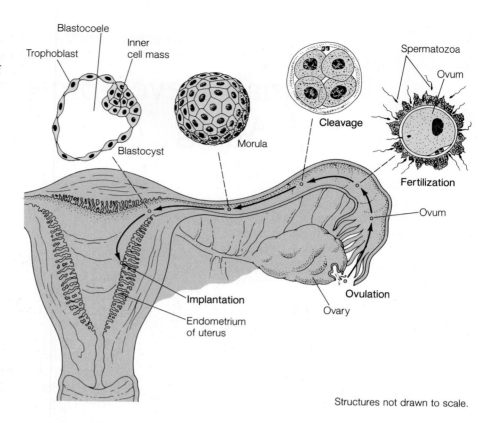

FIGURE 17–1 Fertilization and Early Embryonic Development Fertilization usually occurs in the upper third of the uterine tube. The zygote begins to divide, but remains in the uterine tube for about three days.

Blastocoele

Trophoblast

Inner cell mass

Spermatozoa

Ovum

Blastocyst

Morula

Cleavage

Fertilization

Ovum

Implantation

Ovulation

Ovary

Endometrium of uterus

Structures not drawn to scale.

Sperm are deposited in the vagina and quickly make their way up the reproductive tract (Figure 17–2). Within a few minutes of ejaculation, they enter the cervical canal; 30 minutes later, they arrive at the junction of the uterine tube and uterus, then travel to the upper reaches of the uterine tube. Along the way, many millions of sperm die, killed by acidic secretions of the vagina and cervix. Only one in a thousand reaches the uterine tube.

Although sperm are motile, the principal driving

FIGURE 17–2 Sperm Transport in the Female Reproductive System Sperm move rapidly up the female reproductive tract principally as a result of contractions in the muscular walls of the uterus and uterine tubes. Notice the rapid decline in sperm number along the way.

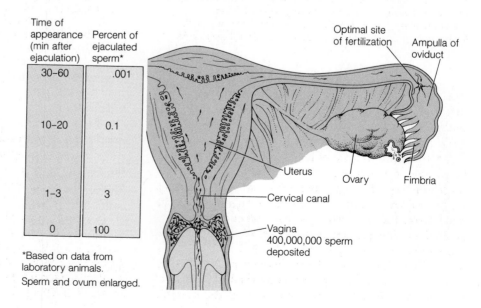

Time of appearance (min after ejaculation)	Percent of ejaculated sperm*
30–60	.001
10–20	0.1
1–3	3
0	100

*Based on data from laboratory animals.
Sperm and ovum enlarged.

Optimal site of fertilization

Ampulla of oviduct

Uterus

Ovary

Fimbria

Cervical canal

Vagina
400,000,000 sperm deposited

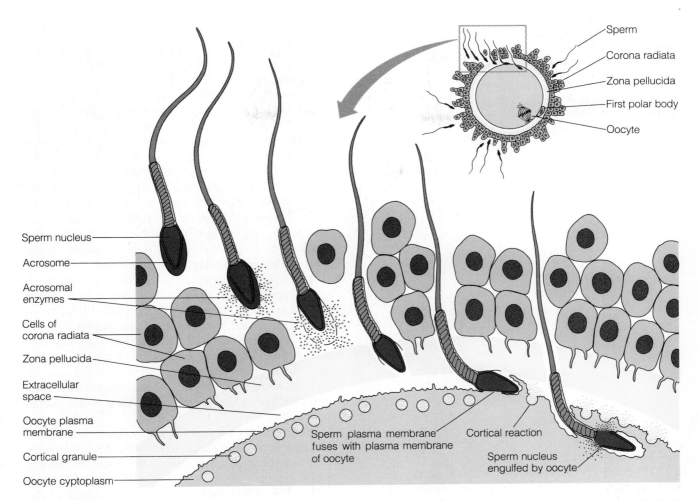

Sperm
Corona radiata
Zona pellucida
First polar body
Oocyte

Sperm nucleus
Acrosome
Acrosomal enzymes
Cells of corona radiata
Zona pellucida
Extracellular space
Oocyte plasma membrane
Cortical granule
Oocyte cyptoplasm

Sperm plasma membrane fuses with plasma membrane of oocyte

Cortical reaction

Sperm nucleus engulfed by oocyte

FIGURE 17–3 Fertilization and Cortical Reaction
After capacitation, the plasma membrane of the sperm and the outer membrane of the acrosome fuse and the membranes break down, releasing enzymes that allow the sperm to penetrate the corona radiata. Sperm digest their way through the zona pellucida via enzymes associated with the inner acrosomal membrane. Sperm are engulfed by the oocyte plasma membrane. Cortical granules are released when the sperm cell contacts the membrane. These granules cause other sperm in contact with the membrane to detach.

force is the muscular contraction of the wall of the uterus and uterine tubes. Some researchers believe that these contractions are stimulated by prostaglandins, hormonelike substances in the semen.

Sperm reach the site of fertilization within an hour or so after ejaculation, but cannot fertilize an ovum until they have been in the female reproductive tract for six to seven hours. This time is required for **capacitation**, the removal of a layer of cholesterol deposited on the plasma membranes of sperm by the secretions of the sex accessory glands. The coat helps to stabilize the sperm plasma membrane and helps protect the sperm as it moves along the female reproductive tract. Removing the coat renders the sperm's

membranes fragile and disruptible, a prerequisite for fertilization.

After capacitation, the plasma membrane over the head of the sperm fuses with the outer membrane of the acrosome, a caplike structure filled with digestive enzymes (Figure 17–3). Tiny openings develop at the points of fusion, allowing acrosomal enzymes to leak out.

As Figure 17–3 shows, the follicle cells around the ovum become elongated and radiate outward like the spokes of a wheel; the cells form the corona radiata (radiating crown). Acrosomal enzymes of the sperm swarming around the oocyte dissolve the extracellular material holding the cells of the corona radiata to-

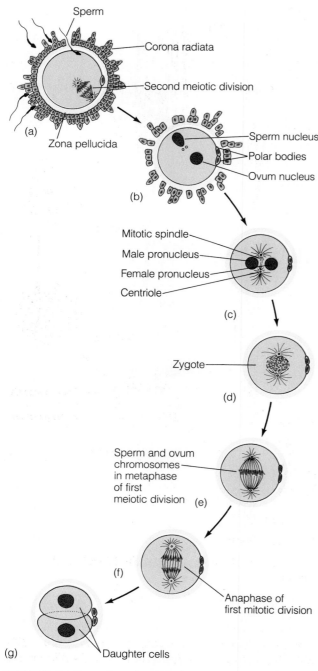

FIGURE 17–4 The Zygote Prepares for Division
(*a*) The sperm contacts the plasma membrane of the oocyte.
Second meiotic division takes place. (*b*) Sperm and oocyte
pronuclei form. (*c*) The pronuclei migrate toward the center
of the cell. Chromosomes condense and a mitotic spindle
forms. (*d*) Chromosomes condense and nuclear membrane
breaks down. (*e*) Metaphase plate is formed. (*f*) Anaphase of
the first meiotic division. (*g*) Two daughter cells form.

gether. After passing through the corona radiata,
sperm digest their way through the zona pellucida, a
gel-like layer surrounding the oocyte, with the aid of
acrosomal enzymes (Figure 17–3).

Sperm cells that traverse the zona pellucida enter
the space between the plasma membrane of the oocyte
and the zona. As a rule, the sperm cell that first comes
in contact with the plasma membrane of the oocyte is
the one that will fertilize it; all other sperm are excluded.[1] Sperm are excluded by at least two mechanisms. The first is called the **fast block to polyspermy**
(many sperm). The fast block to polyspermy occurs
when the sperm cell contacts the plasma membrane of
the oocyte, triggering membrane depolarization. For
reasons not well understood, depolarization blocks
other sperm from fusing with the oocyte.

Sperm are also excluded by the **slow block to
polyspermy**. Sperm contact triggers the release of
cortical granules, membrane-bound vesicles lying beneath the plasma membrane of the oocyte (Figure
17–3). Enzymes released from the lysosomelike cortical granules cause the zona pellucida to harden and
may block other sperm from reaching the oocyte.
Cortical secretions may also cause sperm that may
have attached to the plasma membrane of the oocyte
to detach.

Sperm contact triggers the second meiotic division,
thus converting the secondary oocyte into an ovum.
Once the sperm nucleus enters, the ovum becomes a
zygote.

Sperm do not pierce the plasma membrane of the
oocyte, but rather are phagocytized by oocytes. Inside
the oocyte, the sperm cell nucleus, containing 23
single-stranded chromosomes, swells. At this stage,
the nuclei of the ovum and sperm are referred to as
the male and female **pronuclei** (Figure 17–4). The
chromosomes in the pronuclei soon begin to duplicate, and the pronuclei move toward the center of the
ovum in preparation for the first mitotic division. A
mitotic spindle assembles in the zygote as the chromosomes duplicate. After duplication, the chromosomes begin to condense, and the membranes of the
pronuclei disintegrate. The spindle fibers attach to the
chromosomes, and the chromosomes line up on the
equatorial plate. The zygote is now ready for the first
of many mitotic divisions.

---

[1]If more than one sperm penetrates, the resulting cell would probably be unable to divide successfully.

PRE-EMBRYONIC DEVELOPMENT

Human development is divided into three stages: pre-embryonic, embryonic, and fetal. This section explores pre-embryonic development.

Pre-embryonic development begins at fertilization and ends at implantation. During this phase, the zygote undergoes rapid cellular division and is converted into a tiny ball of cells, the **morula**, not much bigger than the fertilized ovum (Figure 17–5). The morula is nourished by secretions from glands in the uterine tubes. Approximately three to four days after ovulation, the morula passes into the uterus.

Fluid soon begins to accumulate in the morula, converting it into a **blastocyst**, a hollow sphere of cells slightly larger than the morula. As Figure 17–5 shows, the blastocyst consists of two parts: a clump of cells, the **inner cell mass**, which will become the embryo, and a ring of flattened cells called the **trophoblast** (meaning "to nourish" the "blastocyst"). The trophoblast gives rise to the embryonic portion of the **placenta**, an organ that supplies nutrients to the growing embryo and removes wastes.

The blastocyst remains unattached in the uterine lumen for about two to three days. During this period, the blastocyst is nourished by secretions produced by uterine glands. If the embryo is to survive, however, it must implant in the wall of the uterus.

Implantation: The Blastocyst Invades

The blastocyst attaches to the uterine lining and digests its way into the endometrium. Called **implantation**, this process begins about six to seven days after ovulation.

Most embryos implant high on the back wall of the uterus. The cells of the trophoblast contact the endometrium and adhere to it, but only if the uterine lining is healthy and properly primed by estrogen and progesterone (Figure 17–6a). If the endometrium is not ready or is "unhealthy"—for example, because of the presence of an IUD or a endometrial infection—the blastocyst will not implant. Blastocysts may also fail to implant if their cells contain genetic mutations. Unable to bind, the blastocyst perishes and will be either resorbed (phagocytized by the cells of the endometrium) or shed during menstruation.

In the region of attachment, cells of the endometrium enlarge. Enzymes released by the trophoblast digest a minute hole in the endometrium, and the

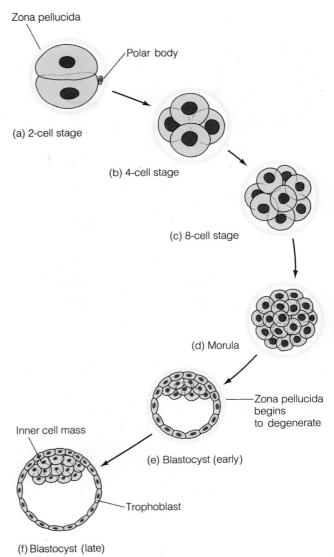

(a) 2-cell stage

(b) 4-cell stage

(c) 8-cell stage

(d) Morula

(e) Blastocyst (early)

(f) Blastocyst (late)

FIGURE 17–5 Formation of the Morula and Blastocyst

blastocyst "bores" its way into the lining (Figure 17–6b). Nutrients released from the digested cells serve as nourishment for the blastocyst, helping sustain it before the placenta forms.

Endometrial cells respond to the invasion of the blastocyst by producing prostaglandins. Prostaglandins increase the development of uterine blood vessels and, therefore, ensure an ample supply of blood and nutrients for the blastocyst. By day 14, the uterine

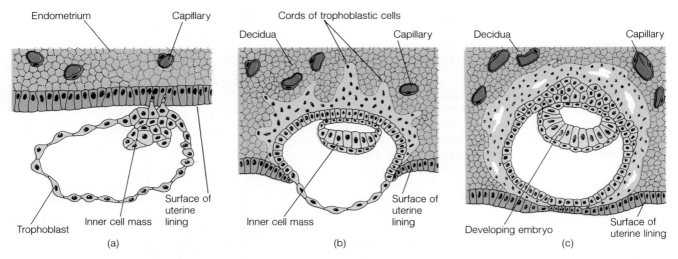

Endometrium   Capillary
Trophoblast   Inner cell mass   Surface of uterine lining
(a)

Cords of trophoblastic cells
Decidua   Capillary
Inner cell mass   Surface of uterine lining
(b)

Decidua   Capillary
Developing embryo   Surface of uterine lining
(c)

FIGURE 17–6 Implantation (*a*) The blastocyst fuses with the endometrial lining. Endometrial cells proliferate forming the decidua. (*b*) The blastocyst digests its way into the endometrium. Cords of trophoblastic cells invade, digesting maternal tissue and providing nutrients for the developing blastocyst. (*c*) The blastocyst soon becomes completely embedded in the endometrium.

endometrium grows over the blastocyst, walling it off from the uterine lumen (cavity).

Placental Development: Ensuring Survival

Placenta is the Latin word for cake. Shaped somewhat like a cake, the placenta forms from maternal and embryonic tissue. Soon after implantation begins, the outer layer of the trophoblast proliferates and begins to invade the endometrium (Figure 17–7). Cavities form among the cells of the outer layer as it digests its way into the uterine lining. During its inward march, this layer of cells severs the walls of maternal blood capillaries. Blood pours out of the capillaries and fills the cavities.

As Figure 17–7c shows, the inner layer of the trophoblast invades some time later, forming fingerlike projections called placental villi. The **placental villi** carry blood vessels, which absorb nutrients from the pools of blood formed in the outer layer of the trophoblast. The blood vessels of the placental villi connect to the developing embryo, providing a route for nutrients to flow from the maternal blood to the embryo.

The placental villi grow and divide, increasing the surface area for absorption. As they grow, they continue to invade the maternal tissue. At the same time, the blood-filled cavities enlarge, forming even bigger pools of maternal blood. Most villi project into the cavities where they absorb nutrients and expel embry-

onic wastes, such as carbon dioxide and urea. Since the walls of the villi are thin, wastes and nutrients can diffuse back and forth between the embryonic and maternal blood with ease. Some villi span the blood-filled cavities and anchor the embryonic portion of the placenta to the maternal tissue.

Besides providing nutrients and getting rid of embryonic wastes, the placenta produces a variety of hormones needed to maintain pregnancy. The placenta, therefore, is a respiratory, nutritive, excretory, and endocrine organ.

Placental Hormones and Their Role in Pregnancy

This section discusses three of the placenta's hormones: human chorionic gonadotropin, estrogen, and progesterone (Table 17–1).

FIGURE 17–7 Placental Formation (right)
(*a*) Invasion of the maternal tissue by the outer layer of the trophoblast. (*b*) Invasion continues. Cavities form. Note the presence of extraembryonic mesoderm from which blood vessels and blood cells will form. (*c*) The inner layer of the trophoblast and blood vessels invade, forming placental villi. (*d*) Fully formed placenta showing rich vascular supply. (*e*) Enlarged view showing the relationship of maternal and fetal blood vessels in the placenta. Note that the fetal blood vessels are in the placental villus, which is bathed in a pool of maternal blood.

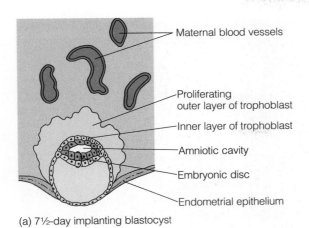

Maternal blood vessels

Proliferating outer layer of trophoblast

Inner layer of trophoblast

Amniotic cavity

Embryonic disc

Endometrial epithelium

(a) 7½-day implanting blastocyst

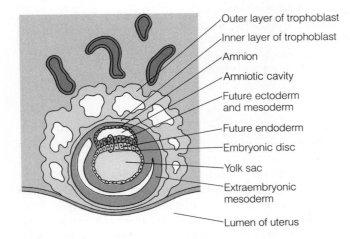

Outer layer of trophoblast

Inner layer of trophoblast

Amnion

Amniotic cavity

Future ectoderm and mesoderm

Future endoderm

Embryonic disc

Yolk sac

Extraembryonic mesoderm

Lumen of uterus

(b) 9-day implanted blastocyst

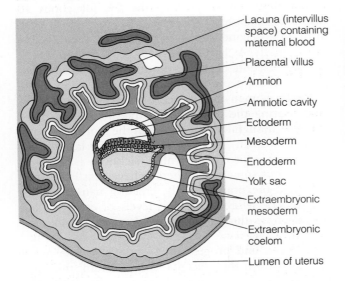

Lacuna (intervillus space) containing maternal blood

Placental villus

Amnion

Amniotic cavity

Ectoderm

Mesoderm

Endoderm

Yolk sac

Extraembryonic mesoderm

Extraembryonic coelom

Lumen of uterus

(c) 16-day embryo

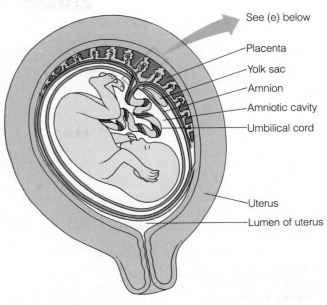

See (e) below

Placenta

Yolk sac

Amnion

Amniotic cavity

Umbilical cord

Uterus

Lumen of uterus

(d) 13-week fetus

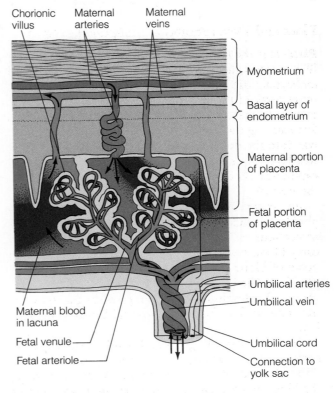

Chorionic villus   Maternal arteries   Maternal veins

Myometrium

Basal layer of endometrium

Maternal portion of placenta

Fetal portion of placenta

Umbilical arteries

Umbilical vein

Umbilical cord

Connection to yolk sac

Maternal blood in lacuna

Fetal venule

Fetal arteriole

(e) Placental structure

TABLE 17–1   Hormones Produced by the Placenta

HORMONE	FUNCTION
Human chorionic gonadotropin	Maintains corpus luteum of pregnancy Stimulates secretion of testosterone by developing testes in XY embryos
Estrogen (also secreted by corpus luteum of pregnancy)	Stimulates growth of myometrium, increasing uterine strength for parturition Helps prepare mammary glands for lactation
Progesterone (also secreted by corpus luteum of pregnancy)	Suppresses uterine contractions to provide quiet environment for fetus Promotes formation of cervical mucus plug to prevent uterine contamination Helps prepare mammary glands for lactation
Human chorionic somatomammotropin	Helps prepare mammary glands for lactation Believed to reduce maternal utilization of glucose so that greater quantities of glucose may be shunted to the fetus
Relaxin (also secreted by corpus luteum of pregnancy)	Softens cervix in preparation for cervical dilation at parturition Loosens connective tissue between pelvic bones in preparation for parturition

Source: Reprinted by permission from Table 20–5, p. 750 of *Human Physiology: From Cells to Systems* by Lauralee Sherwood. Copyright © 1989 by West Publishing Company. All rights reserved.

FIGURE 17–8   Blood Levels of Placental Hormones
Human chorionic gonadotropin levels peak in the second month of pregnancy, then drop off by the end of the third month. Levels of estrogen and progesterone, produced chiefly by the placenta, continue to rise.

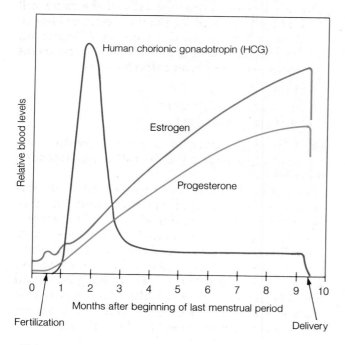

The very first hormone released by the placenta is HCG or human chorionic gonadotropin (Chapter 16). HCG is an LH-like hormone produced by the embryo early in pregnancy. HCG prevents the corpus luteum (CL) from deteriorating and stimulates estrogen and progesterone production by the CL for about 10 weeks. As shown in Figure 17–8, HCG levels peak during the second month of pregnancy, then decline fairly rapidly. As HCG levels fall, the CL degenerates, and the placenta begins to produce estrogen and progesterone, supplanting the CL.

Many women experience nausea (morning sickness) during the first two to three months of pregnancy. Although it often occurs in the morning hours, in some women "morning sickness" may last all day. The exact cause of morning sickness is not yet known, but some researchers believe that HCG may stimulate the brain directly, creating nausea. Other researchers believe that high levels of estrogen and progesterone during pregnancy are responsible for nausea.

Estrogen and progesterone—first from the ovary and then from the placenta—have a number of important functions. These hormones, for example, stimulate the growth of the uterine endometrium. Estrogen by itself stimulates growth of the smooth mus-

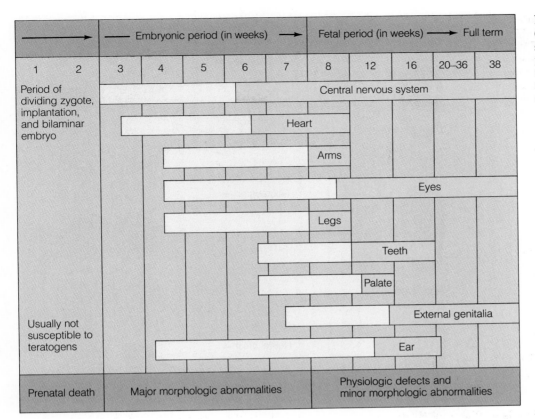

FIGURE 17–9 Organogenesis The shaded area indicates the periods most sensitive to teratogenic agents (agents that can cause birth defects).

Chart labels:

- Embryonic period (in weeks) ⟶
- Fetal period (in weeks) ⟶ Full term
- Weeks: 1 2 3 4 5 6 7 8 12 16 20–36 38
- Period of dividing zygote, implantation, and bilaminar embryo
- Central nervous system
- Heart
- Arms
- Eyes
- Legs
- Teeth
- Palate
- External genitalia
- Ear
- Usually not susceptible to teratogens
- Prenatal death
- Major morphologic abnormalities
- Physiologic defects and minor morphologic abnormalities

cle cells of the myometrium, allowing the uterus to expand many times its original size during pregnancy. Additional muscle helps to expel the baby at birth.

Progesterone helps calm the uterine musculature during pregnancy, preventing premature expulsion of the embryo and fetus. Progesterone also stimulates mucus production by the cervix. Mucus plugs the cervical canal, preventing bacteria from entering.

The Amnion Forms

As the placenta begins to form, the inner cell mass (ICM) of the blastocyst undergoes some remarkable changes of its own. As Figure 17–7 shows, early in development a layer of cells separates from the ICM to form the **amnion**. A small cavity, the **amniotic cavity**, forms between the ICM and the amnion. The amniotic cavity fills with a watery fluid known as **amniotic fluid**, which helps protect the baby during development.[2]

[2]A portion of the amniotic fluid is produced by the fetus. The rest apparently comes from the amniotic membranes.

EMBRYONIC AND FETAL DEVELOPMENT

After the amnion forms, the cells of the inner cell mass differentiate, forming the three germ cell layers: ectoderm, mesoderm, and endoderm (Chapter 6). The formation of the three primary germ layers marks the beginning of **embryonic development**.

Embryonic Development: Differentiation Begins

The primary germ layers of the embryo give rise to the organs of the body in a process called **organogenesis** (Figure 17–9).

One of the first events of organogenesis is the formation of the spinal cord and brain, which are produced from ectoderm (Table 17–2). Early in embryonic development, the ectoderm invaginates (folds inward) and forms a groove, the **neural groove**, which runs the length of the embryo. The neural groove deepens and eventually closes off, forming the **neural tube** (Figure 17–10). The walls of the neural tube thicken, eventually forming the spinal cord. Anteriorly, the neural tube expands to form the brain.

FIGURE 17–10   Formation of the Spinal Cord (a) A cross section of a 17-day embryo showing the relationship between the three embryonic tissues and the amnion and yolk sac. (b) The formation of the neural groove from ectoderm. (c) The neural groove deepens. (d) The neural tube forms. Note the presence of the neural crest, ectodermal cells that give rise to nerves.

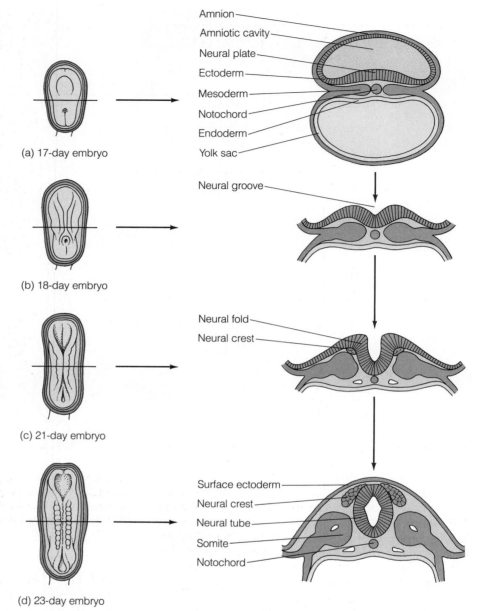

(a) 17-day embryo

(b) 18-day embryo

(c) 21-day embryo

(d) 23-day embryo

Amnion
Amniotic cavity
Neural plate
Ectoderm
Mesoderm
Notochord
Endoderm
Yolk sac

Neural groove

Neural fold
Neural crest

Surface ectoderm
Neural crest
Neural tube
Somite
Notochord

The nerves that attach to the spinal cord (spinal nerves) and brain (cranial nerves) develop from ectodermal cells lying alongside the neural tube (Figure 17–10). These cells sprout processes that extend into the body and attach to organs, muscle, bone, and skin, among others.

The middle germ layer, the mesoderm, gives rise to muscle, cartilage, bone, and other structures. Much of the mesoderm first aggregates in blocks, called the **somites**, situated alongside the neural tube (Figure 17–11). The somites form the vertebrae (the backbone) and the muscles of the neck and trunk. Mesoderm lateral to the somites becomes the dermis of the skin, connective tissue, and the bones and muscles of the limbs.

The endoderm, the "lowermost" germ layer of the inner cell mass, forms a large pouch under the embryo called the **yolk sac** (Figure 17–12). In human embryos, the blood cells first form in the wall of the yolk sac. The primitive germ cells, called **primordial**

TABLE 17-2	End Products of Embryonic Germ Layers
Ectoderm	Epidermis
	Hair, nails, sweat glands
	Brain and spinal cord
	Cranial and spinal nerves
	Retina, lens, and cornea of eye
	Inner ear
	Epithelium of nose, mouth, and anus
	Enamel of teeth
Mesoderm	Dermis
	All muscles of the body
	Cartilage
	Bone
	Blood
	All other connective tissue
	Blood vessels
	Reproductive organs
	Kidneys
Endoderm	Lining of the digestive system
	Lining of the respiratory system
	Urethra and urinary bladder
	Gallbladder
	Liver and pancreas
	Thyroid gland
	Parathyroid gland
	Thymus

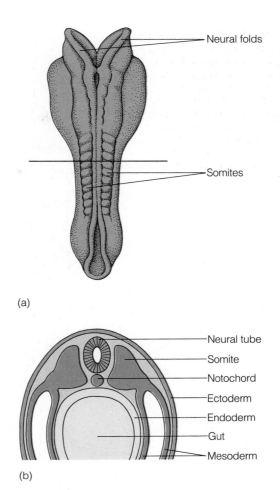

FIGURE 17-11   Somites   (a) Blocks of mesoderm lying lateral to the neural tube in the embryo give rise to the vertebrae of the spinal column. (b) A cross section through the embryo showing the relationship between the somites and the neural tube.

germ cells, originate here as well. These cells migrate from the wall of the yolk sac via ameboid action to the developing testes and ovaries. Here they become spermatogonia or oogonia, depending on the sex of the embryo. The uppermost part of the yolk sac becomes the lining of the intestinal tract.

Fetal Development

As shown in Figure 17-9, organogenesis is well on its way by the end of the eighth week of development. All of the organ systems have begun to develop, and some—like the urinary system and the circulatory system—are fully operational.

Fetal development begins in the ninth week and ends at birth. It involves two basic processes: (1) continued organ development and growth and (2) changes in body proportions-—for example, elongation of the limbs.

The fetus grows rapidly during the fetal period, increasing from about 2.5 centimeters (1 inch) to 35 to 50 centimeters (14 to 21 inches) and increasing in weight from 1 gram to 3000 to 4000 grams. During this period, the fetus also undergoes considerable change in physical appearance.

Fetal Circulation

Figure 17-13 illustrates the fetal circulatory system. Fetal circulation is similar to adult circulation with the exception of several bypasses that divert blood around various organs. As the figure illustrates, in the fetus, blood travels to and from the placenta in the

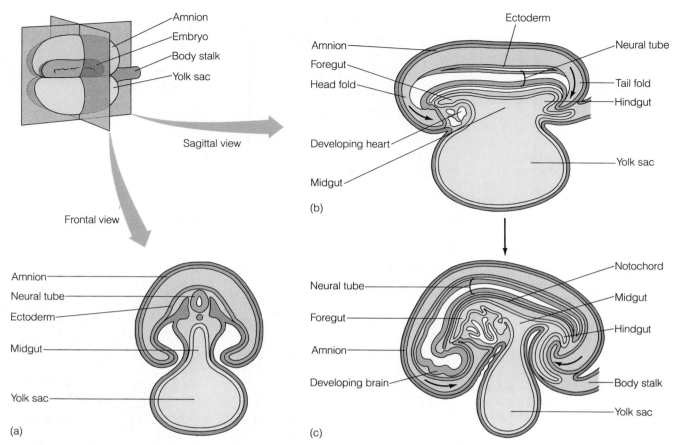

FIGURE 17–12 The Yolk Sac (*a*) A cross section of the embryo. The yolk sac forms from embryonic endoderm. (*b*) A longitudinal section showing how the upper end of the yolk sac forms the embryonic gut, (*c*) which will become the lining of the intestinal tract.

umbilical cord. The umbilical cord contains two **umbilical arteries** and a single **umbilical vein**. The umbilical vein carries oxygen- and nutrient-rich blood from the placenta to the fetus. Some of this blood flows into and through the fetal liver, as shown in Figure 17–13. From the liver, the blood passes into the inferior vena cava and on to the heart. Much of the blood, however, bypasses the liver in the **ductus venosus**, a shunt that connects from the umbilical vein to the inferior vena cava.

Blood from the inferior vena cava flows into the right atrium of the heart. In an adult, the blood follows one path into the right ventricle. In the fetus, however, blood follows two paths. Part of the blood flows from the right atrium into the right ventricle. From the right ventricle, it is pumped to the developing lungs via the pulmonary artery, supplying the

growing tissue with oxygen and nutrients and picking up waste products of cellular metabolism. The rest of the blood in the right ventricle is shunted through a hole in the interatrial septum, known as the **foramen ovale**. Blood entering the left atrium passes into the left ventricle, then is delivered to the head and the rest of the body through the aorta and its branches.

The foramen ovale closes at birth in most infants, establishing the adult circulation pattern (Chapter 8). In some children, however, the foramen ovale remains open and must be corrected surgically. Babies born with an open foramen ovale often appear bluish, due to the lack of oxygen in their blood.

Figure 17–13 also illustrates a third shunt, the ductus arteriosus. The **ductus arteriosus** lies between the pulmonary artery and the aorta and diverts blood from the lungs. The ductus arteriosus and the fora-

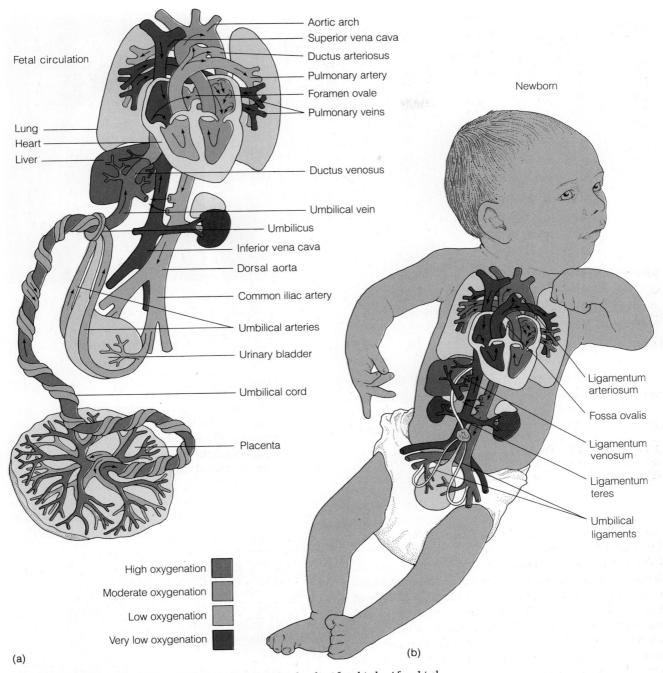

Fetal circulation

Aortic arch
Superior vena cava
Ductus arteriosus
Pulmonary artery
Foramen ovale
Pulmonary veins

Lung
Heart
Liver

Ductus venosus

Umbilical vein

Umbilicus

Inferior vena cava

Dorsal aorta

Common iliac artery

Umbilical arteries

Urinary bladder

Umbilical cord

Placenta

High oxygenation
Moderate oxygenation
Low oxygenation
Very low oxygenation

(a)

Newborn

Ligamentum arteriosum

Fossa ovalis

Ligamentum venosum

Ligamentum teres

Umbilical ligaments

(b)

FIGURE 17–13 Fetal Circulation (*a*) Before birth. (*b*) After birth. After birth the umbilical arteries and umbilical vein in the fetus will shrivel and disappear.

men ovale help to reduce blood flow to the lungs. The lungs will need this blood when the baby is born and the lungs start operating, but during fetal development the lungs' requirement for blood is small.

Blood pumped to the body through the aorta and its branches returns to the heart via the inferior vena cava. Fetal blood, however, must also recirculate to the placenta. A return route, shown in Figure 17–13,

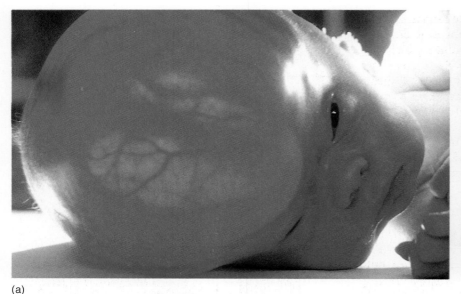

(a)

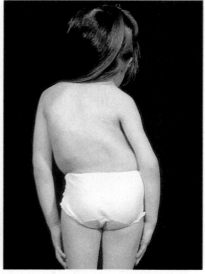

(b)

FIGURE 17–14 Common Birth Defects (*a*) Hydrocephalus, or "water on the brain," is caused by an enlargement of the brain's ventricles. (*b*) Scoliosis, a lateral curvature of the spine.

is provided by the umbilical arteries. The **umbilical arteries** are branches of the large arteries of the legs (the iliac arteries) and carry blood to the placenta where it is recharged with nutrients and cleansed of much of its waste.

At birth, the fetal circulation pattern is replaced by the adult pattern. The umbilical cord is tied off and cut, and the umbilical arteries and umbilical veins inside the fetus shrivel and disappear. The ductus venosus also vanishes, and the foramen ovale closes.

Ectopic Pregnancy

Ectopic pregnancy or **tubal pregnancy** occurs when a fertilized ovum develops outside the uterus, usually in the uterine tube. The zygote implants in the lining of the uterine tube. Placental development damages the uterine tubes, resulting in internal bleeding and constant, severe abdominal pain. The uterine tube cannot sustain the embryo, and pregnancy cannot continue. Surgery is required to remove the embryo.

About 1 in every 200 pregnancies is ectopic. Usually diagnosed within the first two months, they occur as a result of congenital defects in the oviducts or scar tissue from a previous infection, which impedes the transport of the blastocyst to the uterus.

Miscarriage and Birth Defects

Embryonic and fetal development are complex processes that can be altered by outside influences. Thirty-one percent of all fertilizations, in fact, end in miscarriage—spontaneous abortion. Two of every three of these miscarriages occur before a woman is even aware that she is pregnant. Why such a high rate? Biologists hypothesize that early miscarriage is nature's way of "discarding" defective embryos. It is an evolutionary mechanism that ensures a healthier population.

Nevertheless, humans experience a surprisingly high rate of **birth defects**, physical or physiological defects in newborns. By various estimates, 10 to 12% of all newborns have some kind of birth defect, ranging from minor biochemical or physiological problems that are not even noticed at birth to gross defects (Figure 17–14).[3] The study of birth defects is called **teratology** (*teratos* is Greek for monster). Birth defects arise from a variety of chemical, biological and physical agents, collectively known as **teratogens**. Table 17–3 lists known teratogens in humans.

The effect of teratogenic agents is related to the time of exposure, the nature of the agent, and the

[3]About 2% of all births show gross malformations.

dose. Consider time first. Since the organ systems develop at different times during embryonic development, the timing of exposure determines which systems are affected by a teratogen. Organ systems are usually most sensitive to potentially harmful agents early in their development, as indicated by the colored bars in Figure 17–9. The central nervous system, for example, begins to develop during the third week of **gestation** or pregnancy. The teeth, palate, and genitalia, however, do not begin to form until about the sixth or seventh week of pregnancy. Exposure to a teratogen during the seventh week would affect the teeth, palate, and genitalia, but might have little effect on the central nervous system.

The nature of the teratogen also influences the outcome of exposure. Some teratogens are broad-acting agents—that is, they affect a variety of developing systems. Others are narrow-spectrum agents, affecting only one system. Ethanol is a broad-acting teratogen found in beer, wine, and hard liquor. Consequently, children who are born to alcoholic mothers—or even women who have consumed one or two drinks early in pregnancy—may have a variety of birth defects, including facial, heart, and skeletal defects. Other teratogenic agents are more selective, acting on one system. Methyl mercury, for instance, damages the central nervous system, creating defects in the brain and spinal cord, but has little effect on other systems.

Finally, consider the dose. Like most toxins, teratogens generally follow a dose-response relationship: the greater the dose, the greater the effect.

One of the best-known teratogens is not a chemical agent, but rather a biological agent, a virus that causes German measles. German measles or rubella is less common and less contagious than ordinary measles. However, if a pregnant woman contracts the disease during the first three months of pregnancy, she has a one in three chance of giving birth to a baby with a serious birth defect. Deafness, cataracts, heart defects, and mental retardation are common defects. Fifteen percent of all babies with congenital rubella syndrome die within one year after birth.

Many birth defects can be avoided. German measles vaccine can be given to girls and may protect them throughout their childbearing years. It is also important to vaccinate boys so they do not spread the virus. Special precautions should be taken to avoid alcohol and other potential teratogens during the first eight weeks of pregnancy when the organ systems are developing and are most vulnerable. Unfortunately,

TABLE 17–3 Known and Suspected Human Teratogens	
KNOWN AGENTS	POSSIBLE OR SUSPECTED AGENTS
Progesterone	Aspirin
Thalidomide	Certain antibiotics
Rubella (German	Insulin
Measles)	Antitubercular drugs
Alcohol	Antihitamines
Irradiation	Barbiturates
	Iron
	Tobacco
	Antacids
	Excess vitamins A and D
	Certain antitumor drugs
	Certain insecticides
	Certain fungicides
	Certain herbicides
	Dioxin
	Cortisone
	Lead

most women do not know they are pregnant for two to four weeks after conception. Because of this, a woman who is trying to get pregnant should control what she eats and drinks.

Good nutrition and a healthy environment are also essential to fetal development. A number of physical and chemical agents in our homes and places of work are potentially toxic to the fetus. Fetal toxins can stunt growth and, in higher quantities, can actually kill a fetus.

MATERNAL CHANGES DURING PREGNANCY

During pregnancy a woman's body undergoes dramatic (largely reversible) changes. The uterus, for example, enlarges considerably, increasing to about 20 times its original weight (excluding the weight of the fetus and placenta). Originally about the size of a fist, the uterus (with its contents) eventually occupies

FIGURE 17–15 Growth of the Uterus during Pregnancy

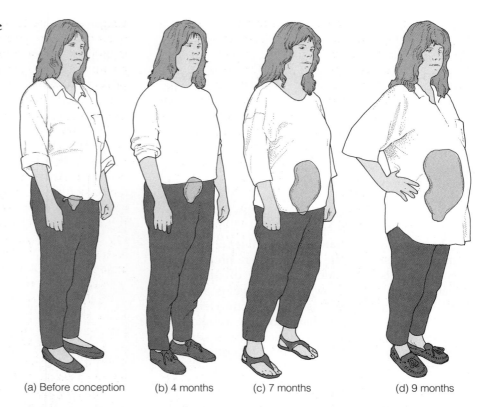

(a) Before conception   (b) 4 months   (c) 7 months   (d) 9 months

most of the pelvic and abdominal cavities. The enlarged uterus pushes up on the diaphragm and rib cage, sometimes making breathing difficult (Figure 17–15). The enlarged uterus also presses against the abdominal organs, such as the stomach. This may cause acid from the stomach to percolate up into the esophagus, creating heartburn. The uterus also pushes out on the abdominal wall, which accentuates the normal curvature of the lower back and strains the ligaments of the spinal column, resulting in back pain in the last month or two of pregnancy.

During pregnancy, the uterus and vagina become engorged with blood. In some women, the rise in vaginal vascularity increases its sensitivity to touch, heightening sexual pleasure. The breasts also enlarge during pregnancy. Glands and ducts grow dramatically and are responsible for the increase in the size of the breasts. These changes help prepare the breasts for milk production or **lactation** (described below).

A woman's blood volume also increases during pregnancy by about 30%. On average women gain about 20 to 30 pounds during pregnancy. Weight gain results from the increase in blood volume, the growth of the fetus and placenta, and the enlargement of the uterus and breasts (Table 17–4).

During pregnancy, a woman is eating, breathing, and excreting (urinating) for two living beings. Thus, three additional changes are seen during pregnancy: an increase in appetite, an increase in the rate of breathing, and an increase in urinary output. A woman produces more urine, but also urinates more frequently, in part, because the uterus also pushes against the bladder, reducing its ability to expand.

TABLE 17–4   Weight Gain during Pregnancy

AVERAGE WEIGHT GAIN	POUNDS
Total	24
Fetus	7–8
Amniotic fluid, placenta, and fetal membranes	4–5
Uterus	2
Blood	6–7
Breasts	2

The embryo and fetus are akin to an internal parasite, acquiring nutrients from its host, sometimes at the host's expense. If the fetus needs calcium, for example, it "robs" its mother's bones. If it needs iron, it generally gets what it needs, even if the mother becomes anemic in the process. This is not to imply that a fetus always acquires sufficient nutrients from its mother. Maternal dietary deficiencies can result in fetal deficiencies. If a mother's diet contains inadequate protein, for example, fetal growth may be stunted. Accordingly, pregnant women are advised to eat carefully during pregnancy—not only for their own nutrition but for the nourishment of their offspring.

After the baby is delivered, the uterus shrinks, and the blood volume returns to normal. The breasts remain engorged and begin producing milk, but if the baby is not breast fed, the breasts will return to normal in a matter of weeks. Many women are able to return to their prepregnancy weight.

CHILDBIRTH AND LACTATION

For most expectant parents, the birth of a child is one of the most exciting events of their lives. Childbirth or labor begins with mild uterine contractions, which increase in intensity and frequency.

Initiating the Contractions

Uterine muscle contractions are stimulated by a change in hormonal levels. During the last few weeks of pregnancy, estrogen levels increase dramatically. Placental estrogen stimulates the production of oxytocin receptors in the smooth muscle cells of the uterus. Oxytocin stimulates uterine muscle contractions. The increase in the number of receptors renders the muscle cells more responsive to oxytocin from the posterior pituitary.

Researchers also believe that high levels of estrogen may negate the quieting influence of placental progesterone. As a result, the uterus begins to contract at irregular intervals. Irregular contractions are called **Braxton Hicks contractions** or, more commonly, false labor, for they begin a month or two before childbirth. Uterine contractions occur more frequently as the due date approaches and have caused many a couple to race to the hospital only to be told to go home and wait for several weeks for the real thing.

More intense and more frequent contractions that occur at the beginning of true labor are thought to be triggered by the fetus itself. The fetus releases small quantities of oxytocin before birth. Fetal oxytocin stimulates contractions in the sensitized uterine musculature. Fetal oxytocin also stimulates the release of prostaglandins by the placenta. Prostaglandins act in concert with fetal oxytocin stimulating powerful and more frequent uterine contractions.

Emotional and physical stress in the mother may also play a role in triggering childbirth. Stress, resulting from uterine contractions and discomfort, is believed to trigger maternal oxytocin release. Maternal oxytocin, in turn, augments muscle contractions that are already underway (Figure 17–16). Thus, a positive feedback mechanism is triggered. As uterine muscle contraction increases, maternal oxytocin is released. This stimulates even stronger contractions, resulting in additional oxytocin release—a cycle that continues until the baby is delivered.

Childbirth also requires the hormone relaxin. **Relaxin** is produced by the CL and the placenta. Released near the end of gestation, relaxin has two effects. First, it softens the fibrocartilage uniting the pubic bones (Chapter 14). This allows the pelvic cavity to widen and thus greatly facilitates childbirth. Relaxin also softens the cervix, allowing it to expand to allow the baby to pass.

Stages of Childbirth

Childbirth occurs in three stages. Stage 1, the **dilation stage**, begins when uterine contractions start and generally lasts 6 to 12 hours, but is sometimes much longer. At the beginning of stage 1, uterine contractions usually last only 30 seconds and may come every half hour or so. However, as time passes, uterine contractions become more frequent and powerful.

Contractions are responsible for the progressive dilation of the relaxin-softened cervix. The infant's head is pushed against the cervix, causing it to thin (Figure 17–17b). By the end of stage 1, the cervix has dilated to about 10 centimeters (4 inches)—about the diameter of a baby's head—in preparation for delivery. Uterine contractions also rupture the amnion, causing it to release a flood of amniotic fluid. This event is commonly called "breaking the water."

The baby is pushed downward by the uterine contractions and descends into the pelvic cavity during the dilation stage. When the head is "locked" in the pelvis, the baby is said to be **engaged**.

Stage 2, the **expulsion stage**, begins after the cervix is dilated to 10 centimeters and the baby is engaged

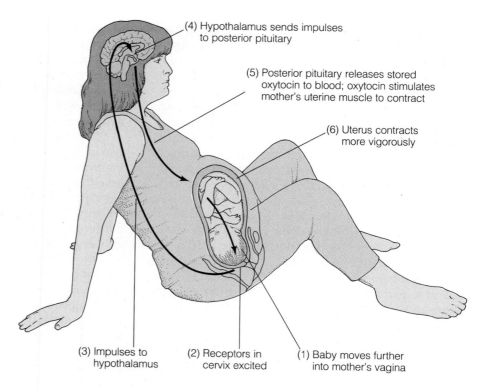

FIGURE 17–16 Oxytocin Positive Feedback Mechanism in Birth The positive feedback mechanism continues to cycle until interrupted by the birth of the baby.

(4) Hypothalamus sends impulses to posterior pituitary

(5) Posterior pituitary releases stored oxytocin to blood; oxytocin stimulates mother's uterine muscle to contract

(6) Uterus contracts more vigorously

(3) Impulses to hypothalamus

(2) Receptors in cervix excited

(1) Baby moves further into mother's vagina

(Figure 17–17c). By the time the cervix is fully dilated, uterine contractions are occurring every 2 or 3 minutes and last 1 to 1.5 minutes. For most women having their first child, 50 to 60 minutes are required for childbirth. If a woman is delivering her second child, only 20 to 30 more minutes of uterine contraction are required to push the infant's head through the cervix and the vagina. The expulsion stage ends when the child is pushed through the vagina into the waiting hands of a doctor or midwife.

To facilitate the delivery, physicians or midwives will often make an incision to widen the vaginal opening. This procedure, called an **episiotomy**, is performed when the baby's head enters the vagina in stage 2 of labor. The incision enlarges the vaginal orifice, prevents tearing, and allows the infant to pass quickly. The incision is stitched up immediately after the baby is born.

Once the baby's head emerges from the vagina, the rest of the body slips out almost instantly. However, the baby remains attached to the placenta via the umbilical cord. Before tying off and cutting the cord, many health care workers wait a minute or two to allow the blood remaining in the placenta to be pumped into the newborn.

Most babies (95%) are delivered head first with their noses pointed toward the mother's tailbone (Fig-

ure 17–17). Occasionally, however, babies may be oriented in other positions. This makes delivery more difficult, time-consuming, and hazardous. The most common alternative delivery is the **breech birth** in which the baby is expelled feet first. Breech births require more time and may cause extreme fatigue in the mother and brain damage in the baby. The umbilical cord sometimes wraps around the baby's neck, cutting off its supply of blood. To avoid complications, breech babies are often delivered by a **cesarean section**, a horizontal incision through the abdomen just at the pubic hair. Cesarean sections may be performed for other reasons as well, such as prolonged labor.

The final stage of delivery is the **placental stage** (Figure 17–17d). The placenta—sometimes called the afterbirth—remains attached to the uterine wall for a short while, then is expelled by uterine contractions, usually within 15 minutes of childbirth. After the placenta is expelled, uterine blood vessels clamp shut, preventing hemorrhage.

The uterus gradually returns to its normal size after delivery. Uterine involution (shrinkage) results from the rapid decline in estrogen and progesterone and is accelerated by oxytocin released by the posterior pituitary during suckling. Complete involution in women who breast feed usually occurs within four

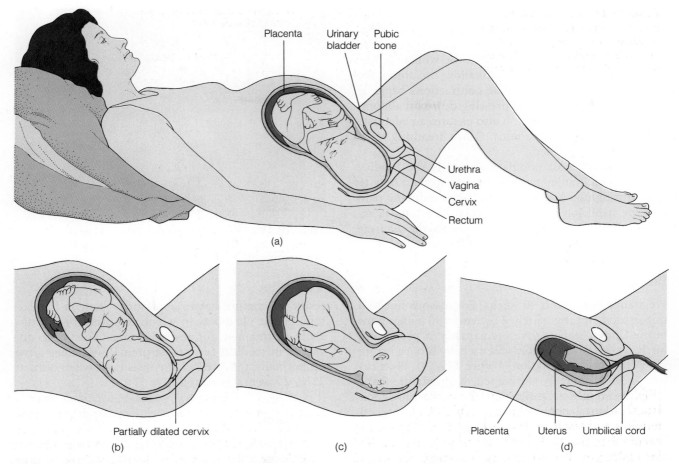

Placenta   Urinary   Pubic
bladder   bone

Urethra
Vagina
Cervix
Rectum

(a)

Partially dilated cervix

(b)

(c)

Placenta   Uterus   Umbilical cord

(d)

FIGURE 17–17   Stages of Labor   (a) Position of the fetus near birth. (b) Dilation stage. Uterine contractions push the fetal head lower in the uterus and cause the relaxin-softened cervix to dilate. (c) Expulsion stage. Fetus is expelled through the cervix and vagina. (d) Placental stage. Placenta is delivered.

weeks of pregnancy. In women who do not, the process usually takes six weeks.

Relieving the Pain of Childbirth

The level of pain a woman feels during childbirth varies greatly and is partly governed by her level of fear and tension. The more tense a woman is, generally the more pain she feels. For this reason, many hospitals provide comfortable birthing rooms and relaxation training. Drugs can also be given to reduce tension and pain, but generally must be given well before the birth of the baby. If given just before delivery, they may affect the baby's breathing.

Painkilling drugs can also be injected into the wall of the vagina with a syringe. Known as a **pudendal block**, this procedure is performed if an episiotomy is likely or if forceps are going to be used to facilitate birth. Many hospitals offer **epidural anesthesia**, commonly called an **epidural**. During this procedure, an anesthetic is injected into the base of a woman's spine. The drug blocks pain in the nerves that supply the lower body, destroying all sensations. If the anesthetic is given late in the first stage of labor, however, its effect may be so strong that a woman is unable to push the baby out. Forceps may be required, but forceps can damage a newborn and should be avoided.

Many couples and health care workers believe that drugs may be harmful to mothers and their babies. This belief and a desire to do things "naturally" have spawned the **natural childbirth** movement. Natural childbirth means different things to different people. In general, it refers to drug-free deliveries. Couples receive training in relaxation and special breathing techniques.

The most popular natural childbirth method today is known as the Lamaze technique, after its founder. Parents who are going to use the Lamaze technique attend childbirth classes in which the woman learns special breathing techniques. Shallow breathing, for example, is used when uterine contractions begin; it keeps the diaphragm from pressing down on abdominal organs, reducing pain. It also ensures an adequate supply of oxygen to the fetus. Other breathing techniques are also learned.

The second most popular method is the Bradley method. The Bradley method teaches women how to relax during uterine contractions, thus reducing pain and the need for drugs. No special breathing is required. The woman helps push, as in the Lamaze method, only near the end of childbirth.

Premature Birth

Pregnancy lasts about 40 weeks. For reasons not well understood, approximately 7 of every 100 babies born in the United States arrive prematurely. A **premature birth**, by definition, occurs when a baby is born before the 37th week of gestation (Figure 17–18). Born before it is completely developed, a premature baby faces several medical problems. One of the more common problems, mentioned in Chapter 9, is **respiratory distress syndrome** or **RDS** (or **hyaline membrane disease**). RDS results from an insufficient amount of surfactant in the infant's lungs. Surfactant lowers surface tension and prevents the alveoli from collapsing. Without treatment, the alveoli cave in, and the baby will die.

Premature babies are also subject to infections, resulting from an inadequate level of antibodies in the blood and also from increased exposure to bacteria during medical treatment. Babies born prematurely are also likely to suffer from jaundice due to the immaturity of the liver. The liver normally removes bilirubin, a yellow pigment formed in the breakdown of hemoglobin from red blood cells. The buildup of bilirubin can result in brain damage.

Premature babies also suffer from cerebral hemorrhaging. Weak blood vessels in the brain may burst and may cause death or permanent brain damage. Premature babies also have trouble digesting fat and must be fed a low-fat diet. Finally, premature infants generally have great difficulty regulating body temperature. Left alone, their body temperatures may stabilize in the low 90s or high 80s. Even a slightly depressed body temperature can cause death. As a

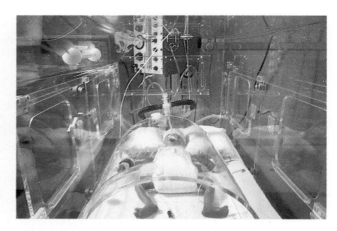

FIGURE 17–18   A Premature Baby   Kept alive through heroic and costly measures, this baby will probably live a normal life.

result, premature babies are usually kept in incubators, plastic chambers in which temperature is artificially maintained, for several weeks. Thanks to advances in medical care, most premature babies grow up normally and live healthy lives even when born at 25 to 26 weeks gestation.

Lactation

A baby emerges into a novel environment at birth. No longer connected to the placental lifeline, it must now breathe to get its oxygen and dispose of carbon dioxide, and it must also find a new way of acquiring food. For most of human evolution, that source was breast milk.

The breasts of a nonpregnant woman consist primarily of fat and connective tissue interspersed with milk-producing glandular tissue. The glands are drained by ducts that lead to the nipple (Figure 17–19). During pregnancy, the ducts and glands proliferate under the influence of placental and ovarian estrogen and progesterone. Milk production is induced by prolactin and a placental hormone.

Prolactin secretion in the mother begins during the fifth week of pregnancy and increases throughout gestation, peaking at birth (Figure 17–20). The placenta also releases a mildly lactogenic (milk-producing) hormone, **human chorionic somatomammotropin (HCS)**. Despite the presence of these lactogenic hormones, milk production does not begin until two or three days after birth. Why? High levels of estrogen and progesterone throughout pregnancy (necessary

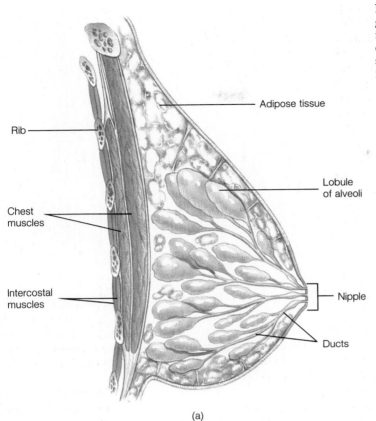

Adipose tissue

Rib

Chest
muscles

Intercostal
muscles

Lobule
of alveoli

Nipple

Ducts

(a)

for breast development) inhibit the action of these hormones, impeding milk production.

Although the breasts do not produce milk for two or three days, they immediately begin producing small quantities of colostrum. **Colostrum** is a fluid rich in protein and lactose, but lacking fat. Colostrum contains antibodies that protect the infant from bacteria as described in Health Note 10–1. A newborn subsists on colostrum for the first few days.

After birth, estrogen and progesterone levels fall, and milk production begins in earnest (Figure 17–20). Thus within two to three days, the breasts start to produce large quantities of milk. Milk production is facilitated by several other hormones, notably growth hormone, cortisol, and parathyroid hormone.

Prolactin secretion is maintained after childbirth by suckling. Suckling causes a surge in prolactin secretion; prolactin levels remain elevated for approximately one hour after each feeding. Each surge stimulates milk production needed for the next feeding. Milk production continues as long as the baby suckles. If nursing is interrupted for three or four days, however, the breasts stop producing milk. In most

women, milk production begins to decline by the seventh to ninth month of lactation, a time at which her baby is beginning to feed on semisolid food.

Prolactin secretion is controlled by a neuroendocrine reflex (Chapter 15). Suckling generates nerve impulses that travel to the hypothalamus, causing the release of **prolactin releasing hormone** (**PRH**). PRH, in turn, stimulates prolactin secretion. When suckling ends, PRH secretion stops.

In order for milk to reach the nipple, it must be actively propelled from the glandular units and through the ducts. This is the function of oxytocin. Oxytocin secretion is also controlled by a neuroendocrine reflex, described in Chapter 15.

INFANCY, CHILDHOOD, ADOLESCENCE, AND ADULTHOOD

Describing the changes that occur from birth to adulthood in one section of one chapter is a little like trying to define the universe in a paragraph. Never-

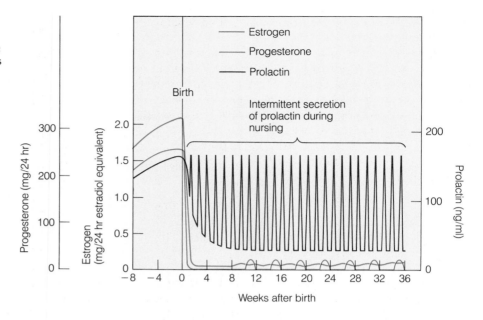

FIGURE 17–20 Hormone Levels Before and After Birth Prolactin, estrogen, and progesterone secretion prior to birth and 36 weeks after birth in a lactating woman.

theless, we will take a look at some milestones along this exciting journey.

Infancy

Infancy lasts from birth to the end of the second year of life. At birth, the infant emerges from the warm, relatively secure uterine home, where its needs were catered to automatically, to the home environment, where it must play a much more active role. The newborn survives its new home because of parental care and because it comes equipped with several built-in reflexes. One of the most important is the **suckling reflex.** Within minutes of birth, the baby begins to suckle the breast (or anything else put in its mouth).

Babies also come equipped with a **rooting reflex,** which helps the infant find the nipple. Pressure from a nipple causes the infant to turn its head toward the stimulus. You can test it by stroking a newborn's cheek. The baby will turn its head toward your finger automatically.

Crying is also a reflex in the newborn. Babies cry reflexively when they are hungry, hurt, or uncomfortable. Crying alerts the parents that something is wrong. Usually, a simple remedy will cure the crying and restore quiet—at least for a while.

During infancy, the child transforms physically and mentally. Although children develop in predictable stages, the timetable varies from one child to the next (Figure 17–21). One infant, for example, may learn

to crawl at five months while another does not crawl until its seventh month. This variability is natural and to be expected.

Within the first year, the child's body weight triples. Even though growth continues at a fairly fast clip, the growth rate begins to fall during infancy. Rapid physical growth is replaced by a growth in sensory, motor, and intellectual capacity.

At birth, the child is fairly immobile, but capable of moving its arms and feet in joy, protest, or play. Next the baby learns to roll over, a maneuver that provides surprising mobility. Next, the infant learns to crawl, often slithering along the floor like a prehistoric creature. During the seventh month, the baby begins to crawl on its hands and knees. In the eleventh or twelfth month, the child stands upright. From an upright position, the child quickly masters the art of walking, and a parent's life is forever changed. Throughout this period, the brain is changing as myelin is laid down on axons.

During infancy, the child also becomes more aware of its environment and more and more interactive with parents and peers. Child development involves important emotional and intellectual achievements. One of the first emotional milestones of normal human development is **bonding,** the establishment of an intimate relationship between the infant and one (or preferably both) of its parents or caregivers. Some psychologists think that bonding begins within an hour or so of birth. The initial contact, they say, is crucial to developing the bond. Others disagree. They think that bonding may require a period of several months.

What is important is for a child to develop a close relationship with a caregiver—to learn that it can be comforted by another human and that it can trust another person explicitly. Successful bonding or attachment is important for later development and may lead to emotional security later in life.

As the brain develops, infants become more curious. They begin to show memory and signs of reasoning. With these changes come improvements in motor control.

Research suggests that intelligence, like motor skills, improves with use. Children are eager explorers, ready to tear into drawers or closets to explore the fascinating objects of the adult world. Parents are encouraged to foster their children's curious nature—to let them explore and to provide opportunities for a rich and varied experience. A child's innate curiosity, then, is a seed a parent can nurture to facilitate intellectual development. Parents are also advised to provide ample stimuli for their children—toys, books, and plenty of personal interaction. Parents should read to infants even before their children can talk, point out and name objects, and talk to infants whenever possible.

Childhood and Adolescence

Childhood lasts from infancy to puberty, the beginning of adolescence. Growth continues throughout childhood, but as in infancy, the rate of growth continues to decline. During childhood, children display increased motor abilities. Language skills improve enormously. Curiosity abounds and mental capacity increases remarkably. In early childhood (and often during infancy), children start to assert their own will. They may resist parental wishes or even disobey parents outright. Developing a will and an identity are important and natural phases of development, and parents should be careful not to quash a child's developing will. That requires skillful parenting, the judicious use of discipline, and, of course, love.

Puberty, sexual maturation, marks the onset of **adolescence**, a sometimes tumultuous period of growth and development that lasts until adulthood (age 18 to 20). For girls, puberty begins about age 10 or 11. For boys, it begins at age 12 or 13. As in all developmental processes, individual variation is expected.

Discussed briefly in Chapter 16, sexual maturation results from the production of sex steroids in both boys and girls. Research suggests that the pituitary,

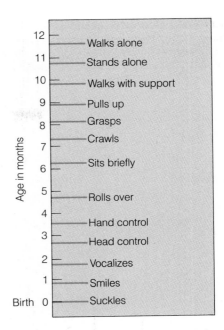

FIGURE 17–21   Motor Development in the Year after Birth

hypothalamus, and gonads begin functioning long before puberty, but the hypothalamus is extremely sensitive to circulating levels of sex steroid. As a result, gonadotropin release is greatly reduced. As an individual ages, however, the hypothalamus becomes less sensitive to the negative feedback influence of the sex steroids. GnRH release increases, and the pituitary output of gonadotropins climbs. This, in turn, increases gonadal hormone production, inducing puberty.

The physical changes induced by ovarian and testicular steroids during puberty were discussed in Chapter 16. Effects on bone were discussed in Chapter 14. Tables 17–5 and 17–6 summarize some of the important physical changes.

Adolescents also undergo a tremendous change in mental capacity. Their minds are now capable of dealing with abstract concepts, and their thinking skills may have improved considerably. Rudiments of critical thinking are apparent. A teenager may, in fact, use the scientific method to test hypotheses. Many teens become obsessively interested in appearance and intent on fitting into their peer group.

Adolescence is a time of emerging identity. That identity may contrast sharply with a parent's. It may also run counter to a parent's desire for the child. All too often, this leads to conflict. Accepting the emerging personality and maintaining a healthy relationship

TABLE 17–5   Physical Development in Adolescent Girls

PHYSICAL CHANGE	AVERAGE AGE WHEN CHANGE BEGINS	AVERAGE AGE WHEN NOTICEABLE CHANGE USUALLY STOPS	REMARKS
Increase in rate of growth	10 to 11	15 to 16	If conspicuous growth fails to begin by 15, consult your physician.
Breast development	10 to 11	13 to 14	Noticeable development of breasts (one of which may begin to "bud" before the other) is usually the first sign of puberty. If changes do not occur by 16, there may be cause for concern.
Emergence of body hair	Pubic hair: 10 to 11 Underarm hair: 12 to 13	Pubic hair: 13 to 14 Underarm hair: 15 to 16	Age at first appearance of body hair extremely variable. Pubic hair usually darkens and thickens as puberty progresses.
Development of sweat glands	12 to 13	15 to 16	The apocrine sweat glands are responsible for increased underarm sweating, which causes a type of body odor not present in children.
Menstruation	11 to 14	15 to 17	Menstruation often begins with extremely irregular periods, but by 17 a regular cycle (3 to 7 days every 28 days) usually becomes evident. If menstruation begins before 10 or has not begun by 17, consult your physician.

Source: From *The American Medical Association Family Medical Guide.* Copyright © 1982 by The American Medical Association. Reprinted by permission of Random House, Inc.

require that parents learn to tolerate differences. Love, support, and understanding are as important now as at any time in human development.

Adulthood

Adulthood begins when adolescence ends. Physical growth stops, but, for many people, social and psychological development continue throughout adult life. Not all people see adulthood as a period of personal growth, however. Many seem to congeal, remaining more or less the same throughout their adult life. They develop fixed patterns of behavior and live according.

In their 20s, many adults begin to show signs of aging. With it comes the gradual loss of youth—wrinkles may appear, weight may increase, gray hairs may develop. Aging is the topic of the next section.

TABLE 17-6   Physical Development in Adolescent Boys

PHYSICAL CHANGE	AVERAGE AGE WHEN CHANGE BEGINS	AVERAGE AGE WHEN NOTICEABLE CHANGE USUALLY STOPS	REMARKS
Increase in rate of growth	12 to 13	17 to 18	If conspicuous growth fails to begin by 15, you should consult your physician.
Enlargement of genitals	Testicles and scrotum:11 to 12 Penis: 12 to 13	16 to 17 15 to 16	As testicles grow, the skin of the scrotum darkens. The penis usually lengthens before it broadens. Ability to ejaculate seminal fluid usually begins about a year after the penis starts to lengthen.
Emergence of body hair	Pubic hair: 11 to 12 Underarm hair: 13 to 15	Pubic hair: 15 to 16 Underarm hair: 16 to 18	Development of hair is extremely variable and largely dependent on genetic inheritance. The spread of hair up the abdomen and onto the chest usually continues into adulthood.
Development of sweat glands	13 to 15	17 to 18	See remarks in the accompanying table for girls.
Voice change	Enlargement of the larynx, or voice box, begins at 13 to 14, and the voice deepens at 14 to 15	16 to 17	Growth of the larynx may make the adam's apple more prominent. The voice may change rapidly or gradually. If childhood voice persists after 16, consult your physician.

Source: From *The American Medical Association Family Medical Guide.* Copyright © 1982 by The American Medical Association. Reprinted by permission of Random House, Inc.

AGING AND DEATH

Aging is part of the life process, but aging is also one of the great mysteries of biology. Some define **aging** as a progressive deterioration of the body's homeostatic responses, a view that has gained wide acceptance in recent years.

The most notable changes occurring with age are physical changes—wrinkling, loss of hair, and stoop-ing. Aging also involves a gradual deterioration of the function of body organs. Sensory impairment is common. Vision and hearing both deteriorate with age. Muscular strength also declines as the number of myofibrils (bundles of contractile filaments) in the muscle decreases. Bones tend to thin, and joints often show signs of wear and tear. Aging is accompanied by a gradual reduction in cardiac output and pulmonary function—a reduction in oxygen absorption and vital capacity (the amount of air that can be inhaled and exhaled). The kidneys show signs of deterioration as well. The number of functional nephrons declines and so does renal function. The immune system declines, becoming less able to respond to new and old antigens. Finally, the nervous system function also declines with age. Memory and reaction time both decline. Although these changes are a part of the aging process, not all of them begin at the same time.

Aging: The Decline in Cell Numbers and Cell Function

The decline in homeostatic systems and general body function results from at least two factors: a decrease in the number of cells in the organs and a decline in the function of existing cells. What causes the decline in cell number?

Cell number is determined by the balance between cell division and cell death. Laboratory experiments show that body cells grown in culture divide a certain number of times, then stop. The number of divisions a cell undergoes is directly related to the age of the organism from which the cell is taken: the older the organism, the fewer divisions possible.

Laboratory studies such as these also show an interesting correlation between the number of divisions in culture and the life span of the species from which the cells were taken. In other words, cells from species with long life spans undergo more divisions than cells from species with short life spans. These data and others suggest that the end of cell division and, therefore, aging, may be genetically programmed.

Why then don't all humans live to be the same age? People no doubt differ genetically. In addition, people live markedly different lifestyles. Their diets differ. Research shows that the better one lives, the longer the life span (see Health Note 17–1). Diet, exercise, and stress management are keys to a healthy life.

Studies show that by manipulating the chemical environment of cells, a researcher can produce more cell divisions than normal. Vitamin E, when given in large quantities, for example, increases the number of cell divisions *in vitro*—that is, it increases the cell's life span. Whether vitamin E will extend a person's life is not yet known.

What causes a deterioration of cellular function? The deterioration in cellular function results from problems arising in the DNA, RNA, and cell proteins. As cells are exposed to natural and anthropogenic (produced by human activities) radiation and other

potentially harmful agents, such as chemicals in the home and workplace environment, they may be damaged and may lose their ability to function normally. As the damage increases, body function deteriorates.

The decline in cell number and the loss of cell function are thought to be the two leading causes of aging. But other researchers think that the process may be largely the result of a gradual decline in immune system function. Physiological deterioration of the immune system may result in disease and even cancer.

Aging and Disease

Many diseases are associated with old age. Osteoporosis, arthritis, atherosclerosis, and cancer are four examples. It is important to note, however, that these diseases are not an inevitable consequence of aging. As previous chapters have shown, the likelihood of developing these diseases can be reduced by exercise, diet, and other lifestyle adjustments. People can age in a healthy manner by living healthy lives.

Sharpening Your Critical Thinking Skills: Are We Living Longer?

"Thanks to improvements in medicine, people are living longer." You have heard this statement dozens of times before in one form or another. The assertion is that advances in medicine are increasing life span. A careful examination of the facts, however, shows that this statement is not entirely true.

Medical scientists measure longevity (how long people live) by a statistic called **life expectancy**. Life expectancy at birth is the number of years, on average, a person lives after he or she is born. Life expectancy at birth has increased dramatically in the past 80 years. In 1900, for example, on average, white American females lived only 50 years. Today, life expectancy is 81 years. For males, a similar trend is observed. In 1900, for example, the life expectancy of a white American male was 47 years; today, it is 74 years.

Unfortunately, these numbers are somewhat deceiving. Most people take them to mean that human beings are actually living longer. In reality, though, something very different is happening. What medicine has chiefly done is dramatically reduce infant death. As a result, more people are living past the first year of life. This results in an increased average life expectancy.

Can We Reverse the Process of Aging?

The search for a "cure" for old age has been frustrating. Numerous treatments have been tried and have failed. There is some encouraging news, however. Scientists recently discovered a protein called **stomatin**, which was first isolated from fibroblasts that had stopped dividing in tissue culture.

Scientists think that stomatin may stop a cell from dividing. If this is true, medical researchers may have discovered a significant clue in the aging puzzle—the trigger that ends cell division. If this protein signals the end of cellular division, scientist believe that there must be a gene that controls its production. If the stomatin gene can be located, it could also be inactivated. A drug or genetically engineered chemical could be injected in people to block the gene. The tissues of that person might continue to regenerate beyond the genetically programmed life span, and an individual's life span could be extended.

Research on ways to reverse the aging process has met with success in another arena as well—slowing down the aging of skin. Scientists have long studied the skin to understand the process of aging, for the changes in skin are similar to those that take place in other aging tissues.

Skin aging is the result of two processes: intrinsic, chronological aging, which may be genetically programmed, and extrinsic, or accumulated environmental, damage resulting, for instance from sunlight.

Scientists question whether anything can be done about intrinsic aging. Research indicates that after a skin cell has lived out its lifetime, membrane receptors in the cell become insensitive to growth factors. Growth factors stimulate DNA replication and cell division.

New research suggests, however, that extrinsic aging, especially that induced by sunlight, is preventable and reversible. Sunlight takes an indirect toll inside the skin cells and fibroblasts of the underlying dermis. Molecules inside various skin cells become electrically excited after exposure to sunlight. The energy they absorb may be released by reacting with other chemicals, thus causing the dissociation of chemical bonds. In this way, plasma membranes can be damaged within a few minutes of exposure to sunlight. Sunlight energy can also be absorbed by oxygen in the tissue, resulting in the formation of oxygen free radicals, highly destructive chemicals that are primarily responsible for extrinsic aging.

In 1988, medical researchers discovered that Retin-A, a derivative of vitamin A used to treat severe cases of acne, could reduce wrinkling of the skin caused by sunlight. Researchers do not know how the drug works. But studies suggest that the drug has several beneficial effects. First, it may stimulate the growth of new blood vessels, which may nurture the regeneration of damaged skin cells. Researchers believe it may regulate genes that play a role in cell growth and differentiation. Retin-A also detoxifies oxygen free radicals in tissues and can inhibit the destruction of collagen. A study of Retin-A that followed patients for 22 months or more after treatment showed that the drug reduces wrinkles, age spots, and roughness for at least 22 months—as long as it is applied regularly.

Despite the discovery of stomatin and Retin-A, there is no evidence that medical scientists can prevent or even retard the overall rate of aging. Numerous experiments have failed. The process of cellular aging, many researchers fear, may be beyond our grasp and may always remain so.

The best hope to live a long and healthy life is to live well: learn ways to reduce stress, eat well, exercise regularly, take alcohol in moderation or not at all, and avoid harmful practices, such as smoking.

To illustrate this point, consider a simplified example: Suppose 10 people are born in a given year on a small island. If 5 die the first year, and the other 5 live until they are 70, the average life expectancy is 35 years. Now suppose, a doctor moves onto the island and is able to reduce the infant death rate so that only 1 child dies out of every 10. The rest of the people live until they are 70. The average life expectancy is now 63 years. In this example, the island's residents haven't conquered aging. People aren't really living longer. All that has happened is that more children are living past the dangerous first year of life.

The same holds true for the United States. A drastic drop in infant death has occurred since the early 1900s (from over 100 per thousand to about 12 per thousand). The increase in life expectancy over the last 80 years is largely the product of decreased infant mortality.

This is not to say that all of the gain in life expectancy results from a decline in infant mortality. Medical advances have increased life expectancy, but these changes are small in comparison to the increase brought about by decreasing infant mortality. About 85% of the increase in life span in the past 90 years is the result of decreased infant mortality.

Death and Dying

Aging results in a deterioration of function that eventually leads to death. Death also results from traumatic injury to the body—severe damage to the brain or a sudden loss of blood, for example.

Although it is difficult to define precisely, death is generally described as the cessation of life. Trouble begins when people try to define death more precisely, especially in cases requiring life-support systems. In 25 states, a person is considered dead when his or her heart stops beating and breathing ceases. In the remaining 25 states, however, death is defined as an irreversible loss of brain function. A woman in a coma, for example, often shows little or no brain activity other than that needed to keep the heart beating and continue breathing. Is this woman alive? In some states, she is; in others, she isn't.

Maintaining a person on life support is extremely costly and can be financially and emotionally crippling to a family. The money spent on maintaining a life could, arguably, be used to save dozens of lives. These issues create an ethical dilemma in our society over euthanasia, a word derived from the Greek meaning "easy death." **Euthanasia**, an act or method of causing death painlessly, may be either passive or active. Passive euthanasia consists of deliberate actions and decisions to withhold treatment that might prolong life. Active euthanasia consists of actions and decisions that actively shorten a person's life—for example, injecting a lethal substance to terminate a patient's life.

Opinions on euthanasia vary widely—from those who oppose all forms of euthanasia to those who support active measures. The controversy will, no doubt, be with us for many years. Individuals who want to save their relatives emotional and legal turmoil, however, can sign a living will declaration. This is a legal document that stipulates conditions under which a person is allowed to die.

ENVIRONMENT AND HEALTH: VIDEO DISPLAY TERMINALS

Computer monitors or video display terminals (VDTs) produce numerous types of radiation with frequencies ranging from X-rays to radio waves. Protective shields prevent most of this radiation from escaping. It's what escapes the shield, however, that has some health officials concerned.

Most scientists agree that small amounts of middle- and high-frequency radiation escaping the protective shielding are not a threat to health. Lower-frequency radiation (radio frequencies) is another story. Laboratory experiments show that electromagnetic fields generated by extremely low frequency radiation can alter fetal development in chicken, rabbits, and swine.

Researchers in Oakland, California, recently published results of a medical study on the incidence of miscarriage in nearly 1600 women. The study showed that women who sit in front of a computer VDT for more than 20 hours a week during the first three months of pregnancy are nearly twice as likely to miscarry as women in similar jobs not using computers.

What causes miscarriage? The researchers suspect that they result from radiation emitted from the VDTs. However, other factors may also be involved. For example, stress or some other factor in the office may be the cause. Only more research will tell. Preferring to err on the conservative side, some scientists advise pregnant women to minimize their exposure to radiation from monitors.

Should the link between radiation from VDTs and miscarriage be substantiated by further research, it would once again illustrate the importance of a healthy environment to overall human health and reproduction.

SUMMARY

FERTILIZATION: A NEW LIFE BEGINS

1. Human life begins at fertilization when the sperm and oocyte unite to form a zygote. Fertilization usually occurs in the upper third of the uterine tube.
2. Sperm deposited in the vagina reach the site of fertilization with the aid of muscular contractions in the walls of the uterus and uterine tube.
3. Sperm bore through the zona pellucida and contact the plasma membrane. The first one to contact the membrane fertilizes the oocyte. Further sperm penetration is blocked.

4. Sperm are engulfed by the oocyte. The chromosomes of the sperm and oocyte duplicate and merge in the center of the cell where mitosis begins.

PRE-EMBRYONIC DEVELOPMENT

5. Human development is divided into three stages: pre-embryonic development, embryonic development, and fetal development.

6. Pre-embryonic development begins at fertilization and ends at implantation. The zygote undergoes rapid cellular division and forms a solid ball of cells, the morula.

7. A cavity forms in the morula, converting it to a blastocyst, a hollow sphere of cells. The blastocyst consists of a clump of cells, the inner cell mass, which will become the embryo, and the trophoblast, which gives rise to the embryonic portion of the placenta.

8. The morula arrives in the uterus approximately three to four days after fertilization and is converted into a blastocyst, which implants two to three days later. The blastocyst contacts the wall of the endometrium, then digests its way into it. The cells it digests provide nourishment.

9. Soon after implantation begins, the trophoblast differentiates into two layers. The outer layer invades the endometrium. Cavities form in this layer and fill with blood.

10. Fingerlike projections of the inner layer invade a little later, carrying embryonic blood vessels. Known as placental villi, the projections are bathed in maternal blood and provide a means of acquiring oxygen and nutrients from the mother and disposing of embryonic wastes.

EMBRYONIC AND FETAL DEVELOPMENT

11. While the placenta forms, a layer of cells from the inner cell mass of the blastocyst separates from it and forms the amnion. The amnion fills with fluid. The amnion enlarges during embryonic and fetal development and eventually surrounds the entire embryo and fetus. The amniotic fluid protects the embryo and fetus during development.

12. After the amnion forms, the cells of the inner cell mass differentiate into the three germ cell layers: ectoderm, mesoderm, and endoderm. The formation of the three primary germ layers marks the beginning of embryonic development.

13. The organs develop from the three basic tissues during organogenesis. Table 17-2 lists the organs and tissues formed from each of the layers.

14. Fetal development begins eight weeks after fertilization. Since most of the organ systems have developed or are under development, fetal development is primarily a period of growth.

15. Blood flows to and from the fetus via the umbilical cord, permitting the discharge of wastes and the acquisition of nutrients in the placenta.

16. The placenta also produces a number of hormones. Human chorionic gonadotropin maintains the corpus luteum during pregnancy. Progesterone and estrogen stimulate uterine growth and the development of the glands and ducts of the breast.

17. About 10 to 12% of all newborns enter the world with some form of birth defect. The study of birth defects is teratology.

18. Birth defects arise from a variety of chemical, biological, and physical agents, known as teratogens. Table 17-3 lists known human teratogens.

19. The effect of teratogenic agents is related to the time of exposure, the nature of the agent, and the dose. A defect is most likely to arise if a woman is exposed to a teratogen during the embryonic period when the organs are forming.

20. Many physical and chemical agents are toxic to the human fetus and when present in sufficient quantities can kill a fetus or retard its growth.

MATERNAL CHANGES DURING PREGNANCY

21. During pregnancy a woman's body undergoes incredible change. The uterus and breasts enlarge considerably, blood volume increases, respiration rate climbs, and urination increases.

CHILDBIRTH AND LACTATION

22. The uterus begins to contract spontaneously in the months prior to birth. These contractions are false labor.

23. True labor consists of more intense and more frequent contractions. They may be caused by the release of small amounts of fetal oxytocin prior to birth. Fetal oxytocin stimulates the release of prostaglandins by the placenta. Oxytocin and prostaglandins stimulate contractions in the sensitized uterine musculature.

24. Emotional and physical stress in the mother may trigger maternal oxytocin release, thus augmenting muscle contractions in progress. Contractions stimulate a positive feedback mechanism. As uterine contraction increases, it causes more maternal oxytocin to be released. This stimulates stronger contractions and more oxytocin release, a cycle that continues until the baby is born.

25. Labor consists of three stages. During stage 1, the dilation phase, uterine contractions push the fetal head lower in the uterus and cause the relaxin-softened cervix to dilate. In stage 2, the expulsion phase, the fetus is expelled through the cervix and vagina. In stage 3, the placental stage, the placenta is delivered.

26. A premature birth occurs when a baby is born before 37 weeks of gestation. Premature babies face numerous medical problems.

27. The breasts of a nonpregnant woman consist primarily

of fat and connective tissue interspersed with milk-producing glandular tissue and ducts.

28. During pregnancy, the glands and ducts proliferate under the influence of placental and ovarian estrogen and progesterone.

29. Milk production is induced by maternal prolactin and human chorionic somatomammotropin, but does not begin until two to three days after birth.

30. Before milk production begins, the breasts produce small quantities of a protein-rich material called colostrum. A newborn can subsist on colostrum for the first few days and derives antibodies from the colostrum that help protect it from bacteria.

31. Suckling causes a surge in prolactin secretion, and prolactin levels remain elevated for approximately one hour after each feeding. Each surge stimulates milk production needed for the next feeding.

INFANCY, CHILDHOOD, ADOLESCENCE, AND ADULTHOOD

32. The newborn survives because of parental care and because it comes equipped with several built-in reflexes, including the suckling reflex, rooting reflex, and crying reflex.

33. During infancy, the child transforms physically and mentally. Children develop in predictable stages. However, the timetable varies considerably.

34. During infancy, sensory, motor, and intellectual capacity increase dramatically.

35. During infancy, children develop emotionally as well. One of the first emotional milestones of normal human development is bonding, the establishment of an intimate relationship between the infant and its caregivers. Successful bonding is important for later development and may facilitate emotional security later in life.

36. Research suggests that intelligence, like motor skills, improves with use. Parents are encouraged to foster their children's curious nature—to let them explore and to provide opportunities for a rich and varied experience.

37. Childhood lasts from infancy to puberty. During childhood, children undergo dramatic improvements in motor skills and use and comprehension of language.

38. In early childhood (and often during infancy), children start to assert their own will. Developing a will and an identity are natural phases of development.

39. Puberty, sexual maturation, marks the onset of adolescence, which lasts until adulthood (age 18 to 20). Sexual maturation results from the production of sex steroids.

40. Adolescence is also a time of emerging identity. Conflict between teenagers and their parents is common during this period, but can be reduced if parents learn to accept the emerging personality and maintain a supportive relationship with their teenagers.

41. Adulthood begins after adolescence ends. Physical growth stops, but, for many people, social and psychological development continue.

AGING AND DEATH

42. Aging is a natural part of the life process. Aging is the progressive deterioration of the body's homeostatic abilities and the gradual deterioration of the function of body organs.

43. These changes result from at least two factors: a decrease in the number of cells in the organs and a decline in the function of existing cells.

44. Death results from aging, traumatic injury, and infectious disease.

EXERCISING YOUR CRITICAL THINKING SKILLS

1. A vitamin manufacturer claims that vitamin E will prolong your life. This conclusion is based on research in animals showing that rats given large doses of vitamin E live longer. Using your critical thinking skills, draft a list of questions you would ask the vitamin manufacturer.

2. Using your critical thinking skills, critique the study of video display terminals and miscarriage presented in this chapter.

TEST OF TERMS

1. The ovum and sperm unite to form a _____ during fertilization, which takes place in the upper third of the _____ _____ .

2. _____ renders sperm membranes fragile, allowing the sperm to release enzymes from the _____ .

3. When a sperm cell contacts the plasma membrane of the oocyte, it triggers an immediate change in the plasma membrane that prevents other sperm from entering. This is called the _____ block to _____ .

4. During pre-embryonic development, the cells multiply first forming a solid ball of cells, the _____ . A cavity soon forms in the morula, converting it to a blastocyst. The blastocyst consists of a clump of cells, the _____ _____ _____ , which gives rise to the embryo, and a ring of flattened cells, the _____ , which will give rise to the embryonic portion of the placenta.

5. The blastocyst attaches and digests its way into the uterine lining in a process called _____ , which begins approximately _____ days after ovulation.

6. The placenta forms from fetal and maternal tissue. Blood vessels invade the forming placenta with the inner layer of the trophoblast. The fingerlike projections into the maternal blood are called _____ .

7. Early in embryonic development, the _____ forms from the cells that will form the embryo proper. It eventually forms a complete fluid-filled sac that surround the fetus.

8. The organs form during embryonic development; this process is called _____ .

9. Ectoderm forms the neural tube, which later develops into the _____ and _____ _____ .

10. Blocks of mesoderm along the neural tube, called _____ , form the muscles of the neck and trunk as well as the _____ .

11. The lining of the intestines comes from embryonic _____ .

12. The umbilical _____ carries blood rich in oxygen and nutrients from the placenta to the fetus. The _____ _____ shunts blood from this vessel to the _____ _____ _____ .

13. The hole in the interatrial septum shunts blood from the _____ atrium to the _____ atrium and is called the _____ .

14. The placental hormone _____ _____ _____ maintains the corpus luteum and stimulates production of _____ and _____ by the CL.

15. The study of birth defects is called _____ . Birth defects result from _____ agents.

16. _____ , a hormone produced by the placenta and ovaries of pregnant women, loosens the connective tissue in the pubic symphysis and softens the cervix in preparation for childbirth.

17. Stage 1 of childbirth is also known as the _____ phase. During this phase, the baby descends into the pelvic cavity, and the _____ dilates.

18. Childbirth in which the baby is born buttocks first is called a _____ birth.

19. The final stage of childbirth is the _____ stage.

20. An anesthetic injected into the vagina during childbirth is called a _____ block.

21. Lactation, the production of milk from the breasts, is stimulated by the maternal hormone, _____ , secreted from the anterior pituitary gland.

22. The protein-rich secretion produced by the breasts during the first few days after birth is called _____ .

23. A baby turns its head toward the nipple that brushes against its cheek. This is called the _____ reflex.

24. One of the first emotional milestones of normal human development is _____ , the establishment of an intimate relationship between the infant and one (or preferably both) of its parents or caregivers.

25. Sexual maturation occurs during _____ .

26. Aging is the progressive deterioration of the body's _____ responses and also involves a gradual deterioration of the function of body organs.

27. _____ _____ at birth is the average life span of an individual.

Answers to the Test of Terms are located in Appendix B.

TEST OF CONCEPTS

1. Describe the process of fertilization in detail. Use drawings to elaborate your points and label all drawings.
2. Define each of the following terms: morula, blastocyst, inner cell mass, trophoblast.
3. Describe where the embryo acquires nutrients before the placenta forms.
4. Describe the formation of the placenta.
5. What are the major functions of the placenta?
6. Describe the flow of blood from the placenta to the fetus and back. What is the role of the various shunts?
7. List and describe the function of the five placental hormones.
8. Describe the changes that occur in pregnant women. Describe the factors that contribute to maternal weight gain during pregnancy.
9. Discuss how the nature of a teratogenic agent and the time of exposure both affect teratogenesis—the production of birth defects.
10. What factors trigger labor?
11. What factors are responsible for cervical dilation during labor?
12. Design a humane experiment to test the two major hypotheses about bonding presented in this chapter. Describe how you would go about the experiment and the special considerations that would be necessary to eliminate bias. Could such an experiment be objective? Could such an experiment be ethical?
13. Describe the various medical problems of newborn infants.
14. Prolactin, which stimulates milk production by the breasts, is present in the mother's blood well before the onset of lactation. It is not until estrogen and progesterone levels drop, however, that milk production begins. Design an experiment to test this hypothesis. Using your critical thinking skills, critique your own experimental design.
15. Describe some of the reflexes of the newborn and their importance to the infant's survival.
16. Define the term *aging* and describe some of the hypotheses that attempt to explain it.
17. Thanks to modern medicine, humans are living longer, says a friend. Using your understanding of average life expectancy data and the reason for an increase in life expectancy in the last hundred years, explain why this statement is not quite true.

SUGGESTED READINGS

Guyton, A. C. 1986. *Textbook of medical physiology.* Philadelphia: Saunders. See Chapter 82 for a well-written description of the physiology of pregnancy and lactation.

Sizer, F. S., and E. N. Whitney. 1988. *Life choices: Health concepts and strategies.* St. Paul, Minn.: West. Filled with practical advice on contraception, childbirth, and parenting.

Tuchmann-Duplessis, H., G. David, and P. Haegel. 1975. *Illustrated human embryology,* vol. 1. *Embryogenesis.* New York: Springer Verlag. Detailed graphical description of embryonic and placental development.

Tuchmann-Duplessis, H. and P. Haegel. 1975. *Illustrated human embryology.* vol. 2. *Organogenesis.* New York: Springer Verlag. Detailed graphical description of organogenesis.

Weiss, R. 1988. Wrestling with wrinkles. *Science News* 134(13): 200—202. Explores the evidence linking Retin-A with wrinkle reduction.

18 Principles of Heredity and Human Inheritance

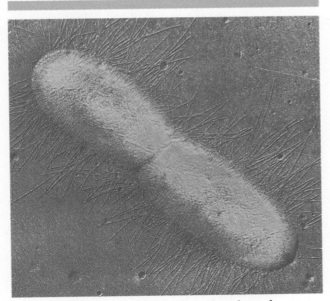

Bacterial fission. Many of the details we have learned about genetics come from studies of bacteria like this common one which is about to divide.

For most of us, life is something of a roller coaster. Our emotions tend to rise and fall with the changing circumstances. Emotional peaks and troughs are quite normal. For some individuals, however, the emotional roller coaster ride is much more intense. Their feelings swing widely. One minute they are in a state of elated overactivity (mania), the next they fall into a state of depressed inactivity. This condition, known as manic-depression, tends to be cyclical and is not related to external events that generally elicit emotional changes.

Manic-depression varies considerably from one person to the next. In some, the bouts last only a short time, but in others the bouts of depression and elation continue for years. When in a state of deep depression, victims sometimes threaten suicide, but their lack of energy often prevents them from following through with their threats. The desire may persist and result in suicide when the victim recovers from depression, however. During mania—the animated and highly energetic state—manic-depressives may act ir-

501

(a)

(b)

FIGURE 18-1 Mendel and His Garden (a) Gregor Mendel worked out some of the basic rules of inheritance in the mid-1800s in his research on garden peas. (b) Mendel's garden at the monastery.

rationally or obnoxiously. Such behavior can ruin social and professional relationships and can lead to financial disaster, especially if it interferes with decision making.

Manic-depression occurs in about 3% of the U.S. population and tends to run in families, suggesting a genetic link. Researchers, in fact, recently located a gene near the tip of chromosome 11 that may predispose its bearers to manic-depression and possibly severe depression.

Chapters 4 and 5 described the human genes and chromosomes, noting how they affect our anatomy, physiology, and even our behavior. This chapter examines another aspect of genetics, notably the inheritance of traits from our parents. We begin with some basic principles and concepts.

PRINCIPLES OF HEREDITY: MENDELIAN GENETICS

Genetics is the study of the structure and function of genes and the transmission of genes from parents to offspring. The formal study of genetics began with the work of a nineteenth-century monk, Gregor Mendel (Figure 18-1). Born in eastern Europe in 1821, Men-del entered an Augustinian monastery in Brno, in what is now Czechoslovakia, at the age of 21.[1] After completing his studies, Mendel enrolled at the University of Vienna, where he studied mathematics, botany, chemistry, and other sciences. Returning to the monastery, Mendel began a series of experiments on garden peas; his studies would reveal several key principles of genetics still relevant to the study of genetics in eukaryotic organisms.

During Mendel's time, many scientists believed that the traits of a child's parents were somehow blended together in the offspring, producing a child with intermediate characteristics. In addition, since the ova of most species were much larger than the sperm, scientists believed that the female had a greater influence on the characteristics of the offspring than the male.

Mendel's studies were designed to answer two questions (Table 18-1). First, were the physical characteristics of an offspring the result of a blending of parental traits? Second, if blending did occur, did both parents contribute equally to the offspring?

Mendel studied 28,000 plants over a 10-year period to avoid problems that might arise from small sample

[1]Brno is now the city of Brünn.

TABLE 18–1 Genetic Traits Mendel Studied

STRUCTURE STUDIED	DOMINANT	RECESSIVE
Seeds	Smooth	Wrinkled
	Yellow	Green
Pods	Full	Constricted
	Green	Yellow
Flowers	Axial (along stems)	Terminal (top of stems)
	Purple	White
Stems	Long	Short

size, a requirement of careful experimentation, as noted in the critical thinking section of Chapter 1. Mendel also repeated his experiments to be certain that his results were reproducible, and he subjected his results to statistical analysis.[2] In fact, Mendel was probably the first to use statistical analysis in biological research.

Do Traits Blend?

Mendel's first discovery in his analysis of inheritance in peas was that the physical characteristics of the parents were not blended to produce offspring of intermediate traits. When he bred a pea with a white flower to a pea with a purple flower, he did not produce pea plants with intermediate pink flowers, but rather plants with purple flowers. Other experiments he performed showed similar results.

Do Parents Contribute Equally to the Traits of Their Offspring?

Mendel's research led him to conclude that each adult plant carries a pair of hereditary factors, which govern the inheritance of a trait. Today, these hereditary factors are known as genes. Genes are carried on the chromosomes. Mendel's findings led him to conclude that the hereditary factors of the mother and father separate during gamete formation and that each gamete contains only one hereditary factor (gene) for each trait. This concept is known as the **principle of**

[2]A study of Mendel's notebooks, however, suggests that he discarded data that did not fit the nice patterns he was getting and only reported data for which he could supply a theoretical explanation.

segregation or Mendel's first law. During fertilization, the gametes of the parents combine, producing an offspring inheriting two genes for each trait—one from each parent. From his studies, Mendel concluded that the contributions of the parents must be equal—in other words, each parent contributes one hereditary factor for each trait.

These conclusions may seem unremarkable to you now, especially with your knowledge of meiosis and gamete formation (Chapter 16). However, it is important to remember that Mendel performed his experiments in the 1850s and 1860s before chromosomes had been discovered! Mendel hypothesized the presence and behavior of genes long before science knew even the most basic facts of cell biology and inheritance.

Dominant and Recessive Traits

Mendel also postulated that hereditary factors (genes) may be either **dominant** or **recessive**. When a dominant and recessive factor were present in a pea, Mendel showed the dominant factor was always expressed. A recessive factor was expressed only when the dominant factor was missing. The dominant and recessive factors today are known as dominant and recessive genes. Dominant genes are designated by a capital letter; recessive genes are signified by a lowercase letter.

In human somatic cells, genes exist in pairs with one member of the pair on each chromosome. Alternatives forms of a gene are called **alleles**. In Mendel's peas, for instance, the gene for flower color is the P gene. The dominant form is P (purple flowers), and the recessive form is p (white flowers). The P and p

genes are both alleles of the flower color gene. For genes with two alleles, such as these, three possible combinations are possible. The first consists of two dominant genes—for example, *PP*. The individual is said to be **homozygous dominant** for that particular trait. The second consists of two recessive genes—in this example, *pp*—and the individual is said to be **homozygous recessive** for that trait. The third occurs when a dominant and recessive gene are present—*Pp*—and the individual is **heterozygous**.

Genotypes and Phenotypes

Mendel determined that blending did not occur in the inheritance of traits, such as flower color. Thus, a pea plant with purple flowers (*PP*) bred with a pea plant with white flowers (*pp*) produced offspring with purple flowers and not pink flowers, which would occur if blending took place. Mendel proposed that the genotype of the purple-flowered offspring of this cross was *Pp*—that is, heterozygous. In such cases, he argued, the recessive hereditary factor (gene) is masked by the dominant hereditary factor (gene). Because of dominance, Mendel noted that the outward appearance of a plant—its **phenotype**—did not always reflect its genetic makeup, or **genotype**. In this example, two different genotypes can lead to the same phenotype (*PP* = purple and *Pp* = purple). As a general rule, a homozygous dominant individual and a heterozygous individual are indistinguishable on the basis of phenotype.

Tracking Genotypes and Phenotypes: The Punnett Square

To track genotypes and phenotypes of breeding experiments, such as those that Mendel performed, geneticists use a relatively simple tool, called the Punnett square (Figure 18–2). To illustrate the process, consider one of the traits that Mendel studied, seed conformation, a gene with two alleles: one that codes for wrinkled seeds (*s*) and the other that codes for smooth seeds (*S*). As indicated by the capital *S*, the smooth-seed allele is dominant.

Suppose that you bred a homozygous dominant (*SS*) plant with a homozygous recessive plant (*ss*). Figure 18–2 lists the genotypes of each parent and lists the possible genotypes of the gametes. To determine the outcome of crossing these two plants, the gametes from one plant are listed along the top of the Punnett square and the gametes from the other are listed along the side. The gametes are then combined,

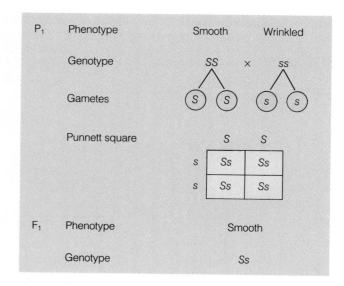

FIGURE 18–2 Mendel's Early Experiments When studying the inheritance of seed conformation, Mendel first crossed homozygous dominant (*SS*) and homozygous recessive (*ss*) plants. The offspring were all heterozygotes (*Ss*).

producing all of the possible genotypes present in the offspring.

As this example shows, a cross between a homozygous dominant individual and a homozygous recessive individual produces only heterozygous offspring. The offspring are phenotypically uniform, and all of them resemble the homozygous dominant parent.

In the language of genetics, when one organism is bred with another to study the transmission of a single trait, the procedure is called a **monohybrid cross**. In a monohybrid cross, the organisms that are bred initially constitute the P_1 generation (P for parents). The first set of offspring constitutes the F_1 generation (first filial generation). In this example, the F_1 offspring could be bred to produce additional offspring, the F_2 generation.

To determine the genotypes and phenotypes of such a cross, the Punnett square can be used. As Figure 18–3 shows, the gametes of the F_1 generation are *S* and *s*. The possible gametes from the mother and father are placed along the top and side of the Punnett square and then combined in the box. As illustrated, this cross produces three different genotypes: *SS*, *Ss*, and *ss* (Figure 18–3). The ratio of genotypes or genotypic ratio of the F_2 generation is 1 *SS*:2 *Ss*:1 *ss*. Thus, if 100 offspring are produced, you would expect 25 of them to be *SS* (homozygous dominant), 50 to be *Ss* (heterozygous), and 25 to be *ss* (homozygous recessive). Because of dominance, this cross yields

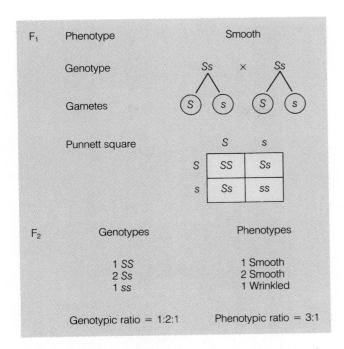

FIGURE 18–3 Monohybrid Cross In Mendel's early work, he crossed offspring from his F_1 generation (*Ss*) and found a variety of genotypes and phenotypes.

only two phenotypes: smooth-seed plants (*SS* and *Ss*) and wrinkled-seed plants (*ss*). The ratio of phenotypes, or phenotypic ratio, is 3 smooth:1 wrinkled.

The Principle of Independent Assortment

In his later experiments, Mendel tracked two traits at the same time, a procedure geneticists refer to as a **dihybrid cross** (Figure 18–4). Consider an example. Peas contain a gene for smooth and wrinkled seeds, the *S* gene. The dominant form is *S* (smooth), and the recessive form is *s* (wrinkled). Peas also contain a gene for seed color; the dominant form (*Y*) yields yellow seeds, and the recessive form (*y*) yields green seeds.

Mendel crossed homozygous dominant plants with smooth, yellow seeds (*SSYY*) with homozygous recessive plants (*ssyy*). As Figure 18–4 shows, the genotype of the F_1 generation was *SsYy*. Consequently, the offspring were all phenotypically identical, and all displayed the dominant characteristics. When Mendel crossed two of the F_1 offspring, however, he got a mixture of phenotypes and genotypes (Figure 18-4). To produce this combination of offspring, Mendel concluded that the hereditary factors *S* and *s* must have segregated independently of *Y* and *y* during ga-

mete formation. Translated into modern terms, this means that the *Y* and *S* genes were on different chromosomes, which separate independently of one another during gamete formation. Because of independent assortment, the gametes produced by the F_1 generation contain all combinations of the alleles in equal proportions: *SY, Sy, sY,* and *sy*. Fertilization occurred at random, giving rise to 16 possible combinations (Figure 18–4).

Mendel's research on dihybrid crosses led him to propose the **principle of independent assortment** or Mendel's second law. Stated briefly, the principle of independent assortment says that the segregation of the alleles of one gene is independent of the segregation of the alleles of another gene during gamete formation.

Mendel may have been the luckiest scientist in the world, for it just so happened that he chose to study seven traits and, fortunately for him, peas have seven pairs of chromosomes and each of the seven traits he studied was on a different chromosome. If two genes under study are located on the same chromosome, they are said to be linked, and they tend to be inherited together. Independent assortment holds only for genes that are on different chromosomes.[3]

MENDELIAN GENETICS IN HUMANS

Mendelian genetics applies to many human traits. Table 18–2 lists a number of human traits and diseases that are carried on our chromosomes and whose genetic expression follows basic Mendelian principles.

Autosomal Recessive Traits

As noted in Chapter 4, human cells contain 23 pairs of chromosomes consisting of two types: the sex chromosomes and the autosomes. The **sex chromosomes** are involved in sex determination. As noted in earlier chapters, two types of sex chromosomes exist, X and Y. Females have two homologous X chromosomes, and males have a nonhomologous pair, consisting of one X and one Y chromosome. The remaining 22 pairs of chromosomes are called the **autosomes**. These chromosomes carry numerous genes that control a wide variety of traits. This section examines several autosomal recessive traits.

[3]Widely separated genes on the same chromosome assort independently if the distance between them is great and if crossing over occurs with great frequency.

FIGURE 18-4 Dihybrid Cross
The dihybrid cross examines the inheritance of two traits. Independent assortment of genes produces a large number of genotypes.

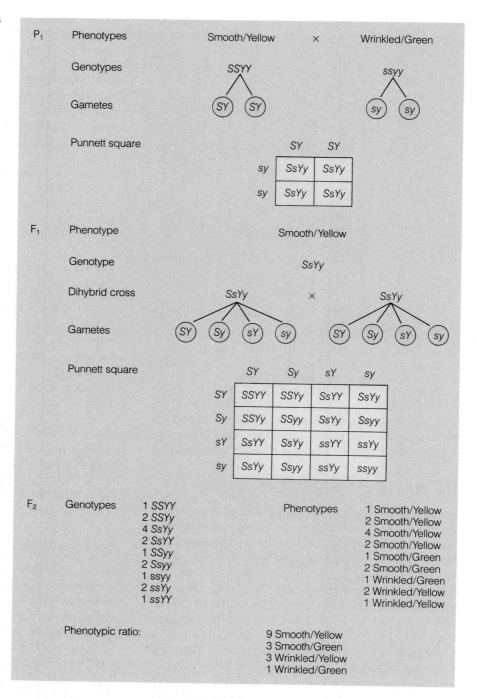

Autosomal recessive traits are expressed only when both recessive alleles are present. Over 600 traits in humans have been identified as autosomal recessive, and another 800 are strongly suspected of being autosomal recessive. Interestingly, most of the autosomal recessive traits cause serious problems.

Three common autosomal recessive traits are sickle-cell anemia, albinism, and cystic fibrosis.

Sickle-Cell Anemia.   Sickle-cell anemia was briefly described in Chapter 8. Occurring chiefly in African Americans and Caucasians of Mediterranean descent,

TABLE 18–2 Traits and Diseases Carried on Human Chromosomes

Autosomal recessive	
Albinism	Lack of pigment in eyes, skin, and hair
Cystic fibrosis	Pancreatic failure, mucus buildup in lungs
Sickle-cell anemia	Abnormal hemoglobin leading to sickle-shaped RBCs that obstruct vital capillaries
Tay-Sachs disease	Improper metabolism of a class of chemicals called gangliosides in nerve cells, resulting in early death
Phenylketonuria	Accumulation of phenylalanine in blood; results in mental retardation
Attached earlobe	Earlobe attached to skin
Hyperextendible thumb	Thumb bends past 45° angle
Autosomal dominant	
Achondroplasia	Dwarfism resulting from a defect in metaphysial plates of forming long bones
Marfan's syndrome	Defect manifest in connective tissue, resulting in excessive growth, aortic rupture
Widow's peak	Hairline comes to a point
Huntington's disease	Progressive deterioration of the nervous system beginning in late twenties or early thirties; results in mental deterioration and early death
Brachydactyly	Disfiguration of hands, shortened fingers

sickle-cell anemia occurs in individuals who are homozygous recessive—that is, individuals who contain both recessive alleles. Sickle-cell anemia leads to complications when oxygen levels in the blood drop—for example, in metabolically active tissues or in capillaries. The defective hemoglobin in red blood cells (RBCs) of individuals with sickle-cell anemia causes the cells to become sickle shaped when exposed to low levels of oxygen. Sickle-shaped RBCs clog capillaries and reduce oxygen flow to brain cells, heart cells, and other organs. As a result, sickle-cell anemia is usually lethal, with most victims dying by their late twenties.

Sickle-cell anemia occurs in people who have two recessive genes for the trait. Individuals who are heterozygous for the trait are said to be **carriers**, because they can pass the gene on to their children. Carriers generally lead relatively normal lives, but are subject to occasional problems. In fact, about one-third of the carriers contain some defective hemoglobin and may suffer from anemia and other complications under certain conditions.

Approximately 1 in every 500 African Americans born in the United States is homozygous recessive, and about 1 of every 12 is a carrier of sickle-cell anemia. The sickle-cell trait protects carriers and homozygotes from malaria, a deadly disease prevalent in African countries from which many black Americans'

ancestors came. Malaria is caused by a microscopic single-celled parasite called *Plasmodium*, which is transmitted from one person to the next by the *Anopheles* mosquito. Inside the body, the parasites invade and colonize the liver where they multiply rapidly. New parasites leave the liver and enter the bloodstream. In the blood, they enter and destroy RBCs. Many of the parasites remain in the liver, however, where they continue to reproduce, periodically releasing new offspring. Unless treated, a victim of malaria suffers from repeated attacks (chills followed by fever). Each attack corresponds with the release of a new batch of *Plasmodia* from the liver.

The hemoglobin in RBCs of carriers and homozygotes is altered by the sickle-cell gene. For reasons not entirely clear, the presence of altered hemoglobin changes the plasma membrane of RBCs, preventing the parasite from entering. Homozygotes and carriers are, therefore, relatively immune to the parasite. In East Africa, where the parasite is common, 45% of the black population are carriers. In the United States, malaria is virtually nonexistent, and the frequency of the sickle-cell allele has decreased considerably.

Albinism. **Albinism** is one of the most common genetic defects known to science. Albinism is an autosomal recessive trait that results in a metabolic deficiency in the production of melanin, the brown

FIGURE 18–5   **Albinism**   Albinism is an autosomal recessive trait. Albino husband and wife from India.

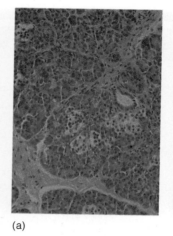
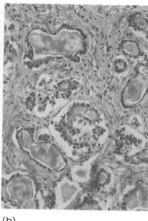

(a)   (b)

FIGURE 18–6   **Pancreas from Patient with Cystic Fibrosis**   (a) Normal pancreas. Cystic fibrosis is a disease caused by an autosomal recessive gene. (b) It results in a blockage of the ducts draining the pancreas, leading to cysts in the pancreatic tissue. The tissue degenerates and is replaced by fibrous connective tissue.

pigment responsible for coloration of the eyes, skin, and hair. Individuals with the disorder may have no melanin at all or may have reduced levels (Figure 18–5). Consequently, the skin of an albino is pale and the hair is white. The eyes are usually pink. Albinism occurs in 1 of every 37,000 Caucasian births, but 1 of every 15,000 blacks. Among the Hopi and Navajo Indians of the desert in the American Southwest, the incidence is 1 in every 200.

Melanin in the skin protects against the effects of ultraviolet light. Its absence in albinos makes them highly susceptible to sunburn and skin cancer. The lack of pigment in the retina of the eyes may result in visual problems and blindness.

Cystic Fibrosis.   Cystic fibrosis is an autosomal recessive disease, which, like sickle-cell anemia, leads to early death. The presence of two recessive genes for the disease leads to problems in sweat glands, mucous glands, and the pancreas. Defective sweat glands release excess amounts of salt, a marker that helps physicians diagnose the disease. In victims of cystic fibrosis, the ducts that drain digestive enzymes from the pancreas into the small intestine become clogged, thus impairing digestion. The buildup of enzymes results in the formation of cysts in the pancreas. Over time, the pancreas begins to degenerate and fibrous tissue replaces glandular tissue (Figure 18–6). Hence the name cystic fibrosis.

Despite adequate nutritional intake, victims of cystic fibrosis often show signs of malnutrition. To enhance digestion, patients are often given powdered or granular extracts of animal pancreases containing di-

gestive enzymes. Massive doses of vitamins and nutrients are also given.

The respiratory systems of most victims of cystic fibrosis produce copious amounts of mucus. Mucus blocks the respiratory passages, making breathing difficult. Victims must be treated several times a day to remove the mucus (Figure 18–7). Despite antibiotic therapy and other treatments, most cystic fibrosis patients live only into their late teens or early twenties.

Cystic fibrosis is one of the most common genetic diseases known to medical science. Surprisingly, 1 of every 25 Caucasians carries a gene for this disease, and approximately 1 of every 2000 Caucasians born in the United States suffers from it. In the black population, the incidence is about 1 in 100,000 or 1 in 150,000.

Autosomal Dominant Traits

Many human traits are **autosomal dominant**—that is, they are carried on the autosomes and are expressed in heterozygotes and homozygote dominants. Nearly 1200 traits in humans have been identified as autosomal dominants, and another 1000 are suspected. The absence of dermal ridges (which give rise to fingerprints), short fingers and toes, freckles, cleft chin, nearsightedness, and drooping eyelids are all autosomal dominant traits. This section discusses three

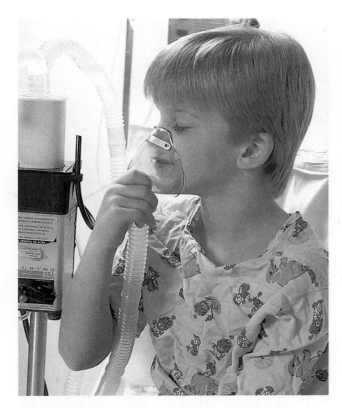

FIGURE 18–7 **Cystic Fibrosis** Inhalants, antibiotics, and special physical therapy techniques are used to treat victims of cystic fibrosis. To remove mucus from the lungs, parents or physical therapists must treat the victim two or three times a day. Pounding on the rib cage with a cupped hand (clopping) loosens mucus.

other examples: widow's peak, achondroplasia, and Marfan's syndrome.

Widow's Peak. Take a moment or two to examine the hairlines of your friends and classmates. You will notice that in some individuals the hairline runs straight across the forehead. In others, it juts forward in the center, forming a "widow's peak" (Figure 18–8). Widow's peak results from an autosomal dominant gene, indicated by W. Since the W allele is dominant, this phenotype is expressed in homozygous dominant individuals (WW) and also heterozygotes (Ww). Individuals with the genotype ww have a continuous hairline.

Achondroplasia. The young man shown in Figure 18–9 suffers from a genetic disease called **achondroplasia**, one form of dwarfism (Chapter 15). Victims of the disease have short, stubby legs and arms, but a

(a)   (b)

FIGURE 18–8 **Widow's Peak** (*a*) Widow's peak is a dominant trait carried on one of the autosomes. (*b*) A straight hairline is a recessive trait. Simple Mendelian genetics can be used to determine the genotype of the offspring.

relatively normal-size body. As a rule, they never grow more than four feet tall.

Achondroplasia afflicts about 1 of every 10,000 children born in the United States and results from an autosomal dominant gene. Most cases are believed to arise from spontaneous mutations, because most children with the condition are born to phenotypically normal parents.

Marfan's Syndrome. President Abraham Lincoln died from an assassin's bullet. More than a hundred years after his death, however, Lincoln has become the subject of scientific controversy. The debate isn't over who killed him or why, but rather over the possibility that Lincoln suffered from a genetic disorder called **Marfan's syndrome**.

Marfan's syndrome is an autosomal dominant disorder that affects the skeletal system, the eye, and the cardiovascular system. It is difficult to diagnose and is usually identified only after its victims die. Flo Hyman, the 1984 star of the U.S. Women's Olympic Volleyball team, and at one time believed to be the best woman volleyball player in the world, had Marfan's syndrome (Figure 18–10). In 1986, Flo was taken out of a game because of ill health. Shortly afterward, she collapsed on the floor and died. Upon autopsy, pathologists found that her aorta had ruptured, killing her instantly.

Marfan's victims are characterized by exceedingly long arms and legs, making them excellent candidates

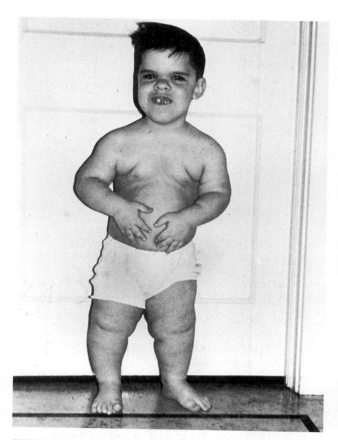

FIGURE 18–9 Achondroplasia Achondroplasia, one form of dwarfism, results from an autosomal dominant trait.

FIGURE 18–10 Marfan's Syndrome Flo Hyman, volleyball player extraordinaire, suffered from Marfan's syndrome, an autosomal dominant trait.

for volleyball and basketball. In normal individuals, the arm span (the distance from fingertip to fingertip) is equal to their height. In Marfan's syndrome, however, an individual's arm span exceeds the height.

Nearsightedness and lens defects are also common among Marfan's patients. The most serious problem, however, is enlargement and weakening of the aortic arch, which results from a weakening of the connective tissue in the wall of the aorta. This causes the aorta to gradually enlarge; if untreated, it will burst.

As for Abe Lincoln, no one really knows for sure whether he had the disorder. Some evidence suggests that he did. Lincoln was tall and lanky, for example, and he wore glasses to correct his vision. Lincoln's great-great-grandfather had the disease, making it possible that Abe had received the Marfan's gene. The critical thinking skills section, however, reminds us of the importance of a close examination of the facts. Lincoln's glasses were to correct farsightedness, not

myopia, a characteristic of victims of the disease. Lincoln also had shown no sign of cardiovascular disease, and his limbs were within the normal dimensions of tall people.

VARIATIONS IN MENDELIAN GENETICS

Mendel presented his work at meetings of the Natural Science Society in Czechoslovakia in 1865 and published his results the following year. Like many ideas ahead of their time, Mendel's conclusions went largely unnoticed. It was not until 1900, 16 years after he died, that his work received the attention it deserved. At that time, the publication of three other studies confirmed Mendel's findings. More studies followed and excitement began to grow. The scientific community came to realize that Mendel's principles pertained to a great many organisms. New research, however, also uncovered additional modes of inheritance and gene expression that are more common than simple dominance.

Incomplete Dominance

One example is a phenomenon called incomplete or partial dominance. **Incomplete dominance** occurs when heterozygous offspring exhibit intermediate phenotypes. In other words, incomplete dominance produces F_1 offspring with phenotypes intermediate to the parental phenotypes. Incomplete dominance occurs in a plant called *Mirabilis*. Figure 18–11 shows a cross between two plants, one with red flowers *(RR)* and one with white flowers *(rr)*. This cross produces offspring with pink flowers, an intermediate phenotype (Figure 18–11). If the *R* gene were completely dominant, you would expect the F_1 offspring to be red. In this case, however, the gene does not exert complete dominance. Incomplete dominance also occurs in a number of human traits.

Multiple Alleles and Codominance

Mendel studied seven characteristics of peas. Each characteristic is determined by a gene with two alleles—that is, two alternative forms. In human genetics, some genes have more than two alleles; such a gene is said to have **multiple alleles**.

Consider blood types (Chapter 10). The gene that controls blood type is called the **I gene** (for isoagglutinin). This gene can exist in one of three biochemically distinct forms. The three alleles are I^A, I^B, and I^O. The *I* gene is carried at one region (or locus) on one pair of chromosomes. Therefore, even though there are three possible alleles in human beings, an individual can only have two of the alleles in his or her genome.

As you may recall, four possible blood types exist: A, B, AB, and O. Table 18–3 lists the four blood types and the six different genotypes that give rise to them. The *I* gene codes for glycoproteins (proteins with carbohydrate attached) that project from the surface of the RBC. The A and B alleles of the *I* gene produce slightly different glycoproteins, the A and B glycoproteins, illustrated in Figure 18–12. The O allele produces no cell surface glycoproteins. Both A and B are dominant genes, and the O allele is recessive. Therefore, in AO individuals, the RBCs contain only type A glycoproteins (Figure 18–12). In BO individuals, the type B glycoproteins are present in the plasma mem-

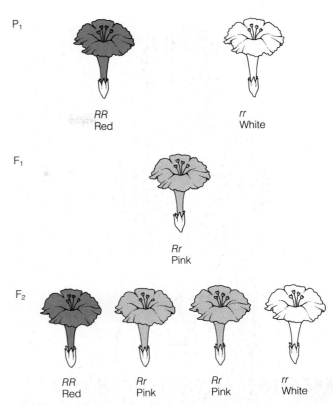

FIGURE 18–11 Incomplete Dominance Incomplete dominance involves two alleles, neither of which is dominant over the other. The result is an intermediate phenotype as shown here in the flower of the plant *Mirabilis*.

brane. When the A and B alleles are both present, the result is a cell surface with both types of surface molecule. The I^A and I^B genes are said to be codominant. **Codominant genes** are expressed fully and equally. The result is a cell with both type A and type B glycoproteins.

TABLE 18–3	ABO Blood Types
GENOTYPES	PHENOTYPES
$I^A I^A$, $I^A I^O$	Type A
$I^B I^B$, $I^B I^O$	Type B
$I^A I^B$	Type AB
$I^O I^O$	Type O

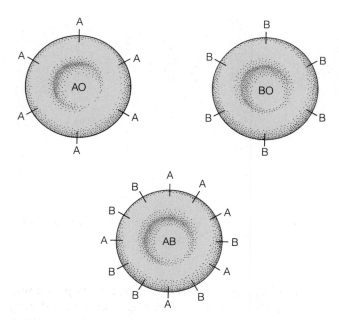

FIGURE 18–12 Codominance Codominant genes are expressed fully when present in the same cell. In blood type AB, both A and B genes produce their characteristic glycoproteins. Type A (genotype AO) cells have only type A glycoprotein, and type B (genotype BO) cells have only type B glycoprotein. Type O cells have neither.

Codominance and incomplete dominance are two exceptions to Mendel's principles. These examples do not negate Mendel's discoveries, but are simply addi-tional modes of gene expression. Mendelian princi-ples, in fact, can even be used to predict the genotypic ratios and patterns of inheritance in instances of in-complete dominance and codominance.

Polygenic Inheritance

A rule of thumb presented in Chapters 4 and 5 is that each gene controls a single trait. In humans, however, most traits are controlled by not one but a number of genes—from a few to hundreds of different genes. Skin color, for example, is controlled by as many as eight genes. This type of inheritance is called **polygenic inheritance**. Polygenic inheritance results in incredible phenotypic variation.

For simplicity, consider an example using two genes for skin color, designated A and B. The geno-type of a black person is *AABB*. The genotype of a Caucasian is *aabb*. Because there are several possible genotypes, skin color exists on a continuum from white to black (Table 18–4 and Figure 18–13).

Polygenic inheritance helps explain variation in skin color. How many gene pairs are involved in these traits, however, remains unknown. Polygenic inherit-ance is also responsible for height, weight, intelli-gence, and a number of behavioral traits. As a rule of thumb, the genotype establishes the range in which a phenotype will fall, but environmental factors deter-mine exactly how much of the potential will be real-ized.

TABLE 18–4 Possible Skin Color Genotypes and Phenotypes with Two Skin Color Genes*

GENOTYPE	PHENOTYPE	NUMBER OF RECESSIVE GENES
AABB	Black	0
AABb	Dark	1
AaBB	Dark	1
AaBb	Mulatto	2
AAbb	Mulatto	2
aaBB	Mulatto	2
Aabb	Light	3
aaBb	Light	3
aabb	White	4

*Skin color probably involves many more genes.

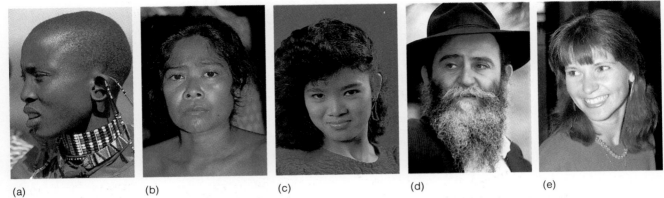

(a)   (b)   (c)   (d)   (e)

FIGURE 18–13   Polygenic Inheritance   Skin color and height are probably determined by at least two genes, resulting in a wide range of phenotypes. (*a*) Black, (*b*) dark, (*c*) mulatto, (*d*) light, and (*e*) white.

Linkage

Mendel found that during gamete formation (meiosis), the seven genes under study were segregated independently. As noted earlier, independent assortment generally occurs only when the genes under study are on different chromosomes. As a rule, if two genes are on the same chromosomes, they do not segregate independently.

In humans there are an estimated 100,000 genes on the 46 chromosomes. Those genes found on the same chromosome tend to be inherited together and are said to be **linked**. To illustrate the concept, imagine that you were studying two traits. We'll designate the dominant form of the first trait as *A* and the recessive allele as *a*. The dominant form of the second trait is designated *B* and the recessive form as *b*. In this example, we will cross a homozygous dominant individual with a homozygous recessive mate (Figure 18–14a). The result is an F_1 generation that is entirely heterozygous (*AaBb*). Notice that four different gametes are produced by the F_1 offspring. If an F_1 heterozygote mates with another heterozygote, the result is an F^2 generation with nine distinct genotypes and four different phenotypes.

If the two genes are on the same chromosome, however, the outcome changes dramatically. As Figure 18–14b shows, the F_1 generation is heterozygous, but produces only two types of gametes, *AB* and *ab*, because the genes are linked. Consequently, only two phenotypes are produced in the F^2 generation.

As a rule, independent assortment does not occur when genes are linked. There is one exception—a phenomenon called **crossing over**. Crossing over was discussed in Chapter 5. Crossing over occurs during meiosis. When the homologous chromosomes unite in prophase I, many of them exchange strands of chromatin. During this process, a segment of one chromosome is exchanged with the corresponding section of the homologous chromosome.

Figure 18–15 presents a hypothetical example. For the sake of illustration, suppose that the exchange involves the section of the chromosome bearing the *A* and *B* genes. Suppose the break occurs between the two genes in the production of female gametes. As illustrated in Figure 18–15, because of crossing over, four distinct gametes can be produced, instead of the two that would be expected.

Crossing over produces additional genetically different gametes. This leads to additional genetic variation in offspring or more genetic combinations. The more variation, the more genotypes in a population. As Chapter 19 points out, variation is essential to evolution. Genetic variants may have characteristics that give one organism an advantage over another.

Crossing over can occur anywhere along the length of a chromosome. However, the greater the distance between two genes on the same chromosome, the more likely it is that a crossover will occur between them.

SEX-LINKED GENES

Sex is determined by the Y chromosome. In humans, the presence of the Y chromosome results in the male

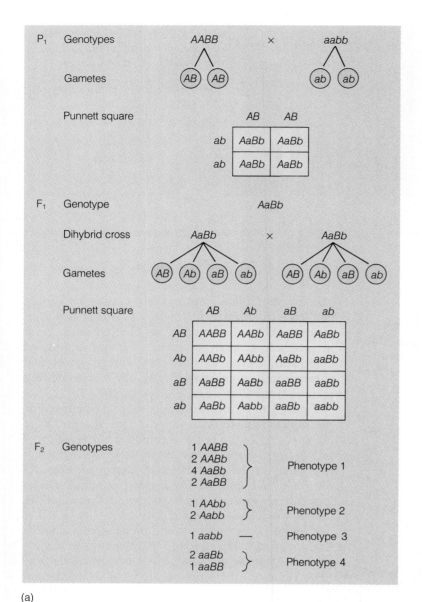

(a)

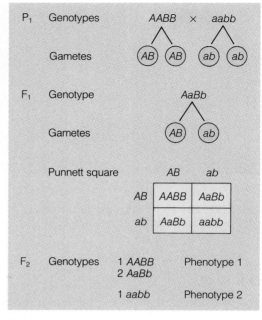

(b)

FIGURE 18–14 Linkage (*a*) Hypothetical dihybrid cross with no linkage. (*b*) Hypothetical dihybrid cross with linkage. Linkage reduces the number of genotypes in the offspring.

phenotype. The gonads in male and female embryos develop identically at first. When the Y chromosome is present, however, the embryonic gonad becomes a testis. That is because the Y chromosome contains a gene *t* (testes-determining gene) responsible for the differentiation of the gonad into a testis.[4] The testis, in turn, produces testosterone and other androgens, which are responsible for the male secondary sex

characteristics (Chapter 16). Genes on the Y chromosome also control spermatogenesis. The absence of the Y chromosome results in the development of an ovary and a female phenotype.

The X and Y chromosomes also carry genes that determine a great many other traits. A gene on a sex chromosome is known as a **sex-linked** gene. Most sex-linked genes known to science are located on the X chromosome and are, therefore, known as X-linked genes. The following sections describe the inheritance of some common sex-linked genes.

[4]Sex determination may require one or more genes.

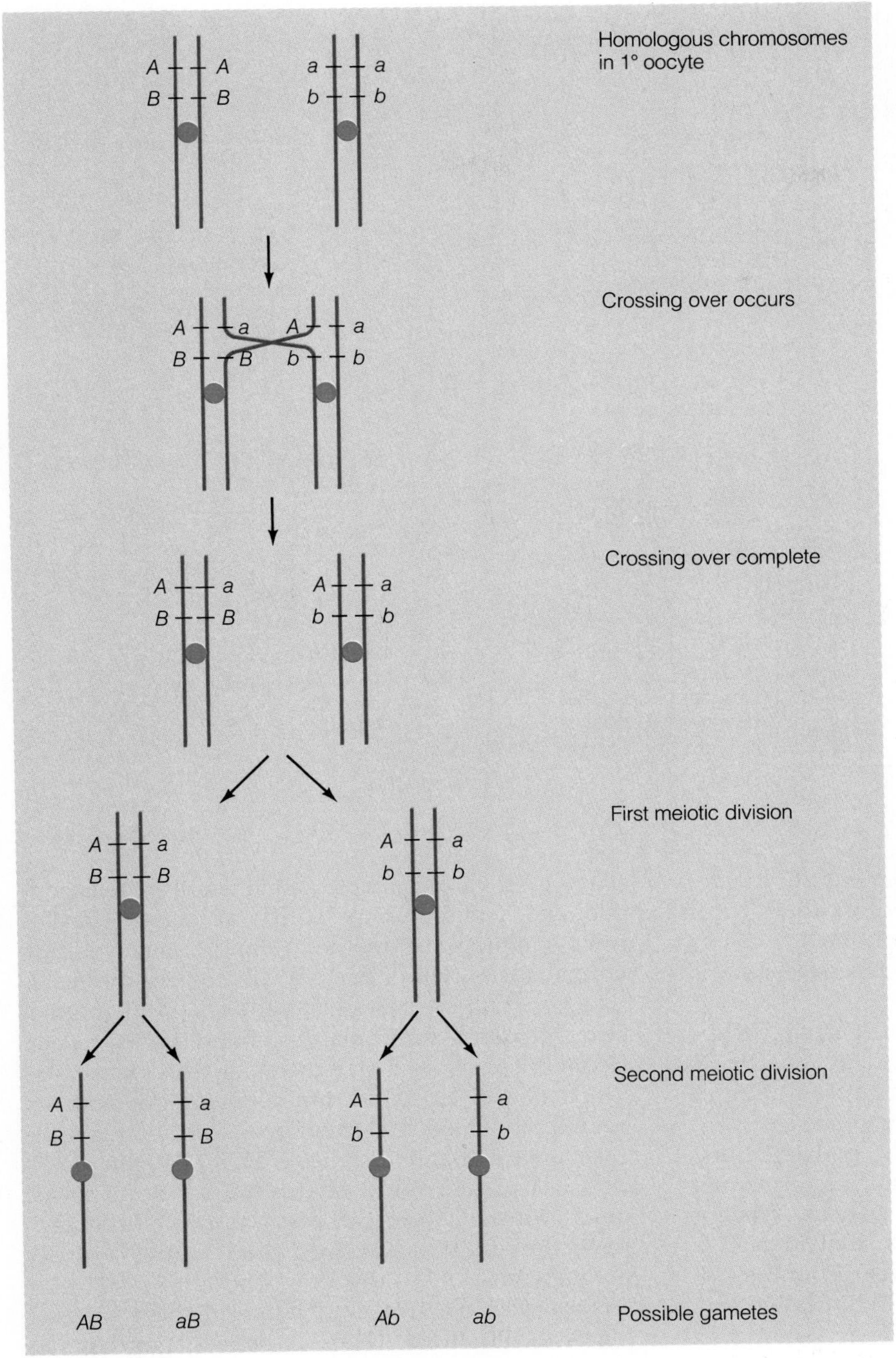
Homologous chromosomes in 1° oocyte
A A B B
a a b b
Crossing over occurs
A a B B
A a b b
Crossing over complete
A a B B
A a b b
First meiotic division
A a B B
A a b b
Second meiotic division
A B
a B
A b
a b
Possible gametes
AB
aB
Ab
ab

FIGURE 18–15 Hypothetical Crossing Over involving a Homologous Pair of Chromosomes Crossing over increases genetic variation in gametes and offspring.

Recessive X-Linked Genes

At least 124 genes have been assigned to the X chromosome; 160 or more are thought to be located there as well. Color blindness, discussed in Chapter 13, for instance, is a recessive trait carried on the X chromosome. In addition, certain forms of hemophilia are caused by a recessive X-linked gene.

In order for a female to display a recessive sex-linked trait, each of her X chromosomes must carry

FIGURE 18–16 Inheritance of Color Blindness Four possible genetic ways a sex-linked recessive gene like color blindness can be passed to offspring.

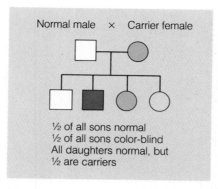

Normal male × Carrier female

½ of all sons normal
½ of all sons color-blind
All daughters normal, but
½ are carriers

(a)

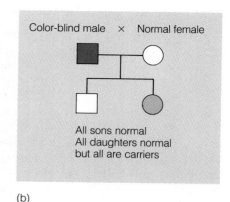

Color-blind male × Normal female

All sons normal
All daughters normal
but all are carriers

(b)

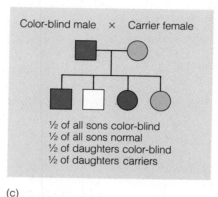

Color-blind male × Carrier female

½ of all sons color-blind
½ of all sons normal
½ of daughters color-blind
½ of daughters carriers

(c)

Normal male × Color-blind female

All sons color-blind
All daughters carriers

(d)

the recessive gene. For males, however, only one recessive gene is required. That is because the Y chromosome is not genetically equivalent to the X chromosome.[5] Thus, in men only one recessive gene is needed to exhibit the trait.

Figure 18–16 shows four possible genetic combinations leading to color blindness. This illustration introduces you to a genetic tracking system used to follow traits in families. The boxes represent men. The circles represent women and the horizontal line linking a box to a circle (□—○) indicates a mating. The offspring are shown below the parents. When a box or square is lightly shaded, the individual is a carrier of the gene, who does not suffer from the disease. A darkly shaded box or circle indicates the person has the disease.

In 18–16a, for example, a man and a woman have four children, two boys and two girls. The woman

(circle) is a carrier of color blindness. Her cells contain one X chromosome with a recessive gene for color blindness and another X chromosome with the normal allele. Half of her ova will contain the recessive gene, and the other half will contain the normal gene. As shown, the woman's husband is not a carrier. He produces sperm with either X or Y chromosomes. When one of his X-bearing sperm unites with an ovum carrying an X chromosome with the recessive gene for color blindness, the result is a daughter (XX) who is a carrier. When one of his X-bearing sperm unites with an ovum carrying a normal X chromosome, the result is a daughter who is neither a carrier nor a victim.

Now what about male children of these parents? Males are produced when a Y-bearing sperm unites with an ovum carrying an X chromosome. If the X chromosome carries the recessive gene for color blindness, the boy is color-blind. If the X chromosome is normal, the boy's color vision is unimpaired. Take a moment to study the other possibilities in Figure 18–16.

[5]The X and Y chromosomes are believed to share few genes, and only part of the Y chromosome is homologous with the X.

Dominant X-Linked Genes

Although recessive X-linked genes are the most common type of sex linkage seen in human beings, there are a few noteworthy examples of dominant X-linked genes. One of the best understood is a disorder with a tongue-twisting name of **hypophosphatemia**—low phosphate levels. This genetic disorder results in a form of rickets or bowleggedness. Rickets usually results from a dietary deficiency of vitamin D or insufficient exposure to sunlight (Chapter 7). Alleviating the dietary deficiency usually solves the problem. In this genetic disease, however, vitamin D cannot reverse the symptoms.

Hypophosphatemia occurs when either the male (XY) or the female (XX) has one X chromosome bearing the dominant gene (X'). Figure 18–17 illustrates the pattern of inheritance when a woman who is heterozygous for the trait mates with a man who does not carry the trait.

Y-Linked Genes

Y-linked genes are those genes found only on the Y chromosome. Genes that affect gonadal differentiation, spermatogenesis, and other secondary sex characteristics are thought to be located on the Y gene. Y-linked genes have a simple but fairly distinct pattern of inheritance. Because only males have Y chromosomes, Y-linked traits only appear in males, and Y-linked genes are transmitted only from fathers to sons. The X and Y chromosomes in men are not homologous, thus each gene on the Y chromosome has only one allele, which is always expressed.

Sex-Limited Genes

Certain autosomal genes are expressed only in one sex or the other and are, therefore, called **sex-limited genes**. Consider several examples. Full beard development occurs only in men. It results from testosterone, which stimulates genes that control hair growth on the face. The hair-growth genes are also present in women, but normally are not activated because testosterone levels in women are low. Breast development in women is another example. Genes in the cells of the breasts are stimulated by female sex steroids. Those same genes are found in the autosomes of men, but are not activated because of the relatively low levels of female sex steroids in men.

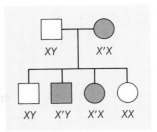

FIGURE 18–17 Inheritance of a Sex-Linked Dominant Gene

Sex-Influenced Genes

Certain autosomal genes behave differently in the different sexes. In one sex, for example, an allele will be dominant; in the other sex, it will be recessive. These genes are known as **sex-influenced genes**. The best-known example is the gene for pattern baldness, which is present in men and women. Pattern baldness is the loss of hair that often begins in a man's twenties. Affected individuals do not go completely bald, but retain a rim of hair on the temples and back of the head. The allele for pattern baldness is dominant in men and is, therefore, expressed in both heterozygous and homozygous individuals. In women, however, the allele is recessive. Women who are homozygous for the trait will exhibit baldness.

CHROMOSOMAL ABNORMALITIES AND GENETIC COUNSELING

A young couple waits in their doctor's office for the results of a genetic test on their unborn baby. Studies of chromosomes from cells removed from the amniotic fluid during amniocentesis, discussed in Chapter 4, have revealed a chromosome abnormality. Abnormal chromosome numbers are surprisingly common in humans, but this is only one of several genetic defects with which physicians and parents must contend. Changes may also occur in the DNA. Such mutations may be beneficial, but many are lethal to cells. Still others can lead to debilitating diseases. Chromosomes can also be torn apart in meiosis, resulting in missing segments. A segment from one chromosome may attach to another, which can also lead to serious medical problems. This section examines some chromosome abnormalities—how they arise and the problems they create.

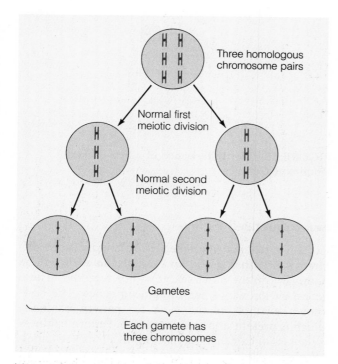

(a)

(b)

FIGURE 18–18 Meiosis and Abnormal Chromosome Numbers (a) A simplified version of meiosis. During the first meiotic division, the homologous pairs line up, then separate, producing daughter cells with one-half the number of chromosomes. In the second meiotic division, the chromosomes line up single file and separate with one chromatid going to each daughter cell. (b) Nondisjunction in the first meiotic division. A chromosome pair may fail to separate during meiosis I, resulting in abnormal gametes. Half are missing a chromosome, and the other half have an extra chromosome.

Abnormal Chromosome Numbers

Alterations in the number of chromosomes result chiefly from errors in meiosis (gamete formation). During meiosis I, homologous chromosomes pair, then separate. One double-stranded (replicated) chromosome migrates to each pole, the other chromosome migrates to the other pole (Figure 18–18a). If a homologous pair fails to separate during meiosis, however, one of the new cells will end up with an extra chromosome (Figure 18–18b). The other cell will be short one chromosome. The failure of homologous chromosomes to separate is called **nondisjunction**.

Nondisjunction can also occur in the second meiotic division. In this division, you may recall, the 23 double-stranded (replicated) chromosomes split apart, with one chromatid going to each daughter cell. If a chromosome fails to separate into its two chromatids, the result is the same as nondisjunction in meiosis I—a daughter cell with an extra chromosome and another daughter cell missing one chromosome.

When a gamete with an extra chromosome unites with a normal gamete, the resulting zygote will contain 47 chromosomes. The zygote may be able to divide, producing an embryo whose cells have an additional chromosome. Thus, instead of the normal 23 chromosome pairs, the cells will contain 22 pairs and one triplet. This condition is called **trisomy** (three bodies).

Gametes with a missing chromosome can unite with normal gametes, producing individuals with 45 chromosomes—22 chromosome pairs and a chromosome singlet. This condition is called **monosomy**.

Aneuploidy: Trisomies and Monosomies. Monosomy and trisomy are collectively referred to as **aneuploidy**. Aneuploidy (literally, "not a true number") has profound effects on human reproduction and de-

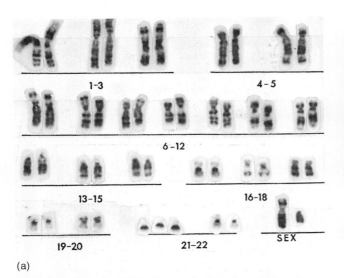

(a)

(b)

FIGURE 18–19 Down Syndrome (a) Karyotype of Down syndrome girl. (b) Notice the distinguishing characteristics described in the text.

velopment. One of every two conceptions is aneuploid. Aneuploidy is responsible for 70% of all early embryonic deaths and 30% of all fetal deaths. Since most aneuploid embryos and fetuses die *in utero*, aneuploidy is also associated with an increased miscarriage rate in older mothers.

Down Syndrome. One of the most common trisomies is **Down syndrome** or **Trisomy 21**.[6] Approximately 1 of every 700 babies born in the United States has Down syndrome. Down syndrome children are typically short and have round moonlike faces (Figure 18–19). Their tongues protrude forward, forcing their mouths open. Down syndrome children also have Mongol-like eyes, which slant upward at the corners, and are mentally retarded with IQs rarely over 70. A significant number of Down syndrome babies die in the first year of infancy from heart defects and respiratory infections. Modern medical care, however, has helped reduce early death, and many Down syndrome patients live to about age 20.

The incidence of Down syndrome (and many other aneuploidies) increases with maternal age. As Figure 18–20 shows, after age 35, a woman's chances of having a Down syndrome baby increase dramatically. For this reason, many pregnant women over 35 opt for amniocentesis. If the test shows that the fetus has

Down syndrome, a couple may decide to have an abortion. Amniocentesis also helps parents for whom abortion is not an option to prepare psychologically for a Down syndrome child and to receive the education to help them care for their child.

Some researchers hypothesize that the rise in Down syndrome results from exposure to radiation or potentially harmful chemicals. All of the primary oocytes that a woman will have during her lifetime are present in her ovaries at birth. Thus, as a woman gets older, the more likely her ovaries will be exposed to some potentially harmful chemical or radiation from natural or human sources. The more exposure to potentially such agents, the more likely a nondisjunction will occur.

Nondisjunction of the Sex Chromosomes. Nondisjunction of the sex chromosomes leads to a variety of genetic disorders. For example, if an ovum with two X chromosomes is fertilized by a Y-bearing sperm, the result is a XXY genotype. This condition, known as **Klinefelter syndrome**, occurs in about 1 of every 700 to 1000 live males born.[7] Although Klinefelter syndrome victims are males, masculinization is incomplete. The victims' external genitalia and testes are unusually small, and about half of the victims develop breasts (Figure 18–21). Spermatogenesis is abnormal

---

[6]Contemporary geneticists generally refer to this as Down syndrome, rather than Down's syndrome.

[7]Nondisjunction can also occur if a nondisjunction occurs in sperm development, resulting in an XY sperm.

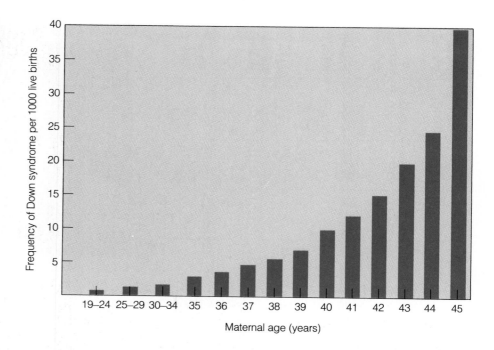

FIGURE 18–20 Incidence of Down Syndrome Babies versus Maternal Age The incidence of Down syndrome rises quickly after maternal age 35.

Frequency of Down syndrome per 1000 live births

Maternal age (years)

and Klinefelter patients are generally sterile.

Another common disorder involving the sex chromosomes is **Turner syndrome**, a monosomy. Turner syndrome results when an ovum lacking its X chromosome is fertilized by a X-bearing sperm. It may also result when a genetically normal ovum is fertilized by a sperm lacking an X or Y chromosome. In either case, the result is an offspring with 22 pairs of autosomes and a single, unmatched X chromosome (X0).

Turner syndrome patients are phenotypically female and are characteristically short with wide chests and a prominent fold of skin on their necks (Figure 18–22). Because the ovaries fail to develop at puberty, Turner syndrome patients are sterile, have low levels of estrogen, and have small breasts.[8] For the most part, Turner syndrome patients lead fairly normal lives. Mental retardation is not associated with the disorder, although some studies suggest that Turner patients are not as capable at numerical skills and spatial perception as genetically normal children. Turner syndrome occurs in 1 of every 10,000 live female births.

Polyploidy. Occasionally, zygotes are formed with a complete extra set of chromosomes. Instead of having

[8]Some estrogen is released from the adrenal cortex, but not enough to permit breast development.

the normal 46 chromosomes, the cells have 69 — that is, 23 triplets rather than 23 pairs. This condition, known as **triploidy**, results when a normal (haploid) gamete combines with a gamete that has twice the normal number of chromosomes (a diploid gamete). A diploid gamete can be produced during meiosis by a complete nondisjunction — that is, a complete failure of chromosome separation during meiosis I or meiosis II.

Triploidy occurs in 1 of every 100 conceptions, but over 99% of all triploid embryos die and are aborted or resorbed. Thus, the incidence of triploidy at birth is fairly small — 1 in 10,000 live births. Most of the triploid offspring that are born die shortly after birth.

Rarer still is tetraploidy. **Tetraploidy** occurs when an individual's cells have two complete sets of chromosomes. In other words, instead of having 46 chromosomes, the cells have 92. Tetraploidy probably arises during the first mitotic division after fertilization. Researchers believe that nuclear division occurs normally, but it is not followed by cytokinesis — division of the cytoplasm. The result is a cell with two complete sets of chromosomes. Although tetraploid cells may divide normally, problems soon arise, and most of the tetraploid embryos are aborted or absorbed. Only rarely is a tetraploid baby born. Tetraploidy and triploidy are collectively referred to as **polyploidy**.

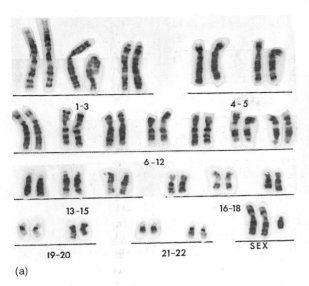

(a)

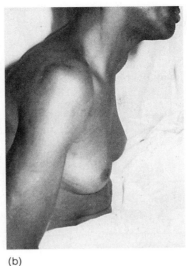

(b)

FIGURE 18–21
Klinefelter Syndrome
(*a*) Karyotype of Klinefelter. Notice the two X chromosomes. (*b*) Breast development in a male with Klinefelter syndrome.

Variations in Chromosome Structure

Variations may also occur in chromosome structure. These most often result from two occurrences: **deletions**, the loss of a piece of chromosome, and **translocations**, breakage followed by reattachment elsewhere.

Deletions.   Most deletions are fairly deleterious to life, and nature eliminates many offspring with deletions fairly early in embryonic life. Nevertheless, there are a number of examples (Table 18–5). One of the more striking is called **Praeder-Willi syndrome**. This condition, characterized by slow infant growth, compulsive eating, and obesity, occurs when one of the arms of chromosome 15 breaks off during gamete formation. Babies born with the syndrome are weak. They have a poor suckling reflex and do not feed well. By age five or six, however, children become compul-

sive eaters. Parents must lock their cupboards and refrigerators. Neighbors must be warned to discourage begging and must keep their garbage cans under lock and key. The urge to eat results in obesity, which often leads to diabetes. If food intake is not restricted, victims will literally eat themselves to death.

Researchers believe that the eating disorder may result from an endocrine imbalance. Praeder-Willi syndrome occurs in an estimated 1 in 10,000 to 1 in 25,000 births.

Translocations.   Translocations occur when a segment of a chromosome breaks off, but reattaches to another site on the same chromosome or to another chromosome. The movement of a segment of a chromosome to another site can upset the delicate balance of gene expression. Translocations, for example, may be the cause of certain forms of leukemia.

TABLE 18–5   Chromosome Deletions

SYNDROME	PHENOTYPE
Wolf-Hirschhorn syndrome	Growth retardation, heart malformations, cleft palate; 30% die within 24 months.
Cri-du-chat syndrome	Infants have catlike cry, some facial anomalies, severe mental retardation.
Wilm's tumor	Kidney tumors, genital and urinary tract abnormalities.
Retinoblastoma	Cancer of eye, increased risk of other cancers.
Praeder-Willi syndrome	Infants weak, slow growth; in children and adults, obesity and compulsive eating.

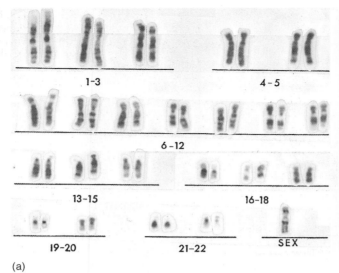

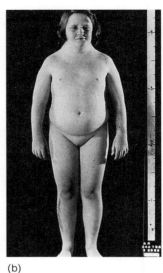

FIGURE 18–22 **Turner Syndrome** (*a*) Karyotype of Turner syndrome. Notice the single X chromosome. (*b*) Characteristic physical features of a Turner syndrome girl.

1-3   4-5

6-12

13-15   16-18

19-20   21-22   SEX

(a)   (b)

Genetic Screening and Genetic Counseling

Thanks to advances in modern medicine, parents can now find out the sex and genetic makeup of their child well before birth. One of the procedures for studying the genome of an unborn child, amniocentesis, was discussed in Chapter 4. Removing cells and amniotic fluid via amniocentesis and studying the fetus's chromosomal makeup and characteristic banding patterns allows doctors to examine hundreds of chromosomal and biochemical disorders before birth, although only a dozen or so are routinely screened.

Amniocentesis increases the risk of spontaneous abortion by about 1%. It also slightly increases the risk of maternal uterine infection. Therefore, this procedure is usually recommended only if (1) a woman is over 35, (2) she has already delivered a baby with a genetic defect, (3) she is a carrier of an X-linked biochemical disorder, or (4) the mother or the father has a known chromosomal abnormality.

Amniocentesis cannot be performed until the 16th week of pregnancy. Prior to this time, there is not enough fluid in the amnion surrounding the fetus, and the needle inserted into the amnion could damage the fetus. Analysis of the fetal cells withdrawn from the amnion requires another 10 to 15 days. If a serious defect is observed, a couple may elect to have an abortion, but abortions this late in pregnancy are more risky than those performed earlier (Chapter 16).

To permit earlier detection of genetic defects, a new procedure, known as **chorionic villus biopsy**, has been developed. As Chapter 17 noted, placental or chorionic villi form from embryonic tissue early in development. Inserting a catheter into the uterus through the vagina allows physicians to remove a small sample of the villi (Figure 18–23). The cells of the villi are then examined for chromosomal abnormalities. If a defect is found, an abortion can be performed earlier—at 8 to 12 weeks gestation, which is safer than an abortion at 16 to 20 weeks. Although chorionic villus biopsy allows for earlier detection, the procedure is not as safe as amniocentesis. Biopsies pose a greater threat to the mother and her fetus.

A Final Note: Chromosomal Abnormalities in Mitochondria

Mitochondria contain about 0.3% of a cell's DNA. The genes on this DNA code for some of the enzymes mitochondria need to produce ATP. Researchers recently discovered a rare form of blindness in humans linked to a defect in the mitochondrial DNA. This confirms suspicions that defective mitochondrial genes can result in genetic defects and suggests an additional mechanism for inheriting genetic diseases.

The defective gene in the mitochondria that leads to blindness codes for a protein required in the first step of ATP production. The absence of this protein results in optic nerve death, producing blindness by age 20. Since mitochondria are passed on by the mother, every one of her children will inherit the defective gene. However, only a small fraction of the children who inherit the defective gene actually go blind. Thus, it is thought that the mutation probably only predisposes people to blindness. By itself, it is

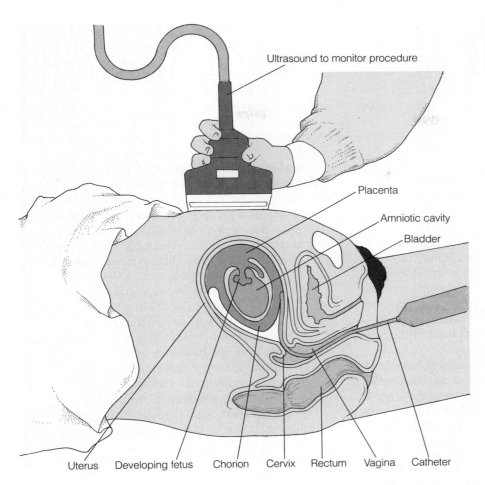

Ultrasound to monitor procedure

Placenta

Amniotic cavity

Bladder

Uterus Developing fetus Chorion Cervix Rectum Vagina Catheter

not sufficient to create blindness. Other factors may contribute to blindness.

Several other rare genetic diseases may also result from mitochondrial DNA defects. Researchers suggest that even some of the more common diseases may be caused by genetic defects in mitochondrial DNA. Some cases of heart, kidney, and central nervous system failure, whose causes are now unknown, may one day prove to be the result of defective mitochondrial DNA.

ENVIRONMENT AND HEALTH: NATURE VERSUS NURTURE

Throughout this book, a central theme has emerged over and over—the idea that humans are influenced profoundly by their environment. Our social, psychological, and physical environments,—even our diet—

have a profound impact on our internal environment and, therefore, our health.

In this section, we examine a controversy that has raged in science for decades: the connection between our genes and our environment. We will look at two basic questions: How do our genes contribute to our personalities? What is the role of our environment in determining our personality and behavior?

At one time, psychologists viewed the baby as a blank slate. A child's personality, they said, develops through interaction with its environment—its parents, friends, teachers, and others. Extensive research suggests, however, that our personalities are also influenced by our genes. Thus, our genes and our environment probably operate together in this matter.

Michael Lewis, a researcher at the Robert Wood Johnson Medical School, studies infant response to stress. His research shows that newborn babies differ markedly in how they respond to the stress of a blood test performed—well before their environment could have affected their personalities. Lewis found that some children wail when poked with a needle during routine blood tests in the first few days of life; others hardly seem to notice. Of the newborns who cry, some quickly dampen their response. Others seem to go on forever.

Lewis notes that a child's reaction to stress at this time is likely to be repeated three months later when the child receives an inoculation. Lewis believes that the difference in response to stress is genetically based and that the inherent differences will persist.

Lewis has also found that babies differ in how they react to frustration. He performed a series of experiments to test infant response to a frustrating situation. Most children responded with anger. Some, however, showed no response at all and others displayed sadness. These innate differences in behavior, occurring too early to stem from differences in upbringing, probably result from genetic differences, says Lewis. They may help account for the profound differences in responses to stress seen in adults.

Psychologist Nathan Fox has performed important studies on shy and extroverted children. His studies show that shy children cling to their mothers when a clown suddenly appears; extroverted children eagerly engage the clown in play. Fox has found that shy children show greater electrical activity in the right part of the brain; extroverted children show a higher level of activity in the left side. Since the children differ at birth in personality and since the personality differences are reflected in sharp differences in brain activity, Fox argues that genetics may be playing a powerful role in personality development.

Allison Rosenberg, a researcher at the National Institutes of Health, has also studied shy and outgoing children. Her work shows marked differences in heart rate and cortisol secretion between these two groups, further supporting the notion that there are inherent physiological differences in children from the outset. These are related to differences in early personality and, very possibly, to differences in genetic makeup.

Infant monkeys display a wide range of personality traits early in life. Research shows that timid animals differ physiologically from their braver counterparts. Stephen Soumi, a researcher at the National Institutes of Health, believes that the individual differences seen in personality and in underlying physiology in response to stress are so pronounced that they must be genetically based. Furthermore, he and colleagues have found that shyness and extroversion tend to run in families—further suggesting a link between behavior early in life and genetics.

Soumi notes, however, that environmental effects, especially events very early in life, can modify genetically programmed behavior. By taking a shy baby monkey and pairing it with an exceptionally nurturant mother, he finds the shy animal can thrive. It develops rapidly and actively explores its environment.

Another example of behavioral modification by environment comes from research on thrill-seeking. Psychologists believe that some individuals are naturally born thrill-seekers. They have labeled them type T people. Researchers believe that thrill-seeking may be genetically based. One theory is that risk takers—the type T or Big T individuals—may be underarousable. They are hard to excite and, therefore, attempt to seek out arousal. At the opposite end of the spectrum are small "t" people, risk avoiders, who are easily arousable. They seek to avoid stimulation.

Some researchers believe that type T personality individuals have an imbalance of a neurotransmitter, known as monoamine oxidase (MO), in the brain. Thrill-seeking supposedly increases the MO levels in the brain, creating a feeling of exhilaration.

Type T behavior can be modified by an individual's upbringing. It can be turned in a negative or positive direction, say some psychologists. A positive direction might lead an individual to play for the Los Angeles Rams. A negative direction might lead that same individual, under different environmental conditions, to gang fighting and crime in the streets of Los Angeles.

Peers, teachers, relatives, ministers, parents, and others comprise the environment; their influence may turn the type T child to healthy, constructive opportunities or unhealthy, destructive ends. A healthy psychological and social environment proves itself a positive asset.

SUMMARY

PRINCIPLES OF HEREDITY: MENDELIAN GENETICS

1. Gregor Mendel, a nineteenth-century monk, derived several important principles of inheritance from his work on garden peas.
2. Mendel first determined that, in peas, traits did not blend as was commonly thought at the time. Mendel also postulated that each adult has two heredity factors (now called genes) for a given trait. The genes separate during gamete formation. This is the principle of segregation.

3. Mendel also postulated that a hereditary factor may be either dominant or recessive. A dominant factor masks a recessive factor. A recessive factor is expressed only when the dominant factor is missing. The dominant and recessive genes are alternative forms of the gene or alleles.

4. Three genetic combinations are possible for a given trait: heterozygous, homozygous dominant, and homozygous recessive.

5. The genetic makeup of an organism is called its genotype. The physical appearance, which is determined by the genotype in its environment, is its phenotype.

6. From his studies, Mendel also concluded that the hereditary factors are separated independently of one another during gamete formation. This is the principle of independent assortment and holds true only for non-linked genes.

MENDELIAN GENETICS IN HUMANS

7. Human cells contain 23 pairs of chromosomes: 22 pairs of autosomes and 1 pair of sex chromosomes. These chromosomes carry dominant and recessive traits, and inheritance of these traits is consistent with Mendel's principles of inheritance, although numerous additional mechanisms are at work in humans.

8. Sickle-cell anemia, cystic fibrosis, and albinism are autosomal recessive traits and are expressed only in homozygous recessive genotypes.

9. Widow's peak, achondroplasia, and Marfan's syndrome are autosomal dominant traits or diseases and are expressed in heterozygous and homozygous dominant genotypes.

VARIATIONS IN MENDELIAN GENETICS

10. Genetic research since Mendel's time has turned up several additional modes of inheritance.

11. One mode is incomplete dominance. Incomplete dominance occurs when an allele exerts only partial dominance, producing intermediate phenotypes.

12. Some genes have more than two possible alleles. Multiple alleles result in more genotypes and phenotypes in a population. Because chromosomes exist in pairs, individuals have only two of the possible alleles.

13. Codominance occurs in multiple allele genes. Codominant genes are expressed fully and equally.

14. Some traits are controlled by many genes. This is called polygenic inheritance.

15. Genes that are found on the same chromosome are said to be linked. If crossing over does not occur, these genes are inherited together.

SEX-LINKED GENES

16. Sex is determined by the sex chromosomes, particularly by the Y chromosome. The sex chromosomes also carry genes that determine other physical traits. A trait determined by a gene on a sex chromosome is a sex-linked trait.

17. Most sex-linked traits occur on the X chromosome. Both dominant and recessive sex-linked traits are present.

CHROMOSOMAL ABNORMALITIES AND GENETIC COUNSELING

18. Abnormalities in the human genome arise from mutations (changes in DNA structure), abnormalities in chromosome number (aneuploidy and polyploidy), and alterations in chromosome structure (deletions and translocations).

19. Alterations in the number of chromosomes result chiefly from errors in gamete formation when chromosomes fail to separate, a process called nondisjunction. An additional chromosome in a sperm or ovum is passed on to the offspring, and the result is a trisomy, an individual with 47 chromosomes.

20. A sperm or ovum missing a chromosome that unites with a normal gamete produces a monosomy, an individual with 45 chromosomes.

21. Nondisjunction of the sex chromosomes during meiosis can result in extra or missing sex chromosomes in offspring.

22. Variations in chromosome structure result from two occurrences: deletions, the loss of a piece of chromosome, and translocations, breakage followed by reattachment elsewhere.

23. Embryos with abnormal chromosome numbers or abnormal chromosome structure are likely to die and to be aborted spontaneously.

ENVIRONMENT AND HEALTH: NATURE VERSUS NURTURE

24. At one time, psychologists thought that a child's personality developed principally through interaction with its environment—its parents, friends, and teachers. New research suggests, however, that our personalities are also influenced by our genes.

The environment and health perspective discusses the role of genes and environment in determining behavior. Discuss the evidence given to support the hypothesis that some early behavioral characteristics result from genes. Using your critical thinking skills, discuss other possible interpretations (if any) of the studies cited in that section.

TEST OF TERMS

1. Many of the basic principles of inheritance arose from the work of the Austrian monk, _____ _____ .

2. During his research, Mendel hypothesized that each adult plant carries a pair of _____ factors, now known as genes. These factors separate during gamete formation, a concept called the principle of _____ .

3. A gene that can only be expressed in the homozygous condition is called a(n) _____ gene.

4. An alternative form of a gene is called a(n) _____ .

5. The genotype Aa is described as _____ .

6. The outward appearance of an organism, or its _____ , is determined by its genotype.

7. A genetic cross in which a single pair of genes is under study is called a(n) _____ cross.

8. The principle of _____ holds true for genes that are not linked.

9. Human cells contain 23 pairs of chromosomes. A female contains a pair of sex chromosomes, _____ , and 22 pairs of _____ .

10. Cystic fibrosis, albinism, and sickle-cell anemia are all _____ genetic disorders. A heterozygote is said to be a(n) _____ of these diseases.

11. Widow's peak, achondroplasia, and Marfan's syndrome are _____ traits or diseases. They occur in the _____ dominant and _____ genotypes.

12. The H gene determines whether hair is curly or straight, but neither gene is dominant. An individual with curly hair mates with an individual with straight hair and the result is an intermediate form, wavy hair. This is an example of _____ .

13. Blood types are determined by the I gene. It has three possible alleles. This phenomenon is called _____ alleles.

14. _____ genes are expressed fully and equally.

15. Skin color and height are probably controlled by two or more genes. They are examples of _____ inheritance.

16. Two genes located on the same chromosome are said to be _____ .

17. A trait carried on a sex chromosome is called a _____ trait.

18. Color blindness results from a _____ gene located on the _____ chromosome.

19. Certain autosomal genes are influenced by sex hormones and are called _____ genes.

20. Monosomy and trisomy are caused by _____ _____ during gamete formation.

21. Down syndrome is also called _____ .

22. Klinefelter and Turner syndrome result from _____ of sex chromosomes.

23. Tetraploidy and triploidy are conditions that are collectively called _____ .

24. Variations in chromosome structure result from _____ , the loss of a piece of chromosome, and _____ , breakage followed by reattachment elsewhere.

25. To permit earlier detection of genetic defects, medical scientists have developed a procedure called _____ _____ biopsy.

Answers to the Test of Terms are located in Appendix B.

TEST OF CONCEPTS

1. Mendel's research was designed to answer two basic questions. What were the questions and what were his findings?

2. Define the following terms: principle of segregation, principle of independent assortment, allele, genome, phenotype, genotype, heterozygous, homozygous, monohybrid cross, and dihybrid cross.

3. Freckles are dominant. A woman with freckles (*Ff*) marries and has a baby with a man without freckles (*ff*). What are the chances that their children will have freckles?

4. Two freckled adults marry and have children. The first baby has no freckles. What are the genotypes of the parents?

5. Attached earlobes (*A*) are dominant over unattached earlobes (*a*).

A woman with freckles and attached earlobes (*FfAa*) marries and wants to have children with a man who has freckles and attached earlobes (*FfAa*). Draw a Punnett square showing the various gametes as well as the genotypes of the offspring. List all possible phenotypes and the genotypes that correspond to them.

6. What is sickle-cell anemia? What causes it? Why can a person be a carrier of the disease but not display outward symptoms?

7. How do incomplete dominance and codominance differ? Give examples of each.

8. Assuming that two genes control height (*A* and *B*), list all of the possible genotypes and indicate the phenotype associated with each.

9. Describe how crossing over works. What impact does it have on the genotype of a person's gametes?

10. Color blindness is a recessive X-linked gene. A color-blind man and his wife have four children. They have two boys and two girls. One boy and one girl are both color-blind, and the other two are normal. What is the genotype of the woman?

11. What is a sex-influenced gene? Give some examples. What criteria would you use to assess whether or not a trait was a sex-influenced trait?

12. Explain how each of the following genetic defects could arise: trisomy 18, monosomy 10, triploidy, and tetraploidy.

SUGGESTED READINGS

Bower, B. 1988. Alcoholism's elusive genes. *Science News* 134(5): 74 — 75, 79. Looks at how alcoholism seems to run in families, and explores the question of whether alcoholism is inherited.

Cummings, M. R. 1988. *Human heredity: Principles and issues.* St. Paul, Minn.: West. Introductory-level coverage of basic human genetics.

Dworetzky, J. P. 1988. *Psychology,* 3d ed. St. Paul, Minn.: West. See Chapter 12 for a discussion of environmental and genetic aspects of intelligence.

Klug, W. S., and M. R. Cummings. 1986. *Concepts of genetics,* 2d ed. Glenville, Ill.: Scott, Foresman and Company. Detailed information on genetics presented in a clear and understandable way.

Maxson, L. R., and C. H. Daugherty. 1989. *Genetics: A human perspective,* 2d ed. Dubuque, Iowa: Wm. C. Brown. Excellent introduction to human genetics.

V

EVOLUTION, ECOLOGY, AND BEHAVIOR

Evolution: Five Billion Years of Change

The Grand Canyon.

Preceding page: Maples in autumn. Deciduous forests are alive with colors in the fall.

T he atoms that make up your body and the bodies of other organisms and the atoms in all nonliving matter around you once existed in cold vast reaches of outer space as a giant cloud of cosmic dust and gas (Figure 19–1). Nearly 5 billion years ago, one of these huge clouds condensed. Thus began a process that led to the formation of our solar system and to life on Earth.

What triggered the condensation, no one knows, but scientists think that a nearby exploding star (supernova) may have been the cause. Dust particles in the center of the condensing cloud compacted rapidly, and the center of the cloud grew hotter. When the mass of material in the center of the cloud—soon to be the sun—reached a critical density, heat and pressure caused small atoms to unite, forming larger atoms. This process, known as **fusion**, releases enormous amounts of energy in the form of heat, light, rays, radio waves, and so on. Material lying outside the center of the cosmic cloud also condensed, forming the planets (Greek for "wanderers"); among them was the Earth.

When it originated, the Earth was a solid mass of rock and ice. Scientists believe that radioactive decay, intense solar heat, and other sources of heat caused the Earth to melt. The Earth became a glowing molten mass suspended in the sky. Water and the lighter elements, such as hydrogen, escaped into the atmosphere.

FIGURE 19–1 **Cosmic Clouds** Outer space is riddled with enormous clouds of dust and gas from which stars and planets are formed.

In the millennia that followed, the Earth gradually cooled and its crust formed. Today, the crust encircles a molten core, a remnant of the planet's fiery past (Figure 19–2). As the Earth cooled, water began to fall from the skies, creating lakes and oceans. Today, the oceans cover nearly 70% of the planet. It is in these oceans, or on their shores, that life began.

This chapter traces the emergence of life from the Earth's beginnings. It describes evolution, the biological process responsible for the rich and varied life-forms that make this planet their home. It concludes with a discussion of human biological and cultural evolution, illustrating the connection between environmental health and human health.

THE EVOLUTION OF LIFE

In the broadest sense, **evolution** is the process in which existing life-forms change to meet changing conditions on the newly formed Earth. Evolution is also the source of new life-forms. Evolution is gener-

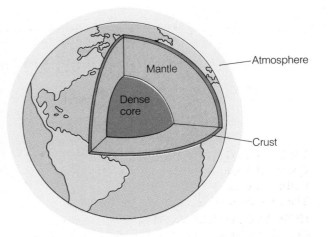

FIGURE 19–2 **The Earth in Cross Section** The molten core is a remnant of the Earth's early history.

ally divided into three phases: chemical evolution, cellular evolution, and the evolution of multicellular organisms.

Chemical Evolution: Life-Forms in the Seas

One of the most intriguing aspects of evolution is how life formed on the newly formed Earth. The early seas, formed by rain, were lifeless bodies of water. The landmasses, so richly carpeted with grasses, trees, and other plants today, were barren rock. How could life form on this lifeless planet?

In 1924, a Russian scientist, A. I. Oparin, suggested that life arose from nonliving matter (Figure 19–3). Briefly stated, Oparin's hypothesis says that molecules in the Earth's primitive atmosphere formed simple organic molecules; these molecules, in turn, became the building blocks of life, forming polymers. Polymers, in turn, combined to form primitive cells.

Scientists believe that the Earth's primitive atmosphere did indeed contain a mixture of water vapor, methane, ammonia, and hydrogen as Oparin had suggested. It also contained lesser amounts of several other gases, such as nitrogen, carbon monoxide, and hydrogen sulfide. The early atmosphere contained no oxygen, however.

The formation of organic molecules from the Earth's primitive atmosphere which Oparin hypothesized, is now called **chemical evolution** and began about 4 billion years ago—about 600 million years after the Earth first formed. As the Earth cooled, rain

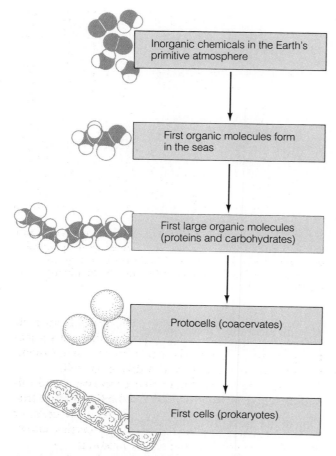

FIGURE 19–3 Oparin's Hypothesis Oparin hypothesized that the organic molecules necessary for life formed from the Earth's primitive atmosphere. Today, this idea is called the theory of chemical evolution and is supported by a considerable body of evidence.

The flow chart shows:
- Inorganic chemicals in the Earth's primitive atmosphere
- First organic molecules form in the seas
- First large organic molecules (proteins and carbohydrates)
- Protocells (coacervates)
- First cells (prokaryotes)

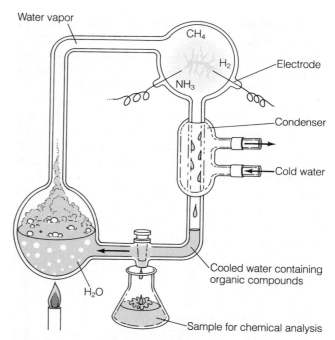

FIGURE 19–4 Miller's Apparatus Graduate student Stanley Miller showed that organic molecules could be produced from the chemical components of the Earth's early atmosphere.

Labels in figure: Water vapor, CH_4, H_2, NH_3, Electrode, Condenser, Cold water, Cooled water containing organic compounds, Sample for chemical analysis, H_2O

washed methane, ammonia, and hydrogen from the sky. Sunlight, heat from volcanoes, or lightning may have energized these molecules, causing them to react with one another. The reactions produced a variety of simple organic molecules, including monosaccharides and amino acids. Many scientists believe that the formation of these chemicals took place in the shallow waters of the seas and, perhaps, even in the atmosphere itself. Some scientists believe that these reactions may also have occurred in and around **geothermal vents**, openings in the ocean floor that emit hot water. Geothermal vents are underwater geysers where conditions may be favorable for the synthesis of organic molecules.

According to the theory of chemical evolution, the organic molecules that first formed began to react

with one another, forming larger molecules, the polymers. Primitive proteins and perhaps even rudimentary RNA or DNA molecules may have formed.

Little by little, the organic molecules necessary for life began to emerge. These molecules, in turn, combined to form aggregates or **protocells**, the precursors of cells, discussed in more detail shortly.

Chemical Evolution: Is It Fact or Fantasy? Although chemical evolution may sound like a science fiction story, a considerable body of research supports this theory.

The first support came from an ingenious American graduate student by the name of Stanley Miller. While studying for his Ph.D. in chemistry at the University of Chicago in the early 1950s, Miller devised an apparatus to test Oparin's hypothesis (Figure 19–4). To a closed, sterilized glass container, Miller added three gases thought to have existed in the Earth's primitive atmosphere: methane (CH_4), ammonia (NH_3), and hydrogen (H_2). A sparking device simulating lightning provided energy. Boiling water created steam that circulated through the apparatus, carrying with it the gases and the products of the reaction.

When Miller turned off the apparatus, some days later, the water had turned brown. An analysis of the contents of this, the first man-made primordial soup, revealed that it contained several important amino acids and several other important organic compounds, including urea and lactic acid.

Miller's experiments, for which he later was awarded a Nobel Prize, were repeated by a number of others. This work has shown that a variety of organic molecules, including the building blocks of DNA and RNA (purines and pyrimidines), can be created abiotically—in the absence of living organisms. This research supports Oparin's contention that the molecular components of the Earth's early atmosphere produced the simple building blocks of life.

Researchers also performed experiments to determine if the organic building blocks produced on Earth could have assembled into larger polymers abiotically. One notable experiment was performed by Sidney Fox from the University of Miami. Fox found that when he heated amino acids in air, the amino acids joined to form small chains, which he called **proteinoids**. Biologists hypothesize that dilute solutions of organic monomers may have been splashed onto hot rocks or lava where the amino acids polymerized. Polymerization may also have occurred on clay particles. Clay particles concentrate amino acids dissolved in water. The amino acids bind to the charged surface of the clay particles and may have united via peptide bonds (Chapter 2).

Nucleic acids and polymers of carbohydrates have also been produced in similar experiments, thus confirming the second step in Oparin's hypothesis. But that left the third and final step, the most phenomenal of all, the formation of living cells.

Cellular Evolution: The First Cells

Additional experiments by Fox suggest one way in which living cells arose. Fox immersed the proteinoids he had produced abiotically in a boiling salt solution, then cooled the solution. This experiment produced tiny globules of protein, **microspheres** (Figure 19–5). Microspheres look like bacteria. They are delimited by a membrane that resembles the plasma membrane of modern cells. Microspheres also grow, and when they reach a critical size, tiny protrusions form and break off, forming new microspheres.

Researchers have also found that mixing dilute solutions of several other polymers produces small globules similar to microspheres. Called **coacervates**, these microscopic globules of variable size and com-

FIGURE 19–5 Microspheres Sidney Fox showed that his abiotically produced proteinoids formed small spherical structures that resemble cells. Could they have acquired genetic material and enzymes to become the first living cells?

position selectively incorporate molecules from their environment, in much the same way that cells selectively transport materials across their plasma membranes. Coacervates grow and divide as well.

Some researchers believe that coacervates and microspheres were the precursors of the Earth's first true cells. Some of the proteins in the precursor cells, or **protocells**, may have had enzymatic properties, allowing cells to synthesize their own products.

Nucleic acids may have also formed during chemical evolution. Taken up by protocells, the nucleic acids may have provided a primitive mechanism of heredity. Research also suggests that RNA molecules may have even acted as enzymes.

The first true cells probably contained primitive enzymes, rudimentary genes, and a selectively permeable membrane. These cells probably absorbed organic molecules, such as glucose, from their environment and may have catabolized them down via anaerobic glycolysis (remember: there is no oxygen in the atmosphere yet). Thus, the early cells are known as **heterotrophic fermenters**. Fermentation is the anaerobic breakdown of glucose (Chapter 3).

Biologists believe that autotrophs (photosynthetic organisms, the "self-feeders") may have arisen from early heterotrophs. Autotrophs are cells that can synthesize their own food. Plants and algae are common modern-day examples. These organisms use carbon dioxide, water, and sunlight energy to produce organic food molecules. Water acts as a source of electrons. The first autotrophs, however, probably lacked chlorophyll and acquired the electrons not from water, but from other molecules, such as hydrogen sulfide or methane. The first autotrophs, then, were probably single-celled **chemosynthetic organisms**, organisms that acquire energy from sources other than sunlight. Some chemosynthetic bacteria can be found today.[1]

Over many millions of years, photosynthesis evolved. The emergence of photosynthesis depended on the evolution of metabolic pathways that produce chlorophyll. The depletion of organic molecules in the shallow waters of the seas may have been the evolutionary driving force behind the emergence of chlorophyll. As the primordial soup became depleted, organisms that could capture sunlight and make their own organic foodstuffs—that is, photosynthetic organisms—probably "enjoyed" an advantage over organisms that were dependent on food from the sea. In the language of evolutionary biology, this is called a **selective advantage**. Organisms that could photosynthesize would be more likely to reproduce and pass on their genes. These organisms formed the foundation of future generations.

The advent of photosynthesis marks a pivotal point in evolution. Besides producing food for plants, photosynthesis produces oxygen. As noted earlier, the early atmosphere of the Earth contained no free oxygen. Oxygen released by photosynthesis was probably toxic to the organisms living at that time. The evolution of photosynthesis, then, may have been the Earth's first global pollution disaster. Undoubtedly, many of the anaerobic organisms living at the time died off, unable to cope with oxygen. Others retreated to oxygen-free environments, such as the mud or sediment beneath lakes and oceans, where their descendants remain today. Still others evolved mechanisms that allowed them to live in the oxygen atmosphere. These organisms were the eukaryotes (organisms with true nuclei).

The process by which eukaryotes are believed to have formed is called **endosymbiotic evolution**. The theory of endosymbiotic evolution states that eukaryotes arose from preexisting cells. All of the cells that existed at the time were prokaryotes (bacteria-like organisms), lacking nuclei and organelles. According to the theory, eukaryotes came into existence when a host cell (perhaps a heterotrophic fermenter) phagocytized a smaller oxygen-respiring bacterium. The bacteria lived inside the host cell, using oxygen. As

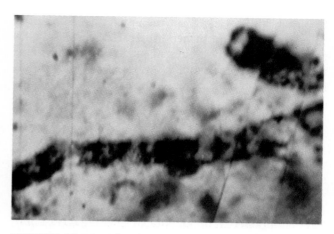

FIGURE 19–6 **The First Prokaryote** A composite photograph of the first prokaryote fossil discovered in western Australia approximately 30 years ago and believed to be about 3.5 billion years old.

atmospheric oxygen levels increased, such unions may have been advantageous. The relationship became hereditary. Thus, when the host cell divided, it passed on its internal partners to the daughter cells.

Endosymbiotic evolution may explain how mitochondria, chloroplasts, and even flagella became cellular organelles. At one time, these organelles may have been free-living organisms that entered into a partnership with other cells. A permanent relationship was established and has persisted for a billion years or more.

Putting It All In Perspective. The first prokaryotes probably arose about 1 billion years after the Earth formed. Imprints of these organisms have been found in 3.5 billion-year-old rock (Figure 19–6). These imprints resemble cyanobacteria—photosynthetic bacteria—present today. The first eukaryotes probably emerged about 2 billion years later, and are seen in rock about 1.1 billion years old. Therefore, for nearly 2 billion years, life on Earth consisted of a variety of prokaryotic organisms. At first, these prokaryotes subsisted on organic molecules produced abiotically. Over time, they developed systems by which they could produce their own food—chemosynthesis and photosynthesis. As photosynthesis became more elaborate and began producing oxygen, the stage was set for the evolution of the eukaryotes. First came bacteria capable of utilizing oxygen. Some of them were incorporated by other organisms, thus becoming internal symbionts—organisms living inside other organisms—and giving rise to the eukaryotic line.

---

[1]On the sea floor, chemosynthetic bacteria live on hydrogen sulfide or natural gas.

Evidence Supporting the Theory of Endosymbiotic Evolution. Evidence for endosymbiotic evolution is circumstantial or indirect. Added together, however, this evidence presents a rather convincing argument for endosymbiotic evolution. Consider the mitochondrion.

Mitochondria, which are thought to have arisen from oxygen-respiring organisms over a billion years ago, contain circular DNA, similar to all bacteria (Chapter 4). Mitochondria also contain ribosomes similar to those found in bacteria, and they contain some of the enzymes needed to produce energy. Mitochondria can divide and contain two membranes. Biologists believe that the outer membrane could have been formed as the bacterium that gave rise to mitochondria was phagocytized during endosymbiotic evolution. The resemblance of mitochondria and bacterium is quite compelling.

The theory of endosymbiotic evolution is also bolstered by the fact that endosymbiosis (one organism living inside another) is a rather common occurrence. Termites, for example, contain a microorganism in their gut that digests the wood termites eat. The microorganism also contains its own internal symbiont—a bacterium that lives inside it. Dozens of other examples exist, suggesting that endosymbiosis is commonplace in the biological world. This evidence alone does not "prove" that the theory is right, but suggests its plausibility.

The Evolution of Multicellular Organisms

The evolution of eukaryotes was a major turning point in the history of life on Earth, opening the door for the evolution of multicellular organisms—plants, animals, and fungi. A complete discussion of the evolution of these organisms is beyond the scope of this book, but a few words are in order.

Evolution of Life in the Sea: A Brief Overview.

As noted earlier, life probably arose in the oceans and remained there for hundreds of millions of years. Aquatic prokaryotes (bacteria) lived in the seas and eventually gave rise to aquatic eukaryotes. A great variety of plants and animals evolved in the salt water of the ocean from the single-celled eukaryotes.

Marine algae were one of the first eukaryotes to form. Numerous types of algae exist today. The earliest forms were probably single-celled organisms that floated freely in the water, absorbing sunlight and carbon dioxide and releasing oxygen into the seas. Today,

(a)

(b)

FIGURE 19–7 Early Marine Plants (*a*) Fucus or rockweed is a form of brown algae. (*b*) Kelp is also a form of brown algae.

single-celled green algae still float aimlessly in the seas. Some single-celled algae evolved the ability to colonize, or form aggregations or colonies of cells. The long green filaments attached to rocks in freshwater streams, for example, are colonies of algae. A variety of marine algae form colonies. Other algae evolved to form multicellular organisms that belong to the plant kingdom, such as kelp and sea lettuce, shown in Figure 19–7. These organisms display some degree of differentiation.

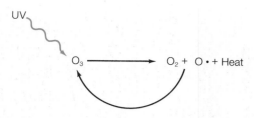

UV

$$O_3 \longrightarrow O_2 + O \cdot + Heat$$

FIGURE 19–8 Action of Ozone Molecules in the Ozone Layer When ultraviolet light strikes an ozone molecule, it causes the molecule to dissociate. The energy absorbed by the molecule is given off as heat, and the ozone molecule reassociates.

A variety of multicellular animals also arose in the sea. One of the first was a jellyfish-like organism, a soft-bodied invertebrate (an animal without a backbone). Mollusks (shellfish) and arthropods (crabs and lobsters) soon followed.

Evolution of Life on the Land. Life remained restricted to the ocean for many millions of years. The emergence of land organisms was prevented until about 400 million years ago by ultraviolet light from the sun. Ultraviolet light is lethal to plants and other organisms.

Today, terrestrial life is protected from ultraviolet light by the **ozone layer**, a thin layer some 20 to 30 miles above the Earth's surface containing the mole-cule **ozone (O_3)**. Early in evolutionary history, there was no ozone layer. Had plants or animals invaded the land, they would have been killed by ultraviolet radiation.

The ozone layer began to form from oxygen released by photosynthesis. Oxygen molecules emitted by photosynthetic marine organisms accumulated in the atmosphere. In the upper atmosphere, they react to form ozone molecules. Over time, the concentration of ozone increased in the upper atmosphere, creating the ozone layer, which filters out 99% of the incoming ultraviolet light (Chapter 21). When ultraviolet light strikes a molecule of ozone, the molecule absorbs the energy and splits (Figure 19–8). Ultraviolet light is then converted to heat in much the same way that microwaves are converted to heat in food.[2]

The formation of the ozone layer permitted the colonization of land by plants about 400 million years ago. The first "land" plants probably evolved from green algae that were splashed onto the rocks by waves. Those algae that could survive periods of dryness may have persisted, eventually giving rise to other land plants, particularly a group called the bryophytes. The **bryophytes** are mosses and other low-lying plants all of which thrive in moist environments (Figure 19–9). Next came the ferns, then conifers

[2]Microwaves deposit their energy in food molecules.

FIGURE 19–9 The First Land Plants Bryophytes, similar to (a) liverworts and (b) mosses, were some of the earliest plants to inhabit land. These short-stalked plants live in moist environments.

(a)

(b)

and a whole host of flowering plants. The success of these plants was no doubt due in part to the evolution of a protective coating (a layer of wax called the cuticle) that reduces evaporative loss, keeping the plant from drying out.

Soon after the land plants evolved, the first animals invaded from the seas. The first to come to feed on the plant matter were probably the **arthropods**, invertebrates with segmented bodies and jointed legs, such as insects and crustaceans. The very first terrestrial arthropods may have been the early ancestors of the scorpion (Figure 19–10). Arthropods have hard outer skeletons that reduce evaporation and also provide support. These features—already acquired in the ocean—made the transition to terrestrial life that much easier.[3] The previous emergence of plants provided a ready supply of food.

Later came a peculiar group of fishes, the lobe-fins. Biologists believe that the lobe-fins actually evolved in fresh water. One group colonized shallow ponds and streams. Some of these fish had lungs that allowed them to breathe out of water. They could also "walk" on dry land using their sturdy fleshy fins. This enabled them to travel from pond to pond in search of food or water when one pond dried up.

Amphibians evolved from the lobe-fins. The earliest amphibians were like the lobe-fins in several key respects—they could walk on land and could breathe air. Amphibians include organisms such as frogs and salamanders, which are dependent on both land and water for their subsistence. Amphibians must return to water to lay their eggs.

Reptiles (snakes, lizards, and turtles) evolved from a group of amphibians. Thanks to a thick scaly outer layer, reptiles can survive on land. One group of reptiles gave rise to the dinosaurs (Figure 19–11). The dinosaurs lived for over 100 million years and have to be considered one of the most successful of all animals that have inhabited the planet. Approximately 60 million years ago, the dinosaurs vanished from the face of the Earth. Many scientists think that asteroids striking the Earth may have created huge dust storms that cooled the Earth's atmosphere. Great clouds of dust may have persisted for several months, creating conditions too cold for dinosaurs to survive. Not all scientists agree, however. Some believe that while the extinction of the dinosaurs was abrupt, it was not as abrupt as those supporting the asteroid theory believe. Other natural processes may have changed the climate, making conditions inimical to the dinosaurs. Over time the dinosaurs died off.

[3]Organisms with characteristics that make them suitable for other environments are said to be preadapted.

FIGURE 19–10   The First Land Animals
Arthropods, perhaps the early ancestors of the scorpion, were the first to invade the land.

Birds and mammals also evolved from reptiles. The earliest mammals were small creatures that lived in trees and were active principally at night. Mammals are characterized by mammary glands and hair. They have the largest brains (relative to their size) of all organisms on Earth. The primates, which include the apes and humans, are a group of mammals of particular interest. The primates evolved from a tree-dwelling mammal that lived among the dinosaurs and persisted long after the giant beasts perished.

HOW EVOLUTION WORKS

Evolution explains how life formed and why it has changed. Evolution explains how life became so diverse, yet why life-forms are so similar.

Human knowledge of the evolutionary history of life on Earth comes chiefly from fossils. **Fossils** consist of the remains (bones) or imprints of organisms that lived on Earth many years ago. For example, dinosaurs that died along muddy river banks in the western United States decayed or were eaten by scavengers. Over time, the bones of the dinosaurs were covered with mud and buried. Rivers changed course, leaving the bones entombed in sediment that often turned to rock. Dinosaurs also left footprints in the mud that hardened, forming stone. These footprints are part of the vast fossil evidence of earlier life-forms

(a)

(b)

(c)

FIGURE 19–11 The Dinosaurs (*a*) Probably one of the most successful species ever to inhabit the land, the dinosaurs lived on Earth for about 100 million years. Why they disappeared still remains a mystery. (*b*) Birds, (*c*) alligators, and crocodiles are modern descendants of the dinosaurs.

(Figure 19–12a). Imprints of leaves also make up part of the fossil record (Figure 19–12b). Some ancient creatures (frogs and insects) were preserved in amber, resins released by ancient trees (Figure 19–12c). Blocks of amber, like plastic blocks containing biological specimens, provide a window to our past. Gas bubbles in amber have even been analyzed to determine the composition of the atmosphere in earlier times.

Genetic Variation: The Raw Material of Evolution

For years, biologists believed that evolution was a slow and gradual process of change, sometimes leading to the emergence of new species. A **species** is a group of organisms that are anatomically and physiologically similar to one another. The members of a species interbreed successfully—that is, they produce viable, reproductively functional offspring.

The gradual changes in populations that occur during evolution take place because of shifts in the genetic composition—the **gene pool**—of a species. These shifts are the result of mutations and natural selection (described in the next section).

Mutations in the genes of organisms occur naturally and randomly. Many of these errors are corrected by the cells, but some remain. Mutations may occur in the body cells (somatic cells) or germ cells. In some instances, mutations in body cells may have no effect on a cell. In other instances, they may so alter cellular function that they kill the cell or greatly impair its function. In still other instances, mutations may make the cell operate more efficiently. In the evolutionary process of multicellular organisms, however, somatic mutations are meaningless, for they cannot be passed to future generations. It is the germ cell mutation that is of importance in evolution.

Like somatic cell mutations, germ cell mutations can be harmful, even lethal. Many human embryos that are aborted spontaneously, for instance, are the result of harmful germ cell mutations in one or both of the parents (Chapter 17).[4] Other germ cell mutations, however, are beneficial to the offspring if they result in characteristics that give the offspring an advantage over other members of the species. It is these mutations that are the raw material of evolution in all species that reproduce via gametes.

---

[4]A mutation in a somatic cell in a morula or blastocyst could also lead to embryonic or fetal death and miscarriage.

A beneficial mutation inherited by an offspring may confer an advantage over others in the population. The advantage it confers thus increases the likelihood that the individual will reproduce. Organisms that do not share the trait are less likely to reproduce or will produce fewer offspring. Organisms that breed successfully form the foundation of future generations. Offspring carrying the advantageous genes, in turn, are more likely to reproduce than others, leaving greater numbers of offspring, thus causing the gene pool to shift over time.

Mutations are the exclusive source of new genes and thus a major source of **variation** in a population—differences in anatomical, physiological, and even behavioral characteristics.[5] In sexually reproducing organisms, variation also results from gene flow and recombination. **Gene flow** is the introduction of new genes to a population when individuals from another population join.[6] **Recombination** is the formation of new combinations of genes, new genotypes, during meiosis and fertilization (Chapter 16). During meiosis, homologous chromosomes pair up and often exchange segments in a process called crossing over (Chapter 18). Crossing over, you may recall, results in new genetic combinations—and, therefore, variation in a population. The random assortment of genes during meiosis also produces new genetic combinations, which may result in phenotypic and behavioral variation.

Genetically based characteristics that increase an organism's chances of passing on its genes are called **adaptations**. Adaptations arise spontaneously in populations from mutations, recombination, and gene flow.

Natural Selection

Sociologist Andrew Schmookler once wrote that evolution employs no author but only an extremely patient editor. By that he meant that evolution is not directed. Mutations arise spontaneously. Recombination, gene flow, and mutations produce new variants. The genetically based characteristics (adaptations) that arise, if beneficial, tend to persist. The species changes over time becoming better suited to its environment. In some cases, a whole new species may evolve. The gene pool shifts because of evolution's

[5]For bacteria and other organisms that reproduce asexually (by dividing or budding), mutation is the only source of variation.

[6]The new genes most likely arose from a mutation.

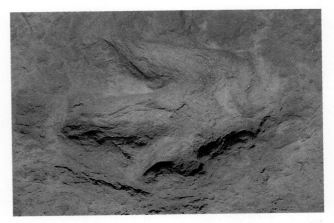

(a)

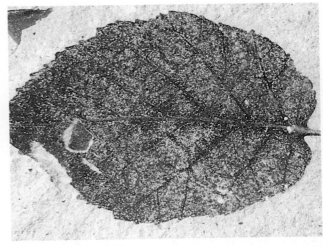

(b)

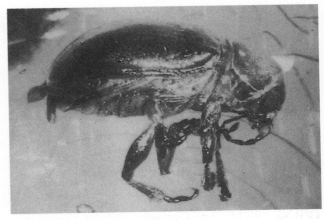

(c)

FIGURE 19–12 Fossils (*a*) Dinosaur tracks made in a Texas streambed about 120 million years ago. (*b*) Imprint of leaf. (*c*) Insect embedded in amber about 40 million years ago.

(a)

patient editor, a process called natural selection. **Natural selection** is nature's way of editing. Charles Darwin, who originated the idea, described natural selection as a process in which slight variations, if useful, are preserved (Figure 19–13). Natural selection is a process by which organisms become better adapted to their environment.

Two principal factors influence a population: **biotic factors**, or other organisms, and **abiotic factors**, the physical and chemical environment (temperature, rainfall, and so on). Abiotic and biotic factors are evolution's chief editors. They influence survival by "selecting" the fittest—those best able to reproduce and pass on their genes. If the conditions change, those organisms that are best adapted to the new conditions tend to remain and pass their genes to subsequent generations. Natural selection, therefore, causes a shift in the frequency of certain genes in the gene pool of a population.

FIGURE 19–13   Charles Darwin   (*a*) Darwin as a young man proposed the theory of evolution by natural selection, helping to solve one of the key puzzles of biology: how species evolve. (*b*) From 1831 to 1836, Darwin made a famous voyage on the *Beagle*, collecting and cataloging thousands of diverse species.

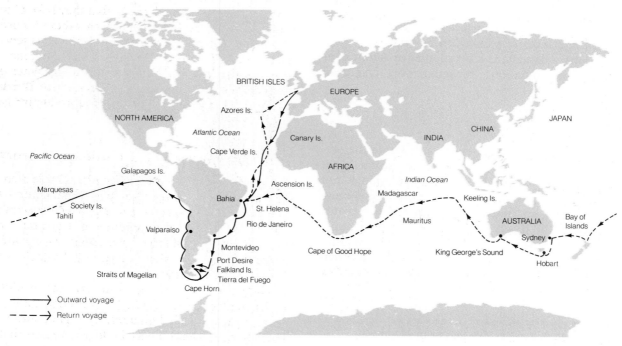

(b)

Charles Darwin was a nineteenth-century British naturalist who is in large part responsible for the concept of evolution through natural selection. In Darwin's time, evolution was widely discussed among naturalists and other scientists, but the mechanism by which it occurred remained an enigma. Darwin dedicated many years to the search for an answer, traveling by ship to South and Central America, cataloging species, and pondering the mechanism responsible for evolutionary change (Figure 19–13b).

According to Stephen Jay Gould, a well-known evolutionary biologist who writes widely on the topic, when Darwin finally arrived at his notion of natural selection, he kept it quiet. Why? According to Gould, the young scientist feared ostracism from the scientific community. Darwin's theory of natural selection opposed the prevalent religious belief (also held by scientists at the time) that God was responsible for the diverse array of living creatures. To protect his career, Darwin remained mute on the point.

In 1859, however, more than 20 years after he had first proposed the idea, Darwin published his concept. What inspired him was not a change in the times, but the knowledge that a contemporary of his, Alfred Russell Wallace, was about to publish a virtually identical theory.[7]

As in the case of Mendel, Darwin's ideas took many years to be understood and accepted. Not until the 1940s, about 60 years after his death, did natural selection become widely known and appreciated.

Darwin used the phrase "survival of the fittest" to describe how natural selection worked.[8] Survival of the fittest is commonly translated as "survival of the strongest." To a biologist, however, **fitness**, strictly speaking, is a measure of reproductive success and, therefore, of the genetic influence an individual has on future generations. By definition, then, the fittest individuals leave the largest number of descendants in future generations. Their influence on the gene pool will be greater than their less-fit contemporaries.

Fitness does not necessarily result from strength or speed. An organism that is better able to hide than its cohorts, for example, is more fit than one that falls victim to predators. An organism that is better able to digest grasses may be more fit than one that cannot.

The Evolution of New Species

Evolution is a process by which species change over time. In the process, they become better adapted to their environment. Evolution can also produce new species; in fact, all of the species alive today have evolved from others. As pointed out earlier, one group of fishes (the lobe-fins) gave rise to amphibians. Two groups of reptiles evolved to form birds and mammals.

Perhaps the most common mechanism of **speciation**, the evolution of a new species, results from geographic isolation. **Geographic isolation** occurs when members of a population of organisms are physically separated by some barrier—mountain ranges, lakes, oceans, rivers, or simply great distance. In their separate and distinctly different habitats, the members of the subpopulations are exposed to different environmental influences. Over time, two or more species may form in a process known as **divergent evolution**. Divergent evolution is responsible for many (possibly most) of all modern species.

If a population is separated long enough, its members undergo mutations that could cause them to lose the ability to interbreed with members of the original population. Biologists then say that they have become **reproductively isolated** from each other. According to some, the first step in isolation is the emergence of behavioral differences. Females detect differences in the courtship pattern. New species emerge when geographical isolation results in reproductive isolation. Table 19–1 lists several causes of reproductive isolation.

A Modern Version of Evolutionary Theory

Darwin's theory of evolution by natural selection can be summarized in three principles. First, natural variations exist in all species. Second, which species survive to reproduce or reproduce at a greater rate is determined by inherited variations (adaptations). Third, natural selection "determines" which organisms survive and reproduce.

For many years, biologists thought that evolutionary changes occurred gradually over many millions of years, a process called **gradualism**. If gradualism does indeed occur, evolutionary biologists reason that the fossil record should contain many intermediate forms

---

[7]Russell held that natural selection was responsible for the evolution of all aspects of life, except the human brain. God, he said, had managed that task. Darwin had received a letter from Wallace asking for comments on Wallace's ideas. Darwin and Wallace even published a joint paper on the concept of natural selection, a paper that received little attention from the scientific community.

[8]Herbert Spencer is thought to have coined the phrase.

TABLE 19–1	Reasons for Reproductive Isolation

Factors that prevent mating
  Mate at different times
  Develop different cues (songs and coloration)
  Develop different courtship behaviors
  Genital incompatibility
Factors that prevent production of viable offspring and/or reproduction
  Sperm cannot reach ova
  Hybrid offspring die in utero or shortly after birth
  Hybrid offspring survive but are sterile
  Hybrid offspring survive but have lower fitness

of plants and animals. During the evolution of birds, for example, one would expect numerous intermediate forms. The fossil record, however, contains only one transitional form in the "evolutionary gap" between reptiles and birds.

As a rule, new species in the fossil record appear rather abruptly, persist for several million years, then vanish as abruptly as they arrived on the scene. Based on this and other data, Stephen Jay Gould of Harvard University and Niles Eldredge of the American Museum of Natural History proposed an alternative hypothesis—that evolution occurs in spurts. According to them, evolution consists of long periods of relatively little change (equilibrium) interspersed (punctuated) with briefer periods of relatively rapid change, although these periods are still many thousands of years long. Their hypothesis is called **punctuated equilibrium**. Proponents of punctuated equilibrium believe that species undergo most of their morphological change when they first diverge from their parent species. Thus, a species will appear rather suddenly, then will undergo little or no change for long periods.

Evidence Supporting Evolution

Evolution by natural selection is one of the central theories of modern biology. Although biologists may argue over some of the details, such as gradualism or punctuated equilibrium, they agree about the basic tenets of the theory. From the outside, this bickering may appear to be evidence that the theory of evolution is on shaky ground. Nothing could be further from the truth. What is disputed is the mechanism, not the fact that it has occurred. The scientific knowledge in support of evolution is rich and varied.

One of the critical thinking rules discussed in Chapter 1 is that hypotheses gain credibility in science when they are supported by considerable amounts of data. Scientific truths (facts) are built on a foundation of consistent discovery. The same is true of the evolutionary theory.

The fossil record yields some of the best supporting evidence for the existence of evolution. Imprints and remains of ancient species in sedimentary rock can be dated using radioactive dating techniques. The oldest known fossils are of simple prokaryotic organisms and are about 3.5 billion years old. In layers of rock deposited later, the fossils become more complex and many new forms arise. The presence of new life-forms in "younger" rock suggests an evolutionary progression.

The commonalities among Earth's many different organisms, called the unity of life and discussed earlier, also lends support to the theory of evolution. Besides biochemical similarities, organisms share many structural similarities. All vertebrates (backboned animals), for example, have similar skeletons, which suggests a common ancestry.

Figure 19–14 illustrates the bone structure of six different vertebrates. As shown, the bones in the wing of a bird, the flipper of a whale, and the arm of a human being are quite similar. These similarities are very likely the result of common ancestry. Structures, such as these, which are thought to have arisen from common ancestors, are known as **homologous structures**.

Structural and functional similarities among species allow scientists to classify organisms into groups. The largest grouping is the kingdom. It includes all organisms united by one or a few common features. For example, the kingdom Animalia includes all multicellular organisms that are heterotrophs—that is, organisms that feed on others.

Within a kingdom, biologists create subcategories, based on anatomical, behavioral, and physiological similarities and differences. The broadest grouping in a kingdom is the **phylum**. In the kingdom Animalia, for example, are many phyla (Table 19–2). Phyla can also be divided into subgroups called **classes**. Classes are broken into **orders**, and orders are broken into

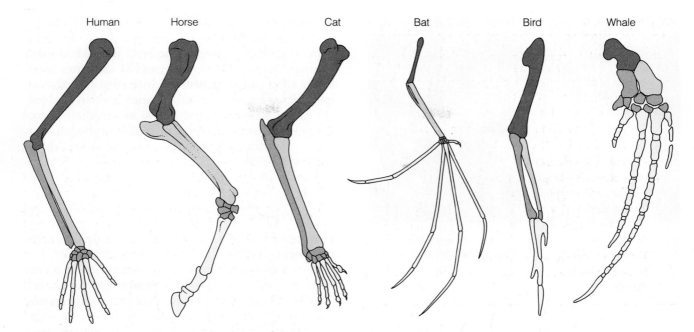

| Human | Horse | Cat | Bat | Bird | Whale |

FIGURE 19–14 **Homologous Structures among Vertebrates** The presence of homologous structures in vertebrates and other groups supports the theory of evolution.

families. Families consist of **genera** (singular, **genus**), and each genus contains one or more **species.**[9] Table 19–2 shows the classification of humans.

Biologists often refer to animals, plants, and other organisms by their common name. Scientific names are also frequently used. The scientific name of a species is unique and often descriptive. It consists of both the genus and the species. The robin, for example, goes by the scientific name *Turdus migratorius* (the migratory thrush). The grizzly bear goes by the name *Ursus horribilis*—the horrible bear. The genus and species are always underlined or in italic type because these words are Latin or Latin derivatives, and foreign words are italicized or underlined when they appear in English.

As noted earlier, the common biochemical makeup of organisms is also evidence that supports the theory of evolution. Organisms as distantly related as roses and rhinos, for example, are made of the same basic biochemicals—ATP, DNA, RNA, and protein. The more closely related two species are, the more similar their biochemistry. Chimpanzees and humans, for instance, share many (about 90%) of the same genes.

Although many species have similar enzymes and other proteins, differences do exist. Hemoglobin, for example, is found in vertebrate red blood cells, but variations in the composition of proteins, such as hemoglobin, are common among vertebrates. These variations are useful tools for evolution. By comparing the amino acid sequence of certain proteins (such as hemoglobin) from different species, evolutionary biologists can determine how related species are. The more differences, the more distant the evolutionary relationship.

TABLE 19–2	Taxonomic Classification of Humans
Kingdom	Animalia
Phylum	Chordata
Class	Mammalia
Order	Primates
Family	Hominidae
Genus	Homo
Species	Sapiens

[9]Here's a good opportunity to use a mnemonic (learning aid). Try this one: *keep putting chocolate on for goodness sake*

FIGURE 19–15   The Tree Shrew   An organism resembling the tree shrew is believed to have been the early ancestor of the primates.

Another line of evidence that supports the theory of evolution comes from embryology. Studies show that the embryos of members of many groups of organisms undergo similar structural development. The embryos of chickens, pigs, and humans—all vertebrates—are remarkably similar during their early stages. This phenomenon is summarized in a biological rule: ontogeny is an approximate recapitulation of phylogeny. **Ontogeny** is the development of an organism starting with fertilization. To say it is an approximate recapitulation means that it approximately repeats the anatomical stages that occur during the evolutionary development of a species, **phylogeny**.

HUMAN EVOLUTION

Humorist Will Cuppy once quipped that "All modern men are descended from a wormlike creature, but it shows more in some people." In reality, humans belong to a group (order) called primates that evolved, not from worms, but from a mammalian insectivore—an insect-eating mammal. This early ancestor of the primates resembled the modern-day tree shrew (Figure 19–15).

Early Primate Evolution

Primates include four basic groups: prosimians (premonkeys), monkeys, apes, and humans. The first primates to evolve from the treeshrew–like creatures were the tree-dwelling **prosimians** (premonkeys). Prosimians include modern-day lemurs and tarsiers, shown in Figure 19–16.

About 38 million years ago, the prosimians gave rise to the **anthropoids**. Anthropoids include the New World monkeys (monkeys of Central and South America), the Old World monkeys (monkeys of Africa and Asia), the great apes, and humans (Figure 19–17). The first anthropoids to evolve were the New World monkeys. As shown in Figure 19–17, the Old World monkeys evolved later.

Apes evolved from the Old World monkeys and are part of a subgroup of the anthropoids called **hominoids**. Approximately 25 million years ago, apes were widely distributed, living in Africa, Europe, and Asia. One group of apes, belonging to the genus *Dryopith-*

FIGURE 19–16   Prosimians   The prosimians (premonkeys) were probably the first primates. The earliest ones probably resembled modern-day (*a*) tarsiers and (*b*) lemurs.

(a)

(b)

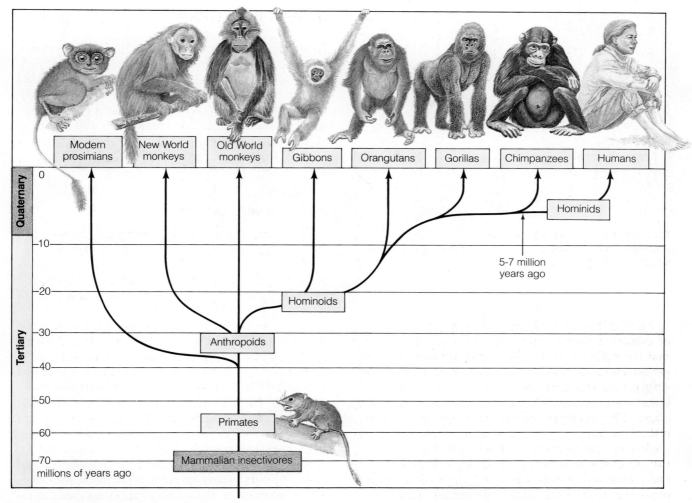

FIGURE 19–17   Evolution of Primates

ecus, is thought to have given rise to the four apes alive today: the gibbons, gorillas, orangutans, and chimpanzees. Archaeological evidence suggests that **dryopithecines** lived in forests. They spent most of their time in the trees, and when they walked on the ground, they probably rambled about on all fours, much like chimpanzees and gorillas.

Dryopithecines may also have been the evolutionary forerunners of the first humanlike creatures, or **hominids** (Figure 19–18). Archaeological evidence suggests that the first hominid was *Australopithecus* ("southern apeman"). The oldest known australopithecine skeleton was unearthed in Africa and is believed to be about 3.5 million years old. Called *Australopithecus afarensis* (*afarensis* means from the Afar re-

gion of Ethiopia), it stood only about three feet high and had a brain only slightly larger than an ape. The shape of its pelvis suggests that *A. afarensis* walked erect.

Many biologists believe *Australopithecus afarensis* gave rise to three additional species: *A. africanus*, *A. boisei*, and *A. robustus* (Figure 19-19a). *Australopithecus afarensis* may also have given rise to the genus *Homo*, the ancestors of modern humans, *Homo sapiens* (Figure 19–18).

The three "younger" australopithecines were all **bipedal**—that is, they walked on two legs. All three species apparently coexisted 1 to 3 million years ago. About 1 million years ago, however, australopithecines disappeared. Why they vanished no one knows.

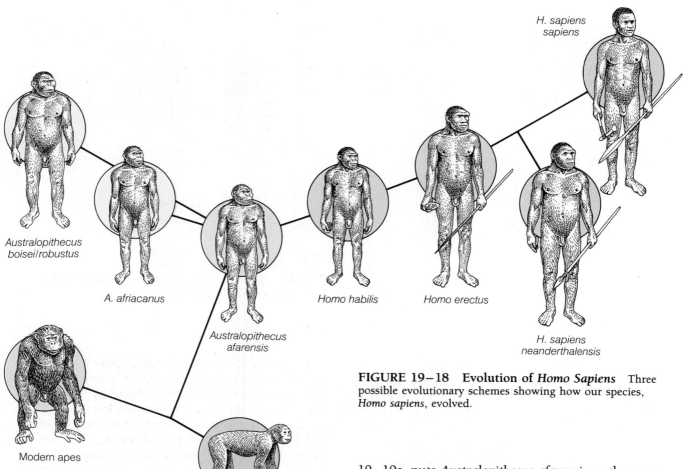

Australopithecus
boisei/robustus

A. afriacanus

Australopithecus
afarensis

Homo habilis

Homo erectus

H. sapiens
sapiens

H. sapiens
neanderthalensis

Modern apes

Dryopithecus

FIGURE 19–18 Evolution of *Homo Sapiens* Three possible evolutionary schemes showing how our species, *Homo sapiens*, evolved.

Evolution of the First Truly Human Primates

The exact origin of the genus to which modern humans belong is in dispute. Many scientists think that a 1.8-million-year-old skeleton found in Tanzania in 1960 by Mary Leakey represents the first member of the genus *Homo*. Mary Leakey and her husband, Louis, called this species *Homo habilis* ("skillful man"). Not all scientists agree with their classification, however. Some contend that this skeleton is merely another member of the genus *Australopithecus*. Recent archaeological discoveries have not yet resolved the issue.

Several schemes have been presented for the evolution of the genus *Homo*. The first, shown in Figure

19–19a, puts *Australopithecus afarensis* as the ancestor of all three *australopithecus* species as well as the genus *Homo*. The second holds that *A. afarensis* first gave rise to *A. africanus* (Figure 19–20b). This species, in turn, gave rise to two other australopithecines as well as the genus *Homo*. Still another scheme proposes that an as yet undiscovered ancestor gave rise to two separate lines (Figure 19–19c).

Two hundred thousand years after the emergence of *Homo habilis* arose the first unmistakable member of the genus *Homo*, called **Homo erectus** ("upright man"). Skeletons of *Homo erectus* appear in geological formations 300,000 to 1.6 million years old in the Old World, starting in Africa and spreading to Europe and Asia. *Homo erectus* stood about five feet tall, used fire, and made more sophisticated tools and weapons than *Homo habilis*. With a brain slightly smaller than ours, *Homo erectus* is believed to be the direct ancestor of **Homo sapiens**, the self-proclaimed "thinking man."

Homo sapiens emerged about 400,000 years ago. *Homo sapiens* consists of two subspecies, **Homo sapi-**

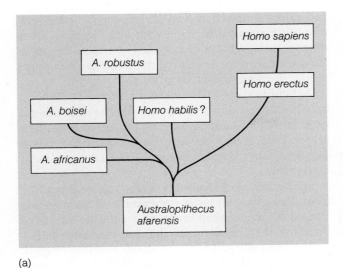

(a)

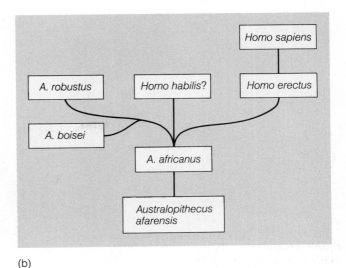

(b)

FIGURE 19–19 Three Possible Schemes for the Evolution of *Homo Sapiens*

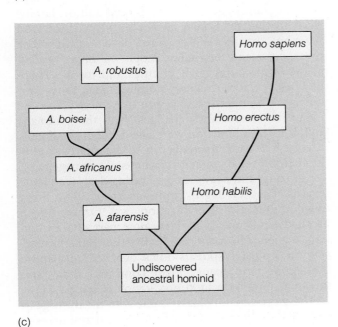

(c)

ens neanderthalensis, the Neanderthals, and *Homo sapiens sapiens*, modern humans. Widely distributed in Europe and Asia, the Neanderthals lived in caves and camps. They gathered fruits, berries, grains, and roots and hunted animals with weapons. They cooked some of their food on fires. Neanderthals stood erect and walked upright. Their brains were about the same size as ours today. Archaeological evidence suggests that they lived in small clans and buried their dead in elaborate rituals.

For unknown reasons, approximately 40,000 years ago the Neanderthals disappeared. Some archaeologists believe that they were replaced by modern humans, the **Cro-Magnons**, the earliest known members of *Homo sapiens sapiens*. Cro-Magnons arose in Africa and invaded Europe and northern Asia, wiping out the Neanderthals or possibly interbreeding with them. Archaeological evidence shows that Cro-Magnons used sophisticated tools and may have had a well-developed language. They lived in caves, which they decorated with elaborate art work.

Over the past 40,000 years, human evolution has provided little noticeable change in the physical appearance of humans. A Cro-Magnon on the streets of Los Angeles, in fact, would probably go unnoticed. Those 40,000 years have not been without change, however, for during this period, *Homo sapiens* has developed a rich and varied culture, complex language, and extraordinarily sophisticated tools.

HUMAN CULTURAL EVOLUTION: AN ENVIRONMENTAL PERSPECTIVE

Human culture has evolved through three phases: hunting and gathering, agricultural, and industrial societies. For most of human history, humans have been hunters and gatherers. In fact, 99% of the time spent

FIGURE 19–20 Hunters and Gatherers Some anthropologists believe that the early hunters and gatherers acquired most of their food by gathering nuts, fruits, berries, roots, and seeds. Hunting may have provided a supplementary food source.

on Earth humans have made their living gathering grains, fruits, nuts, berries, and roots and hunting animals (Figure 19–20). Recent archaeological evidence suggests that our earliest ancestors were also scavengers, eating animals that died from natural causes.

Based on studies of modern-day hunters and gatherers, whom some anthropologists call our "living ancestors," early hunters and gatherers must have had a profound knowledge of the environment. They knew how to find water in deserts and had an intimate knowledge of edible and medicinal plants. They were more adept at locating edible insects and grubs than the best trained field biologists today.

Some studies of our "living ancestors" suggest that their lives were not as hard as we often imagine. In many locales, they did not live in constant danger of starvation and did not spend a great deal of time finding food. Their lives were often leisurely and healthy and free from disease.

Hunters and gatherers lived in relative harmony with nature, taking what they needed and generally causing little ecological disruption—especially compared to modern humans. Their lack of environmental damage can be attributed to three basic facts of life: (1) many hunters and gatherers were nomadic, wandering in search of food and a favorable climate, (2) their numbers were small, and (3) their tools were primitive.

New archaeological research suggests that many groups of hunters and gatherers grew their own food and raised animals to feed their people. Some may even have engaged in trade with other groups. Over time, more and more of our early ancestors began to grow their own food, slowly giving rise to a new form of life, the agricultural society.

Agricultural societies emerged between 10,000 and 6,000 years ago. In the moist rain forests of Southeast Asia, farmers cleared small jungle plots to raise their crops. They grew a variety of vegetables and raised pigs and other domesticated animals to supplement their crops.

Seed crops originated in a wider region extending from China west to India and eastern Africa. Farmers cleared forests to plant crops, and with the advent of the plow, they began to till the rich grassland soils.

The plow allowed for larger fields and higher grain production. Farming, which had more or less supplied the immediate needs of a farmer and his family, began to change. Now a farmer could produce food for many families. With this development arose towns, cities and the trades. People no longer needed on the farm congregated in cities and towns. There they began trades, making clothing, pots, tools, and weapons. Commerce had begun in earnest.

Several important changes in the human-environment interaction were also evident as agriculture grew. First, humans began to drastically modify the natural environment. Poor soil management, overgrazing, and heavy timber cutting destroyed large regions. The rich Tigris-Euphrates river valley, often called the cradle of civilization, was turned into a parched landscape. This area, now part of Iran and Iraq, was once lush and productive, but human intervention brought about its quick demise. Land abuse from poor farm management continues today and threatens long-term food production (Chapter 21).

The second major change came with the upsurge in commerce. Commerce demands natural resources—metals, energy, and stone. These materials came from the outlying countryside. The towns and cities, therefore, drew heavily on the surrounding land, often causing considerable damage.

The third change, possibly the most fundamental of all, was the severance of the link between humans and nature. Hunters and gatherers were part of nature, taking from the land and living within the bounds of nature. Agriculturalists, however, were driven by a desire to command nature, or to harness or control its forces. No longer did humans see themselves as a part of the environment. Instead, they began to see themselves as separate from the environ-

FIGURE 19–21   Power Plant Spewing Out Toxic Fumes

ment, as its masters. This change in attitude followed humans into the next phase of cultural evolution, the industrial society. With the advent of trade and commerce in agricultural societies, humans began to see the natural world in a different light and began to regard it as a source of wealth.

The industrial society is a recent occurrence in human history. In fact, if the Earth's history were condensed to a one-year-long movie, the Industrial Revolution would occur in the last half-second of the film. The **Industrial Revolution**, the advent of mechanized production, began in England in the 1700s and in the United States in the 1800s. Machines took over manual production, but machines required energy—enormous amounts of it—and produced enormous amounts of air and water pollution (Figure 19–21).

Mechanization swept the farms, too. As a result, still fewer people were needed to raise food. Cities grew. Pollution increased. Streams, once rich in fish, turned putrid with the stench of human and factory wastes. The countryside was ravaged to provide energy and materials for factories. Pollution and species extinction were two of the principal environmental problems that arose during the Industrial Revolution.

Changes occurring during the Agricultural and Industrial Revolutions planted the seeds of a dramatic increase in human population. Between 1850 and 1990, human population increased from 1 billion to over 5.25 billion. The rise in population occurred principally for two reasons: (1) modern medicines that lowered infant mortality and (2) improved sanitation that stopped the spread of disease in the increasingly crowded urban environment.

Humans' relationship with nature grew even more strained. Philosophers and economists argued that people must seek power over nature. Control became the byword. Survival, as many saw it, required complete domination of nature. Today, that domination continues, but global environmental problems are a warning sign that we have pushed too far. Some current problems, such as global warming and ozone depletion, threaten to alter global climate and bring widespread species extinction (Chapter 21). Some ecologists believe that widespread social and economic disruption will occur because we have exceeded the Earth's ability to support humans. In a sense, our health and survival on the planet depend on reestablishing a balance with nature.

ENVIRONMENT AND HEALTH: THE PESTICIDE TREADMILL

After World War II, chemical pesticides became the cornerstone of modern agriculture. Pesticides kill potentially harmful insects and other organisms that destroy crops. Each year an estimated 40% of the U.S. harvest is lost to insects, rodents, birds, and other "pests," resulting in billions of dollars worth of losses.

To combat weeds and animal pests, farmers, homeowners, and others spray 2.5 million metric tons of pesticides on the land each year. These chemicals cause a number of problems. First, many broad-spectrum pesticides kill beneficial insects that control pests naturally. Spider mites were once only a minor crop pest in California, but the heavy use of chemical pesticides to control them has inadvertently killed the mites' natural predators, species that were more sensitive to spraying than the mites. As a result, spider mites today cause twice as much damage in California as all other insect pests combined.

A second problem of great concern is **genetic resistance**. A small proportion (about 5%) of any insect population is naturally resistant to chemical pesticides. Pesticide resistance results from genetic variation—the presence of a mutation that permits the insects to resist particular chemical sprays. Therefore, when farmers spray their fields with chemical pesticides, they kill the nonresistant insects. Surviving the onslaught, however, is a small population of genetically resistant insects. These insects, in turn, breed and produce a new population that is resistant to spraying. Chemical pesticides used to control insects artificially therefore select for resistant insects.

Controversy over Antibiotics in Meat

Livestock growers began adding antibiotics to cattle and pig feed over 30 years ago to protect animals confined to pens from disease as they were being fattened for market (see the figure). Antibiotics help control disease, which can run rampant under crowded conditions, but the drugs had an unanticipated effect that is still unexplained: they accelerated the rate of body growth. That meant that farmers could turn a higher profit.

Not surprisingly, today 70% of all cattle and 90% of all veal calves and pigs are reared on feed laced with penicillin or tetracycline. Nearly half of the antibiotics sold in the United States, in fact, are used for livestock feed.

The addition of antibiotics to feed has been sharply criticized by microbiologists and health officials. Why? They fear that widespread antibiotic use could help foster the evolution of superstrains of bacteria that are immune to antibiotics. In an editorial in the *New England Journal of Medicine*, Tufts University microbiologist Stuart Levy noted, "Every animal . . . taking an antibiotic . . . becomes a factory producing resistant strains" of bacteria. Resistant bacteria, in turn, could transfer their resistance to other bacteria, creating highly lethal strains that could infect humans. This is called the crossover effect.

Scientists are also concerned that some resistant bacteria in cattle, such as *Salmonella*, could be transmitted directly to people in meat or milk. The effects could be grave.

Despite these concerns, efforts to reduce the use of antibiotics in feed have been soundly defeated. In 1978, for instance, the Food and

Feedlot Cattle are housed in feedlots like this one to fatten up before slaughter. Concentrated confinement facilitates the spread of disease.

Drug Administration, which controls food additives such as antibiotics, proposed cutting back on penicillin and tetracycline use in feeds. Livestock producers, feed producers, and the multimillion-dollar drug industry, however, fought vigorously. They argued that microbiologists' concerns were not proven.

New evidence in recent years, however, confirms the medical suspicions. Dr. Thomas O'Brien of the Harvard University School of Medicine, for example, published a study in the *New England Journal of Medicine* in 1982, showing that bacteria that commonly infect humans and other animals share genetic information quite freely through the exchange of plasmids, tiny segments of DNA separate from the bacteria's chromosomes. O'Brien argued that drug resistance could be transferred easily from bacterium to bacterium.

Another study by researchers at the Centers for Disease Control, published in *Science* in 1984, confirmed the suspicion that antibiotic-resistant bacteria could be transferred directly from meat to men and women. This research showed that most of the outbreaks of antibiotic-

resistant *Salmonella* in the previous decade could be traced to meat from animals that were fed antibiotic-treated grains. The research showed that 20 to 30% of the *Salmonella* outbreaks involved antibiotic-resistant strains. About 4.2% of the people contracting antibiotic-resistant bacteria died, compared to only 0.2% of the victims of normal bacteria.

Proponents of antibiotics believe that the link between antibiotics and human disease is still weak and that further research is needed. Even if these findings are substantiated by further research, though, proponents believe that the benefits of using antibiotics outweigh the potential health effects. Banning antibiotics or cutting back on their use could have enormous economic impacts that must be weighed against sickness and loss of life. Any way you look at it, it's a tough issue and it's not going to go away. Health experts hope that the United States, like Europe, which strictly limited the use of antibiotics in animal feed in the early 1970s, will find the political will to end this potentially dangerous activity.

To kill them, farmers must apply more chemical pesticide or switch to an alternative form. However, a small segment of the new genetically resistant population is usually genetically resistant to either the higher dose or the new chemical preparation. That group survives the spraying and breeds, producing an even more resistant population, starting an ever-escalating cycle that some people refer to as the "pesticide treadmill".

The pesticide treadmill is a result of genetic variation. It warns farmers that they can never win in the battle against pests. Thanks to genetic variation in a population, no matter what pesticide they use, there will always be a resistant strain. Ever more powerful and ever more frequent applications at higher doses will be necessary just to stay even. Not surprisingly, today entomologists have identified over 450 species of insects that are resistant to one or more chemical pesticides. Twenty of the worst pests are now resistant to all types of insecticide. A few decades ago, farmers in Central America applied pesticides about 8 times a year; today, 30 to 40 applications are the norm on any given field.

One of the rules of critical thinking is to examine the big picture. Unfortunately, when it comes to pest controls, farmers and chemical pesticide manufacturers have failed to do so. They insist on waging a war against insects that cannot be won. They fail to take into account the existence of genetic variation and the selection process that is at work in their fields.

Knowledge of these phenomena suggests alternative strategies, of which there are many. A simple practice of **crop rotation**, for example, can hold pest populations to manageable levels. Practiced years ago on many farms, but now largely abandoned, crop rotation requires farmers to alternate the crops they plant in a given field. One year corn might be grown in the field; the next year, beans, and the third year, alfalfa. This practice helps maintain soil nutrient levels and also effectively cuts down on pests. Why?

Insect pest populations increase in farm fields as the season progresses. A field of corn or wheat provides an unlimited supply of food for these tiny creatures. At the end of the season, the insects lay their eggs in the soil or in organic material on the surface. If the same crop is planted the next year, the offspring (from eggs laid by insects from the previous year) will have an abundant food supply once again. The insect population increases dramatically, causing considerable damage. If a different crop is planted, however, the insect pest population will likely decline. Continued rotation helps hold down pests year after year

without costly and potentially harmful pesticide applications.

Farmers can also plant several crops in the same field, a practice called **heteroculture**. A field planted with corn and peanuts, for example, can reduce corn borers (insects that can devastate corn crops) by as much as 80%.

Chemical resistance is a rather common occurrence in the modern world. Weeds may become resistant to herbicides; even microorganisms develop resistance to antibiotics, which is one reason many physicians prescribe antibiotics judiciously to patients. They fear, and rightly so, that the more antibiotics we use in society, the more likely a resistant strain will emerge. Antibiotics are used in livestock feed to protect closely housed animals and to stimulate growth; Health Note 19–1 discusses the problems that can arise from this common practice.

Heavy pesticide use accelerates the pesticide treadmill. A little knowledge of genetics and evolution proves handy.

SUMMARY

THE EVOLUTION OF LIFE

1. All atoms and molecules in our solar system came from the same source, an enormous cloud of cosmic dust and gas that gave rise to the Earth and the sun.
2. The evolution of life can be divided into three phases: chemical evolution, cellular evolution, and the evolution of multicellular organisms.
3. Chemical evolution, scientists hypothesize, is a process that began about 4 billion years ago. At that time, the Earth's primitive atmosphere contained a mixture of water vapor and gases. During chemical evolution, these molecules gave rise to small organic molecules, which, in turn, gave rise to organic polymers—small proteins and nucleic acids.
4. Organic polymers gave rise to aggregates called coacervates and microspheres, the precursors of the cell. Aggregates of polymers exhibit structural and functional characteristics of living organisms.
5. Cellular evolution is the evolutionary development of cells from cell precursors. The first true cells probably contained primitive enzymes, rudimentary genes, and selectively permeable membranes. These cells probably derived nourishment from organic molecules they absorbed from their environment and are thought to have been heterotrophic fermenters.
6. Autotrophs, organisms capable of synthesizing their own food, arose next. The earliest autotrophs were probably chemosynthetic bacteria that acquired energy from chemicals in the environment.

7. With the evolution of chlorophyll, photosynthetic autotrophs arose.

8. Oxygen produced by photosynthesis was toxic to the organisms living at that time in an oxygen-free environment. Many of the anaerobic organisms died off, unable to cope with oxygen. Others retreated to oxygen-free environments. Still others evolved ways to survive the oxygen atmosphere.

9. The theory of endosymbiotic evolution states that eukaryotes arose from prokaryotes when a host cell (perhaps a heterotrophic fermenter) acquired an internal symbiotic partner (possibly a smaller oxygen-respiring bacterium). As atmospheric oxygen levels increased, these unions persisted and, in time, the relationship became hereditary.

10. Prokaryotes emerged about 3.5 billion years ago, and eukaryotes evolved about 1.1 billion years ago. The evolution of eukaryotes opened the door for the evolution of multicellular organisms.

11. A variety of multicellular plants and animals evolved from single-celled eukaryotes in the oceans. As the ozone layer developed, life on land became possible.

12. The first species to invade the land were plants that thrive in moist environments. Soon after the land plants invaded, animals came ashore.

HOW EVOLUTION WORKS

13. Evolution has produced a great diversity of organisms. These organisms arose from a common ancestor and, therefore, share many biochemical similarities. Common biochemistry is evidence of the conservative nature of evolution.

14. Changes in populations that occur during evolution take place because of shifts in the gene pool.

15. Genetic variation in a species arises from mutations, gene flow, and recombination. Genetic variation may result in the emergence of adaptations—anatomical, physiological, and behavioral characteristics that confer a selective advantage on certain offspring, giving them a better chance of reproducing and passing the genes on to future generations.

16. Beneficial traits are preserved in a population by natural selection, the preservation of favorable characteristics by the environment. Abiotic and biotic factors are the agents of natural selection.

17. Natural selection results in organisms that are better adapted to their environment and can lead to the evolution of new species.

18. Geographical isolation is one of the most common mechanisms by which new species arrive. Geographical isolation occurs when two subpopulations of the same species become separated. Subject to different environmental conditions, the populations may evolve independently. Over time, they may become reproductively isolated—that is, they become unable to interbreed.

When this occurs, two different species are said to have formed.

19. Many evolutionary biologists believe that evolution occurs in spurts. According to the theory of punctuated equilibrium, long periods of relatively little change (equilibrium) are punctuated by briefer periods of relatively rapid change, many thousands of years long.

20. The scientific knowledge in support of evolution is rich and varied. The fossil record, anatomical similarities in groups of organisms, the common biochemical makeup of organisms, and similar embryological development among many groups of organisms all support the theory.

HUMAN EVOLUTION

21. Humans belong to a group (order) called primates that evolved from a mammalian insectivore that resembled the modern-day tree shrew.

22. The first primates to evolve were the tree-dwelling prosimians (premonkeys). Modern-day prosimians include the lemurs and tarsiers.

23. The prosimians gave rise to the New World and Old World monkeys. Apes evolved from the Old World monkeys. Humans probably evolved from a genus of apes called *Dryopithecus*.

24. The dryopithecines gave rise to the first hominid, *Australopithecus* ("southern apeman"). The oldest known australopithecine skeleton was unearthed in Africa and is believed to be about 3.5 million years old. It belongs to a group called *Australopithecus afarensis*.

25. *A. afarensis* may also have given rise to the genus *Homo*, the ancestors of modern humans.

26. The exact origin of the genus to which we belong is in dispute, however. Many scientists think that a 1.8-million-year-old skeleton found in Tanzania was a member of the genus *Homo*. It was called *Homo habilis* ("skillful man").

27. Two hundred thousand years after the emergence of *Homo habilis* arose the first unmistakable member of the genus *Homo*, called *Homo erectus* ("upright man"). Skeletons of *Homo erectus* appear in geological formations 300,000 to 1.6 million years old in the Old World, starting in Africa and spreading to Europe and Asia.

28. With a brain slightly smaller than ours, *Homo erectus* is believed to be the direct ancestor of *Homo sapiens*.

29. *Homo sapiens* emerged about 400,000 years ago and consists of two subspecies: *Homo sapiens sapiens* and *Homo sapiens neanderthalensis*, the Neanderthals. The Neanderthals (now extinct) lived in caves and camps in Europe and Asia.

30. Approximately 40,000 years ago the Neanderthals disappeared. Some archaeologists believe that they were replaced by modern humans, the Cro-Magnons, the earliest known members of *Homo sapiens sapiens*.

31. The Cro-Magnons arose in Africa and invaded Europe and northern Asia, perhaps wiping out the Neanderthals or possibly interbreeding with them.
32. Over the past 40,000 years, human evolution has provided little noticeable change in the physical appearance of humans. During that period, however, *Homo sapiens* has developed a rich and varied culture, complex language, and extraordinarily sophisticated tools.

HUMAN CULTURAL EVOLUTION: AN ENVIRONMENTAL PERSPECTIVE

33. Human culture has evolved through three phases: hunting and gathering, agricultural, and industrial societies.
34. During that transition, humans have sought ways to control the environment, sometimes with disastrous consequences.
35. Today, our attempts to dominate nature continue. But current global environmental problems are warning signs that we may have pushed too far, that our society is out of balance with the natural world. This upset threatens our own existence.

ENVIRONMENT AND HEALTH: THE PESTICIDE TREADMILL

36. Evolution can be witnessed today in modern agriculture. Chemical pesticides used to control insects artificially select resistant organisms.
37. Farmers who spray their fields to kill insects leave behind a genetically resistant subpopulation that breeds and repopulates farm fields. A second application at a higher dose or an application of another pesticide leaves behind another subset that often becomes a pest, forcing farmers to try higher doses or still another pesticide. This escalation in the war against pests is called the pesticide treadmill.
38. Chemical resistance is a rather common occurrence in the modern world. Weeds can become resistant to herbicides, and even microorganisms develop resistance to antibiotics.

⑄ EXERCISING YOUR CRITICAL THINKING SKILLS

A researcher at a major university makes a new discovery that will allow farmers to plant the same crop in their fields year after year without depleting the vital nutrients in the soil. You are asked by the press to comment on the discovery. Is it a good idea? What problems might arise from this discovery?

TEST OF TERMS

1. Many scientists believe that life arose from the chemicals found in the Earth's primitive atmosphere. The process in which organic chemicals formed from chemicals in the atmosphere is called _____ evolution.
2. Organic molecules combined to form polymers, and they, in turn, gave rise to the first primitive cells, _____ .
3. Research chemist Stanley Miller demonstrated the plausibility of Oparin's hypothesis, showing that organic molecules could be produced _____ in conditions similar to those thought to have existed early in the Earth's history.
4. Microspheres and _____ are small spherical structures exhibiting some characteristics of life and containing organic polymers.
5. These first true cells may have absorbed glucose and broken it down by anaerobic glycolysis and are, therefore, called _____ .
6. The first autotrophic organisms were probably _____ bacteria that acquired electrons needed to synthesize their food from chemicals, such as hydrogen sulfide, in the environment.
7. The evolution of chemical mechanisms allowing cells to synthesize _____ permitted the evolution of photosynthesis.
8. An organism that is better adapted to its environment than its cohorts is said to have a _____
_____ over others.
9. The evolution of photosynthesis was a turning point in evolutionary history. One of its chief products, _____ , changed the course of evolutionary history and may have been a major stimulus for the evolution of _____ .
10. The evolution of cellular organelles, such as the mitochondrion and chloroplast, is explained by the theory of _____ evolution.
11. Life first evolved in the seas, but with the development of the _____ , plants were able to invade the land.
12. Animals invaded the land some time after plants. The first to invade were probably the _____ ,

organisms with jointed legs and segmented bodies.

13. A(n) _____ is a group of organisms that are structurally and functionally similar and produce viable, reproductively competent offspring.

14. Mutations create genetic _____ in a population—that is, differences in physical or functional characteristics.

15. A genetically based characteristic that increases an organism's chances of passing on its genes is a(n) _____ .

16. _____ _____ is a process in which slight variations, if useful, are preserved.

17. _____ is a measure of reproductive success and of the genetic influence an individual has on future generations.

18. The evolution of a new species, _____ , occurs when _____ isolation leads to _____ isolation.

19. The process in which one species gives rise to several others that occupy different environments is called _____ _____ .

20. According to the theory of _____ _____ , long periods of relatively little change are broken up by briefer periods of relatively rapid change, still lasting many thousands of years.

21. Structures such as the forelimb of a human and the wing of a bird, which are thought to have arisen from common ancestors, are called _____ structures.

22. The broadest grouping in a kingdom is the _____ . Organ-

isms, however, are generally referred to by their common name or by their _____ and _____ .

23. _____ resistance in insects exposed to pesticides is responsible for the phenomenon called the pesticide treadmill.

24. Humans evolved from an early ape, _____ , which may have given rise to the genus, _____ .

25. Louis and Mary Leakey discovered what may be the first member of the genus *Homo*, which they called *Homo* _____ .

26. _____ _____ is believed to be the direct ancestor of *Homo sapiens*, modern humans.

Answers to the Test of Terms are located in Appendix B.

TEST OF CONCEPTS

1. In 1924, a Russian scientist, A. I. Oparin, suggested that life arose from nonliving matter. Explain his hypothesis and discuss the research supporting it.

2. Describe how the first cells may have arisen during evolution. What critical requirements must have been met for life to begin?

3. If life formed abiotically in the shallow waters of the sea, would you expect the process to be occurring now? Why or why not?

4. Describe the theory of endosymbiotic evolution and the evidence supporting it.

5. The development of photosynthesis and the emergence of eukary-

otes were pivotal events in evolution. Why?

6. Numerous plants and animals evolved in the sea before life emerged on land. Why?

7. A critic of evolution says, "Life forms are too diverse for them to have come from a common ancestor." How would you respond?

8. What is meant by the phrase "the conservative nature of evolution"?

9. Over the course of the Earth's history, mountain ranges rise and are gradually worn down. Explain what might happen to a population of a species that is geographically isolated by the emergence of a new mountain range.

10. How do random mutations in germ cells contribute to the evolutionary process?

11. Define the following terms: adaptation, variation, natural selection, biotic factors, abiotic factors, selective advantage, and fitness.

12. Discuss the following statement: Natural selection is nature's editor.

13. Discuss the evidence supporting the theory of evolution.

14. How do crop rotation and heteroculture keep pest populations in check?

15. You are on an archaeological dig and discover parts of a skeleton. How would you go about determining the age of the skeleton and its genus and species?

SUGGESTED READINGS

Audesirk, G. and T. Audesirk. 1989. *Biology: Life on Earth,* 2d ed. New York: Macmillan. See Chapters 13–15 for clear and readable discussions of evolution and the history of life on Earth.

Campbell, N. A. 1988. *Biology.* Menlo Park, Calif.: Benjamin/Cummings. See Chapters 20–23 for detailed and highly readable coverage of evolution.

Chiras, D. D. 1991. *Environmental science: Action for a sustainable future,* 3d ed. Redwood City, Calif.: Benjamin/Cummings. Chapter 17 discusses genetic resistance in pests and alternatives to chemical pesticides.

Gould, S. J. 1977. *Ever since Darwin: Reflections in natural history.* New York: Norton. Excellent collection of essays on evolution.

_____ . 1977. *The panda's thumb: More reflections in natural history.* New York: Norton. Excellent collection of essays on evolution.

_____ . 1989. An asteroid to die for. *Discover* 10(10): 60–65. Extraordinary piece outlining the evidence for the punctuated equilibrium hypothesis.

Levine, N. D. 1989. Roundtable: Evolution and extinction. *Bioscience* 39(1): 38. Part of a debate on the issue of preservation of species and the impact of extinction on evolution. See also Myers.

Lovejoy, C. O. 1988. Evolution of human walking. *Scientific American* 259(5): 118–25. Examination of similarities among the pelvises and legs of modern humans and our primate ancestors, providing clues to the evolution of walking.

Myers, N. 1989. Roundtable: Extinction past and present. *Bioscience* 39(1): 39. Part of a debate on the issue of preservation of species and the impact of extinction on evolution. See also Levine.

Nathans, J. 1989. The genes for color vision. *Scientific American* 260(2): 42–49. Provides new insights into the evolution of color vision.

20 Ecology: The Economy of Nature

Canada geese bask in the sun.

M ost of us live our lives seemingly apart from nature. We make our homes in cities and towns, surround ourselves with concrete and steel, and inadvertently drown out the sound of birds with our noise. The closest many of us get to nature is a romp with the family dog on the grass in our backyards.

Raymond Dasmann, a world-renowned ecologist, once wrote that despite what many of us may think, a human apart from nature is an abstraction. No such thing exists. Our lives depend on nature and the environment. The clothes we wear, our morning coffee, and even the breakfast cereal we eat are all products of nature. So is the oxygen we breathe. Annually, one-half of the oxygen in the atmosphere is replenished by plants and algae. Without them, humans and other species could not survive. Trees, grasses, and other plants also provide other free services. They protect watersheds near our homes, preventing flooding. Swamps help purify the water we drink. Birds control insect populations, and predators control rodent populations. Nature "serves" us well. Thus, although we may have isolated ourselves from nature, we have not emancipated ourselves. The ties that bind us cannot be broken.

This chapter examines **ecology**, the study of living organisms and the web of relationships that binds them together in nature. Ecology takes as its domain the entire living world. Ecologists study how organ-

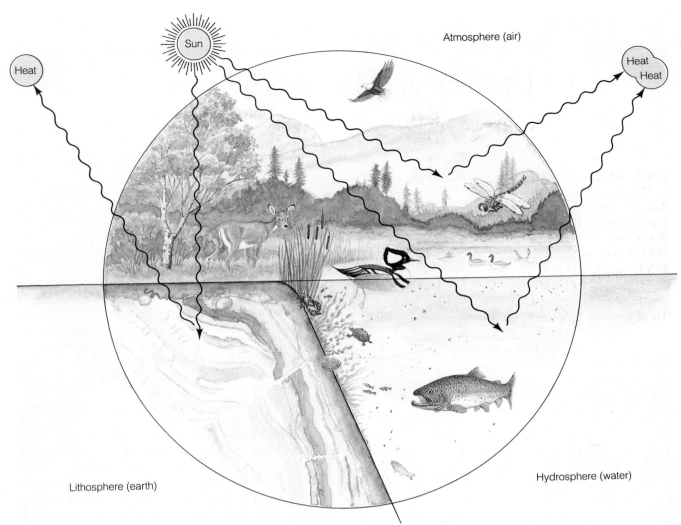

FIGURE 20-1   The Biosphere   Life exists at the intersection of land, air, and water.

isms interact with one another and how they interact with the abiotic, or nonliving, components of the environment.

PRINCIPLES OF ECOLOGY: ECOSYSTEM STRUCTURE

Ecology, like all science, is a body of knowledge and a process of inquiry that seeks to understand the mysteries of nature. A great many lessons can be learned by studying nature and applying what we learn to human society.

Ecology, however, probably ranks as one of the most misused words in the English language. Banners proclaim "Save Our Ecology." Speakers argue that "our ecology is in danger," and others talk about the "eco-

logical movement." These common uses of the word *ecology* are incorrect. Why? Ecology is a scientific field of inquiry. It is not synonymous with the word *environment*. It does not mean the web of interactions in the environment. We can save our ecology department and ecology textbooks, but we cannot save our ecology. Our ecology is not in danger, our environment is. You cannot join the ecology movement, but would be a welcome addition to the environmental movement.

The Biosphere

Life exists at the intersection of the land, air, and water. The region that supports life is called the **biosphere** (Figure 20-1). The biosphere extends from the bottom of the ocean to the tops of the highest mountains. Life, however, is rare at the extremes of

FIGURE 20–2   A View of the Earth from Outer Space

The Biomes and Aquatic Life Zones

Viewed from outer space, the Earth resembles a giant jigsaw puzzle, consisting of large landmasses and vast expanses of ocean (Figure 20–2). The landmasses, or continents, can be divided into large subregions or **biomes** (Figure 20–3). A biome is a region characterized by a distinct climate and a characteristic type of plant and animal life adapted to it.

Figure 20–3 shows a biome map of the world. As illustrated, the North American continent contains seven biomes, five of which are discussed here. Starting in the frozen north is the **tundra**, a region of long, cold winters and very short growing seasons (Figure 20–4a). The rolling terrain of the tundra supports grasses, mosses, lichen, wolves, musk-oxen, and other animals adapted to the bitter cold. Trees cannot grow on the tundra because of the short growing season and because the subsoil (permafrost) remains frozen all year round, preventing deep root growth necessary for trees. Immediately south of the tundra lies the **taiga**, the northern coniferous forests. The milder climate and longer growing season result in a greater diversity and abundance of plant and animal life than are found on the tundra. Evergreen trees, bears, wolverines, and moose are characteristic species (Figure 20–4b). East of the Mississippi lies the **temperate deciduous forest** biome, characterized by even warmer climate and more abundant rainfall (Figure 20–4c). Broad-leafed trees make their home in this biome. Opossum, black bear, squirrels, and foxes are characteristic animal species. West of the Mississippi lies the **grassland** biome (Figure 20–4d). Inadequate rainfall and periodic drought prevent trees from flourishing here, except near rivers and streams. Over the centuries deep-rooted grasses have evolved on the plains. These grasses tolerate periodic drought and can withstand grazing. White-tailed deer and coyotes are characteristic animal species. In the Southwest, where even less rain falls, is the **desert** biome (Figure 20–4e). Contrary to what many people think, the desert often contains a rich assortment of life—both plants and animals—uniquely adapted to the aridity and heat. Cacti, mesquite trees, rattlesnakes, and a variety of lizards all make their home in this seemingly inhospitable environment.

The oceans can also be divided into subregions, known as **aquatic life zones**. Aquatic life zones are the aquatic equivalent of biomes. Like their land-based counterparts, each of these regions has a distinct en-

the biosphere because conditions for survival at the fringes are marginal.

If the Earth were the size of an apple, the biosphere would be about the thickness of its skin. This thin zone of life is referred to as a **closed system**—much like a sealed terrarium. A closed system receives no materials from the outside.[1] The only external contribution is sunlight. Sunlight powers virtually all life-processes and is, unbeknownst to many, also a major source of energy in modern industrial societies. The energy released by the combustion of coal, oil, and natural gas, which we use to light our homes and run our factories, for example, owes its origin to sunlight that fell on the Earth several hundred million years ago.

Because the Earth is a closed system, all materials necessary for life must be recycled over and over again. The carbon dioxide you exhale, for instance, may be used by a plant during photosynthesis next week halfway across the world. The water in your morning coffee may have evaporated from a pond in India last month.

[1]Cosmic dust settles on the Earth, but virtually all of the materials necessary for life come from the Earth itself.

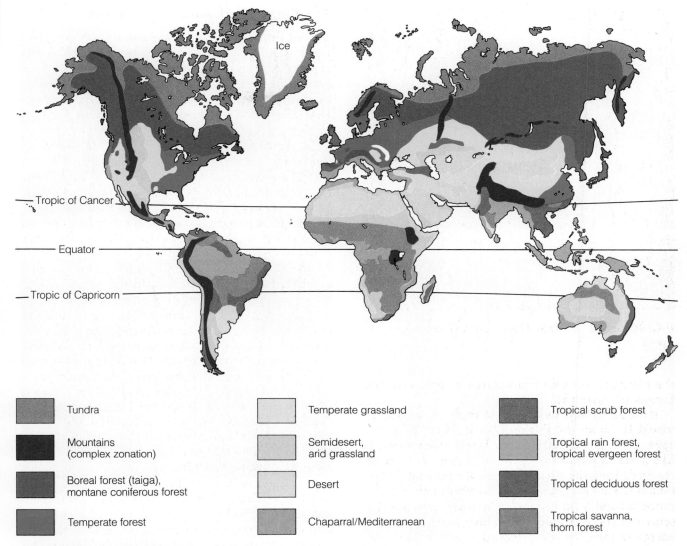

Ice

Tropic of Cancer

Equator

Tropic of Capricorn

Tundra	Temperate grassland	Tropical scrub forest
Mountains (complex zonation)	Semidesert, arid grassland	Tropical rain forest, tropical evergeen forest
Boreal forest (taiga), montane coniferous forest	Desert	Tropical deciduous forest
Temperate forest	Chaparral/Mediterranean	Tropical savanna, thorn forest

FIGURE 20–3   The Biomes

vironment and characteristic plant and animal life. Four major aquatic life zones exist: coral reefs, estuaries (the mouths of rivers where fresh and salt water mix), the deep ocean, and the continental shelf.

Ecosystems

The biosphere is a global **ecological system** or **ecosystem**. An ecosystem consists of organisms and their environment. Innumerable interactions are possible within an ecosystem. For the sake of convenience, ecologists often limit their study to a small ecological system—for example, a pond, a rotting log, or a grassy meadow. Even in these small ecosystems, the number of organisms present and the number of possible interactions can be astounding.

Abiotic Components.   Reduced to a minimum, ecosystems consist of two basic components: abiotic and biotic. The **abiotic components** are the physical and chemical factors that are needed for life, including sunlight, precipitation, temperature, and nutrients. The **biotic components** are the organisms that live in an ecosystem.

FIGURE 20–4 North American Biomes (*a*) Tundra; (*b*) taiga; (*c*) temperate deciduous forest; (*d*) grassland; (*e*) desert.

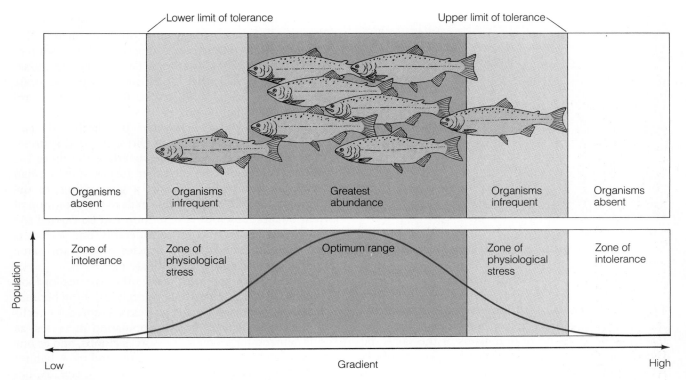

FIGURE 20–5 Range of Tolerance Organisms tolerate a range of conditions but thrive within an optimum range.

The abiotic conditions within a biome often vary. To survive in any given biome, then, organisms must be able to tolerate a range of conditions. The range of conditions in which an organism is adapted is called its **range of tolerance** (Figure 20- 5). As Figure 20–5 shows, organisms do best in the optimum range. Outside of that are the **zones of physiological stress**, where survival and reproduction may be possible, but not optimal. Outside of these zones are the **zones of intolerance**, where life is not possible.

Humans often alter the physical and chemical nature of their environment. If conditions shift drastically, species present in the ecosystem may be eliminated. Dams, for example, create lakes in streambeds, changing the water temperature, water flow, and other abiotic factors. As a result, many fish that lived in the stream perish. Native razorback suckers and other species that once lived in the Colorado River are now endangered species (in danger of extinction) because of the large dams constructed along the river. These dams release extremely cold water from the bottom of huge reservoirs, changing the water temperature in the river and wiping out most of the native fish.

Although species are sensitive to all of the abiotic factors in their environment, one factor often turns out to be more important than others in regulating growth. This factor, which ultimately regulates growth of a population, is called a **limiting factor**.

In freshwater lakes and rivers, for example, dissolved phosphate is a limiting factor. Phosphate is needed by plants and algae for growth, but phosphate concentrations are naturally low. Plant and algal growth is held in check. When phosphate is added to a body of water, however, plants and algae proliferate. Algae often form dense surface mats, blocking sunlight (Figure 20–6). As a result, plants rooted on the bottom may die, and oxygen levels in the deeper waters may decline, killing fish and other aquatic organisms.

Phosphates are added to lakes and rivers from several sources. One of the major sources is the sewage treatment plant. The phosphates released by these plants come from detergents and human waste.

On land, precipitation tends to be the limiting factor. At any given temperature, the more moisture that falls, the richer the plant and animal life.

FIGURE 20-6 Algal Bloom This pond is choked with algae due to the abundance of plant nutrients from human sources.

Biotic Components.   Within biomes and aquatic life zones, organisms of a given species often occupy specific regions. A group of organisms of the same species occupying a specific region constitutes a **population**. The members of a population may remain together throughout all or much of the year. Bighorn sheep, for example, remain in herds throughout the year. Other organisms, such as the grizzly bear, are solitary animals, for the most part, keeping to themselves except for mating. In any given ecosystem, several populations exist together and form a **community**, a network of plants, animals, and microorganisms. Organisms in a community are part of an interdependent web of life and interact in many ways.

Habitat and Niche

If asked to give a brief description of yourself, you probably would begin by describing the place you live. You would then probably discuss the work you do, the friends you have, and other important relationships that describe your place in society. A biologist would do much the same when describing an organism. He or she would start with a description of the place an organism lives—that is, its **habitat**. Next, he or she would describe how the organism "fits" into the ecosystem, or its **ecological niche** or simply **niche**. An organism's niche includes its habitat and all of the relationships that exist between that organism and its environment. This includes what an organism eats, what eats it, its range of tolerance for various environmental factors, and other important factors.

Organisms in a community live in the same habitat, but most of them occupy quite different niches. This phenomenon minimizes competition for resources and is, no doubt, a condition favored by natural selection. The fact that organisms occupy separate niches provides for a wider use of an ecosystem's resources, especially food.

Niches do overlap somewhat. In other words, two species may feed on some of the same food sources. The more two species' niches overlap, the more the species compete with each other. Just as in the human economy, competition can be a good thing. In the biological world, competition leads to the evolution of new adaptations. But competition can be carried too far. When niches overlap considerably, one species is usually eliminated. If two species occupy the same niche, competition will eliminate one.

Humans compete with many other species for food and living space. The competition is anything but fair, however, because humans possess an incredible technological advantage. As our population increases, as our demand for food and resources climbs, and as our technological power increases, more and more habitat for wild animals and plants will be lost. Already, because of overfishing, dozens of world fisheries have been depleted by commercial fishing interests. Seals and other fish-eating species will perish. Currently, at least one species becomes extinct every day, largely because of human activities.

ECOSYSTEM FUNCTION

Life on land and in the Earth's waters is possible principally because of the **producers**—algae and plants that absorb sunlight and use its energy to synthesize organic foodstuffs from atmospheric carbon dioxide and water via photosynthesis.[2] These organic molecules are used by the producers themselves but also provide nourishment for the rest of the biological world. As a result, producers form the foundation of the living world.

Organisms that are nourished by the food produced by the plants and algae are known as **consumers**. Ecologists place consumers in four general categories, depending on the type of food they eat. Some, such as deer, elk, and cattle, feed directly on plants and are called **herbivores**. Others, such as wolves,

[2]Some producers are chemosynthetic organisms; that is, organisms capable of using methane and hydrogen sulfide as a source of energy to produce organic molecules.

FIGURE 20−7 A Simplified Food Chain Two grazer food chains are shown here. On the top is a terrestrial grazer food chain, and on the bottom is an aquatic grazer food chain.

feed on grazers and other animals and are known as **carnivores**. Humans and a great many other animal species subsist on a mixed diet of plants and animals and are known as **omnivores**. The final group feeds on waste—animal waste or the remains of plants and animals. These organisms are called **detrivores**. (Remember that detritus is waste material.)

Food Chains and Food Webs

Biological communities consist of numerous food chains. A **food chain** is a series of organisms, each one feeding on the organism preceding it (Figure 20−7).

All organisms in the community are members of one or more food chains. Two types of food chains exist: grazer and decomposer.

Grazer food chains begin with plants and algae, which are consumed by herbivores or **grazers**. **Decomposer food chains** begin with dead material—either animal wastes (feces) or the dead remains of plants and animals (Figure 20−8a). In ecosystems, decomposer and grazer food chains are linked (Figure 20−8b). Thus, waste from the grazer food chain enters the decomposer food chain where it is returned to the environment for reuse. Nutrients liberated by the decomposer food chain enter the soil and water and

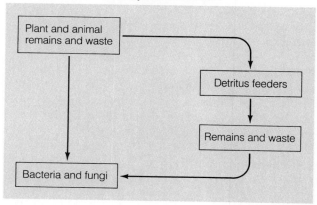

Decomposer food chain

(a)

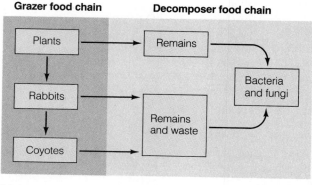

Grazer food chain   **Decomposer food chain**

(b)

FIGURE 20–8   Food Chains   (*a*) A decomposer food chain. (*b*) A Grazer food chain and a decomposer food chain showing the connection between the two.

are incorporated into plants, thus entering the grazer food chain.

Food chains exist only on the pages of textbooks; in a community, all food chains are part of a much more complex network of feeding interactions called the **food web**. Food webs present a complete picture of the feeding relationships in any given ecosystem.

The Flow of Energy and Matter through Ecosystems

Food chains are a conduit for the flow of energy and nutrients through the environment. Consider energy first. Sunlight energy is captured by plants and algae and used to produce organic food molecules. Sunlight energy is stored in the chemical bonds of these molecules. In the food chain, organic molecules pass from

one organism to the next. In animals, glucose and other organic molecules originally produced by plants are broken down to release energy to power numerous cellular functions. But not all of the energy in food molecules is captured by the grazers; as we saw in Chapter 3, nearly two-thirds of the energy contained in glucose molecules is lost as heat. This heat is radiated into space and lost from Earth. Eventually, all organisms die and are decomposed. During decomposition, much of the remaining energy still locked in the covalent bonds is released. Thus, all of the sunlight energy that is absorbed by the plants is eventually released into the environment as heat. Sunlight energy, therefore, flows unidirectionally through the food chain. It cannot be recycled.

In contrast, nutrients flow cyclically—that is, they are recycled over and over again. Nutrients found in the soil, air, and water are incorporated into plants and algae, then passed along the food chain. Nutrients in the food chain reenter the environment in two principal ways: through the decomposition of dead organisms and the decomposition of waste. When a plant or animal dies, for instance, it decomposes. Bacteria responsible for decomposition release nutrients contained in dead plants and animals. These nutrients enter the water or soil for reuse. Other nutrients reenter the environment through animal waste. The feces of a rhinoceros, for example, are broken down by bacteria, which liberate carbon dioxide and minerals.

One way or another, all nutrients eventually make their way back to the environment. Each new generation of organisms, therefore, relies on the recycling of material in the biosphere.

Trophic Levels and Ecological Pyramids

Ecologists break down a food chain according to an organism's position or **trophic level** (literally, "feeding" level). The producers are the base of the grazer food chain and are, therefore, members of the first trophic level. The grazers are members of the second trophic level. Carnivores that feed on grazers are members of the third trophic level and so on.

Most terrestrial food chains are limited to three or four trophic levels. Longer terrestrial food chains are rare. The reason for this is that food chains generally do not have a large enough producer base to support many levels of consumers. Put another way, the longer the food chain, the less food is available for the top-level consumers. Why?

Plants absorb only a small portion of the sunlight that strikes the earth—only 1 to 2%. They use this to

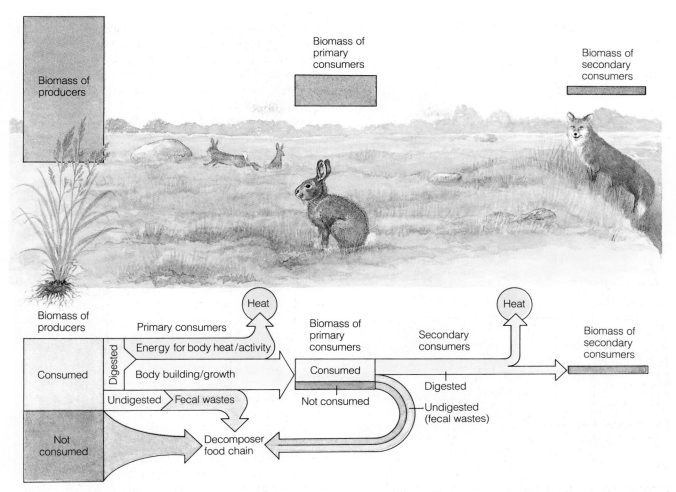

FIGURE 20−9   Flow of Energy and Biomass through a Food Chain   Not all biomass from one trophic level ends up in the next level for reasons shown in the diagram and discussed in the text. Note that all energy is lost as heat.

produce organic matter or biomass. Technically, **biomass** is the dry weight of living material in an ecosystem. The biomass at the first trophic level is the raw material for the second trophic level. The biomass at the second trophic level is the raw material for the third trophic level and so on.

As Figure 20−9 shows, not all of the biomass produced by plants is converted to grazer biomass. At least three reasons account for the incomplete or partial transfer of biomass from one trophic level to the next. First, some of the plant material, such as the roots, is not eaten. Second, not all of the material that the grazers eat is digested. Third, some of the digested material is broken down to produce energy and heat and, therefore, cannot be used to build biomass in the grazers (Figure 20−9). As a rule, only about 5 to 20%

of the biomass at any one trophic level can be passed to the next.

When plotted on graph paper, the biomass at the various trophic levels forms a pyramid, the **biomass pyramid** (Figure 20−10). Since biomass contains energy (stored in the covalent bonds), one can also plot the chemical energy present in the various trophic levels. When plotted, the energy contained in the biomass at each trophic level forms an **energy pyramid**. In most food chains, the number of organisms also decreases with each trophic level, forming a **pyramid of numbers**.

Knowledge of ecological pyramids helps explain why people in many Third World countries generally subsist on a diet of grains (corn, rice, or wheat) rather than meat. Figure 20−11 illustrates two food chains.

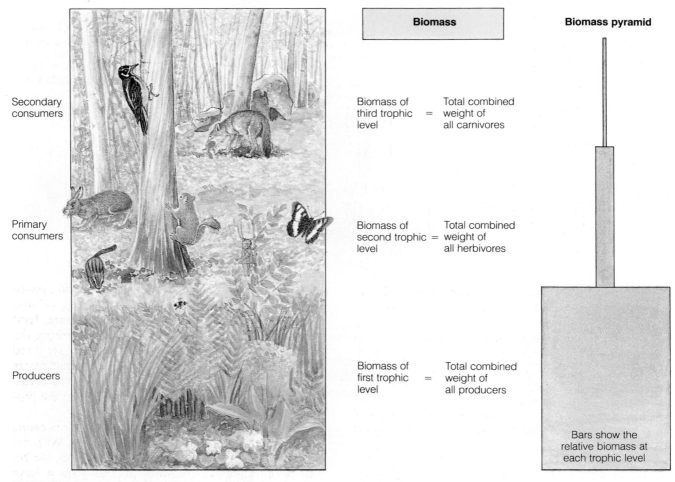

| Biomass | | Biomass pyramid |

Secondary consumers

Biomass of third trophic level = Total combined weight of all carnivores

Primary consumers

Biomass of second trophic level = Total combined weight of all herbivores

Producers

Biomass of first trophic level = Total combined weight of all producers

Bars show the relative biomass at each trophic level

FIGURE 20–10 Biomass Pyramid In most food chains, biomass decreases from one trophic level to the next higher one.

In the grain-human food chain, more biomass is available for humans from a given amount of grain than in the grain-cow-human chain. In this somewhat simplified example, 20,000 kilocalories of corn can feed 10 people for a day. However, if that corn is fed to a cow, and the beef is fed to humans, only one person can subsist on the 20,000 kilocalories. In the grain-cow-human food chain, the 20,000 kilocalories fed to the cow produces only 2000 kilocalories of food, barely enough for one person (assuming 10% transfer of biomass). The shorter the food chain, the more food is available to top-level consumers.

The human population increases by 89 million people a year. Feeding these people poses an enormous challenge. The most efficient food source will be crops—such as corn, rice, and wheat—fed directly to people. It is obviously far less efficient to feed corn and other grains to cattle and other livestock, which are then slaughtered for human consumption. People, however, must receive adequate protein, as noted in Chapter 7. Protein can be supplied by fish, meat, and a mixture of legumes. (For more on feeding the world's hungry, see Chapter 21.)

Nutrient Cycles

Nature has its own economy: a system of exchange, competition, and cooperation. It is an economy driven by the sun. The economy of nature is dependent on the circular flow of many nutrients through food webs and a continual inflow of solar energy. Nutrients flow from the environment through food webs

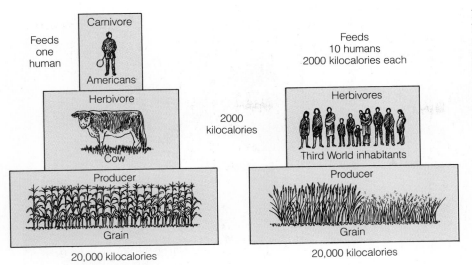

FIGURE 20–11 Energy
Pyramids in Two Food Chains
(*a*) The typical American diet. Note
that the 20,000 kilocalories of corn
fed to cows feeds only one American.
(*b*) In the shorter food chain in the
Third World nations, the same
20,000 kilocalories will feed 10
people directly.

and are released back into the environment. This circular flow constitutes a **nutrient cycle** or **biogeochemical cycle**.

Nutrient cycles can be divided broadly into two phases: the environmental phase and the organismic phase (Figure 20–12). In the **environmental phase**, a nutrient exists in the air, water, or soil—or sometimes in two or more of them simultaneously. In the **organismic phase**, nutrients are found in the members of the community, the plants, animals, and microorganisms.

Nutrient cycles ensure the availability of building materials for future generations. Our lives depend on them. However, a great many human activities can disrupt these cycles. Such actions can profoundly influence the survival of a species. In this chapter, we look at two of the most important nutrient cycles—the carbon and nitrogen cycles—and the ways they are being altered.

The Carbon Cycle.

The carbon cycle is illustrated in Figure 20–13 in simplified form. We begin with free carbon dioxide. In the environmental phase of the cycle, carbon dioxide resides in two reservoirs or sinks: the atmosphere and surface waters (the oceans, lakes, and rivers). Carbon dioxide is absorbed by plants and algae from these reservoirs and thus enters the organismic phase of the cycle.

As noted earlier, carbon dioxide is the raw material of photosynthesis, which plants and algae use to produce organic food materials. These materials then travel along the food chain from one trophic level to the next.

Carbon dioxide reenters the environmental phase by two routes. First, it is given off during cellular energy production in the organisms of the grazer food chain (Chapter 3). Second, it is released during the decomposer food chain. In the decomposer food chain, organic wastes and the remains of dead plants and animals are decomposed by bacteria and fungi. Bacteria and fungi liberate carbon dioxide in the process.

For many tens of thousands of years, our ancestors lived in harmony with nature (Chapter 19). With the advent of the Industrial Revolution, however, we began to interfere with natural processes on a large scale. One of the victims of our technological development has been the global carbon cycle. The widespread combustion of fossil fuels (which releases carbon dioxide) and rampant deforestation (which reduces carbon dioxide uptake) have seriously altered the cycle.

Before the Industrial Revolution, global carbon dioxide production more or less equaled carbon dioxide absorption by plants and algae. Today, about 5.2 billion tons of carbon dioxide are added to the atmosphere each year. Three quarters of the increase results from the combustion of fossil fuels; the remaining quarter stems from deforestation. Trees absorb enormous amounts of carbon dioxide. Therefore, as forests are cleared, the amount of atmospheric carbon dioxide increases. Making matters worse, many forests are burned after cutting, further adding to the global carbon dioxide levels in the atmosphere.

In the past hundred years, global atmospheric carbon dioxide levels have increased about 25%. In the

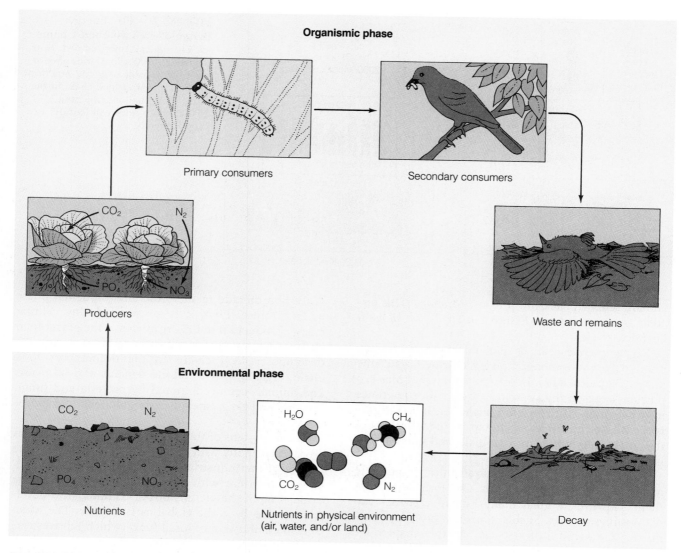

Organismic phase

Primary consumers

Secondary consumers

Producers

CO_2 N_2 PO_4 NO_3

Waste and remains

Environmental phase

CO_2 N_2 PO_4 NO_3

Nutrients

H_2O CH_4 CO_2 N_2

Nutrients in physical environment
(air, water, and/or land)

Decay

FIGURE 20–12   General Structure of a Nutrient Cycle   Nutrients exist in the organismic and environmental phase and cycle back and forth between them.

atmosphere, carbon dioxide traps heat escaping from Earth and reradiates the heat to the Earth's surface. As carbon dioxide levels increase, global temperature may rise dramatically. Such a rise could have devastating effects on global climate. It could shift rainfall patterns, destroy agricultural production in many regions, and wipe out thousands of species. Rising global temperature may cause glaciers and the polar ice caps to melt, raising the sea level and flooding many low-lying coastal regions (For more on the greenhouse effect, see Chapter 21.)

The Nitrogen Cycle.   Nitrogen is an element essential to many important biological molecules, including amino acids, DNA, and RNA. The Earth's atmosphere is a major source of nitrogen, but atmospheric nitrogen is in the form of nitrogen gas (N_2), which is unusable to all but a few organisms. Because of this, atmospheric nitrogen must be converted to a usable form—either nitrate or ammonia.

The conversion process, known as **nitrogen fixation**, takes place in the soil by two principal routes. As Figure 20–14 shows, the roots of leguminous plants

FIGURE 20—13
The Carbon Cycle

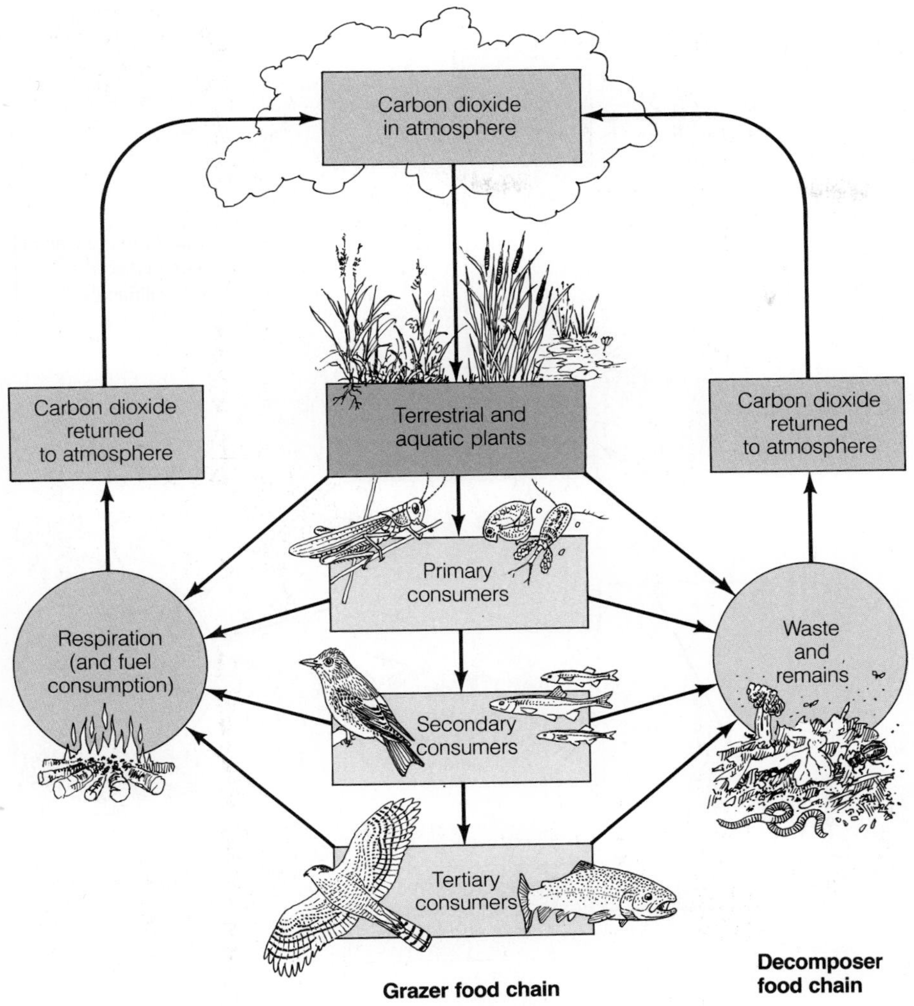

Carbon dioxide
in atmosphere

Carbon dioxide
returned
to atmosphere

Terrestrial and
aquatic plants

Carbon dioxide
returned
to atmosphere

Primary
consumers

Respiration
(and fuel
consumption)

Secondary
consumers

Waste
and
remains

Tertiary
consumers

Grazer food chain

**Decomposer
food chain**

(peas, beans, clover, alfalfa, vetch, and others) contain small swellings, called **root nodules**. Inside the nodules are symbiotic bacteria that convert atmospheric nitrogen to ammonia. Ammonia is also produced by bacteria that live in the soil.

Once ammonia is produced, other soil bacteria convert it to nitrate. Nitrates are incorporated by plants and used to make amino acids and nucleic acids. All consumers ultimately receive the nitrogen they require from plants.

The decay of animal waste and the remains of dead plants and animals returns ammonia to the soil for reuse (Figure 20–14). Ammonia can also be converted to nitrate and reused. Some nitrate, however, may be converted to nitrite and then to nitrous oxide (N_2O) by denitrifying bacteria, as illustrated on the

left of Figure 20–14. Nitrous oxide is converted to nitrogen and released into the atmosphere.

Humans alter the nitrogen cycle in three ways: (1) by applying excess nitrogen-containing fertilizer on land, much of which ends up in waterways, (2) by disposing of nitrogen-rich municipal sewage in waterways, and (3) by raising cattle in feedlots adjacent to waterways. These activities all increase the concentration of nitrogen in the soil or water, upsetting the ecological balance. Excess fertilizer applied to farm fields, for example, often ends up in neighboring lakes and rivers. Even though sewage treatment plants remove much of the nitrogen from human waste, large quantities are dumped in our waterways. Nitrogen, like phosphate, is a plant nutrient. It stimulates growth of aquatic plants and causes rivers and lakes to

FIGURE 20–14   The Nitrogen Cycle   Nitrogen in the atmosphere is converted to NH_3 (ammonia) by bacteria and cyanobacteria in soil. Ammonia is converted to nitrates and taken up by plants.

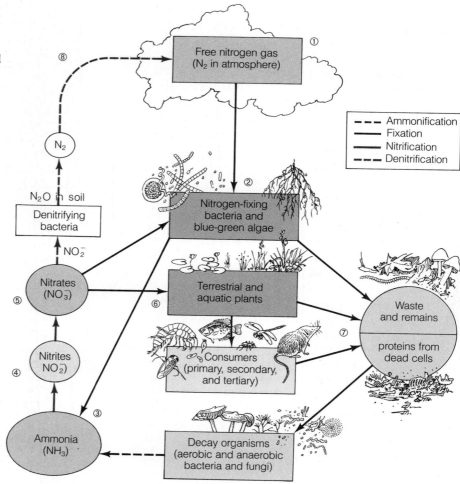

become congested with dense mats of vegetation. Water bodies become unnavigable. Sunlight penetration to deeper levels is impaired, and oxygen levels in deeper waters may decrease. In the fall, when aquatic plants die and decay, oxygen levels may fall further, killing aquatic life.

ECOSYSTEM BALANCE AND IMBALANCE

This book has pointed out that human health is dependent on homeostasis at several levels. First, our health requires internal balance. A breakdown of homeostasis can lead to disease and death. Human health is also dependent on homeostasis in the environment. A breakdown of the Earth's homeostatic systems could have profound effects on our lives. Global climate change is a superb and frightening example. Balance is also required in the environments closer to

home—our homes, our communities, and our states. As numerous examples in this book have shown, a healthy physical, chemical, social, and psychological environment is crucial to preserving our own health.

Like our bodies, the environment possesses numerous homeostatic mechanisms that help regulate constancy, or ecosystem balance. **Ecosystem balance** is a dynamic equilibrium. In balanced ecosystems, for example, populations grow and decline in natural cycles. But from year to year, they remain more or less the same.

Ecosystem Stability: Population Growth/Environmental Resistance

Ecosystem balance is the result of opposing forces that act on individual populations (Figure 20–15). The first set of forces are those that cause populations to grow, the **growth factors**. Favorable weather, ample food supplies, and a high reproductive rate are three of many factors that can cause a population to increase. The second set are those that cause populations to decline, the **reduction factors**. Adverse weather, lack of food, and excess predation, for instance, can cause populations to decline. Reduction factors collectively constitute **environmental resistance**.

Numerous growth and reduction factors operate simultaneously in any given ecosystem. As Figure 20–15 also shows, growth and reduction factors have two components: abiotic and biotic. To understand how these work, consider a simplified example.

In wet years, grasses and other plants of the Midwest grow rapidly. Mice and other rodents thrive on the abundance of food. Inevitably, their populations increase. The increase in rodents may result in an increase in the size of the hawk and owl populations, which feed on rodents. The more food that is available, the more owl and hawk young that will survive.

In this chain of events, the favorable weather conditions result in an increase in the populations of mice and other rodents, which leads to an increase in owls and hawks. If the winter is particularly harsh, however, the mouse population will probably decline. The owl and hawk populations will also decline as a result of the reduction in prey, thus restoring the ecological balance. The system is going through natural oscillations, which are common in ecosystems and part of the dynamic equilibrium known as ecosystem balance.

This simplified example illustrates the mechanisms by which slight imbalances are corrected. In this case, the increase in owl and hawk populations is an example of a **biotic** reduction factor that will help achieve ecosystem stability. Adverse weather could also reduce the mouse population and is an **abiotic** reduction factor that offsets the rise in the rodent population and helps reduce the size of the hawk and owl populations.

Density-Independent and Density-Dependent Factors

Ecosystem balance is a complex issue, but is basically the net result of factors that increase and decrease the size of constituent populations. Ecologists group the components of environmental resistance into two broad categories: density-independent factors and density-dependent factors.

Density-independent factors are those events that limit the growth of populations irrespective of

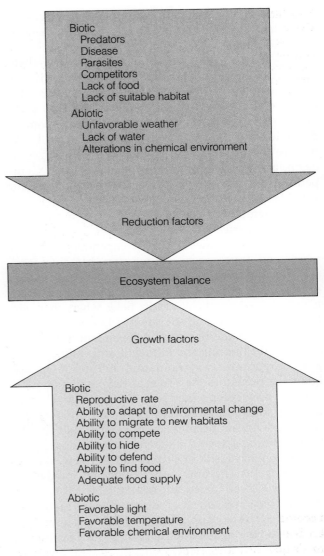

FIGURE 20–15 Ecosystem Balance Ecosystem balance is achieved by a complex balancing act brought about by biotic and abiotic growth and reduction factors.

Source: From *Environmental Science,* Second Edition, by Daniel D. Chiras. Copyright © 1988, the Benjamin/Cummings Publishing Company, Menlo Park, Calif. Reprinted by permission.

their size or density (number of organisms per acre). Drought, heat waves, cold spells, tornadoes, floods, and storms are good examples. A freeze in the usually warm tropics, for example, kills most members of a population of native butterflies, no matter what the population density.

Density-dependent factors are those conditions or events whose influence in controlling population size increases or decreases depending on the density of a population. Ecologists recognize four major density-dependent factors: competition, predation, parasitism, and disease.

Take competition, for example. Suppose a field supports 1000 mice, which are eaten by hawks. If only one hawk feeds on the mice, the supply will be sufficient for it and its offspring. If, however, 100 additional hawks set up residence in the field, the competition for the limited supply of mice would be so intense that some hawks would inevitably starve to death, assuming no other food were available. Competition for food, therefore, is a density-dependent variable. Competition can occur within a species, as in this example, or between species—for example, if house cats started feeding on the mice.

Population increases intensify competition in animals when food, water, cover, nesting sites, breeding dens, and space are limited. Competition eliminates the weak or unskillful hunters. Less skillful hunters, for example, may be forced to inhabit nearby fields where there are fewer mice. Death from starvation is likely.

Predation is another density-dependent factor. **Predation** is the hunting and killing of other animals, the **prey**. Organisms that hunt and kill for food are known as **predators**. As a rule, as the population density of a prey species increases, the percentage of organisms killed by predators increases, possibly because individuals become easier to find and attack. Furthermore, increased competition in the dense prey population may result in a larger number of weakened organisms that become easy targets for predators. Prey may also be forced into less suitable habitat and may become weakened or diseased and, therefore, more likely to be captured and killed by a predator. Ecologists have also found that when the density of a prey population increases, predators that eat a variety of prey tend to shift their attention to the most concentrated ones.

Parasites are another density-dependent factor. Parasites, such as tapeworms, feed on the bodies of other living organisms. As a rule, though, they do so without killing their hosts. Parasites spread from organism to organism in a population, and the higher the population density, the more readily they spread. In Colorado in the late 1970s, bighorn sheep offspring died in record numbers because of a parasite called the lungworm. Lungworms weaken sheep, making them more prone to pneumonia and other diseases. Under normal (uncrowded) conditions, the lungworm causes little harm, but because humans have come to occupy much of the bighorn sheep's habitat, the population density on the remaining habitat in-creased. Lungworm eggs are deposited in the sheep's feces. Because they were pressed for space, females picked up more and more eggs as they graze for food. Record numbers of parasites were found in the females; the parasitic worms passed through the placenta and infected fetuses. Too weak to withstand the parasite, newborns perished in record numbers. Fortunately, wildlife biologists found out about the problem and have captured sheep, treated them, and relocated them in suitable habitat in the Rockies.

Infectious diseases also increase when population density increases. The incidence of infectious diseases in humans, for example, increases with increasing population density. This was demonstrated in World War I when large numbers of American soldiers, who were crowded in the barracks and trenches, died from the influenza virus. Because organisms that cause infectious diseases are often transmitted by contact, through food, and by animals (insects), an increase in population density results in a increase in the spread of disease.

Density-independent and density-dependent factors operate together on a given population. For example, disease and predation may wipe a species out during a hard winter. Heavy rains may flood land and reduce the population of a species afflicted with a parasite, causing the population to decline sharply.

Succession

Changes in the density-independent and density-dependent factors are commonplace. The ability of an ecosystem to recover from minor changes such as these gives it a measure of stability. Not all changes can be quickly reversed, however. Ecosystem balance, in other words, can be drastically upset by natural forces, such as volcanoes, or by human forces. In such instances, ecosystems may recover, but the recovery is quite slow. A deciduous forest that is cleared and then planted with crops, for example, will return to forest if abandoned, but the process will require about 70 years (Figure 20–16). Long-term recovery such as this is called secondary succession.

Succession, in general, is a process of sequential change in which one community is replaced by another until a mature or climax ecosystem is formed. Two types of succession can be seen: primary and secondary.

Primary succession occurs where no biotic community previously existed—for example, when deep-sea volcanoes form islands or when glaciers retreat exposing barren rock. On volcanic islands, a rich

Abandoned farmland			10–30 years
0–1 years			Established pine forest
Crabgrass colonizes first			
1–3 years			30–70 years
Tall grass/ herbaceous plants			Hardwoods invade
3–10 years			70+ years
Pines invade			Hardwood climax forest
			Succession complete

FIGURE 20–16 Secondary Succession Occurring on an Abandoned Eastern Farm

tropical paradise can form on the barren rock, but it takes tens of thousands of years. Seeds for plants may be carried to the islands by waves or may be dropped by birds. Over time, the plants take root, then spread to cover the entire island. Along the way, new species may evolve (Chapter 19). The process of primary succession on rock exposed by the retreat of glaciers is shown in Figure 20–17.

Secondary succession occurs where a biotic community previously existed, but was destroyed by natural forces or human actions. Figure 20–16 illustrates secondary succession on an abandoned eastern farm field.

As a rule, secondary succession occurs more rapidly than primary succession because the soil is already present. In primary succession, soil must be formed, and soil formation is a rather slow process.

Secondary succession proceeds until a mature or climax ecosystem develops, unless the land is disturbed again. Secondary succession may not always be successful. In Vietnam, for example, tropical forests were destroyed by U.S. forces who sprayed Agent Orange, a herbicide, from planes and helicopters. The trees died quickly, and the forests were invaded by some hardy species that some scientists think may prevent mature forests from reestablishing. As another example, clear-cutting large tracts of tropical forests may also prevent or severely retard secondary succession. As explained in Chapter 21, clear-cutting may expose the land to intense sunlight, baking the soil to a hard bricklike consistency and preventing plants from taking root. Clear-cutting can also result in serious soil erosion, making full recovery difficult, if not impossible.

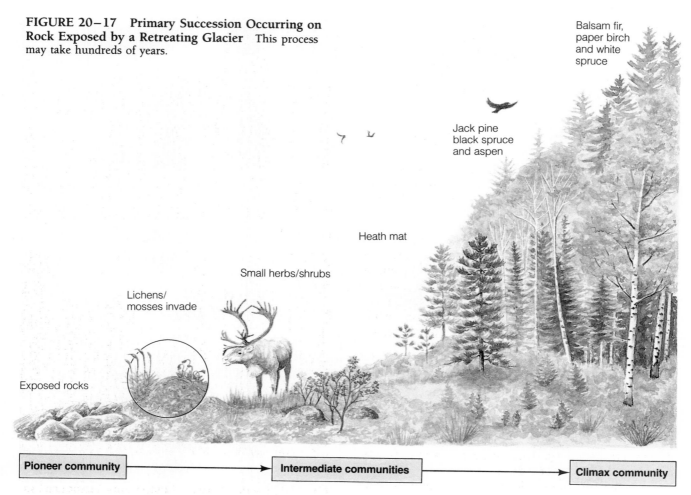

FIGURE 20–17 Primary Succession Occurring on Rock Exposed by a Retreating Glacier This process may take hundreds of years.

Balsam fir, paper birch and white spruce

Jack pine black spruce and aspen

Heath mat

Small herbs/shrubs

Lichens/ mosses invade

Exposed rocks

Pioneer community → **Intermediate communities** → **Climax community**

Species Diversity and Ecosystem Balance

What makes an ecosystem stable? Ecosystem stability is the result of the interplay of growth and reduction factors, nature's homeostatic mechanisms. Ecosystem stability may also result from species diversity. Generally speaking, **species diversity** is a measure of the number of species in a given ecosystem. The more species there are, the more diverse an ecosystem is. Some ecologists believe that species diversity leads to ecosystem stability. They support this contention with the following facts. First, extremely stable ecosystems, such as tropical rain forests, are characterized by a remarkably high species diversity. Unstable ecosystems, such as the tundra, contain populations that tend to oscillate widely. These ecosystems have a much lower diversity. Scientists also know that intentionally simplifying an ecosystem (removing species) tends to make it unstable. Farm fields are a good example. Monotype crops are more vulnerable to pests and disease than heteroculture crops containing sev-

eral different crops planted in the same field. Both monotype and heteroculture crops are more vulnerable than nature's heteroculture grasslands, which contain many species of plants and animals.

To understand how diversity leads to stability, consider the simple models shown in Figure 20–18. As illustrated, in the simplified ecosystem, fewer species and fewer connections exist. The loss of one species in this system would be much more noticeable than the loss of one species from the much more diverse ecosystem in Figure 20–18b. For example, the loss of one producer in a simplified ecosystem would have a more dramatic effect on that ecosystem than the loss of a producer in an ecosystem where several are present.

Not all ecologists agree. Some say that diverse ecosystems such as tropical rain forests are stable not because of their diversity, but because of their uniform climate. Biological stability is a product of a constant climate. Diversity may be an outcome of the

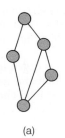
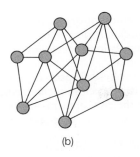

(a)   (b)

FIGURE 20–18 Diversity and Stability (*a*) In a simplified ecosystem, the loss of one species generally has a more profound impact on the food web than (*b*) the loss of one species in a more complex ecosystem.

climate, which provides ample food for a great variety of species. Instability in the tundra is a result not of the low level of diversity, but of the volatile climate.

Despite differences of opinion, some generalizations can be drawn from the evidence. The most important is that intentionally reducing species diversity can make ecosystems unstable.

ENVIRONMENT AND HEALTH: THE CUMULATIVE IMPACT

Humans alter the environment in two principal ways: (1) by altering the abiotic components and (2) by altering the biotic components.

Consider some examples of ways we tamper with abiotic factors. Chemical pollutants from factories, and even our lawns and gardens, enter streams and lakes, altering the chemical environment and killing fish and other aquatic organisms. Hot water is another kind of pollutant that is released from power plants and factories into lakes and streams where it alters the physical environment of the stream, killing aquatic life. Carbon dioxide and other air pollutants may be increasing global atmospheric temperature, causing a dramatic shift in the abiotic conditions on planet Earth. Overgrazing and cutting forests often reduce rainfall. The list goes on.

These intrusions have numerous indirect impacts on the biotic components of the environment, the plants, animals, and microorganisms. Human populations can have direct effects as well. Overfishing, for example, wipes out valuable members of the food chain. Introducing foreign species to an ecosystem can cause chaos. Foreign competitors or predators, for instance, can wipe out native plants and animals (Figure 20–19). Eliminating predators can have equally

(a)

(b)

FIGURE 20–19 Foreign Competitors Introduced to the United States (*a*) Kudzu grows wild in the south, covering fields and homes. (*b*) Water hyacinth, introduced from South America, proliferates in southern waters, choking out native plants and making navigation impossible.

disruptive effects. The elimination of wolves, mountain lions, and bears in the Kaibab plateau on the north rim of the Grand Canyon in the early 1900s, for instance, may have resulted in an explosion in the local deer population from 4000 to 100,000 by 1924. The dramatic increase was followed by widespread environmental damage caused by overgrazing and an equally dramatic crash in the deer population. Approximately 60,000 deer died in the following winters.

All organisms have an impact on their environment; virtually all living things alter the abiotic and biotic components of the ecosystem they live in. But the impact of human populations is different from other organisms for several key reasons. First, our advanced technological development has given us incredible power to alter the face of the Earth. Second, the human population has reached unprecedented numbers. Today, over 5.25 billion people live on the planet. Even small acts are now beginning to add up and may be putting the planet—and our own population—at risk. The next chapter examines the evidence behind this assertion and describes ways to reduce human impact and build a sustainable future. The key to success, many experts believe, is restoring global ecosystem balance and building a relationship with nature that does not endanger the homeostatic mechanisms that maintain the planet's health—and the health and well-being of the millions of species that make this planet their home.

SUMMARY

PRINCIPLES OF ECOLOGY: ECOSYSTEM STRUCTURE

1. Most of us see our lives as apart from nature. But in truth our lives depend heavily on our environment for food, oxygen, clothing, materials, and much else.
2. Ecology is the study of ecosystems—organisms and their environment and the many interactions that exist between abiotic and biotic components of ecosystems.
3. The living "skin" of the planet is called the biosphere. It extends from the bottom of the ocean to the tops of the highest mountains.
4. The biosphere is a closed system in which materials are recycled over and over again. The only outside contribution is sunlight, which powers all biological processes.
5. The Earth's surface is divided into large biological regions, or biomes, each with a characteristic climate and characteristic plant and animal life.
6. The oceans may also be divided into biological regions, which are called aquatic life zones.
7. The biosphere is a global ecological system or ecosystem. An ecosystem is a community of organisms, their environment, and all of their interactions. Ecosystems consist of biotic and abiotic components.
8. Organisms are adapted to a range of conditions in the ecosystem in which they live. This is called the range of tolerance. The conditions vary naturally from day to day and season to season, but as long as they remain within the optimal zone, organisms will thrive.
9. Human activities can alter the abiotic and biotic conditions in an ecosystem, causing considerable harm.
10. In any ecosystem, one factor tends to limit growth and is, therefore, called a limiting factor.
11. A group of organisms of the same species living in a specific region constitutes a population. Two or more populations occupying that region constitute a community.
12. The physical space a species occupies is called its habitat. A species's niche includes its habitat and its place in the environment—its position in the food chain, its range of tolerance, and so on.

ECOSYSTEM FUNCTION

13. Virtually all life on Earth depends on the producers, organisms that synthesize organic materials from sunlight, carbon dioxide, and water. The major producers are the plants and algae.
14. Organisms dependent on producers and other organisms for food are called consumers. Four types of consumers are present: herbivores, carnivores, omnivores, and detrivores.
15. All organisms are part of food chains. A food chain represents the feeding relationships in an ecosystem. Food chains that begin with plants that are consumed by grazers (herbivores) are known as grazer food chains. Those that begin with animal waste or the dead remains of plants, animals, and microorganisms are called decomposer food chains. Food chains are simplified components of larger networks, called food webs, which represent a truer picture of the feeding relationships in an ecosystem.
16. Food chains and food webs are a conduit for the one-way flow of energy through an ecosystem. They also are a channel for the circular flow (recycling) of minerals and other nutrients through the ecosystem.
17. The position of an organism in a food chain is called its trophic level. In a grazer food chain, plants (producers) are on the first trophic level; herbivores are on the second; carnivores are on the third.
18. Most food chains are limited to three or four trophic levels because only 5 to 20% of the biomass is transferred from one trophic level to the next. This reduces the amount of biomass available to the higher-level organisms. The important implication is that the lower an organism feeds in a food chain, the more food is potentially available.
19. Nutrients flow through the environment in nutrient cycles. Nutrients in the environment enter the organismic phase through the producers. The nutrients are then shunted through the food chain (organismic phase of the cycle) and eventually reenter the environment (environmental phase) where they generally are available for reuse.
20. Humans can interrupt nutrient cycles locally and globally. The nitrogen cycle, for example, is often flooded with nitrogen from human sources (fertilizer, sewage, and animal waste) in towns, cities, and rural communities.

21. The global carbon cycle is flooded with carbon dioxide from the worldwide consumption of fossil fuels. Deforestation also increases atmospheric carbon dioxide.

ECOSYSTEM BALANCE AND IMBALANCE

22. Human health is dependent on homeostasis at two levels: within the environment and within ourselves. A healthy physical, chemical, social, and psychological environment is crucial to a healthy body.
23. The environment contains numerous homeostatic mechanisms that help to create a balanced environment.
24. A balanced environment is one in which populations remain more or less stable over long periods. It experiences regular change, but remains fairly constant over time.
25. Ecosystem balance is the net result of a variety of biotic and abiotic growth and reduction factors. They increase and decrease population, one offsetting the other, in a constant balancing act, thus maintaining ecosystem balance.
26. Ecosystem balance is a complex issue, but is basically the net result of factors that increase and decrease the size of constituent populations.
27. Environmental resistance plays a major role in controlling populations. Environmental resistance consists of factors that fall into two broad categories: density-independent factors and density-dependent factors.
28. Density-independent factors are those events that limit the growth of populations irrespective of their size or density (number of organisms per acre), such as drought, heat waves, cold spells, and storms.
29. Density-dependent factors are those conditions or events whose influence in controlling population size increases or decreases depending on the density of the population: competition, predation, parasitism, and disease.
30. Density-independent and density-dependent factors operate together on a given population.
31. Ecosystems can recover from minor disturbances rather easily, but more significant changes may require more time. In some cases, damage may be so severe that recovery is impossible.
32. The reestablishment of a community in an ecosystem disturbed by humans or natural causes is called secondary succession.
33. Succession, in general, refers to a series of changes in an ecosystem, in which one community replaces another until a mature or climax ecosystem is produced.
34. Ecosystem stability results from the interplay of growth and reduction factors. It may also result from high species diversity or possibly abiotic factors, such as climate.

ENVIRONMENT AND HEALTH: THE CUMULATIVE IMPACT

35. Humans alter the environment in two principal ways: (1) by altering the abiotic components and (2) by altering the biotic components.
36. All organisms have an impact on their environment. The impact of human populations, however, is different from other organisms because of our technological power and because our population has reached unprecedented numbers. Even small acts are now beginning to add up and may be putting the planet—and our own population—at risk.

EXERCISING YOUR CRITICAL THINKING SKILLS

Imagine that you work for your state's department of natural resources and are asked to comment on the proposed introduction of a deer from a remote part of the Soviet Union. The rationale behind the introduction is that it would enhance hunting in the state. How would you evaluate the proposal? What ecological criteria would you rely on to determine if it was a good idea?

TEST OF TERMS

1. _____ is the study of ecosystems, which includes biotic and abiotic components and the numerous interactions that exist between them.
2. The region of the planet that supports life is called the _____ .
3. A _____ is a region characterized by a distinct climate and distinct plants and animals.
4. An ecosystem consists of two basic components: _____ and
5. The range of conditions an organism can survive is called the _____ - _____ .
6. One factor is often more important than others in regulating growth

in an ecosystem. It is, therefore, called a _____ _____.

7. A group of like organisms occupying a specific region constitutes a(n) _____. In any given ecosystem, several populations exist together and form a(n) _____.

8. An ecological _____ includes the habitat of an organism and its place in the ecosystem.

9. Plants and algae are also called _____, because they synthesize organic molecules necessary for all other organisms, which are collectively referred to as _____.

10. Animals that eat only plants are called grazers or _____; animals that eat plants and animals are called _____; and organisms that consume waste and the dead remains of animals and plants are _____.

11. Two types of food chains exist: _____ and _____.

12. Nutrients are recycled in ecosystems, but all sunlight energy is eventually dissipated into space as _____.

13. The position of an organism in the food chain is called its _____ level.

14. _____ is the dry weight of the living material in an ecosystem.

15. Each nutrient cycle consists of two basic parts: the _____ phase and the _____ phase.

16. The conversion of nitrogen gas to nitrate and ammonia by root nodules and soil bacteria is called _____ _____.

17. Ecosystem balance or stability results from the interplay of two sets of factors: _____ factors, which tend to increase the size of populations, and _____ factors, which tend to decrease the size of populations.

18. Competition and predation are two _____-dependent components of environmental resistance.

19. _____ is a process of sequential change in which one community is replaced by another until a mature or climax ecosystem is formed. When this process occurs on a newly formed island, it is called _____ _____.

Answers to the Test of Terms are located in Appendix B.

TEST OF CONCEPTS

1. Humans are a part of nature. Do you agree or disagree with this statement? Support your answer.

2. Define the term *ecology* and give examples of its proper and improper use.

3. The Earth is a closed system. What are the implications of this statement?

4. Define the following terms: biosphere, biome, aquatic life zone, and ecosystem.

5. Define the term *range of tolerance*. Using your knowledge of ecology, give some examples of ways humans alter the abiotic and biotic conditions of certain organisms, and describe the potential consequences of such actions.

6. Describe ways that humans alter their own range of tolerance for various environmental factors.

7. What is a limiting factor? Give some examples.

8. Define the following terms: habitat, niche, producer, consumer, trophic level, food chain, and food web.

9. Explain why biomass at one trophic level is lower than the biomass at the next lower trophic level.

10. Outline the flow of carbon dioxide through the carbon cycle, and describe ways that humans can impact the carbon cycle.

SUGGESTED READINGS

Chiras, D. D. 1991. *Environmental science: Action for a sustainable future,* 3d ed. Redwood City, Calif.: Benjamin/Cummings. Chapters 3 and 4 present this material in much more detail.

_____. 1992. *Healing the Earth.* Washington, D.C.: Island Press. Author's description of ways to build a sustainable society.

Ehrlich, P. R., and J. Roughgarden. 1987. *The science of ecology.* New York: Macmillan. Good advanced-level introductory textbook.

McPhee, J. 1989. *The control of nature.* New York: Farrar Straus Giroux. Fascinating account of human efforts to control nature and the many ways our attempts backfire, creating worse problems.

Owen, O. S., and D. D. Chiras. 1990. *Natural resource conservation: An ecological approach,* 5th ed. New York: Macmillan. See Chapter 2 for a discussion of ecology.

Perry, D. A., M. P. Amaranthus, J. G. Borchers, S. L. Borchers, and R. E. Brainerd. 1989. Bootstrapping in ecosystems. *Bioscience* 39(4): 230–37. Examines the productivity and stability of biological systems and the effect of outside disturbances.

21

Environmental Issues: Population, Pollution, and Resources

Bridger Mountains, Montana. We must choose between a clean and healthy environment and a polluted, dangerous world.

Nature has its own economy—a complex system of exchange, competition, and co-operation that is, in many ways, like our own economy. Like the human economy, the economy of nature is driven by the sun; the economy of nature has its producers and consumers. The avenues of commerce are food chains and food webs. In many ways, however, the economy of nature and the massive multibillion-dollar human economy are at odds, like two gears in a huge machine, running in opposite directions. The small wheel of the human economy is growing bigger and more powerful and is now beginning to tear apart the larger one that drives it. After two hundred years of industrial growth and significant gains in human welfare in many countries, the human economy is threatening the workings of nature and the health of the planet. Not just human existence is in danger, but the existence of all life. The threat to the environment is commonly referred to as the "environmental crisis."

To many people, the phrase "environmental crisis" may seem like an exaggeration. To them, such a proclamation seems unwarranted. But a close examination of trends in population growth, pollution, and resource use and depletion yields a different view. Many experts who have studied the trends in fact believe that the problems are so severe as to constitute a crisis. Dr. Jay Hair, president of the National Wildlife Federation, a zoologist and former professor of zoology at North Carolina State University, in fact, believes that humanity has at best 10 or 15 years to come to grips with the menacing trends. If not, our future 100 years down the road will be far from optimum.

This chapter outlines current trends in population, pollution, and resource use and describes several of the most pressing environmental problems. It also discusses a variety of solutions, for individuals, corporations, and governments. The chapter concludes with a discussion of ways to steer a sustainable course, a kind of harmony between humans and the environment that many experts believe is needed to ensure long-term ecological stability on the planet—a condition essential for a healthy human existence.

OVERSHOOTING THE EARTH'S CARRYING CAPACITY

The environmental problems you hear about on the news and read about in the paper occur in the rich, industrialized nations as well as in the poor, nonindustrialized nations. Although the problems vary from one nation to another, they all have a common root. In a phrase, environmental problems result from people living beyond the means of the environment. In the language of ecologists, human society is exceeding the Earth's carrying capacity.

Carrying capacity is the number of organisms an ecosystem can support indefinitely—that is, the number it can support on a sustainable basis. Carrying capacity is limited by at least three factors: (1) food production, (2) resource supply, and (3) the environment's ability to assimilate pollution.

Food and resource supplies are primary determinants of population size, a fact that is as true for humans as it is for all other organisms. As noted in the last chapter, an increase in food availability can boost reproductive success in natural populations, resulting in an increase in population size. A shortage of food has the opposite effect (Figure 21–1).

As the global human population grows, nations are finding it more and more difficult to meet the demands for food. Starvation abounds in the Third

FIGURE 21–1 Food Shortages Overpopulation, political turmoil, and mismanagement of farmland are chief causes of scenes such as this one. Children die by the thousands every day in similar camps.

World, and millions of people perish each year from malnutrition and starvation and diseases worsened by hunger (Chapter 7).

Human populations require a great many other resources, such as fuel, fiber, and building materials. Those resources that are finite, such as oil, natural gas, and minerals, are known as **nonrenewable resources**. Resources that replenish themselves via natural biological and geological processes are called **renewable resources**. Wind, hydropower, trees, and fish are examples.

Natural resources, those materials from the environment essential to human well-being, are also vital to other species. Consequently, many resource scientists have expanded the definition of natural resources. In this new light, **natural resources** are those materials from the environment essential to all of Earth's organisms. The mutual dependence on natural resources creates many apparent environmental "con-

FIGURE 21–2 **Air Pollution** Studies show that air pollutants in some locales can increase respiratory disease, such as emphysema and lung cancer. Air pollution is aesthetically unappealing, and levels that do not affect human health can cause considerable environmental damage.

through nutrient cycles. Feces, for instance, are broken down by soil bacteria, which release carbon dioxide, nitrogen, and other nutrients into the environment for reuse. Nutrient cycles can be severely altered by human action. Too much animal waste in an aquatic environment, for example, may shift the concentration of nitrogen and phosphorus, starting a series of changes that result in rapid growth of algae and aquatic plants; algae and plants decay in the fall, consuming oxygen in the process and killing fish.

Human populations produce enormous amounts of wastes. In the United States, a world's leading producer of waste, human sewage, garbage from our homes, hazardous wastes from factories, pollution from automobiles and homes, and wastes from agriculture and mining amount to 50,000 pounds per year for every man, woman, and child, according to a study by the Conservation Foundation. The U.S. population represents only 6% of the world population, but produces about 25% of the world's pollution.

The environment can dilute some of this waste to harmless levels. It can break down other wastes, rendering them harmless, but in many locations, the environment's assimilative capacity is being overtaxed. In lakes, rivers, and oceans, pollution is taking a toll on fish and wildlife. Discarded plastic fishnets, six-pack yokes, and other plastic garbage kill an estimated 100,000 marine mammals, such as seals, each year. In U.S. waters, tighter regulations have reduced the outflow of noxious pollutants from factories and sewage treatment plants, helping to achieve cleaner waters. Nevertheless, many lakes and rivers have shown little improvement over the past decade. Many of these bodies of water contain hazardous levels of pollutants that are being washed from urban lawns, golf courses, parking lots, and farm fields or are washed from the sky by rain and snow. These pollutants have offset many of the gains in pollution control of our factories and sewage treatment plants.

The overtaxing of the assimilative capacity of the atmosphere is also evident around most cities with populations over 50,000. These pockets of air and water pollution pose a threat to human health and a threat to a great many species that make this planet their home (Figure 21–2). Today, over 110 million Americans (nearly half of our population) live in air deemed harmful to human health.

One of the "advances" of modern industrial society was the advent of synthetic organic chemicals, such as the insecticide DDT and plastics. Many of these products cannot be broken down by natural processes—or

flicts," a kind of competition between humans and nonhuman species that we nearly always win.

Human existence today depends on our ability to draw sustenance from a finite natural world, but our continued existence depends, in large part, on our ability to shift to renewable resources and to refrain from destroying the natural regenerative systems that provide a continual supply of them. Today, many renewable resources are badly abused. Heavy cutting of tropical forests, overhunting, and overfishing all diminish the renewable resource base upon which all life depends.

The final determinant of carrying capacity is the capacity of the environment to assimilate and degrade pollution. In natural ecosystems, wastes are usually diluted to harmless levels. Wastes are recycled

are degraded very slowly—because naturally occurring bacteria are not equipped with the enzymes needed to decompose these new molecules. As a result, DDT, PCBs, and a number of other toxic organic chemicals often persist in the environment for decades, accumulating in the tissues of organisms, impairing reproduction, and threatening their future.

The environmental crisis is multifaceted. In many of the poor, nonindustrialized countries, it results from too many people and not enough food. In other countries, the problem results from too many people depleting both the renewable and nonrenewable resource base, a trend that will lead to major resource shortages in your lifetime. In other countries, such as the United States, pollution production exceeds the environment's assimilative capacity. Residents are, quite literally, drowning in a sea of pollution. In all three instances, the problem is basically the same: humans are exceeding the carrying capacity. Correcting that imbalance is one way of alleviating the problems.

OVERPOPULATION: PROBLEMS AND SOLUTIONS

Overpopulation occurs when populations exceed the carrying capacity of the environment. The world human population is currently 5.25 billion and is increasing at a rate of 1.8% per year. Although the growth rate may seem small, if the current rate of growth continues, world population could reach 10 billion people by the year 2030—in 40 years.

Growth Rate and Doubling Time

The annual **growth rate** of the world population is calculated by the following formula:

$$\text{growth rate} = \text{birth rate} - \text{death rate}$$

The world birth rate is equal to 28 per 1000. In other words, 28 children are born for every 1000 people living on the Earth each year. The global death rate is 10 per 1000. To calculate the growth rate, you simply subtract the death rate from the birth rate. The global growth rate is equal to 28/1000 - 10/1000, or 18 per 1000. This means that 18 new residents are added to the planet each year for every 1000 people alive. To convert this to a percentage, simply multiply by 100. The growth rate is equal to 18/1000 x 100, or 1.8%.

Growth rates are important measures of population dynamics, but can be somewhat deceiving: low growth rates appear inconsequential. For this reason, **demographers**, scientists who study populations, use the growth rate to calculate the **doubling time**, the time it takes a population to double. Doubling time is calculated by a simple equation:

$$\text{doubling time} = 70/\text{growth rate (in percent)}$$
$$= 70/1.8$$
$$= 39 \text{ years}$$

The figure 70 in this equation is a demographic constant. The global growth rate is an average of all countries and, thus, masks regional growth differences. For example, in the United States, the growth rate is currently about 0.7%. In other words, 7 new residents are added to the population each year for every 1000 people alive.[1] Our present population of 248 million people will double in 100 years if current growth continues. In contrast, in Africa the growth rate is 2.9%, yielding an alarming doubling time of 24 years!

Today, the most rapid growth is occurring on three continents: Africa, Asia, and Latin America (Table 21–1). The slowest growth is occurring in Europe where the population is growing at a rate of 0.3% per year. At this rate, Europe's population will double in about 230 years.

For most people, overpopulation is a problem of the Third World. Many Third World nations are plagued with too many people and not enough food to go around. Actually, overpopulation is a problem in all countries. It is as serious in the United States as it is in Bangladesh.

The reason for this is the fairly high standard of living in the rich, industrialized countries. Because of this, citizens of these nations have an enormous impact on the environment. One baby born in the United States will use 25 to 40 times as many resources as an Indian baby. The 248 million Americans cause as much damage as 6 to 10 billion people in the Third World.

The Dangers of Exponential Growth

The human population has not always been so large nor has it always grown so rapidly. As Figure 21–3a shows, it was not until the last two hundred years that

---

[1] Population growth in any country is the result not only of the difference between birth rate and death rate, but also of the net migration—that is, the number of people entering or leaving the country. In the United States, about 40% of our annual growth results from legal and illegal immigration.

TABLE 21-1   Growth Rate and Doubling Time

REGION	GROWTH RATE (%)	DOUBLING TIME (YEARS)
World	1.8	39
Developed countries	0.6	117
Less-developed countries	2.1	33
Africa	2.9	24
Asia	1.9	38
North America	0.7	100
Latin America	2.1	33
Europe	0.3	230
Soviet Union	1.0	68
Oceania	1.2	56

global human population began to skyrocket, and then only because of better sanitation, improvements in medicine, and advances in technology (Chapter 20).

The graph of human population growth is J-shaped or exponential. What is exponential growth? To understand this simple, yet important concept, consider an example. Suppose your parents invested $1000 in a savings account at 10% interest the day you were born. Suppose also that the interest your money earned was applied to the balance, so that it too earned interest. Your bank account would grow exponentially. **Exponential growth** occurs any time a value, such as population size, grows by a fixed percentage with the interest (increase) being applied to the base amount.

Exponential growth is deceptive. At first, the growth in your account seems small. At 10% interest, the $1000 will double in 7 years, yielding $2000 (Figure 21-3b). It will double again in another 7 years, yielding $4000. The next doubling, occurring when you are 21 years old, will give you $8000. By age 42, the account will have grown to $64,000. But by age 49, it would hold $128,000. At age 56, you'd have over $250,000. If you waited 7 more years, the account would grow to over $500,000. In 7 more years, at 70 years of age, you'd be a millionaire, but probably too old to enjoy the money!

What is deceptive about exponential growth is that it takes off so slowly, even though the rate is constant. In this example, it took your account 49 years to grow from $1000 to $128,000. In the following 21 years, the account grew by $900,000. Thus, once the base amount reaches a certain level, each doubling yields incredible gains.

Exponential growth of population, resource demand, and pollution are at the heart of the environmental crisis. The human population will increase by about 5 billion in the next 40 years if current growth continues, but it has taken over 3 million years to reach the current size (Table 21-2). Pollution and

TABLE 21-2   Estimates of Global Population Growth

POPULATION SIZE	YEAR	TIME REQUIRED TO DOUBLE
1 billion	1850	All of human history
2 billion	1930	80 years
4 billion	1975	45 years
8 billion (projected)	2017	42 years

FIGURE 21-3 Exponential Growth of the World Population (a) World population. (b) Exponential growth of a bank account starting with $1000 at 10% interest.

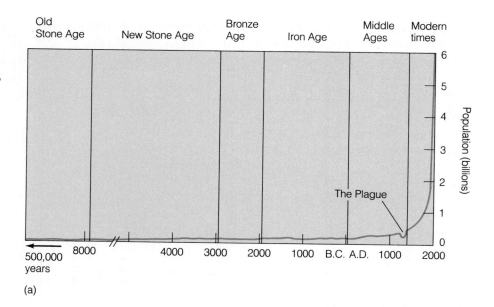

(a)

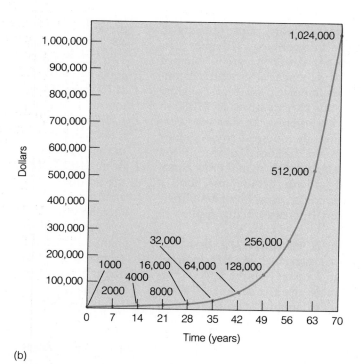

(b)

resource demand also grow exponentially, and therein lies the concern of many environmentalists. A resource with a billion-year life span at the current rate of consumption will last under 500 years if demand increases at 5% per year. Thus, even though many resource supplies seem adequate now, in a few doublings, demand will likely outstrip supply.

Too Many People, Reproducing Too Quickly

The human population problem can be summed in a phrase: too many people, reproducing too quickly. In many countries, large segments of the population are living in extreme poverty. For many of these people, there is not enough food to eat. People live in makeshift homes or on the streets (Figure 21–4). Today, an estimated 700 to 800 million people are malnourished and severely undernourished. About 2 billion people are on the edge of poverty with barely enough to eat. Thus, nearly three of every five people on the planet are living in subhuman conditions. Forty million people, about half of them children, die of starvation and diseases worsened by hunger each year.

Humans are living beyond the carrying capacity of the environment now. Further growth will only worsen problems. Each year 89 million new residents are added to the world population. Approximately 9 of every 10 of these people are born in the Third World, where hunger and poverty have become a way of life (Figure 21–5). In African nations, such as Kenya, where the human population is expected to double in 17 years, efforts aimed at keeping up the substandard food supplies will put enormous strains on the economy and the environment. Making improvements in the human diet to reduce starvation and persistent hunger seems impossible. In this struggle, many believe that nature will restore a more equitable balance. Millions of people will die unless something is done to curb population growth and increase food supply.

Steps to Solve World Hunger

There is no easy answer to world hunger. Solving such a complex issue requires a complex, multifaceted response (Table 21–3).

Slowing the Rate of Growth. One of the most important steps is to reduce population growth. Most world leaders agree that greatly reducing the rate of

FIGURE 21–4   Poverty Abounds   About three-fifths of the world's people live in poverty. Nearly one of every five people on this planet does not have enough to eat. Many live in makeshift shelters.

increase will help Third World countries produce the food they need and help improve the human condition. Reducing the rate of growth will also help industrial countries reduce their demand for resources and help reduce pollution and habitat destruction as well (Figure 21–6).

Reducing Numbers through Attrition.   Judging from the state of the environment, today's population of 5.25 billion people is well beyond the carrying capacity. Slowing the growth of human population, then, will not be sufficient. Over the next 50 years, it may be necessary to reduce the size of the human population, a measure that will help permit human society to live within the Earth's carrying capacity. The principal means of reducing population size is

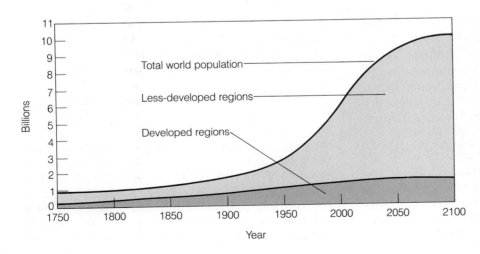

FIGURE 21–5   Growth of Human Population   The graph shows the relative size of the populations in the less-developed regions and the developed regions and projected growth. Ninety percent of the growth is occurring in the Third World. Today, about one of every five people in the world lives in an industrialized country. By 2100, if current growth rates continue, the ratio will be about 1 to 10.

TABLE 21-3 Some Solutions to Help Alleviate World Hunger

Reduce population growth
Reduce soil erosion
Reduce desertification
Reduce farmland conversion
Improve yield through better strains
Improve yield through better soil management
Improve yield through fertilization
Improve yield through better pest control
Reduce spoilage and pest damage after harvest
Use native animals for meat production
Tap farmland reserves available in some countries

FIGURE 21-6 Habitat Destruction Wetlands are a valuable natural resource but have been destroyed by home building and other activities.

through attrition—reducing the birth rate so that it falls below the death rate.

Reducing the Total Fertility Rate. Reducing the population size of a country, indeed the world, is not as difficult as many might believe. Already, a dozen or so countries in Europe have reached a stage of extremely slow growth, no growth, or even "negative" growth (shrinkage), as shown in Table 21-4. The populations of West Germany and Hungary, for example, are declining slightly. The decline in population size is achieved, in large part, because couples are having smaller families.

The total number of children a woman has in her lifetime is called the **total fertility rate**. For a population to remain stable in countries, such as ours, the total fertility rate must be maintained at 2.1.[2] This means that each woman of reproductive age has, on average, 2.1 children. This level of fertility is called the **replacement-level fertility**. It is the number of children that will replace a couple when they die. A replacement-level fertility rate of 2.1 means that every 10 couples in a population must produce 21 children to maintain a steady population size. The additional child accounts for typical mortality and for childless couples.

When the total fertility rate is below the replacement-level fertility, a population will decline, but only if there is no net immigration—that is, there are no newcomers from other countries. In the United States, the total fertility rate has been below replacement-level fertility since 1972 (Figure 21-7). Despite this important development, the U.S. population continues to grow. Why?

One of the reasons for growth is the steady inflow of illegal and legal immigrants—about a million people each year. Growth also occurs because of the age structure of our population. Figure 21-8 is a **population histogram** or **population profile** of the United States in 1960 and 1988. It shows the number or percentage of males and females in each age group. As you can see, the profile for 1960 is bottom-heavy.

[2]Assuming there is no immigration into the country.

There are more people in the lower age groups than in the upper ones. Many of the people in the lower age groups are products of the baby boom era, that is, they were children born after World War II.

After the war, America's prospects looked bright. Judging from the rise in total fertility to well over 3, people must have been extremely optimistic and happy to be done with the brutal war. The large postwar families caused the U.S. population to swell (Figure 21–9).

Today, the baby boomers are having children of their own, and even though couples are having far fewer children than their parents did, the reproductive age group is larger today than it was after the war. Growth in U.S. population today, therefore, occurs not because women are having more children, but because more women are having children.

If immigration quotas are not increased and if the total fertility rate remains the same, the U.S. population profile should become more boxlike or stationary (Figure 21–10). At this point, the population will cease growing, reaching a steady state called **zero population growth**.

Worldwide the population profile today looks a lot like the U.S. population profile did shortly after the war. The profile is triangular or expansive. Today, 35% of the world's people are under the age of 15. Soon, they will be entering the reproductive age group and will start having families. Unless these children curb

TABLE 21–4	Nations Experiencing Slow Growth or No Growth	
NATION	GROWTH RATE (%)
Hungary	-0.2
West Germany	-0.1
Denmark	0.0
Italy	0.0
Austria	0.1
Belgium	0.1
Bulgaria	0.1
East Germany	0.1
Greece	0.1
Luxembourg	0.1
Czechoslovakia	0.2
Norway	0.2
Sweden	0.2
United Kingdom	0.2
Finland	0.3
Portugal	0.3
Spain	0.3
Switzerland	0.3
France	0.4
Netherlands	0.4

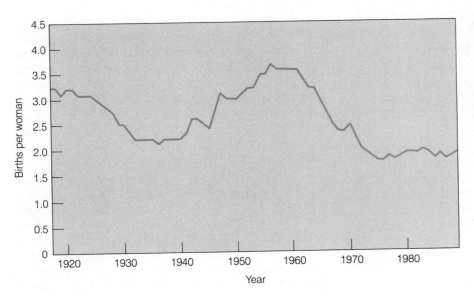

FIGURE 21–7   Total Fertility Rate in the United States   Since 1972, the total fertility rate in the United States has been below replacement-level fertility. Yet overall population growth continues because of immigration and because of the large number of women having children.

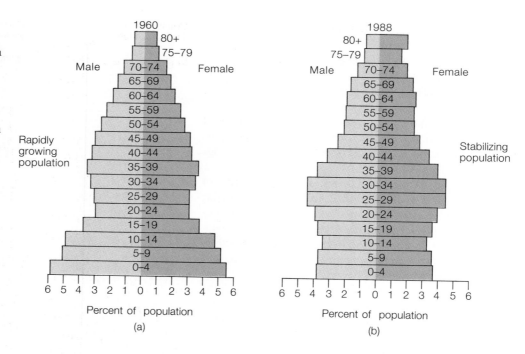

FIGURE 21–8
Population Profiles for the United States (*a*) In 1960, the U.S. population was growing. The broad base of the population profile reflects this fact. (*b*) In 1988, growth was slowing, and the population was beginning to stabilize; the profile is becoming more boxlike.

1960

Male   Female

80+
75–79
70–74
65–69
60–64
55–59
50–54
45–49
40–44
35–39
30–34
25–29
20–24
15–19
10–14
5–9
0–4

Rapidly growing population

6 5 4 3 2 1 0 1 2 3 4 5 6

Percent of population

(a)

1988

Male   Female

80+
75–79
70–74
65–69
60–64
55–59
50–54
45–49
40–44
35–39
30–34
25–29
20–24
15–19
10–14
5–9
0–4

Stabilizing population

6 5 4 3 2 1 0 1 2 3 4 5 6

Percent of population

(b)

their own reproduction, they will cause massive population growth in the next four decades.

RESOURCE DEPLETION: ERODING THE PROSPECTS OF ALL ORGANISMS

The human population depends on a variety of resources for its survival and well-being: forests, soil, water, minerals, and oil. Today, however, many of these resources are in danger. According to one estimate, 5 billion trees are cut down every day, and worldwide the rate of timber harvest now exceeds regrowth. Productive soils are being eroded by careless farming practices, and once-productive farmland is turning to desert at an alarming rate. Water for drinking and irrigation is being withdrawn from wells faster than it can be replaced, creating regional shortages that are bound to worsen in the near future. Minerals are being extracted in ever-increasing quantities, and many vital mineral supplies will last no more than four decades. All of these problems are serious in and of themselves; when taken together, they create an environmental crisis. Each of these

FIGURE 21–9 The Baby Boom Effect This figure follows the babies born from 1955 to 1959 through the year 2010. Population profiles provide a means of projecting future trends in populations.

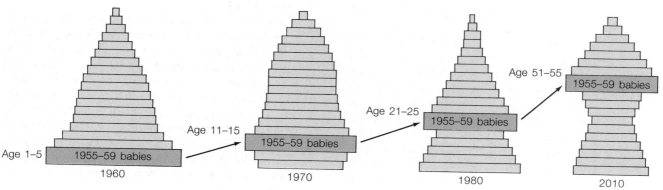

Age 1–5 1955–59 babies
1960

Age 11–15 1955–59 babies
1970

Age 21–25 1955–59 babies
1980

Age 51–55 1955–59 babies
2010

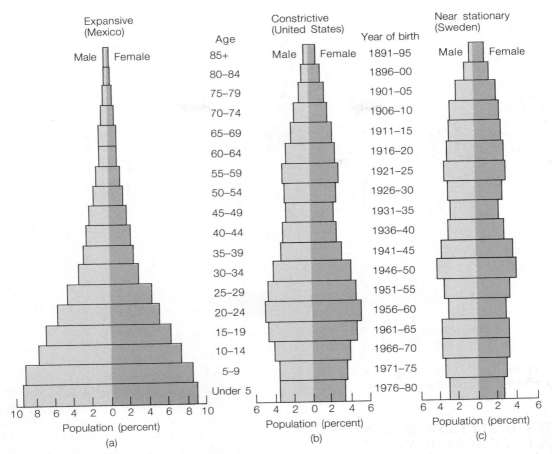

FIGURE 21–10 Three Possible Population Profiles (*a*) Mexico's population profile is typical of a rapidly growing population. (*b*) The population profile of the United States reflects its much slower growth. (*c*) Sweden's population profile is near stationary.

problems is solvable by responsible actions, but actions must begin soon. The next section describes the problems and discusses some of the solutions.

Destroying the World's Forests

At one time, the tropical rain forests covered a region about the size of the United States (Figure 21–11). Today, at least one-third of those forests are gone. Because of an exponential increase in population size and timber harvest, most of the remaining forests could be gone by the year 2000. Along with them, thousands of species—perhaps as many as a million, many of them unknown to science—would vanish.

The decline of the world's forests is not limited to the tropics. In the United States, for example, 45% of the forested land present when the eastern seaboard was first colonized has vanished. In the Pacific Northwest, 90 to 95% of the old-growth forests, consisting of trees from 250 to 1000 years of age, have been cut (Figure 21–12).

Since 1920, however, the overall rate of timber harvest in the United States, has more or less equaled the rate of regeneration. That is due, in large part, to the reforestation of abandoned farmland in the East and more stringent policies in national forests. The rolling hills of Connecticut, for example, once cleared for farming, now support healthy young forests. Expected increases in the demand for timber and wood products, however, could shift the balance once again.

Since global deforestation exceeds reforestation and since population and demand for resources con-

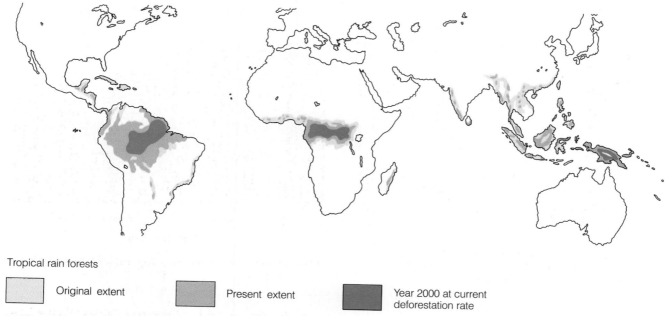

Tropical rain forests

☐ Original extent   ☐ Present extent   ■ Year 2000 at current deforestation rate

FIGURE 21–11   **Tropical Rain Forests**   This map shows the steady decline in tropical forests and projected decline for the year 2000.

tinue to climb, the fate of the world's forests is dim. In your lifetime, many tropical forests will be destroyed. Consequently, many conservationists and environmentalists are urging sharp cutbacks in current de-

FIGURE 21–12   **Old-Growth Forest**   (*a*) These ancient trees in Washington are protected in Olympic National Park. (*b*) But 95% of the old-growth stands, home to the endangered spotted owl, have been cut, and much of the rest of the old-growth forest on private land and on state and federal land is slated for harvest.

(a)   (b)

mand and other strategies that will ease the pressure on forests. Smaller homes, paper recycling, widespread tree planting, and better forest management are four ways to help avert a shortage of timber in coming years and help protect remaining forests. Increasing the rate of paper recycling by 30% in the United States would save an estimated 350 million trees every year.

Saving forests is not just a matter of ensuring a steady supply of wood and wood products. Forest conservation also protects the habitat of wild species, helps purify air, and reduces carbon dioxide buildup in the atmosphere. It protects watersheds, reduces soil erosion, and protects recreational opportunities as well. Tropical forests are a source of new medicines and potentially new food plants—plants that could help feed the world's people.

Destroying tropical forests also disrupts native cultures, forcing people who have lived off the land for centuries to move to town and cities. Protecting forests is a way of protecting these cultures and peoples. New efforts are now underway to save the forests by developing markets for sustainable products, such as nuts and fruits. Studies suggest that sustainable harvest of the forest is much more profitable than timber harvest and cattle ranching.

FIGURE 21–13 **Soil Erosion** Millions of acres of farmland are destroyed each year because of poor land management that leads to severe soil erosion.

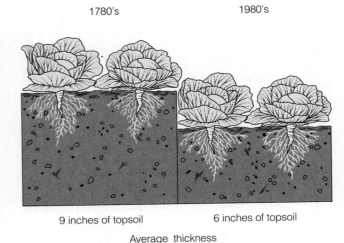

1780's   1980's

9 inches of topsoil   6 inches of topsoil

Average thickness

FIGURE 21–14 **American Topsoil—Then and Now**

Depleting Soils

Soil erosion, like deforestation, is also a worldwide phenomenon. Although soil erosion occurs naturally, it is often greatly accelerated by human activities, such as farming, construction, and mining (Figure 21–13). In the past hundred years, about one-third of the fertile topsoil on American farmland has been eroded away by water and wind, resulting in large part from poor land management (Figure 21–14). Each year, an estimated 2–3 billion tons of topsoil are eroded by wind and water from U.S. farmland. That's enough topsoil to fill a line of dump trucks 3500 miles long. Over one-half of America's once-rich farmland is eroding faster than it can be replaced.

Worldwide, soil erosion is estimated to be about 25 billion tons each year. Annual erosion rates are 18 to 100 times greater than the annual renewal rate, the rate at which soil is reformed (the average renewal rate is 500 years, but it ranges from 200 to 1000 years). Erosion of soil by wind and water robs enough topsoil each year to fill a train that would encircle the Earth at the equator 150 times.

The annual loss of topsoil causes farmers to retire millions of acres of once-productive farmland and, in many cases, to cut down forests or plow up grassland to replace what has been lost. Approximately 25% of the tropical rain forests cut down each year is leveled to replace farmland destroyed by human activities.

Making matters worse, millions of acres of farmland are lost to urban and suburban sprawl, highway construction, and other human activities (Figure 21–15). In the United States, 7000 acres of actual or potential farmland, pasture, and rangeland are lost every day. That's equivalent to 2.5 million acres a year, or a strip 0.6 miles wide extending from New York City to San Francisco. Overgrazing and poor land management are destroying millions of acres of farmland as well. Worldwide, an area the size of Ohio becomes desert each year, principally because of overgrazing by

FIGURE 21–15 **Farmland Conversion** Cities are often surrounded by excellent farmland—soils that drain well and are flat and highly productive. Many of the features that make soils suitable for farmland also make them suitable for building. This scene, unfortunately, is becoming all too common as urban areas expand.

FIGURE 21-16 Desertification Poor land management and overgrazing are turning valuable farmland and rangeland into desert, further compounding global food production problems. The rangeland behind the fence has been properly grazed, ensuring its productivity. The land on the right has been overgrazed and destroyed.

livestock and poor land management practices (Figure 21-16).

These figures paint a rather grim picture for the long-term future of food production here and abroad, especially when viewed against the inevitable increase in human population. But trend need not be destiny. The loss of topsoil can be stopped, and soils can be replenished, but such efforts will require a reduction in population growth and worldwide conservation measures. Table 21-3 lists additional measures to reduce the loss of productive farmland and rangeland. For these actions to be effective, society must begin work soon.

Depleting Global Water Supplies

Figure 21-17 is a map of areas in the United States that will face water shortage in the near future. Regional water shortages result because too many people are drawing on limited water supplies. Long-term prospects here and abroad appear dim. Between 1975 and 2000, irrigated agriculture worldwide is expected to double to meet the rising demand for food. Industry's demand for water is expected to increase twentyfold. By the end of the century, water demand is expected to exceed supply in at least 30 countries.

Figure 21-18 shows an enormous **aquifer**, a porous underground zone that contains water, known as the **Ogallala aquifer**. It supplies irrigation and drinking water for farms in Nebraska, Kansas, Colorado,

Oklahoma, and Texas. But severe **groundwater overdraft**—withdrawals that exceed natural replenishment—may put an end to irrigated farmland in these states. Texas alone is expected to lose half of its irrigated farmland by 2000.

Reducing the mining of groundwater is essential to meeting future needs. That will require strict conservation efforts, such as lining irrigation ditches with concrete or using pipes rather than open ditches to transport water to fields to reduce evaporation. More efficient sprinklers, computerized systems that monitor soil moisture so that farmers know exactly how much irrigation is required, and other measures can also help.

Depleting Global Mineral Supplies

Metals, such as steel and aluminum, are produced from the Earth's mineral deposits. More than a hundred minerals, worth billions of dollars to the global economy, are traded on the world market. Several dozen of these minerals are so important to modern society that if they were suddenly no longer available or were no longer available at a reasonable price, the U.S. economy, like those of other industrialized countries, would be brought to a standstill.

At least 18 economically important minerals will fall in short supply in the next 40 years, even if countries expand their recycling programs. Silver, mercury, lead, sulfur, tin, tungsten, and zinc are all candidates. Even if new discoveries and new technologies make it possible to extract five times the currently known reserves, this group will be 80% depleted on or before 2040.

The End of Oil

Oil is the lifeblood of modern society. In the United States, for example, oil supplies 43% of our annual energy demand. But the supply of oil is finite. The proven global reserves of oil (the amount known to exist and to be economically recoverable) are about 900 billion barrels. Although that may sound like a lot of oil, it will only last about 40 years *at the current rate of consumption*. The undiscovered global reserves (oil thought to exist and to be economically recoverable) amount to 525 billion barrels. That's enough oil for another 25 years at the current rate of consumption.

Unfortunately, except for two brief periods, global energy use has risen 5% per year since 1860—that's a doubling of demand every 14 years for over 100 years.

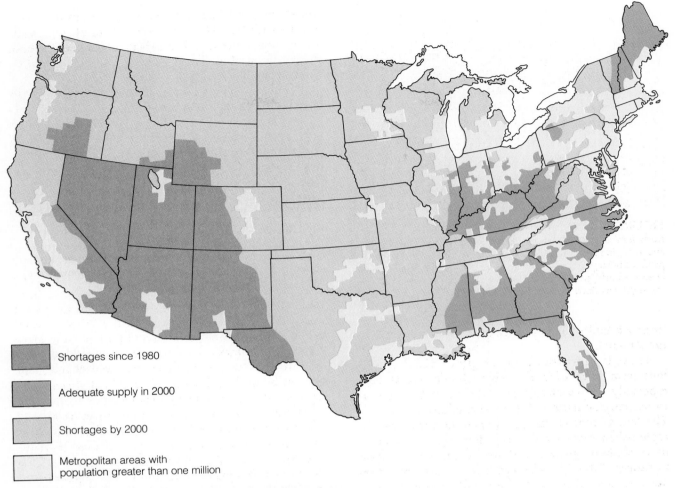

FIGURE 21–17   Water-Short Regions in the United States

Should this increase continue, oil's modest 65-year life span would be sharply reduced. Many experts believe that global demand for oil will exceed the supply early in the next century, creating a new oil crisis that could cripple the world economy. World oil prices are projected to begin to climb in the 1990s, once again reaching $25–$35 per barrel by 1995.

The outlook for oil in the United States is even grimmer. By various estimates, only about 100 billion barrels remain in U.S. territories, enough to last only 17 years at the current rate of consumption. The United States has become increasingly dependent on foreign oil in recent years, and that dependence is bound to grow. Unless a clean, economical substitute is found—and soon—our country could face a severe economic crisis.

Clearly, the time is running out for oil. Most of you will see the end of oil within your lifetime. It is time to reduce population growth, become much more efficient in our use of energy, and tap renewable fuels. The immediate need is for a source of energy to replace oil, which is used chiefly in home heating and transportation. Alternative fuels and energy conservation represent our biggest hopes.

Ethanol produced from corn and wheat could be used to power automobiles and trucks (Figure 21–19). Grown on special fuel farms, ethanol could virtually replace gasoline, but a shift to ethanol fuel in the United States will reduce food output, possibly reducing U.S. exports.

The efficient use of oil and oil products, such as gasoline, diesel fuel, and home heating oil, despite

FIGURE 21–18 The Ogallala Aquifer Map of the Ogallala aquifer. The water of the Ogallala is replenished very slowly and is currently being withdrawn much faster than it can be replaced. Farmers' wells are running dry, and eventually many irrigated farms will go out of business.

FIGURE 21–19 Ethanol as a Replacement for Oil Ethanol produced from sugar cane, corn, and other crops could fuel trucks, buses, and cars in the future, but farmland currently used to produce food will have to be converted to fuel farms, further impairing food production.

common misconceptions, can help us stretch current oil supplies considerably. Storm doors, storm windows, and added insulation coupled with measures to cut air inflow (infiltration) can reduce home energy consumption by 30 to 50%. More efficient cars, such as the Geo and Honda CVCC, already use fuel two times more efficiently than the average new car (Figure 21–20). Even higher mileage is possible. The Japanese, for example, have a car that gets 98 miles per gallon on the highway and seats four people! Mass transit is four times more efficient than the automobile and could also help us stretch our oil supplies. Alternative fuels, such as oil shale, a rock that when heated gives off an oily substance that can be refined to make gasoline, can also ease the crunch, but their supplies are small in comparison to global oil demand, and producing oil from shale is costly and inefficient (Figure 21–21). Oil shale development

would cause considerable air pollution in the West where the deposits are located and would destroy large tracts of winter habitat vital to deer, elk, and other wildlife.

Energy efficiency is one of the keys to meeting future demands. To give you an idea of the underlying potential of conservation, consider just one fact. Amory Lovins, an expert on energy efficiency, calculates that U.S. electrical demand could be cut by 75% by using energy-efficient light bulbs, motors, and other technologies *that are currently available* (Figure 21–22).

Renewable energy will also provide much of the fuel we need in years to come. The renewable energy base is enormous and, in some cases, already can be tapped quickly and inexpensively. According to one estimate, nonrenewable energy sources (coal, oil, natural gas, and so on) still in the ground would provide the equivalent of 8.8 trillion barrels of oil. Renewable energy could provide 10 times that amount of energy—every year!

POLLUTION: FOULING OUR NEST

All organisms produce waste; that is an inescapable fact of life. Humans, however, are by far the most prolific generators of waste on the planet. Today, our waste is overwhelming nutrient cycles, poisoning other species (and ourselves), and destroying the

FIGURE 21–20 Energy-Efficient Automobile The Geo is the top mileage car in the United States. It gets 58 miles per gallon on the highway and 53 in the city. Further improvements in gasoline mileage are possible. The average new car in the United States gets only about 27.5 miles per gallon.

FIGURE 21–21 Oil Shale This sedimentary rock contains an organic material called kerogen. When heated, oil shale releases its kerogen, forming shale oil. Shale oil can be refined to produce gasoline, kerosene, jet fuel, and other products, much like crude oil. Shale oil production, however, is expensive, limited by supply, and environmentally harmful.

planetary homeostasis. This section recaps four of the most serious waste problems.

Global Warming

Carbon dioxide is produced during cellular respiration and the combustion of any organic material, especially fuel. In some respects, atmospheric carbon dioxide is like a prescription drug; it is beneficial at low levels, but potentially harmful at higher concentrations. In normal concentrations in the Earth's atmosphere, carbon dioxide has a warming effect on the planet. It acts like the glass in a greenhouse, trapping heat escaping from the Earth and radiating it back to the surface. Carbon dioxide is, therefore, also known as a **greenhouse gas**. A little bit of carbon dioxide is necessary; too much could lead to overheating. As a result of industrialization, which has been fueled principally by fossil fuels, global atmospheric carbon dioxide levels have been increasing for over a hundred years. Today, atmospheric carbon dioxide levels are 25% higher than they were a century ago and are continuing to rise at a rate of about 1.5% per year.

Several other pollutants also contribute to global warming. Methane, chlorofluorocarbons, nitrous oxide, and even water vapor all radiate heat back to the Earth, causing the atmosphere to heat up. Methane is released from manure; humankind's nearly 1 billion cattle annually release about 73 million metric tons of

this gas. Methane production has increased 435% in the last century. Termites excrete methane as well, and termites now thrive in the deforested tropics. Today, methane levels are increasing at a rate of about 1% per year. What is so alarming about this trend is that methane is 20 times more effective at trapping heat than carbon dioxide.

Chlorofluorocarbons or CFCs are known for their effect on the ozone layer and are released from spray

FIGURE 21–22 Energy-Efficient Light Bulb Many new light bulbs use only 25% of the energy a standard light bulb requires. Although they cost more, they last as long as 10 standard bulbs and can save $20 to $40 in electricity over their lifetime.

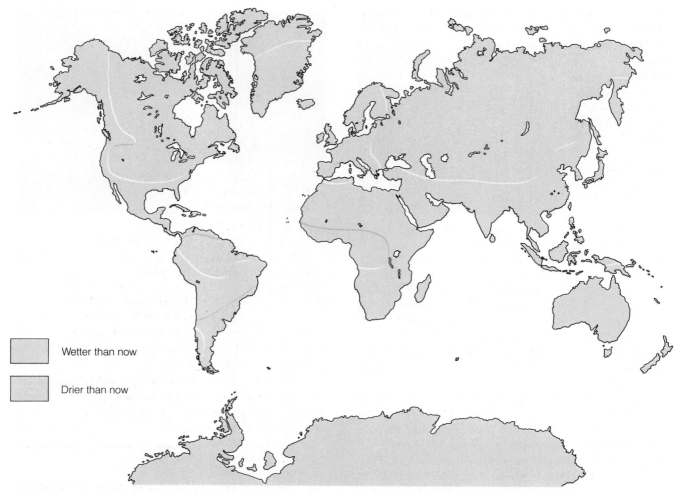

FIGURE 21–23   Projected Rainfall Patterns Resulting from Global Warming

cans, refrigerators, air conditioners, and freezers.[3] One class of CFCs is used to clean circuit boards used in computers and other electronic equipment. In the atmosphere, CFCs also trap heat, contributing to the greenhouse effect.

Because of the dramatic increase in the release of greenhouse gases, many atmospheric scientists believe that the global temperature may rise dramatically in the coming decades, causing extraordinary and dangerous shifts in climate. A graph of global temperature indicates a warming trend, but some scientists think that the apparent rise may result from normal climatic variation (for more on this subject, see the Point/Counterpoint).

[3]CFCs used in spray cans were banned in the United States and several other countries in the late 1970s.

Impacts of Global Warming.   The consequences of global warming could be substantial. According to computer models based on the present rate of increase in greenhouse gases, global temperatures could be 2–5°C hotter within 40 years. The models suggest that global rainfall patterns will shift dramatically. The Midwest and much of the western United States will be drier and hotter than they are today. Many midwestern farmers will be driven out of business. The southern United States and Pacific Coast may be wetter but probably also hotter (Figure 21–23). In Dallas, Texas, the number of days over 90°C, the models suggest, will increase from 17 today to 78 by 2030. In Washington, D.C., the number of days over 90°C will increase from 36 to 87.

Food shortages and rising food prices will impact the U.S. economy in profound ways. Local farm communities in the Midwest will be hardest hit as agriculture shifts to northern states. But this shift to northern regions will probably not offset the loss of production in the Midwest, in large part, because northern soils are not as rich as those in the Midwest.

A rise in global temperature is expected to melt glaciers and the polar ice caps. Warmer temperatures will also expand the volume of the seas. In the past 50 years, sea level has risen 10 to 12 centimeters (4 to 6 inches). By 2050, the models suggest that the sea level will rise 50 to 100 centimeters (2 to 3 feet) because of global warming.

In the United States, about one-half of the population lives within 50 miles of the ocean. Rising sea levels will flood many low-lying regions. Storm surges will cause more damage because they can move farther inland. Expensive dikes and levees would be needed to protect cities, such as Miami and New Orleans. Other coastal cities will have to be rebuilt on higher ground, but relocation will be astronomically expensive. The effects of rising sea level will be particularly hard felt in Bangladesh and other Asian countries with extensive lowland rice paddies. Approximately 17% of Bangladesh would be reclaimed by the sea.

The loss of land surface will increase competition for existing land. And rising temperatures will greatly impact many species. Some may adapt or find new habitat; many plants, however, will be wiped out by the relatively rapid increase in temperature. Forests will dry out and die off. The incidence of forest fires may increase, further adding to global warming.

What Can Be Done? While scientists and politicians debate global warming, the Earth appears to be getting hotter. Five of the hottest years in the past hundred years, in fact, have occurred since 1980. Can anything be done to stave off global warming?

The answer is a resounding yes, but changes will have to be dramatic and swift. Sharp reductions in fossil fuel consumption through energy efficiency, conservation, and the use of alternative sources of fuel will be required. Global reforestation, most experts agree, is a must. But for each family of four in the United States, six acres of fast-growing trees would have to be planted to offset the carbon dioxide the family will produce during its life. Additional strategies are shown in Table 21–5.

On average, each gallon of gasoline you consume in your automobile produces 5 pounds of carbon dioxide. Every hour you watch television results in the production of 0.64 pounds. Your frost-free refrigera-

TABLE 21–5 Measures to Reduce Global Warming
Reduce population growth rate
Switch from coal-fired and oil-fired power plants to natural gas, which produces much less carbon dioxide per unit of electricity
Implement the technologies that burn coal more efficiently
Expand cogeneration—processes that trap waste heat and put it to good use
Boost automobile efficiency
Expand mass transit
Develop alternative liquid fuels for the transportation sector
Improve the efficiency of industry
Make new and existing homes more energy-efficient through insulation, weather stripping, storm doors, and storm windows
Build many new homes that use solar energy for space heating
Reduce global deforestation
Begin a massive global reforestation effort
Phase out all CFCs and halons soon
Reduce consumption of unnecessary items
Expand recycling efforts

tor accounts for nearly 13 pounds of carbon dioxide a day (Table 21–6). The lesson from this is that individuals are, in part, responsible for global warming. We can help reduce the problem by using energy more efficiently. Using mass transit, recycling, insulating our homes, and a great many other strategies can help individuals cut back on carbon dioxide pollution (Table 21–7).

Acid Deposition

In the Adirondack Mountains of New York, hundreds of lakes are dying. Fish have vanished as the lakes turned acidic. A similar phenomenon is occurring in southeastern Canada and in Sweden and Norway, where dying lakes number in the thousands.

The acids falling from the skies are produced from two atmospheric pollutants, sulfur dioxide and nitrogen dioxide, arising chiefly from the combustion of fossil fuels: coal, oil, and natural gas. In the atmosphere, these gases, or **acid precursors**, combine with

GLOBAL WARMING IS REAL

Stephen H. Schneider

Stephen H. Schneider is currently the head of the Interdisciplinary Climate Systems Section at the National Center for Atmospheric Research. His research interests include climatic change; global warming; and climatic modeling of human impacts on climate.

It is already established beyond doubt by observations of the Earth that atmospheric constituents, such as water vapor, clouds, carbon dioxide (CO_2), methane, (CH_4), nitrous oxide (N_2O), and chlorofluorocarbons (CFCs), trap heat escaping from the Earth's surface, causing the greenhouse effect. Likewise, it is virtually certain that an unprecedented 25% increase in CO_2 and 100% increase in CH_4 over the past 150 years have resulted from increased use of fossil fuels and expanded deforestation. It is also well accepted that the buildup of these gases has trapped two extra watts of radiative energy averaged over every square meter of Earth. What then is the basis of the debate over global warming?

First of all, translating 2 watts/m² of heating into X degrees of temperature rise requires calculations that are based on not-yet-verified assumptions about how clouds, soils, forests, ice, and oceans will respond to this heating. These factors could change in ways that feed back on the heating, either reducing it, as global warming critics like to point out, or enhancing it, as most present climate models project. Such models predict that the past hundred years should have experienced some 1°C warming from the extra 2 watts/m² of heating, provided all other factors were constant—a dubious prospect. Already, half a dozen national and international assessment bodies have suggested that if the CO_2, CH_4, and CFC trends continue, global warming of some 1.5–4.5°C can be expected in the next century. Already, 0.5± 0.2°C of global warming have been observed from 1890 to 1990.

Some critics note that there is not a perfect match between decade-to-decade fluctuations in climate and the buildup of greenhouse gases. Unfortunately, such critics often fail to mention that no knowledgeable scientist would ever expect such agreement, since fluctuations of several tenths of a degree Celsius per decade occur naturally. Thus, it is illogical to look for a decade-by-decade match. Only long-term, global trends can verify or deny that a warming trend due to the buildup of greenhouse gases has been detected. A high level of certainty will take 10 to 20 years more to establish, not a few years as some critics of immediate changes argue. Moreover, waiting for such scientific certainty is not cost-free because the Earth could then be forced to adapt to a greater amount and rate of climate change than if we acted now.

Over the past 10 years, half a dozen government-sponsored assessments have agreed that there is a better than 50% chance that current trends in population growth, fossil fuel use, and land utilization practices will cause climatic changes of 2°C or more in the next century. Moreover, the rate of projected human-induced changes is some 10 times greater than the long-term rate of natural, global climate changes.

Critics suggest that three to five more years of testing is needed before taking action. While this sounds prudent, such testing will not provide definitive answers on climatic change or its implications for ecosystems, forestry, agriculture, water supplies, human health, sea level, and severe storms.

I'm not a planetary gambler. I'd prefer to slow down the rate of buildup of greenhouse gases rather than gamble that things may work out all right in the end. Study after study has shown that the best way to start to reduce the buildup rate of greenhouse gases is to make cost-effective improvements, such as controlling population growth, increasing the efficiency of energy use, and virtually eliminating the production of CFCs. Yale economist William Nordhaus, a critic of severe cuts in CO_2 emissions, has argued nevertheless that modest cuts in CO_2 emissions would, at present, yield economic benefits in excess of the costs. And his calculations did not even include the free extras: less acid rain, less air pollution, a lower balance of payments deficit from importing foreign oil, and lower long-term energy costs of manufactured goods.

To me, reducing CO_2 and other measures is a kind of climate change "insurance" that pays other dividends. We need to eliminate the billions of dollars of government subsidies to inefficient current fossil fuel uses and deforestation practices and move toward lower real costs and an environmentally more stable society. Political rhetoric about uncertainty only commits the future to greater risks.

TOO EARLY TO TELL

S. Fred Singer

S. Fred Singer, professor of environmental sciences at the University of Virginia, has served as deputy assistant administrator of the Environmental Protection Agency and as the first director of the U.S. weather satellite program in the Department of Commerce. An atmospheric and space physicist, he developed satellites and instruments and predicted the increase of atmospheric methane due to human activities.

Greenhouse warming (GW) has emerged as the issue of the 1990s. Wide acceptance of the Montreal Protocol, which reduces the manufacture of chlorofluorocarbons (CFCs), considered a threat to the stratospheric ozone layer, has encouraged environmental activists to call for similar controls on carbon dioxide (CO_2). They have expressed disappointment with the White House for not supporting immediate action on CO_2. Should the United States assume "leadership" in this campaign, or would it be more prudent to first assure through scientific research that the problem is both real and urgent?

The scientific base for GW includes some facts, lots of uncertainty, and just plain ignorance. What is needed are more observations, better theories, and more extensive calculations. There is consensus about an increase in greenhouse gases (CO_2, CFCs, methane, nitrous oxide, ozone) in the Earth's atmosphere. There is some uncertainty about their rate of generation and their rate of removal. There is major uncertainty and disagreement about whether this increase has caused a change in the climate during the last hundred years; many observations do not fit the theory. There is also major disagreement in the scientific community about predicted changes from further increases in greenhouse gases; the models used to calculate future climate are not yet refined enough to simulate nature. As a consequence, we cannot be sure whether the next century will bring a warming that is negligible or significant. Finally, even if there is a warming and associated climate changes, it is debatable whether the consequences will be good or bad; likely, we will get some of each.

Has the observed increase of greenhouse gases in the last decades had an effect on climate? The data are ambiguous to say the least. Advocates of immediate action profess to see a global warming of about 0.5°C since 1880 and point to record temperatures experienced in the 1980s. Others tend to be more cautious; they call attention to the fact that the strongest increase occurred *before* the major rise in greenhouse gas concentration; it was followed by a quarter-century decrease, between 1940 and 1965. Since then temperatures have begun to climb again. Some researchers consider the warming observed before 1940 to be a recovery from the "Little Ice Age" that prevailed from 1600 to about 1850.

We can sum up our conclusions in a simple message: *The scientific base for a greenhouse warming is too uncertain to justify drastic action at this time.* There is little risk in delaying policy responses to this century-old problem, since there is every expectation that scientific understanding will be substantially improved within a few years. Instead of panicky and premature actions, we will then be able to apply specific remedies as necessary. That is not to say that steps cannot be taken now; indeed, many kinds of energy conservation and efficiency increases make economic sense even *without* the threat of greenhouse warming.

Drastic, precipitous—and especially, unilateral—steps to delay the putative greenhouse impacts can cost jobs and prosperity, without being effective. Yale economist William Nordhaus, one of the few who has been trying to deal quantitatively with the economics of the greenhouse effect, has pointed out that "...those who argue for strong measures to slow greenhouse warming have reached their conclusion without any discernible analysis of the costs and benefits . . ." It would be prudent to complete the ongoing and recently expanded research so that we will know what we are doing before we act. "Look before you leap" may still be good advice.

SHARPENING YOUR CRITICAL THINKING SKILLS

1. Summarize the major points made by each author. Are there areas where they are looking at close to the same data but reaching different conclusions? If so, how is this possible?
2. Using your critical thinking skills, analyze each essay. What flaws do you see in the logic, if any?

TABLE 21–6	Carbon Dioxide Production from Common Activities	
ELECTRICAL APPLIANCES	POUNDS OF CARBON DIOXIDE ADDED TO ATMOSPHERE*	
Color television	0.64	per hour
Steam iron	0.85	per hour
Vacuum cleaner	1.70	per hour
Air conditioner, room	4.00	per hour
Toaster oven	12.80	per hour
Ceiling fan	4.00	per day
Refrigerator, frost-free	12.80	per day
Waterbed heater	24.00	per day
with thermostat	12.80	per day
Clothes dryer	10.00	per load
Dishwasher	2.60	per load
Toaster	0.12	per use
Microwave oven	0.25	per 5-minute use
Coffeemaker	0.50	per brew

*At room temperature and sea level, every pound of carbon dioxide occupies 8.75 cubic feet, about half the size of a refrigerator.

Source: Copyright © 1990 by the National Wildlife Federation. Reprinted from the February/March 1990 issue of *National Wildlife.*

water and oxygen to form sulfuric and nitric acids and are deposited from the atmosphere, a phenomenon called **acid deposition**. Rain and snow wash acids from the sky; they constitute wet deposition. Fog and clouds carry heavy loads of acids that are deposited on trees. Particulates in the atmosphere may also attach to airborne acids as well, then fall to the Earth, forming **dry deposition**.

In the United States and Europe, acid deposition has been growing worse in the past three decades. A map of the United States, for example, shows that the region affected by acid deposition is growing larger and that the level of acidity, measured by pH, is increasing (Figure 21–24). Today, acid deposition is common downwind from virtually any major population center. It comes from power plants, automobiles, factories, and even our own homes.

Acid deposition changes the pH of lakes and streams, killing fish and other aquatic organisms. Acids on land dissolve toxic minerals, such as aluminum, from the soil and wash them into water bodies. Aluminum causes the gills of fish to clog with mucus, literally suffocating the fish. Extensive acidification of lakes and rivers is putting resort owners out of business in the Northeast, upper Midwest, and southern Canada.

Figure 21–25 shows areas most susceptible to acid deposition. These regions are generally mountainous and contain soils with little capacity to neutralize acids. Acids that fall on the land quickly wash to nearby lakes and streams.

Acids can also damage crops and trees. The damage may be direct or indirect. Direct damage results when acids damage growing buds or leaves. Indirect damage occurs when acids alter soil chemistry, often inhibiting soil bacteria necessary for nutrient recycling. Acids may leach important minerals from the soil, resulting in slower plant growth.

Massive forest diebacks are occurring throughout the world, and some scientists believe they may be the result of acidic rainfall and acidic fog that blanket forests (Figure 21–26). Acid deposition also damages buildings, statues, and other man-made structures. Estimated damage in the United States is about $5 billion per year.

Reducing the deposition of acids will require a dramatic reduction in the release of sulfur dioxide and nitrogen dioxide from power plants, factories, and automobiles. Smokestack scrubbers, for instance, trap sulfur dioxide escaping from power plants, removing up to 95% of the gas. Installing and operating a scrubber is rather expensive, and some utilities have objected to this strategy. Using low-sulfur coal can also help, but eastern coal producers find this strategy objectionable, since their coal is especially high in sulfur. Because of these and other objections, progress in reducing acid deposition has been painfully slow. Energy efficiency could help cut energy use at tremendous savings to individuals and businesses, but this strategy has been largely overlooked.

Stratospheric Ozone Depletion

Encircling the Earth, 20 to 30 miles above its surface, is a layer of the atmosphere, the **ozone layer**, containing a slightly elevated level of ozone gas. As noted in Chapter 19, the ozone layer shields the Earth from

TABLE 21–7   Individual Actions That Can Reduce Global Warming

Automobile energy savings:
   Buy energy-efficient vehicles
   Reduce unnecessary driving
   Car pool, take mass transit, walk, or bike to work
   Combine trips
   Keep your car tuned and your tires inflated to the proper level
   Drive at the speed limit
Home energy savings:
   Increase your attic insulation to R30 or R38
   Caulk and weather-strip your house
   Add storm windows and insulated curtains
   Install an automatic thermostat
   Turn the thermostat down a few degrees in winter and wear warmer clothing
   Replace furnace filters when needed
   Lower water heater setting to 120–130° F.
   Insulate water heater and pipes, install a water heater insulation blanket, and repair or replace all
      leaky faucets
   Take shorter showers
   Use cold water as much as possible
   Avoid unnecessary appliances
   Buy energy-efficient appliances
   Use low-energy light bulbs
Reducing waste and resource consumption:
   Recycle at home and at work
   Avoid products with excessive packaging
   Reuse shopping bags
   Refuse bags for single items
   Use a diaper service instead of disposable diapers
   Reduce consumption of throwaways
   Donate used items to Goodwill, Disabled American Veterans, the Salvation Army, or other charities
   Buy durable items
   Give environmentally sensitive gifts

potentially harmful ultraviolet light coming from the sun. During evolution, in fact, the generation of the ozone layer probably "allowed" the colonization of land by plants and animals.

Today, however, the ozone layer is being gradually destroyed by chemicals released into the atmosphere by modern society. The most potent destroyer of ozone is a class of chemicals called chlorofluorocarbons (CFCs).[4] Used as spray can propellants in some countries, refrigerants, blowing agents for plastic foam, and cleansing agents, CFCs are highly stable

molecules. Scientists, in fact, selected them for use in spray cans in large part because of their lack of chemical reactivity.

In the early 1970s, however, two U.S. scientists made a shocking discovery—CFCs were broken down by sunlight. CFCs released on the Earth's surface enter the atmosphere and gradually diffuse into the upper atmosphere. Here, the scientists hypothesized that the breakdown products would react with and destroy ozone molecules, removing them from the layer.

This hypothesis sparked a round of controversy. Dire predictions were made, suggesting massive declines in the ozone layer. One study, in fact, suggested

[4]Nitric oxide (NO) from high-flying jet airplanes, such as the SST, can also destroy the ozone layer.

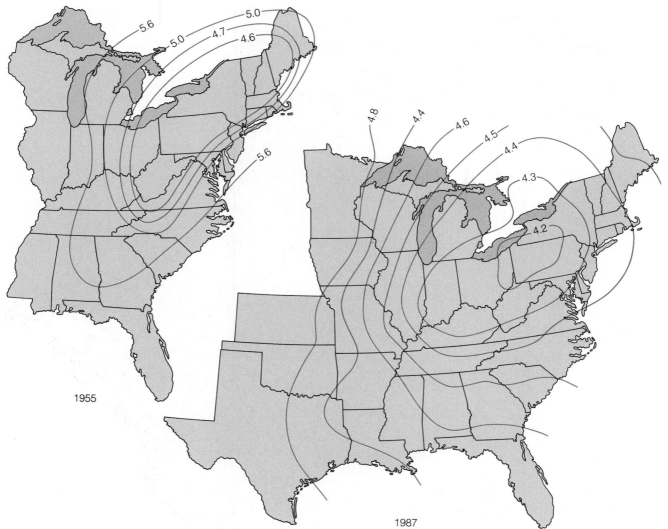

FIGURE 21–24   Extent and Strength of Acid Rain in the United States in 1955 and 1988

a 16% decline in ozone by the year 2000. What frightened scientists and a great many others was a projection that each 1% decline in ozone would increase skin cancer by at least 2% (some think it may be 4%). A 16% decline would increase the skin cancer rate by 32%. Even though skin cancer is not, as a rule, highly lethal, this dramatic increase would result in tremendous escalation in treatment costs and would kill 4000 to 12,000 people each year in the United States alone. Worldwide, the death toll would be staggering. Ozone depletion could be harmful to a great many other species, especially plants. Increased levels of ultraviolet light, for example, can damage plants.

Other scientists projected smaller declines. Still others thought that the fears were unfounded. Estimates of ozone depletion came and went over the years, but scientists were hard pressed to detect any measurable decline, in large part, because ozone levels fluctuate naturally from year to year. In 1988, however, a panel of a hundred atmospheric scientists reviewing 20 years of satellite data on ozone levels put the controversy to rest for many people. The panel announced that the ozone layer was indeed on the decline. Over North America, ozone levels had fallen 2% to 3% (Figure 21–27). Much larger declines were evident at the poles. Each year over Antarctica a giant

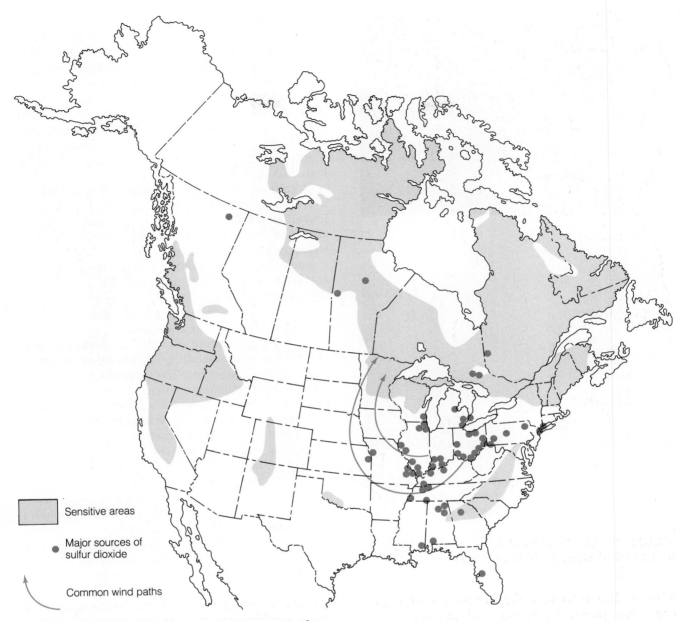

FIGURE 21–25 Acid-Sensitive Regions in North America and Major Sources of Acid Precursors

Sensitive areas

Major sources of sulfur dioxide

Common wind paths

hole about the size of the United States forms in the ozone layer. It appears in late winter and early spring; in this massive hole, ozone levels are about 50% below normal. Studies strongly suggested that one of the main reasons for the decline in the ozone layer was the accumulation of chlorofluorocarbons.

In a show of solidarity in September 1988, 23 nations signed an agreement to cut back on the produc-

tion of many ozone-destroying chemicals. By 1999, the treaty would achieve a 50% reduction in CFC manufacture and release. Experts objected, saying that a 50% cut would still result in a 10% decline in the ozone layer. Deeper cuts were needed. Because of these concerns and because of efforts on the part of the CFC industry to find what they hope will be safe substitutes, CFCs could be eliminated entirely by 2000.

(a)

(b)

FIGURE 21-26   Forest Die-off (*a*) Ghostly remains of trees killed by acid deposition. (*b*) Brown patches are trees killed by acid deposition in New England.

Hazardous Wastes

Hazardous wastes are products of industry as well as our homes that are highly toxic to humans and other organisms (Figure 21-28). In the United States each year, factories create an estimated 260 to 280 million metric tons of hazardous waste—or over a ton for every man, woman, and child. Other industrialized countries produce enormous amounts of hazardous waste as well. In fact, according to one estimate, each pound of material produced by industry results in 20 pounds of hazardous waste.

For many years, hazardous wastes in the United States ended up in abandoned warehouses; in rivers,

streams, and lakes; in leaky landfills that contaminate groundwater; in fields and forests; and along highways (Figure 21-28). Ill-conceived and irresponsible waste disposal has left a legacy of polluted groundwater, contaminated land, human disease, well closures, habitat destruction, fish kills, and difficult cleanups. The U.S. Office of Technology Assessment estimates that there are about 10,000 contaminated sites in the United States that pose a serious threat to human health, and that cleaning them up will cost at least $100 billion.

Despite legislation that provides billions of dollars for cleanup, only about 30 of the 1000 sites slated for immediate cleanup have been addressed. A decade after Americans realized that they had a hazardous waste problem, many experts agree that we are in for a much longer, more difficult battle than once was anticipated. Lee Thomas, former head of the Environmental Protection Agency (EPA), contends that "There are far more sites that are far more difficult to deal with than anybody ever anticipated."

Cleaning up existing sites will no doubt continue for many decades. Handling current waste, however, requires action now. In 1976 Congress passed legislation aimed at eliminating illegal and improper waste disposal. A reporting network was established to trace hazardous wastes from cradle to grave. The EPA also established standards for waste disposal sites and began issuing permits for approved sites. Dumping elsewhere was illegal. Unfortunately, implementation has been slow. Too many loopholes exist in the legislation, allowing much hazardous waste to escape control. New amendments closed many of the loopholes and tightened the net on hazardous wastes. Because of improvements in hazardous waste management, less toxic waste is being improperly discharged. Regardless, much more needs to be done; the problem is far from solved. A study in California in 1987 illustrates why. In that year alone, California businesses produced 2.5 million metric tons of hazardous waste. Where did it end up? The answer may prove shocking: (1) 65% was discharged into the ocean, lakes, and rivers; (2) 27% was injected into deep wells; (3) 4.3% went to public sewers and sewage treatment plants; (4) 1.8% went to treatment, storage, and disposal facilities; and (5) 1.3% escaped into the air.

One of the results of the legislation has been a dramatic increase in hazardous waste disposal costs. To avoid the cost of disposing of wastes, some companies have turned to foreign countries with lax or

FIGURE 21–27
Decline in the Ozone Layer

−2.3%

−3.0%
Average change

−1.7%

64°N

33°N

40°N

30°N

Vancouver

Montreal

Toronto
Chicago   New York

Los Angeles

Mexico City

nonexistent laws. Here, critics worry, wastes will not be disposed of properly.

Solving all waste problems requires a three-step process. The first line of defense is to reduce hazardous waste production in factories. Simple changes in industrial processes can cut back on waste production and can save companies sizable amounts of money. Hazardous waste generation can also be reduced by reusing or recycling products. In some cases, waste products are perfectly usable in other processes. Discarding them is like throwing away leftover lumber at a building site. Recycling networks are springing up all over the country to put our hazardous waste to good use. The second line of attack is detoxification and stabilization. Some wastes can be detoxified—made less toxic—by bacteria or sunlight. Incineration is also an alternative for detoxifying organic wastes; however, toxic waste incinerators are expensive and do not always burn cleanly, and even when operating efficiently they often release toxic emissions. Unburned residual matter from incinerators must be disposed of properly. High-temperature furnaces can operate at stationary waste disposal sites, on ships, or on mobile trailers.

Hazardous wastes that cannot be broken down can be stored permanently in landfills lined with synthetic liners and thick, impermeable layers of clay. Although landfills are the cheapest option, they have many dis-

FIGURE 21–28 Hazardous Wastes Barrels of toxic chemicals abandoned in the country are part of a legacy of improper waste disposal that threatens humans and other species.

advantages, the most noteworthy being their low popularity among citizens and the worry that all landfills, no matter how well built, will inevitably leak. Even a small leak can contaminate large quantities of groundwater. Cleaning a polluted aquifer is costly and time-consuming.

ENVIRONMENT AND HEALTH: BUILDING A SUSTAINABLE FUTURE

The trends and problems discussed in the previous sections add up to a problem of epic proportions. Solving them will require tougher laws and regulations, reductions in population growth, more efficient technologies, and changes in human behavior. Ultimately, however, many people believe that much broader societal changes are required. Even public opinion polls suggest that many citizens think that, to solve our problems, we must redirect society. One suggestion is that we build a **sustainable society**, a society that lives within the carrying capacity of the environment. Living within the limits of nature implies achieving a population size and a way of life that does not exceed the planet's ability to supply food and other resources and does not exceed the planet's ability to handle wastes. It is a society that manages its affairs in a way that protects the Earth's homeostatic mechanisms.

Foundations of a Sustainable Society

For much of human history, our ancestors lived in a sustainable relationship with nature. They took little and then principally what they needed to survive. They did not waste materials. They relied on renewable resources—plants and animals—for food and clothing. Unbeknownst to them, they were participants in the Earth's elaborate recycling mechanisms—the nutrient cycles discussed in Chapter 20. What our ancestors threw away was recycled and made available for reuse. Sustainability also arose from the fact that for much of our early history human populations were small, held in check through disease and other mechanisms.

Interestingly, these very same factors are responsible for the persistence of ecosystems. Ecosystems for the most part remain stable and will persist thousands of years, barring major disaster or human intervention, because of four operational principles: conservation, recycling, renewable resources, and population control.

In contrast, modern society is wasteful. The modern human economy, in fact, is built on waste, much of it resulting from the disposable products that flood the markets in developed countries. Modern society recycles only a fraction of the materials it uses. Despite the recent upsurge in recycling in the United States, for example, we still recycle only 13% of our municipal garbage. By various estimates, 50 to 90% of our municipal garbage could be recycled or composted. American society, like many others, also relies heavily on nonrenewable resources, such as oil, natural gas, and minerals. We tap only a small fraction of our renewable resource potential. Acid deposition and global warming, two of the major pollution problems we now face, result from pollutants released from fossil fuel combustion. On a final note, in contrast to "natural" populations, human society has, for years, pretty much ignored the imperatives of population control.

Building a sustainable society, say its proponents, will require a shift to a pattern seen in nature. It will require that we become more efficient in our use of resources, seriously engage in recycling, shift to renewable resources wherever possible, and stabilize—if not decrease—world population.

Building a sustainable society will also depend on a profound change in attitudes. The prevalent frontier notions of unlimited resources and human dominance over nature, say proponents, must be replaced with a new ethic, a sustainable-Earth ethic. A **sustainable-Earth ethic** is based on three tenets: (1) the world has a limited supply of resources ("there's not always more"), which must be shared with all living things; (2) humans are a part of nature and subject to its rules; and (3) nature is not something to conquer, but rather a force we must learn to cooperate with. New attitudes will not come easily, but are needed to strike a balance with nature.

Building a sustainable society will require a lifetime of commitment on the part of businesses, governments, and individuals. Given the course of modern society, it means taking a 180-degree turn. By comparison, the moon landing will seem like a weekend fix-it project.

How Do We Build a Sustainable Society?

To build a sustainable society, we must make profound changes. Some may come as a result of legal and legislative actions, such as new laws and regulations that force us to rely more heavily on conservation, recycling, and renewable resources and to control population size to reach a stable, sustainable level. Technological innovations will also help us become a sustainable society. For example, a new line of compact fluorescent light bulbs is now available. These bulbs fit into a regular light socket and use one-fourth as much energy as a regular incandescent bulb (Figure 21–22). Photovoltaics, thin silicon wafers that produce electricity from sunlight, could also help us make the transition to solar energy. Water-sensing computers can help farmers apply appropriate amounts of water to crops to prevent overwatering. Improvements in automobile efficiency can help us cut back on fossil fuel consumption and will help us clean up the air. But new technologies are not the only, or even the most important, answer. Many existing technologies simply need to be installed. Insulation, weather stripping, flow restrictors for shower-heads and faucets, and efficient appliances—already on the market—can help us make tremendous inroads into waste.

Many proponents also believe that we will eventually need to reduce consumption and alter our lifestyles. Taking shorter showers, shaving without the water running, turning off lights, and hundreds of other small actions on the part of individuals, when combined with similar actions by millions of other people, can result in significant cuts in resource demand.

The days of profligate resource use are quickly coming to an end. Ending the waste and pollution—and soon—is essential if we are to protect nature, the Earth, and ourselves. The balance of nature is, after all, the balance that sustains us. We risk upsetting it at our own peril. Protecting the planet is the ultimate form of health insurance.

SUMMARY

1. The economy of nature and the human economy are at odds. Today, that "conflict" manifests itself in a global environmental crisis characterized by overpopulation, resource depletion, and pollution.

OVERSHOOTING THE EARTH'S CARRYING CAPACITY

2. Although environmental problems vary from one nation to another, they are all linked by a common phenomenon: human populations are exceeding the carrying capacity of the environment.
3. Carrying capacity is the number of organisms an ecosystem can support indefinitely.
4. Carrying capacity is limited by food and resource supplies and by the capacity of the environment to assimilate or destroy waste products of organisms.
5. The human population sustains itself with a mix of renewable and nonrenewable resources. Long-term survival, in large part, depends on a shift to renewable resources.

OVERPOPULATION: PROBLEMS AND SOLUTIONS

6. Overpopulation occurs any time an organism overshoots its carrying capacity. It is manifest in shortages of food or other resources or excessive pollution, sometimes a combination of all three.
7. The world population is currently about 5.25 billion and is growing at a rate of 1.8% per year, giving a doubling time of about 40 years.
8. Growth rate is calculated by subtracting the death rate from the birth rate and multiplying by 100 to convert to a percentage.
9. Today, the most rapid growth in human population is occurring in the Third World, in parts of Africa, Asia, and Latin America. Here, resource depletion and food shortages are the most common problems.
10. Many experts believe that the industrialized countries are also overpopulated; judging from the quality of our air, water, and soils, we are exceeding the Earth's carrying capacity.

11. Because of our excessive resource demand, in fact, an average American has 25 to 40 times as much impact on the environment as a resident of a Third World country.

12. The human population has not always been so large. The rapid growth in numbers has occurred in the last two hundred years thanks to better sanitation, disease control, and modern technology.

13. The human population is growing exponentially. Exponential growth occurs any time a value, such as population size, grows by a fixed percentage with the interest (increase) being applied to the base amount.

14. Exponential growth can be deceptive. Growth in absolute numbers begins slowly, then escalates sharply as the base amount increases.

15. Exponential growth of population, resource demand, and pollution are at the heart of all environmental problems.

16. The human population problem can be summed up by a phrase: too many people, reproducing too quickly.

17. The size of the human population and the rapid rate of growth result in resource shortages and excessive pollution. Approximately one-fifth of the world population lives in extreme poverty without adequate food or shelter; another two-fifths live on the border of poverty, barely able to get enough to eat. Further growth only worsens the problems.

18. Solving overpopulation requires a decline in population growth rate, perhaps followed by a humane decrease in population size. A decline in total fertility rate, the number of children a woman will have over her lifetime, to replacement-level fertility, the rate at which parents replace themselves, is required to reduce the rate of population growth.

RESOURCE DEPLETION: ERODING THE PROSPECTS OF ALL ORGANISMS

19. The human population requires a variety of renewable and nonrenewable resources for survival, and many of these resources are being mismanaged. Some are even in danger of depletion.

20. Worldwide, forests are being cut faster than they can regenerate.

21. Forests are important sources of wood and wood products, but also provide wildlife habitat, protect watersheds, and help regenerate oxygen and remove pollutants, such as carbon dioxide.

22. To prevent the destruction of the world's forests, conservationists and environmentalists urge reductions in current demand, tree planting, recycling, and other strategies that will ease the pressure on our forests. Increasing the rate of paper recycling by 30% in the United States would save an estimated 350 million trees every year.

23. Worldwide, soils are being eroded from rangeland and farmland at an unsustainable rate. Making matters worse, millions of acres of farmland are being destroyed by human encroachment. Soil erosion, human encroachment, and other factors that destroy good productive soils threaten the long-term prospects for food production.

24. Reducing the rate of population growth and soil conservation measures can help ensure an adequate supply of soil, but efforts must begin soon and must be widespread.

25. Because of regional overpopulation, many areas suffer water shortages. The rise in population and the subsequent rise in demand are likely to cause further shortages over the coming decades.

26. Reducing growth and strict water conservation measures must be at the forefront of strategies aimed at reducing shortages.

27. Several dozen metals are vital to modern economies—so important, in fact, that if they were to suddenly fall in short supply or become too expensive, modern industrial societies such as ours would come to a standstill.

28. At least 18 commercially important metals will fall into short supply in the next 20 years. The more efficient use of metals, recycling, and a decline in demand (resulting from a reduction in population growth and individual consumption) are needed to help offset inevitable future shortages.

29. Oil supplies, like metal supplies, are finite. At best, the world has 65 years of oil remaining at the current rate of consumption. A rise in consumption will cut the oil supply sharply, resulting in shortages within the next 10 years.

30. Again, by reducing population growth, using oil much more efficiently, and seeking alternative fuel supplies, most likely renewable energy resources, we can help make a smooth transition to a sustainable society.

POLLUTION: FOULING OUR NEST

31. All organisms produce waste, but humans are by far the most prolific generators of waste on the planet. Today, our waste is overwhelming nutrient cycles, poisoning other species (and ourselves), and destroying the planet's homeostatic mechanisms.

32. One of the most serious threats from pollution comes from carbon dioxide. Global atmospheric concentrations of carbon dioxide have increased 22% in the past hundred years in large part due to the combustion of fossil fuels and deforestation.

33. Carbon dioxide is a greenhouse gas, trapping heat in the Earth's atmosphere. A little carbon dioxide is important, but too much will increase the planet's surface temperature, altering its climate, shifting rainfall pat-

terns and agricultural zones, flooding low-lying regions, and destroying many species that cannot adapt to the sudden change in temperature.

34. Chlorofluorocarbons and methane are also greenhouse gases that are on the increase.

35. Many scientists believe that global warming has begun. To slow it down or stop it, they recommend massive reforestation projects and dramatic improvements in efficiency of fossil fuel combustion. Alternative fuels and reductions in population growth can also help.

36. Sulfur dioxide and nitrogen dioxide are two gaseous pollutants released from power plants, factories, automobiles, and other sources. In the atmosphere, they are converted to sulfuric and nitric acid, respectively.

37. Acids fall from the sky in wet and dry deposition processes. Acids alter the pH of lakes and streams, killing fish and other organisms. They also leach toxic metals that kill fish from soils. They destroy trees and crops and deface buildings, costing society billions of dollars a year.

38. Pollution control devices called scrubbers, which are now in use on some power plants, can help remove sulfur dioxide from smokestack gases, but do not stop nitrogen dioxide. Stopping acid rain requires a multifaceted approach, but conservation, coal cleaning, and pollution control devices are the three prominent strategies.

39. The ozone layer encircles the Earth, trapping ultraviolet light. It is being destroyed by chlorofluorocarbons or CFCs (and other pollutants).

40. A panel of atmospheric scientists laid to rest much of the debate on ozone depletion in 1988 when they concluded that the ozone layer was indeed being eroded by CFCs. This led to an international agreement to reduce CFC production by 50% by the year 1999. Fearing this will still result in a significant decline in the ozone layer, numerous countries are now negotiating a complete phase-out of CFC production by 2000.

41. Hazardous wastes are materials produced by factories and our own homes that are toxic to humans and other organisms. Decades of improper hazardous waste disposal have left a legacy of human disease and contamination, which, in the United States, will cost at least $100 billion to clean up, and probably more.

42. Despite a growing awareness of the problem and tighter regulations, hazardous waste continues to be disposed of improperly. To help solve the problem, a three-tiered approach is needed. The first approach is to reduce hazardous waste production in the first place by process redesign, reuse, recycling, and reductions in demand for products that generate toxic wastes. The second approach is to stabilize or detoxify wastes. Finally, the remaining materials must be properly disposed of, for example, in secured landfills.

ENVIRONMENT AND HEALTH: BUILDING A SUSTAINABLE FUTURE

43. Solving our problems will require tougher laws and tighter regulations, but may also require a profound change in the society in which we live. We must build a sustainable society.

44. A sustainable society is built on four operating principles: conservation, recycling, renewable resources, and population control.

45. It is based on respect for nature and a willingness to cooperate with natural processes. Building a sustainable society is the ultimate form of health insurance.

EXERCISING YOUR CRITICAL THINKING SKILLS

1. Read an article in an environmental magazine on a topic of your interest. Summarize the key points made. Using your critical thinking skills, analyze the arguments.

2. Read an article on the same topic in a business magazine. Summarize the key points made. Do the authors agree or disagree on most of the key points? Why? Analyze the second article using your critical thinking skills.

TEST OF TERMS

1. The _____ _____ is the number of organisms an ecosystem can support on a sustainable basis.

2. Wind, sunlight, and trees are examples of _____ natural resources.

3. _____ occurs when populations exceed the carrying capacity of the environment.

4. To calculate the _____ _____ of a human popu-

lation, divide 70 by the annual _____ _____ .

5. _____ growth occurs any time a value, such as population size, grows by a fixed percentage with the interest (increase) being applied to the base amount.

6. The total number of children a woman is expected to have during her lifetime is the _____ _____ rate. The number of children a couple must have to replace themselves is called _____ _____ fertility.

7. In a population, _____ _____ growth occurs when the birth rate and death rate are equal and there is no net migration.

8. A porous underground layer that is saturated with water is called a(n) _____ .

9. Carbon dioxide, methane, and chlorofluorocarbons are all considered _____ gases because they contribute to global warming.

10. Sulfur dioxide and nitrogen dioxide are considered to be _____ _____ because they are converted in the atmosphere to sulfuric and nitric acid, respectively.

11. Rain and snow are a form of acid deposition called _____ deposition.

Answers to the Test of Terms are located in Appendix B.

TEST OF CONCEPTS

1. In what ways are the economy of nature and the human economy working in opposite directions? In your estimation, is this a serious problem and, if so, how can it be reduced or eliminated?

2. Human populations in the industrialized and nonindustrialized countries are overshooting the Earth's carrying capacity. Do you agree or disagree with this statement? Be sure to describe what is meant by overshooting the carrying capacity.

3. The global human population is growing at a rate of 1.8% per year, so there's nothing to worry about. Do you agree or disagree with this statement? Support your position.

4. Why is overpopulation as much a problem in the United States as it is in Bangladesh?

5. The U.S. population fell below replacement-level fertility in 1972, yet the U.S. population continues to increase. Why?

6. Describe key trends in the use of forests, soils, water, minerals, and oil that suggest humanity is on an unsustainable course. List and discuss solutions to each of the problems.

7. Given trends in natural resource use, many experts believe that global population growth must stop. Do you agree or disagree? Why? Is your position supported by scientific fact or based more on a general feeling?

8. Describe the cause and impacts of each of the following: global warming, acid deposition, stratospheric ozone depletion, and hazardous wastes.

9. Make a list of solutions for each of the problems you identified in question 8. How could conservation (efficiency), recycling, renewable resources, and population control factor into the solutions?

10. Explain what is meant by the following statement: A sustainable society is based on a design from nature.

11. Outline the operational and ethical principles of a sustainable society. Using your critical thinking skills, determine how they differ from the principles of modern society.

SUGGESTED READINGS

Brown, L. R., et al. 1990. *State of the world.* Washington, D.C.: Worldwatch Institute. An annual publication outlining global environmental problems and sustainable solutions.

Chiras, D. D. 1990. *Beyond the fray: Reshaping America's environmental movement.* Boulder, Colo.: Johnson. A critique of the American environmental response with suggestions on how to bring about more effective change.

_____ . 1991. *Environmental science: Action for a sustainable future,* 3d ed. Redwood City, Calif.: Benjamin/Cummings. In-depth coverage of population, resource, and pollution problems.

_____ . 1992. *Healing the Earth.* Washington, D. C.: Island Press. Discussion of steps needed to build a sustainable society.

Diamond, J. 1990. Playing dice with megadeath. *Discover* 11(4): 54–59. Disturbing article on species extinction.

Flavin, C., R. Piltz and C. Nichols. 1990. *Sustainable energy.* Washington, D.C.: Renew America. Excellent overview of renewable energy technologies.

Ruckelshaus, W. D. 1989. Toward a sustainable world. *Scientific American* 261: 166–75. Good overview of some of the requirements of a sustainable society. The September issue is dedicated to the environment and is worthwhile reading.

PERIODIC TABLE OF THE ELEMENTS

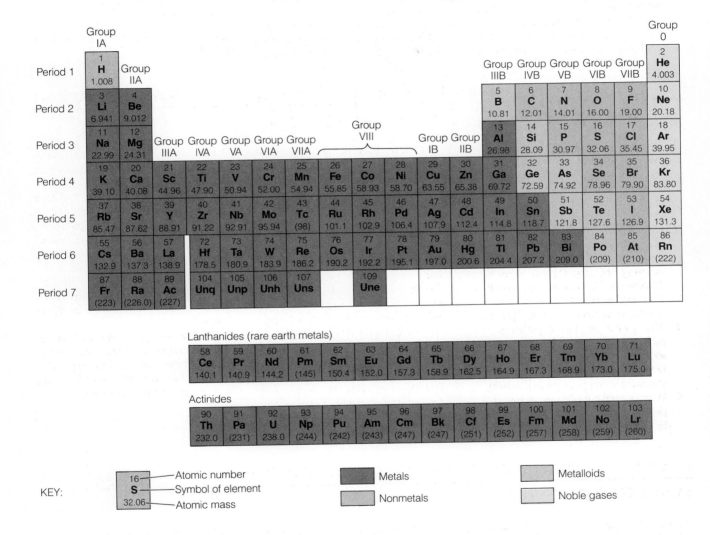

Lanthanides (rare earth metals)

58	59	60	61	62	63	64	65	66	67	68	69	70	71
Ce	**Pr**	**Nd**	**Pm**	**Sm**	**Eu**	**Gd**	**Tb**	**Dy**	**Ho**	**Er**	**Tm**	**Yb**	**Lu**
140.1	140.9	144.2	(145)	150.4	152.0	157.3	158.9	162.5	164.9	167.3	168.9	173.0	175.0

Actinides

90	91	92	93	94	95	96	97	98	99	100	101	102	103
Th	**Pa**	**U**	**Np**	**Pu**	**Am**	**Cm**	**Bk**	**Cf**	**Es**	**Fm**	**Md**	**No**	**Lr**
232.0	(231)	238.0	(244)	(242)	(243)	(247)	(247)	(251)	(252)	(257)	(258)	(259)	(260)

KEY:

16 ——— Atomic number
S ——— Symbol of element
32.06 ——— Atomic mass

Metals

Nonmetals

Metalloids

Noble gases

APPENDIX B
ANSWERS TO THE TEST OF TERMS

CHAPTER 1

1. biosphere; 2. Homeostasis; 3. insulin; 4. emphysema; 5. Asexual; 6. variation; 7. metabolism; 8. Irritability; 9. natural selection; 10. Culture; 11. knowledge; 12. scientific method; 13. Hypotheses; 14. theory.

CHAPTER 2

1. matter; 2. Elements; 3. electrons, protons, and neutrons; 4. nucleus, positive; 5. atomic number; 6. neutrons; 7. Radiation; 8. ion; 9. covalent; 10. hydrogen; 11. anabolic, catabolic; 12. coupled; 13. ATP, 14. chemical equilibrium; 15. carbon, hydrogen, covalent; 16. glycogen; 17. hydrogen, hydroxide; 18. acid, base; 19. acids, greater; 20. buffer; 21. monosaccharides; 22. fatty acid, glycerol; 23. phosphoglyceride; 24. Cholesterol; 25. polymers, peptide; 26. Enzymes; 27. primary, tertiary; 28. nucleic acids, nucleotides; 29. inorganic phosphate, energy.

CHAPTER 3

1. 3.5; 2. bacteria; 3. Cells; 4. nucleus or chromosomes, DNA; 5. cytoskeleton, enzymes; 6. metabolic pathway; 7. active, specific; 8. catalysts, allosteric; 9. feedback; 10. double, phospholipid, integral; 11. selectively permeable; 12. glycoprotein, Golgi complex, RER; 13. diffusion; 14. carrier; 15. phagocytosis; 16. osmotic; 17. isotonic, hypotonic, swell; 18. envelope, pores; 19. chromatin, chromosomes; 20. Ribosomal RNA, protein; 21. mitochondrion; 22. RER, messenger RNA; 23. Golgi complex; 24. lysosome, food vacuole; 25. flagellum, 9, 9 + 2; 26. cilia, basal body; 27. glycolysis, 2, ATP or pyruvate; 28. cellular respiration, 38; 29. electron transport system, electrons, citric acid; 30. glycolysis.

CHAPTER 4

1. cell division, interphase; 2. S, 46, 2; 3. gene; 4. mitosis, cytokinesis; 5. contact inhibition; 6. two, histone; 7. centromere, anaphase; 8. amniocentesis, karyotype; 9. mitotic spindle; 10. metaphase; 11. microfilamentous network; 12. metastasis, secondary; 13. transformation; 14. carcinogen; 15. ozone, chlorofluorocarbons.

CHAPTER 5

1. hydrogen, double helix; 2. nucleotide; 3. thymine, guanine; 4. transcription, translation; 5. codon; 6. tRNA; 7. Complementary base; 8. structural, operon; 9. operon, promoter, repressor, operator; 10. heterochromatin; 11. Enhancers; 12. introns, exons; 13. oncogenes; 14. protooncogenes.

CHAPTER 6

1. ectoderm, mesoderm, endoderm; 2. extracellular material; 3. connective, nervous, muscle, epithelium; 4. exocrine; 5. epidermis, epithelial; 6. dense connective; 7. fibroblast; 8. hyaline cartilage; 9. fibrocartilage; 10. central, canaliculi, osteocytes; 11. compact; 12. plasma; 13. platelets, red blood cells; 14. smooth, actin, myosin; 15. neuron; 16. organ system; 17. set point; 18. reflexes; 19. integration; 20. effectors; 21. hormone, paracrine; 22. circadian; 23. suprachiasmatic; 24. jet lag.

CHAPTER 7

1. macronutrients; 2. carbohydrates, lipids; 3. Aerobic, lipids; 4. fat; 5. essential, complete; 6. fiber, colon; 7. Vitamins, fat-soluble vitamins, water-soluble vitamins; 8. trace minerals; 9. amylase, lipase; 10. taste buds; 11. pharynx; 12. peristalsis; 13. gastroesophageal; 14. chyme; 15. pepsin, pepsinogen; 16. pyloric; 17. liver, bile salts; 18. sodium bicarbonate, digestive enzymes; 19. circular folds, villi, microvilli; 20. lymphatic.

CHAPTER 8

1. pulmonary, systemic; 2. right atrium; 3. aorta, elastic; 4. semilunar; 5. atrioventricular; 6. sinoatrial, right atrium; 7. atrioventricular bundle; 8. tunica intima, tunica media, tunica adventitia; 9. Elastic fibers; 10. systolic; 11. capillaries, venules; 12. Varicose veins; 13. edema; 14. valves; 15. plasma; 16. antibodies; 17. RBC or red blood cell, hemoglobin; 18. red bone; 19. erythropoietin, RBCs; 20. porphyrin; 21. Anemia; 22. neutrophils, monocytes; 23. lymphocyte; 24. Leukemia; 25. platelet; 26. Fibrinogen; 27. Plasmin; 28. interstitial fluid or tissue fluid; 29. Lymph nodes; 30. Carbon monoxide.

CHAPTER 9

1. Emphysema; 2. pharynx; 3. trachea, bronchi; 4. bronchioles; 5. mucous; 6. surfactant; 7. dust; 8. vocal cords; 9. olfactory membrane; 10. hemoglobin, bicarbonate; 11. brain stem, breathing center; 12. inspiration, diaphragm, intercostal; 13. expiration; 14. chronic bronchitis.

CHAPTER 10

1. capsid; 2. reverse transcriptase; 3. binary fission; 4. inflammatory response; 5. Pyrogens; 6. Interferons; 7. antigen or allergen; 8. immunocompetence; 9. humoral; 10. memory; 11. immunoglobulins; 12. neutralization; 13. agglutination; 14. cytotoxic or killer T cell; 15. vaccine, active.

CHAPTER 11

1. urinary, renal; 2. ureters; 3. urethra; 4. cortex; 5. renal pelvis; 6. glomerulus, renal tubule; 7. afferent, Bowman's capsule or Bowman's space; 8. podocytes; 9. peritubular capillaries, tubular secretion; 10. internal sphincter; 11. dialysis; 12. ADH or antidiuretic hormone, increases; 13. aldosterone; 14. nephrotoxin.

CHAPTER 12

1. central, peripheral; 2. autonomic; 3. dendrites; 4. myelin sheath; 5. terminal boutons, neurotransmitters; 6. resting potential, -60; 7. sodium ions, action potential; 8. spinal nerves, ventral; 9. interneuron or association; 10. primary motor; 11. gyri, sulci; 12. primary sensory, central sulcus; 13. association; 14. limbic system; 15. cerebellum; 16. hypothalamus, nuclei.

CHAPTER 13

1. pain; 2. Receptors, encapsulated receptors; 3. Merkel discs; 4. Pacinian; 5. Meissner's corpuscle; 6. muscle spindle; 7. Golgi tendon organs; 8. Habituation; 9. special senses; 10. taste buds, papillae; 11. olfactory, bipolar; 12. sclera, cornea; 13. pigmented, choroid, ciliary body; 14. retina, cones, night; 15. ganglion; 16. fovea centralis, optic disc; 17. lens, ciliary body, cataracts; 18. vitreous humor; 19. Glaucoma, aqueous humor; 20. refraction, change velocity; 21. extrinsic eye; 22. nearsightedness, elongated eyeball, strong lens; 23. radial keratotomy; 24. astigmatism; 25. Rhodopsin; 26. sex-linked; 27. external auditory, tympanic membrane or eardrum; 28. ossicles, oval window; 29. eustachian or auditory tube; 30. ampullae, semicircular, endolymph; 31. maculae; 32. organ of Corti, vestibular; 33. conduction; 34. temporary threshold.

CHAPTER 14

1. vertebrae, axial; 2. appendicular; 3. long, diaphysis, compact, marrow; 4. spongy; 5. synovial, synovial, the joint capsule, tendons; 6. flexion, extension; 7. arthroscope; 8. osteoarthritis; 9. Rheumatoid, synovial; 10. prosthesis; 11. hyaline cartilage, primary center; 12. osteoclasts, osteoblasts; 13. epiphyseal plate; 14. calcitonin, thyroid, par-

athormone; 15. osteoporosis; 16. nuclei, striated; 17. endomysium; 18. epimysium; 19. myofibril; 20. sarcomere; A, I, 21. calcium, sarcoplasmic, troponin, actin; 22. cross bridges; 23. acetylcholine, T tubules; 24. Muscle fatigue, lactic acid; 25. twitch, all-or-none; 26. wave, motor; 27. slow-twitch, myoglobin; 28. fast-twitch, myosin ATPase; 29. anabolic steroids.

CHAPTER 15

1. hormone, target; 2. receptors; 3. trophic; 4. steroids, amines; 5. second messenger, ATP, adenylate cyclase, phosphates; 6. two-step, genes or DNA; 7. pituitary, hypothalamus; 8. anterior pituitary, releasing, neurosecretory, hypothalamus; 9. Growth, hypertrophy, amino acids; 10. Hyposecretion; 11. ACTH; 12. gonadotropins; 13. prolactin, suckling; 14. neuroendocrine; 15. neuroendocrine, oxytocin, ADH; 16. thyroxin, calcitonin; 17. thyroglobulin; 18. Goiter; 19. parathyroid hormone or parathormonec, parathyroid, calcitonin, thyroid; 20. Insulin, pancreas (or islets of Langerhans or beta cells), glucose, glycogen; 21. gluconeogenesis; 22. insulin-dependent or early-onset, autoimmune; 23. late-onset or insulin-independent, obesity, diet; 24. adrenalin, noradrenalin; 25. adrenal cortex, gluconeogenesis, immune; 26. Aldosterone, adrenal cortex.

CHAPTER 16

1. scrotum, seminiferous; 2. epididymis, vas deferens; 3. sex accessory; 4. germinal, spermatogonia, primary spermatocytes, secondary spermatocytes; 5. spermatids, 23; 6. acrosome; 7. interstitial cells, testosterone, luteinizing hormone or interstitial cell stimulating hormone; 8. foreskin, circumcision; 9. erectile; 10. Erection, semen; 11. uterus, uterine tubes; 12. vagina, cervical; 13. labia majora; 14. clitoris; 15. ovulation, menstrual; 16. tertiary, zona pellucida; 17. first, first polar; 23, double; 18. corpus luteum, estrogen, progesterone,; 19. follicular, estrogen, FSH; 20. endometrium, menstruation; 21. premenstrual syndrome; 22. menopause, estrogen; 23. vasectomy, uterine tubes, tubal ligation; 24. estrogen, progesterone; 25. pap smear; 26. intrauterine device; 27. diaphragm, spermicidal; 28. condom; 29. rhythm or natural; 30. temperature, cervical; 31. sexually transmitted diseases; 32. Nonspecific urethritis; 33. Herpes.

CHAPTER 17

1. zygote, uterine tube or Fallopian tube; 2. Capacitation, acrosome; 3. fast, polyspermy; 4. morula, inner cell mass, trophoblast; 5. implantation, six or seven; 6. placental villi; 7. amnion; 8. organogenesis; 9. brain, spinal cord; 10. somites, vertebrae (or backbone); 11. endoderm; 12. vein, ductus venosus, inferior vena cava; 13. right, left, foramen ovale; 14. human chorionic gonadotropin, estrogen, progesterone; 15. teratology, teratogenic; 16. Relaxin; 17. dilation, cervix; 18. breech; 19. placental; 20. pudendal; 21. prolactin; 22. colostrum; 23. rooting; 24. bonding; 25. puberty; 26. homeostatic; 27. Life expectancy.

CHAPTER 18

1. Gregor Mendel; 2. hereditary, segregation; 3. recessive; 4. allele; 5. heterozygous; 6. phenotype; 7. monohybrid; 8. independent assortment; 9. XX, autosomes; 10. autosomal recessive, carrier; 11. autosomal dominant, homozygous, heterozygous; 12. incomplete dominance; 13. multiple; 14. Codominant; 15. polygenic; 16. linked; 17. sex-linked; 18. recessive, X; 19. sex-influenced; 20. nondisjunction; 21. trisomy; 22. nondisjunction; 23. polyploidy; 24. deletions, translocations; 25. chorionic villus.

CHAPTER 19

1. chemical; 2. protocells; 3. abiotically; 4. coacervates; 5. heterotrophic fermenters; 6. chemosynthetic; 7. chlorophyll; 8. selective advantage; 9. oxygen, eukaryotes; 10. endosymbiotic; 11. ozone layer; 12. arthropods; 13. species; 14. variation; 15. adaptation; 16. Natural selection; 17. Fitness; 18. speciation, geographic, reproductive; 19. divergent evolution; 20. punctuated equilibrium; 21. homologous; 22. phylum, genus, species; 23. Genetic; 24. *Dryopithecus, Australopithecus*; 25. *habilis*; 26. *Homo erectus*.

CHAPTER 20

1. Ecology; 2. biosphere; 3. biomes; 4. abiotic, biotic; 5. range of tolerance; 6. limiting factor; 7. population, community; 8. niche; 9. producers, consumers; 10. herbivores, omnivores, detrivores; 11. grazer, decomposer; 12. heat; 13. trophic; 14. Biomass; 15. organismic, environmental; 16. nitrogen fixation; 17. growth, reduction; 18. density; 19. Succession, primary succession.

CHAPTER 21

1. carrying capacity; 2. renewable; 3. Overpopulation; 4. doubling time, growth rate; 5. Exponential; 6. total fertility, replacement-level; 7. zero population; 8. aquifer; 9. greenhouse; 10. acid precursors; 11. wet.

THE METRIC SYSTEM

In the United States, the metric system is a system of measurement used principally by scientists. In our day-to-day lives, though, most Americans use the English system of measurement—miles, inches, feet, pounds, tons, and so on.

Many countries like New Zealand use the metric system for weights and other measures. If you travel abroad, road signs will list distance in terms of kilometers rather than miles. Linear measurements will be given in meters and centimeters instead of yards, feet, and inches. The weight of objects will be expressed in kilograms or grams instead of pounds and ounces.

It can be confusing if you don't know how to convert from one system of weights and measures to another. The lists below show some of the most common units you will encounter in biology and other sciences and compare them to their English equivalents.

Most Common English Units and the Corresponding Metric Units

	ENGLISH UNIT	METRIC UNIT
Weight	tons	metric tons
	pounds	kilograms
	ounces	grams
Length	miles	kilometers
	yards	meters
	inches	centimeters
Square Measure	acres	hectares
	square miles	square kilometers
Volume	quarts and gallons	liters
	fluid ounces	milliliters

Converting English Units to Metric Units

	ENGLISH UNIT		METRIC UNIT
Weight	1 ton	=	2000 pounds
	1 ton	=	0.9 metric tons
	1 pound	=	0.454 kilograms
	1 ounce	=	28.35 grams
Length	*1 mile	=	1.6 kilometers
	1 yard	=	0.9 meters
	*1 inch	=	2.54 centimeters
Square Measure	1 acre	=	0.4 hectares
	1 square mile	=	2.59 square kilometers
Volume	1 quarts	=	0.95 liters
	*1 gallons	=	3.78 liters
	1 fluid ounce	=	29.58 milliliters

*Most useful conversions to know

Converting Metric Units to English Units

	ENGLISH UNIT		METRIC UNIT
Weight	*1 metric ton	=	2204 pounds
	1 metric ton	=	1.1 tons
	*1 kilogram	=	2.2 pounds
	1 gram	=	28.35 ounces
Length	*1 kilometer	=	.6 miles
	1 meter	=	1.1 yards
	1 centimeter	=	.39 inches
Square Measure	*1 hectare	=	2.47 acres
	1 square kilometer	=	0.386 square miles
Volume	1 liter	=	1.057 quarts
	1 liter	=	.26 gallons
	1 milliliter	=	0.0338 fluid ounces

*Most useful conversions to know

GLOSSARY

Abiotic factors Physical and chemical components of an organism's environment.

Accommodation Change in the shape of the lens caused by contraction or relaxation of the smooth muscle of the ciliary body. Through accommodation, the lens adjusts the degree to which incoming light rays are bent, permitting objects to be focused on the retina.

Acetylcholine Neurotransmitter substance in the central and peripheral nervous systems of humans.

Acetylcholinesterase Enzyme that destroys the neurotransmitter acetylcholine in the synaptic cleft.

Achondroplasia Genetic disease that results from an autosomal dominant gene. Individuals with the disease have short legs and arms, but a relatively normal body size.

Acid deposition Deposition of sulfuric and nitric acids in the atmosphere onto the Earth's surface. Damages buildings, lakes, streams, crops, and forests. Acids come from sulfur dioxide and nitrogen dioxide produced during the combustion of fossil fuels.

Acid precursors Sulfur dioxide and nitrogen dioxide gases that combine with water and oxygen to form sulfuric and nitric acids in the atmosphere.

Acrosome Enzyme-filled cap over the head of a sperm. Helps the sperm dissolve its way through the corona radiata and zona pellucida.

Actin microfilaments Contractile filaments made of protein and found in cells as part of the cytoskeleton. Especially abundant in the microfilamentous network beneath the plasma membrane and in muscle cells.

Action potential Recording of electrical change in membrane potential when a neuron is stimulated.

Active immunity Immune resistance gained when an antigen is introduced into the body either naturally or through vaccination.

Active site Small pocket or indentation on an enzyme where the chemical reaction occurs. Its shape corresponds to the molecule(s) undergoing reaction.

Active transport Movement of molecules across membranes using protein molecules and energy supplied by ATP. Moves molecules and ions from regions of low to high concentration.

Adaptation Genetically based characteristic that increases an organism's chances of passing on its genes.

Adaptive radiation Process in which one species gives rise to many others that occupy different environments. Also known as divergent evolution.

Adenylate cyclase Enzyme bound to the inner surface of the plasma membrane. Linked to plasma membrane hormone receptors. Responsible for the conversion of ATP to cyclic AMP (a second messenger).

Adipose tissue Type of loose connective tissue containing numerous fat cells. Important storage area for lipids.

Adolescence Period of human life from puberty until adulthood. Characterized by sexual maturity.

Adrenal cortex Outer portion of the adrenal gland. Produces a variety of steroid hormones, including cortisol and aldosterone.

Adrenal gland Endocrine gland located on top of the kidney. Consists of two parts: adrenal cortex and medulla, each with separate functions.

Adrenal medulla Inner portion of the adrenal glands. Produces epinephrine (adrenalin) and norepinephrine (noradrenalin).

Adrenalin (epinephrine) Hormone secreted under stress. Contributes to the fight-or-flight response by increasing heart rate, shunting blood to muscles, increasing blood glucose levels, and other functions.

Adrenocorticotropic hormone (ACTH) Polypeptide hormone produced by the anterior pituitary. Stimulates the cells of the adrenal cortex, causing them to synthesize and release its hormones, especially glucocorticoids.

Aerobic exercise Exercise, such as swimming, that does not deplete muscle oxygen. Excellent for strengthening the heart and for losing weight.

Age-structure diagram (population histogram) Graphical representation of the number or percentage of males and females in various age groups in a population.

Agglutination Clumping of antigens that occurs when antibodies bind to several antigens.

Aging Inevitable and progressive deterioration of the body's function, especially its homeostatic mechanisms.

AIDS Acquired immune deficiency syndrome. Fatal disease caused by the HIV virus, which attacks T helper cells, greatly reducing the body's ability to fight infection.

Albinism Genetic disease resulting in a lack of pigment in the eyes or the eyes, skin, and hair. An autosomal recessive trait.

Aldosterone Steroid released by the adrenal cortex in response to a decrease in blood pressure, blood volume, and osmotic concentration. Acts principally on the kidney.

Alleles Alternative form of a gene.

Allergen Antigen that stimulates an allergic response.

Allergy Extreme overreaction to some antigens, such as pollen or foods. Characterized by sneezing, mucus production, and itchy eyes.

Allosteric site Region of an enzyme where products of metabolic pathways bind, changing the shape of the active site. In some enzymes this prevents substrates from binding to the active site; in others, it allows them to bind. Thus, allosteric sites can either turn on or turn off enzymes.

Alveoli Tiny, thin-walled sacs in the lung where oxygen and carbon dioxide are exchanged between the blood and the air.

Ameboid motion Cellular locomotion common in single-celled organisms and some cells in the human body. The cells send out slender cytoplasmic projections that attach to the substrate "ahead" of the cell. The cytoplasm flows into the projections, or pseudopodia, advancing the organism.

Amniocentesis Procedure whereby physicians extract cells and fluid from the amnion surrounding the fetus. The cells are examined for genetic defects, and the fluid is studied biochemically.

Amnion Layer of cells that separates from the inner cell mass of the embryo and eventually forms a complete sac around the fetus.

Amniotic fluid Liquid in the amniotic cavity surrounding the embryo and fetus during development. Helps protect the fetus.

Ampulla Enlarged area of each semicircular canal that houses receptor cells for movement.

Amylase Enzyme in saliva that helps break down starch molecules.

Anabolic steroids Synthetic androgen hormones that promote muscle development.

Analogous structures Anatomical structures that function similarly but differ in structure; for example, the wing of a bird and the wing of an insect.

Anaphase Phase of mitosis during which the chromatids of each chromosome begin to uncouple and are pulled in opposite directions with the aid of the mitotic apparatus.

Androgen-binding protein Cytoplasmic receptor protein that binds to and concentrates testosterone within the Sertoli cell. Production of ABP is stimulated by FSH.

Androgens Male sex steroids such as testosterone produced principally by the testes.

Anemia Condition characterized by an insufficient number of red blood cells in the blood or insufficient hemoglobin. Often caused by insufficient iron intake.

Aneuploidy Describes a genetic condition in which there is an abnormal number of chromosomes.

Aneurysm Ballooning of the arterial wall caused by a degeneration of the tunica media.

Anoxia Lack of oxygen.

Antagonistic Refers to hormones or muscles that exert opposite effects.

Anterior pituitary Major portion of the pituitary gland, which is controlled by hypothalamic hormones. Produces seven protein and polypeptide hormones.

Anthropoids Monkeys, the great apes, and humans.

Antibodies Proteins produced by immune system cells that destroy or inactivate antigens, including pollen, bacteria, yeast, and viruses.

Anticodon loop Part of the transfer RNA molecule that bears three bases that bind to the three bases of the codon on messenger RNA.

Anticodon Sequence of three bases found on the transfer RNA molecule.

Aligns with the codon on messenger RNA and helps control the sequence of amino acids inserted into the growing protein.

Antidiuretic hormone (ADH) Hormone released by the posterior pituitary. Increases the permeability of the distal convoluted tubule and collecting tubules, increasing water reabsorption.

Antigens Any substance that is detected as foreign by an organism and elicits an immune response. Most antigens are proteins and large molecular weight carbohydrates.

Anvil (incus) One of three bones of the middle ear that helps transmit sound waves to the receptor for sound in the inner ear.

Aorta Largest artery in the body; carries the oxygenated blood away from the heart and delivers it to the rest of the body through many branches.

Appendicular skeleton Bones of the arms, legs, shoulders, and pelvis. Contrast with axial skeleton.

Aquatic life zones Subregions of the ocean characterized by a distinct plant and animal life.

Aqueous humor Liquid in the anterior and posterior chambers of the eye.

Aquifer Porous underground zone containing water.

Arteries Vessels that transport blood away from the heart.

Arteriole Smallest of all arteries; usually drains into capillaries.

Arthropods Invertebrates with segmented bodies and jointed legs.

Arthroscope Device used to examine internal joint injuries.

Asexual reproduction Reproductive strategy common in single-celled organisms, such as the amoeba. Reproduction occurs by cell division.

Association cortex Area of the brain where integration occurs.

Association neurons Nerve cells that receive input from many sensory neurons and help process them, ultimately carrying impulses to nearby multipolar neurons.

Aster Array of microtubules found in the cell in association with the spindle fibers during cell division.

Asthma Respiratory disease resulting from an allergic response. Allergens cause histamine to be released in the lungs. Histamine causes the air-carrying ducts (bronchioles) to constrict, cutting down airflow and making breathing difficult.

Astigmatism Unequal curvature of the cornea (sometimes the lens) that distorts vision.

Atom Smallest particles of matter that can be achieved by ordinary chemical means, consisting of protons, neutrons, and electrons.

Atomic mass units Unit used to measure atomic weight. One atomic mass unit is 1/12 the weight of a carbon atom.

Atomic weight Average mass of the atoms of a given element, measured in atomic mass units.

Atrioventricular bundle Tract of modified cardiac muscle fibers that conduct the pacemaker's impulse into the ventricular muscle tissue.

Atrioventricular node (AV node) Knot of tissue located in the right ventricle. Picks up the electrical signal arriving from the atria and transmits it down the atrioventricular bundle.

Atrioventricular valves Valves between the atria and ventricles.

Auditory (eustachian) tube Collapsible tube that joins the nasopharynx and middle ear cavities and helps equalize pressure in the middle ear.

Auricle (or pinna) Skin-covered cartilage portion of the outer ear.

Autocrines Chemical substances produced by cells, which affect the function of the cells producing them.

Autoimmune reaction Immune response directed at one's own cells.

Autonomic nervous system That part of the nervous system not under voluntary control.

Autosomal dominant trait Trait that is carried on the autosomes and is expressed in heterozygotes and homozygote dominants.

Autosomal recessive trait Trait that is carried on the autosomes and is expressed only when both recessive genes are present.

Autosomes All human chromosomes except the sex chromosomes.

Axial skeleton The skull, vertebral column, and rib cage. Contrast with appendicular skeleton.

Axon Long, unbranched process attached to the nerve cell body of a neuron. Transports bioelectric impulses away from the cell body.

Basal body Organelle located at the base of the cilium and flagellum. Consists of nine sets of microtubules arranged in a circle. Each "set" contains three microtubules.

Basilar membrane Membrane that supports the organ of Corti in the cochlea.

Benign tumor Abnormal cellular proliferation. Unlike a malignant tumor, the cells in a benign tumor stop growing after a while, and the tumor remains localized.

Beta cells Insulin-producing cells of the islets of Langerhans in the pancreas.

Bile Fluid produced by the liver and stored and concentrated in the gallbladder.

Bile salts Steroids produced by the liver, stored in the gallbladder, and released into the small intestine where they emulsify fats, a step necessary for enzyme digestion.

Binary fission Bacterial cellular division.

Bioelectric impulse Nerve impulse resulting from the influx of sodium ions along the plasma membrane of a neuron.

Biomass The dry weight of living material in an ecosystem.

Biomass pyramid Diagram of the amount of biomass at each trophic level in an ecosystem or, more commonly, a food chain.

Biome Terrestrial region characterized by a distinct climate and a characteristic plant and animal life.

Biorhythms (biological cycles) Naturally fluctuating physiological process.

Biosphere Region on Earth that supports life. Exists at the junction of the atmosphere, lithosphere, and hydrosphere.

Biotic factor Biological components of ecosystems.

Bipedal Refers to the ability to walk on two legs.

Birth control Any method or device that prevents conception and birth.

Birth control pill Pill generally taken to inhibit ovulation. The most commonly used pills contain both synthetic estrogen and progesterone.

Birth defects Physical or physiological defects in newborns. Caused by a variety of biological, chemical, and physical factors.

Blastocyst Hollow sphere of cells formed from the morula. Consists of the inner cell mass and the trophoblast.

Blood Specialized form of connective tissue. Consists of white blood cells, platelets, red blood cells, and plasma.

Blood clot Mass of fibrin containing platelets, red blood cells, and other cells. Forms in walls of damaged blood vessels, halting the efflux of blood.

B-lymphocytes Type of lymphocyte that transforms into a plasma cell when exposed to antigens.

Bonding Process in which an infant establishes an intimate relationship with one or both of its parents or caretakers. Essential for emotional health.

Bone (organ) Structure comprised of bone tissue. Provides internal support, protects organs, and helps maintain blood calcium levels.

Bone (tissue) Tissue consisting of a calcified extracellular material with numerous cells (osteocytes) embedded in it.

Bowman's capsule Cup-shaped end of the nephron that participates in glomerular filtration.

Bowman's space Cavity between the inner and outer layer of Bowman's capsule.

Brain stem Part of the brain that consists of the medulla and pons. Houses structures, such as the breathing control center and reticular activating systems, that control many basic body functions.

Braxton-Hicks contractions Contractions that begin a month or two before childbirth. Also known as false labor.

Breathing center Aggregation of nerve cells in the brain stem that controls breathing.

Breech birth Delivery of a baby feet first.

Bronchi Ducts that convey air from the trachea to the bronchioles and alveoli.

Bronchioles Smallest ducts in the lungs. Their walls are largely made of smooth muscle that contracts and relaxes, regulating the flow of air into the lung.

Bulimia Eating disorder characterized by recurrent binge eating followed by vomiting.

Calcitonin (thyrocalcitonin) Polypeptide hormone produced by the thyroid gland that inhibits osteoclasts and stimulates osteoblasts to produce bone, thus lowering blood calcium levels.

Canal of Schlemm Network of channels located at the junction of the sclera and cornea that drains aqueous humor from the anterior chamber of the eye.

Canaliculi Tiny canals in compact bone that provide a route for nutrients and wastes to flow to and from the osteocytes.

Cancer Disease characterized by the uncontrollable replication of cells.

Capacitation Process in which the outer protective coat of the sperm is dissolved away in the female reproductive tract. Makes the membrane fragile and disruptible, allowing the acrosome to break down and release its enzymes.

Capillaries Tiny vessels in body tissues whose walls are composed of a flattened layer of cells that allow water and other molecules to flow freely into and out of the tissue fluid.

Capillary bed Branching network of capillaries supplied by arterioles and drained by venules.

Capsid Protein coat of a virus.

Capsomere Globular proteins that make up the capsid of viruses.

Carbohydrate Organic compound consisting of carbon, hydrogen, and oxygen. A structural component of plant cells, it is used principally as a source of energy in animal cells.

Carbonic anhydrase Enzyme found in red blood cells that catalyzes the conversion of carbon dioxide to carbonic acid.

Carcinogens Cancer-causing agents.

Cardiac muscle Type of muscle found in the walls of the heart; it is striated and involuntary.

Carnivores Meat eaters; organisms that feed on grazers and other animals.

Carrier proteins Class of proteins that transport smaller molecules and ions across the plasma membrane of the cell. Are involved in facilitated diffusion.

Carriers Individuals who carry a gene for a particular trait that can be passed on to their children, but who do not express the trait.

Carrying capacity The number of organisms an ecosystem can support on a sustainable basis.

Cartilage Type of specialized connective tissue. Found in joints on the articular surfaces of bones and other locations.

Catalyst Class of compounds that speed up chemical reactions. Although they take an active role in the process, they are left unchanged by the reaction. Thus, they can be used over and over again. *See also* Enzymes.

Cataracts Disease of the eye resulting in cloudy spots on the lens (and sometimes the cornea) that cause cloudy vision.

Cell body Part of the nerve cell that contains the nucleus and other cellular organelles; the center of chemical synthesis.

Cell culture Glass bottle or shallow dish containing nutrient medium and designed to permit cells to grow in the laboratory under controlled conditions.

Cell cycle Repeating series of events in the lives of many cells. Consists of two principal parts: interphase and cellular division.

Cell division Process by which the nucleus and the cytoplasm of a cell are split, creating two daughter cells. Consists of mitosis and cytokinesis.

Cellular respiration The complete breakdown of glucose in the cell, producing carbon dioxide and water. Comprised of four separate but interconnected parts: glycolysis, the intermediate reaction, the citric acid cycle, and the electron transport system.

Central nervous system The brain and spinal chord.

Centriole Organelle consisting of a ring of microtubules, arranged in nine sets of three. Structurally identical to basal bodies, but associated with the spindle apparatus. Gives rise to the basal body in ciliated cells.

Centromere Region on each chromatid that joins with the centromere of its sister chromatid.

Cerebellum Structure of the brain that lies blow the cerebral cortex. It has many important functions including synergy.

Cerebral cortex Outer layer of each cerebral hemisphere, consisting of many multipolar neurons and nerve cell fibers.

Cerebral hemisphere Convoluted mass of nervous tissue located above the deeper structures, such as the hypothalamus and limbic system. Home of consciousness, memory, and sensory perception; originates much conscious motor activity.

Cervical cap Birth control device consisting of a small cup that fits over the tip of the cervix.

Cervix Lowermost portion of the uterus; it protrudes into the vagina.

Cesarean section Delivery of a baby via an incision through the abdominal and uterine walls.

Chemical evolution Formation of organic molecules from inorganic molecules early in the history of the Earth.

Chemiosmosis Process responsible for most ATP production in the electron transport system of cells.

Chemosynthetic organisms Cells that lack chlorophyll and acquire electrons from inorganic molecules.

Childhood Period of human life that lasts from infancy to puberty.

Chlamydia Bacterium that causes nonspecific urethritis, a type of sexually transmitted disease.

Chlorofluorocarbons (CFCs) Chemical substances often used as spray can propellants (outside the United States and several other countries) and refrigerants that drift to the upper stratosphere and dissociate. Chlorine released by CFCs reacts with ozone, thus eroding the ozone layer.

Cholecystokinin (CCK) Hormone produced by cells of the duodenum when chyme is present. Causes the gallbladder to contract, releasing bile.

Chordae tendineae Tendinous chords that anchor the atrioventricular valves to the inner walls of the ventricles.

Chorionic villus biopsy Medical procedure to detect genetic defects; involves removing a small portion of the villi and then examining it for chromosomal abnormalities.

Choroid Middle layer of the eye that absorbs stray light and supplies nutrients to the eye.

Chromatid Strand of the chromosome consisting of DNA and protein.

Chromatin Long, threadlike fibers containing DNA and protein in the nucleus.

Chromosome-to-pole fibers Microtubules of the spindle that extend from the centriole to the chromosome where they attach. They play a crucial role in separating the double-stranded chromosomes during mitosis.

Chronic bronchitis Persistent irritation of the bronchi, which causes a mucus buildup, coughing, and difficulty breathing.

Chyme Liquified food in the stomach.

Ciliary body Portion of the middle layer of the eye that is located near the lens. Contains smooth muscle that constricts, thus helping control the shape of the lens and permitting the eye to focus.

Circadian rhythm Biorhythm that occurs on a daily cycle.

Circulatory system Organ system consisting of the heart, blood vessels, and blood.

Circumcision Operation to remove the foreskin of the penis. Generally performed on newborns.

Cisterna Channels of the endoplasmic reticulum and Golgi.

Citric acid Six-carbon compound produced in the first reaction of the citric acid or Krebs cycle. Formed when oxaloacetate reacts with acetyl Coenzyme A.

Class Subgroup of a phylum.

Clitoris Small knot of tissue located where the labia minora meet. Consists of erectile tissue.

Cloning Technique of genetic engineering whereby many copies of a gene are produced.

Closed system System that receives no materials from the outside.

Coacervates Microscopic globules that selectively incorporate molecules from their environment. Coacervates existing on Earth over 3 billion years ago may have given rise to the earliest cells.

Cochlea Sensory organ of the inner ear that houses the receptor for hearing.

Codominant Refers to two equally expressed alleles.

Codon Three adjacent bases in the messenger RNA that code for a single amino acid.

Collecting tubules Tubules in the kidney into which nephrons drain. They converge and drain into the renal pelvis.

Color blindness Condition that occurs in individuals who have a deficiency of certain cones. The most common form involves difficulty in distinguishing between red and green.

Colostrum Protein-rich product of the breast, produced for two to three days immediately after delivery.

Community All of the plants, animals, and microorganisms in an ecosystem.

Compact bone Dense bony tissue in the outer portion of all bones.

Complement Group of blood proteins that circulate in the blood in an inactive state until the body is invaded by bacteria; then they help destroy the bacteria.

Complementary base pairing Unalterable coupling of purine adenine to pyrimidine thymine, and purine guanine to pyrimidine cytosine. Responsible for the accurate transmission of genetic information from parent to offspring.

Condom Birth control device, consisting of a thin latex rubber sheath or other material that is rolled onto the erect penis. Prevents sperm from entering the vagina and helps prevent the spread of sexually transmitted diseases.

Conduction deafness Loss of hearing that occurs when the conduction of sound waves to the inner ear is impaired. May be caused by ruptured eardrum or damage to ossicles.

Cone Type of photoreceptor that operates in bright light; is responsible for color vision.

Connective tissue One of the primary tissues. It contains cells and varying amounts of extracellular material and holds cells together, forming tissues and organs.

Connective tissue proper Name referring to loose and dense connective tissue; supports and joins various body structures.

Consumers Organisms that eat plants and algae (producers) and other consumers.

Contact inhibition Cessation of growth that results when two or more cells contact each other. A feature of normal cells but absent in cancer cells.

Contraceptive Any measure that helps prevent fertilization and pregnancy.

Convergence Inward turning of the eyes to focus on a nearby object.

Corepressor Molecule that binds to a repressor protein, allowing it to bind to the operator site, which, in turn, shuts down the structural genes by blocking RNA polymerase.

Cornea Clear part of the wall of the eye continuous with the sclera; allows light into the interior of the eye.

Coronary bypass surgery Surgical technique used to reestablish blood flow to the heart muscle by grafting a vein to shunt blood around a clogged coronary artery.

Corpus luteum (CL) Structure formed from the ovulated follicle in the ovary; produces estrogen and progesterone.

Cortical granules Secretory vesicles lying beneath the plasma membrane of the oocyte that are released when a sperm cell contacts the oocyte membrane. They block additional sperm from fertilizing the ovum.

Cortisol Glucocorticoid hormone that increases blood glucose by stimulating gluconeogenesis. Also stimulates protein breakdown in muscle and bone.

Cowper's gland Smallest of the sex accessory glands; empties into the urethra.

Cranial nerves Nerves arising from the brain and brain stem.

Creatine phosphate High-energy molecule in muscle.

Cristae Folds formed by the inner membrane of the mitochondrion.

Cro-Magnons Earliest known members of *Homo sapiens sapiens*.

Cross bridges Part of the myosin molecule that attaches to and pulls actin molecules inward causing the sarcomere to shorten.

Crossing over Exchange of chromatin by homologous chromosomes during prophase I. Results in considerably more genetic variation in gametes and offspring.

Crystallin Protein inside the lens that may denature, causing cataracts.

Culture The ideas, customs, skills, and arts of a given people in a given time that can change—or evolve—over time.

Cushing's disease Disease that results from pharmacologic doses of cortisone usually administered for rheumatoid arthritis or allergies.

Cyclic AMP Nucleotide derived from ATP. Its synthesis is stimulated when protein and polypeptide hormones bind to the plasma membrane of cells. In the cytoplasm, it activates protein kinase, which, in turn, activates other enzymes.

Cyclosporine Drug used to suppress graft rejection.

Cystic fibrosis Autosomal recessive disease that leads to problems in sweat glands, mucus glands, and the pancreas. Pancreas may become blocked, thus reducing the flow of digestive enzymes to the small intestine. Mucus buildup in the lungs makes breathing difficult.

Cytokinesis Cytoplasmic division brought about by the contraction of a microfilamentous network lying beneath the plasma membrane at the midline. Usually begins when the cell is in late anaphase or early telophase.

Cytoplasm Material occupying the cytoplasmic compartment of a cell. Consists of a semifluid substance, the cytosol, containing many dissolved substances, and formed elements, the organelles.

Cytoskeleton A network of protein tubules in the cytoplasmic compartment of a cell. Attaches to many organelles and enzyme molecules and thus helps organize cellular activities, increasing efficiency.

Cytotoxic cells Type of T cell (T-lymphocyte) that attacks and kills virus-infected cells, parasites, fungi, and tumor cells.

Daughter cells Cells produced during cell division.

Decibel Unit used to measure the intensity of sound.

Decomposer food chain Series of organisms that feed on organic wastes and the dead remains of other organisms.

Defibrillation Procedure to stop fibrillation (erratic electrical activity) of the heart.

Deletion Loss of a piece of a chromosome.

Demographer Scientist who studies populations.

Dendrite Short, highly branched fiber that carries impulses to the nerve cell body.

Dense connective tissue Type of connective tissue that consists primarily of densely packed fibers, such as those found in ligaments and tendons.

Deoxyribonuclease Pancreatic enzyme that breaks RNA and DNA into shorter chains.

Depth perception Ability to judge the relative position of objects in our visual field.

Dermis Layer of dense irregular connective tissue that binds the epidermis to underlying structures.

Desert Biome characterized by low rainfall and a hot climate. Contains organisms well adapted to these conditions.

Detrivores Organisms that feed on animal waste or the remains of plants and animals.

Diabetes insipidus Condition caused by lack of ADH. Main symptoms are polydipsia (excessive drinking) and polyuria (excessive urination).

Diabetes mellitus Insulin disorder either resulting from insufficient insulin production or decreased sensitivity of target cells to insulin. Results in elevated blood glucose levels unless treated.

Dialysis Procedure used to treat patients whose kidneys have failed. Blood is removed from the body and pumped through an artificial filter that removes impurities.

Diaphragm (birth control) Birth control devise consisting of a rubber cup that fits over the end of the cervix. Used in conjunction with spermicidal jelly or cream.

Diaphragm (muscle) Dome-shaped muscle that separates the abdominal and thoracic cavities.

Diaphysis Shaft of the long bones. Consists of an outer layer of compact bone and an inner marrow cavity.

Diastolic pressure The pressure at the moment the heart relaxes. The lower of the two blood pressure readings.

Differentiation Structural and functional divergence from the common cell line. Occurs during embryonic development.

Dihybrid cross Procedure where one plant is bred with another to study two traits.

Diplopia Double vision. May occur when the eyes fail to move synchronously.

Distal convoluted tubule Section of the nephron that connects the loop of Henle to the collecting tubule. Site of tubular reabsorption.

Divergent evolution Process in which organisms evolve in different directions due to exposure to different environmental influences.

Diverticulitis Expansion of the large intestine due to obstruction.

DNA polymerase Enzyme that helps align the nucleotides and join the phosphates and sugar molecules in a newly forming DNA strand.

Dominant Adjective used in genetics to refer to an allele that is always expressed in heterozygotes. Designated by a capital letter.

Dorsal root (of a spinal nerve) Inlet for sensory nerve fibers to the spinal cord.

Double helix Describes the helical structure formed by two polynucleotide chains making up the DNA molecule.

Doubling time Time it takes a population to double.

Down syndrome Genetic disorder caused by an additional chromosome 21 that results in distinctive facial characteristics and mental retardation. Also known as Trisomy 21.

Dryopithecus Genus of apelike creatures that is thought to have given rise to the gibbons, gorillas, orangutans, and chimpanzees.

Ductus arteriosus Shunt that lies between the pulmonary artery and the aorta, helping divert blood from the lungs.

Ductus venosus Shunt that connects directly from the umbilical vein to the inferior vena cava.

Duodenum First portion of the small intestine; site where most food digestion and absorption takes places.

Dust cell Cell found in and around the alveoli; phagocytizes particulate matter that has entered the lung.

E. coli Common bacterium that lives in the large intestine of humans and digests leftover glucose and other materials from food. Used in much genetic research.

Ecological niche An organism's habitat and all of the relationships that exist between that organism and its environment.

Ecological system (ecosystem) System consisting of organisms and their environment and all of the interactions that exist between these components.

Ecology Study of living organisms and the web of relationships that binds them together in the economy of nature. The study of ecosystems.

Ecosystem balance Dynamic equilibrium in ecosystems. Maintained by the interplay of growth and reduction factors.

Ectoderm One of the three types of cells that emerges in human embryonic development. Gives rise to the skin and associated structures, including the eyes.

Edema Swelling resulting from the buildup of fluid in the tissues.

Effector General term for any organ or gland that is controlled by the nervous system.

Ejaculation Ejection of semen from the male reproductive tract.

Elastic arteries Arteries that contain numerous elastic fibers interspersed among the smooth muscle cells of the tunica media.

Elastic cartilage Type of cartilage containing many elastic fibers found in regions where support and flexibility are required.

Electron Highly energetic particle carrying a negative charge that orbits the nucleus of an atom.

Electron cloud A region surrounding the nucleus of an atom where electrons orbit.

Electron transport system Series of protein molecules in the inner membrane of the mitochondrion that pass electrons from the citric acid cycle from one to another, eventually donating them to oxygen. During their journey along this chain of proteins, the electrons lose energy, which is used to make ATP. *See also* Chemiosmosis.

Elements Purest form of matter; substances that cannot be separated into different substances by chemical means.

Emphysema Progressive, debilitating disease that destroys tiny air sacs in the lung (alveoli). The fastest growing cause of death in the United States.

Endocrine glands Glands of internal secretion. They produce hormones that are secreted into the bloodstream.

Endocrine system Numerous, small, hormone-producing glands scattered throughout the body.

Endocytosis Process by which cells engulf solid particles, bacteria, viruses, and even other cells.

Endoderm One of the three types of cells that emerges during embryonic development. Gives rise to the intestinal tract and associated glands.

Endolymph Fluid inside the semicircular canals that deflects the cupula, signaling rotational movement of the head and body.

Endometrium Uterine endothelium or lining.

Endoplasmic reticulum Branched network of channels found throughout the cytoplasm of many cells. Formed from flattened sheets of membrane derived from the nuclear membrane.

Endosymbiotic evolution Theory that accounts for the development of the first eukaryotes. Says that free-living bacteria-like organisms were engulfed by other cells and became internal symbionts. Internal symbionts later became the organelles of eukaryotes.

Endothelium Single-celled lining of blood vessels.

Energy pyramid Diagram of the amount of energy at various trophic levels in a food chain or ecosystem.

Enhancer Segment of DNA that increases the activity of nearby genes several hundred times.

Envelope Protective membrane of some viruses; lies outside the capsid.

Enzymes Special proteins that participate in chemical reactions in the body, greatly accelerating their rate.

Epidermis Outermost layer of the skin that protects underlying tissues from drying out and from bacteria and viruses.

Epididymal duct Duct within the epididymis; site where sperm are stored until ejaculation.

Epididymis Storage site of sperm. Located on the testis, it consists of a long, tortuous duct.

Epiglottis Flap of tissue that closes off the trachea during swallowing.

Epiphyseal plate Band of cartilage cells between the shaft of the bone and the epiphysis. Allows for bone growth.

Epiphysis Expanded end of the long bones.

Episiotomy Surgical incision that runs from the vaginal opening toward the rectum. Enlarges the vaginal opening, easing childbirth.

Epithelium One of the primary tissues. Forms linings and external coatings of organs.

Erectile tissue Spongy tissue of the penis that fills with blood during sexual excitement, making the penis turgid.

Erythropoietin Hormone produced by the kidney when oxygen levels decline. Stimulates red blood cell production in the bone marrow.

Esophagus Muscular tube that transports food to the stomach.

Essential amino acid One of nine amino acids that must be provided in the human diet.

Euchromatin Metabolically active chromatin.

Evolution Process that leads to structural and functional changes in species, making them better able to survive in their environment; also

leads to the formation of new species. Results from natural genetic variation and environmental conditions that select for organisms best suited to their environment.

Exhalation Expulsion of air from the lungs.

Exocrine gland Gland of external secretion; empties its contents into ducts.

Exocytosis Process by which cells release materials stored in secretory vesicles. The reverse of endocytosis.

Exon Expressed segment of DNA.

Experiment Test performed to prove or disprove a hypothesis.

Exponential growth Type of growth that occurs when a value grows by a fixed percentage and the increase is applied to the base amount.

Extension Movement of a body part (limbs, fingers, and toes) that opens a joint.

External auditory canal Channel that directs sound waves to the eardrum.

External genitalia External portion of the female reproductive system consisting of the clitoris, labia minor, and labia majora.

External sphincter of the bladder Voluntary muscular valve that controls urine release under conscious control. Formed by a flat band of muscle that forms the floor of the pelvic cavity.

Extrinsic eye muscles Six muscles located outside the eye that are responsible for eye movement.

Facilitated diffusion Process in which carrier proteins shuttle molecules across plasma membranes. The molecules move in response to concentration gradients.

Feces Semisolid material containing undigested food, bacteria, ions, and water; produced in the large intestine.

Feedback mechanism Method of control in which the product of a reaction or process regulates the reaction or process. Both positive and negative feedbacks exist in biological systems.

Fenestrae Minute openings in capillary walls that help permit movement of molecules to and from the capillary.

Fermentation Process occurring in eukaryotic cells in the absence of oxygen, during which pyruvic acid is converted to lactic acid. Also occurs in prokaryotes.

Fertilization Union of sperm and ovum.

Fiber Any of the indigestible polysaccharides in fruits, vegetables, and grains.

Fibrillation Cardiac muscle spasms occurring during heart attacks due to a loss of synchronized electrical signals.

Fibrin Fibrous protein produced from fibrinogen, a soluble plasma protein. Helps form blood clots.

Fibrinogen Protein in plasma that forms fibrin.

Fibroblast Connective tissue cell, found in loose and dense connective tissues that produces collagen, elastic fibers, and a gelatinous extracellular material; responsible for repairing damage created by cuts or tears to connective tissue.

Fibrocartilage Type of cartilage whose extracellular matrix consists of numerous bundles of collagen fibers. Principally found in the intervertebral disks.

Fimbriae Fingerlike projections of the end of the oviduct that sweep the oocyte into the oviduct.

First polar body Cast-off nuclear material produced during the first meiotic division during oogenesis.

Fitness Measure of reproductive success of an organism and, therefore, the genetic influence an individual has on future generations.

Flagellum Long, whiplike extension of the plasma membrane of certain protozoans and sperm cells in humans. Used for motility.

Flexion Movement of a limb, finger, or toe that involves closing a joint.

Follicle (ovary) Structure found in the ovary. Each follicle contains an oocyte and one or more layers of follicle cells that are derived from the loose connective tissue of the ovary surrounding the follicle.

Follicle (thyroid) Structure found in the thyroid gland. Consists of an outer layer of cuboidal cells surrounding thyroglobulin.

Follicle-stimulating hormone (FSH) Gonadotropic hormone that promotes gamete formation in both men and women.

Food chain Series of organisms in an ecosystem in which each organism feeds on the organism preceding it.

Food vacuole Membrane-bound vacuole in a cell containing material engulfed by the cell.

Food web All of the connected food chains in an ecosystem.

Foramen ovale Hole in the interatrial septum of the embryonic heart that diverts blood from the right atrium to the left atrium, reducing the flow of blood to the pulmonary arteries and lungs.

Foreskin Sheath of skin that covers the glans penis.

Fossil Remains or imprints of organisms that lived on Earth many years ago, usually embedded in rocks or sediment.

Fovea centralis Tiny spot in the center of the macula of the eye that contains only cones. Objects are focused onto the fovea for sharp vision.

Fusion Joining of two atoms, which releases large amounts of energy.

Gallbladder Sac on the underside of the liver that stores and concentrates bile.

Gastrin Stomach hormone that stimulates HCl production and release by the gastric glands.

Gastroesophageal sphincter Ring of muscle located in the lower esophagus that opens when food arrives, allowing food to pass into the stomach, and then closes to keep food and stomach acid from percolating upward.

Gene Segment of the DNA that controls cell structure and function.

Gene flow Introduction of new genes into a population when new individuals join the population.

Gene pool All the genes of all of the members of a population or species.

Genera Plural of genus.

Genome Genes of an organism.

Genotype Genetic makeup of an organism.

Genus Subgroup of a family of organisms.

Geographic isolation Physical separation of a population by some barrier. Sometimes results in reproductive isolation and the formation of new species.

Germinal epithelium Germ cells in the wall of the seminiferous tubule that give rise to sperm.

Gestation The period of pregnancy.

Glans penis Slightly enlarged tip of the penis.

Glaucoma Disease of the eye caused by pressure resulting from a buildup of aqueous humor in the anterior chamber.

Glomerular filtration Movement of materials out of the glomeruli into Bowman's capsule in the kidney.

Glomerulus Tuft of capillaries that make up part of the nephron; site of glomerular filtration.

Glucagon Hormone released by the pancreas that stimulates the breakdown of glycogen in liver and muscle and the release of glucose molecules, thus increasing blood levels of glucose.

Glucocorticoids Group of steroid hormones produced by the adrenal cortex that stimulate gluconeogenesis.

Gluconeogenesis Synthesis of glucose from fatty acids and amino acids. Takes place in the liver where amino acids and fatty acids are stored.

Glycogenolysis Breakdown of glycogen, releasing glucose.

Glycolysis Metabolic pathway in the cytoplasm of the cell, during which glucose is split in half, forming two molecules of pyruvic acid. The energy released during the reaction is used to generate two molecules of ATP.

Glycoproteins Proteins that have carbohydrate attached to them.

Goiter Condition in which the thyroid gland enlarges due to lack of dietary iodide.

Golgi complex Organelle consisting of a series of flattened membranes that form channels. It sorts and chemically modifies molecules and repackages its proteins into secretory vesicles.

Golgi tendon organs Special receptors found in tendons that respond to stretch. Also known as neurotendinous organs.

Gonadotropin General term for FSH and LH, which are produced by the anterior pituitary and target male and female gonads.

Gonadotropin-releasing hormone (GnRH) Hormone produced by the hypothalamus that controls the release of FSH (ICSH in males) and LH.

Gonorrhea Sexually transmitted disease caused by a bacterium.

Gray matter Gray, outermost region of the cerebral cortex.

Grazer Herbivorous organism.

Grazer food chain Food chain beginning with plants and grazers (herbivores).

Greenhouse gas Gas, such as carbon dioxide and chlorofluorocarbons, that traps heat escaping from the Earth and radiates it back to the surface.

Growth factor Any biotic or abiotic factor that causes a population to grow.

Growth hormone A protein hormone produced by the anterior pituitary that stimulates cellular growth in the body, causing cellular hypertrophy and hyperplasia. Its major targets are bone and muscle.

Growth rate (of a population) Determined by subtracting the death rate from the birth rate.

Habitat Place in which an organism lives.

Habituation Condition where sensory receptors stop generating impulses, even though a stimulus is still present.

Hammer (malleus) One of three bones of the middle ear. Abuts the tympanic membrane and helps transmit sound from the eardrum to the inner ear.

Helper cell Type of T-lymphocyte that stimulates the proliferation of T and B cells when antigen is present.

Heme group Subunit of the hemoglobin molecule. Consists of a porphyrin ring and a central iron ion to which oxygen binds.

Hemoglobin Protein molecules inside red blood cells; binds to oxygen.

Hemophilia Disease caused by a gene defect occurring on the Y chromosome. Results in absence of certain blood-clotting factors.

Herbicide Any chemical applied to crops to control weeds.

Herbivore Any organism that feeds directly on plants. Also known as a grazer.

Herpes One of the most common sexually transmitted diseases; caused by a virus.

Heterochromatin Inactive chromatin that is slightly coiled or compacted in the interphase nucleus.

Heterotrophic fermenter Evolutionarily probably one of the first cells. Absorbed glucose from the environment and broke it down by anaerobic glycolysis.

Heterozygous Adjective describing a genetic condition in which an individual contains one dominant and one recessive gene in a gene pair.

High-density lipoproteins (HDLs) Complexes of lipid and protein that transport cholesterol to the liver for destruction.

Histamine Potent vasodilator released by certain cells in the body during allergic reactions.

Histone Globular protein thought to play a role in regulating the genes.

Homeostasis A condition of stability or equilibrium within any biological or social system. Achieved through a variety of automatic mechanisms that compensate for internal and external changes.

Hominid First humanlike creatures.

Hominoids Subgroup of anthropoids.

Homo sapiens neanderthalensis The Neanderthals. Subspecies of *Homo sapiens*.

Homo sapiens sapiens Species of modern humans that emerged about 400,000 years ago.

Homologous structures Structures thought to have arisen from a common origin.

Homozygous Adjective describing a genetic condition marked by the presence of two identical alleles for a given gene.

Hormone Chemical substance produced in one part of the body that travels to another where it elicits a response.

Human chorionic gonadotropin (HCG) Hormone produced by the embryo that stimulates the corpus luteum to produce estrogen.

Humoral immunity Immune reaction that protects the body primarily against viruses and bacteria in the body fluids via antibodies produced by plasma cells.

Hyperglycemia High blood glucose levels.

Hyperopia (farsightedness) Condition that occurs when the eyeball is too short or the lens is too weak, resulting in poor focus on nearby objects.

Hypertension   High blood pressure.

Hypertonic   Adjective describing a solution with a higher solute concentration than the cell's cytoplasm, causing the cell to shrivel.

Hypothalamus   Structure in the brain located beneath the thalamus. It consists of many aggregations of nerve cells and controls a variety of autonomic functions aimed at maintaining homeostasis.

Hypothesis   Tentative and testable explanation for a phenomenon or observation.

Hypotonic   Adjective describing a solution with a solute concentration lower than the cell's cytoplasm, resulting in a swelling of the cell.

I gene   Gene that controls blood type through the synthesis of glycoproteins on the plasma membrane of the red blood cell.

Immune system   Diffuse system consisting of trillions of cells that circulate in the blood and lymph and take up residence in the lymphoid organs, such as the spleen, thymus, lymph nodes, and tonsils, as well as other body tissues. Helps protect the body against foreign cells, such as bacteria and viruses, and protects against cancer cells.

Immunity   Term referring to the resistance of the body to infectious disease.

Immunocompetence   Process in which lymphocytes mature and become capable of responding to specific antigens.

Immunoglobulins   Antibodies.

Implantation   Process in which the blastocyst embeds in the uterine lining.

Impotency   Inability of a male to achieve an erection.

In vitro   Term referring to any procedure carried out in a test tube or petri dish, such as *in vitro* fertilization.

Incomplete dominance   Partial dominance. Occurs when an allele exerts only partial dominance over another allele, resulting in an intermediate trait.

Incontinence   Inability to control urination.

Induced abortion   Deliberate expulsion of a fetus or embryo.

Inducer   Chemical substance that activates inducible genes.

Inducible operon   Set of genes that remain inactive until needed. Activated by inducers.

Infectious mononucleosis   White blood cell disorder caused by a virus. Characterized by a rapid increase of monocytes and lymphocytes.

Inferior vena cava   Large vein that empties deoxygenated blood from the body below the heart into the right atrium of the heart.

Infertility   Inability to conceive; can be due to problems in either the male or the female.

Inflammatory response   Response to tissue damage including an increase in blood flow, the release of chemical attractants, which draw monocytes to the scene, and an increase in the flow of plasma into a wound.

Inhalation   Process of air being drawn into the lungs.

Inhibin   Substance produced by the seminiferous tubules that inhibits the production of FSH by the anterior pituitary.

Inhibiting hormone   Hormone from the hypothalamus that inhibits the release of hormones from the anterior pituitary.

Initiator codon   Codon found on a messenger RNA strand that marks where protein synthesis begins.

Inner cell mass   Cells of the blastocyst that become the embryo and amnion.

Insulin   Hormone that stimulates the uptake of glucose by body cells, especially muscle and liver cells.

Stimulates the synthesis of glycogen in liver and muscle cells.

Insulin-dependent diabetes
Type of diabetes that can only be treated with injections of insulin. May be caused by an autoimmune reaction. Also known as early-onset diabetes.

Insulin-independent diabetes
Type of diabetes that often occurs in obese people. In most patients, it can be controlled by diet. Also know as late-onset diabetes.

Integral protein   Large protein molecules in the lipid bilayer of the plasma membrane.

Integration   Process of making sense of various nervous inputs so that a meaningful response can be achieved.

Intercostal muscles   Short, powerful muscles that lie between the ribs. Involved in inspiration and active exhalation.

Interferon   Protein released from cells infected by viruses that stops the replication of viruses in other cells.

Interleukin 2   Chemical released by helper cells that activates T and B cells, stimulating cell division.

Internal sphincter (of the bladder)
Involuntary muscular valve that relaxes reflexively, releasing urine. Formed by a smooth muscle in the neck of the bladder at the junction of the bladder and the urethra.

Internodes   Segments of the axon between nodes of Ranvier.

Interphase   Period of cellular activity occurring between cell divisions. Synthesis and growth occur in preparation for cell division.

Interstitial cells   Cells located in the loose connective tissue between the seminiferous tubules of the testes. Produce testosterone.

Interstitial cell stimulating hormone (ICSH)   Luteinizing hormone in males. Regulates testosterone secretion.

Interstitial fluid   Fluid surrounding cells in body tissues. Provides a path through which nutrients, gases, and wastes can travel between the capillary and the cells.

Intervertebral disks Shock-absorbing material between the bones of the spine.

Intrauterine device (IUD) Birth control device that consists of a small plastic or metal object with a string attached that is inserted into the uterus through the cervix. Prevents implantation.

Intron Segment of DNA that is not expressed. Lies between exons (expressed segments).

Ion Atom that has gained or lost one or more electrons. May be either positively or negatively charged.

Ionic bond Weak bond that forms between oppositely charged ions.

Iris Colored segment of the middle layer of the eye visible through the cornea.

Irritability Ability to perceive and respond to stimuli.

Islets of Langerhans Group of endocrine cells found in the pancreas that produce insulin and glucagon.

Isotonic Having the same solute concentration as a cell or body fluid.

Isotope Alternative form of an atom; differs from other atoms in the number of neutrons in the nucleus.

Joint capsule Connective tissue that connects to the opposing bones of a joint and forms the synovial cavity. The inner layer of the joint capsule produces synovial fluid.

Kidney Organ that rids the body of wastes and plays a key role in regulating the chemical constancy of blood.

Klinefelter syndrome Genetic disorder that results from a XXY genotype.

Labia majora Outer folds of skin of the external genitalia in women.

Labia minora Inner folds of the external genitalia in women.

Labor The process or period of childbirth.

Lactation Milk production in the breasts.

Laparoscope Instrument used to examine internal organs through small openings made in the skin and underlying muscle.

Larynx Rigid but hollow cartilaginous structure that houses the vocal cords and participates in swallowing.

Lens Transparent structure that lies behind the iris and in front of the vitreous humor. Focuses light on the retina.

Leukemia Cancer of white blood cells.

Leukocytosis An increase in the concentration of white blood cells, which often occurs during a bacterial or viral infection.

Life expectancy Average length of time a person will live.

Ligament Connective tissue structure that runs from bone to bone, located alongside and sometimes inside the joint. Offers support for joints.

Limbic system Array of structures in the brain that work in concert with centers of the hypothalamus. Site of instincts and emotions.

Limiting factor One factor that is most important in regulating growth in an ecosystem.

Lipase Enzyme that removes some of the fatty acids from the glycerol molecule, forming a monoglyceride. Produced by the salivary glands and the pancreas.

Lipid Commonly known as fats. Water-insoluble organic molecules that provide energy to body cells, help insulate the body from heat loss, and serve as precursors in the synthesis of certain hormones. A principal component of the plasma membrane.

Liposuction Technique used to remove subcutaneous fat.

Liver Organ located in the abdominal cavity that performs many functions essential to homeostasis. It stores glucose and fats, synthesizes some key blood proteins, stores iron and certain vitamins, detoxifies certain chemicals, and plays an important role in digestion by producing bile.

Long bones Bones of the skeleton that form parts of the extremities.

Loose connective tissue Type of connective tissue that serves primarily as a packing material. Contains many cells among a loose network of collagen and elastic fibers, especially cells that help protect the body from foreign organisms.

Low-density lipoproteins (LDLs) Complexes of protein and lipid that transport cholesterol, depositing it in body tissues.

Lungs Two large saclike organs in the thoracic cavity where the blood and air exchange carbon dioxide and oxygen.

Luteinizing hormone (LH) Hormone that stimulates gonadal hormone production. In men, LH stimulates the production of testosterone, the male sex steroid. In women, LH stimulates estrogen secretion.

Lymph Fluid contained in the lymphatic vessels. Similar to tissue fluid, but also contains white blood cells and may contain large amounts of fat.

Lymph node Small nodular organ interspersed along the course of the lymphatic vessels. Serves as a filter for lymph.

Lymphatic system Network of vessels that drains extracellular fluid from body tissues and returns it to the circulatory system.

Lymphocyte Type of white blood cell. *See also* B-lymphocyte and T-lymphocyte.

Lymphoid organs Organs, such as the spleen and thymus, that belong to the lymphatic system.

Lymphokine Chemical released by suppressor T cells that inhibits the division of B and T cells.

Lysosome Membrane-bound organelle that contains enzymes. Responsible for the breakdown of material that enters the cell by endocytosis. Also destroys aged or malfunctioning cellular organelles.

Lysozyme Enzyme produced in saliva that dissolves the cell wall of bacteria, killing them.

Macronutrients Nutrients required in relatively large amounts by organisms. Includes water, proteins, carbohydrates, and lipids.

Macrophage Phagocytic cell derived from monocytes that resides in loose connective tissues and helps guard tissues against bacterial and viral invasion.

Macula lutea Region of the retina located lateral to the optic disc where cones are most abundant.

Maculae Receptor organs in the saccule and utricle that play a role in position sense.

Malignant tumor Structure resulting from uncontrollable cellular growth. Cells often spread to other parts of the body.

Marfan's syndrome Autosomal dominant genetic disorder that affects the skeletal system, the eye, and the cardiovascular system.

Marrow cavity Cavity inside a bone containing either red or yellow marrow.

Mast cell Cell found in many tissues, especially in the connective tissue surrounding blood vessels. Contains large granules containing histamine.

Matrix Extracellular material found in cartilage. Also the material in the inner compartment of the mitochondrion.

Matter Anything that has mass and occupies space.

Medulla Term referring to the central portion of some organs; for example, the adrenal medulla.

Megakaryocyte Large cell found in bone marrow that produces platelets.

Meiosis Type of cell division that occurs in the gonads during the formation of gametes. Requires two cellular divisions (meiosis I and meiosis II). In humans, it reduces the chromosome number from 46 to 23.

Meiosis I First meiotic division.

Meiosis II Second meiotic division.

Meissner's corpuscle Encapsulated sensory receptor thought to respond to light touch.

Membranous epithelium Refers to any sheet of epithelium that forms a continuous lining on organs.

Memory cells T or B cells produced after antigen exposure. They form a reserve force that responds rapidly to antigen during subsequent exposure.

Menopause End of the reproductive function (ovulation) in women. Usually occurs between the ages of 45 and 55.

Menstrual cycle Recurring series of events in the reproductive functions of women. Characterized by dramatic changes in ovarian and pituitary hormone levels and changes in the uterine lining that prepare the uterus for implantation.

Menstruation Process in which the endometrium is sloughed off, resulting in bleeding. Occurs approximately once every month.

Merkel disk Light touch receptor. Consists of dendrites that end on cells in the epidermis.

Mesoderm One of the three types of cells that emerge in human embryonic development. Lies in the middle of the forming embryo and forms muscle, bone, and cartilage.

Messenger RNA (mRNA) Type of RNA that carries genetic information needed to synthesize proteins to the cytoplasm of a cell.

Metabolic pathway Series of linked chemical reactions in which the product of one reaction becomes the reactant in another reaction.

Metabolism Sum total of all of the chemical reactions that occur in an organism.

Metaphase Stage of cellular division in which chromosomes line up in the center of the cell.

Metarterioles Arterioles that serve as circulatory short cuts, connecting arterioles with venules in a capillary bed. Also known as thoroughfare channels.

Metastasis Spread of cancerous cells throughout the body, through the lymph vessels and circulatory system or directly through tissue fluid.

Microfilament Solid fiber consisting of contractile proteins that is found in cells in a dense network under the plasma membrane. Forms part of the cytoskeleton.

Micronutrients Nutrients required in small quantities. They include two broad groups, vitamins and minerals.

Microspheres Small globules consisting of protein that may have been precursors of the first cells. Also known as proteinoids.

Microsurgery Type of surgery performed under dissecting microscopes. Used to reconnect axons, blood vessels, and other small structures.

Microtubules Hollow protein tubules in the cytoplasm of cells that form part of the cytoskeleton. Also form spindles.

Microvilli Tiny projections of the plasma membranes of certain epithelial cells that increase the surface area for absorption.

Middle ear Portion of the ear located within a bony cavity in the temporal bone of the skull. Houses the ossicles.

Mineralocorticoids Group of steroid hormones produced by the adrenal cortex. Involved in electrolyte or mineral salt balance.

Mitochondrion Membrane-bound organelle where the bulk of cellular energy production occurs in eukaryotic cells. Houses the citric acid cycle and electron transport system.

Mitosis   Term referring specifically to the division of a cell's nucleus. Consists of four stages: prophase, metaphase, anaphase, and telophase.

Mitotic spindle   Array of microtubules constructed in the cytoplasm during prophase. Microtubules of the mitotic spindle connect to the chromosomes and help draw them apart during mitosis.

Monocyte   White blood cell that phagocytizes bacteria and viruses in body tissues.

Monohybrid cross   Procedure in which one plant is bred with another to study the inheritance of a single trait.

Monosomy   Genetic condition caused by a missing chromosome.

Morning sickness   Nausea that often occurs in the first two to three months of pregnancy.

Morula   Solid ball of cells produced from the zygote by numerous cellular divisions.

Motor unit   Muscle fibers supplied by a single axon and its branches.

Mucus   Thick, slimy material produced by the lining of the respiratory tract and parts of the digestive tract. Moistens and protects them.

Multipolar neuron   Motor neuron found in the central nervous system. Contains a prominent, multiangular cell body and several dendrites.

Muscle fiber   Long, unbranched, multinucleated cell found in skeletal muscle.

Muscle spindles   Stretch receptors found in skeletal muscle. Also known as neuromuscular spindles.

Muscle tone   Inherent firmness of muscle, resulting from contraction of muscle fibers during periods of inactivity.

Muscular artery   Any one of the main branches of the aorta. Tunica media consists primarily of smooth muscle cells.

Mutation   Technically, a change in the DNA caused by chemical and physical agents. Also refers to a wide range of chromosomal defects.

Myelin sheath   Layer of fatty material coating the axons of many neurons in the central and peripheral nervous systems. Formed by Schwann cells.

Myofibril   Bundle of contractile myofilaments in skeletal muscle cells.

Myoglobin   Cytoplasmic protein in muscle cells that binds to oxygen.

Myometrium   Uterine smooth muscle.

Myopia (nearsightedness)   Visual condition that results when the eyeball is slightly elongated or the lens is too strong. In the uncorrected eye, light rays from distant images come into focus in front of the retina.

Myosin   Protein filament found in many cells in the microfilamentous network. Also found in muscle cells.

Myosin ATPase   Enzyme found in the myosin cross bridges that splits ATP during muscle contraction.

Naked nerve ending   Unmodified dendritic ending of the sensory neurons. Responsible for at least three sensations: pain, temperature, and light touch.

Natural childbirth   Childbirth without the use of drugs.

Natural selection   Evolutionary process in which environmental abiotic and biotic factors "weed" out the less fit—those organisms not as well adapted to the environment as their counterparts.

Nephron   Filtering unit in the kidney. Consists of a glomerulus and renal tubule.

Nerve   Bundle of nerve fibers. May consist of axons, dendrites, or both. Carries information to and from the central nervous system.

Nerve deafness   Loss of hearing resulting from nerve or brain damage.

Nervous tissue   One of the primary tissues. Found in the nervous system and consists of two types of cells: conducting cells (neurons) and supportive cells.

Neural groove   Ectodermal groove that forms early in embryonic development and runs the length of the embryo, later forming the neural tube.

Neural tube   Tube of ectoderm that arises from the neural groove and will become the spinal cord.

Neuroendocrine reflex   A reflex involving the endocrine and nervous systems.

Neuron   Highly specialized cell that generates and transmits bioelectric impulses from one part of the body to another.

Neurosecretory neurons   Specialized nerve cells of the hypothalamus and posterior pituitary that produce and secrete hormones.

Neurotransmitter   Chemical substance released from the terminal ends (terminal boutons) of axons when a bioelectric impulse arrives. May stimulate or inhibit the next neuron.

Neutron   Uncharged particle in the nucleus of the atom.

Neutrophil   Type of white blood cell that phagocytizes bacteria and cellular debris.

Nicotinamide adenine dinucleotide (NAD)   Electron acceptor molecule that shuttles energetic electrons from glycolysis, the transition reaction, and the citric acid cycle to the electron transport system.

Nitrogen fixation   Process in which bacteria and a few other organisms convert atmospheric nitrogen to nitrate or ammonia, forms usable by plants.

Node of Ranvier   Small gap in the myelin sheath of an axon; located between segments formed by Schwann cells. Responsible for saltatory conduction.

Nondisjunction   Failure of a chromosome pair or chromatids of a double-stranded chromosome to separate during mitosis or meiosis.

Nonspecific urethritis (NSU) One of the most common sexually transmitted diseases. Caused by several different bacteria.

Noradrenaline (norepinephrine) Hormone produced by adrenal medulla and secreted under stress. Contributes to the fight-or-flight response.

Nuclear envelope Double membrane delimiting the nucleus.

Nuclear pores Minute openings in the nuclear envelope that allow materials to pass to and from the nucleus.

Nucleoli Temporary structures in the nuclei of cells during interphase. Regions of the DNA that are active in the production of RNA.

Nucleus (atom) Dense, center region of the atom that contains neutrons and protons.

Nucleus (cell) Cellular organelle that contains the genetic information that controls the structure and function of the cell.

Nutrient cycle Circular flow of nutrients from the environment through the various food chains back into the environment.

Olfactory membrane Receptor for smell; found in the roof of the nasal cavity.

Olfactory nerve Nerve that transmits impulses from the olfactory membrane to the brain.

Omnivores Organisms that feed on both plants and animals.

Ontogeny Development of an organism starting with fertilization.

Oogenesis Production of ova.

Oogonium Germ cell in ovary that contains 46 double-stranded chromosomes. Forms primary oocytes.

Operator site Region of the DNA molecule adjacent to the structural genes that acts as a switch to turn the operon on or off.

Operon Functional unit of the DNA of bacteria. Consists of structural and regulatory genes.

Optic disk Site in the retina where the optic nerve exits. Also known as the blind spot.

Optic nerve Nerve that carries impulses from the retina to the brain.

Order Taxonomic term that refers to a subgroup of a class.

Organ Discrete structure that carries out specialized functions.

Organ of Corti Receptor for sound; located in the inner ear within the cochlea.

Organ system Group of organs that participate in a common function.

Organogenesis Organ formation during embryonic development.

Osmosis Diffusion of water across a selectively permeable membrane.

Osmotic pressure Force that drives water across a selectively permeable membrane. Created by differences in solute concentrations.

Ossicles Three small bones inside the middle ear that transmit vibrations created by sound waves to the organ of Corti.

Osteoarthritis Degenerative joint disease caused by wear and tear that impairs movement of joints.

Osteoblast Bone-forming cell; secretes collagen.

Osteoclast Cell that digests the extracellular material of bone. Stimulated by the parathyroid hormone.

Osteocyte Bone cell derived from osteoblasts that has been surrounded by calcified extracellular material.

Osteoporosis Degenerative disease resulting in the deterioration of bone. Due to inactivity in men and women and loss of ovarian hormones in postmenopausal women.

Outer ear External portion of the ear.

Oval window Membrane-covered opening in the cochlea where vibrations are transmitted from the stirrup to the fluid within the cochlea.

Ovary Female gonad; produces ova.

Overpopulation Condition in which a species has exceeded the carrying capacity of the environment.

Ovulation Release of the oocyte from ovary. Stimulated by hormones from the anterior pituitary.

Ovum Germ cell containing 23 single-stranded chromosomes. Produced during the second meiotic division.

Oxaloacetate Four-carbon compound of the citric acid cycle. It is involved in the very first reaction of the cycle and is regenerated during the cycle.

Oxytocin Hormone from the posterior pituitary hormone. Stimulates contraction of the smooth muscle of the uterus and smooth-muscle–like cells surrounding the glandular units of the breast.

Ozone O_3. Molecule produced in the stratosphere (upper layer of the atmosphere) from molecular oxygen. Helps screen out incoming ultraviolet light. *See also* Ozone layer.

Ozone layer Region of the atmosphere located approximately 12 to 30 miles above the Earth's surface where ozone molecules are produced. Helps protect the Earth from ultraviolet light.

Pacinian corpuscle Large encapsulated nerve ending that is located in the deeper layers of the skin and near body organs. Responds to pressure.

Pancreas Organ found in the abdominal cavity under the stomach, nestled in a loop formed by the first portion of the small intestine. Produces enzymes needed to digest foodstuffs in the small intestine and hormones that regulate blood glucose levels.

Pap smear Procedure in which cells are retrieved from the cervical canal to be examined for the presence of cancer.

Papillae Small protrusions on the upper surface of the tongue. Some papillae contain taste buds.

Paracrines   Chemicals released by cells that elicit a response in nearby regions.

Parasympathetic division (of the autonomic nervous system)   Portion of the autonomic nervous system responsible for a variety of involuntary functions.

Parathyroid glands   Endocrine glands located on the posterior surface of the thyroid gland in the neck. Produce parathyroid hormone.

Parathyroid hormone (PTH)   Hormone that helps regulate blood calcium levels. Stimulates osteoclasts to digest bone, thus raising blood calcium levels. Also known as parathormone.

Parturition   Childbirth.

Passive immunity   Temporary protection from antigen (bacteria and others) produced by the injection of immunoglobulins.

Penis   Male organ of copulation.

Pepsin   Enzyme released by the gastric glands of the stomach. Breaks down proteins into large peptide fragments.

Pepsinogen   Inactive form of pepsin.

Perforin   Chemical released by cytotoxic cells that destroys bacteria. Binds to plasma membrane of target cells, forming pores that make the target cells leak and die.

Perichondrium   Connective tissue layer surrounding most types of cartilage. Contains blood vessels that supply nutrients to cartilage cells.

Periodic table of elements   Table that lists elements by ascending atomic number. Also lists other vital statistics of each element.

Peripheral nervous system   Portion of the nervous system consisting of the cranial and spinal nerves and receptors.

Peristalsis   Involuntary contractions of the smooth muscles in the wall of the esophagus, stomach, and intestines, which propel food along the digestive tract.

Peritubular capillaries   Capillaries that surround nephrons. They pick up water, nutrients, and ions from the renal tubule, thus helping maintain the osmotic concentration of the blood.

Permanent threshold shift   Permanent hearing loss caused by repeated exposure to noise. Results from damage to hair cells of the organ of Corti.

Pharynx   Chamber that connects the oral cavity with the esophagus.

Phenotype   Outward appearance of an organism.

Phonation   Production of sound.

Photoreceptors   Modified nerve cells that respond to light. Located in the retina of humans and other animals.

Photosynthesis   Process in plants and algae in which sunlight is used to produce organic molecules from carbon dioxide from the atmosphere and water.

Phylogeny   Evolutionary development of a species.

Phylum   Largest groups in a kingdom.

Pineal gland   Small gland located in the brain that secretes a hormone thought to help control the biological clock.

Pituitary gland   Small pea-sized gland located beneath the brain in the sella turcica. It produces numerous hormones and consists of two main subdivisions: anterior and posterior pituitary.

Placenta   Organ produced from maternal and embryonic tissue. Supplies nutrients to the growing embryo and fetus and removes fetal wastes.

Plasma   Extracellular fluid of blood. Comprises about 55% of the blood.

Plasma cell   Cell produced from B-lymphocytes (B cells); synthesizes and releases antibodies.

Plasma membrane   Outer layer of the cell. Consists of lipid and protein and controls the movement of materials into and out of the cell.

Plasmids   Small circular strands of DNA found in bacterial cytoplasm separate from the main DNA.

Plasmin   Enzyme in the blood that helps dissolve blood clots.

Plasminogen   Inactive form of plasmin.

Plasmodium   Single-celled parasite responsible for malaria.

Platelet   Cell fragment produced from megakaryocytes in the red bone marrow. Plays a key role in blood clotting.

Podocyte   Type of cell forming the inner lining of Bowman's capsule of the nephron. Part of the filtration mechanism in the glomerulus.

Polar body   Discarded nuclear material produced during meiosis I and meiosis II of oogenesis.

Pole-to-pole fibers   Type of microtubule found in the spindle. Extend from one centriole to the other.

Polygenic inheritance   Transmission of traits that are controlled by more than one gene.

Polyploidy   Term referring to a genetic disorder caused by an abnormal number of chromosomes. Includes tetraploidy and triploidy.

Polyribosome   Also known as polysome. Organelle formed by several ribosomes attached to a single messenger RNA. Synthesizes proteins used inside the cell.

Population   Group of like organisms occupying a specific region.

Portal system   Arrangement of blood vessels in which a capillary bed drains to a vein, which drains to another capillary bed.

Posterior chamber   Posterior portion of the anterior cavity of the eye.

Posterior pituitary   Neuroendocrine gland that consists of neural tissue and releases two hormones, oxytocin and antidiuretic hormone.

Precapillary sphincters   Tiny rings of smooth muscle that surround the capillaries arising from the metarterioles.

Premature birth Birth of a baby before 37 weeks of gestation.

Premenstrual syndrome (PMS) Condition that occurs in some women in the days before menstruation normally begins. Characterized by a variety of symptoms such as irritability, depression, fatigue, headaches, bloating, swelling, and tenderness of breasts, joint pain, and tension.

Premotor area Region of the brain in front of the primary motor area. Controls muscle contraction and other less voluntary actions (playing a musical instrument).

Presbyopia Visual impairment caused by aging. Lens becomes stiffer, making it more difficult to focus on nearby objects.

Primary center of ossification Region in the interior of a cartilage mass that first becomes bone.

Primary follicle Structure in the ovary consisting of a primary oocyte and a complete single layer of cuboidal follicle cells.

Primary motor area Ridge of tissue in front of a central groove (the central sulcus). Controls voluntary motor activity.

Primary oocyte Germ cell produced from oogonium in the ovary. Undergoes the first meiotic division.

Primary response Immune response elicited when an antigen first enters the body.

Primary sensory area Region of the brain located just behind the central sulcus. The point of destination for many sensory impulses traveling from the body into the spinal cord and up to the brain.

Primary spermatocyte Cell produced from spermatogonium in the seminiferous tubule. Will undergo first meiotic division.

Primary succession Process of sequential change in which one community is replaced by another. Occurs where no biotic community has existed before.

Primary tissue One of major tissue types, including epithelial, connective, muscle, and nervous tissue.

Primary tumor Cancerous growth that gives rise to cells that spread to other regions of the body.

Primates An order of the kingdom Animalia. Includes prosimians (premonkeys), monkeys, apes, and humans.

Primordial follicle Structure in the ovary that consists of a primary oocyte surrounded by a layer of flattened follicle cells. Gives rise to the primary follicle.

Primordial germ cells Cells that originate in the wall of the yolk sac and eventually become either spermatogonia or oogonia.

Principle of independent assortment Mendel's second law. Hereditary factors are segregated independently during gamete formation. Occurs only when genes are on different chromosomes.

Principle of segregation Mendel's first law, which states that hereditary factors separate during gamete formation.

Producers Generally refers to organisms that can synthesize their own foodstuffs. Major producers are the algae and plants that absorb sunlight and use its energy to synthesize organic foodstuffs from water and carbon dioxide.

Prolactin Protein hormone under control of the hypothalamus. In humans, it is responsible for milk production by the glandular units of the breast.

Promoter Region of the operon between the regulator gene and operator site. Binds to RNA polymerase.

Pronuclei Name of the ovum and sperm cell nuclei shortly after fertilization occurs. Each contains 23 chromosomes.

Prophase First phase of mitosis during which chromosomes condense, the nuclear membrane disappears, and the spindle forms.

Proprioception Sense of body and limb position.

Prosimians Premonkeys; tarsiers and lemurs.

Prostaglandins Group of chemical substances that have a variety of functions. Act on nearby cells.

Prostate gland Sex accessory gland that is located near the neck of the bladder and empties into the urethra. Produces fluid that is added to the sperm during ejaculation.

Proteinoids Spherical structures composed of small amino acid chains formed when amino acids are heated in air. May have been an early precursor of the first cells.

Proton Subatomic particle found in the nucleus of the atom. Each proton carries a positive charge.

Proto-oncogenes Genes in cells that, when mutated, lead to cancerous growth.

Puberty Period of sexual maturation in humans.

Pulmonary circuit (or circulation) Short circulatory loop that supplies blood to the lungs and transports it back to the heart.

Pulmonary veins Veins that carry oxygenated blood from the lungs to the left atrium.

Punctuated equilibrium Hypothesis explaining how evolutionary change occurs. States that long periods of relatively little change are broken up by briefer periods of relatively rapid evolution.

Pupil Opening in the iris that allows light to penetrate deeper into the eye.

Purine Type of nitrogenous base found in DNA nucleotides. Consists of two fused rings.

Purkinje fiber Modified cardiac muscle fiber that conducts bioelectric impulses to individual heart muscle cells.

Pus Liquid emanating from a wound. Contains plasma, many dead neutrophils, dead cells, and bacteria.

Pyloric sphincter Ring of smooth muscle cells in the lower portion of the stomach where it joins the duodenum. Serves as a gate valve. Opens periodically after a meal, releasing spurts of chyme (liquified, partially digested food) into the small intestine.

Pyramid of numbers Diagram of the number of organisms at various trophic levels in a food chain or ecosystem.

Pyrimidine One of two types of nitrogen base found in DNA nucleotides. Consists of one ring.

Pyrogen Chemical released primarily from macrophages that have been exposed to bacteria and other foreign substances. Responsible for fever.

Radial keratotomy Procedure to correct nearsightedness. Numerous, small superficial incisions are made in the cornea, flattening it and reducing its refractive power.

Radioactivity Tiny bursts of energy or particles emitted from the nucleus of some unstable atoms. Results from excess neutrons in the nuclei of some atoms.

Radionuclide Radioactive isotope of an atom.

Range of tolerance Range of conditions in which an organism is adapted.

Receptor Any structure that responds to internal or external changes. Three types of receptors are found in the body: encapsulated, nonencapsulated (naked nerve endings), and specialized (e.g., the retina and semicircular canals).

Recessive Term describing an allele of a gene that is expressed when the dominant factor is missing.

Recombinant DNA technology Procedure in which scientists take segments of DNA from an organism and combine them with DNA from other organisms.

Recombination Process of crossing over during meiosis, resulting in new genetic combinations.

Red blood cells (RBCs) Enucleated cells in blood that transport oxygen in the bloodstream.

Red bone marrow Tissue found in the marrow cavity of bones. Site of blood cell and platelet production.

Reduction factor Any of the factors that cause populations to decline.

Reflex Automatic response to a stimulus. Mediated by the nervous system.

Refraction Bending of light.

Regulator gene Gene that codes for the synthesis of repressor protein in an operon.

Relaxin Hormone produced by the corpus luteum and the placenta. It is released near the end of pregnancy and softens the cervix and the fibrocartilage uniting the pubic bones, thus facilitating birth.

Releasing hormone Any of a group of hormones that stimulates the release of other hormones by the anterior pituitary.

Renal pelvis Hollow chamber inside the kidney. Receives urine from the collecting tubules and empties into the ureter.

Renal tubule That portion of the nephron where urine is produced.

Renewable resources Resources that replenish themselves via natural biological and geological processes, such as wind, hydropower, trees, fish, and wildlife.

Replacement-level fertility Number of children that will replace a couple when they die.

Repressible operon Operon whose genes remain active unless turned off. Found in bacteria and may be present in eukaryotes as well.

Repressor protein Protein produced by a regulator gene. Binds to a region of the DNA molecule (the operator site) adjacent to the structural genes. Blocks RNA polymerase from transcribing structural genes.

Reproductive isolation Condition in which two groups of similar organisms derived from the same parent stock lose the ability to interbreed. Often due to geographic isolation.

Respiratory distress syndrome Disease of premature babies that results from an insufficient amount of surfactant in the infant's lungs, causing alveoli to collapse. Also known as hyaline membrane disease.

Resting potential Minute voltage differential across the membrane of neurons. Also known as the membrane potential.

Restriction endonuclease Enzyme used in recombinant DNA technology. Cuts off segments of the DNA molecule for cloning and splicing.

Reticular activating system (RAS) Region of the medulla that receives nerve impulses from neurons transmitting information to and from the brain. Impulses are transmitted to the cortex, alerting it.

Retina Innermost, light-sensitive layer of the eye. Consists of an outer pigmented layer and an inner layer of nerve cells and photoreceptors (rods and cones).

Reverse transcriptase Enzyme that allows the production of DNA from strands of viral RNA.

Rheumatoid arthritis Type of arthritis in which the synovial membrane of the joint becomes inflamed and thickens. Results in pain and stiffness in joints. Thought to be an autoimmune disease.

Rhodopsin (visual purple) Pigment contained in the rods.

Rhythm method Birth control method in which a couple abstains from sexual intercourse around the time of ovulation. Also known as the natural method.

Ribosomal RNA (rRNA) RNA produced at the nucleolus. Combines with protein to form the ribosome.

Ribosome Cellular organelle consisting of two subunits, each made of protein and ribosomal RNA. Plays an important part in protein synthesis.

RNA polymerase   Enzyme that helps align and join the nucleotides in a replicating RNA molecule.

Rod   Type of photoreceptor in the eye. Provides for vision in dim light.

Root nodule   Swelling in the roots of certain plants (legumes) containing nitrogen-fixing bacteria.

Rough endoplasmic reticulum (RER)   Ribosome-coated endoplasmic reticulum. Produces lysosomal enzymes and proteins for use outside the cell.

Saccule   Membranous sac located inside the vestibule. Contains a receptor for movement and body position.

Salivary gland   Any of several exocrine glands situated around the oral cavity. Produces saliva.

Saltatory conduction   Conduction of a bioelectric impulse down a myelinated neuron from node to node.

Sarcomere   Functional unit of the muscle cell. Consists of the myofilaments, actin, and myosin.

Sarcoplasmic reticulum   Term given to the smooth endoplasmic reticulum of a skeletal muscle fiber. Stores and releases calcium ions essential for muscle contractions.

Schwann cell   Type of neuroglial cell or supportive cell in the nervous system. Responsible for the formation of the myelin sheath.

Science   Body of knowledge on the workings of the world and a method of accumulating knowledge. *See also* Scientific method.

Scientific method   Deliberate, systematic process of discovery. Begins with observation and measurement. From observations, hypotheses are generated and tested. This leads to more observation and measurement that supports or refutes the original hypothesis.

Sclera   Outermost layer of the eye.

Scrotum   Skin-covered sac containing the testes.

Sebum   Oil excreted by sebaceous glands onto the surface of the skin.

Second messenger mechanism Describes how protein hormones and others effect intracellular change by binding to a receptor, activating adenylate cyclase, which leads to the production of cyclic AMP. Cyclic AMP, the second messenger, then activates a cytoplasmic enzyme, protein kinase, which activates or inactivates other enzymes.

Secondary center of ossification Region of bone formation that occurs in the ends (epiphyses) of bones.

Secondary response   Generally, a powerful, swift immune system response occurring the second time an antigen enters the body. Much faster than the primary response.

Secondary sex characteristics Distinguishing features of men and women resulting from the sex steroids. In men, includes facial hair growth and deeper voices. In women, includes breast development and fatty deposits in the hips and other regions.

Secondary succession   Process of sequential change in which one community is replaced by another. It occurs where a biotic community previously existed, but was destroyed by natural forces or human actions.

Secondary tumor   Cancerous growth formed by cells arising from a primary tumor.

Secretin   Hormone produced by the cells of the duodenum. Stimulates the pancreas to release sodium bicarbonate.

Secretory vesicles   Membrane-bound vesicles containing protein (hormones or enzymes) produced by the endoplasmic reticulum and packaged by the Golgi complex of some cells. They fuse with the membrane, releasing their contents by exocytosis.

Selective permeability   Control of what moves across the plasma membrane of a cell.

Semen   Fluid containing sperm and secretions of the secondary sex glands.

Semicircular canal   Sensory organ of the inner ear. Houses the receptors that detect body position and movement.

Semilunar valve   Type of valve lying between the ventricles and the arteries that conduct blood away from the heart.

Seminal vesicles   Sex accessory glands that empty into the vas deferens. Produce the largest portion of ejaculate.

Seminiferous tubule   Sperm-producing tubule in the testis.

Sertoli cell   Cell in the germinal epithelium of the seminiferous tubule. Houses spermatogenic cells as they develop.

Sex accessory gland   One of several glands that produce secretions that are added to sperm during ejaculation.

Sex chromosomes   X and Y chromosomes that help determine the sex of an individual.

Sex-linked trait   Trait produced by a gene carried on a sex chromosome.

Sex steroid   Steroid hormones produced principally by the ovaries (in women) and testes (in men). Help regulate secretion of gonadotropins and determine secondary sex characteristics.

Sexually transmitted diseases (venereal diseases)   Infections that are transmitted by sexual contact.

Sickle-cell anemia   Genetic disease common in African Americans that results in abnormal hemoglobin in red blood cells, causing cells to become sickle shaped when exposed to low oxygen levels. Sickling causes cells to block capillaries, restricting blood flow to tissues.

Sinoatrial node   The heart's pacemaker. Located in the wall of the right atrium, it sends timed impulses to the heart muscle, thus synchronizing muscle contractions.

Skeletal muscle Muscle that is generally attached to the skeleton and causes body parts to move.

Skeleton Internal support of humans and other animals. Consists of bones joined together at joints.

Sliding filament mechanism Sliding of actin filaments toward the center of a sarcomere, causing muscle contraction.

Smooth endoplasmic reticulum (SER) Endoplasmic reticulum without ribosomes. Produces phosphoglycerides used to make the plasma membrane. Performs a variety of different functions in different cells.

Smooth muscle Involuntary muscle that lacks striations. Found around circulatory system vessels and in the walls of such organs as the stomach, uterus, and intestines.

Somite Block of mesoderm that gives rise to the vertebrae, muscles of the neck, and trunk.

Special sense Vision, hearing, taste, smell, and balance.

Speciation Formation of new species resulting from geographic isolation.

Species Group of organisms that is structurally and functionally similar. When members of the group breed, they produce viable, reproductively competent offspring. Also a subgroup of a genus.

Spermatogenesis Formation of sperm in the seminiferous tubules.

Spermatogonia Sperm-producing cells in the periphery of the germinal epithelium of the seminiferous tubules.

Spermatozoan Sperm.

Spinal nerve Nerve that arises from the spinal cord.

Spongy bone Type of bony tissue inside most bones. Consists of an irregular network of bone spicules.

Sterilization Procedure to render a man or woman sterile or infertile. In men, the method is generally a va-

sectomy; in women, it is usually tubal ligation.

Stirrup (stapes) One of three bones of the middle ear that conducts vibrations from the eardrum to the inner ear.

Structural gene Any gene of an operon that codes for the production of enzymes and other proteins.

Subatomic particles Electrons, pro- tons, and neutrons. Particles that can be separated from an atom by physical means.

Substrate Molecule that fits into the active site of an enzyme.

Succession Process of sequential change in which one community is replaced by another until a mature or climax ecosystem is formed. *See also* Primary succession and Secondary succession.

Sulcus Indented region or groove in the cerebral cortex between ridges.

Suppressor cell Cell of the immune system that shuts down the immune reaction as the antigen begins to disappear.

Suprachiasmatic nucleus Clump of nerve cells in the hypothalamus. Thought to play a major role in coordinating several key functions and several other control centers. Sometimes called the "master clock."

Surfactant Detergent-like substance produced by the lungs. Dissolves in the thin watery lining of the alveoli; helps reduce surface tension, keeping the alveoli from collapsing.

Suspensory ligament Zonular fibers that connect the lens to the ciliary body.

Sustainable-Earth ethic Ethic based on three tenets: (1) the world has a limited supply of resources that must be shared with all living things, (2) humans are a part of nature and subject to its rules, and (3) nature is not something to conquer, but rather a force we must learn to cooperate with.

Sustainable society Society that lives within the carrying capacity of the environment.

Sympathetic division Division of the autonomic nervous system that is responsible for many functions, especially those involved in the fight-or-flight response.

Synapse Juncture of two neurons.

Synaptic cleft Gap between an axon and the dendrite or effector (e.g., gland or muscle) it supplies.

Synergy Coordination of the workings of antagonistic muscle groups.

Synovial fluid Lubricating liquid inside joint cavities. Produced by the synovial membrane.

Synovial membrane Inner layer of the joint capsule.

Syphilis Potentially serious, sexually transmitted disease caused by a bacterium.

Systemic circulation System of blood vessels that transports blood to and from the body and heart, excluding the lungs.

Systolic pressure Peak pressure at the moment the ventricles contract. The higher of the two numbers in a blood pressure reading.

Taiga The northern coniferous forests biome.

Taste bud Receptor for taste principally found in the surface epithelium and certain papillae of the tongue.

T cell *See* T-lymphocyte.

Telophase Final stage of mitosis in which the nuclear envelope reforms from vesicles and the chromosomes uncoil.

Temperate deciduous forest Biome that in the United States lies east of the Mississippi River and is characterized by broad-leafed trees.

Temporary threshold shift Temporary loss of hearing after being exposed to a noisy environment.

Tendons Connective tissue structures that generally attach muscles to bones.

Teratogen Chemical, biological, or physical agent that causes birth defects.

Teratology Study of birth defects.

Terminal boutons Small swellings on the terminal fibers of axons. They lie close to the membranes of the dendrites of other axons or the membranes of the effectors, and transfer bioelectric impulses from one cell to another.

Terminator codon Codon found on each strand of messenger RNA that marks where protein synthesis should end.

Testes Male gonads. They produce sex steroids and sperm.

Testosterone Male sex hormone that stimulates sperm formation and is responsible for secondary sex characteristics, such as facial hair growth and muscle growth.

Tetraploidy Condition in which an individual is endowed with two complete sets of chromosomes. Instead of having 46 chromosomes, he or she has 92.

Theories Principles of science—the broader generalizations about the world and its components. Theories are supported by considerable scientific research.

Thoracic duct Duct carrying lymph to the circulatory system. Empties into the large veins at the base of the neck.

Thoroughfare channel Vessel that connects the arterioles with the venules, thus allowing blood to bypass a capillary bed. Also known as a metarteriole.

Thyroid gland U- or H-shaped gland located in the neck on either side of the trachea just below the larynx. Produces three hormones: thyroxin, triiodothyronine, and calcitonin.

Thyroid-stimulating hormone (TSH) Hormone produced by the pituitary gland. Stimulates production and release of thyroxine and triiodothyronine by the thyroid gland.

Thyroxin Hormone produced by the thyroid gland that accelerates the rate of mitochondrial glucose catabolism in most body cells and also stimulates cellular growth and development.

Tissue Component of the body from which organs are made. Consists of cells and extracellular material (fluid, fibers, and so on).

T-lymphocyte Type of lymphocyte responsible for cell-mediated immunity. Attacks foreign cells, virus-infected cells, and cancer cells directly. Also known as T cell.

Total fertility rate Number of children a woman is expected to have during her lifetime.

Trachea Duct that leads from the pharynx to the lungs.

Transcription RNA production on a DNA template.

Transfer RNA (tRNA) Small RNA molecules that bind to amino acids in the cytoplasm and deliver them to specific sites on the messenger RNA.

Transformation Conversion of a normal cell to a cancerous one.

Transition reaction Part of cellular respiration in which one carbon is cleaved from pyruvic acid, forming a two-carbon compound, which reacts with Coenzyme A. The resulting chemical enters the citric acid cycle.

Translation Synthesis of protein on a messenger RNA template.

Translocation Process in which a segment of a chromosome breaks off but reattaches to another site on the same chromosome or another one.

Triiodothyronine Hormone produced by the thyroid gland. Nearly identical in function to thyroxin.

Triploidy Genetic disorder in which cells have 69 chromosomes instead of 46.

Trisomy Genetic condition characterized by the presence of one extra chromosome.

Trophic hormones Hormones that stimulate the production and secretion of other hormones. Also known as tropic hormones.

Trophic level Feeding level in a food chain.

Trophoblast Outer ring of cells of the blastocyst that form the embryonic portion of the placenta.

TSH-releasing hormone (TSH-RH) Hormone secreted by the posterior lobe of the pituitary gland. Stimulates thyroxin secretion by the thyroid gland.

T tubules (transverse tubules) Invaginations of the plasma membrane of skeletal muscle fibers that conduct an impulse to the interior of the cell.

Tubal ligation Sterilization procedure in women. Uterine tubes are cut, preventing sperm and ova from uniting.

Tubular reabsorption Process in which nutrients are transported out of the nephron into the peritubular capillaries.

Tubular secretion Process in which wastes are transported from the peritubular capillaries into the nephron.

Tumor Mass of cells derived from a single cell that has begun to divide. In malignant tumors, the cells divide uncontrollably and often release clusters of cells or single cells that spread in the blood and lymphatic systems to other parts of the body. Benign tumors grow to a certain size, then stop.

Tundra Northernmost biome with long, cold winters and a short growing season.

Turner syndrome Genetic disorder in which an offspring contains 22 pairs of autosomes and a single, unmatched X chromosome. Phenotypically female.

Twitch Single muscle fiber contraction.

Tympanic membrane (eardrum) Membrane between the external auditory canal and middle ear that oscillates when struck by sound waves.

Type I diabetes Form of diabetes that occurs mainly in young people and results from an insufficient amount of insulin production and release. Brought on by damage to insulin-producing cells of the pancreas. Also known as early-onset diabetes.

Type II diabetes Form of diabetes that occurs chiefly in older individuals (around age of 40) and results from a loss of tissue responsiveness to insulin. Also known as late-onset diabetes.

Type II alveolar cell Cell found in the lining of the alveoli. Produces surfactant.

Umbilical artery One of two arteries in the umbilical cord that carries blood from the embryo to the placenta.

Umbilical vein Vein in the umbilical cord that carries blood from the placenta to the fetus.

Ureter Hollow, muscular tube that transports urine by peristaltic contractions from the kidney to the urinary bladder.

Urethra Narrow tube that transports urine from the urinary bladder to the outside of the body. In males, it also conducts sperm and semen to the outside.

Urinary bladder Hollow, distensible organ with muscular walls that stores urine. Drained by the urethra.

Urine Fluid containing various wastes that is produced in the kidney and excreted out of the urinary bladder.

Uterus Organ that houses and nourishes the developing embryo and fetus.

Utricle Membranous sac containing a receptor for body position and movement. Located inside the vestibule of the inner ear.

Vaccine Preparation containing dead or weakened bacteria and viruses that, when injected in the body, elicits an immune response. *See also* Active immunity.

Vagina Tubular organ that serves as a receptacle for sperm and provides a route for delivery of the baby at birth.

Vagus nerve Nerve that terminates in the stomach wall and stimulates HCl production by cells in the gastric glands.

Variation Genetically based differences in physical or functional characteristics within a population.

Varicose vein Vein whose wall balloons out because the flow of blood downstream is obstructed.

Vas deferens Duct that carries sperm from the testis to the urethra. Contracts during ejaculation.

Vasectomy Contraceptive procedure in men in which the vas deferens is cut and the free ends sealed to prevent sperm from entering the urethra during ejaculation.

Vasopressin Also known as antidiuretic hormone, which in high concentrations increases blood pressure.

Vena cava One of two large veins that empty into the right atrium of the heart.

Vein Type of blood vessel that carries blood to the heart.

Venule Smallest of all veins. Empties into capillary networks.

Villi Fingerlike projections of the lining of the small intestine that increase the surface area for absorption.

Virus Nonliving entity consisting of a nucleic acid—either DNA or RNA—core surrounded by a protein coat, the capsid. Viruses are cellular parasites, invading cells and taking over their metabolic machinery to reproduce.

Vitamin Any of a diverse group of organic compounds. Essential to many metabolic reactions.

Vitreous humor Gelatinous material found in the posterior cavity of the eye.

Vocal cords Elastic ligaments inside the larynx that vibrate as air is expelled from the lungs, generating sound.

White blood cells (WBCs) Cells of the blood formed in the bone marrow. Principally involved in fighting infection.

White matter The portion of the brain and spinal cord that appears white to the naked eye. Consists primarily of white, myelinated nerve fibers.

Yellow marrow Inactive marrow of bones in adults containing fat. Formed from red marrow.

Yolk sac Embryonic pouch formed from endoderm. Site of early formation of red blood cells and germ cells.

Zero population growth Condition in which a population stops growing.

Zona pellucida Band of material surrounding the oocyte.

Zonular fibers Thin fibers that attach the lens to the ciliary body.

Zygote Cell produced by a sperm and ovum during fertilization. Contains 46 chromosomes.

INDEX

CREDITS

PHOTOGRAPHS

PART I   p. 1, © Richard D. Poe/Visuals Unlimited

CHAPTER 1   Opener p. 2, © Polaroid-Desiderio/Visuals Unlimited; Fig. 1–2a, p. 4, © Frank Hanna/Visuals Unlimited; Fig. 1–2b, p. 4, © Peter K. Ziminski/Visuals Unlimited; Health Note 1–1, p. 9, © Will & Deni McIntyre/Science Source/Photo Researchers; Fig. 1–6a, p. 10, © Manfred Kage/Peter Arnold; Fig. 1–6b, p. 10, © M. Moore/Visuals Unlimited; Fig. 1–7a, p. 10, © Carolina Biological/Visuals Unlimited; Fig. 1–7b, p. 10, © Carolina Biological/Visuals Unlimited; Fig. 1–8, p. 11, © G. Kirtley-Perkins/Visuals Unlimited; Fig. 1–9, p. 12, © David Gucwa/Visuals Unlimited; Fig. 1–10a, p. 13, © Grant Heilman/Grant Heilman Photography; Fig. 1–10b, p. 13, © Grant Heilman/Grant Heilman Photography; Fig. 1–12, p. 15, © Jim Harvey/Visuals Unlimited; Fig. 1–13a, p. 17, © William Banaszewski/Visuals Unlimited; Fig. 1–13b, p. 17, © John D. Cunningham/Visuals Unlimited

CHAPTER 2   Opener p. 21, © George Musil/Visuals Unlimited; Fig. 2–1, p. 22, © SIU/Visuals Unlimited; Fig. 2–2, p. 22, © Richard Treptow/Visuals Unlimited; Fig. 2–10, p. 30, © AP/Wide World Photos; Fig. 2–11, p. 31, © Steve McCutcheon/Visuals Unlimited; Fig. 2–13, p. 32, © Walt Anderson/Visuals Unlimited; Fig. 2–15, p. 34, © Bill Beatty/Visuals Unlimited; Fig. 2–19, p. 37, © John D. Cunningham/Visuals Unlimited; Fig. 2–21a, p. 39, © Carolina Biological/Visuals Unlimited; Fig. 2–21b, p. 39, © Veronika Burmeister/Visuals Unlimited; Fig. 2–22, p. 39, © Nicholas Desciose/Photo Researchers; Health Note 2–1, p. 40, © SIU/Visuals Unlimited; Fig. 2–23, p. 41, © John D. Cunningham/Visuals Unlimited; Fig. 2–24a, p. 41, © Carolina Biological/Visuals Unlimited; Fig. 2–24b, p. 41, © William Ober/Visuals Unlimited; Fig. 2–30a, p. 45, © Stanley Flegler/Visuals Unlimited; Fig. 2–30b, p. 45, © Stanley Flegler/Visuals Unlimited

PART II   p. 53, © Carolina Biological/Visuals Unlimited

CHAPTER 3   Opener p. 54, © Michael Gabridge/Visuals Unlimited; Fig. 3–2a, p. 56, © Ralph A. Slepecky/Visuals Unlimited; Fig. 3–2b, p. 56, © K. G. Murti/Visuals Unlimited; Fig. 3–4a, p. 59, © M. Schliwa/Visuals Unlimited; Health Note 3–1, p. 64, © University of Colorado Health Sciences Center; Fig. 3–11a, p. 69, © K. G. Murti/Visuals Unlimited; Fig. 3–11b, p. 69, © Lester Bergman & Associates, Inc.; Fig. 3–12, p. 69, © K. G. Murti/Visuals Unlimited; Fig. 3–13a, p. 72, © G. Musil/Visuals Unlimited; Fig. 3–14, p. 73, © William Banaszewski/Visuals Unlimited; Fig. 3–15, p. 73, © Jon Turk/Visuals Unlimited; Fig. 3–16b, p. 74, © K. G. Murti/Visuals Unlimited; Fig. 3–16c, p. 74, © David M. Phillips/Visuals Unlimited; Fig. 3–18b, p. 75, © David M. Phillips/Visuals Unlimited; Fig. 3–19a, p. 76, © David M. Phillips/Visuals Unlimited; Fig. 3–21a, p. 77, © David M. Phillips/Visuals Unlimited; Fig. 3–21b, p. 77, © David M. Phillips/Visuals Unlimited; Fig. 3–22a, p. 78, © Ellen R. Dirksen/Visuals Unlimited; Fig. 3–22b, p. 78, © David M. Phillips/Visuals Unlimited; Fig. 3–22c, p. 78, © David M. Phillips/Visuals Unlimited; Fig. 3–22d, p. 78, © David M. Phillips/Visuals Unlimited; Fig. 3–27, p. 83, © M. Long/Visuals Unlimited

CHAPTER 4   Opener p. 88, © John D. Cunningham/Visuals Unlimited; Fig. 4–3b, p. 92, Science VU/Visuals Unlimited; Fig. 4–4b, p. 93, © M. Coleman/Visuals Unlimited; Fig. 4–5a, p. 94, © SIU/Visuals Unlimited; Fig. 4–6a, p. 95, © Carolina Biological/Visuals Unlimited; Fig. 4–6b, p. 95, © David M. Phillips/Visuals Unlimited; Fig. 4–7a, p. 96, © Michael Abbey/Visuals Unlimited; Fig. 4–7b (left), p. 96, © John D. Cunningham/Visuals Unlimited; Fig. 4–7b (right), p. 96, © John D. Cunningham/Visuals Unlimited; Fig. 4–7c, p. 97, © John D. Cunningham/Visuals Unlimited; Fig. 4–7d, p. 97, © John D. Cunningham/Visuals Unlimited; Fig. 4–7e, p. 97, © John D. Cunningham/Visuals Unlimited; Fig. 4–10b, p. 99, © David M. Phillips/Visuals Unlimited; Fig. 4–11, p. 99, © Cytographics, Inc./Visuals Unlimited; Health Note 4–1, p. 102, © Carolina Biological/Visuals Unlimited

CHAPTER 5   Opener p. 107, © NIH/Visuals Unlimited; Fig. 5–12a, p. 119, © Ralph Slepecky/Visuals Unlimited; Fig. 5–12b, p. 119, © K. G. Murti/Visuals Unlimited; Fig. 5–16, p. 128, © Dana Richter/Visuals Unlimited

CHAPTER 20   Opener p. 556, © Ron Spomer/Visuals Unlimited; Fig. 20–2, p. 558, Science VU-NASA/Visuals Unlimited; Fig. 20–4a, p. 560, © Steve McCutcheon/Visuals Unlimited; Fig. 20–4b, p. 560, © Albert J. Copley/Visuals Unlimited; Fig. 20–4c, p. 560, © William J. Weber/Visuals Unlimited; Fig. 20–4d, p. 560, © Ron Spomer/Visuals Unlimited; Fig. 20–4e, p. 560, © John D. Cunningham/Visuals Unlimited; Fig. 20–6, p. 562, © Richard Thom/Visuals Unlimited; Fig. 20–19a, p. 575, © Richard Ashley/Visuals Unlimited; Fig. 20–19b, p. 575, © G. Prance/Visuals Unlimited

CHAPTER 21   Opener p. 579, © Daniel D. Chiras; Fig. 21–1, p. 580, © AP/Wide World Photos; Fig. 21–2, p. 581, © Jerome Wyckoff/Visuals Unlimited; Fig. 21–4, p. 585, © Sylvan Wittwer/Visuals Unlimited; Fig. 21–6, p. 586, © Frank M. Hanna/Visuals Unlimited; Fig. 21–12a, p. 590, © Daniel D. Chiras; Fig. 21–12b, p. 590, © William Grenfell/Visuals Unlimited; Fig. 21–13, p. 591, © William Banaszewski/Visuals Unlimited; Fig. 21–15, p. 591, © Science VU-SCS/Visuals Unlimited; Fig. 21–16, p. 592, © Dennis Paulson/Visuals Unlimited; Fig. 21–19, p. 594, Oak Ridge National Laboratories; Fig. 21–20, p. 595, Chevrolet Division, General Motors; Fig. 21–21, p. 595, © Science VU-API/Visuals Unlimited; Fig. 21–22, p. 595, General Electric; Fig. 21–26a, p. 604, © Joe McDonald/Visuals Unlimited; Fig. 21–26b, p. 604, © Tim Perkins/Visuals Unlimited; Fig. 21–28, p. 606, © Dana Richter/Visuals Unlimited

ARTWORK

Wayne Clark   Figure 8–20, p. 211; Figure 9–7a, p. 229; Figure 11–3, p. 284; Figure 12–5a,b,d, p. 310; Figure 16–1, p. 433; Figure 16–7, p. 440; Figure 16–10, p. 443; Figure 16–12a, p. 445; Figure 16–18, p. 452; Figure 17–6 (layout), p. 474

Cyndie C. H.-Wooley   Figure 1–5, p. 7; Figure 2–34, p. 49; Figure 4–5b, p. 94; Figure 6–13, p. 147; Figure 6–14, pp. 148–49; Figure 7–5, p. 174; Figure 7–6, p. 175; Figure 8–1, p. 188; Figure 8–12a, p. 202; Figure 8–25, p. 215; Figure 9–1, p. 223; Figure 9–3, p. 225; Health Note figure, p. 226; Figure 11–2, p. 283; Figure 13–2, p. 342; Figure 13–6, p. 345; Figure 13–26, p. 364; Figure 15–1, p. 402; Figure 15–18, p. 414; Figure 16–2, p. 434 (layout); Figure 16–9 bottom, p. 442 (layout); Figure 17–1, p. 470 (layout); Figure 17–2, p. 470 (layout); Figure 17–13, p. 481; Figure 17–15, p. 484; Figure 17–16, p. 486; Figure 17–17, p. 487; Figure 18–23, p. 523

Darwen and Vally Hennings   Figure 1–1, p. 3; Figure 1–3, p. 5; Figure 6–2, p. 136; Figure 6–3, p. 137; Figure 6–5a art, b, p. 139; Figure 7–7a, p. 176; Figure 7–8a, p. 177; Figure 8–2, p. 189; Figure 8–4a,b, p. 191; Figure 8–5, p. 192; Figure 8–7, p. 198; Figure 9–9a, p. 230; Figure 9–15a, p. 235; Figure 10–12, p. 259; Figure 12–11, p. 318; Figure 12–12, p. 319; Figure 12–13, p. 320; Figure 12–14, p. 321; Table 12–1, pp. 322–23; Figure 12–15, p. 324; Figure 12–16, p. 325; Figure 12–17, p. 327; Figure 12–18, p. 328; Figure 12–19, p. 329; Figure 13–1, p. 339; Figure 13–3, p. 343; Figure 13–4, p. 344; Figure 13–5, p. 344; Figure 13–8, pp. 348–49; Figure 13–10, p. 350; Figure 13–13, p. 352; Figure 13–14, p. 352; Figure 13–19, p. 356; Figure 13–21b, p. 359; Figure 13–22b,c, p. 360; Figure 13–25, p. 363; Figure 14–1, p. 373; Figure 14–2b,c, p. 374; Figure 14–3, p. 375; Figure 14–4, p. 375; Figure 14–5, p. 376; Figure 14–9a, p. 380; Figure 14–10, p. 381; Figure 14–15, p. 385; Figure 14–17, p. 387; Figure 15–8a,b, p. 407; Figure 15–26, p. 421; Figure 19–17, p. 545; Figure 19–18, p. 546; Figure 20–1, p. 557; Figure 20–7, p. 563; Figure 20–9, p. 565; Figure 20–10, p. 566; Figure 20–11, p. 567; Figure 20–13, p. 569; Figure 20–14, p. 570; Figure 20–16, p. 573; Figure 20–17, p. 574

Georg Klatt (layouts only)   Figure 1–11, p. 14; Figure 2–3, p. 23; Figure 2–4, p. 24; Figure 2–5, p. 25; Figure 2–6, p. 25; Figure 2–7, p. 28; Figure 2–8, p. 26; Figure 2–9, p. 29; Figure 2–12, p. 31; Figure 2–14, p. 33; Figure 2–16, p. 34; Figure 2–17, p. 35; Figure 2–18, p. 36; Figure 2–20, p. 38; Figure 2–25, p. 42; Figure 2–26, p. 42; Figure 2–27, p. 43; Figure 2–28, p. 43; Figure 2–29, p. 44; Figure 2–31, p. 46; Figure 2–32, p. 47; Figure 2–33, p. 48; Figure 3–6, p. 61; Figure 3–7, p. 61; Figure 3–26, p. 83; Figure 4–1, p. 89; Figure 4–2, p. 91; Figure 4–3a, p. 92; Figure 5–1, p. 108; Figure 5–2, p. 109; Figure 5–3, p. 110; Figure 5–4, p. 111; Figure 5–5, p. 112; Figure 5–6, p. 113; Figure 5–8, p. 115; Figure 5–9, p. 116; Figure 5–10, p. 117; Figure 5–11, p. 118; Figure 5–13, p. 120; Health Note figure, p. 123; Figure 8–19, p. 210; Figure 10–1, p. 247; Figure 10–2, pp. 248–49; Figure 10–3b, p. 250; Figure 10–5, p. 251; Figure 10–7 art, p. 253; Figure 10–15, p. 262; Figure 15–3, p. 403; Figure 15–6, p. 405; Figure 15–21, p. 415; Figure 19–8, p. 536; Figure 20–18, p. 575; Figure 21–8, p. 588; Figure 21–9, p. 588; Figure 21–10, p. 589

Sandra McMahon   Figure 13–7, p. 346; Figure 14–18, p. 388; Figure 17–19, p. 489

Elizabeth Morales-Denney   Figure 1–11, p. 14 (final rendering); Figure 2–3, p. 23 (final rendering); Figure 2–4, p. 24 (final rendering); Figure 2–5, p. 25 (final rendering); Figure 2–6, p. 25 (final rendering); Figure 2–7, p. 28 (final rendering); Figure 2–8, p. 29 (final rendering); Figure 2–9, p. 29 (final rendering); Figure 2–12, p. 31 (final rendering); Figure 2–14, p. 33 (final inking/pasteup); Figure 2–16, p. 34 (final rendering); Figure 2–17, p. 35 (final rendering); Figure 2–18, p. 36 (final rendering); Figure 2–20, p. 38 (final rendering); Figure 2–25, p. 42 (final rendering); Figure 2–26, p. 42 (final rendering); Figure 2–27, p. 43 (final rendering); Figure 2–28, p. 43 (final rendering); Figure 2–29, p. 44 (final rendering); Figure 2–31, p. 46 (final rendering); Figure 2–32, p. 47 (final rendering); Figure 2–33, p. 48 (final rendering); Figure

3–3, p. 57; Figure 3–4b, p. 59; Figure 3–6, p. 61 (final rendering); Figure 3–7, p. 61 (final rendering); Figure 3–8, p. 63; Figure 3–13b, p. 72; Figure 3–16a, p. 74; Figure 3–17, p. 75; Figure 3–18a, p. 75; Figure 3–19b, p. 76; Figure 3–26, p. 83 (final rendering); Figure 4–1, p. 89 (final rendering); Figure 4–2, p. 91 (final rendering); Figure 4–3a, p. 92 (final rendering); Figure 4–8, p. 98; Figure 4–9, p. 98; Figure 5–1, p. 108 (final rendering); Figure 5–2, p. 109 (final rendering); Figure 5–3, p. 110 (final rendering); Figure 5–4, p. 111 (final rendering); Figure 5–5, p. 112 (final rendering); Figure 5–6, p. 113 (final rendering); Figure 5–8, p. 115 (final rendering); Figure 5–9, p. 116 (final rendering); Figure 5–10, p. 117 (final rendering); Figure 5–11, p. 118 (final rendering); Figure 5–13, p. 120 (final rendering); Health Note figure, p. 123 (final rendering); Figure 8–19, p. 210 (final rendering); Figure 10–1, p. 247 (final rendering); Figure 10–2, pp. 248–49 (final rendering); Figure 10–3b, p. 250 (final rendering); Figure 10–5, p. 251 (final rendering); Figure 10–7 art, p. 253 (final rendering); Figure 10–15, p. 262 (final rendering); Figure 15–3, p. 403 (final rendering); Figure 15–6, p. 405 (final rendering); Figure 15–21, p. 415 (final rendering); Figure 19–8, p. 536 (final rendering); Figure 20–18, p. 575 (final rendering); Figure 21–8, p. 588 (final rendering); Figure 21–9, p. 588 (final rendering); Figure 21–10, p. 589 (final rendering)

Rolin Graphics Figure 10–14, p. 261; Figure 13–16, p. 355; Figure 14–21, p. 392; Figure 14–23, p. 393; Figure 15–4, p. 404; Figure 16–11, p. 444; Figure 16–22b, p. 456; Figure 17–8, p. 476

John and Judy Waller Figure 1–4, p. 6; Figure 2–14, p. 33 (airbrushing); Figure 3–1, p. 55; Figure 3–5, p. 60; Figure 3–9, p. 66; Figure 3–10, p. 68; Figure 3–20, p. 77; Figure 3–23, p. 79; Figure 3–24, p. 81; Figure 3–25, p. 82; Figure 4–4a, p. 93; Figure 4–7, pp. 96–97; Figure 4–10a, p. 99; Figure 4–12, p. 101; Figure 5–7, p. 113; Figure 5–14, p. 121; Figure 5–15, p. 125; Figure 6–1, p. 135; Figure 6–4, p. 138; Figure 6–8d, p. 141; Figure 6–9c, p. 142; Figure 6–10, p. 143; Health Note figure, p. 145; Figure 6–15, p. 150; Figure 6–16, p. 151; Figure 6–17, p. 153; Figure 6–18, p. 154; Figure 6–19, p. 154; Figure 6–20, p. 155; Figure 6–21, p. 156; Figure 6–22, p. 156; Figure 7–1, p. 164; Figure 7–3, p. 166; Figure 7–8b, p. 177; Figure 7–9, p. 178; Figure 7–10, p. 178; Figure 8–3, p. 190; Figure 8–6b, p. 194; Figure 8–8b, p. 199; Figure 8–10, p. 200; Figure 8–12b, p. 202; Figure 8–14, p. 204; Figure 8–15, p. 205; Figure 8–16a, p. 206; Figure 8–17, p. 207;

Figure 8–23, p. 213; Figure 8–26, p. 216; Figure 9–5, p. 227; Figure 9–10, p. 231; Figure 9–11, p. 231; Figure 9–12, p. 232; Figure 9–13, p. 232; Figure 9–16, p. 238; Figure 9–17, p. 239; Figure 9–18, p. 242; Figure 10–8, p. 254; Figure 10–9, p. 256; Figure 10–10, p. 257; Figure 10–13, p. 260; Table 10–2, p. 263; Figure 10–16, p. 264; Figure 10–17, p. 265; Figure 10–18 left, p. 267; Figure 10–19, p. 273; Figure 10–21, p. 274; Figure 11–4b, p. 285; Figure 11–5b, p. 286; Figure 11–6, p. 287; Figure 11–8a, p. 289; Figure 11–9, p. 290; Figure 11–10, p. 292; Figure 11–12, p. 296; Figure 11–13, p. 297; Figure 11–14, p. 300; Figure 12–3, p. 308; Figure 12–4b, p. 309; Figure 12–6, p. 311; Figure 12–7, p. 312; Figure 12–8, p. 313; Figure 12–9, p. 314; Figure 12–10b, p. 316; Figure 13–12, p. 351; Figure 13–15, p. 353; Figure 13–20, p. 357, Figure 13–21a, p. 359; Figure 13–22a, p. 360; Table 13–5, p. 365; Figure 14–11, p. 381; Figure 14–13, p. 382; Figure 14–19, p. 389; Figure 14–20, p. 390; Figure 14–22, p. 392; Figure 15–2, p. 403; Figure 15–5, p. 404; Figure 15–7, p. 406; Figure 15–8c, p. 407; Figure 15–11, p. 409; Figure 15–12, p. 410; Figure 15–13, p. 411; Figure 15–15, p. 412; Figure 15–16, p. 412; Figure 15–17, p. 413; Figure 15–22, p. 416; Figure 15–24, p. 417; Figure 15–25, p. 418; Figure 16–2, p. 434 (final rendering); Figure 16–3b, p. 435; Figure 16–4, pp. 436–37; Figure 16–5, p. 438; Figure 16–6, p. 439; Figure 16–8, p. 441; Figure 16–9 top, p. 442; Figure 16–9 bottom, p. 442 (final rendering); Figure 16–13, p. 446; Figure 16–14, p. 449; Figure 16–16, p. 451; Figure 16–17, p. 452; Figure 16–26, p. 458; Figure 16–27, p. 458; Figure 17–1, p. 470 (final rendering); Figure 17–2, p. 470 (final rendering); Figure 17–3, p. 471; Figure 17–4, p. 472; Figure 17–5, p. 473; Figure 17–6, p. 474 (final rendering); Figure 17–7, p. 475; Figure 17–9, p. 477; Figure 17–10, p. 478; Figure 17–11, p. 479; Figure 17–12, p. 480; Figure 17–20, p. 490; Figure 17–21, p. 491; Figure 18–2, p. 504; Figure 18–3, p. 505; Figure 18–4, p. 506; Figure 18–11, p. 511; Figure 18–12, p. 512; Figure 18–14, p. 514; Figure 18–15, p. 515; Figure 18–16, p. 516; Figure 18–17, p. 517; Figure 18–18, p. 518; Figure 18–20, p. 520; Figure 19–2, p. 531; Figure 19–3, p. 532; Figure 19–4, p. 532; Figure 19–14, p. 543; Figure 19–19, p. 547; Figure 20–3, p. 559; Figure 20–5, p. 561; Figure 20–8, p. 564; Figure 20–12, p. 568; Figure 20–15, p. 571; Figure 21–3, p. 584; Figure 21–5, p. 585; Figure 21–7, p. 587; Figure 21–11, p. 590; Figure 21–14, p. 591; Figure 21–17, p. 593; Figure 21–18, p. 594; Figure 21–23, p. 596; Figure 21–24, p. 602; Figure 21–25, p. 603; Figure 21–27, p. 605

BIOLOGICAL ROOTS

One of the major challenges of a biology course is mastering new terminology. As noted in the Study Skills section of this book, one of the best ways of learning and remembering technical terms is to first learn their component parts—that is, their roots. But studying common Latin and Greek roots for biological terms *before* you begin your study can also help. Spend a few minutes early in your course studying these common roots. They will not only help you understand new terms, but also make them easier to remember.

a-, an- [Gk. *an-*, not, without, lacking]: anaerobic, abiotic, anemia

ad- [L. *ad-*, toward, to]: adrenalin

amphi- [Gk. *amphi-*, two, both, both sides of]: amphibian

ana- [Gk. *ana-*, up, up against]: anaphase, anabolic, anatomy

andro- [Gk. *andros*, an old man]: androgen

anti- [Gk. *anti-*, against, opposite, opposed to]: antibiotic, antibody, antigen, antidiuretic hormone

arthro- [Gk. *arthron*, a joint]: arthropod, arthritis

auto- [Gk. *auto-*, self, same]: autoimmune, autotroph

bi-, bin- [L. *bis*, twice; *bini*, two-by-two]: binary fission, binocular vision, bicarbonate

bio- [Gk. *bios*, life]: biology, biomass, biome, biosphere, biotic

blasto-, -blast [Gk. *blastos*, sprout; pertains to embryo]: blastula, trophoblast, osteoblast

broncho- [Gk. *bronchos*, windpipe]: bronchus, bronchi, bronchiole, bronchitis

carb- [L. *carbo*, coal]: carbon, carbohydrate

carcino- [Gk. *karkin*, a crab, cancer]: carcinogen

cardio- [Gk. *kardia*, heart]: cardiac, myocardium, electrocardiogram

cat- [Gk. *kata*, down, downward]: catabolic

chloro- [Gk. *chloros*, green]: chlorophyll, chloroplast, chlorine

chromo- [Gk. *chroma*, color]: chromosome, chromatin

coelo-, -coel [Gk. *koilos*, hollow, cavity]: coelom

com-, con-, col-, cor-, co- [L. *cum*, with, together]: coenzyme, covalent

cranio- [Gk. *kranios*, L. *cranium*, skull]: cranial, cranium

cuti- [L. *cutis*, skin]: cutaneous, cuticle

cyto-, -cyte [Gk. *kytos*, vessel or container; now, "cell"]: cytoplasm, cytokinesis, erythrocyte, leucocyte

de- [L. *de-*, away, off; removal, separation]: deciduous, decomposer, dehydration

derm-, dermato- [Gk. *derma*, skin]: dermis, epidermis, ectoderm, endoderm, mesoderm

di- [Gk. *dis*, twice, two, double]: disaccharide, dioxide

dia- [Gk. through, passing through, thorough, thoroughly]: diabetes, dialysis, diaphragm

diplo- [Gk. *diploos*, two-fold]: diploid

eco- [Gk. *oikos*, house, home]: ecology, ecosystem, economy

ecto- [Gk. *ektos*, outside]: ectoderm

endo- [Gk. *endon*, within]: endoderm, endometrium

epi- [Gk. *epi*, on, upon, over]: epidermis, epididymis, epiglottis, epithelium

equi- [L. *aequus*, equal]: equilibrium

eu- [Gk. *eus*, good; *eu*, well, true]: eukaryote

ex-, exo-, ec, e- [Gk., L. out, out of, from, beyond]: emission, ejaculation, excretion, exergonic, exhale, exocytosis, exoskeleton

extra- [L. outside of, beyond]: extracellular, extraembryonic

-fer [L. *ferre*, to bear]: fertile, fertilizaton, conifer

gam-, gameto- [Gk. gamos, marriage; now usually in reference to gametes (sex cells)]: gamete

gastro- [Gk. *gaster*, stomach]: gastric, gastrin, gastrovascular cavity

gen- [Gk. *gen*, born, produced by; Gk. *genos*, race, kind; L. *genus, generare*, to beget]: polygenic, genotype, geneology, glycogen, pyrogen, heterogenous

gluco, glyco- [Gk. *glykys*, sweet; now pertaining to sugar]: glucose, glycogen, glycolysis, glycoprotein

hemo-, hemato-, -hemia, -emia [Gk. *haima*, blood]: hematology, hemoglobin, hemophilia

hepato- [Gk. *hepar, hepat-*, liver]: hepatitis, hepatic portal system